A Course in
Mathematical
Methods
for Physicists

A Course in Mathematical Methods for Physicists

Russell L. Herman

CRC Press
Taylor & Francis Group
Boca Raton London New York

CRC Press is an imprint of the
Taylor & Francis Group, an **informa** business

CRC Press
Taylor & Francis Group
6000 Broken Sound Parkway NW, Suite 300
Boca Raton, FL 33487-2742

© 2014 by Taylor & Francis Group, LLC
CRC Press is an imprint of Taylor & Francis Group, an Informa business

No claim to original U.S. Government works

Printed on acid-free paper
Version Date: 20130830

International Standard Book Number-13: 978-1-4665-8467-9 (Paperback)

Library of Congress Cataloging-in-Publication Data

Herman, Russell L., author.
  A course in mathematical methods for physicists / Russell L. Herman.
      pages cm
  Includes bibliographical references and index.
  ISBN 978-1-4665-8467-9 (hardcover : alk. paper)
  1. Mathematical physics--Textbooks. I. Title.

QC20.H485 2014
530.15--dc23   2013030827

Visit the Taylor & Francis Web site at
http://www.taylorandfrancis.com

and the CRC Press Web site at
http://www.crcpress.com

Dedicated to those students who have endured previous versions of A COURSE IN MATHEMATICAL METHODS FOR PHYSICISTS *and to those about to embark on the journey.*

Special thoughts go to my wife, Ann, and to EliJacob, Arianna, Shoshana, Nathan, and Avi.

Contents

Prologue

"All science is either physics or stamp collecting."—Ernest Rutherford (1871–1937)

Introduction

THIS IS A BOOK ON MATHEMATICAL METHODS SEEN IN PHYSICS and aimed at undergraduate students who have completed a year-long introductory course in physics. The intent of the course is to introduce students to many of the mathematical techniques useful in their undergraduate science education long before they are exposed to more focused topics in physics.

Most texts on mathematical physics are encyclopedic works which can never be covered in one semester and are often presented as a list of seemingly unrelated topics with some examples from physics inserted to highlight the connection of the particular topic to the real world. The point of these excursions is to introduce the student to a variety of topics and not to delve into it with as much rigor as one would find in some mathematics courses. Most of these topics have equivalent semester-long courses that go into the details and proofs of the main conjectures in that topic. Students may decide to later enroll in such courses during their undergraduate or graduate study. Often the relevance to physics must be found in more advanced courses in physics when the particular methods are used for specialized applications.

So, why not teach the methods in topics courses as they are needed? Part of the reason is that going into the details can take away from the global view of the course. Students often get lost in the mathematical details, as the proverbial tree can be lost in a forest of trees. Many of the mathematical techniques used in one course can be found in other courses. Collecting these techniques in one place, such as a course in mathematical methods for physicists or engineers, can provide a uniform background for students entering later courses on specialized topics. Repeated exposure to standard methods also helps to ingrain these methods. Furthermore, in such a course as this, the student first sees both the physical and mathematical connections between different fields. Instructors can use this course as an opportunity to show students how the physics curriculum ties together what otherwise might appear to be a group of seemingly different courses.

This originated as an introduction to topics in mathematical physics, introduced using the physics of oscillations and waves. It was based upon a one-semester junior-level course in mathematical physics taught at the University of North Carolina, Wilmington, and originally set in book form in 2005. The notes were later modified and used for several semesters before being turned into the current text.

The typical topics covered in a course on mathematical methods in physics are vector analysis, vector spaces, linear algebra, complex variables, power series, ordinary and partial differential equations, Fourier series, Laplace and Fourier transforms, Sturm–Liouville Theory, special functions and possibly other more advanced topics, such as tensors, group theory, the calculus of variations, or approximation techniques. We will cover many of these topics, but at times we will do so under the guise of exploring specific physical problems. In particular, we will introduce many of these topics in the context of the physics of oscillations and waves.

In the course of turning a set of course notes designed for one semester into a book, this book now contains material that cannot be covered in one semester. It was designed, unlike other texts on mathematical methods, to begin in one dimension and later extend the basic ideas to higher dimensions. The core topics that can be covered in a semester might come from the core of Chapters 2, 3, 5, 6, 7, 8, 9, and 11. Several sections of these chapters can be omitted for a broad overview covering the standard topics. Depending on the background of the students, a typical course might start with a quick review of some topics in the first chapter and then proceeding to visit applications of differential equations in Chapter 2 and some linear algebra and applications in Chapter 3. Chapters 5 and 6 introduce the ideas of Fourier analysis for both trigonometric and special functions. Chapter 7 gives quick coverage of complex analysis, which is useful for some parts of the transforms chapter. Vector analysis and curvilinear coordinates are then introduced so that students are ready to solve higher-dimensional problems in the final chapter. The optional topics consist of nonlinear dynamics topics, the calculus of variations, and the thread of seeing Green's functions throughout the text. Students can return to these topics as they progress through their more advanced classes.

What Is Mathematical Physics?

WHAT DO YOU THINK WHEN YOU HEAR the phrase "mathematical physics"? If one does a search on Google, one finds in Wikipedia, the following:

"Mathematical physics is an interdisciplinary field of academic study in between mathematics and physics, aimed at studying and solving problems inspired by physics within a mathematically rigorous framework. Although mathematical physics and theoretical physics are related, these two notions are often distinguished. Mathematical physics emphasizes the mathematical rigor of the same type as found in mathematics while theoretical physics emphasizes the links to actual observations and experimental physics that often requires the theoretical physicists to use heuristic, intuitive, and approximate arguments. Arguably, mathematical physics is closer to mathematics, and theoretical physics is closer to physics.

Because of the required rigor, mathematical physicists often deal with questions that theoretical physicists have considered as solved for decades. However, the mathematical physicists can sometimes (but neither commonly nor easily) show that the previous solution was incorrect.

Quantum mechanics cannot be understood without a good knowledge of mathematics. It is not surprising, then, that its developed version under the name of quantum field theory is one of the most abstract, mathematically-based areas of physical sciences, being backward-influential to mathematics. Other subjects researched by mathematical physicists include operator algebras, geometric algebra, noncommutative geometry, string theory, group theory, statistical mechanics, random fields etc."

However, we will not adhere to the rigor suggested by this definition of mathematical physics, but will aim more toward the theoretical physics approach. Thus, this course is more about mathematical methods in physics. With this approach in mind, the course will be designed as a study of physical topics leading to the use of standard mathematical techniques. However, we should keep in mind Freeman Dyson's (b. 1923) words,

> "For a physicist mathematics is not just a tool by means of which phenomena can be calculated, it is the main source of concepts and principles by means of which new theories can be created." from *Mathematics in the Physical Sciences Mathematics in the Physical Sciences*, **Scientific American**, 211(3), September 1964, pp. 129–146.

It has not always been the case that we had to think about the differences between mathematics and physics. Until about a century ago, people did not view physics and mathematics as separate disciplines. The Greeks did not separate the subjects, but developed an understanding of the natural sciences as part of their philosophical systems. Later, many of the big-name physicists and mathematicians actually worked in both areas only to be placed in these categories through historical hindsight. People like Newton and Maxwell made just as many contributions to mathematics as they did to physics while trying to investigate the workings of the physical universe. Mathematicians such as Gauss, Leibniz, and Euler made their share of contributions to physics.

Mathematics and physics are intimately related.

In the 1800s, the climate changed. The study of symmetry led to group theory, problems of convergence of the trigonometric series used by Fourier and others led to the need for rigor in analysis, the appearance of non-Euclidean geometries challenged the millennia-old Euclidean geometry, and the foundations of logic were challenged shortly after the turn of the century. This led to a whole population of mathematicians interested in abstracting mathematics and putting it on a firmer foundation without much attention to applications in the real world. This split is summarized by Freeman Dyson:

> "I am acutely aware of the fact that the marriage between mathematics and physics, which was so enormously fruitful in past centuries, has recently ended in divorce." from Missed Opportunities, 1972. (Gibbs Lecture)

In the meantime, many mathematicians were interested in applying and extending their methods to other fields, such as physics, chemistry, biology, and economics. These applied mathematicians have helped to mediate the divorce. Likewise, over the past century, a number physicists with a strong bent toward mathematics have emerged as mathematical physicists. So, Dyson's report of a divorce might be premature.

Some of the most important fields at the forefront of physics are steeped in mathematics. Einstein's general Theory of Relativity, a theory of gravitation, involves a good dose of differential geometry. String Theory is also highly mathematical, delving well beyond the topics in this book. While we will not get into these areas in this course, I would hope that students reading this book at least get a feel for the need to maintain the needed balance between mathematics and physics.

An Overview of the Course

It is difficult to list all of the topics needed to study a subject like String Theory. However, it is safe to say that a good grasp of the topics in this and more advance books on mathematical physics would help. A solid background in complex analysis, differential geometry, Lie groups and algebras, and variational principles should provide a good start.

ONE OF THE PROBLEMS WITH COURSES in mathematical physics and some of the courses taught in mathematics departments is that students do not always see the tie with physics. In this class we hope to enable students to see the mathematical techniques needed to enhance their future studies in physics. We will try not to provide the mathematical topics devoid of physical motivation. We will instead introduce the methods studied in this course while studying one underlying theme from physics. For example, most of the topics can be shown to have direct ties to oscillations and waves. Even though this theme is not the only possible collection of applications seen in physics, it is one of the most pervasive and has proven to be at the center of the revolutions of twentieth-century physics.

In this section we provide an overview of the course in terms of the theme of oscillations where appropriate. Besides making this connection throughout the text, we have also ordered the topics covered a bit differently than found in other common books on mathematical methods in physics. Many of the students taking such a course at the intermediate undergraduate level might have just completed their third semester of calculus, multivariate calculus. So, they are not quite comfortable with vector analysis. However, this is where many texts begin. So, in order to ease into higher-dimensional problems, we have put off three-dimensional problems until much further into the book. We begin instead with one-dimensional problems by solving ordinary differential equations and systems of differential equations. Later we look at the solution of partial differential equations. We will first solve the one-dimensional wave equation and one-dimensional heat equation. This will lead to the study of Fourier series.

We will see throughout the book that in the background there is some vector space, and an understanding of this idea requires some familiarity with linear algebra. Thus, we will discuss finite and infinite dimensional vectors spaces at an introductory level. This will lead to generalizations of Fourier series and provide tools for studying problems in higher dimensions. The culmination of these methods will be the solutions of problems typically encountered in upper-level courses in physics. The specific layout of the book is summarized below.

1. Introduction and Review

In this chapter we review some of the key computational tools that are seen in the first two courses in calculus and we recall some of the basic formulae for elementary functions. Then, we provide a short overview of the basic physics background, which will be useful in this course. It is meant to be a reference.

We also cover a few added topics not typically seen in many calculus classes, both here and later in the text. For example, we discuss the use of hyperbolic function substitution which seems to be used in physics texts more than in mathematics texts. Also, we describe the tabular method for doing integration by parts and differentiating under the integral, a trick often described by Feynman as one of his methods of doing integrals quickly. While we do cover key infinite series in this review chapter, a more extensive discussion of sequences and series is left for the Appendix.

As the aim of this course is to introduce techniques useful in exploring the basic physics concepts in more detail through computation, we also provide an overview of how one can use mathematical tables and computer algebra systems to help with the tedious tasks often encountered in solving physics problems.

What Do I Need to Know from Calculus?

What I Need from My Intro Physics Class?

Using Technology and Tables

Dimensional Analysis

2. Free Fall and Harmonic Oscillators

A major theme throughout this book is that of oscillations, starting with simple vibrations and ending with the vibrations of membranes, electromagnetic fields, and the electron wave function. We begin the first chapter by studying the simplest type of oscillation, simple harmonic motion. We look at examples of a mass on a spring, LRC circuits, and oscillating pendula. These examples lead to constant coefficient differential equations, whose solutions we study along the way. We further consider the effects of damping and forcing in such systems, which are key ingredients in understanding the qualitative behavior of oscillating systems. Another important topic in physics is that of a nonlinear system. We touch upon such systems in a few places in the text as we develop the tools needed in these first chapters.

Even before introducing differential equations for solving problems involving simple harmonic motion, we first look at differential equations for simpler examples, beginning with a discussion of free fall and terminal velocity. As you have been exposed to simple differential equations in your calculus class, we need only review some of the basic methods needed to solve standard applications in physics.

More complicated physical problems involve coupled systems. In fact, the problems in this chapter can be formulated as linear systems of differential equations. Such systems can be posed using matrices and the solutions are then obtained by solving eigenvalue problems, which is treated in the following chapter.

Other techniques for studying such problems described by differential equations involve power series methods, and Laplace and other integral transforms. These ideas are explored later in the book when we move on to exploring partial differential equations in higher dimensions.

We also touch on numerical solutions of differential equations, as not all problems can be solved analytically. We discuss Euler's Method, higher-order Taylor Methods, and Runge–Kutta Methods. These are sufficient for a simple understanding of how some computer applications carry out numerical solutions of differential equations. We apply these techniques to several problems, such as free falling bodies, the two-body problem, and the Friedmann Equation.

Free Fall

First-Order Differential Equations

Simple Harmonic Oscillator

Second-Order Differential Equations

LRC Circuits

Damped Oscillations

Forced Oscillations

Numerical Solutions of Differential Equations

Linear Systems

3. Linear Algebra

One of the most important mathematical topics in physics is linear algebra. Nowadays, the linear algebra course in most mathematics departments has evolved into a hybrid course covering matrix manipulations and some basics from vector spaces. However, it is seldom the case that applications, especially from physics, are covered. In this chapter we introduce vector spaces, linear transformations and view matrices as representations of linear transformations. The main theorem of linear algebra is the spectral theorem, which means studying eigenvalue problems. Essentially, when can a given operator (or matrix representation) be diagonalized? As operators act between vector spaces, it is useful to understand the concepts of both finite and infinite dimensional vector spaces and linear transformations on them.

The mathematical basis of much of physics relies on an understanding of both finite and infinite dimensional vector spaces. Linear algebra is important in the study of ordinary and partial differential equations, Fourier analysis, quantum mechanics, and general relativity. We return to this idea throughout the text. In this chapter we introduce the concepts of vector spaces, linear transformations, and eigenvalue problems. We also show how techniques from linear algebra are useful in solving coupled linear systems of differential equations. Later we see how much of what we do in physics involves linear transformations on vector spaces and eigenvalue problems.

Finite Dimensional Vector Spaces

Linear Transformations

Eigenvalue Problems

Matrix Formulation of Linear Systems

Applications of Linear Systems

Diagonalization and Linear Systems

4. **Nonlinear Dynamics**

An important area of research and interesting behaviors in the past several decades has been that of nonlinear dynamics and chaos. This is a huge area, and we only touch on some basic topics such as the idea of stability for nonlinear systems. The key is to linearize the system near equilibrium points and determine the local behavior of solutions. A standard physical model is that of the nonlinear pendulum. However, more complicated behaviors can arise as seen with the Duffing oscillator, and such behaviors can also arise in some nonlinear circuits.

Autonomous First-Order Equations

The Logistic Equation

Bifurcations of First-Order Equations

Nonlinear Pendulum

Stability of Fixed Points

Nonlinear Population Models

Limit Cycles

Nonautonomous Nonlinear Systems

5. **The Harmonics of Vibrating Strings**

The oscillations that we study here are solutions of the one-dimensional wave equation. A key example is provided by the finite length vibrating string. We study traveling wave solutions and look at techniques for solving the wave equation. The standard technique is to use separation of variables, turning the solution of a partial differential equation into the solution of several ordinary differential equations. The resulting general solution is written as an infinite series of sinusoidal functions, leading us to the study of Fourier series, which in turn provides the basis for studying the spectral content of complex signals.

In the meantime, we also introduce the heat, or diffusion, equation as another example of a generic one-dimensional partial differential equation exploiting the methods of this chapter. These problems begin our study of initial-boundary value problems, which pervade upper-level physics courses, especially in electromagnetic theory and quantum mechanics. At the end of the chapter we indicate that the solutions can be written in terms of special functions, called Green's functions.

Harmonics and Vibrations

The 1D Heat Equation

The 1D Wave Equation

Introduction to Fourier Trigonometric Series

Sine and Cosine Series

Finite Length Strings

The Gibbs Phenomenon

6. Special Functions and the Space in Which They Live

In our studies of systems in higher dimensions, we encounter a variety of new solutions of boundary value problems. These collectively are referred to as Special Functions and have been known for a long time. They appear later in the undergraduate curriculum and we cover several important examples. At the same time, we see that these special functions may provide bases for infinite dimensional function spaces. Understanding these functions spaces goes a long way toward understanding generalized Fourier theory, differential equations, and applications in electrodynamics and quantum theory.

In order to fully appreciate the special functions typically encountered in solving problems in higher dimensions, we develop Sturm–Liouville Theory with some further excursion into the theory of infinite dimensional vector spaces.

Infinite Dimensional Function Spaces

Classical Orthogonal Polynomials

Legendre Polynomials

Gamma Function

Bessel Functions

Sturm–Liouville Eigenvalue Problems

7. Complex Representations of Functions

Another simple example, useful later for studying electromagnetic waves, is the infinite one-dimensional string. We begin with the solution of the finite length string, which consists of an infinite sum over a discrete set of frequencies, or a Fourier series. Allowing for the string length to get large will turn the infinite sum into a sum over a continuous set of frequencies. Such a sum is now an integration and the resulting integrals are defined as Fourier transforms. Fourier transforms are useful in representing analog signals and localized waves in electromagnetism and quantum mechanics. Such integral transforms are explored in the next chapter. However, useful results can be obtained only after first introducing complex variable techniques.

So, we spend some time exploring complex variable techniques and introducing the calculus of complex functions. In particular, we become comfortable manipulating complex expressions and learn how to use contour methods to aid in the computation of integrals. We can apply these techniques to solving some special problems. We explore dispersion relations, relations between frequency and wave number for wave propagation, and the computation of complicated integrals such as those encountered in computing induced current using Faraday's Law.

Complex Representations of Waves

Complex Numbers

Complex Valued Functions

Complex Differentiation

Harmonic Functions and Laplace's Equation

Complex Integration

8. Transforms of the Wave and Heat Equations

For problems defined on an infinite interval, solutions are no longer given in terms of infinite series. They can be represented in terms of integrals, which are associated with integral transforms. We explore Fourier and Laplace transform methods for solving both ordinary and partial differential equations. By transforming our equations, we are led to simpler equations in transform space. We apply these methods to ordinary differential equations modeling forced oscillations and to the heat and wave equations.

Transform Theory

Exponential Fourier Transform

The Dirac Delta Function

The Convolution Operation

The Laplace Transform

Solution of Initial Value Problems

The Inverse Laplace Transform

9. Vector Analysis and Electromagnetic Waves

One of the major theories is that of electromagnetism. In this chapter we will recall Maxwell's Equations and use vector identities and vector theorems to derive the wave equation for electromagnetic waves. This requires us to recall some vector calculus from Calculus III. In particular, we review vector products, gradients, divergence, curl, and standard vector identities useful in physics. In the next chapter we solve the resulting wave equation for some physically interesting systems.

In preparation for solving problems in higher dimensions, we pause to look at generalized coordinate systems and the transformation of gradients and other differential operators in these new systems. This will be useful in the final chapter for solving problems in other geometries. The idea of using generalized coordinates leads to a discussion of tensors and their applications.

> Vector Analysis
>
> Maxwell's Equations
>
> Electromagnetic Waves
>
> Curvilinear Coordinates
>
> Tensors

10. Extrema and Variational Calculus

Before moving on to the higher-dimensional wave equations, we pause to pursue another standard topic in mathematical methods of physics: the calculus of variations. This chapter is about extrema principles. We begin with searching for minima and maxima of functions. However, many problems are related to the search for functions that lead to extreme values of certain integrals. In physics, these are related to Fermat's Principle of Least Time, or the Principle of Least Resistance, or the Principle of Least Action. Also, minimal surfaces and geodesics come from the exploration of such optimization problems. In this chapter we discuss some of the conditions, such as the Euler–Lagrange Equations, to solve standard problems like the brachistochrone problem or finding the path of shortest distance between two points on the surface of a sphere.

> Stationary and Extreme Values of Functions
>
> The Calculus of Variations
>
> Hamilton's Principle
>
> Geodesics

11. Problems in Higher Dimensions

Having studied one-dimensional oscillations, we are now be prepared to move on to higher-dimensional applications. These involve heat flow and vibrations in different geometries, primarily using the method of separation of variables. We apply these methods to the solution of the wave and heat equations in higher dimensions.

Another major equation of interest that you will encounter in upper-level physics is the Schrödinger Equation. We introduce this equation and explore solution techniques obtaining the relevant special functions involved in describing the wavefunction for a hydrogenic electron. We conclude the chapter with a summary of Green's function techniques.

Vibrations of Rectangular Membranes

Vibrations of a Kettle Drum

Laplace's Equation in 3D

Three Dimensional Cake Baking

Laplace's Equation and Spherical Symmetry

Schrödinger's Equation in Spherical Coordinates

Green's Functions for Partial Differential Equations

Tips for Students

GENERALLY, SOME TOPICS IN THE COURSE might seem difficult the first time through, especially not having had the upper-level physics at the time the topics are introduced. However, like all topics in physics, you will understand many of the topics at deeper levels as you progress through your studies. It will become clear that the more adept one becomes in the mathematical background, the better one's understanding of the physics.

You should read through this set of notes and then listen to the lectures. As you read the notes, be prepared to fill in the gaps in derivations and calculations. This is not a spectator sport, but a participatory adventure. Discuss the difficult points with others and your instructor. Work on problems as soon as possible. These are not problems that you can do the night before they are due. This is true of all physics classes. Feel free to go back and reread your old calculus and physics texts.

Note to Instructors

THIS BOOK IS BASED UPON A FOUR-CREDIT, ONE-SEMSETER COURSE on mathematical methods in physics typically taken by students just beginning their intermediate physics courses. It is assumed that the students have taken three semesters of calculus and may be currently enrolled in courses in differential equations or linear algebra. The first chapter is treated mostly as a review of calculus with some emphasis on the techniques used in physics. It is typically covered in only a few lectures. The real start of the course begins with Chapter 2, an introduction to differential equations with an emphasis on applications in physics and aimed at solving oscillation problems. We then introduce finite dimensional vector spaces and matrix representations in Chapter 3. This leads to using eigenvalue problems to study linear systems of differential equations, but also to provide some background to general vector spaces that will be used later to move on to infinite dimensional function spaces and special functions.

Fourier series are introduced in Chapter 5. Keeping with the theme of the book, one-dimensional partial differential equations are introduced through the solution of the heat and wave equations using the method of separation of variables. This provides a reason for studying Fourier series in physics.

In Chapter 6, Fourier series are generalized from a trigonometric basis to bases involving other functions, such as Legendre polynomials and Bessel functions. In that chapter we describe such series as the expansions of functions in certain function spaces over a basis of orthogonal functions.

It is natural to seek such expansions over infinite intervals, which lead to the Fourier transform. Before doing that, we first need to know how to compute complex integrals. So, in Chapter 7 we introduce complex numbers, complex functions, and derivatives and integrals of complex functions. Then, in Chapter 8 we discuss integral transforms with emphasis on Fourier and Laplace transforms.

Now we are ready to move on to higher dimensions. With the goal of solving higher-dimensional problems in physics that are described by partial differential equations, we provide the background needed to derive the wave equation for the electromagnetic field. This requires knowing the vector operations of divergence, gradient, and curl. We also discuss curvilinear coordinates in preparation for solving problems in geometries other than Cartesian.

In Chapter 11 we are now prepared to solve some typical problems in higher dimensions. In particular, we introduce students to problems they will see in their senior year, introducing spherical harmonics and solving the hydrogen problem.

There have been topics added to the original course as it was taught. Adding more content, this book now contains much more material than can be covered in a one-semester course. It now contains some discussion of numerical solutions of differential equations, the calculus of variations, and more nonlinear dynamics. Depending upon the background of the students and the goals of each individual instructor, one can spend less time on the first couple of chapters, or leave out some sections as one progresses through the course. Of course, this book has been written so that it can be read. It is not meant to be a reference book, and I think students would get more from the book by reading it and working through the examples.

Acknowledgments

MOST, IF NOT ALL, OF THE IDEAS AND EXAMPLES are not my own. These notes are a compendium of topics and examples that I have used in teaching not only mathematical physics, but also numerous courses in physics and applied mathematics. Some of the notions even extend back to when I first learned them in courses I had taken. In particular, I should note that many of the seeds for how I see such a course go back to what I learned about mathematical physics from Jerrold Franklin at Temple University and Fourier analysis and integral transforms from Abdul J. Jerri at Clarkson University. Some references to specific topics are included within the book, while other useful references are provided in the bibliography for further study.

I would also like to express my gratitude to the many students who have found typos, or suggested sections needing more clarity in the core set of notes upon which this book was based. This applies not only to the set of notes used in my mathematical physics course, but also to my other sets of notes, *An Introduction to Fourier and Complex Analysis with Application to the Spectral Analysis of Signals* and *A Second Course in Ordinary Differential Equations: Dynamical Systems and Boundary Value Problems* with which this text has some significant overlap.

Supplementary Material

Some materials will be provided at a site for the book as long as I maintain the site. The url is http://www.russherman.com/cmmp.

MATLAB is a registered trademark of The Math Works, Inc. For product information, please contact:
The MathWorks, Inc.
3 Apple Hill Drive
Natick, MA 01760-2098 USA
Tel: 508 647 7000
Fax: 508-647-7001
E-mail: info@mathworks.com
Web: www.mathworks.com

1
Introduction and Review

"Ordinary language is totally unsuited for expressing what physics really asserts, since the words of everyday life are not sufficiently abstract. Only mathematics and mathematical logic can say as little as the physicist means to say."—Bertrand Russell (1872–1970)

BEFORE WE BEGIN OUR STUDY OF MATHEMATICAL METHODS in physics, perhaps we should review some things from your past classes. You definitely need to know something before taking this class. It is assumed that you have taken Calculus and are comfortable with differentiation and integration. You should also have taken some introductory physics class, preferably the calculus-based course. Of course, you are not expected to know every detail from these courses. However, there are some topics and methods that will come up, and it would be useful to have a handy reference to what it is you should know.

Most importantly, you should still have your introductory physics and calculus texts to which you can refer throughout the course. Looking back on that old material, you will find that it appears easier than when you first encountered the material. That is the nature of learning mathematics and physics. Your understanding is continually evolving as you explore topics more in depth. It does not always sink in the first time you see it.

In this chapter we will give a quick review of these topics. We will also mention a few new things that might be interesting. This review is meant to make sure that everyone is at the same level before moving on to new topics.

1.1 What Do I Need to Know from Calculus?

1.1.1 Introduction

THERE ARE TWO MAIN TOPICS IN CALCULUS: derivatives and integrals. You learned that derivatives are useful in providing rates of change in either time or space. Integrals provide areas under curves, and also are useful in providing other types of sums over continuous bodies, such as lengths, areas, volumes, moments of inertia, or flux integrals. In physics, one can look at graphs of position versus time and the slope (derivative) of such a function gives the velocity. (See Figure 1.1.) By plotting velocity versus time, you can either look at the derivative to obtain acceleration, or you can look at the area under the curve and get the displacement:

$$x = \int_{t_0}^{t} v \, dt. \tag{1.1}$$

This is shown in Figure 1.2.

Of course, you need to know how to differentiate and integrate the given functions. Even before getting into differentiation and integration, you need to have a bag of functions useful in physics. Common functions are the polynomial and rational functions. You should be fairly familiar with these. Polynomial functions take the general form

$$f(x) = a_n x^n + a_{n-1} x^{n-1} + \cdots + a_1 x + a_0, \tag{1.2}$$

where $a_n \neq 0$. This is the form of a polynomial of degree n. Rational functions, $f(x) = \frac{g(x)}{h(x)}$, consist of ratios of polynomials. Their graphs can exhibit vertical and horizontal asymptotes.

Next are the exponential and logarithmic functions. The most common are the natural exponential and the natural logarithm. The natural exponential is given by $f(x) = e^x$, where $e \approx 2.718281828\ldots$. The natural logarithm is the inverse to the exponential, denoted by $\ln x$. (One needs to be careful, because some mathematics and physics books use log to mean natural exponential, whereas many of us were first trained to use this notation to mean the common logarithm, which is the "log base 10." Here we will use $\ln x$ for the natural logarithm.)

The properties of the exponential function follow from the basic properties for exponents. Namely, we have

$$
\begin{aligned}
e^0 &= 1, & (1.3)\\
e^{-a} &= \frac{1}{e^a}, & (1.4)\\
e^a e^b &= e^{a+b}, & (1.5)\\
(e^a)^b &= e^{ab}. & (1.6)
\end{aligned}
$$

The relation between the natural logarithm and natural exponential is given by

$$y = e^x \Leftrightarrow x = \ln y. \tag{1.7}$$

Some common logarithmic properties are

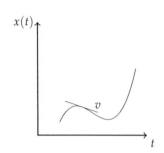

Figure 1.1: Plot of position versus time.

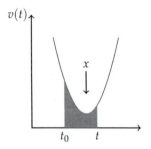

Figure 1.2: Plot of velocity versus time.

Exponential properties.

Logarithmic properties.

$$\ln 1 = 0, \tag{1.8}$$
$$\ln \frac{1}{a} = -\ln a, \tag{1.9}$$
$$\ln(ab) = \ln a + \ln b, \tag{1.10}$$
$$\ln \frac{a}{b} = \ln a - \ln b, \tag{1.11}$$
$$\ln \frac{1}{b} = -\ln b. \tag{1.12}$$

We will see applications of these relations as we progress through the course.

1.1.2 Trigonometric Functions

ANOTHER SET OF USEFUL FUNCTIONS are the trigonometric functions. These functions have probably plagued you since high school. They have their origins as far back as the building of the pyramids. Typical applications in your introductory math classes probably have included finding the heights of trees, flag poles, or buildings. It was recognized a long time ago that similar right triangles have fixed ratios of any pair of sides of the two similar triangles. These ratios only change when the non-right angles change.

Thus, the ratio of two sides of a right triangle only depends upon the angle. Because there are six possible ratios (think about it!), there are six possible functions. These are designated as sine, cosine, tangent and their reciprocals (cosecant, secant, and cotangent). In your introductory physics class, you really only needed the first three. You also learned that they are represented as the ratios of the opposite to hypotenuse, adjacent to hypotenuse, etc. Hopefully, you have this down by now.

You should also know the exact values of these basic trigonometric functions for the special angles $\theta = 0, \frac{\pi}{6}, \frac{\pi}{3}, \frac{\pi}{4}, \frac{\pi}{2}$, and their corresponding angles in the second, third, and fourth quadrants. This becomes internalized after much use, but we provide these values in Table 1.1 just in case you need a reminder.

θ	$\cos\theta$	$\sin\theta$	$\tan\theta$
0	1	0	0
$\frac{\pi}{6}$	$\frac{\sqrt{3}}{2}$	$\frac{1}{2}$	$\frac{\sqrt{3}}{3}$
$\frac{\pi}{3}$	$\frac{1}{2}$	$\frac{\sqrt{3}}{2}$	$\sqrt{3}$
$\frac{\pi}{4}$	$\frac{\sqrt{2}}{2}$	$\frac{\sqrt{2}}{2}$	1
$\frac{\pi}{2}$	0	1	undefined

Table 1.1: Table of Trigonometric Values

The problems students often have using trigonometric functions in later courses stem from using, or recalling, identities. We will have many an occasion to do so in this class as well. What is an identity? It is a relation that holds true all the time. For example, the most common identity for trigonometric functions is the Pythagorean identity

$$\sin^2 \theta + \cos^2 \theta = 1. \tag{1.13}$$

This holds true for every angle θ! An even simpler identity is

$$\tan \theta = \frac{\sin \theta}{\cos \theta}. \tag{1.14}$$

Other simple identities can be derived from the Pythagorean identity. Dividing the identity by $\cos^2 \theta$, or $\sin^2 \theta$, yields

$$
\begin{aligned}
\tan^2 \theta + 1 &= \sec^2 \theta, \tag{1.15} \\
1 + \cot^2 \theta &= \csc^2 \theta. \tag{1.16}
\end{aligned}
$$

Several other useful identities stem from the use of the sine and cosine of the sum and difference of two angles. Namely, we have that

Sum and difference identities.

$$
\begin{aligned}
\sin(A \pm B) &= \sin A \cos B \pm \sin B \cos A, \tag{1.17} \\
\cos(A \pm B) &= \cos A \cos B \mp \sin A \sin B. \tag{1.18}
\end{aligned}
$$

Note that the upper (lower) signs are taken together.

Example 1.1. *Evaluate* $\sin \frac{\pi}{12}$.

$$
\begin{aligned}
\sin \frac{\pi}{12} &= \sin \left(\frac{\pi}{3} - \frac{\pi}{4} \right) \\
&= \sin \frac{\pi}{3} \cos \frac{\pi}{4} - \sin \frac{\pi}{4} \cos \frac{\pi}{3} \\
&= \frac{\sqrt{3}}{2} \frac{\sqrt{2}}{2} - \frac{\sqrt{2}}{2} \frac{1}{2} \\
&= \frac{\sqrt{2}}{4} \left(\sqrt{3} - 1 \right). \tag{1.19}
\end{aligned}
$$

Double angle formulae.

The double angle formulae are found by setting $A = B$:

$$
\begin{aligned}
\sin(2A) &= 2 \sin A \cos B, \tag{1.20} \\
\cos(2A) &= \cos^2 A - \sin^2 A. \tag{1.21}
\end{aligned}
$$

Using Equation (1.13), we can rewrite Equation (1.21) as

$$
\begin{aligned}
\cos(2A) &= 2 \cos^2 A - 1, \tag{1.22} \\
&= 1 - 2 \sin^2 A. \tag{1.23}
\end{aligned}
$$

These, in turn, lead to the half angle formulae. Solving for $\cos^2 A$ and $\sin^2 A$, we find that

Half angle formulae.

$$
\begin{aligned}
\sin^2 A &= \frac{1 - \cos 2A}{2}, \tag{1.24} \\
\cos^2 A &= \frac{1 + \cos 2A}{2}. \tag{1.25}
\end{aligned}
$$

Example 1.2. *Evaluate* $\cos\frac{\pi}{12}$. *In the previous example, we used the sum/difference identities to evaluate a similar expression. We could have also used a half angle identity. In this example, we have*

$$
\begin{aligned}
\cos^2\frac{\pi}{12} &= \frac{1}{2}\left(1+\cos\frac{\pi}{6}\right) \\
&= \frac{1}{2}\left(1+\frac{\sqrt{3}}{2}\right) \\
&= \frac{1}{4}\left(2+\sqrt{3}\right).
\end{aligned}
\tag{1.26}
$$

So, $\cos\frac{\pi}{12} = \frac{1}{2}\sqrt{2+\sqrt{3}}$. *This is not the simplest form and is called a nested radical. In fact, if we proceeded using the difference identity for cosines, then we would obtain*

$$
\cos\frac{\pi}{12} = \frac{\sqrt{2}}{4}(1+\sqrt{3}).
$$

So, how does one show that these answers are the same?

Let's focus on the factor $\sqrt{2+\sqrt{3}}$. *We seek to write this in the form* $c+d\sqrt{3}$. *Equating the two expressions and squaring, we have*

$$
\begin{aligned}
2+\sqrt{3} &= (c+d\sqrt{3})^2 \\
&= c^2+3d^2+2cd\sqrt{3}.
\end{aligned}
\tag{1.27}
$$

In order to solve for c and d, it would seem natural to equate the coefficients of $\sqrt{3}$ *and the remaining terms. We obtain a system of two nonlinear algebraic equations:*

$$
c^2+3d^2 = 2, \tag{1.28}
$$

$$
2cd = 1. \tag{1.29}
$$

Solving the second equation for $d=1/2c$, *and substituting the result into the first equation, we find*

$$
4c^4-8c^2+3=0.
$$

This fourth-order equation has four solutions:

$$
c = \pm\frac{\sqrt{2}}{2}, \pm\frac{\sqrt{6}}{2}
$$

and

$$
b = \pm\frac{\sqrt{2}}{2}, \pm\frac{\sqrt{6}}{6}.
$$

Thus,

$$
\begin{aligned}
\cos\frac{\pi}{12} &= \frac{1}{2}\sqrt{2+\sqrt{3}} \\
&= \pm\frac{1}{2}\left(\frac{\sqrt{2}}{2}+\frac{\sqrt{2}}{2}\sqrt{3}\right) \\
&= \pm\frac{\sqrt{2}}{4}(1+\sqrt{3})
\end{aligned}
\tag{1.30}
$$

It is useful at times to know when one can reduce square roots of such radicals, called denesting. More generally, one seeks to write $\sqrt{a+b\sqrt{q}}=c+d\sqrt{q}$. Following the procedure in this example, one has $d=\frac{b}{2c}$ and

$$
c^2 = \frac{1}{2}\left(a\pm\sqrt{a^2-qb^2}\right).
$$

As long as a^2-qb^2 is a perfect square, there is a chance to reduce the expression to a simpler form.

and

$$\cos \frac{\pi}{12} = \frac{1}{2}\sqrt{2 + \sqrt{3}}$$

$$= \pm \frac{1}{2}\left(\frac{\sqrt{6}}{2} + \frac{\sqrt{6}}{6}\sqrt{3}\right)$$

$$= \pm \frac{\sqrt{6}}{12}(3 + \sqrt{3}). \tag{1.31}$$

Of the four solutions, two are negative and we know that the value of the cosine for this angle must be positive. The remaining two solutions are actually equal! A quick computation will verify this:

$$\frac{\sqrt{6}}{12}(3 + \sqrt{3}) = \frac{\sqrt{3}\sqrt{2}}{12}(3 + \sqrt{3})$$

$$= \frac{\sqrt{2}}{12}(3\sqrt{3} + 3)$$

$$= \frac{\sqrt{2}}{4}(\sqrt{3} + 1). \tag{1.32}$$

We could have bypassed this situation by requiring that the solutions for b and c were not simply proportional to $\sqrt{3}$ like they are in the second case.

Product identities

Finally, another useful set of identities are the product identities. For example, if we add the identities for $\sin(A + B)$ and $\sin(A - B)$, the second terms cancel and we have

$$\sin(A + B) + \sin(A - B) = 2 \sin A \cos B.$$

Thus, we have that

$$\boxed{\sin A \cos B = \frac{1}{2}(\sin(A + B) + \sin(A - B)).} \tag{1.33}$$

Similarly, we have

$$\boxed{\cos A \cos B = \frac{1}{2}(\cos(A + B) + \cos(A - B))} \tag{1.34}$$

and

$$\boxed{\sin A \sin B = \frac{1}{2}(\cos(A - B) - \cos(A + B)).} \tag{1.35}$$

Know the above boxed identities!

These boxed equations are the most common trigonometric identities. They appear often and should just roll off your tongue.

We will also need to understand the behaviors of trigonometric functions. In particular, we know that the sine and cosine functions are periodic. They are not the only periodic functions, as we shall see. [Just visualize the teeth on a carpenter's saw.] However, they are the most common periodic functions.

Periodic functions.

A periodic function $f(x)$ satisfies the relation

$$f(x + p) = f(x), \quad \text{for all } x$$

for some constant p. If p is the smallest such number, then p is called the period. Both the sine and cosine functions have period 2π. This means that the graph repeats its form every 2π units. Similarly, $\sin bx$ and $\cos bx$ have the common period $p = \frac{2\pi}{b}$. We will make use of this fact in later chapters.

Related to these are the inverse trigonometric functions. For example, $f(x) = \sin^{-1} x$, or $f(x) = \arcsin x$. Inverse functions give back angles, so you should think

$$\theta = \sin^{-1} x \quad \Leftrightarrow \quad x = \sin \theta. \tag{1.36}$$

Also, you should recall that $y = \sin^{-1} x = \arcsin x$ is only a function if $-\frac{\pi}{2} \leq x \leq \frac{\pi}{2}$. Similar relations exist for $y = \cos^{-1} x = \arccos x$ and $\tan^{-1} x = \arctan x$.

Once you think about these functions as providing angles, then you can make sense out of more complicated looking expressions, like $\tan(\sin^{-1} x)$. Such expressions often pop up in evaluations of integrals. We can untangle this in order to produce a simpler form by referring to Expression (1.36). $\theta = \sin^{-1} x$ is simple an angle whose sine is x. Knowing that the sine is the opposite side of a right triangle divided by its hypotenuse, one just draws a triangle in this proportion as shown in Figure 1.3. Namely, the side opposite the angle has length x and the hypotenuse has length 1. Using the Pythagorean Theorem, the missing side (adjacent to the angle) is simply $\sqrt{1 - x^2}$. Having obtained the lengths for all three sides, we can now produce the tangent of the angle as

$$\tan(\sin^{-1} x) = \frac{x}{\sqrt{1 - x^2}}.$$

In Feynman's *Surely You're Joking Mr. Feynman!*, Richard Feynman (1918–1988) discussed his invention of his own notation for both trigonometric and inverse trigonometric functions as the standard notation did not make sense to him.

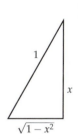

Figure 1.3: $\theta = \sin^{-1} x \Rightarrow \tan \theta = \frac{x}{\sqrt{1-x^2}}$

1.1.3 Hyperbolic Functions

SO, ARE THERE ANY OTHER FUNCTIONS that are useful in physics? Actually, there are many more. However, you have probably not seen many of them to date. We will see by the end of the semester that there are many important functions that arise as solutions of some fairly generic, but important, physics problems. In your calculus classes you have also seen that some relations are represented in parametric form. However, there is at least one other set of elementary functions which you should already know about. These are the hyperbolic functions. Such functions are useful in representing hanging cables, unbounded orbits, and special travelling waves called solitons. They also play a role in special and general relativity.

Hyperbolic functions are actually related to the trigonometric functions, as we shall see after a little bit of complex function theory. For now, we just want to recall a few definitions and identities. Just as all of the trigonometric functions can be built from the sine and the cosine, the hyperbolic functions can be defined in terms of the hyperbolic sine and hyperbolic cosine (shown in Figure 1.4):

Solitons are special solutions to some generic nonlinear wave equations. They typically experience elastic collisions and play special roles in a variety of fields in physics, such as hydrodynamics and optics. A simple soliton solution is of the form

$$u(x, t) = 2\eta^2 \, \text{sech}^2 \, \eta(x - 4\eta^2 t).$$

Hyperbolic functions; We will later see the connection between the hyperbolic and trigonometric functions in Chapter 7.

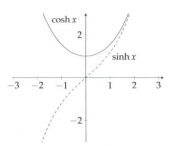

Figure 1.4: Plots of $\cosh x$ and $\sinh x$. Note that $\sinh 0 = 0$, $\cosh 0 = 1$, and $\cosh x \geq 1$.

$$\sinh x = \frac{e^x - e^{-x}}{2}, \tag{1.37}$$

$$\cosh x = \frac{e^x + e^{-x}}{2}. \tag{1.38}$$

There are four other hyperbolic functions. These are defined in terms of the above functions similar to the relations between the trigonometric functions. We have

$$\tanh x = \frac{\sinh x}{\cosh x} = \frac{e^x - e^{-x}}{e^x + e^{-x}}, \tag{1.39}$$

$$\operatorname{sech} x = \frac{1}{\cosh x} = \frac{2}{e^x + e^{-x}}, \tag{1.40}$$

$$\operatorname{csch} x = \frac{1}{\sinh x} = \frac{2}{e^x - e^{-x}}, \tag{1.41}$$

$$\coth x = \frac{1}{\tanh x} = \frac{e^x + e^{-x}}{e^x - e^{-x}}. \tag{1.42}$$

There are also a whole set of identities similar to those for the trigonometric functions. For example, the Pythagorean identity for trigonometric functions, $\sin^2 \theta + \cos^2 \theta = 1$, is replaced by the identity

$$\cosh^2 x - \sinh^2 x = 1.$$

This is easily shown by simply using the definitions of these functions. This identity is also useful for providing a parametric set of equations describing hyperbolae. Letting $x = a \cosh t$ and $y = b \sinh t$, one has

$$\frac{x^2}{a^2} - \frac{y^2}{b^2} = \cosh^2 t - \sinh^2 t = 1.$$

Hyperbolic identities.

A list of commonly needed hyperbolic function identities is given by the following:

$$\cosh^2 x - \sinh^2 x = 1, \tag{1.43}$$

$$\tanh^2 x + \operatorname{sech}^2 x = 1, \tag{1.44}$$

$$\cosh(A \pm B) = \cosh A \cosh B \pm \sinh A \sinh B, \tag{1.45}$$

$$\sinh(A \pm B) = \sinh A \cosh B \pm \sinh B \cosh A, \tag{1.46}$$

$$\cosh 2x = \cosh^2 x + \sinh^2 x, \tag{1.47}$$

$$\sinh 2x = 2 \sinh x \cosh x, \tag{1.48}$$

$$\cosh^2 x = \frac{1}{2}(1 + \cosh 2x), \tag{1.49}$$

$$\sinh^2 x = \frac{1}{2}(\cosh 2x - 1). \tag{1.50}$$

Note the similarity with the trigonometric identities. Other identities can be derived from these.

There also exist inverse hyperbolic functions and these can be written in terms of logarithms. As with the inverse trigonometric functions, we begin with the definition

$$y = \sinh^{-1} x \quad \Leftrightarrow \quad x = \sinh y. \tag{1.51}$$

The aim is to write y in terms of x without using the inverse function. First, we note that

$$x = \frac{1}{2}\left(e^y - e^{-y}\right). \tag{1.52}$$

Next we solve for e^y. This is done by noting that $e^{-y} = \frac{1}{e^y}$ and rewriting the previous equation as

$$0 = (e^y)^2 - 2xe^y - 1. \tag{1.53}$$

This equation is in quadratic form which we can solve using the quadratic formula as

$$e^y = x + \sqrt{1 + x^2}.$$

(There is only one root as we expect the exponential to be positive.)

The final step is to solve for y,

$$y = \ln\left(x + \sqrt{1 + x^2}\right). \tag{1.54}$$

The inverse hyperbolic functions care given by

$$
\begin{aligned}
\sinh^{-1} x &= \ln\left(x + \sqrt{1 + x^2}\right), \\
\cosh^{-1} x &= \ln\left(x + \sqrt{x^2 - 1}\right), \\
\tanh^{-1} x &= \frac{1}{2}\ln\frac{1 + x}{1 - x}.
\end{aligned}
$$

1.1.4 Derivatives

Now that we know some elementary functions, we seek their derivatives. We will not spend time exploring the appropriate limits in any rigorous way. We are only interested in the results. We provide these in Table 1.2. We expect that you know the meaning of the derivative and all of the usual rules, such as the product and quotient rules.

Function	Derivative
a	0
x^n	nx^{n-1}
e^{ax}	ae^{ax}
$\ln ax$	$\frac{1}{x}$
$\sin ax$	$a\cos ax$
$\cos ax$	$-a\sin ax$
$\tan ax$	$a\sec^2 ax$
$\csc ax$	$-a\csc ax \cot ax$
$\sec ax$	$a\sec ax \tan ax$
$\cot ax$	$-a\csc^2 ax$
$\sinh ax$	$a\cosh ax$
$\cosh ax$	$a\sinh ax$
$\tanh ax$	$a\operatorname{sech}^2 ax$
$\operatorname{csch} ax$	$-a\operatorname{csch} ax \coth ax$
$\operatorname{sech} ax$	$-a\operatorname{sech} ax \tanh ax$
$\coth ax$	$-a\operatorname{csch}^2 ax$

Table 1.2: Table of Common Derivatives (a is a constant)

Also, you should be familiar with the Chain Rule. Recall that this rule tells us that if we have a composition of functions, such as the elementary functions above, then we can compute the derivative of the composite function. Namely, if $h(x) = f(g(x))$, then

$$\frac{dh}{dx} = \frac{d}{dx}\left(f(g(x))\right) = \frac{df}{dg}\bigg|_{g(x)}\frac{dg}{dx} = f'(g(x))g'(x). \tag{1.55}$$

Example 1.3. *Differentiate* $H(x) = 5\cos\left(\pi\tanh 2x^2\right).$

This is a composition of three functions: $H(x) = f(g(h(x)))$, *where* $f(x) = 5\cos x$, $g(x) = \pi\tanh x$, *and* $h(x) = 2x^2$. *Then the derivative becomes*

$$
\begin{aligned}
H'(x) &= 5\left(-\sin\left(\pi\tanh 2x^2\right)\right)\frac{d}{dx}\left(\left(\pi\tanh 2x^2\right)\right)\\
&= -5\pi\sin\left(\pi\tanh 2x^2\right)\operatorname{sech}^2 2x^2\frac{d}{dx}\left(2x^2\right)\\
&= -20\pi x\sin\left(\pi\tanh 2x^2\right)\operatorname{sech}^2 2x^2. \tag{1.56}
\end{aligned}
$$

1.1.5   Integrals

INTEGRATION IS TYPICALLY A BIT HARDER. Imagine being given the last result in Expression (1.56) and having to figure out what was differentiated in order to get the given function. As you may recall from the Fundamental Theorem of Calculus, the integral is the inverse operation to differentiation:

$$\int \frac{df}{dx}\,dx = f(x) + C. \tag{1.57}$$

It is not always easy to evaluate a given integral. In fact, some integrals are not even doable! However, you learned in calculus that there are some methods that could yield an answer. While you might be happier using a computer algebra system, such as Maple or WolframAlpha.com, or a fancy calculator, you should know a few basic integrals and know how to use tables for some of the more complicated ones. In fact, it can be exhilarating when you can do a given integral without reference to a computer or a Table of Integrals. However, you should be prepared to do some integrals using what you have been taught in calculus. We will review a few of these methods and some of the standard integrals in this section.

First of all, there are some integrals you are expected to know without doing any work. These integrals appear often and are just an application of the Fundamental Theorem of Calculus to the previous Table 1.2. The basic integrals that students should know off the top of their heads are given in Table 1.3.

These are not the only integrals you should be able to do. We can expand the list by recalling a few of the techniques you learned in calculus: the Method of Substitution, Integration by Parts, integration using partial fraction decomposition, and trigonometric integrals, and trigonometric substitution. There are also a few other techniques that you had not seen before. We will look at several examples.

Example 1.4. *Evaluate* $\int \frac{x}{\sqrt{x^2+1}}\, dx$.

When confronted with an integral, you should first ask if a simple substitution would reduce the integral to one you know how to do.

The ugly part of this integral is the $x^2 + 1$ under the square root. So, we let $u = x^2 + 1$.

Noting that when $u = f(x)$, we have $du = f'(x)\, dx$. For our example, $du = 2x\, dx$.

Looking at the integral, part of the integrand can be written as $x\, dx = \frac{1}{2}u\, du$. Then, the integral becomes

$$\int \frac{x}{\sqrt{x^2+1}}\, dx = \frac{1}{2} \int \frac{du}{\sqrt{u}}.$$

The substitution has converted our integral into an integral over u. Also, this integral is doable! It is one of the integrals we should know. Namely, we can write it as

$$\frac{1}{2} \int \frac{du}{\sqrt{u}} = \frac{1}{2} \int u^{-1/2}\, du.$$

This is now easily finished after integrating and using the substitution variable to give

$$\int \frac{x}{\sqrt{x^2+1}}\, dx = \frac{1}{2} \frac{u^{1/2}}{\frac{1}{2}} + C = \sqrt{x^2+1} + C.$$

Note that we have added the required integration constant, and that the derivative of the result easily gives the original integrand (after employing the Chain Rule).

Function	Indefinite Integral
a	ax
x^n	$\frac{x^{n+1}}{n+1}$
e^{ax}	$\frac{1}{a}e^{ax}$
$\frac{1}{x}$	$\ln x$
$\sin ax$	$-\frac{1}{a}\cos ax$
$\cos ax$	$\frac{1}{a}\sin ax$
$\sec^2 ax$	$\frac{1}{a}\tan ax$
$\sinh ax$	$\frac{1}{a}\cosh ax$
$\cosh ax$	$\frac{1}{a}\sinh ax$
$\text{sech}^2 ax$	$\frac{1}{a}\tanh ax$
$\sec x$	$\ln\lvert \sec x + \tan x\rvert$
$\frac{1}{a+bx}$	$\frac{1}{b}\ln(a+bx)$
$\frac{1}{a^2+x^2}$	$\frac{1}{a}\tan^{-1} ax$
$\frac{1}{\sqrt{a^2-x^2}}$	$\frac{1}{a}\sin^{-1} ax$
$\frac{1}{\sqrt{x^2-a^2}}$	$\frac{1}{a}\sec^{-1} ax$

Table 1.3: Table of Common Integrals

Often we are faced with definite integrals, in which we integrate between two limits. There are several ways to use these limits. However, students often forget that a change of variables generally means that the limits have to change.

Example 1.5. *Evaluate $\int_0^2 \frac{x}{\sqrt{x^2+1}}\,dx$.*

This is the previous example but with integration limits added. We proceed as before. We let $u = x^2 + 1$. As x goes from 0 to 2, u takes values from 1 to 5. So, this substitution gives

$$\int_0^2 \frac{x}{\sqrt{x^2+1}}\,dx = \frac{1}{2}\int_1^5 \frac{du}{\sqrt{u}} = \sqrt{u}\Big|_1^5 = \sqrt{5} - 1.$$

When you become proficient at integration, you can bypass some of these steps. In the next example we try to demonstrate the thought process involved in using substitution without explicitly using the substitution variable.

Example 1.6. *Evaluate $\int_0^2 \frac{x}{\sqrt{9+4x^2}}\,dx$*

As with the previous example, one sees that the derivative of $9 + 4x^2$ is proportional to x, which is in the numerator of the integrand. Thus, a substitution would give an integrand of the form $u^{-1/2}$. So, we expect the answer to be proportional to $\sqrt{u} = \sqrt{9 + 4x^2}$. The starting point is therefore

$$\int \frac{x}{\sqrt{9+4x^2}}\,dx = A\sqrt{9+4x^2},$$

where A is a constant to be determined.

We can determine A through differentiation since the derivative of the answer should be the integrand. Thus,

$$\frac{d}{dx}A(9+4x^2)^{\frac{1}{2}} = A(9+4x^2)^{-\frac{1}{2}}\left(\frac{1}{2}\right)(8x)$$
$$= 4xA(9+4x^2)^{-\frac{1}{2}}. \tag{1.58}$$

Comparing this result with the integrand, we see that the integrand is obtained when $A = \frac{1}{4}$. Therefore,

$$\int \frac{x}{\sqrt{9+4x^2}}\,dx = \frac{1}{4}\sqrt{9+4x^2}.$$

We now complete the integral

$$\int_0^2 \frac{x}{\sqrt{9+4x^2}}\,dx = \frac{1}{4}[5-3] = \frac{1}{2}.$$

The function

$$\mathrm{gd}(x) = \int_0^x \frac{dx}{\cosh x} = 2\tan^{-1}e^x - \frac{\pi}{2}$$

is called the Gudermannian and connects trigonometric and hyperbolic functions. This function was named after Christoph Gudermann (1798-1852), but introduced by Johann Heinrich Lambert (1728-1777), who was one of the first to introduce hyperbolic functions.

Example 1.7. *Evaluate $\int \frac{dx}{\cosh x}$.*

This integral can be performed by first using the definition of $\cosh x$, followed by a simple substitution:

$$\int \frac{dx}{\cosh x} = \int \frac{2}{e^x + e^{-x}}\,dx$$
$$= \int \frac{2e^x}{e^{2x}+1}\,dx. \tag{1.59}$$

Now, we let $u = e^x$ and $du = e^x dx$. Then,

$$\int \frac{dx}{\cosh x} = \int \frac{2}{1+u^2}\,du$$
$$= 2\tan^{-1}u + C$$
$$= 2\tan^{-1}e^x + C. \tag{1.60}$$

Integration by Parts

When the Method of Substitution fails, there are other methods you can try. One of the most used is the Method of Integration by Parts. Recall the Integration by Parts Formula:

Integration by Parts Formula.

$$\int u\,dv = uv - \int v\,du. \qquad (1.61)$$

The idea is that you are given the integral on the left and you can relate it to an integral on the right. Hopefully, the new integral is one you can do, or at least it is an easier integral than the one you are trying to evaluate.

However, you are not usually given the functions u and v. You have to determine them. The integral form that you really have is a function of another variable, say x. Another form of the Integration by Parts Formula can be written as

$$\int f(x)g'(x)\,dx = f(x)g(x) - \int g(x)f'(x)\,dx. \qquad (1.62)$$

This form is a bit more complicated in appearance, though it is clearer than the u-v form as to what is happening. The derivative has been moved from one function to the other. Recall that this formula was derived by integrating the product rule for differentiation. (See your calculus text.)

These two formulae can be related using the differential relations

Note: Often in physics one needs to move a derivative between functions inside an integrand. The key: Use integration by parts to move the derivative from one function to the other under an integral.

$$
\begin{aligned}
u = f(x) &\quad \rightarrow \quad du = f'(x)\,dx, \\
v = g(x) &\quad \rightarrow \quad dv = g'(x)\,dx.
\end{aligned}
\qquad (1.63)
$$

This also gives a method for applying the Integration by Parts Formula.

Example 1.8. *Consider the integral $\int x \sin 2x\,dx$. We choose $u = x$ and $dv = \sin 2x\,dx$. This gives the correct left side of the Integration by Parts Formula. We next determine v and du:*

$$du = \frac{du}{dx}dx = dx,$$

$$v = \int dv = \int \sin 2x\,dx = -\frac{1}{2}\cos 2x.$$

We note that one usually does not need the integration constant. Inserting these expressions into the Integration by Parts Formula, we have

$$\int x \sin 2x\,dx = -\frac{1}{2}x\cos 2x + \frac{1}{2}\int \cos 2x\,dx.$$

We see that the new integral is easier to do than the original integral. Had we picked $u = \sin 2x$ and $dv = x\,dx$, then the formula still works, but the resulting integral is not easier.

For completeness, we finish the integration. The result is

$$\int x \sin 2x\,dx = -\frac{1}{2}x\cos 2x + \frac{1}{4}\sin 2x + C.$$

As always, you can check your answer by differentiating the result, a step students often forget to do. Namely,

$$\frac{d}{dx}\left(-\frac{1}{2}x\cos 2x + \frac{1}{4}\sin 2x + C\right) = -\frac{1}{2}\cos 2x + x\sin 2x + \frac{1}{4}(2\cos 2x)$$
$$= x\sin 2x. \tag{1.64}$$

So, we do get back the integrand in the original integral.

We can also perform integration by parts on definite integrals. The general formula is written as

Integration by Parts for Definite Integrals.

$$\boxed{\int_a^b f(x)g'(x)\,dx = f(x)g(x)\Big|_a^b - \int_a^b g(x)f'(x)\,dx. \qquad (1.65)}$$

Example 1.9. *Consider the integral*

$$\int_0^\pi x^2 \cos x\,dx.$$

This will require two integrations by parts. First, we let $u = x^2$ and $dv = \cos x$. Then,

$$du = 2x\,dx, \quad v = \sin x.$$

Inserting into the Integration by Parts Formula, we have

$$\int_0^\pi x^2 \cos x\,dx = x^2 \sin x\Big|_0^\pi - 2\int_0^\pi x\sin x\,dx$$
$$= -2\int_0^\pi x\sin x\,dx. \tag{1.66}$$

We note that the resulting integral is easier than the given integral, but we still cannot do the integral off the top of our head (unless we look at Example 1.3!). So, we need to integrate by parts again. (Note: In your calculus class you may recall that there is a tabular method for carrying out multiple applications of the formula. We will show this method in the next example.)

We apply integration by parts by letting $U = x$ and $dV = \sin x\,dx$. This gives $dU = dx$ and $V = -\cos x$. Therefore, we have

$$\int_0^\pi x\sin x\,dx = -x\cos x\Big|_0^\pi + \int_0^\pi \cos x\,dx$$
$$= \pi + \sin x\Big|_0^\pi$$
$$= \pi. \tag{1.67}$$

The final result is

$$\int_0^\pi x^2 \cos x\,dx = -2\pi.$$

There are other ways to compute integrals of this type. First of all, there is the Tabular Method to perform integration by parts. A second method is to use differentiation of parameters under the integral. We will demonstrate this using examples.

Example 1.10. *Compute the integral $\int_0^\pi x^2 \cos x \, dx$ using the Tabular Method.*

First we identify the two functions under the integral, x^2 and $\cos x$. We then write the two functions and list the derivatives and integrals of each, respectively. This is shown in Table 1.4. Note that we stopped when we reached zero in the left column.

Next, one draws diagonal arrows, as indicated, with alternating signs attached, starting with $+$. The indefinite integral is then obtained by summing the products of the functions at the ends of the arrows along with the signs on each arrow:

$$\int x^2 \cos x \, dx = x^2 \sin x + 2x \cos x - 2 \sin x + C.$$

To find the definite integral, one evaluates the antiderivative at the given limits.

$$
\begin{aligned}
\int_0^\pi x^2 \cos x \, dx &= \left[x^2 \sin x + 2x \cos x - 2 \sin x \right]_0^\pi \\
&= (\pi^2 \sin \pi + 2\pi \cos \pi - 2 \sin \pi) - 0 \\
&= -2\pi. \hspace{3cm} (1.68)
\end{aligned}
$$

Actually, the Tabular Method works even if a zero does not appear in the left column. One can go as far as possible, and if a zero does not appear, then one needs only to integrate, if possible, the product of the functions in the last row, adding the next sign in the alternating sign progression. The next example shows how this works.

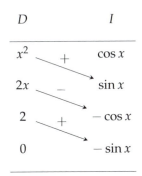

Table 1.4: Tabular Method

Example 1.11. *Use the Tabular Method to compute $\int e^{2x} \sin 3x \, dx$.*

As before, we first set up the table as shown in Table 1.5.

Putting together the pieces, noting that the derivatives in the left column will never vanish, we have

$$\int e^{2x} \sin 3x \, dx = (\frac{1}{2} \sin 3x - \frac{3}{4} \cos 3x)e^{2x} + \int (-9 \sin 3x) \left(\frac{1}{4} e^{2x} \right) \, dx.$$

The integral on the right is a multiple of the one on the left, so we can combine them:

$$\frac{13}{4} \int e^{2x} \sin 3x \, dx = (\frac{1}{2} \sin 3x - \frac{3}{4} \cos 3x)e^{2x},$$

or

$$\int e^{2x} \sin 3x \, dx = (\frac{2}{13} \sin 3x - \frac{3}{13} \cos 3x)e^{2x}.$$

Table 1.5: Tabular Method—Showing a
Non-terminating Example

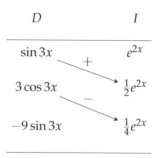

Differentiation under the Integral

Another method that one can use to evaluate this integral is to differentiate under the integral sign. This is mentioned in Richard Feynman's memoir *Surely You're Joking, Mr. Feynman!*. In the book, Feynman recounts using this "trick" to be able to do integrals that his MIT classmates could not do. This is based on a theorem found in Advanced Calculus texts. Readers unfamiliar with partial derivatives should be able to grasp their use in the following example.

Theorem 1.1. *Let the functions $f(x, t)$ and $\frac{\partial f(x,t)}{\partial x}$ be continuous in both t, and x, in the region of the (t, x) plane which includes $a(x) \le t \le b(x)$, $x_0 \le x \le x_1$, where the functions $a(x)$ and $b(x)$ are continuous and have continuous derivatives for $x_0 \le x \le x_1$. Defining*

$$F(x) \equiv \int_{a(x)}^{b(x)} f(x, t)\, dt,$$

then

$$\frac{dF(x)}{dx} = \left(\frac{\partial F}{\partial b}\right)\frac{db}{dx} + \left(\frac{\partial F}{\partial a}\right)\frac{da}{dx} + \int_{a(x)}^{b(x)} \frac{\partial}{\partial x} f(x, t)\, dt$$

$$= f(x, b(x))\, b'(x) - f(x, a(x))\, a'(x) + \int_{a(x)}^{b(x)} \frac{\partial}{\partial x} f(x, t)\, dt.$$

$$(1.69)$$

for $x_0 \le x \le x_1$. This is a generalized version of the Fundamental Theorem of Calculus.

In the next examples we show how we can use this theorem to bypass integration by parts.

Example 1.12. *Use differentiation under the integral sign to evaluate $\int xe^x\, dx$. First, consider the integral*

$$I(x, a) = \int e^{ax}\, dx = \frac{e^{ax}}{a}.$$

Then,

$$\frac{\partial I(x, a)}{\partial a} = \int xe^{ax}\, dx.$$

So,

$$
\begin{aligned}
\int x e^{ax}\, dx &= \frac{\partial I(x,a)}{\partial a} \\
&= \frac{\partial}{\partial a}\left(\int e^{ax}\, dx\right) \\
&= \frac{\partial}{\partial a}\left(\frac{e^{ax}}{a}\right) \\
&= \left(\frac{x}{a} - \frac{1}{a^2}\right) e^{ax}.
\end{aligned}
\tag{1.70}
$$

Evaluating this result at $a = 1$, we have

$$
\int x e^x\, dx = (x-1)e^x.
$$

The reader can verify this result by employing the previous methods or by just differentiating the result.

Example 1.13. *We will do the integral $\int_0^\pi x^2 \cos x\, dx$ once more. First, consider the integral*

$$
\begin{aligned}
I(a) &\equiv \int_0^\pi \cos ax\, dx \\
&= \left.\frac{\sin ax}{a}\right|_0^\pi \\
&= \frac{\sin a\pi}{a}.
\end{aligned}
\tag{1.71}
$$

Differentiating the integral $I(a)$ twice with respect to a gives

$$
\frac{d^2 I(a)}{da^2} = -\int_0^\pi x^2 \cos ax\, dx.
\tag{1.72}
$$

Evaluation of this result at $a = 1$ leads to the desired result. Namely,

$$
\begin{aligned}
\int_0^\pi x^2 \cos x\, dx &= \left.-\frac{d^2 I(a)}{da^2}\right|_{a=1} \\
&= \left.-\frac{d^2}{da^2}\left(\frac{\sin a\pi}{a}\right)\right|_{a=1} \\
&= \left.-\frac{d}{da}\left(\frac{a\pi \cos a\pi - \sin a\pi}{a^2}\right)\right|_{a=1} \\
&= \left.-\left(\frac{a^2\pi^2 \sin a\pi + 2a\pi \cos a\pi - 2\sin a\pi}{a^3}\right)\right|_{a=1} \\
&= -2\pi.
\end{aligned}
\tag{1.73}
$$

Trigonometric Integrals

Other types of integrals that you will see often are trigonometric integrals. In particular, integrals involving powers of sines and cosines. For odd powers, a simple substitution will turn the integrals into simple powers.

Example 1.14. *For example, consider*

$$\int \cos^3 x \, dx.$$

This can be rewritten as

$$\int \cos^3 x \, dx = \int \cos^2 x \cos x \, dx.$$

Integration of odd powers of sine and cosine.

Let $u = \sin x$. Then, $du = \cos x \, dx$. Since $\cos^2 x = 1 - \sin^2 x$, we have

$$
\begin{aligned}
\int \cos^3 x \, dx &= \int \cos^2 x \cos x \, dx \\
&= \int (1 - u^2) \, du \\
&= u - \frac{1}{3} u^3 + C \\
&= \sin x - \frac{1}{3} \sin^3 x + C.
\end{aligned}
\tag{1.74}
$$

A quick check confirms the answer:

$$
\begin{aligned}
\frac{d}{dx} \left(\sin x - \frac{1}{3} \sin^3 x + C \right) &= \cos x - \sin^2 x \cos x \\
&= \cos x (1 - \sin^2 x) \\
&= \cos^3 x.
\end{aligned}
\tag{1.75}
$$

Even powers of sines and cosines are a little more complicated, but doable. In these cases we need the half angle formulae (1.24) and (1.25).

Integration of even powers of sine and cosine.

Example 1.15. *As an example, we will compute*

$$\int_0^{2\pi} \cos^2 x \, dx.$$

Substituting the half angle formula for $\cos^2 x$, we have

$$
\begin{aligned}
\int_0^{2\pi} \cos^2 x \, dx &= \frac{1}{2} \int_0^{2\pi} (1 + \cos 2x) \, dx \\
&= \frac{1}{2} \left(x - \frac{1}{2} \sin 2x \right) \Big|_0^{2\pi} \\
&= \pi.
\end{aligned}
\tag{1.76}
$$

Recall that RMS averages refer to the root mean square averages. This is computed by first computing the average, or mean, of the square of some quantity. Then one takes the square root. Typical examples are RMS voltage, RMS current, and the average energy in an electromagnetic wave. AC currents oscillate so fast that the measured value is the RMS voltage.

We note that this result appears often in physics. When looking at root mean square averages of sinusoidal waves, one needs the average of the square of sines and cosines. Recall that the average of a function on interval $[a, b]$ is given as

$$f_{\text{ave}} = \frac{1}{b - a} \int_a^b f(x) \, dx. \tag{1.77}$$

So, the average of $\cos^2 x$ over one period is

$$\frac{1}{2\pi} \int_0^{2\pi} \cos^2 x \, dx = \frac{1}{2}. \tag{1.78}$$

The root mean square is then found by taking the square root, $\frac{1}{\sqrt{2}}$.

Trigonometric Function Substitution

Another class of integrals typically studied in calculus are those involving the forms $\sqrt{1-x^2}$, $\sqrt{1+x^2}$, or $\sqrt{x^2-1}$. These can be simplified through the use of trigonometric substitutions. The idea is to combine the two terms under the radical into one term using trigonometric identities. We will consider some typical examples.

Example 1.16. *Evaluate $\int \sqrt{1-x^2}\,dx$.*
Since $1 - \sin^2\theta = \cos^2\theta$, we perform the sine substitution

$$x = \sin\theta, \quad dx = \cos\theta\,d\theta.$$

Then,

$$\begin{aligned}
\int \sqrt{1-x^2}\,dx &= \int \sqrt{1-\sin^2\theta}\,\cos\theta\,d\theta \\
&= \int \cos^2\theta\,d\theta.
\end{aligned} \tag{1.79}$$

Using the last example, we have

$$\int \sqrt{1-x^2}\,dx = \frac{1}{2}\left(\theta - \frac{1}{2}\sin 2\theta\right) + C.$$

However, we need to write the answer in terms of x. We do this by first using the double angle formula for $\sin 2\theta$ and $\cos\theta = \sqrt{1-x^2}$ to obtain

$$\int \sqrt{1-x^2}\,dx = \frac{1}{2}\left(\sin^{-1}x - x\sqrt{1-x^2}\right) + C.$$

Similar trigonometric substitutions result for integrands involving $\sqrt{1+x^2}$ and $\sqrt{x^2-1}$. The substitutions are summarized in Table 1.6. The simplification of the given form is then obtained using trigonometric identities. This can also be accomplished by referring to the right triangles shown in Figure 1.5.

Form	Substitution	Differential
$\sqrt{a^2-x^2}$	$x = a\sin\theta$	$dx = a\cos\theta\,d\theta$
$\sqrt{a^2+x^2}$	$x = a\tan\theta$	$dx = a\sec^2\theta\,d\theta$
$\sqrt{x^2-a^2}$	$x = a\sec\theta$	$dx = a\sec\theta\tan\theta\,d\theta$

Table 1.6: Standard Trigonometric Substitutions

In any of these computations, careful attention must be paid to simplifying the radical. This is because

$$\sqrt{x^2} = |x|.$$

For example, $\sqrt{(-5)^2} = \sqrt{25} = 5$. For $x = \sin\theta$, one typically specifies the domain $-\pi/2 \le \theta \le \pi/2$. In this domain we have $|\cos\theta| = \cos\theta$.

Example 1.17. *Evaluate $\int_0^2 \sqrt{x^2+4}\,dx$.*
Let $x = 2\tan\theta$. Then, $dx = 2\sec^2\theta\,d\theta$ and

$$\sqrt{x^2+4} = \sqrt{4\tan^2\theta + 4} = 2\sec\theta.$$

So, the integral becomes

$$\int_0^2 \sqrt{x^2+4}\,dx = 4\int_0^{\pi/4} \sec^3\theta\,d\theta.$$

Figure 1.5: Geometric relations used in trigonometric substitution.

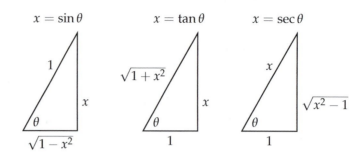

$$x = \sin\theta \qquad x = \tan\theta \qquad x = \sec\theta$$

One has to recall, or look up,

$$\int \sec^3\theta\, d\theta = \frac{1}{2}\left(\tan\theta\sec\theta + \ln|\sec\theta + \tan\theta|\right) + C.$$

This gives

$$
\begin{aligned}
\int_0^2 \sqrt{x^2 + 4}\, dx &= 2\left[\tan\theta\sec\theta + \ln|\sec\theta + \tan\theta|\right]_0^{\pi/4} \\
&= 2\left(\sqrt{2} + \ln|\sqrt{2} + 1| - (0 + \ln(1))\right) \\
&= 2(\sqrt{2} + \ln(\sqrt{2} + 1)).
\end{aligned}
\tag{1.80}
$$

Example 1.18. *Evaluate $\int \frac{dx}{\sqrt{x^2-1}}$, $x \geq 1$.*
In this case one needs the secant substitution. This yields

$$
\begin{aligned}
\int \frac{dx}{\sqrt{x^2 - 1}} &= \int \frac{\sec\theta\tan\theta\, d\theta}{\sqrt{\sec^2\theta - 1}} \\
&= \int \frac{\sec\theta\tan\theta\, d\theta}{\tan\theta} \\
&= \int \sec\theta\, d\theta \\
&= \ln(\sec\theta + \tan\theta) + C \\
&= \ln(x + \sqrt{x^2 - 1}) + C.
\end{aligned}
\tag{1.81}
$$

Example 1.19. *Evaluate $\int \frac{dx}{x\sqrt{x^2-1}}$, $x \geq 1$.*
Again we can use a secant substitution. This yields

$$
\begin{aligned}
\int \frac{dx}{x\sqrt{x^2 - 1}} &= \int \frac{\sec\theta\tan\theta\, d\theta}{\sec\theta\sqrt{\sec^2\theta - 1}} \\
&= \int \frac{\sec\theta\tan\theta}{\sec\theta\tan\theta}\, d\theta \\
&= \int d\theta = \theta + C = \sec^{-1}x + C.
\end{aligned}
\tag{1.82}
$$

Hyperbolic Function Substitution

Even though trigonometric substitution plays a role in the calculus program, students often see hyperbolic function substitution used in physics courses. The reason might be because hyperbolic function substitution is sometimes simpler. The idea is the same as for trigonometric substitution. We use an identity to simplify the radical.

Example 1.20. *Evaluate* $\int_0^2 \sqrt{x^2 + 4}\, dx$ *using the substitution* $x = 2 \sinh u$.

Because $x = 2 \sinh u$, *we have* $dx = 2 \cosh u\, du$. *Also, we can use the identity* $\cosh^2 u - \sinh^2 u = 1$ *to rewrite*

$$\sqrt{x^2 + 4} = \sqrt{4 \sinh^2 u + 4} = 2 \cosh u.$$

The integral can be now be evaluated using these substitutions and some hyperbolic function identities,

$$
\begin{aligned}
\int_0^2 \sqrt{x^2 + 4}\, dx &= 4 \int_0^{\sinh^{-1} 1} \cosh^2 u\, du \\
&= 2 \int_0^{\sinh^{-1} 1} (1 + \cosh 2u)\, du \\
&= 2 \left[u + \frac{1}{2} \sinh 2u \right]_0^{\sinh^{-1} 1} \\
&= 2 \left[u + \sinh u \cosh u \right]_0^{\sinh^{-1} 1} \\
&= 2 \left(\sinh^{-1} 1 + \sqrt{2} \right).
\end{aligned}
\tag{1.83}
$$

In Example 1.17 we used a trigonometric substitution and found

$$\int_0^2 \sqrt{x^2 + 4} = 2(\sqrt{2} + \ln(\sqrt{2} + 1)).$$

This is the same result since $\sinh^{-1} 1 = \ln(1 + \sqrt{2})$.

Example 1.21. *Evaluate* $\int \frac{dx}{\sqrt{x^2 - 1}}$ *for* $x \geq 1$ *using hyperbolic function substitution.*

This integral was evaluated in Example 1.19 using the trigonometric substitution $x = \sec \theta$ *and the resulting integral of* $\sec \theta$ *had to be recalled. Here we will use the substitution*

$$x = \cosh u, \quad dx = \sinh u\, du, \quad \sqrt{x^2 - 1} = \sqrt{\cosh^2 u - 1} = \sinh u.$$

Then,

$$
\begin{aligned}
\int \frac{dx}{\sqrt{x^2 - 1}} &= \int \frac{\sinh u\, du}{\sinh u} \\
&= \int du = u + C \\
&= \cosh^{-1} x + C \\
&= \frac{1}{2} \ln(x + \sqrt{x^2 - 1}) + C, \quad x \geq 1.
\end{aligned}
\tag{1.84}
$$

This is the same result as we had obtained previously, but this derivation was a little cleaner.

Also, we can extend this result to values $x \leq -1$ *by letting* $x = -\cosh u$. *This gives*

$$\int \frac{dx}{\sqrt{x^2 - 1}} = \frac{1}{2} \ln(x + \sqrt{x^2 - 1}) + C, \quad x \leq -1.$$

Combining these results, we have shown

$$\int \frac{dx}{\sqrt{x^2 - 1}} = \frac{1}{2} \ln(|x| + \sqrt{x^2 - 1}) + C, \quad x^2 \geq 1.$$

We have seen in the previous example that the use of hyperbolic function substitution allows us to bypass integrating the secant function in Example 1.19 when using trigonometric substitutions. In fact, we can use hyperbolic substitutions to evaluate integrals of powers of secants. Comparing Examples 1.19 and 1.21, we consider the transformation $\sec\theta = \cosh u$. The relation between differentials is found by differentiation, giving

$$\sec\theta\tan\theta\,d\theta = \sinh u\,du.$$

Since

$$\tanh^2\theta = \sec^2\theta - 1,$$

we have $\tan\theta = \sinh u$, and therefore

$$d\theta = \frac{du}{\cosh u}.$$

In the next example we show how useful this transformation is.

Evaluation of $\int\sec\theta\,d\theta$.

Example 1.22. *Evaluate $\int\sec\theta\,d\theta$ using hyperbolic function substitution.*
From the discussion in the last paragraph, we have

$$
\begin{aligned}
\int\sec\theta\,d\theta &= \int du \\
&= u + C \\
&= \cosh^{-1}(\sec\theta) + C
\end{aligned}
\tag{1.85}
$$

We can express this result in the usual form using the logarithmic form of the inverse hyperbolic cosine,

$$\cosh^{-1}x = \ln(x + \sqrt{x^2 - 1}).$$

The result is

$$\int\sec\theta\,d\theta = \ln(\sec\theta + \tan\theta).$$

This example was fairly simple using the transformation $\sec\theta = \cosh u$. Another common integral that arises often is integrations of $\sec^3\theta$. In a typical calculus class, this integral is evaluated using integration by parts. However, that leads to a tricky manipulation that is a bit scary the first time it is encountered (and probably upon several more encounters.) In the next example, we will show how hyperbolic function substitution is simpler.

Evaluation of $\int\sec^3\theta\,d\theta$.

Example 1.23. *Evaluate $\int\sec^3\theta\,d\theta$ using hyperbolic function substitution.*
First, we consider the transformation $\sec\theta = \cosh u$ with $d\theta = \frac{du}{\cosh u}$. Then,

$$\int\sec^3\theta\,d\theta = \int\frac{du}{\cosh u}.$$

This integral was done in Example 1.7, leading to

$$\int\sec^3\theta\,d\theta = 2\tan^{-1}e^u + C.$$

While correct, this is not the form usually encountered. Instead, we make the slightly different transformation $\tan\theta = \sinh u$. *Since* $\sec^2\theta = 1 + \tan^2\theta$, *we find* $\sec\theta = \cosh u$. *As before, we find*

$$d\theta = \frac{du}{\cosh u}.$$

Using this transformation and several identities, the integral becomes

$$
\begin{aligned}
\int \sec^3\theta\, d\theta &= \int \cosh^2 u\, du \\
&= \frac{1}{2}\int (1 + \cosh 2u)\, du \\
&= \frac{1}{2}\left(u + \frac{1}{2}\sinh 2u\right) \\
&= \frac{1}{2}(u + \sinh u \cosh u) \\
&= \frac{1}{2}\left(\cosh^{-1}(\sec\theta) + \tan\theta\sec\theta\right) \\
&= \frac{1}{2}(\sec\theta\tan\theta + \ln(\sec\theta + \tan\theta)). \quad (1.86)
\end{aligned}
$$

There are many other integration methods, some of which we will visit in other parts of the book, including partial fraction decomposition and numerical integration.

1.1.6 Geometric Series

INFINITE SERIES OCCUR OFTEN in mathematics and physics. Two series which occur often are the geometric series and the binomial series. We will discuss these next.

Geometric series are fairly common and will be used throughout the book. You should learn to recognize them and work with them.

A geometric series is of the form

$$\sum_{n=0}^{\infty} ar^n = a + ar + ar^2 + \ldots + ar^n + \ldots. \quad (1.87)$$

Here a is the first term and r is called the ratio. It is called the ratio because the ratio of two consecutive terms in the sum is r.

Example 1.24. *For example,*

$$1 + \frac{1}{2} + \frac{1}{4} + \frac{1}{8} + \ldots$$

is an example of a geometric series. We can write this using summation notation,

$$1 + \frac{1}{2} + \frac{1}{4} + \frac{1}{8} + \ldots = \sum_{n=0}^{\infty} 1\left(\frac{1}{2}\right)^n.$$

Thus, $a = 1$ is the first term and $r = \frac{1}{2}$ is the common ratio of successive terms. Next, we seek the sum of this infinite series, if it exists.

The sum of a geometric series, when it exists, can easily be determined. We consider the nth partial sum:

$$s_n = a + ar + \ldots + ar^{n-2} + ar^{n-1}. \tag{1.88}$$

Now, multiply this equation by r.

$$rs_n = ar + ar^2 + \ldots + ar^{n-1} + ar^n. \tag{1.89}$$

Subtracting these two equations, while noting the many cancelations, we have

$$
\begin{aligned}
(1-r)s_n &= (a + ar + \ldots + ar^{n-2} + ar^{n-1}) \\
&\quad -(ar + ar^2 + \ldots + ar^{n-1} + ar^n) \\
&= a - ar^n \\
&= a(1 - r^n). \tag{1.90}
\end{aligned}
$$

Thus, the nth partial sums can be written in the compact form

$$s_n = \frac{a(1 - r^n)}{1 - r}. \tag{1.91}$$

The sum, if it exists, is given by $S = \lim_{n \to \infty} s_n$. Letting n get large in the partial sum (1.91), we need only evaluate $\lim_{n \to \infty} r^n$. From the special limits in the Appendix we know that this limit is zero for $|r| < 1$. Thus, we have

---

Geometric Series

The sum of the geometric series exists for $|r| < 1$ and is given by

$$\sum_{n=0}^{\infty} ar^n = \frac{a}{1 - r}, \qquad |r| < 1. \tag{1.92}$$

The reader should verify that the geometric series diverges for all other values of r. Namely, consider what happens for the separate cases $|r| > 1$, $r = 1$ and $r = -1$.

Next, we present a few typical examples of geometric series.

Example 1.25. $\sum_{n=0}^{\infty} \frac{1}{2^n}$

In this case we have that $a = 1$ and $r = \frac{1}{2}$. Therefore, this infinite series converges and the sum is

$$S = \frac{1}{1 - \frac{1}{2}} = 2.$$

Example 1.26. $\sum_{k=2}^{\infty} \frac{4}{3^k}$

In this example we first note that the first term occurs for $k = 2$. It sometimes helps to write out the terms of the series,

$$\sum_{k=2}^{\infty} \frac{4}{3^k} = \frac{4}{3^2} + \frac{4}{3^3} + \frac{4}{3^4} + \frac{4}{3^5} + \ldots.$$

Looking at the series, we see that $a = \frac{4}{9}$ and $r = \frac{1}{3}$. Since $|r| < 1$, the geometric series converges. So, the sum of the series is given by

$$S = \frac{\frac{4}{9}}{1 - \frac{1}{3}} = \frac{2}{3}.$$

Example 1.27. $\sum_{n=1}^{\infty} \left(\frac{3}{2^n} - \frac{2}{5^n} \right)$

Finally, in this case we do not have a geometric series, but we do have the difference of two geometric series. Of course, we need to be careful whenever rearranging infinite series. In this case it is allowed [1]. Thus, we have

[1] A rearrangement of terms in an infinite series is allowed when the series is absolutely convergent. (See Appendix A.)

$$\sum_{n=1}^{\infty} \left(\frac{3}{2^n} - \frac{2}{5^n} \right) = \sum_{n=1}^{\infty} \frac{3}{2^n} - \sum_{n=1}^{\infty} \frac{2}{5^n}.$$

Now we can add both geometric series to obtain

$$\sum_{n=1}^{\infty} \left(\frac{3}{2^n} - \frac{2}{5^n} \right) = \frac{\frac{3}{2}}{1 - \frac{1}{2}} - \frac{\frac{2}{5}}{1 - \frac{1}{5}} = 3 - \frac{1}{2} = \frac{5}{2}.$$

Geometric series are important because they are easily recognized and summed. Other series which can be summed include special cases of Taylor series and *telescoping series*. Next, we show an example of a telescoping series.

Example 1.28. $\sum_{n=1}^{\infty} \frac{1}{n(n+1)}$ *The first few terms of this series are*

$$\sum_{n=1}^{\infty} \frac{1}{n(n+1)} = \frac{1}{2} + \frac{1}{6} + \frac{1}{12} + \frac{1}{20} + \dots.$$

It does not appear that we can sum this infinite series. However, if we used the partial fraction expansion

$$\frac{1}{n(n+1)} = \frac{1}{n} - \frac{1}{n+1},$$

then we find that the kth partial sum can be written as

$$
\begin{aligned}
s_k &= \sum_{n=1}^{k} \frac{1}{n(n+1)} \\
&= \sum_{n=1}^{k} \left(\frac{1}{n} - \frac{1}{n+1} \right) \\
&= \left(\frac{1}{1} - \frac{1}{2} \right) + \left(\frac{1}{2} - \frac{1}{3} \right) + \dots + \left(\frac{1}{k} - \frac{1}{k+1} \right). \quad (1.93)
\end{aligned}
$$

We see that there are many cancellations of neighboring terms, leading to the series collapsing (like a retractable telescope) to something manageable:

$$s_k = 1 - \frac{1}{k+1}.$$

Taking the limit as $k \to \infty$, we find $\sum_{n=1}^{\infty} \frac{1}{n(n+1)} = 1$.

Example 1.29. *The Partition Function*

A common occurrence of geometric series is a series of exponentials. An example of this occurs in statistical mechanics. Statistical mechanics is the branch of physics which explores the thermodynamic behavior of systems containing a large number of particles. An important tool is the partition function, Z. This function is the sum of terms, $e^{-\epsilon_n/kT}$, over all possible quantum states of the system. Here, ϵ_n is the energy of the nth state, T the temperature, and k is Boltzmann's constant. Given Z, one can compute macroscopic quantities, such as the average energy,

$$< E >= -\frac{\partial \ln Z}{\partial \beta},$$

where $\beta = 1/kT$.

For the case of the quantum harmonic oscillator, the energy states are given by $\epsilon_n = \left(n + \frac{1}{2}\right)\hbar\omega$, where ω is the angular frequency, $\hbar$ is Planck's constant divided by 2π, and $n = 0, 1, 2, \ldots$. The partition function is then computed as

$$
\begin{aligned}
Z &= \sum_{n=0}^{\infty} e^{-\beta\epsilon_n} \\
&= \sum_{n=0}^{\infty} e^{-\beta\left(n+\frac{1}{2}\right)\hbar\omega} \\
&= e^{-\beta\hbar\omega/2} \sum_{n=0}^{\infty} e^{-\beta n\hbar\omega}.
\end{aligned}
\tag{1.94}
$$

In some texts the energy is taken as $\epsilon_n = n\hbar\omega$, $n = 0, 1, 2, \ldots$, by choosing the zero point energy as $\epsilon_0 = 0$. Planck's constant is $h = 6.626068 \times 10^{-34}$ m^{2}kg/s.

The terms in the last sum are really powers of an exponential,

$$e^{-\beta n\hbar\omega} = \left(e^{-\beta\hbar\omega}\right)^n.$$

So,

$$Z = e^{-\beta\hbar\omega/2} \sum_{n=0}^{\infty} \left(e^{-\beta\hbar\omega}\right)^n.$$

This is a geometric series, which can be summed as long as $e^{-\beta\hbar\omega} < 1$. Thus,

$$Z = \frac{e^{-\beta\hbar\omega/2}}{1 - e^{-\beta\hbar\omega}}.$$

Multiplying the numerator and denominator by $e^{\beta\hbar\omega/2}$, we have

$$Z = \frac{1}{e^{\beta\hbar\omega/2} - e^{-\beta\hbar\omega/2}} = \left(2\sinh\beta\hbar\omega/2\right)^{-1}.$$

1.1.7   Power Series

Actually, what are now known as Taylor and Maclaurin series were known long before they were named. James Gregory (1638-1675) has been recognized for discovering Taylor series, which were later named after Brook Taylor (1685–1731). Similarly, Colin Maclaurin (1698–1746) did not actually discover Maclaurin series, but the name was adopted because of his particular use of series.

ANOTHER EXAMPLE OF AN INFINITE SERIES that the student has encountered in previous courses is the power series. Examples of such series are provided by Taylor and Maclaurin series.

A power series expansion about $x = a$ with coefficient sequence c_n is given by $\sum_{n=0}^{\infty} c_n(x-a)^n$. For now we will consider all constants to be real numbers with x in some subset of the set of real numbers.

Consider the following expansion about $x = 0$:

$$\sum_{n=0}^{\infty} x^n = 1 + x + x^2 + \dots. \qquad (1.95)$$

We would like to make sense of such expansions. For what values of x will this infinite series converge? Until now we did not pay much attention to which infinite series might converge. However, this particular series is already familiar to us. It is a geometric series. Note that each term is gotten from the previous one through multiplication by $r = x$. The first term is $a = 1$. So, from Equation (1.92), we have that the sum of the series is given by

$$\sum_{n=0}^{\infty} x^n = \frac{1}{1-x}, \qquad |x| < 1.$$

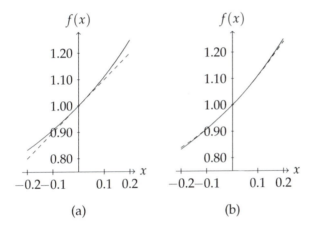

(a)   (b)

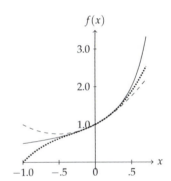

Figure 1.6: (a) Comparison of $\frac{1}{1-x}$ (solid) to $1 + x$ (dashed) for $x \in [-0.2, 0.2]$. (b) Comparison of $\frac{1}{1-x}$ (solid) to $1 + x + x^2$ (dashed) for $x \in [-0.2, 0.2]$.

In this case we see that the sum, when it exists, is a simple function. In fact, when x is small, we can use this infinite series to provide approximations to the function $(1-x)^{-1}$. If x is small enough, we can write

$$(1-x)^{-1} \approx 1 + x.$$

In Figure 1.6a we see that for small values of x, these functions do agree.

Of course, if we want better agreement, we select more terms. In Figure 1.6b we see what happens when we do so. The agreement is much better. But extending the interval, we see in Figure 1.7 that keeping only quadratic terms may not be good enough. Keeping the cubic terms gives better agreement over the interval.

Finally, in Figure 1.8 we show the sum of the first 21 terms over the entire interval $[-1, 1]$. Note that there are problems with approximations near the endpoints of the interval, $x = \pm 1$.

Such polynomial approximations are called Taylor polynomials. Thus, $T_3(x) = 1 + x + x^2 + x^3$ is the third-order Taylor polynomial approximation of $f(x) = \frac{1}{1-x}$.

With this example we have seen how useful a series representation might be for a given function. However, the series representation was a simple

Figure 1.7: Comparison of $\frac{1}{1-x}$ (solid) to $1 + x + x^2$ (dashed) and $1 + x + x^2 + x^3$ (dotted) for $x \in [-1.0, 0.7]$.

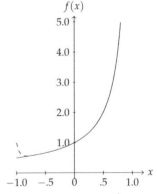

Figure 1.8: Comparison of $\frac{1}{1-x}$ (solid) to $\sum_{n=0}^{20} x^n$ for $x \in [-1, 1]$.

geometric series, which we already knew how to sum. Is there a way to begin with a function and then find its series representation? Once we have such a representation, will the series converge to the function with which we started? For what values of x will it converge? These questions can be answered by recalling the definitions of Taylor and Maclaurin series.

A Taylor series expansion of $f(x)$ about $x = a$ is the series

Taylor series expansion.

$$f(x) \sim \sum_{n=0}^{\infty} c_n (x - a)^n, \qquad (1.96)$$

where

$$c_n = \frac{f^{(n)}(a)}{n!}. \qquad (1.97)$$

Note that we use $\sim$ to indicate that we have yet to determine when the series may converge to the given function. A special class of series consists of those Taylor series for which the expansion is about $x = 0$. These are called Maclaurin series.

Maclaurin series expansion.

A Maclaurin series expansion of $f(x)$ is a Taylor series expansion of $f(x)$ about $x = 0$, or

$$f(x) \sim \sum_{n=0}^{\infty} c_n x^n, \qquad (1.98)$$

where

$$c_n = \frac{f^{(n)}(0)}{n!}. \qquad (1.99)$$

Example 1.30. *Expand $f(x) = e^x$ about $x = 0$.*

We begin by creating a table. In order to compute the expansion coefficients, c_n, we will need to perform repeated differentiations of $f(x)$. So, we provide a table for these derivatives. Then, we only need to evaluate the second column at $x = 0$ and divide by $n!$.

n	$f^{(n)}(x)$	$f^{(n)}(0)$	c_n
0	e^x	$e^0 = 1$	$\frac{1}{0!} = 1$
1	e^x	$e^0 = 1$	$\frac{1}{1!} = 1$
2	e^x	$e^0 = 1$	$\frac{1}{2!}$
3	e^x	$e^0 = 1$	$\frac{1}{3!}$

Next, we look at the last column and try to determine a pattern so that we can write the general term of the series. If there is only a need to get a polynomial approximation, then the first few terms may be sufficient. In this case, the pattern is obvious: $c_n = \frac{1}{n!}$. So,

$$e^x \sim \sum_{n=0}^{\infty} \frac{x^n}{n!}.$$

Example 1.31. *Expand $f(x) = e^x$ about $x = 1$.*

Here we seek an expansion of the form $e^x \sim \sum_{n=0}^{\infty} c_n(x-1)^n$. We could create a table like the previous example. In fact, the last column would have values of the form $\frac{e}{n!}$. (You should confirm this.) However, we will make use of the Maclaurin series expansion for e^x and get the result quicker. Note that $e^x = e^{x-1+1} = ee^{x-1}$. Now, apply the known expansion for e^x:

$$e^x \sim e\left(1 + (x-1) + \frac{(x-1)^2}{2} + \frac{(x-1)^3}{3!} + \ldots\right) = \sum_{n=0}^{\infty} \frac{e(x-1)^n}{n!}.$$

Example 1.32. *Expand $f(x) = \frac{1}{1-x}$ about $x = 0$.*

This is the example with which we started our discussion. We can set up a table in order to find the Maclaurin series coefficients. We see from the last column of the table that we get back the geometric series (1.95).

n	$f^{(n)}(x)$	$f^{(n)}(0)$	c_n
0	$\frac{1}{1-x}$	1	$\frac{1}{0!} = 1$
1	$\frac{1}{(1-x)^2}$	1	$\frac{1}{1!} = 1$
2	$\frac{2(1)}{(1-x)^3}$	$2(1)$	$\frac{2!}{2!} = 1$
3	$\frac{3(2)(1)}{(1-x)^4}$	$3(2)(1)$	$\frac{3!}{3!} = 1$

So, we have found

$$\frac{1}{1-x} \sim \sum_{n=0}^{\infty} x^n.$$

We can replace $\sim$ by equality if we can determine the range of x-values for which the resulting infinite series converges. We will investigate such convergence shortly.

Series expansions for many elementary functions arise in a variety of applications. Some common expansions are provided in Table 1.7.

We still need to determine the values of x for which a given power series converges. The first five of the above expansions converge for all reals, but the others only converge for $|x| < 1$.

We consider the convergence of $\sum_{n=0}^{\infty} c_n(x-a)^n$. For $x = a$, the series obviously converges. Will it converge for other points? One can prove

Theorem 1.2. *If $\sum_{n=0}^{\infty} c_n(b-a)^n$ converges for $b \neq a$, then $\sum_{n=0}^{\infty} c_n(x-a)^n$ converges absolutely for all x satisfying $|x-a| < |b-a|$.*

This leads to three possibilities:

1. $\sum_{n=0}^{\infty} c_n(x-a)^n$ may only converge at $x = a$.

2. $\sum_{n=0}^{\infty} c_n(x-a)^n$ may converge for all real numbers.

Table 1.7: Common Mclaurin Series Expansions

Series Expansions You Should Know		
e^x	$= 1 + x + \dfrac{x^2}{2} + \dfrac{x^3}{3!} + \dfrac{x^4}{4!} + \cdots$	$= \displaystyle\sum_{n=0}^{\infty} \dfrac{x^n}{n!}$
$\cos x$	$= 1 - \dfrac{x^2}{2} + \dfrac{x^4}{4!} - \cdots$	$= \displaystyle\sum_{n=0}^{\infty} (-1)^n \dfrac{x^{2n}}{(2n)!}$
$\sin x$	$= x - \dfrac{x^3}{3!} + \dfrac{x^5}{5!} - \cdots$	$= \displaystyle\sum_{n=0}^{\infty} (-1)^n \dfrac{x^{2n+1}}{(2n+1)!}$
$\cosh x$	$= 1 + \dfrac{x^2}{2} + \dfrac{x^4}{4!} + \cdots$	$= \displaystyle\sum_{n=0}^{\infty} \dfrac{x^{2n}}{(2n)!}$
$\sinh x$	$= x + \dfrac{x^3}{3!} + \dfrac{x^5}{5!} + \cdots$	$= \displaystyle\sum_{n=0}^{\infty} \dfrac{x^{2n+1}}{(2n+1)!}$
$\dfrac{1}{1-x}$	$= 1 + x + x^2 + x^3 + \cdots$	$= \displaystyle\sum_{n=0}^{\infty} x^n$
$\dfrac{1}{1+x}$	$= 1 - x + x^2 - x^3 + \cdots$	$= \displaystyle\sum_{n=0}^{\infty} (-x)^n$
$\tan^{-1} x$	$= x - \dfrac{x^3}{3} + \dfrac{x^5}{5} - \dfrac{x^7}{7} + \cdots$	$= \displaystyle\sum_{n=0}^{\infty} (-1)^n \dfrac{x^{2n+1}}{2n+1}$
$\ln(1+x)$	$= x - \dfrac{x^2}{2} + \dfrac{x^3}{3} - \cdots$	$= \displaystyle\sum_{n=1}^{\infty} (-1)^{n+1} \dfrac{x^n}{n}$

3. $\sum_{n=0}^{\infty} c_n (x-a)^n$ converges for $|x-a| < R$ and diverges for $|x-a| > R$.

Interval and radius of convergence.

The number R is called the radius of convergence of the power series and $(a - R, a + R)$ is called the interval of convergence. Convergence at the endpoints of this interval must be tested for each power series.

In order to determine the interval of convergence, one needs only to note that when a power series converges, it does so absolutely. So, we need only test the convergence of $\sum_{n=0}^{\infty} |c_n(x-a)^n| = \sum_{n=0}^{\infty} |c_n||x-a|^n$. This is easily done using either the ratio test or the nth root test. We first identify the nonnegative terms $a_n = |c_n||x-a|^n$, using the notation from (Appendix A Section A.4). Then we apply one of the convergence tests.

For example, the nth Root Test gives the convergence condition for $a_n = |c_n||x-a|^n$,

$$\rho = \lim_{n \to \infty} \sqrt[n]{a_n} = \lim_{n \to \infty} \sqrt[n]{|c_n|}|x-a| < 1.$$

Because $|x - a|$ is independent of n, we can factor it out of the limit and divide the value of the limit to obtain

$$|x - a| < \left(\lim_{n \to \infty} \sqrt[n]{|c_n|} \right)^{-1} \equiv R.$$

Thus, we have found the radius of convergence, R.

Similarly, we can apply the Ratio Test:

$$\rho = \lim_{n \to \infty} \frac{a_{n+1}}{a_n} = \lim_{n \to \infty} \frac{|c_{n+1}|}{|c_n|}|x-a| < 1.$$

Again, we rewrite this result to determine the radius of convergence:

$$|x - a| < \left(\lim_{n\to\infty} \frac{|c_{n+1}|}{|c_n|} \right)^{-1} \equiv R.$$

Example 1.33. *Find the radius of convergence of the series $e^x = \sum_{n=0}^{\infty} \frac{x^n}{n!}$.*

Because there is a factorial, we will use the Ratio Test.

$$\rho = \lim_{n\to\infty} \frac{|n!|}{|(n+1)!|} |x| = \lim_{n\to\infty} \frac{1}{n+1} |x| = 0.$$

Because $\rho = 0$, it is independent of $|x|$ and thus the series converges for all x. We also can say that the radius of convergence is infinite.

Example 1.34. *Find the radius of convergence of the series $\frac{1}{1-x} = \sum_{n=0}^{\infty} x^n$.*

In this example we will use the nth Root Test.

$$\rho = \lim_{n\to\infty} \sqrt[n]{1} |x| = |x| < 1.$$

Thus, we find that we have absolute convergence for $|x| < 1$. Setting $x = 1$ or $x = -1$, we find that the resulting series do not converge. So, the endpoints are not included in the complete interval of convergence.

In this example we could have also used the Ratio Test. Thus,

$$\rho = \lim_{n\to\infty} \frac{1}{1} |x| = |x| < 1.$$

We have obtained the same result as when we used the nth Root Test.

Example 1.35. *Find the radius of convergence of the series $\sum_{n=1}^{\infty} \frac{3^n (x-2)^n}{n}$.*

In this example, we have an expansion about $x = 2$. Using the nth Root Test we find that

$$\rho = \lim_{n\to\infty} \sqrt[n]{\frac{3^n}{n}} |x - 2| = 3|x - 2| < 1.$$

Solving for $|x - 2|$ in this inequality, we find $|x - 2| < \frac{1}{3}$. Thus, the radius of convergence is $R = \frac{1}{3}$ and the interval of convergence is $\left(2 - \frac{1}{3}, 2 + \frac{1}{3} \right) = \left(\frac{5}{3}, \frac{7}{3} \right)$.

As for the endpoints, we first test the point $x = \frac{7}{3}$. The resulting series is $\sum_{n=1}^{\infty} \frac{3^n (\frac{1}{3})^n}{n} = \sum_{n=1}^{\infty} \frac{1}{n}$. This is the harmonic series, and thus it does not converge. Inserting $x = \frac{5}{3}$, we get the alternating harmonic series. This series does converge. So, we have convergence on $[\frac{5}{3}, \frac{7}{3})$. However, it is only conditionally convergent at the left endpoint, $x = \frac{5}{3}$.

Example 1.36. *Find an expansion of $f(x) = \frac{1}{x+2}$ about $x = 1$.*

Instead of explicitly computing the Taylor series expansion for this function, we can make use of an already known function. We first write $f(x)$ as a function of $x - 1$, as we are expanding about $x = 1$; that is, we are seeking a series whose terms are powers of $x - 1$.

This expansion is easily done by noting that $\frac{1}{x+2} = \frac{1}{(x-1)+3}$. Factoring out a 3, we can rewrite this expression as a sum of a geometric series. Namely, we use the expansion for

$$\begin{aligned} g(z) &= \frac{1}{1+z} \\ &= 1 - z + z^2 - z^3 + \dots, \end{aligned} \quad (1.100)$$

and then we rewrite $f(x)$ as

$$
\begin{aligned}
f(x) &= \frac{1}{x+2} \\
&= \frac{1}{(x-1)+3} \\
&= \frac{1}{3[1+\frac{1}{3}(x-1)]} \\
&= \frac{1}{3}\frac{1}{1+\frac{1}{3}(x-1)}.
\end{aligned}
\tag{1.101}
$$

Note that $f(x) = \frac{1}{3}g(\frac{1}{3}(x-1))$ for $g(z) = \frac{1}{1+z}$. So, the expansion becomes

$$
f(x) = \frac{1}{3}\left[1 - \frac{1}{3}(x-1) + \left(\frac{1}{3}(x-1)\right)^2 - \left(\frac{1}{3}(x-1)\right)^3 + \ldots\right].
$$

This can further be simplified as

$$
f(x) = \frac{1}{3} - \frac{1}{9}(x-1) + \frac{1}{27}(x-1)^2 - \ldots.
$$

Convergence is easily established. The expansion for $g(z)$ converges for $|z| < 1$. So, the expansion for $f(x)$ converges for $|-\frac{1}{3}(x-1)| < 1$. This implies that $|x-1| < 3$. Putting this inequality in interval notation, we have that the power series converges absolutely for $x \in (-2,4)$. Inserting the endpoints, one can show that the series diverges for both $x = -2$ and $x = 4$. You should verify this!

Example 1.37. *Prove Euler's Formula: $e^{i\theta} = \cos\theta + i\sin\theta$.*

As a final application, we can derive Euler's Formula,

Euler's Formula, $e^{i\theta} = \cos\theta + i\sin\theta$, is an important formula and is used throughout the text.

$$
e^{i\theta} = \cos\theta + i\sin\theta,
$$

where $i = \sqrt{-1}$. We naively use the expansion for e^x with $x = i\theta$. This leads us to

$$
e^{i\theta} = 1 + i\theta + \frac{(i\theta)^2}{2!} + \frac{(i\theta)^3}{3!} + \frac{(i\theta)^4}{4!} + \ldots.
$$

Next we note that each term has a power of i. The sequence of powers of i is given as $\{1, i, -1, -i, 1, i, -1, -i, 1, i, -1, -i, \ldots\}$. See the pattern? We conclude that

$$
i^n = i^r, \text{ where } r = \text{ remainder after dividing } n \text{ by 4.}
$$

This gives

$$
e^{i\theta} = \left(1 - \frac{\theta^2}{2!} + \frac{\theta^4}{4!} - \ldots\right) + i\left(\theta - \frac{\theta^3}{3!} + \frac{\theta^5}{5!} - \ldots\right).
$$

We recognize the expansions in the parentheses as those for the cosine and sine functions. Thus, we end with Euler's Formula.

We further derive relations from this result, that will be important for our next studies. From Euler's Formula we have that for integer n,

$$
e^{in\theta} = \cos(n\theta) + i\sin(n\theta).
$$

We also have

$$e^{in\theta} = \left(e^{i\theta}\right)^n = (\cos\theta + i\sin\theta)^n.$$

Equating these two expressions, we are led to de Moivre's Formula, named after Abraham de Moivre (1667–1754),

$$(\cos\theta + i\sin\theta)^n = \cos(n\theta) + i\sin(n\theta). \qquad (1.102)$$

This formula is useful for deriving identities relating powers of sines or cosines to simple functions. For example, if we take $n = 2$ in Equation (1.102), we find

$$\cos 2\theta + i\sin 2\theta = (\cos\theta + i\sin\theta)^2 = \cos^2\theta - \sin^2\theta + 2i\sin\theta\cos\theta.$$

Looking at the real and imaginary parts of this result leads to the well-known double angle identities

$$\cos 2\theta = \cos^2\theta - \sin^2\theta, \quad \sin 2\theta = 2\sin\theta\cos\theta.$$

Replacing $\cos^2\theta = 1 - \sin^2\theta$ or $\sin^2\theta = 1 - \cos^2\theta$ leads to the half angle formulae:

$$\cos^2\theta = \frac{1}{2}(1 + \cos 2\theta), \quad \sin^2\theta = \frac{1}{2}(1 - \cos 2\theta).$$

We can also use Euler's Formula to write sines and cosines in terms of complex exponentials. We first note that due to the fact that the cosine is an even function and the sine is an odd function, we have

$$e^{-i\theta} = \cos\theta - i\sin\theta.$$

Combining this with Euler's Formula, we have that

$$\cos\theta = \frac{e^{i\theta} + e^{-i\theta}}{2}, \quad \sin\theta = \frac{e^{i\theta} - e^{-i\theta}}{2i}.$$

We finally note that there is a simple relationship between hyperbolic functions and trigonometric functions. Recall that

$$\cosh x = \frac{e^x + e^{-x}}{2}.$$

If we let $x = i\theta$, then we have that $\cosh(i\theta) = \cos\theta$ and $\cos(ix) = \cosh x$. Similarly, we can show that $\sinh(i\theta) = i\sin\theta$ and $\sin(ix) = -i\sinh x$.

1.1.8 Binomial Expansion

ANOTHER SERIES EXPANSION WHICH OCCURS often in examples and applications is the binomial expansion. This is simply the expansion of the expression $(a + b)^p$ in powers of a and b. We will investigate this expansion first for nonnegative integer powers p and then derive the expansion

Here we see elegant proofs of well-known trigonometric identities:

$$\cos 2\theta = \cos^2\theta - \sin^2\theta, \quad (1.103)$$
$$\sin 2\theta = 2\sin\theta\cos\theta,$$
$$\cos^2\theta = \frac{1}{2}(1 + \cos 2\theta),$$
$$\sin^2\theta = \frac{1}{2}(1 - \cos 2\theta).$$

Trigonometric functions can be written in terms of complex exponentials:

$$\cos\theta = \frac{e^{i\theta} + e^{-i\theta}}{2},$$
$$\sin\theta = \frac{e^{i\theta} - e^{-i\theta}}{2i}.$$

Hyperbolic functions and trigonometric functions are intimately related:

$$\cos(ix) = \cosh x,$$
$$\sin(ix) = -i\sinh x.$$

The binomial expansion is a special series expansion used to approximate expressions of the form $(a + b)^p$ for $b \ll a$, or $(1 + x)^p$ for $|x| \ll 1$.

for other values of p. While the binomial expansion can be obtained using Taylor series, we will provide a more intuitive derivation to show that

$$(a+b)^n = \sum_{r=0}^{n} C_r^n a^{n-r} b^r, \qquad (1.104)$$

where the C_r^n are called the *binomial coefficients*.

Let's list some of the common expansions for nonnegative integer powers:

$$
\begin{aligned}
(a+b)^0 &= 1, \\
(a+b)^1 &= a+b, \\
(a+b)^2 &= a^2 + 2ab + b^2, \\
(a+b)^3 &= a^3 + 3a^2 b + 3ab^2 + b^3, \\
(a+b)^4 &= a^4 + 4a^3 b + 6a^2 b^2 + 4ab^3 + b^4, \\
&\quad \cdots
\end{aligned}
\qquad (1.105)
$$

We now look at the patterns of the terms in the expansions. First, we note that each term consists of a product of a power of a and a power of b. The powers of a are decreasing from n to 0 in the expansion of $(a+b)^n$. Similarly, the powers of b increase from 0 to n. The sums of the exponents in each term is n. So, we can write the $(k+1)$st term in the expansion as $a^{n-k} b^k$. For example, in the expansion of $(a+b)^{51}$, the 6th term is $a^{51-5} b^5 = a^{46} b^5$. However, we do not yet know the numerical coefficients in the expansion.

Let's list the coefficients for the above expansions:

$$
\begin{array}{llccccccccc}
n = 0: & & & & & 1 & & & & \\
n = 1: & & & & 1 & & 1 & & & \\
n = 2: & & & 1 & & 2 & & 1 & & \\
n = 3: & & 1 & & 3 & & 3 & & 1 & \\
n = 4: & 1 & & 4 & & 6 & & 4 & & 1
\end{array}
\qquad (1.106)
$$

This pattern is the famous Pascal's triangle.[2] There are many interesting features of this triangle. But we will first ask how each row can be generated.

We see that each row begins and ends with a one. The second term and next to last term have a coefficient of n. Next we note that consecutive pairs in each row can be added to obtain entries in the next row. For example, we have for rows $n = 2$ and $n = 3$ that $1 + 2 = 3$ and $2 + 1 = 3$:

$$
\begin{array}{lccccccc}
n = 2: & & 1 & & & 2 & & & 1 \\
& & & \searrow & \swarrow & & \searrow & \swarrow & \\
n = 3: & 1 & & 3 & & & 3 & & 1
\end{array}
\qquad (1.107)
$$

With this in mind, we can generate the next several rows of our triangle:

$$
\begin{array}{llccccccccc}
n = 3: & & & & 1 & & 3 & & 3 & & 1 & \\
n = 4: & & & 1 & & 4 & & 6 & & 4 & & 1 \\
n = 5: & & 1 & & 5 & & 10 & & 10 & & 5 & & 1 \\
n = 6: & 1 & & 6 & & 15 & & 20 & & 15 & & 6 & & 1
\end{array}
\qquad (1.108)
$$

[2] Pascal's triangle is named after Blaise Pascal (1623–1662). While such configurations of numbers were known earlier in history, Pascal published them and applied them to probability theory.

Pascal's triangle has many unusual properties and a variety of uses:

- Horizontal rows add to powers of 2.

- The horizontal rows are powers of 11 (1, 11, 121, 1331, etc.).

- Adding any two successive numbers in the diagonal 1-3-6-10-15-21-28... results in a perfect square.

- When the first number to the right of the 1 in any row is a prime number, all numbers in that row are divisible by that prime number. The reader can readily check this for the $n = 5$ and $n = 7$ rows.

- Sums along certain diagonals lead to the Fibonacci sequence. These diagonals are parallel to the line connecting the first 1 for $n = 3$ row and the 2 in the $n = 2$ row.

So, we use the numbers in row $n = 4$ to generate entries in row $n = 5$: $1 + 4 = 5, 4 + 6 = 10$. We then use row $n = 5$ to get row $n = 6$, etc.

Of course, it would take a while to compute each row up to the desired n. Fortunately, there is a simple expression for computing a specific coefficient. Consider the kth term in the expansion of $(a + b)^n$. Let $r = k - 1$, for $k = 1, \ldots, n + 1$. Then this term is of the form $C_r^n a^{n-r} b^r$. We have seen that the coefficients satisfy

$$C_r^n = C_r^{n-1} + C_{r-1}^{n-1}.$$

Actually, the binomial coefficients, C_r^n, have been found to take a simple form,

$$C_r^n = \frac{n!}{(n-r)! r!} \equiv \binom{n}{r}.$$

This is nothing other than the combinatoric symbol for determining how to choose n objects r at a time. In the binomial expansions, this makes sense. We have to count the number of ways that we can arrange r products of b with $n - r$ products of a. There are n slots to place the b's. For example, the $r = 2$ case for $n = 4$ involves the six products: *aabb, abab, abba, baab, baba*, and *bbaa*. Thus, it is natural to use this notation.

So, we have found that

$$(a + b)^n = \sum_{r=0}^{n} \binom{n}{r} a^{n-r} b^r. \qquad (1.109)$$

Now consider the geometric series $1 + x + x^2 + \ldots$. We have seen that such this geometric series converges for $|x| < 1$, giving

$$1 + x + x^2 + \ldots = \frac{1}{1 - x}.$$

But, $\frac{1}{1-x} = (1 - x)^{-1}$. This is a binomial to a power, but the power is not an integer.

It turns out that the coefficients of such a binomial expansion can be written similar to the form in Equation(1.109). This example suggests that our sum may no longer be finite. So, for p a real number, $a = 1$ and $b = x$, we generalize Equation(1.109) as

$$(1 + x)^p = \sum_{r=0}^{\infty} \binom{p}{r} x^r \qquad (1.110)$$

and see if the resulting series makes sense. However, we quickly run into problems with the coefficients in the series.

Consider the coefficient for $r = 1$ in an expansion of $(1 + x)^{-1}$. This is given by

$$\binom{-1}{1} = \frac{(-1)!}{(-1-1)! 1!} = \frac{(-1)!}{(-2)! 1!}.$$

But what is $(-1)!$? By definition, it is

$$(-1)! = (-1)(-2)(-3) \cdots .$$

Andreas Freiherr von Ettingshausen (1796–1878) was a German mathematician and physicist who in 1826 introduced the notation $\binom{n}{r}$. However, the binomial coefficients were known by the Hindus centuries beforehand.

This product does not seem to exist! But with a little care, we note that

$$\frac{(-1)!}{(-2)!} = \frac{(-1)(-2)!}{(-2)!} = -1.$$

So, we need to be careful not to interpret the combinatorial coefficient literally. There are better ways to write the general binomial expansion. We can write the general coefficient as

$$
\begin{aligned}
\left(\begin{array}{c} p \\ r \end{array} \right) &= \frac{p!}{(p-r)!r!} \\
&= \frac{p(p-1)\cdots(p-r+1)(p-r)!}{(p-r)!r!} \\
&= \frac{p(p-1)\cdots(p-r+1)}{r!}.
\end{aligned}
\tag{1.111}
$$

With this in mind, we now state the theorem:

General Binomial Expansion

The general binomial expansion for $(1+x)^p$ is a simple generalization of Equation (1.109). For p real, we have the following binomial series:

$$(1+x)^p = \sum_{r=0}^{\infty} \frac{p(p-1)\cdots(p-r+1)}{r!}x^r, \quad |x| < 1. \tag{1.112}$$

Often in physics we only need the first few terms for the case that $x \ll 1$:

$$(1+x)^p = 1 + px + \frac{p(p-1)}{2}x^2 + O(x^3). \tag{1.113}$$

The factor $\gamma = \left(1 - \frac{v^2}{c^2}\right)^{-1/2}$ is important in special relativity. Namely, this is the factor relating differences in time and length measurements by observers moving relative inertial frames. For terrestrial speeds, this gives an appropriate approximation.

Example 1.38. *Approximate* $\gamma = \dfrac{1}{\sqrt{1-\frac{v^2}{c^2}}}$ *for* $v \ll c$.

For $v \ll c$, the first approximation is found by inserting $v/c = 0$. Thus, one obtains $\gamma = 1$. This is the Newtonian approximation and does not provide enough of an approximation for terrestrial speeds. Thus, we need to expand γ in powers of v/c.

First, we rewrite γ as

$$\gamma = \frac{1}{\sqrt{1-\frac{v^2}{c^2}}} = \left[1 - \left(\frac{v}{c}\right)^2\right]^{-1/2}.$$

Using the binomial expansion for $p = -1/2$, we have

$$\gamma \approx 1 + \left(-\frac{1}{2}\right)\left(-\frac{v^2}{c^2}\right) = 1 + \frac{v^2}{2c^2}.$$

Example 1.39. *Time Dilation Example*

The average speed of a large commercial jet airliner is about 500 mph. If you flew for an hour (measured from the ground), then how much younger would you be than if you had not taken the flight, assuming these reference frames obeyed the postulates of special relativity?

This is the problem of time dilation. Let Δt be the elapsed time in a stationary reference frame on the ground and $\Delta \tau$ be that in the frame of the moving plane. Then from the Theory of Special Relativity, these are related by

$$\Delta t = \gamma \Delta \tau.$$

The time differences would then be

$$\begin{aligned}
\Delta t - \Delta \tau &= (1 - \gamma^{-1})\Delta t \\
&= \left(1 - \sqrt{1 - \frac{v^2}{c^2}}\right)\Delta t.
\end{aligned} \tag{1.114}$$

The plane speed, 500 mph, is roughly 225 m/s and $c = 3.00 \times 10^8$ m/s. Because $V \ll c$, we would need to use the binomial approximation to get a nonzero result.

$$\begin{aligned}
\Delta t - \Delta \tau &= \left(1 - \sqrt{1 - \frac{v^2}{c^2}}\right)\Delta t \\
&= \left(1 - \left(1 - \frac{v^2}{2c^2} + \ldots\right)\right)\Delta t \\
&\approx \frac{v^2}{2c^2}\Delta t \\
&= \frac{(225)^2}{2(3.00 \times 10^8)^2}(1\,h) = 1.01\,\text{ns}.
\end{aligned} \tag{1.115}$$

Thus, you have aged 1 ns less than if you did not take the flight.

Example 1.40. Small differences in large numbers: *Compute* $f(R,h) = \sqrt{R^2 + h^2} - R$ *for* $R = 6378.164$ km *and* $h = 1.0$ m.

Inserting these values into a scientific calculator, one finds that

$$f(6378164, 1) = \sqrt{6378164^2 + 1} - 6378164 = 1 \times 10^{-7}\,\text{m}.$$

In some calculators, one might obtain 0; in other calculators, or computer algebra systems like Maple, one might obtain other answers. What answer do you get, and how accurate is your answer?

The problem with this computation is that $R \gg h$. Therefore, the computation of $f(R,h)$ depends on how many digits the computing device can handle. The best way to get an answer is to use the binomial approximation. Writing $h = Rx$, or $x = \frac{h}{R}$, we have

$$\begin{aligned}
f(R,h) &= \sqrt{R^2 + h^2} - R \\
&= R\sqrt{1 + x^2} - R \\
&\simeq R\left[1 + \frac{1}{2}x^2\right] - R \\
&= \frac{1}{2}Rx^2 \\
&= \frac{1}{2}\frac{h}{R^2} = 7.83926 \times 10^{-8}\,\text{m}.
\end{aligned} \tag{1.116}$$

Of course, you should verify how many digits should be kept in reporting the result.

In the next examples, we generalize this example. Such general computations appear in proofs involving general expansions without specific numerical values given.

Example 1.41. *Obtain an approximation to $(a + b)^p$ when a is much larger than b, denoted by $a \gg b$.*

If we neglect b, then $(a + b)^p \simeq a^p$. How good an approximation is this? This is where it would be nice to know the order of the next term in the expansion. Namely, what is the power of b/a of the first neglected term in this expansion?

In order to do this, we first divide out a as

$$(a + b)^p = a^p \left(1 + \frac{b}{a}\right)^p.$$

Now we have a small parameter, $\frac{b}{a}$. According to what we have seen earlier, we can use the binomial expansion to write

$$\left(1 + \frac{b}{a}\right)^n = \sum_{r=0}^{\infty} \binom{p}{r} \left(\frac{b}{a}\right)^r. \tag{1.117}$$

Thus, we have a sum of terms involving powers of $\frac{b}{a}$. Since $a \gg b$, most of these terms can be neglected. So, we can write

$$\left(1 + \frac{b}{a}\right)^p = 1 + p\frac{b}{a} + O\left(\left(\frac{b}{a}\right)^2\right).$$

Here we used $O()$, big-Oh notation, to indicate the size of the first neglected term.

Summarizing, we have

$$
\begin{aligned}
(a + b)^p &= a^p \left(1 + \frac{b}{a}\right)^p \\
&= a^p \left(1 + p\frac{b}{a} + O\left(\left(\frac{b}{a}\right)^2\right)\right) \\
&= a^p + pa^p\frac{b}{a} + a^p O\left(\left(\frac{b}{a}\right)^2\right). \tag{1.118}
\end{aligned}
$$

Therefore, we can approximate $(a + b)^p \simeq a^p + pba^{p-1}$, with an error on the order of $b^2 a^{p-2}$. Note that the order of the error does not include the constant factor from the expansion. We could also use the approximation that $(a + b)^p \simeq a^p$, but it is not typically good enough in applications because the error in this case is of the order ba^{p-1}.

Example 1.42. *Approximate $f(x) = (a + x)^p - a^p$ for $x \ll a$.*

In an earlier example we computed $f(R, h) = \sqrt{R^2 + h^2} - R$ for $R = 6378.164$ km and $h = 1.0$ m. We can make use of the binomial expansion to determine the behavior of similar functions in the form $f(x) = (a + x)^p - a^p$. Inserting the binomial expression into $f(x)$, we have as $\frac{x}{a} \to 0$ that

$$f(x) = (a + x)^p - a^p$$

$$= a^p \left[\left(1 + \frac{x}{a} \right)^p - 1 \right]$$

$$= a^p \left[\frac{px}{a} + O\left(\left(\frac{x}{a} \right)^2 \right) \right]$$

$$= O\left(\frac{x}{a} \right) \quad as \ \frac{x}{a} \to 0. \tag{1.119}$$

This result might not be the approximation that we desire. So, we could back up one step in the derivation to write a better approximation as

$$(a + x)^p - a^p = a^{p-1}px + O\left(\left(\frac{x}{a} \right)^2 \right) \quad as \ \frac{x}{a} \to 0.$$

We now use this approximation to compute $f(R, h) = \sqrt{R^2 + h^2} - R$ *for* $R = 6378.164$ km *and* $h = 1.0$ m *in the earlier example. We let* $a = R^2$, $x = 1$ *and* $p = \frac{1}{2}$. *Then, the leading order approximation would be of order*

$$O\left(\left(\frac{x}{a} \right)^2 \right) = O\left(\left(\frac{1}{6378164^2} \right)^2 \right) \sim 2.4 \times 10^{-14}.$$

Thus, we have

$$\sqrt{6378164^2 + 1} - 6378164 \approx a^{p-1}px,$$

where

$$a^{p-1}px = (6378164^2)^{-1/2}(0.5)1 = 7.83926 \times 10^{-8}.$$

This is the same result we had obtained before. However, we also have an estimate of the size of the error, and this might be useful in indicating how many digits we should trust in the answer.

So far, this is enough to get started in the course. We will recall other topics as we need them. For example, we will discuss the method of partial fraction decomposition when we discuss terminal velocity in the next chapter and when we cover applications of the Laplace transform later in the book.

1.2   What I Need from My Intro Physics Class?

So, WHAT DO YOU NEED TO KNOW about physics? You should be comfortable with common terms from mechanics and electromagnetism. In some cases, we will review specific topics. However, it would be helpful to review some topics from your introductory and modern physics texts.

As you may recall, your study of physics began with the simplest systems. You first studied motion for point masses. You were then introduced to the concepts of position, displacement, velocity, and acceleration. You studied motion first in one dimension and even then could only do problems in which the acceleration was constant, or piecewise constant. You looked at horizontal motion and then vertical motion, in terms of free fall. Finally, you moved into two dimensions and considered projectile motion.

Some calculus was introduced and you learned how to represent vector quantities.

We then asked, "What causes a change in the state of motion of a body?" This led to a discussion of forces. The types of forces encountered were the weight, the normal force, tension, the force of gravity, and then centripetal forces. You might have also seen spring forces, which we will see shortly, leading to oscillatory motion, which is one of the underlying themes in this book.

Next, you found out that there are well-known conservation principles for energy and momentum. In these cases, you were led to the concepts of work, kinetic energy, and potential energy. You found out that even when mechanical energy is not conserved, you could account for the missing energy as the work done by nonconservative forces. Momentum becomes important in collision problems or when looking at impulses.

With these basic ideas under your belt, you proceeded to study more complicated systems. Looking at extended bodies, most notably rigid bodies, led to the study of rotational motion. You found out that there are rotational analogs to all of the previously discussed concepts for point masses. For example, there are the natural analogs of rotational velocity and acceleration: angular displacement, angular velocity, and angular acceleration. The cause of angular acceleration is called the torque. The analog to mass is the moment of inertia.

The next level of complication, which sometimes is not covered, are bulk systems. One can study fluids, solids, and gases. These can be investigated by looking at things like mass density, pressure, volume, and temperature. This leads to the study of thermodynamics, in which one studies the transfer of energy between a system and its surroundings. This involves the relationship between the work done on the system, the heat energy added to a systems and its change in internal energy, and entropy. One common example is the diffusion of heat energy, which will be seen in the form of the heat equation.

Bulk systems can also suffer deformations when a force per area is applied. This leads to the idea that small deformations can result in the propagation of energy throughout the system in the form of waves. We will later explore this wave motion in several systems by solving the wave equation.

The second course in physics is spent on electricity and magnetism, leading to electromagnetic waves. You first learned about charges and charge distributions, electric fields, and electric potentials. Then you found out that moving charges produce magnetic fields and are affected by external magnetic fields. Furthermore, changing magnetic fields produce currents. This can all be summarized by Maxwell's equations, which we will recall later in the course. These equations, in turn, predict the existence of electromagnetic waves.

Depending on how far you delved into the book, you may have seen excursions into optics and the impact that trying to understand the existence of electromagnetic waves has had on the development of so-called "modern physics." For example, in trying to understand what medium electromagnetic waves might propagate through, Einstein proposed an answer that completely changed the way we understand the nature of space and time. In trying to understand how ionized gases radiate and interact with matter, Einstein and others followed a path that led them to quantum mechanics and new challenges in the understanding of reality.

So, that is the introductory physics course in a nutshell. In fact, these topics, combined with a good course on modern physics, is most of undergraduate physics. The rest is detail, which you will explore in your other courses as you progress toward a degree in physics.

1.3 Technology and Tables

As you progress through the course, you will often have to compute integrals and derivatives by hand. However, many readers know that some of the tedium can be alleviated using computers, or even looking up what you need in tables. In some cases, you might even find applets online that can quickly give you the answers you seek.

You also need to be comfortable in doing many computations by hand. This is necessary, especially in your early studies, for several reasons. For example, you should try to evaluate integrals by hand when asked to do them. This reinforces the techniques, as outlined earlier. It exercises your brain in much the same way that you might jog daily to exercise your body. The more comfortable you are with derivations and evaluations, the easier it is to follow future lectures without getting bogged down by the details, wondering how your professor got from step A to step D. You can always use a computer algebra system, or a Table of Integrals, to check your work.

Problems can arise when depending purely on the output of computers, or other "black boxes." Once you have a firm grasp of the techniques and a feeling as to what answers should look like, then you can feel comfortable with what the computer gives you. Sometimes, Computer Algebra Systems (CAS) like Maple can give you strange looking answers and sometimes even wrong answers. Also, these programs cannot do every integral or solve every differential equation that you ask them to do. Even some of the simplest looking expressions can cause computer algebra systems problems. Other times, you might even provide wrong input, leading to erroneous results.

Another source of indefinite integrals, derivatives, series expansions, etc., is a Table of Mathematical Formulae. There are several good books that have been printed. Even some of these have typos in them, so you need to be careful. However, it may be worth the investment to have such a book in your personal library. Go to the library, or the bookstore, and look at some of these tables to see how useful they might be.

There are plenty of online resources as well. For example, there is the Wolfram Integrator at http://integrals.wolfram.com/ as well as the recent http://www.wolframalpha.com/. There is also a wealth of information at

the following sites: http://www.sosmath.com/,
http://www.math2.org/, http://mathworld.wolfram.com/, and
http://functions.wolfram.com/.

While these resources are useful for problems that have analytical solutions, at some point you will need to realize that most problems in texts, especially those from a few decades ago, are mostly aimed at solutions that either have nice analytical solutions or have solutions that can be approximated using pencil and paper.

More and more you will see problems that need to be solved numerically. While most of this book (97%) stresses the traditional methods used for determining the exact or approximate behavior of systems based upon solid mathematical methods, there are times when a basic understanding of computational methods is useful. Therefore, we will occasionally discuss some numerical methods related to the subject matter in the text. In particular, we will discuss some methods of computational physics such as the numerical solution of differential equations and fitting data to curves. Applications will be discussed that can only be solved using these methods.

There are many programming languages and software packages that can be used to determine numerical solutions to algebraic equations or differential equations. For example, CAS (Computer Algebra Systems) such as Maple and Mathematica are available. Open source packages such as Maxima, which has been around for a while, Mathomatic, and the SAGE Project, do exist as alternatives. One can use built-in routines and do some programming. The main features are that they can produce symbolic solutions. Generally, they are slow in generating numerical solutions.

For serious programming, one can use standard programming languages like FORTRAN, C, and its derivatives. Recently, Python has become an alternative and much accepted resource as an open source programming language and is useful for doing scientific computing using the right packages.

Also, there is MATLAB®. MATLAB was developed in the 1980s as a Matrix Laboratory and for a long time was the standard outside "normal" programming languages to handle nonsymbolic solutions in computational science. Similar open source clones have appeared, such as Octave. Octave can run most MATLAB files and some of its own. Other clones of MATLAB are SciLab, Rlab, FreeMat, and PyLab.

In this text there are some snippets provided of Maple and MATLAB routines. Most of the text does not rely on these; however, the MATLAB snippets should be relatively readable to anyone with some knowledge of computer packages, or easy to pass to the open source clones, such as Octave. Maple routines are not so simple but may be translatable to other packages with a little effort. However, the theory lying behind the use of any of these routines is described in the text and the text can be read without explicit understanding of the particular computer software.

1.4 Appendix: Dimensional Analysis

In the first chapter of the introductory physics text, you were introduced to dimensional analysis. Dimensional analysis is useful for recalling particular relationships between variables by looking at the units involved, independent of the system of units employed, though most of the time you have used SI, or MKS, units in most of your physics problems.

There are certain basic units—length, mass, and time. By the second semester of physics you found out that you could add charge to the list. We can represent these as [L], [M], [T], and [C]. Other quantities typically have units that can be expressed in terms of these basic units, which are called derived units. For example, the units of acceleration are given by $[L]/[T]^2$ and units of mass density are given by $[M]/[L]^3$.

Slightly more complicated units arise for force. Because $F = ma$, the units of force are

$$[F] = [m][a] = [M]\frac{[L]}{[T]^2}.$$

Similarly, units of magnetic field can be found, though with a little more effort. Recall that $F = qvB \sin\theta$ for a charge q moving with speed v through a magnetic field B at an angle of θ; $\sin\theta$ has no units. So, the units of magnetic field strength, [B], can be derived from the units of force, charge, and speed:

$$\begin{aligned}
[B] &= \frac{[F]}{[q][v]} \\
&= \frac{\frac{[M][L]}{[T]^2}}{[C]\frac{[L]}{[T]}} \\
&= \frac{[M]}{[C][T]}.
\end{aligned} \tag{1.120}$$

Now, assume that you did not know how B depended on F, q, and v, but you knew the units of all the quantities. Can you figure out the relationship between the variables? We could write

$$[B] = [F]^\alpha [q]^\beta [v]^\gamma$$

and solve for the exponents by inserting the dimensions that we know. Thus, we have

$$[M][C]^{-1}[T]^{-1} = \left([M][L][T]^{-2}\right)^\alpha [C]^\beta \left([L][T]^{-1}\right)^\gamma.$$

Right away we can see that $\alpha = 1$ and $\beta = -1$ by looking at the powers of [M] and [C], respectively. Thus,

$$\begin{aligned}
[M][C]^{-1}[T]^{-1} &= [M][L][T]^{-2}[C]^{-1}\left([L][T]^{-1}\right)^\gamma \\
&= [M][C]^{-1}[L]^{1+\gamma}[T]^{-2-\gamma}.
\end{aligned}$$

We see that picking $\gamma = -1$ balances the exponents and gives the correct relation

$$[B] = [F][q]^{-1}[v]^{-1}.$$

An important theorem at the heart of dimensional analysis is the Buckingham Π Theorem. In essence, this theorem tells us that physically meaningful equations in n variables can be written as an equation involving $n - m$ dimensionless quantities, where m is the number of dimensions used. The importance of this theorem is that one can actually compute useful quantities without even knowing the exact form of the equation!

The Buckingham Π Theorem was introduced by Edgar Buckingham (1867–1940) in 1914. Let q_i be n physical variables that are related by

$$f(q_1, q_2, \ldots, q_n) = 0. \tag{1.121}$$

Assuming that m dimensions are involved, we let π_i be $k = n - m$ dimensionless variables. Then, Equation (1.121) can be rewritten as a function of these dimensionless variables as

$$F(\pi_1, \pi_2, \ldots, \pi_k) = 0, \tag{1.122}$$

where the π_i's can be written in terms of the physical variables as

$$\pi_i = q_1^{k_1} q_2^{k_2} \cdots q_n^{k_n}, \quad i = 1, \ldots, k. \tag{1.123}$$

Well, this is the first really new concept (apart from some mathematical tricks) in the book and it is probably a mystery as to its importance. It also seems a bit abstract. However, this is the basis for some of the proverbial "back of the envelope calculations" that you might have heard about. So, let's see how it can be used.

Example 1.43. *Using dimensional analysis to obtain the period of a simple pendulum.*

Let's consider the period of a simple pendulum; for example, a point mass hanging on a massless string. The period, T, of the pendulum's swing could depend upon the string length, ℓ, the mass of the "pendulum bob," m, and gravity in the form of the acceleration due to gravity, g. These are the q_i's in the theorem.

We have four physical variables ($n = 4$). The only units involved are those of length, mass, and time. So, $m = 3$. This means that there is $k = n - m = 1$ dimensionless variable, call this variable π.

According to the Buckingham Pi Theorem, there must be an equation of the form

$$F(\pi) = 0$$

in terms of the dimensionless variable

$$\pi = \ell^{k_1} m^{k_2} T^{k_3} g^{k_4}.$$

We just need to find the k_i's.

We could find the exponents by inspection, or we could find them systematically by writing out the dimensions of each factor and force π can be dimensionless. Thus, the dimensions of π are given in terms of the dimensions of the other variables as

$$
\begin{aligned}
[\pi] &= [\ell]^{k_1}[m]^{k_2}[T]^{k_3}[g]^{k_4} \\
&= [L]^{k_1}[M]^{k_2}[T]^{k_3}\left(\frac{[L]}{[T]^2}\right)^{k_4} \\
&= [L]^{k_1+k_4}[M]^{k_2}[T]^{k_3-2k_4}.
\end{aligned}
\tag{1.124}
$$

π will be dimensionless when

$$
\begin{aligned}
k_1 + k_4 &= 0, \\
k_2 &= 0, \\
k_3 - 2k_4 &= 0.
\end{aligned}
\tag{1.125}
$$

This is a linear homogeneous system of three equations and four unknowns. We can satisfy these equations by setting $k_1 = -k_4$, $k_2 = 0$, and $k_3 = 2k_4$. Therefore, we have

$$
\pi = \ell^{-k_4}T^{2k_4}g^{k_4} = \left(\ell^{-1}T^2 g\right)^{k_4}.
$$

k_4 is arbitrary, so we can pick the simplest value, $k_4 = 1$. Then, $\pi = \ell^{-1}T^2 g$ and

$$
F\left(\frac{T^2 g}{\ell}\right) = 0.
$$

Assuming that this equation has one zero, z, which must be verified by other means, we have that

$$
\frac{gT^2}{\ell} = z = const.
$$

Thus, we have determined that the period is independent of the mass and proportional to the square root of the length. The constant can be determined by experiment as $z = 4\pi^2$. Thus,

$$
T = 2\pi\sqrt{\frac{\ell}{g}}.
$$

Example 1.44. *Estimating the energy of an atomic bomb.*

A more interesting example was provided by Sir Geoffrey Taylor in 1941 for determining the energy release of an atomic bomb. Let's assume that the energy is released in all directions from a single point. Possible physical variables are the time since the blast, t; the energy, E; the distance from the blast, r; the atmospheric density, ρ; and the atmospheric pressure, p. We have five physical variables ($n = 5$) and only three units ($m = 3$). So, there should be $k = n - m = 2$ dimensionless quantities. Let's determine these.

Energy release in the first atomic bomb.

We set

$$
\pi = E^{k_1} t^{k_2} r^{k_3} p^{k_4} \rho^{k_5}.
$$

Inserting the respective units, we find that

$$
\begin{aligned}
[\pi] &= [E]^{k_1}[t]^{k_2}[r]^{k_3}[p]^{k_4}[\rho]^{k_5} \\
&= \left([M][L]^2[T]^{-2}\right)^{k_1}[T]^{k_2}[L]^{k_3}\left([M][L]^{-1}[T]^{-2}\right)^{k_4}\left([M][L]^{-3}\right)^{k_5} \\
&= [M]^{k_1+k_4+k_5}[L]^{2k_1+k_3-k_4-3k_5}[T]^{-2k_1+k_2-2k_4}.
\end{aligned}
\tag{1.126}
$$

Figure 1.9: A photograph of the first atomic bomb test. This image was found at http://www.atomicarchive.com.

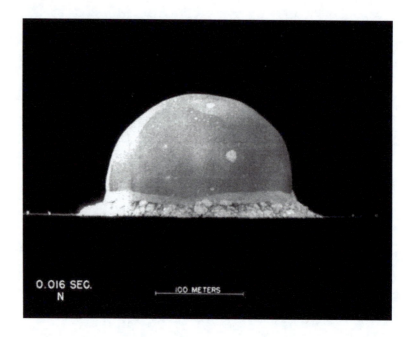

Note: You should verify the units used. For example, the units of force can be found using $F = ma$ and work (energy) is force times distance. Similarly, you need to know that pressure is force per area.

For π to be dimensionless, we have to solve the system:

$$k_1 + k_4 + k_5 = 0,$$
$$2k_1 + k_3 - k_4 - 3k_5 = 0,$$
$$-2k_1 + k_2 - 2k_4 = 0. \tag{1.127}$$

This is a set of three equations and five unknowns. The only way to solve this system is to solve for three unknowns in terms of the remaining two. (In linear algebra, one learns how to solve this using matrix methods.) Let's solve for $k_1, k_2,$ and k_5 in terms of k_3 and k_4. The system can be written as

$$k_1 + k_5 = -k_4,$$
$$2k_1 - 3k_5 = k_4 - k_3,$$
$$2k_1 - k_2 = -2k_4. \tag{1.128}$$

These can be solved by solving for k_1 and k_4 using the first two equations and then finding k_2 from the last one. Solving this system yields

$$k_1 = -\frac{1}{5}(2k_4 + k_3) \quad k_2 = \frac{2}{5}(3k_4 - k_3) \quad k_5 = \frac{1}{5}(k_3 - 3k_4).$$

We have the freedom to pick values for k_3 and k_4. Two independent sets of simple values are obtained by picking one variable as zero and the other as one. This will give the following two cases:

Case I. $k_3 = 1$ and $k_4 = 0$.

In this case, we then have $k_1 = -\frac{1}{5}$, $k_2 = -\frac{2}{5}$, and $k_5 = \frac{1}{5}$. This gives

$$\pi_1 = E^{-1/5}t^{-2/5}r\rho^{1/5} = r\left(\frac{\rho}{Et^2}\right)^{1/5}.$$

Case II. $k_3 = 0$ and $k_4 = 1$.

In this case, we then have $k_1 = -\frac{2}{5}$, $k_2 = \frac{6}{5}$, *and* $k_5 = -\frac{3}{5}$.

$$\pi_2 = E^{-2/5} t^{6/5} p \rho^{-3/5} = p \left(\frac{t^6}{\rho^3 E^2} \right)^{1/5}.$$

Thus, we have that the relation between the energy and the other variables is of the form

$$F \left(r \left(\frac{\rho}{E t^2} \right)^{1/5}, p \left(\frac{t^6}{\rho^3 E^2} \right)^{1/5} \right) = 0.$$

Of course, this is not enough to determine the explicit equation. However, Taylor was able to use this information to get an energy estimate.

Note that π_1 *is dimensionless. It can be represented as a function of the dimensionless variable* π_2. *So, assuming that* $\pi_1 = h(\pi_2)$, *we have that*

$$h(\pi_2) = r \left(\frac{\rho}{E t^2} \right)^{1/5}.$$

Note that for $t = 1$ *second, the energy is expected to be huge, so* $\pi_2 \approx 0$. *Thus,*

$$r \left(\frac{\rho}{E t^2} \right)^{1/5} \approx h(0).$$

Simple experiments suggest that $h(0)$ *is of order one, so*

$$r \approx \left(\frac{E t^2}{\rho} \right)^{1/5}.$$

In 1947, Taylor applied his earlier analysis to movies of the first atomic bomb test in 1945 and his results were close to the actual values. How can one do this? You can find pictures of the first atomic bomb test with a superimposed length scale online similar to Figure 1.9.

We can rewrite the above result to get the energy estimate:

$$E \approx \frac{r^5 \rho}{t^2}.$$

As an exercise, you can estimate the radius of the explosion at the given time and determine the energy of the blast in so many tons of TNT. (See the Challenge problem.)

Problems

1. Prove the following identities using only the definitions of the trigonometric functions, the Pythagorean identity, or the identities for sines and cosines of sums of angles.

    a. $\cos 2x = 2\cos^2 x - 1$.

    b. $\sin 3x = A \sin^3 x + B \sin x$, for what values of A and B?

    c. $\sec\theta + \tan\theta = \tan\left(\frac{\theta}{2} + \frac{\pi}{4} \right)$.

2. Determine the exact values of

   a. $\sin \dfrac{\pi}{8}$.

   b. $\tan 15^0$.

   c. $\cos 105^0$.

3. Denest the following, if possible.

   a. $\sqrt{3 - 2\sqrt{2}}$.

   b. $\sqrt{1 + \sqrt{2}}$.

   c. $\sqrt{5 + 2\sqrt{6}}$.

   d. $\sqrt[3]{\sqrt{5} + 2} - \sqrt[3]{\sqrt{5} - 2}$.

   e. Find the roots of $x^2 + 6x - 4\sqrt{5} = 0$ in simplified form.

4. Determine the exact values of

   a. $\sin\left(\cos^{-1} \dfrac{3}{5}\right)$.

   b. $\tan\left(\sin^{-1} \dfrac{x}{7}\right)$.

   c. $\sin^{-1}\left(\sin \dfrac{3\pi}{2}\right)$.

5. Do the following:

   a. Write $(\cosh x - \sinh x)^6$ in terms of exponentials.

   b. Prove $\cosh(x - y) = \cosh x \cosh y - \sinh x \sinh y$ using the exponential forms of the hyperbolic functions.

   c. Prove $\cosh 2x = \cosh^2 x + \sinh^2 x$.

   d. If $\cosh x = \dfrac{13}{12}$ and $x < 0$, find $\sinh x$ and $\tanh x$.

   e. Find the exact value of $\sinh(\operatorname{arccosh} 3)$.

6. Prove that the inverse hyperbolic functions are the following logarithms:

   a. $\cosh^{-1} x = \ln\left(x + \sqrt{x^2 - 1}\right)$.

   b. $\tanh^{-1} x = \dfrac{1}{2} \ln \dfrac{1 + x}{1 - x}$.

7. Write the following in terms of logarithms:

   a. $\cosh^{-1} \frac{4}{3}$.

   b. $\tanh^{-1} \frac{1}{2}$.

   c. $\sinh^{-1} 2$.

8. Solve the following equations for x:

   a. $\cosh(x + \ln 3) = 3$.

   b. $2\tanh^{-1} \frac{x-2}{x-1} = \ln 2$.

   c. $\sinh^2 x - 7\cosh x + 13 = 0$.

9. Compute the following integrals:

a. $\int x e^{2x^2} dx$.

b. $\int_0^3 \dfrac{5x}{\sqrt{x^2+16}} dx$.

c. $\int x^3 \sin 3x \, dx$. (Do this using integration by parts, the Tabular Method, and differentiation under the integral sign.)

d. $\int \cos^4 3x \, dx$.

e. $\int_0^{\pi/4} \sec^3 x \, dx$.

f. $\int e^x \sinh x \, dx$.

g. $\int \sqrt{9-x^2} \, dx$

h. $\int \dfrac{dx}{(4-x^2)^2}$, using the substitution $x = 2\tanh u$.

i. $\int_0^4 \dfrac{dx}{\sqrt{9+x^2}}$, using a hyperbolic function substitution.

j. $\int \dfrac{dx}{1-x^2}$, using the substitution $x = \tanh u$.

k. $\int \dfrac{dx}{(x^2+4)^{3/2}}$, using the substitutions $x = 2\tan\theta$ and $x = 2\sinh u$.

l. $\int \dfrac{dx}{\sqrt{3x^2-6x+4}}$.

10. Find the sum for each of the series:

a. $5 + \frac{25}{7} + \frac{125}{49} + \frac{625}{343} + \cdots$.

b. $\sum_{n=0}^{\infty} \dfrac{(-1)^n 3}{4^n}$.

c. $\sum_{n=2}^{\infty} \dfrac{2}{5^n}$.

d. $\sum_{n=-1}^{\infty} (-1)^{n+1} \left(\dfrac{e}{\pi} \right)^n$.

e. $\sum_{n=0}^{\infty} \left(\dfrac{5}{2^n} + \dfrac{1}{3^n} \right)$.

f. $\sum_{n=1}^{\infty} \dfrac{3}{n(n+3)}$.

g. What is $0.5\overline{69}$?

11. A superball is dropped from a 2.00 m height. After it rebounds, it reaches a new height of 1.65 m. Assuming a constant coefficient of restitution, find the (ideal) total distance the ball will travel as it keeps bouncing.

12. Here are some telescoping series problems:

a. Verify that

$$\sum_{n=1}^{\infty} \dfrac{1}{(n+2)(n+1)} = \sum_{n=1}^{\infty} \left(\dfrac{n+1}{n+2} - \dfrac{n}{n+1} \right).$$

b. Find the nth partial sum of the series $\sum_{n=1}^{\infty} \left(\dfrac{n+1}{n+2} - \dfrac{n}{n+1} \right)$ and use it to determine the sum of the resulting *telescoping* series.

c. Sum the series $\sum_{n=1}^{\infty} \left[\tan^{-1} n - \tan^{-1}(n+1) \right]$ by first writing the Nth partial sum and then computing $\lim_{N \to \infty} s_N$.

13. Determine the radius and interval of convergence of the following infinite series:

a. $\sum_{n=1}^{\infty} (-1)^n \dfrac{(x-1)^n}{n}$.

b. $\sum_{n=1}^{\infty} \dfrac{x^n}{2^n n!}$.

c. $\sum_{n=1}^{\infty} \dfrac{1}{n} \left(\dfrac{x}{5} \right)^n$.

d. $\sum_{n=1}^{\infty} (-1)^n \dfrac{x^n}{\sqrt{n}}$.

14. Use the partition function Z for the quantum harmonic oscillator to find the average energy, $\langle E \rangle$.

15. Find the Taylor series centered at $x = a$ and its corresponding radius of convergence for the given function. In most cases, you need not employ the direct method of computation of the Taylor coefficients.

a. $f(x) = \sinh x, \, a = 0$.

b. $f(x) = \sqrt{1+x}, \, a = 0$.

c. $f(x) = \ln \dfrac{1+x}{1-x}, \, a = 0$.

d. $f(x) = xe^x, \, a = 1$.

e. $f(x) = \dfrac{1}{\sqrt{x}}, \, a = 1$.

f. $f(x) = x^4 + x - 2, \, a = 2$.

g. $f(x) = \dfrac{x-1}{2+x}, \, a = 1$.

16. Consider Gregory's expansion

$$\tan^{-1} x = x - \frac{x^3}{3} + \frac{x^5}{5} - \cdots = \sum_{k=0}^{\infty} \frac{(-1)^k}{2k+1} x^{2k+1}.$$

a. Derive Gregory's expansion using the definition

$$\tan^{-1} x = \int_0^x \frac{dt}{1+t^2},$$

expanding the integrand in a Maclaurin series, and integrating the resulting series term by term.

b. From this result, derive Gregory's series for π by inserting an appropriate value for x in the series expansion for $\tan^{-1} x$.

17. In the event that a series converges uniformly, one can consider the derivative of the series to arrive at the summation of other infinite series.

a. Differentiate the series representation for $f(x) = \frac{1}{1-x}$ to sum the series $\sum_{n=1}^{\infty} nx^n$, $|x| < 1$.

b. Use the result from part a to sum the series $\sum_{n=1}^{\infty} \frac{n}{5^n}$.

c. Sum the series $\sum_{n=2}^{\infty} n(n-1)x^n$, $|x| < 1$.

d. Use the result from part c to sum the series $\sum_{n=2}^{\infty} \frac{n^2 - n}{5^n}$.

e. Use the results from this problem to sum the series $\sum_{n=4}^{\infty} \frac{n^2}{5^n}$.

18. Evaluate the integral $\int_0^{\pi/6} \sin^2 x \, dx$ by doing the following:

a. Compute the integral exactly.

b. Integrate the first three terms of the Maclaurin series expansion of the integrand and compare with the exact result.

19. Determine the next term in the time dilation example, 1.39. That is, find the $\frac{v^4}{c^2}$ term and determine a better approximation to the time difference of 1 ns.

20. Evaluate the following expressions at the given point. Use your calculator or your computer (such as Maple). Then use series expansions to find an approximation of the value of the expression to as many places as you trust.

a. $\dfrac{1}{\sqrt{1+x^3}} - \cos x^2$ at $x = 0.015$.

b. $\ln \sqrt{\dfrac{1+x}{1-x}} - \tan x$ at $x = 0.0015$.

c. $f(x) = \dfrac{1}{\sqrt{1+2x^2}} - 1 + x^2$ at $x = 5.00 \times 10^{-3}$.

d. $f(R,h) = R - \sqrt{R^2 + h^2}$ for $R = 1.374 \times 10^3$ km and $h = 1.00$ m.

e. $f(x) = 1 - \dfrac{1}{\sqrt{1-x}}$ for $x = 2.5 \times 10^{-13}$.

21. Use dimensional analysis to derive a possible expression for the drag force F_D on a soccer ball of diameter D moving at speed v through air of density ρ and viscosity μ. [Hint: Assuming viscosity has units $\frac{[M]}{[L][T]}$, there are two possible dimensionless combinations: $\pi_1 = \mu D^\alpha \rho^\beta v^\gamma$ and $\pi_2 = F_D D^\alpha \rho^\beta v^\gamma$. Determine α, β, and γ for each case, and interpret your results.]

22. **Challenge:** Read the section on dimensional analysis. In particular, look at the results of Example 1.44. Using measurements in/on Figure 1.9, obtain an estimate of the energy of the blast in tons of TNT. Explain your work. Does your answer make sense? Why?

2

Free Fall and Harmonic Oscillators

"Mathematics began to seem too much like puzzle solving. Physics is puzzle solving, too, but of puzzles created by nature, not by the mind of man."—Maria Goeppert-Mayer (1906–1972)

2.1 Free Fall

IN THIS CHAPTER WE WILL STUDY some common differential equations that appear in physics. We will begin with the simplest types of equations and standard techniques for solving them We will end this part of the discussion by returning to the problem of free fall with air resistance. We will then turn to the study of oscillations, which are modeled by second-order differential equations.

Let us begin with a simple example from introductory physics. Recall that free fall is the vertical motion of an object solely under the force of gravity. It has been experimentally determined that an object near the surface of the Earth falls at a constant acceleration in the absence of other forces, such as air resistance. This constant acceleration is denoted by $-g$, where g is called the acceleration due to gravity. The negative sign is an indication that we have chosen a coordinate system in which up is positive.

Free fall example.

We are interested in determining the position, $y(t)$, of the falling body as a function of time. From the definition of free fall, we have

$$\ddot{y}(t) = -g. \tag{2.1}$$

Note that we will occasionally use a dot to indicate time differentiation. This notation is standard in physics and we will begin to introduce you to this notation, though at times we might use the more familiar prime notation to indicate spatial differentiation, or general differentiation.

Differentiation with respect to time is often denoted by dots instead of primes.

In Equation (2.1) we know g. It is a constant. Near the Earth's surface, it is about 9.81 m/s^2 or 32.2 ft/s^2. What we do not know is $y(t)$. This is our first differential equation. In fact, it is natural to see differential equations appear in physics often through Newton's Second Law, $F = ma$, as it plays an important role in classical physics. We will return to this point later.

So, how does one solve the differential equation in Equation (2.1)? We do so by using what we know about calculus. It might be easier to see when we put in a particular number instead of g. You might still be getting used to the fact that some letters are used to represent constants. We will come back to the more general form after we see how to solve the differential equation.

Consider

$$\ddot{y}(t) = 5. \tag{2.2}$$

Recalling that the second derivative is just the derivative of a derivative, we can rewrite this equation as

$$\frac{d}{dt}\left(\frac{dy}{dt}\right) = 5. \tag{2.3}$$

This tells us that the derivative of dy/dt is 5. Can you think of a function whose derivative is 5? (Do not forget that the independent variable is t.) Yes, the derivative of $5t$ with respect to t is 5. Is this the only function whose derivative is 5? No! You can also differentiate $5t + 1$, $5t + \pi$, $5t - 6$, etc. In general, the derivative of $5t + C$ is 5, where C is an arbitrary integration constant.

So, Equation (2.2) can be reduced to

$$\frac{dy}{dt} = 5t + C. \tag{2.4}$$

Now we ask if you know a function whose derivative is $5t + C$. Well, you might be able to do this one in your head, but we just need to recall the Fundamental Theorem of Calculus, which relates integrals and derivatives. Thus, we have

$$y(t) = \frac{5}{2}t^2 + Ct + D, \tag{2.5}$$

where D is a second integration constant.

Equation (2.5) gives the solution to the original differential equation. That means that when the solution is placed into the differential equation, both sides of the differential equation give the same expression. You can always check your answer to a differential equation by showing that your solution satisfies the equation. In this case we have

$$\ddot{y}(t) = \frac{d^2}{dt^2}\left(\frac{5}{2}t^2 + Ct + D\right) = \frac{d}{dt}(5t + C) = 5.$$

Therefore, Equation (2.5) gives the general solution of the differential equation.

We also see that there are two arbitrary constants, C and D. Picking any values for these gives a whole family of solutions. As we will see, the equation $\ddot{y}(t) = 5$ is a linear second-order ordinary differential equation. The general solution of such an equation always has two arbitrary constants.

Let's return to the free fall problem. We solve it the same way. The only difference is that we can replace the constant 5 with the constant $-g$. So, we find that

$$\frac{dy}{dt} = -gt + C, \tag{2.6}$$

and

$$y(t) = -\frac{1}{2}gt^2 + Ct + D. \qquad (2.7)$$

Once you get down the process, it only takes a line or two to solve.

There seems to be a problem. Imagine dropping a ball that then undergoes free fall. We just determined that there are an infinite number of solutions for the position of the ball at any time! Well, that is not possible. Experience tells us that if you drop a ball, you expect it to behave the same way every time. Or does it? Actually, you could drop the ball from anywhere. You could also toss it up or throw it down. So, there are many ways you can release the ball before it is in free fall producing many different paths, $y(t)$. That is where the constants come in. They have physical meanings.

If you set $t = 0$ in the equation, then you have that $y(0) = D$. Thus, D gives the initial position of the ball. Typically, we denote initial values with a subscript. So, we will write $y(0) = y_0$. Thus, $D = y_0$.

That leaves us to determine C. It appears at first in Equation (2.6). Recall that $\frac{dy}{dt}$, the derivative of the position, is the vertical velocity, $v(t)$. It is positive when the ball moves upward. We will denote the initial velocity $v(0) = v_0$. Inserting $t = 0$ in Equation (2.6), we find that $\dot{y}(0) = C$. This implies that $C = v(0) = v_0$.

Putting this all together, we have the physical form of the solution for free fall as

$$y(t) = -\frac{1}{2}gt^2 + v_0 t + y_0. \qquad (2.8)$$

Doesn't this equation look familiar? Now we see that the infinite family of solutions consists of free fall resulting from initially dropping a ball at position y_0 with initial velocity v_0. The conditions $y(0) = y_0$ and $\dot{y}(0) = v_0$ are called the initial conditions. A solution of a differential equation satisfying a set of initial conditions is often called a particular solution. Specifying the initial conditions results in a unique solution.

So, we have solved the free fall equation. Along the way we have begun to see some of the features that will appear in the solutions of other problems that are modeled with differential equations. Throughout the book we will see several applications of differential equations. We will extend our analysis to higher dimensions, in which case we will be faced with so-called partial differential equations, that involve the partial derivatives of functions of more that one variable.

But are we done with free fall? Not at all! We can relax some of the conditions that we have imposed. We can add air resistance. We will visit this problem later in this chapter after introducing some more techniques. We can also provide a horizontal component of motion, leading to projectile motion.

Finally, we should also note that free fall at constant g only takes place near the surface of the Earth. What if a tile falls off the shuttle far from the surface of the Earth? It will also fall toward the Earth. Actually, the tile also has a velocity component in the direction of the motion of the shuttle. So,

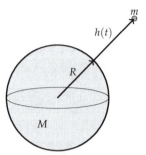

Figure 2.1: Free fall far from the Earth from a height $h(t)$ from the surface.

Here $G = 6.6730 \times 10^{-11}$ m^{3}kg^{-1}s^{-2} is the Universal Gravitational Constant, $M = 5.9736 \times 10^{24}$ kg, and $R = 6,371$ km are the Earth's mass and mean radius, respectively. For $h << R$, $GM/R^2 \approx g$.

it would not necessarily take radial path downward. For now, let's ignore that component.

To look at this problem in more detail, we need to go to the origins of the acceleration due to gravity. This comes out of Newton's Law of Gravitation. Consider a mass m at some distance $h(t)$ from the surface of the (spherical) Earth. Letting M and R be the Earth's mass and radius, respectively, Newton's Law of Gravitation states that

$$ma = F$$
$$m\frac{d^2h(t)}{dt^2} = -G\frac{mM}{(R+h(t))^2}. \tag{2.9}$$

Thus, we arrive at a differential equation

$$\frac{d^2h(t)}{dt^2} = -\frac{GM}{(R+h(t))^2}. \tag{2.10}$$

This equation is not as easy to solve. We will leave it as a homework exercise for the reader.

2.2 First-Order Differential Equations

BEFORE MOVING ON, WE FIRST DEFINE an n-th-order ordinary differential equation. It is an equation for an unknown function $y(x)$ that expresses a relationship between the unknown function and its first n derivatives. One could write this generally as

$$F(y^{(n)}(x), y^{(n-1)}(x), \ldots, y'(x), y(x), x) = 0. \tag{2.11}$$

Here, $y^{(n)}(x)$ represents the nth derivative of $y(x)$.

An initial value problem consists of the differential equation plus the values of the first $n - 1$ derivatives at a particular value of the independent variable, say x_0:

$$y^{(n-1)}(x_0) = y_{n-1}, \quad y^{(n-2)}(x_0) = y_{n-2}, \quad \ldots, \quad y(x_0) = y_0. \tag{2.12}$$

A linear nth-order differential equation takes the form

$$a_n(x)y^{(n)}(x) + a_{n-1}(x)y^{(n-1)}(x) + \ldots + a_1(x)y'(x) + a_0(x)y(x)) = f(x). \tag{2.13}$$

If $f(x) \equiv 0$, then the equation is said to be homogeneous; otherwise it is called nonhomogeneous.

Typically, the first differential equations encountered are first-order equations. A first-order differential equation takes the form

$$F(y', y, x) = 0. \tag{2.14}$$

There are two common first-order differential equations for which one can formally obtain a solution. The first is the separable case and the second is

nth-order ordinary differential equation

Initial value problem.

Linear nth-order differential equation

Homogeneous and nonhomogeneous equations.

First-order differential equation.

a first-order equation. We indicate that we can formally obtain solutions, as one can display the needed integration that leads to a solution. However, the resulting integrals are not always reducible to elementary functions, nor does one obtain explicit solutions when the integrals are doable.

2.2.1 Separable Equations

A FIRST-ORDER EQUATION IS SEPARABLE if it can be written in the form

$$\frac{dy}{dx} = f(x)g(y). \tag{2.15}$$

Special cases result when either $f(x) = 1$ or $g(y) = 1$. In the first case the equation is said to be autonomous.

The general solution to Equation (2.15) is obtained in terms of two integrals:

Separable equations.

$$\boxed{\int \frac{dy}{g(y)} = \int f(x)\,dx + C, \tag{2.16}}$$

where C is an integration constant. This yields a 1-parameter family of solutions to the differential equation corresponding to different values of C. If one can solve Equation (2.16) for $y(x)$, then one obtains an explicit solution. Otherwise, one has a family of implicit solutions. If an initial condition is given as well, then one might be able to find a member of the family that satisfies this condition, which is often called a particular solution.

Example 2.1. $y' = 2xy$, $y(0) = 2$.
Applying Equation (2.16), one has

$$\int \frac{dy}{y} = \int 2x\,dx + C.$$

Integrating yields

$$\ln|y| = x^2 + C.$$

Exponentiating, one obtains the general solution,

$$y(x) = \pm e^{x^2+C} = Ae^{x^2}.$$

Here we have defined $A = \pm e^C$. Because C is an arbitrary constant, A is an arbitrary constant. Several solutions in this 1-parameter family are shown in Figure 2.2.

Next, one seeks a particular solution satisfying the initial condition. For $y(0) = 2$, one finds that $A = 2$. So, the particular solution satisfying the initial condition is $y(x) = 2e^{x^2}$.

Example 2.2. $yy' = -x$. *Following the same procedure as in the last example, one obtains:*

$$\int y\,dy = -\int x\,dx + C \Rightarrow y^2 = -x^2 + A, \quad where \quad A = 2C.$$

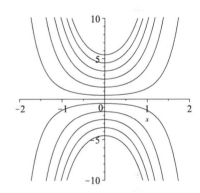

Figure 2.2: Plots of solutions from the 1-parameter family of solutions of Example 2.1 for several initial conditions.

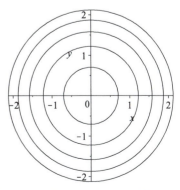

Figure 2.3: Plots of solutions of Example 2.2 for several initial conditions.

Thus, we obtain an implicit solution. Writing the solution as $x^2 + y^2 = A$, we see that this is a family of circles for $A > 0$ and the origin for $A = 0$. Plots of some solutions in this family are shown in Figure 2.3.

2.2.2 Linear First-Order Equations

THE SECOND TYPE OF FIRST-ORDER EQUATION encountered is the linear first-order differential equation in the standard form

$$y'(x) + p(x)y(x) = q(x). \tag{2.17}$$

In this case one seeks an integrating factor, $\mu(x)$, which is a function that one can multiply through the equation making the left side a perfect derivative, and thus obtaining

$$\frac{d}{dx}[\mu(x)y(x)] = \mu(x)q(x). \tag{2.18}$$

The integrating factor that works is $\mu(x) = \exp(\int^x p(\xi)\,d\xi)$. One can derive $\mu(x)$ by expanding the derivative in Equation (2.18),

$$\mu(x)y'(x) + \mu'(x)y(x) = \mu(x)q(x), \tag{2.19}$$

and comparing this equation to the one obtained from multiplying Equation (2.17) by $\mu(x)$:

$$\mu(x)y'(x) + \mu(x)p(x)y(x) = \mu(x)q(x). \tag{2.20}$$

Note that these last two equations would be the same if the second terms were the same. Thus, we will require that

$$\frac{d\mu(x)}{dx} = \mu(x)p(x).$$

Integrating factor.

This is a separable first-order equation for $\mu(x)$ whose solution is the integrating factor:

$$\mu(x) = \exp\left(\int^x p(\xi)\,d\xi\right). \tag{2.21}$$

Equation (2.18) is now easily integrated to obtain the general solution to the linear first-order differential equation:

$$y(x) = \frac{1}{\mu(x)}\left[\int^x \mu(\xi)q(\xi)\,d\xi + C\right]. \tag{2.22}$$

Example 2.3. $xy' + y = x, \quad x > 0, y(1) = 0.$

One first notes that this is a linear first-order differential equation. Solving for y', one can see that the equation is not separable. Furthermore, it is not in the standard form (Equation 2.17). So, we first rewrite the equation as

$$\frac{dy}{dx} + \frac{1}{x}y = 1. \tag{2.23}$$

Noting that $p(x) = \frac{1}{x}$, we determine the integrating factor

$$\mu(x) = \exp\left[\int^x \frac{d\xi}{\xi}\right] = e^{\ln x} = x.$$

Multiplying Equation (2.23) by $\mu(x) = x$, we actually get back the original equation! In this case, we have found that $xy' + y$ must have been the derivative of something to start. In fact, $(xy)' = xy' + x$. Therefore, the differential equation becomes

$$(xy)' = x.$$

Integrating, one obtains

$$xy = \frac{1}{2}x^2 + C,$$

or

$$y(x) = \frac{1}{2}x + \frac{C}{x}.$$

Inserting the initial condition into this solution, we have $0 = \frac{1}{2} + C$. Therefore, $C = -\frac{1}{2}$. Thus, the solution of the initial value problem is

$$y(x) = \frac{1}{2}(x - \frac{1}{x}).$$

We can verify that this is the solution. Because $y' = \frac{1}{2} + \frac{1}{2x^2}$, we have

$$xy' + y = \frac{1}{2}x + \frac{1}{2x} + \frac{1}{2}\left(x - \frac{1}{x}\right) = x.$$

Also, $y(1) = \frac{1}{2}(1 - 1) = 0$.

Example 2.4. $(\sin x)y' + (\cos x)y = x^2$.

Actually, this problem is easy if you realize that the left-hand side is a perfect derivative. Namely,

$$\frac{d}{dx}((\sin x)y) = (\sin x)y' + (\cos x)y.$$

But, we will go through the process of finding the integrating factor for practice.

First, we rewrite the original differential equation in standard form. We divide the equation by $\sin x$ to obtain

$$y' + (\cot x)y = x^2 \csc x.$$

Then, we compute the integrating factor as

$$\mu(x) = \exp\left(\int^x \cot \xi \, d\xi\right) = e^{\ln(\sin x)} = \sin x.$$

Using the integrating factor, the standard form equation becomes

$$\frac{d}{dx}((\sin x)y) = x^2.$$

Integrating, we have

$$y \sin x = \frac{1}{3}x^3 + C.$$

So, the solution is

$$y(x) = \left(\frac{1}{3}x^3 + C\right)\csc x.$$

There are other first-order equations that one can solve for closed-form solutions. However, many equations are not solvable, or one is simply interested in the behavior of solutions. In such cases, one turns to direction fields or numerical methods. We will return to a discussion of the qualitative behavior of differential equations and numerical solutions of ordinary differential equations later in the chapter.

2.2.3   Terminal Velocity

NOW WE RETURN TO FREE FALL. What if there is air resistance? We first need to model the air resistance. As an object falls faster and faster, the drag force becomes greater. So, this resistive force is a function of the velocity. There are a couple of standard models that people use to test this. The idea is to write $F = ma$ in the form

$$m\ddot{y} = -mg + f(v), \tag{2.24}$$

where $f(v)$ gives the resistive force and mg is the weight. Recall that this applies to free fall near the Earth's surface. Also, for it to be resistive, $f(v)$ should oppose the motion. If the body is falling, then $f(v)$ should be positive. If it is rising, then $f(v)$ would have to be negative to indicate the opposition to the motion.

One common determination derives from the drag force on an object moving through a fluid. This force is given by

$$f(v) = \frac{1}{2}CA\rho v^2, \tag{2.25}$$

where C is the drag coefficient, A is the cross-sectional area, and ρ is the fluid density. For laminar flow, the drag coefficient is constant.

Unless you are into aerodynamics, you do not need to get into the details of the constants. So, it is best to absorb all the constants into one to simplify the computation. So, we will write $f(v) = bv^2$. The differential equation including drag can then be rewritten as

$$\dot{v} = kv^2 - g, \tag{2.26}$$

where $k = b/m$. Note that this is a first-order equation for $v(t)$. It is separable too!

Formally, we can separate the variables and integrate over time to obtain

$$t + K = \int^v \frac{dz}{kz^2 - g}. \tag{2.27}$$

This is the first use of Partial Fraction Decomposition. We will explore this method further in the section on Laplace Transforms.

(Note: We used an integration constant of K because C is the drag coefficient in this problem.) If we can do the integral, then we have a solution for v. In fact, we can do this integral. You need to recall another common method of integration, which we have not reviewed yet. Do you remember Partial Fraction Decomposition? It involves factoring the denominator in the integral. In the simplest case, there are two linear factors in the denominator and the integral is rewritten:

$$\int \frac{dx}{(x-a)(x-b)} = \frac{1}{b-a} \int \left[\frac{1}{x-a} - \frac{1}{x-b} \right] dx. \qquad (2.28)$$

The new integral now has two terms which can be readily integrated.

In order to factor the denominator in the current problem, we first have to rewrite the constants. We let $\alpha^2 = g/k$ and write the integrand as

$$\frac{1}{kz^2 - g} = \frac{1}{k} \frac{1}{z^2 - \alpha^2}. \qquad (2.29)$$

Now we use a partial fraction decomposition to obtain

$$\frac{1}{kz^2 - g} = \frac{1}{2\alpha k} \left[\frac{1}{z-\alpha} - \frac{1}{z+\alpha} \right]. \qquad (2.30)$$

Now, the integrand can be easily integrated giving

$$t + K = \frac{1}{2\alpha k} \ln \left| \frac{v-\alpha}{v+\alpha} \right|. \qquad (2.31)$$

Solving for v, we have

$$v(t) = \frac{1 - Be^{2\alpha kt}}{1 + Be^{2\alpha kt}} \alpha, \qquad (2.32)$$

where $B \equiv e^K$. B can be determined using the initial velocity.

There are other forms for the solution in terms of a tanh function, which the reader can determine as an exercise. One important conclusion is that for large times, the ratio in the solution approaches -1. Thus, $v \to -\alpha = -\sqrt{\frac{g}{k}}$ as $t \to \infty$. This means that the falling object will reach a constant terminal velocity.

As a simple computation, we can determine the terminal velocity. We will take an 80-kg skydiver with a cross-sectional area of about 0.093 m^2. (The skydiver is falling head first.) Assume that the air density is a constant 1.2 kg/m^3 and the drag coefficient is $C = 2.0$. We first note that

$$v_{\text{terminal}} = -\sqrt{\frac{g}{k}} = -\sqrt{\frac{2mg}{CA\rho}}.$$

So,

$$v_{\text{terminal}} = -\sqrt{\frac{2(70)(9.8)}{(2.0)(0.093)(1.2)}} = 78\text{m/s}.$$

This is about 175 mph, which is slightly higher than the actual terminal velocity of a skydiver with arms and feet fully extended. One would need a more accurate determination of C and A for a more realistic answer. Also, the air density varies along the way.

2.3   Simple Harmonic Oscillator

THE NEXT PHYSICAL PROBLEM OF INTEREST is that of simple harmonic motion. Such motion comes up in many places in physics and provides a generic first approximation to models of oscillatory motion. This is the beginning of a major thread running throughout this course. You have seen simple harmonic motion in your introductory physics class. We will review SHM (or SHO in some texts) by looking at springs and pendula (the plural of pendulum). We will use this as our jumping board into second-order differential equations and later see how such oscillatory motion occurs in AC circuits.

2.3.1   Mass-Spring Systems

WE BEGIN WITH THE CASE of a single block on a spring as shown in Figure 2.4. The net force in this case is the restoring force of the spring given by Hooke's Law,

$$F_s = -kx,$$

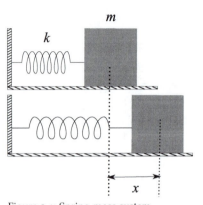

Figure 2.4: Spring-mass system.

where $k > 0$ is the spring constant. Here, x is the elongation, or displacement of the spring from equilibrium. When the displacement is positive, the spring force is negative and when the displacement is negative the spring force is positive. We have depicted a horizontal system sitting on a frictionless surface. A similar model can be provided for vertically oriented springs. However, you need to account for gravity to determine the location of equilibrium. Otherwise, the oscillatory motion about equilibrium is modeled the same.

From Newton's Second Law, $F = m\ddot{x}$, we obtain the equation for the motion of the mass on the spring:

$$m\ddot{x} + kx = 0. \tag{2.33}$$

Dividing by the mass, this equation can be written in the form

$$\ddot{x} + \omega^2 x = 0, \tag{2.34}$$

where

$$\omega = \sqrt{\frac{k}{m}}.$$

This is the generic differential equation for simple harmonic motion.

We will later derive solutions of such equations in a methodical way. For now we note that two solutions of this equation are given by

$$x(t) = A \cos \omega t,$$
$$x(t) = A \sin \omega t, \tag{2.35}$$

where ω is the angular frequency, measured in rad/s, and A is called the amplitude of the oscillation.

The angular frequency is related to the frequency by

$$\omega = 2\pi f,$$

where f is measured in cycles per second, or Hertz. Furthermore, this is related to the period of oscillation, the time it takes the mass to go through one cycle:

$$T = 1/f.$$

2.3.2 Simple Pendulum

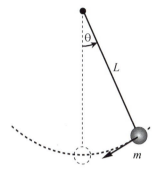

Figure 2.5: A simple pendulum consists of a point mass m attached to a string of length L. It is released from an angle θ_0.

THE SIMPLE PENDULUM consists of a point mass m hanging on a string of length L from some support. (See Figure 2.5.) One pulls the mass back to some starting angle, θ_0, and releases it. The goal is to find the angular position as a function of time.

There are a couple of possible derivations. We could either use Newton's Second Law of Motion, $F = ma$, or its rotational analog in terms of torque, $\tau = I\alpha$. We will use the former only to limit the amount of physics background needed.

There are two forces acting on the point mass. The first is gravity. This points downward and has a magnitude of mg, where g is the standard symbol for the acceleration due to gravity. The other force is the tension in the string. In Figure 2.6 these forces and their sum are shown. The magnitude of the sum is easily found as $F = mg\sin\theta$ using the addition of these two vectors.

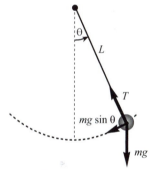

Figure 2.6: There are two forces acting on the mass: the weight mg and the tension T. The net force is found to be $F = mg\sin\theta$.

Now, Newton's Second Law of Motion tells us that the net force is the mass times the acceleration. So, we can write

$$m\ddot{x} = -mg\sin\theta.$$

Next, we need to relate x and θ. x is the distance traveled, which is the length of the arc traced out by the point mass. The arclength is related to the angle, provided the angle is measured in radians. Namely, $x = r\theta$ for $r = L$. Thus, we can write

$$mL\ddot{\theta} = -mg\sin\theta.$$

Canceling the masses, this then gives us the nonlinear pendulum equation

$$L\ddot{\theta} + g\sin\theta = 0. \tag{2.36}$$

We note that this equation is of the same form as the mass-spring system. We define $\omega = \sqrt{g/L}$ and obtain the equation for simple harmonic motion,

$$\ddot{\theta} + \omega^2\theta = 0.$$

There are several variations of Equation (2.36) that will be used in this text. The first one is the linear pendulum. This is obtained by making a small angle approximation. For small angles we know that $\sin\theta \approx \theta$. Under this approximation, Equation (2.36) becomes

$$L\ddot{\theta} + g\theta = 0. \tag{2.37}$$

Linear and nonlinear pendulum equation.

The equation for a compound pendulum takes a similar form. We start with the rotational form of Newton's Second Law: $\tau = I\alpha$. Noting that the torque due to gravity acts at the center of mass position ℓ, the torque is given by $\tau = -mg\ell\sin\theta$. Because $\alpha = \ddot{\theta}$, we have $I\ddot{\theta} = -mg\ell\sin\theta$. Then, for small angles $\ddot{\theta} + \omega^2\theta = 0$, where $\omega = \frac{mg\ell}{I}$. For a simple pendulum, we let $\ell = L$ and $I = mL^2$, and obtain $\omega = \sqrt{g/L}$.

2.4   Second-Order Linear Differential Equations

IN THE LAST SECTION WE SAW how second order differential equations naturally appear in the derivations for simple oscillating systems. In this section we will look at more general second-order linear differential equations.

Second-order differential equations are typically harder than first-order. In most cases, students are only exposed to second-order linear differential equations. A general form for a *second order linear differential equation* is given by

$$a(x)y''(x) + b(x)y'(x) + c(x)y(x) = f(x). \tag{2.38}$$

One can rewrite this equation using operator terminology. Namely, one first defines the differential operator $L = a(x)D^2 + b(x)D + c(x)$, where $D = \frac{d}{dx}$. Then Equation (2.38) becomes

$$Ly = f. \tag{2.39}$$

The solutions of linear differential equations are found by making use of the linearity of L. Namely, we consider the *vector space*[1] consisting of real-valued functions over some domain. Let f and g be vectors in this function space. L is a *linear operator* if for two vectors f and g and scalar a, we have that

a. $L(f + g) = Lf + Lg$

b. $L(af) = aLf$

One typically solves Equation (2.38) by finding the general solution of the homogeneous problem

$$Ly_h = 0$$

and a particular solution of the nonhomogeneous problem

$$Ly_p = f.$$

Then, the general solution of Equation (2.38) is simply given as $y = y_h + y_p$. This is true because of the linearity of L. Namely,

$$\begin{aligned} Ly &= L(y_h + y_p) \\ &= Ly_h + Ly_p \\ &= 0 + f = f. \end{aligned} \tag{2.40}$$

There are methods for finding a particular solution of a nonhomogeneous differential equation. These methods range from pure guessing, to the Method of Undetermined Coefficients, the Method of Variation of Parameters, or Green's functions. We will review these methods later in this chapter.

Determining solutions to the homogeneous problem $Ly_h = 0$ is not always easy. However, many now-famous mathematicians and physicists

[1] We assume that the reader has been introduced to concepts in linear algebra. Later in the text we will recall the definition of a vector space and see that linear algebra is in the background of the study of many concepts in the solution of differential equations.

have studied a variety of second-order linear equations and they have saved us the trouble of finding solutions to the differential equations that often appear in applications. We will encounter many of these in the following chapters. We will first begin with some simple homogeneous linear differential equations.

Linearity is also useful in producing the general solution of a homogeneous linear differential equation. If y_1 and y_2 are solutions of the homogeneous equation, then the *linear combination* $y = c_1 y_1 + c_2 y_2$ is also a solution of the homogeneous equation. In fact, if y_1 and y_2 are *linearly independent*,[2] then $y = c_1 y_1 + c_2 y_2$ is the general solution of the homogeneous problem.

Linear independence can also be established by looking at the Wronskian of the solutions. For a second-order differential equation the Wronskian is defined as

$$W(y_1, y_2) = y_1(x) y_2'(x) - y_1'(x) y_2(x). \tag{2.41}$$

The solutions are linearly independent if the Wronskian is not zero.

2.4.1 *Constant Coefficient Equations*

THE SIMPLEST SECOND-ORDER DIFFERENTIAL EQUATIONS are those with constant coefficients. The general form for a homogeneous constant coefficient second order linear differential equation is given as

$$ay''(x) + by'(x) + cy(x) = 0, \tag{2.42}$$

where a, b, and c are constants.

Solutions to Equation (2.42) are obtained by making a guess of $y(x) = e^{rx}$. Inserting this guess into Equation (2.42) leads to the characteristic equation

$$ar^2 + br + c = 0. \tag{2.43}$$

Namely, we compute the derivatives of $y(x) = e^{rx}$, to get $y(x) = re^{rx}$, and $y(x) = r^2 e^{rx}$. Inserting into Equation (2.42), we have

$$0 = ay''(x) + by'(x) + cy(x) = (ar^2 + br + c)e^{rx}.$$

Because the exponential is never zero, we find that $ar^2 + br + c = 0$.

The roots of this equation, r_1, r_2, in turn lead to three types of solutions, depending upon the nature of the roots. In general, we have two linearly independent solutions, $y_1(x) = e^{r_1 x}$ and $y_2(x) = e^{r_2 x}$, and the general solution is given by a linear combination of these solutions,

$$y(x) = c_1 e^{r_1 x} + c_2 e^{r_2 x}.$$

For two real distinct roots, we are done. However, when the roots are real, but equal, or complex conjugate roots, we need to do a little more work to obtain usable solutions.

[2] A set of functions $\{y_i(x)\}_{i=1}^n$ is a linearly independent set if and only if

$$c_1 y_1(x) + \ldots + c_n y_n(x) = 0$$

implies $c_i = 0$, for $i = 1, \ldots, n$.

For $n = 2$, $c_1 y_1(x) + c_2 y_2(x) = 0$. If y_1 and y_2 are linearly dependent, then the coefficients are not zero and $y_2(x) = -\frac{c_1}{c_2} y_1(x)$ and is a multiple of $y_1(x)$.

The characteristic equation for $ay'' + by' + cy = 0$ is $ar^2 + br + c = 0$. Solutions of this quadratic equation lead to solutions of the differential equation.

Two real, distinct roots, r_1 and r_2, give solutions of the form

$$y(x) = c_1 e^{r_1 x} + c_2 e^{r_2 x}.$$

Example 2.5. $y'' - y' - 6y = 0$ $y(0) = 2, y'(0) = 0$.

The characteristic equation for this problem is $r^2 - r - 6 = 0$. The roots of this equation are found as $r = -2, 3$. Therefore, the general solution can be quickly written down:

$$y(x) = c_1 e^{-2x} + c_2 e^{3x}.$$

Note that there are two arbitrary constants in the general solution. Therefore, one needs two pieces of information to find a particular solution. Of course, we have the needed information in the form of the initial conditions.

One also needs to evaluate the first derivative

$$y'(x) = -2c_1 e^{-2x} + 3c_2 e^{3x}$$

in order to attempt to satisfy the initial conditions. Evaluating y and y' at $x = 0$ yields

$$
\begin{aligned}
2 &= c_1 + c_2 \\
0 &= -2c_1 + 3c_2
\end{aligned}
\tag{2.44}
$$

These two equations in two unknowns can readily be solved to give $c_1 = 6/5$ and $c_2 = 4/5$. Therefore, the solution of the initial value problem is obtained as $y(x) = \frac{6}{5}e^{-2x} + \frac{4}{5}e^{3x}$.

Repeated roots, $r_1 = r_2 = r$, give solutions of the form

$$y(x) = (c_1 + c_2 x)e^{rx}.$$

In the case when there is a repeated real root, one has only one solution, $y_1(x) = e^{rx}$. The question is how does one obtain the second linearly independent solution? Because the solutions should be independent, we must have that the ratio $y_2(x)/y_1(x)$ is not a constant. So, we guess the form $y_2(x) = v(x)y_1(x) = v(x)e^{rx}$. (This process is called the Method of Reduction of Order.)

For constant coefficient second-order equations, we can write the equation as

$$(D - r)^2 y = 0,$$

where $D = \frac{d}{dx}$. We now insert $y_2(x) = v(x)e^{rx}$ into this equation. First we compute

$$(D - r)ve^{rx} = v'e^{rx}.$$

Then,

$$0 = (D - r)^2 ve^{rx} = (D - r)v'e^{rx} = v''e^{rx}.$$

So, if $y_2(x)$ is to be a solution to the differential equation, then $v''(x)e^{rx} = 0$ for all x. So, $v''(x) = 0$, which implies that

$$v(x) = ax + b.$$

So,

$$y_2(x) = (ax + b)e^{rx}.$$

Without loss of generality, we can take $b = 0$ and $a = 1$ to obtain the second linearly independent solution, $y_2(x) = xe^{rx}$. The general solution is then

$$y(x) = c_1 e^{rx} + c_2 xe^{rx}.$$

Example 2.6. $y'' + 6y' + 9y = 0$.

In this example we have $r^2 + 6r + 9 = 0$. There is only one root, $r = -3$. From the above discussion, we easily find the solution $y(x) = (c_1 + c_2 x)e^{-3x}$.

When one has complex roots in the solution of constant coefficient equations, one needs to look at the solutions

$$y_{1,2}(x) = e^{(\alpha \pm i\beta)x}.$$

We make use of Euler's Formula (See Chapter 6 for more on complex variables)

$$e^{i\beta x} = \cos \beta x + i \sin \beta x. \qquad (2.45)$$

Then, the linear combination of $y_1(x)$ and $y_2(x)$ becomes

$$
\begin{aligned}
Ae^{(\alpha + i\beta)x} + Be^{(\alpha - i\beta)x} &= e^{\alpha x}\left[Ae^{i\beta x} + Be^{-i\beta x} \right] \\
&= e^{\alpha x}\left[(A + B)\cos \beta x + i(A - B)\sin \beta x \right] \\
&\equiv e^{\alpha x}(c_1 \cos \beta x + c_2 \sin \beta x). \qquad (2.46)
\end{aligned}
$$

Thus, we see that we have a linear combination of two real, linearly independent solutions, $e^{\alpha x}\cos \beta x$ and $e^{\alpha x}\sin \beta x$.

Complex roots, $r = \alpha \pm i\beta$, give solutions of the form

$$y(x) = e^{\alpha x}(c_1 \cos \beta x + c_2 \sin \beta x).$$

Example 2.7. $y'' + 4y = 0$.

The characteristic equation in this case is $r^2 + 4 = 0$. The roots are pure imaginary roots, $r = \pm 2i$, and the general solution consists purely of sinusoidal functions, $y(x) = c_1 \cos(2x) + c_2 \sin(2x)$, as $\alpha = 0$ and $\beta = 2$.

Example 2.8. $y'' + 2y' + 4y = 0$.

The characteristic equation in this case is $r^2 + 2r + 4 = 0$. The roots are complex, $r = -1 \pm \sqrt{3}i$ and the general solution can be written as

$$y(x) = \left[c_1 \cos(\sqrt{3}x) + c_2 \sin(\sqrt{3}x) \right] e^{-x}.$$

Example 2.9. $y'' + 4y = \sin x$.

This is an example of a nonhomogeneous problem. The homogeneous problem was actually solved in Example 2.7. According to the theory, we need only seek a particular solution to the nonhomogeneous problem and add it to the solution of the last example to get the general solution.

The particular solution can be obtained by purely guessing, making an educated guess, or using the Method of Variation of Parameters. We will not review all of these techniques at this time. Due to the simple form of the driving term, we will make an intelligent guess of $y_p(x) = A \sin x$ and determine what A needs to be. Inserting this guess into the differential equation gives $(-A + 4A)\sin x = \sin x$. So, we see that $A = 1/3$ works. The general solution of the nonhomogeneous problem is therefore $y(x) = c_1 \cos(2x) + c_2 \sin(2x) + \frac{1}{3}\sin x$.

The three cases for constant coefficient linear second-order differential equations are summarized below.

Classification of Roots of the Characteristic Equation
for Second-Order Constant Coefficient ODEs

---

1. **Real, distinct roots** r_1, r_2. In this case, the solutions corresponding to each root are linearly independent. Therefore, the general solution is simply $y(x) = c_1 e^{r_1 x} + c_2 e^{r_2 x}$.

2. **Real, equal roots** $r_1 = r_2 = r$. In this case, the solutions corresponding to each root are linearly dependent. To find a second linearly independent solution, one uses the *Method of Reduction of Order*. This gives the second solution as $x e^{rx}$. Therefore, the general solution is found as $y(x) = (c_1 + c_2 x) e^{rx}$.

3. **Complex conjugate roots** $r_1, r_2 = \alpha \pm i\beta$. In this case, the solutions corresponding to each root are linearly independent. Making use of Euler's identity, $e^{i\theta} = \cos(\theta) + i\sin(\theta)$, these complex exponentials can be rewritten in terms of trigonometric functions. Namely, one has that $e^{\alpha x} \cos(\beta x)$ and $e^{\alpha x} \sin(\beta x)$ are two linearly independent solutions. Therefore, the general solution becomes $y(x) = e^{\alpha x}(c_1 \cos(\beta x) + c_2 \sin(\beta x))$.

As we have seen, one of the most important applications of such equations is in the study of oscillations. Typical systems are a mass on a spring, or a simple pendulum. For a mass m on a spring with spring constant $k > 0$, one has from Hooke's Law that the position as a function of time, $x(t)$, satisfies the equation

$$m\ddot{x} + kx = 0.$$

This constant coefficient equation has pure imaginary roots ($\alpha = 0$), and the solutions are simple sine and cosine functions, leading to simple harmonic motion.

2.5   LRC Circuits

ANOTHER TYPICAL PROBLEM OFTEN ENCOUNTERED in a first-year physics class is that of an LRC series circuit. This circuit is pictured in Figure 2.7. The resistor is a circuit element satisfying Ohm's Law. The capacitor is a device that stores electrical energy and an inductor, or coil, that stores magnetic energy.

The physics for this problem stems from Kirchoff's Rules for circuits. Namely, the sum of the drops in electric potential are set equal to the rises in electric potential. The potential drops across each circuit element are given by

1. Resistor: $V = IR$.

2. Capacitor: $V = \frac{q}{C}$.

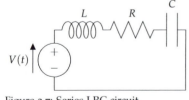

Figure 2.7: Series LRC circuit.

3. Inductor: $V = L\frac{dI}{dt}$.

Furthermore, we need to define the current as $I = \frac{dq}{dt}$. where q is the charge in the circuit. Adding these potential drops, we set them equal to the voltage supplied by the voltage source, $V(t)$. Thus, we obtain

$$IR + \frac{q}{C} + L\frac{dI}{dt} = V(t).$$

Because both q and I are unknown, we can replace the current by its expression in terms of the charge to obtain

$$L\ddot{q} + R\dot{q} + \frac{1}{C}q = V(t).$$

This is a second-order equation for $q(t)$.

More complicated circuits are possible by looking at parallel connections, or other combinations, of resistors, capacitors, and inductors. This will result in several equations for each loop in the circuit, leading to larger systems of differential equations. An example of another circuit setup is shown in Figure 2.8. This is not a problem that can be covered in the first-year physics course. One can set up a system of second-order equations and proceed to solve them. We will see how to solve such problems in the next chapter.

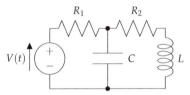

Figure 2.8: Parallel LRC circuit.

2.5.1 Special Cases

IN THIS SECTION WE WILL LOOK AT SPECIAL CASES that arise for the series LRC circuit equation. These include RC circuits, solvable by first-order methods and LC circuits, leading to oscillatory behavior.

Case I. RC Circuits

We first consider the case of an RC circuit in which there is no inductor. Also, we will consider what happens when one charges a capacitor with a DC battery ($V(t) = V_0$) and when one discharges a charged capacitor ($V(t) = 0$) as shown in Figures 2.9 and 2.12, respectively.

For charging a capacitor, we have the initial value problem

$$R\frac{dq}{dt} + \frac{q}{C} = V_0, \quad q(0) = 0. \tag{2.47}$$

This equation is an example of a linear first order equation for $q(t)$. However, we can also rewrite it and solve it as a separable equation, as V_0 is a constant. We will do the former only as another example of finding the integrating factor.

We first write the equation in standard form:

$$\frac{dq}{dt} + \frac{q}{RC} = \frac{V_0}{R}. \tag{2.48}$$

The integrating factor is then

$$\mu(t) = e^{\int \frac{dt}{RC}} = e^{t/RC}.$$

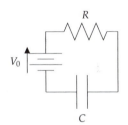

Figure 2.9: RC Circuit for charging.

Charging a capacitor.

Thus,

$$\frac{d}{dt}\left(qe^{t/RC}\right) = \frac{V_0}{R}e^{t/RC}. \tag{2.49}$$

Integrating, we have

$$qe^{t/RC} = \frac{V_0}{R}\int e^{t/RC}\,dt = CV_0 e^{t/RC} + K. \tag{2.50}$$

Note that we introduced the integration constant K. Now divide out the exponential to get the general solution:

$$q = CV_0 + Ke^{-t/RC}. \tag{2.51}$$

(If we had forgotten the K, we would not have gotten a correct solution for the differential equation.)

Next, we use the initial condition to get the particular solution. Namely, setting $t = 0$, we have that

$$0 = q(0) = CV_0 + K.$$

So, $K = -CV_0$. Inserting this into the solution, we have

$$q(t) = CV_0(1 - e^{-t/RC}). \tag{2.52}$$

Now we can study the behavior of this solution. For large times, the second term goes to zero. Thus, the capacitor charges up, asymptotically, to the final value of $q_0 = CV_0$. This is what we expect, because the current is no longer flowing over R and this just gives the relation between the potential difference across the capacitor plates when a charge of q_0 is established on the plates.

Figure 2.10: The charge as a function of time for a charging capacitor with $R = 2.00$ kΩ, $C = 6.00$ mF, and $V_0 = 12$ V.

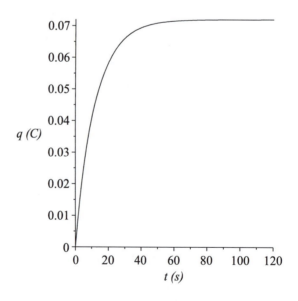

Let's put in some values for the parameters. We let $R = 2.00$ kΩ, $C = 6.00$ mF, and $V_0 = 12$ V. A plot of the solution is given in Figure 2.10. We see that the charge builds up to the value of $CV_0 = 0.072$ C. If we use a smaller resistance, $R = 200\ \Omega$, we see in Figure 2.11 that the capacitor charges to the same value, but much faster.

The rate at which a capacitor charges, or discharges, is governed by the time constant, $\tau = RC$. This is the constant factor in the exponential. The larger it is, the slower the exponential term decays. If we set $t = \tau$, we find that

Time constant, $\tau = RC$.

$$q(\tau) = CV_0(1 - e^{-1}) = (1 - 0.3678794412\ldots)q_0 \approx 0.63q_0.$$

Thus, at time $t = \tau$, the capacitor has almost charged to two thirds of its final value. For the first set of parameters, $\tau = 12$ s. For the second set, $\tau = 1.2$ s.

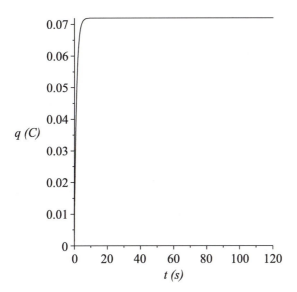

Figure 2.11: The charge as a function of time for a charging capacitor with $R = 200\ \Omega$, $C = 6.00$ mF, and $V_0 = 12$ V.

Now, let's assume the capacitor is charged with charge $\pm q_0$ on its plates. If we disconnect the battery and reconnect the wires to complete the circuit as shown in Figure 2.12, the charge will then move off the plates, discharging the capacitor. The relevant form of the initial value problem becomes

Discharging a capacitor.

$$R\frac{dq}{dt} + \frac{q}{C} = 0, \quad q(0) = q_0. \tag{2.53}$$

This equation is simpler to solve. Rearranging, we have

$$\frac{dq}{dt} = -\frac{q}{RC}. \tag{2.54}$$

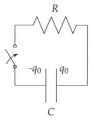

Figure 2.12: RC Circuit for discharging.

This is a simple exponential decay problem, which one can solve using separation of variables. However, by now you should know how to immediately write down the solution to such problems of the form $y' = ky$. The solution is

$$q(t) = q_0 e^{-t/\tau}, \quad \tau = RC.$$

We see that the charge decays exponentially. In principle, the capacitor never fully discharges. That is why you are often instructed to place a shunt across a discharged capacitor to fully discharge it.

In Figure 2.13 we show the discharging of the two previous RC circuits. Once again, $\tau = RC$ determines the behavior. At $t = \tau$ we have

$$q(\tau) = q_0 e^{-1} = (0.3678794412\ldots)q_0 \approx 0.37q_0.$$

So, at this time, the capacitor only has about a third of its original value.

Figure 2.13: The charge as a function of time for a discharging capacitor with $R = 2.00$ kΩ (solid) or $R = 200$ Ω (dashed), and $C = 6.00$ mF, and $q_0 = 0.072$ C.

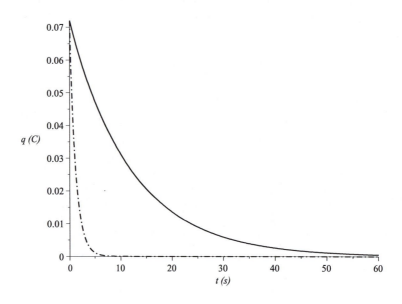

Case II. LC Circuits

LC Oscillators.

Another simple result comes from studying LC circuits. We will now connect a charged capacitor to an inductor as shown in Figure 2.14. In this case, we consider the initial value problem

$$L\ddot{q} + \frac{1}{C}q = 0, \quad q(0) = q_0, \dot{q}(0) = I(0) = 0. \tag{2.55}$$

Dividing out the inductance, we have

$$\ddot{q} + \frac{1}{LC}q = 0. \tag{2.56}$$

This equation is a second-order, constant coefficient equation. It is of the same form as the ones for simple harmonic motion of a mass on a spring or the linear pendulum. So, we expect oscillatory behavior. The characteristic equation is

$$r^2 + \frac{1}{LC} = 0.$$

The solutions are

$$r_{1,2} = \pm \frac{i}{\sqrt{LC}}.$$

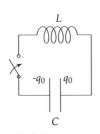

Figure 2.14: An LC circuit.

Thus, the solution of Equation (2.56) is of the form

$$q(t) = c_1 \cos(\omega t) + c_2 \sin(\omega t), \quad \omega = (LC)^{-1/2}. \tag{2.57}$$

Inserting the initial conditions yields

$$q(t) = q_0 \cos(\omega t). \tag{2.58}$$

The oscillations that result are understandable. As the charge leaves the plates, the changing current induces a changing magnetic field in the inductor. The stored electrical energy in the capacitor changes to stored magnetic energy in the inductor. However, the process continues until the plates are charged with opposite polarity and then the process begins in reverse. The charged capacitor then discharges and the capacitor eventually returns to its original state and the whole system repeats this over and over.

The frequency of this simple harmonic motion is easily found. It is given by

$$f = \frac{\omega}{2\pi} = \frac{1}{2\pi} \frac{1}{\sqrt{LC}}. \tag{2.59}$$

This is called the tuning frequency because of its role in tuning circuits.

Example 2.10. *Find the resonant frequency for* $C = 10\mu F$ *and* $L = 100mH$.

$$f = \frac{1}{2\pi} \frac{1}{\sqrt{(10 \times 10^{-6})(100 \times 10^{-3})}} = 160\,\text{Hz}.$$

Of course, this is an ideal situation. There is always resistance in the circuit, even if only a small amount from the wires. So, we really need to account for resistance, or even add a resistor. This leads to a slightly more complicated system in which damping will be present.

2.6 Damped Oscillations

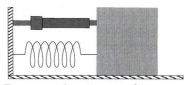

Figure 2.15: A spring-mass-damper system has a damper added that can absorb some of the energy of the oscillations and is modeled with a term proportional to the velocity.

AS WE HAVE INDICATED, simple harmonic motion is an ideal situation. In real systems we often have to contend with some energy loss in the system. This leads to the damping of the oscillations. This energy loss could be in the spring, in the way a pendulum is attached to its support, or in the resistance to the flow of current in an LC circuit. The simplest models of resistance are the addition of a term proportional to first derivative of the dependent variable. Thus, our three main examples with damping added look like:

$$m\ddot{x} + b\dot{x} + kx = 0. \tag{2.60}$$

$$L\ddot{\theta} + b\dot{\theta} + g\theta = 0. \tag{2.61}$$

$$L\ddot{q} + R\dot{q} + \frac{1}{C}q = 0. \tag{2.62}$$

These are all examples of the general constant coefficient equation

$$ay''(x) + by'(x) + cy(x) = 0. \tag{2.63}$$

We have seen that solutions are obtained by looking at the characteristic equation $ar^2 + br + c = 0$. This leads to three different behaviors, depending on the discriminant in the quadratic formula:

$$r = \frac{-b \pm \sqrt{b^2 - 4ac}}{2a}.$$
(2.64)

We will consider the example of the damped spring. Then we have

$$r = \frac{-b \pm \sqrt{b^2 - 4mk}}{2m}.$$
(2.65)

For $b > 0$, there are three types of damping.

Damped oscillator cases: Overdamped, critically damped, and underdamped.

I. Overdamped, $b^2 > 4mk$

In this case, we obtain two real roots. Because this is Case I for constant coefficient equations, we have that

$$x(t) = c_1 e^{r_1 t} + c_2 e^{r_2 t}.$$

We note that $b^2 - 4mk < b^2$. Thus, the roots are both negative. So, both terms in the solution exponentially decay. The damping is so strong that there is no oscillation in the system.

II. Critically Damped, $b^2 = 4mk$

In this case, we obtain one real root. This is Case II for constant coefficient equations and the solution is given by

$$x(t) = (c_1 + c_2 t)e^{rt},$$

where $r = -b/2m$. Once again, the solution decays exponentially. The damping is just strong enough to hinder any oscillation. If it were any weaker, the discriminant would be negative and we would need the third case.

III. Underdamped, $b^2 < 4mk$

Figure 2.16: A plot of underdamped oscillation given by $x(t) = 2e^{0.15t} \cos 3t$. The dashed lines are given by $x(t) = \pm 2e^{0.15t}$, indicating the bounds on the amplitude of the motion.

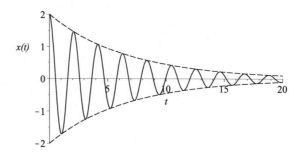

In this case, we have complex conjugate roots. We can write $\alpha = -b/2m$ and $\beta = \sqrt{4mk - b^2}/2m$. Then the solution is

$$x(t) = e^{\alpha t}(c_1 \cos \beta t + c_2 \sin \beta t).$$

These solutions exhibit oscillations due to the trigonometric functions, but we see that the amplitude may decay in time due the overall factor of $e^{\alpha t}$ when $\alpha < 0$. Consider the case that the initial conditions give $c_1 = A$ and $c_2 = 0$. (When is this?) Then, the solution, $x(t) = Ae^{\alpha t}\cos\beta t$, looks like the plot in Figure 2.16.

2.7   Forced Systems

ALL OF THE SYSTEMS PRESENTED at the beginning of the last section exhibit the same general behavior when a damping term is present. An additional term can be added that might cause even more complicated behavior. In the case of LRC circuits, we have seen that the voltage source makes the system nonhomogeneous. It provides what is called a source term. Such terms can also arise in the mass-spring and pendulum systems. One can drive such systems by periodically pushing the mass, or having the entire system moved, or impacted by an outside force. Such systems are called forced, or driven.

Typical systems in physics can be modeled by nonhomogeneous second-order equations. Thus, we want to find solutions of equations of the form

$$Ly(x) = a(x)y''(x) + b(x)y'(x) + c(x)y(x) = f(x). \qquad (2.66)$$

As noted in Section 2.4, one solves this equation by finding the general solution of the homogeneous problem,

$$Ly_h = 0$$

and a particular solution of the nonhomogeneous problem,

$$Ly_p = f.$$

Then, the general solution of Equation (2.38) is simply given as $y = y_h + y_p$.

So far, we only know how to solve constant coefficient, homogeneous equations. So, by adding a nonhomogeneous term to such equations, we will need to find the particular solution to the nonhomogeneous equation.

We could guess a solution, but that is not usually possible without a little bit of experience. So, we need some other methods. There are two main methods. In the first case, the Method of Undetermined Coefficients, one makes an intelligent guess based on the form of $f(x)$. In the second method, one can systematically develop the particular solution. We will come back to the Method of Variation of Parameters and will also introduce the powerful machinery of Green's functions later in this section.

2.7.1   Method of Undetermined Coefficients

LET'S SOLVE A SIMPLE DIFFERENTIAL EQUATION highlighting how we can handle nonhomogeneous equations.

Example 2.11. *Consider the equation*

$$y'' + 2y' - 3y = 4. \qquad (2.67)$$

The first step is to determine the solution of the homogeneous equation. Thus, we solve

$$y_h'' + 2y_h' - 3y_h = 0. \qquad (2.68)$$

The characteristic equation is $r^2 + 2r - 3 = 0$. The roots are $r = 1, -3$. So, we can immediately write the solution

$$y_h(x) = c_1 e^x + c_2 e^{-3x}.$$

The second step is to find a particular solution of Equation (2.67). What possible function can we insert into this equation such that only a 4 remains? If we try something proportional to x, then we are left with a linear function after inserting x and its derivatives. Perhaps a constant function you might think. $y = 4$ does not work. But, we could try an arbitrary constant, $y = A$.

Let's see. Inserting $y = A$ into (2.67), we obtain

$$-3A = 4.$$

Ah ha! We see that we can choose $A = -\frac{4}{3}$ and this works. So, we have a particular solution, $y_p(x) = -\frac{4}{3}$. This step is done.

Combining the two solutions, we have the general solution to the original non-homogeneous equation (2.67). Namely,

$$y(x) = y_h(x) + y_p(x) = c_1 e^x + c_2 e^{-3x} - \frac{4}{3}.$$

Insert this solution into the equation and verify that it is indeed a solution. If we had been given initial conditions, we could now use them to determine the arbitrary constants.

Example 2.12. *What if we had a different source term? Consider the equation*

$$y'' + 2y' - 3y = 4x. \qquad (2.69)$$

The only thing that would change is the particular solution. So, we need a guess.

We know a constant function does not work by the previous example. So, let's try $y_p = Ax$. Inserting this function into Equation (2.69), we obtain

$$2A - 3Ax = 4x.$$

Picking $A = -4/3$ would get rid of the x terms, but will not cancel everything. We still have a constant remaining. So, we need something more general.

Let's try a linear function, $y_p(x) = Ax + B$. Then we get, after substitution into Equation (2.69),

$$2A - 3(Ax + B) = 4x.$$

Equating the coefficients of the different powers of x on both sides, we find a system of equations for the undetermined coefficients:

$$
\begin{aligned}
2A - 3B &= 0 \\
-3A &= 4.
\end{aligned}
\qquad (2.70)
$$

These are easily solved to obtain

$$A = -\frac{4}{3}$$
$$B = \frac{2}{3}A = -\frac{8}{9}. \qquad (2.71)$$

So, the particular solution is

$$y_p(x) = -\frac{4}{3}x - \frac{8}{9}.$$

This gives the general solution to the nonhomogeneous problem as

$$y(x) = y_h(x) + y_p(x) = c_1 e^x + c_2 e^{-3x} - \frac{4}{3}x - \frac{8}{9}.$$

There are general forms that you can guess based upon the form of the driving term, $f(x)$. Some examples are given in Table 2.1. More general applications are covered in a standard text on differential equations. However, the procedure is simple. Given $f(x)$ in a particular form, you make an appropriate guess up to some unknown parameters, or coefficients. Inserting the guess leads to a system of equations for the unknown coefficients. Solve the system and you have the solution. This solution is then added to the general solution of the homogeneous differential equation.

$f(x)$	Guess
$a_n x^n + a_{n-1}x^{n-1} + \cdots + a_1 x + a_0$	$A_n x^n + A_{n-1}x^{n-1} + \cdots + A_1 x + A_0$
ae^{bx}	Ae^{bx}
$a \cos \omega x + b \sin \omega x$	$A \cos \omega x + B \sin \omega x$

Table 2.1: Forms Used in the Method of Undetermined Coefficients

Example 2.13. *Solve*

$$y'' + 2y' - 3y = 2e^{-3x}. \qquad (2.72)$$

According to the above, we would guess a solution of the form $y_p = Ae^{-3x}$. Inserting our guess, we find

$$0 = 2e^{-3x}.$$

Oops! The coefficient, A, disappeared! We cannot solve for it. What went wrong?

The answer lies in the general solution of the homogeneous problem. Note that e^x and e^{-3x} are solutions to the homogeneous problem. So, a multiple of e^{-3x} will not get us anywhere. It turns out that there is one further modification of the method. If the driving term contains terms that are solutions of the homogeneous problem, then we need to make a guess consisting of the smallest possible power of x times the function which is no longer a solution of the homogeneous problem. Namely, we guess $y_p(x) = Axe^{-3x}$ and differentiate this guess to obtain the derivatives $y'_p = A(1 - 3x)e^{-3x}$ and $y''_p = A(9x - 6)e^{-3x}$.

Inserting these derivatives into the differential equation, we obtain

$$[(9x - 6) + 2(1 - 3x) - 3x]Ae^{-3x} = 2e^{-3x}.$$

Comparing coefficients, we have

$$-4A = 2.$$

So, $A = -1/2$ and $y_p(x) = -\frac{1}{2}xe^{-3x}$. Thus, the solution to the problem is

$$y(x) = \left(2 - \frac{1}{2}x\right)e^{-3x}.$$

Modified Method of Undetermined Coefficients

In general, if any term in the guess $y_p(x)$ is a solution of the homogeneous equation, then multiply the guess by x^k, where k is the smallest positive integer such that no term in $x^k y_p(x)$ is a solution of the homogeneous problem.

2.7.2 Periodically Forced Oscillations

A SPECIAL TYPE OF FORCING is periodic forcing. Realistic oscillations will dampen and eventually stop if left unattended. For example, mechanical clocks are driven by compound or torsional pendula, and electric oscillators are often designed with the need to continue for long periods of time. However, they are not perpetual motion machines and will need a periodic injection of energy. This can be done systematically by adding periodic forcing. Another simple example is the motion of a child on a swing in the park. This simple damped pendulum system will naturally slow down to equilibrium (stopped) if left alone. However, if the child pumps energy into the swing at the right time, or if an adult pushes the child at the right time, then the amplitude of the swing can be increased.

There are other systems, such as airplane wings and long bridge spans, in which external driving forces might cause damage to the system. A well-known example is the wind-induced collapse of the Tacoma Narrows Bridge due to strong winds. Of course, if one is not careful, the child in the previous example might get too much energy pumped into the system, causing a similar failure of the desired motion.

While there are many types of forced systems, and some fairly complicated, we can easily get to the basic characteristics of forced oscillations by modifying the mass-spring system by adding an external, time-dependent, driving force. Such a system satisfies the equation

$$m\ddot{x} + b\dot{(x)} + kx = F(t), \tag{2.73}$$

where m is the mass, b is the damping constant, k is the spring constant, and $F(t)$ is the driving force. If $F(t)$ is of simple form, then we can employ the Method of Undetermined Coefficients. Because the systems we have considered so far are similar, one could easily apply the following to pendula or circuits.

As the damping term only complicates the solution, we will consider the simpler case of undamped motion and assume that $b = 0$. Furthermore, we

The Tacoma Narrows Bridge opened in Washington State (U.S.) in mid-1940. However, in November of the same year, the winds excited a transverse mode of vibration, which eventually (in a few hours) led to large amplitude oscillations and then collapse.

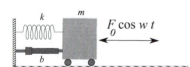

Figure 2.17: An external driving force is added to the spring-mass-damper system.

will introduce a sinusoidal driving force, $F(t) = F_0 \cos \omega t$ in order to study periodic forcing. This leads to the simple, periodically driven mass on a spring system

$$m\ddot{x} + kx = F_0 \cos \omega t. \qquad (2.74)$$

In order to find the general solution, we first obtain the solution to the homogeneous problem,

$$x_h = c_1 \cos \omega_0 t + c_2 \sin \omega_0 t,$$

where $\omega_0 = \sqrt{\frac{k}{m}}$. Next, we seek a particular solution to the nonhomogeneous problem. We will apply the Method of Undetermined Coefficients.

A natural guess for the particular solution would be to use $x_p = A \cos \omega t + B \sin \omega t$. However, recall that the guess should not be a solution of the homogeneous problem. Comparing x_p with x_h, this would hold if $\omega \neq \omega_0$. Otherwise, one would need to use the Modified Method of Undetermined Coefficients as described in the previous section. So, we have two cases to consider.

Dividing through by the mass, we solve the simple driven system,

$$\ddot{x} + \omega_0^2 x = \frac{F_0}{m} \cos \omega t.$$

Example 2.14. *Solve* $\ddot{x} + \omega_0^2 x = \frac{F_0}{m} \cos \omega t$, *for* $\omega \neq \omega_0$.

In this case, we continue with the guess $x_p = A \cos \omega t + B \sin \omega t$. *Because there is no damping term, one quickly finds that* $B = 0$. *Inserting* $x_p = A \cos \omega t$ *into the differential equation, we find that*

$$\left(-\omega^2 + \omega_0^2\right) A \cos \omega t = \frac{F_0}{m} \cos \omega t.$$

Solving for A, we obtain

$$A = \frac{F_0}{m(\omega_0^2 - \omega^2)}.$$

The general solution for this case is thus

$$x(t) = c_1 \cos \omega_0 t + c_2 \sin \omega_0 t + \frac{F_0}{m(\omega_0^2 - \omega^2)} \cos \omega t. \qquad (2.75)$$

Example 2.15. *Solve* $\ddot{x} + \omega_0^2 x = \frac{F_0}{m} \cos \omega_0 t$.

In this case, we need to employ the Modified Method of Undetermined Coefficients. So, we make the guess $x_p = t(A \cos \omega_0 t + B \sin \omega_0 t)$. *As there is no damping term, one finds that* $A = 0$. *Inserting the guess into the differential equation, we find that*

$$B = \frac{F_0}{2m\omega_0},$$

or the general solution is

$$x(t) = c_1 \cos \omega_0 t + c_2 \sin \omega_0 t + \frac{F_0}{2m\omega} t \sin \omega t. \qquad (2.76)$$

The general solution to the problem is thus

$$x(t) = c_1 \cos \omega_0 t + c_2 \sin \omega_0 t + \begin{cases} \frac{F_0}{m(\omega_0^2 - \omega^2)} \cos \omega t, & \omega \neq \omega_0, \\ \frac{F_0}{2m\omega_0} t \sin \omega_0 t, & \omega = \omega_0. \end{cases} \qquad (2.77)$$

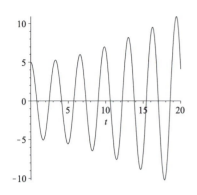

Figure 2.18: Plot of

$$x(t) = 5\cos 2t + \frac{1}{2}t\sin 2t,$$

a solution of $\ddot{x} + 4x = 2\cos 2t$ showing resonance.

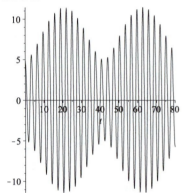

Figure 2.19: Plot of

$$x(t) = \frac{1}{249}\left(2045\cos 2t - 800\cos \frac{43}{20}t\right),$$

a solution of $\ddot{x} + 4x = 2\cos 2.15t$.

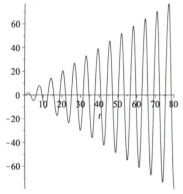

Figure 2.20: Plot of

$$x(t) = t\sin 2t,$$

a solution of $\ddot{x} + x = 2\cos t$.

Special cases of these solutions provide interesting physics, which can be explored by the reader in the homework. In the case that $\omega = \omega_0$, we see that the solution tends to grow as t gets large. This is what is called a resonance. Essentially, one is driving the system at its natural frequency. As the system is moving to the left, one pushes it to the left. If it is moving to the right, one is adding energy in that direction. This forces the amplitude of oscillation to continue to grow until the system breaks. An example of such an oscillation is shown in Figure 2.18.

In the case that $\omega \neq \omega_0$, one can rewrite the solution in a simple form. Let's choose the initial conditions that $c_1 = -F_0/(m(\omega_0^2 - \omega^2)), c_2 = 0$. Then one has (see Problem 21)

$$x(t) = \frac{2F_0}{m(\omega_0^2 - \omega^2)}\sin\frac{(\omega_0 - \omega)t}{2}\sin\frac{(\omega_0 + \omega)t}{2}. \qquad (2.78)$$

For values of ω near ω_0, one finds that the solution consists of a rapid oscillation, due to the $\sin\frac{(\omega_0+\omega)t}{2}$ factor, with a slowly varying amplitude, $\frac{2F_0}{m(\omega_0^2-\omega^2)}\sin\frac{(\omega_0-\omega)t}{2}$. The reader can investigate this solution.

This slow variation is called a beat, and the beat frequency is given by $f = \frac{|\omega_0 - \omega|}{4\pi}$. In Figure 2.19 we see that the high-frequency oscillations are contained by the lower beat frequency, $f = \frac{0.15}{4\pi}$ s. This corresponds to a period of $T = 1/f \approx 83.7$ Hz, which looks about right from the figure.

Example 2.16. *Solve $\ddot{x} + x = 2\cos\omega t$, $x(0) = 0$, $\dot{x}(0) = 0$, for $\omega = 1, 1.15$. For each case, we need the solution of the homogeneous problem,*

$$x_h(t) = c_1\cos t + c_2\sin t.$$

The particular solution depends on the value of ω.

For $\omega = 1$, the driving term, $2\cos\omega t$, is a solution of the homogeneous problem. Thus, we assume

$$x_p(t) = At\cos t + Bt\sin t.$$

Inserting this into the differential equation, we find $A = 0$ and $B = 1$. So, the general solution is

$$x(t) = c_1\cos t + c_2\sin t + t\sin t.$$

Imposing the initial conditions, we find

$$x(t) = t\sin t.$$

This solution is shown in Figure 2.20.

For $\omega = 1.15$, the driving term, $2\cos\omega t$, is not a solution of the homogeneous problem. Thus, we assume

$$x_p(t) = A\cos 1.15t + B\sin 1.15t.$$

Inserting this into the differential equation, we find $A = -\frac{800}{129}$ and $B = 0$. So, the general solution is

$$x(t) = c_1\cos t + c_2\sin t - \frac{800}{129}\cos t.$$

Imposing the initial conditions, we find

$$x(t) = \frac{800}{129} \left(\cos t - \cos 1.15t \right).$$

This solution is shown in Figure 2.21. The beat frequency in this case is the same as with Figure 2.19.

2.7.3 Method of Variation of Parameters

A MORE SYSTEMATIC WAY to find particular solutions is through the use of the Method of Variation of Parameters. The derivation is a little detailed and the solution is sometimes messy, but the application of the method is straightforward if you can do the required integrals. We will first derive the needed equations and then do some examples.

We begin with the nonhomogeneous equation. Let's assume it is of the standard form

$$a(x)y''(x) + b(x)y'(x) + c(x)y(x) = f(x). \tag{2.79}$$

We know that the solution of the homogeneous equation can be written in terms of two linearly independent solutions, which we will call $y_1(x)$ and $y_2(x)$:

$$y_h(x) = c_1 y_1(x) + c_2 y_2(x).$$

Replacing the constants with functions, then we no longer have a solution to the homogeneous equation. Is it possible that we could stumble across the right functions with which to replace the constants and somehow end up with $f(x)$ when inserted into the left side of the differential equation? It turns out that we can.

So, let's assume that the constants are replaced with two unknown functions, which we will call $c_1(x)$ and $c_2(x)$. This change in the parameters is where the name of the method derives. Thus, we are assuming that a particular solution takes the form

$$y_p(x) = c_1(x)y_1(x) + c_2(x)y_2(x). \tag{2.80}$$

If this is to be a solution, then insertion into the differential equation should make the equation hold. To do this we will first need to compute some derivatives.

The first derivative is given by

$$y_p'(x) = c_1(x)y_1'(x) + c_2(x)y_2'(x) + c_1'(x)y_1(x) + c_2'(x)y_2(x). \tag{2.81}$$

Next we will need the second derivative. But, this will yield eight terms. So, we will first make a simplifying assumption. Let's assume that the last two terms add to zero:

$$c_1'(x)y_1(x) + c_2'(x)y_2(x) = 0. \tag{2.82}$$

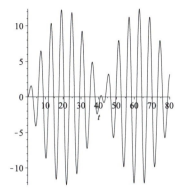

Figure 2.21: Plot of

$$x(t) = \frac{800}{129} \left(\cos t - \cos \frac{23}{20}t \right),$$

a solution of $\ddot{x} + x = 2\cos 1.15t$.

We assume the nonhomogeneous equation has a particular solution of the form

$$y_p(x) = c_1(x)y_1(x) + c_2(x)y_2(x).$$

It turns out that we will get the same results in the end if we did not assume this. The important thing is that it works!

Under the assumption, the first derivative simplifies to

$$y_p'(x) = c_1(x)y_1'(x) + c_2(x)y_2'(x).$$ (2.83)

The second derivative now only has four terms:

$$y_p'(x) = c_1(x)y_1''(x) + c_2(x)y_2''(x) + c_1'(x)y_1'(x) + c_2'(x)y_2'(x).$$ (2.84)

Now that we have the derivatives, we can insert the guess into the differential equation. Thus, we have

$$\begin{aligned} f(x) = {}& a(x)\left[c_1(x)y_1''(x) + c_2(x)y_2''(x) + c_1'(x)y_1'(x) + c_2'(x)y_2'(x)\right] \\ & + b(x)\left[c_1(x)y_1'(x) + c_2(x)y_2'(x)\right] \\ & + c(x)\left[c_1(x)y_1(x) + c_2(x)y_2(x)\right]. \end{aligned}$$ (2.85)

Regrouping the terms, we obtain

$$\begin{aligned} f(x) = {}& c_1(x)\left[a(x)y_1''(x) + b(x)y_1'(x) + c(x)y_1(x)\right] \\ & + c_2(x)\left[a(x)y_2''(x) + b(x)y_2'(x) + c(x)y_2(x)\right] \\ & + a(x)\left[c_1'(x)y_1'(x) + c_2'(x)y_2'(x)\right]. \end{aligned}$$ (2.86)

Note that the first two rows vanish because y_1 and y_2 are solutions of the homogeneous problem. This leaves the equation

$$f(x) = a(x)\left[c_1'(x)y_1'(x) + c_2'(x)y_2'(x)\right],$$

which can be rearranged as

$$c_1'(x)y_1'(x) + c_2'(x)y_2'(x) = \frac{f(x)}{a(x)}.$$ (2.87)

In summary, we have assumed a particular solution of the form

$$y_p(x) = c_1(x)y_1(x) + c_2(x)y_2(x).$$

This is only possible if the unknown functions $c_1(x)$ and $c_2(x)$ satisfy the system of equations

$$\begin{aligned} c_1'(x)y_1(x) + c_2'(x)y_2(x) &= 0 \\ c_1'(x)y_1'(x) + c_2'(x)y_2'(x) &= \frac{f(x)}{a(x)}. \end{aligned}$$ (2.88)

It is standard to solve this system for the derivatives of the unknown functions and then present the integrated forms. However, one could just as easily start from this system and solve the system for each problem encountered.

Margin notes:

In order to solve the differential equation $Ly = f$, we assume

$$y_p(x) = c_1(x)y_1(x) + c_2(x)y_2(x),$$

for $Ly_{1,2} = 0$. Then, one need only solve a simple system of equations (2.88).

System (2.88) can be solved as

$$c_1'(x) = -\frac{fy_2}{aW(y_1, y_2)},$$

$$c_1'(x) = \frac{fy_1}{aW(y_1, y_2)},$$

where $W(y_1, y_2) = y_1y_2' - y_1'y_2$ is the Wronskian. We use this solution in the next section.

Example 2.17. *Find the general solution of the nonhomogeneous problem:* $y'' - y = e^{2x}$.

The general solution to the homogeneous problem $y_h'' - y_h = 0$ *is*

$$y_h(x) = c_1 e^x + c_2 e^{-x}.$$

In order to use the Method of Variation of Parameters, we seek a solution of the form

$$y_p(x) = c_1(x)e^x + c_2(x)e^{-x}.$$

We find the unknown functions by solving the system in (2.88), which in this case becomes

$$\begin{aligned} c_1'(x)e^x + c_2'(x)e^{-x} &= 0 \\ c_1'(x)e^x - c_2'(x)e^{-x} &= e^{2x}. \end{aligned} \tag{2.89}$$

Adding these equations we find that

$$2c_1' e^x = e^{2x} \rightarrow c_1' = \frac{1}{2}e^x.$$

Solving for $c_1(x)$ *we find*

$$c_1(x) = \frac{1}{2}\int e^x\, dx = \frac{1}{2}e^x.$$

Subtracting the equations in the system yields

$$2c_2' e^{-x} = -e^{2x} \rightarrow c_2' = -\frac{1}{2}e^{3x}.$$

Thus,

$$c_2(x) = -\frac{1}{2}\int e^{3x}\, dx = -\frac{1}{6}e^{3x}.$$

The particular solution is found by inserting these results into y_p:

$$\begin{aligned} y_p(x) &= c_1(x)y_1(x) + c_2(x)y_2(x) \\ &= (\frac{1}{2}e^x)e^x + (-\frac{1}{6}e^{3x})e^{-x} \\ &= \frac{1}{3}e^{2x}. \end{aligned} \tag{2.90}$$

Thus, we have the general solution of the nonhomogeneous problem as

$$y(x) = c_1 e^x + c_2 e^{-x} + \frac{1}{3}e^{2x}.$$

Example 2.18. *Now consider the problem:* $y'' + 4y = \sin x$.

The solution to the homogeneous problem is

$$y_h(x) = c_1 \cos 2x + c_2 \sin 2x. \tag{2.91}$$

We now seek a particular solution of the form

$$y_h(x) = c_1(x)\cos 2x + c_2(x)\sin 2x.$$

We let $y_1(x) = \cos 2x$ and $y_2(x) = \sin 2x$, $a(x) = 1$, $f(x) = \sin x$ in system (2.88):

$$c_1'(x) \cos 2x + c_2'(x) \sin 2x = 0$$
$$-2c_1'(x) \sin 2x + 2c_2'(x) \cos 2x = \sin x. \qquad (2.92)$$

Now, use your favorite method for solving a system of two equations and two unknowns. In this case, we can multiply the first equation by $2 \sin 2x$ and the second equation by $\cos 2x$. Adding the resulting equations will eliminate the c_1' terms. Thus, we have

$$c_2'(x) = \frac{1}{2} \sin x \cos 2x = \frac{1}{2}(2\cos^2 x - 1) \sin x.$$

Inserting this into the first equation of the system, we have

$$c_1'(x) = -c_2'(x) \frac{\sin 2x}{\cos 2x} = -\frac{1}{2} \sin x \sin 2x = -\sin^2 x \cos x.$$

These can easily be solved:

$$c_2(x) = \frac{1}{2} \int (2\cos^2 x - 1) \sin x \, dx = \frac{1}{2}(\cos x - \frac{2}{3}\cos^3 x).$$

$$c_1(x) = -\int \sin^x \cos x \, dx = -\frac{1}{3} \sin^3 x.$$

The final step in getting the particular solution is to insert these functions into $y_p(x)$. This gives

$$
\begin{aligned}
y_p(x) &= c_1(x)y_1(x) + c_2(x)y_2(x) \\
&= (-\frac{1}{3}\sin^3 x) \cos 2x + (\frac{1}{2}\cos x - \frac{1}{3}\cos^3 x) \sin x \\
&= \frac{1}{3} \sin x. \qquad (2.93)
\end{aligned}
$$

So, the general solution is

$$y(x) = c_1 \cos 2x + c_2 \sin 2x + \frac{1}{3} \sin x. \qquad (2.94)$$

2.7.4 Initial Value Green's Functions

IN THIS SECTION WE WILL INVESTIGATE the solution of initial value problems involving nonhomogeneous differential equations using Green's functions. Our goal is to solve the nonhomogeneous differential equation

$$a(t)y''(t) + b(t)y'(t) + c(t)y(t) = f(t), \qquad (2.95)$$

subject to the initial conditions

$$y(0) = y_0, \quad y'(0) = v_0.$$

Because we are interested in initial value problems, we will denote the independent variable as a time variable, t.

Equation (2.95) can be written compactly as

$$L[y] = f,$$

where L is the differential operator

$$L = a(t)\frac{d^2}{dt^2} + b(t)\frac{d}{dt} + c(t).$$

The solution is formally given by

$$y = L^{-1}[f].$$

The inverse of a differential operator is an integral operator, which we seek to write in the form

$$y(t) = \int G(t,\tau)f(\tau)\,d\tau.$$

The function $G(t,\tau)$ is referred to as the kernel of the integral operator and is called the Green's function.

$G(t,\tau)$ is called a Green's function.

The history of the Green's function dates back to 1828, when George Green published work in which he sought solutions of Poisson's equation $\nabla^2 u = f$ for the electric potential u defined inside a bounded volume with specified boundary conditions on the surface of the volume. He introduced a function now identified as what Riemann later coined the "Green's function." In this section we will derive the initial value Green's function for ordinary differential equations. Later in the book we will return to boundary value Green's functions and Green's functions for partial differential equations.

George Green (1793–1841), a British mathematical physicist who had little formal education and worked as a miller and a baker, published *An Essay on the Application of Mathematical Analysis to the Theories of Electricity and Magnetism* in which he not only introduced what is now known as Green's function, but he also introduced potential theory and Green's Theorem in his studies of electricity and magnetism. Recently his paper was posted at arXiv.org, arXiv:0807.0088.

In the previous section we solved nonhomogeneous equations like Equation (2.95) using the Method of Variation of Parameters. Letting

$$y_p(t) = c_1(t)y_1(t) + c_2(t)y_2(t), \tag{2.96}$$

we found that we have to solve the system of equations

$$c_1'(t)y_1(t) + c_2'(t)y_2(t) = 0,$$
$$c_1'(t)y_1'(t) + c_2'(t)y_2'(t) = \frac{f(t)}{q(t)}. \tag{2.97}$$

This system is easily solved to give

$$c_1'(t) = -\frac{f(t)y_2(t)}{a(t)\left[y_1(t)y_2'(t) - y_1'(t)y_2(t)\right]},$$
$$c_2'(t) = \frac{f(t)y_1(t)}{a(t)\left[y_1(t)y_2'(t) - y_1'(t)y_2(t)\right]}. \tag{2.98}$$

We note that the denominator in these expressions involves the Wronskian of the solutions to the homogeneous problem, which is given by the determinant

$$W(y_1,y_2)(t) = \begin{vmatrix} y_1(t) & y_2(t) \\ y_1'(t) & y_2'(t) \end{vmatrix}.$$

When $y_1(t)$ and $y_2(t)$ are linearly independent, then the Wronskian is not zero and we are guaranteed a solution to the above system.

So, after an integration, we find the parameters as

$$
\begin{aligned}
c_1(t) &= -\int_{t_0}^{t} \frac{f(\tau)y_2(\tau)}{a(\tau)W(\tau)}\, d\tau, \\
c_2(t) &= \int_{t_1}^{t} \frac{f(\tau)y_1(\tau)}{a(\tau)W(\tau)}\, d\tau,
\end{aligned}
\tag{2.99}
$$

where t_0 and t_1 are arbitrary constants to be determined from the initial conditions.

Therefore, the particular solution of Equation (2.95) can be written as

$$
y_p(t) = y_2(t) \int_{t_1}^{t} \frac{f(\tau)y_1(\tau)}{a(\tau)W(\tau)}\, d\tau - y_1(t) \int_{t_0}^{t} \frac{f(\tau)y_2(\tau)}{a(\tau)W(\tau)}\, d\tau.
\tag{2.100}
$$

We begin with the particular solution Equation (2.100) of the nonhomogeneous differential equation (2.95). This can be combined with the general solution of the homogeneous problem to give the general solution of the nonhomogeneous differential equation:

$$
y_p(t) = c_1 y_1(t) + c_2 y_2(t) + y_2(t) \int_{t_1}^{t} \frac{f(\tau)y_1(\tau)}{a(\tau)W(\tau)}\, d\tau - y_1(t) \int_{t_0}^{t} \frac{f(\tau)y_2(\tau)}{a(\tau)W(\tau)}\, d\tau.
\tag{2.101}
$$

However, an appropriate choice of t_0 and t_1 can be found so that we need not explicitly write out the solution to the homogeneous problem, $c_1 y_1(t) + c_2 y_2(t)$. However, setting up the solution in this form will allow us to use t_0 and t_1 to determine particular solutions that satisfy certain homogeneous conditions. In particular, we will show that Equation (2.101) can be written in the form

$$
y(t) = c_1 y_1(t) + c_2 y_2(t) + \int_{0}^{t} G(t,\tau)f(\tau)\, d\tau,
\tag{2.102}
$$

where the function $G(t,\tau)$ will be identified as the Green's function.

The goal is to develop the Green's function technique to solve the initial value problem

$$
a(t)y''(t) + b(t)y'(t) + c(t)y(t) = f(t), \quad y(0) = y_0, \quad y'(0) = v_0.
\tag{2.103}
$$

We first note that we can solve this initial value problem by solving two separate initial value problems. We assume that the solution of the homogeneous problem satisfies the original initial conditions:

$$
a(t)y_h''(t) + b(t)y_h'(t) + c(t)y_h(t) = 0, \quad y_h(0) = y_0, \quad y_h'(0) = v_0.
\tag{2.104}
$$

We then assume that the particular solution satisfies the problem

$$
a(t)y_p''(t) + b(t)y_p'(t) + c(t)y_p(t) = f(t), \quad y_p(0) = 0, \quad y_p'(0) = 0.
\tag{2.105}
$$

Because the differential equation is linear, then we know that

$$
y(t) = y_h(t) + y_p(t)
$$

is a solution of the nonhomogeneous equation. Also, this solution satisfies the initial conditions:

$$y(0) = y_h(0) + y_p(0) = y_0 + 0 = y_0,$$

$$y'(0) = y_h'(0) + y_p'(0) = v_0 + 0 = v_0.$$

Therefore, we need only focus on finding a particular solution that satisfies homogeneous initial conditions. This will be done by finding values for t_0 and t_1 in Equation (2.100) that satisfy the homogeneous initial conditions, $y_p(0) = 0$ and $y_p'(0) = 0$.

First, we consider $y_p(0) = 0$. We have

$$y_p(0) = y_2(0) \int_{t_1}^0 \frac{f(\tau)y_1(\tau)}{a(\tau)W(\tau)} \, d\tau - y_1(0) \int_{t_0}^0 \frac{f(\tau)y_2(\tau)}{a(\tau)W(\tau)} \, d\tau. \qquad (2.106)$$

Here, $y_1(t)$ and $y_2(t)$ are taken to be any solutions of the homogeneous differential equation. Let's assume that $y_1(0) = 0$ and $y_2 \neq (0) = 0$. Then, we have

$$y_p(0) = y_2(0) \int_{t_1}^0 \frac{f(\tau)y_1(\tau)}{a(\tau)W(\tau)} \, d\tau. \qquad (2.107)$$

We can force $y_p(0) = 0$ if we set $t_1 = 0$.

Now, we consider $y_p'(0) = 0$. First we differentiate the solution and find that

$$y_p'(t) = y_2'(t) \int_0^t \frac{f(\tau)y_1(\tau)}{a(\tau)W(\tau)} \, d\tau - y_1'(t) \int_{t_0}^t \frac{f(\tau)y_2(\tau)}{a(\tau)W(\tau)} \, d\tau, \qquad (2.108)$$

as the contributions from differentiating the integrals will cancel. Evaluating this result at $t = 0$, we have

$$y_p'(0) = -y_1'(0) \int_{t_0}^0 \frac{f(\tau)y_2(\tau)}{a(\tau)W(\tau)} \, d\tau. \qquad (2.109)$$

Assuming that $y_1'(0) \neq 0$, we can set $t_0 = 0$.

Thus, we have found that

$$
\begin{aligned}
y_p(x) &= y_2(t) \int_0^t \frac{f(\tau)y_1(\tau)}{a(\tau)W(\tau)} \, d\tau - y_1(t) \int_0^t \frac{f(\tau)y_2(\tau)}{a(\tau)W(\tau)} \, d\tau \\
&= \int_0^t \left[\frac{y_1(\tau)y_2(t) - y_1(t)y_2(\tau)}{a(\tau)W(\tau)} \right] f(\tau) \, d\tau.
\end{aligned} \qquad (2.110)
$$

This result is in the correct form and we can identify the temporal, or initial value, Green's function. So, the particular solution is given as

$$y_p(t) = \int_0^t G(t, \tau) f(\tau) \, d\tau, \qquad (2.111)$$

where the initial value Green's function is defined as

$$G(t, \tau) = \frac{y_1(\tau)y_2(t) - y_1(t)y_2(\tau)}{a(\tau)W(\tau)}.$$

We summarize

Solution of IVP Using the Green's Function

The solution of the initial value problem,

$$a(t)y''(t) + b(t)y'(t) + c(t)y(t) = f(t), \quad y(0) = y_0, \quad y'(0) = v_0,$$

takes the form

$$y(t) = y_h(t) + \int_0^t G(t, \tau) f(\tau) \, d\tau, \tag{2.112}$$

where

$$G(t, \tau) = \frac{y_1(\tau)y_2(t) - y_1(t)y_2(\tau)}{a(\tau)W(\tau)} \tag{2.113}$$

is the Green's function and y_1, y_2, y_h are solutions of the homogeneous equation satisfying

$$y_1(0) = 0, y_2(0) \neq 0, y_1'(0) \neq 0, y_2'(0) = 0, y_h(0) = y_0, y_h'(0) = v_0.$$

Example 2.19. *Solve the forced oscillator problem*

$$x'' + x = 2\cos t, \quad x(0) = 4, \quad x'(0) = 0.$$

We first solve the homogeneous problem with nonhomogeneous initial conditions:

$$x_h'' + x_h = 0, \quad x_h(0) = 4, \quad x_h'(0) = 0.$$

The solution is easily seen to be $x_h(t) = 4\cos t$.

Next, we construct the Green's function. We need two linearly independent solutions, $y_1(x)$, $y_2(x)$, to the homogeneous differential equation satisfying different homogeneous conditions, $y_1(0) = 0$ and $y_2'(0) = 0$. The simplest solutions are $y_1(t) = \sin t$ and $y_2(t) = \cos t$. The Wronskian is found as

$$W(t) = y_1(t)y_2'(t) - y_1'(t)y_2(t) = -\sin^2 t - \cos^2 t = -1.$$

Because $a(t) = 1$ in this problem, we compute the Green's function,

$$\begin{aligned}
G(t, \tau) &= \frac{y_1(\tau)y_2(t) - y_1(t)y_2(\tau)}{a(\tau)W(\tau)} \\
&= \sin t \cos \tau - \sin \tau \cos t \\
&= \sin(t - \tau). \tag{2.114}
\end{aligned}$$

Note that the Green's function depends on $t - \tau$. While this is useful in some contexts, we will use the expanded form when carrying out the integration.

We can now determine the particular solution of the nonhomogeneous differential equation. We have

$$\begin{aligned}
x_p(t) &= \int_0^t G(t, \tau) f(\tau) \, d\tau \\
&= \int_0^t (\sin t \cos \tau - \sin \tau \cos t)(2\cos \tau) \, d\tau \\
&= 2\sin t \int_0^t \cos^2 \tau \, d\tau - 2\cos t \int_0^t \sin \tau \cos \tau \, d\tau \\
&= 2\sin t \left[\frac{\tau}{2} + \frac{1}{2}\sin 2\tau \right]_0^t - 2\cos t \left[\frac{1}{2}\sin^2 \tau \right]_0^t \\
&= t\sin t. \tag{2.115}
\end{aligned}$$

Therefore, the solution of the nonhomogeneous problem is the sum of the solution of the homogeneous problem and this particular solution: $x(t) = 4\cos t + t\sin t$.

2.8 Cauchy–Euler Equations

ANOTHER CLASS OF SOLVABLE LINEAR DIFFERENTIAL EQUATIONS that is of interest are the Cauchy–Euler type of equations, also referred to in some books as Euler's equation. These are given by

$$ax^2 y''(x) + bxy'(x) + cy(x) = 0. \tag{2.116}$$

Note that in such equations, the power of x in each of the coefficients matches the order of the derivative in that term. These equations are solved in a manner similar to the constant coefficient equations.

One begins by making the guess $y(x) = x^r$. Inserting this function and its derivatives,

$$y'(x) = rx^{r-1}, \qquad y''(x) = r(r-1)x^{r-2},$$

into Equation (2.116), we have

$$[ar(r-1) + br + c]x^r = 0.$$

Because this must be true for all x in the problem domain, we obtain the characteristic equation

$$ar(r-1) + br + c = 0. \tag{2.117}$$

The solutions of Cauchy–Euler Equations can be found using the characteristic equation $ar(r-1) + br + c = 0$.

Just like the constant coefficient differential equation, we have a quadratic equation and the nature of the roots again leads to three classes of solutions. If there are two real, distinct roots, then the general solution takes the form $y(x) = c_1 x^{r_1} + c_2 x^{r_2}$.

For two real, distinct roots, the general solution takes the form

$$y(x) = c_1 x^{r_1} + c_2 x^{r_2}.$$

Example 2.20. *Find the general solution:* $x^2 y'' + 5xy' + 12y = 0$.

As with the constant coefficient equations, we begin by writing the characteristic equation. Doing a simple computation,

$$
\begin{aligned}
0 &= r(r-1) + 5r + 12 \\
&= r^2 + 4r + 12 \\
&= (r+2)^2 + 8, \\
-8 &= (r+2)^2,
\end{aligned}
\tag{2.118}
$$

one determines that the roots are $r = -2 \pm 2\sqrt{2}i$. *Therefore, the general solution is*
$$y(x) = \left[c_1 \cos(2\sqrt{2}\ln|x|) + c_2 \sin(2\sqrt{2}\ln|x|) \right] x^{-2}$$

Deriving the solution for Case 2 for the Cauchy–Euler Equations works in the same way as the second for constant coefficient equations, but it is a bit messier. First note that for the real root, $r = r_1$, the characteristic equation has to factor as $(r - r_1)^2 = 0$. Expanding, we have

$$r^2 - 2r_1 r + r_1^2 = 0.$$

The general characteristic equation is

$$ar(r - 1) + br + c = 0.$$

Dividing this equation by a and rewriting, we have

$$r^2 + (\frac{b}{a} - 1)r + \frac{c}{a} = 0.$$

Comparing equations, we find

$$\frac{b}{a} = 1 - 2r_1, \qquad \frac{c}{a} = r_1^2.$$

So, the Cauchy–Euler Equation for this case can be written in the form

$$x^2 y'' + (1 - 2r_1)xy' + r_1^2 y = 0.$$

Now we seek the second linearly independent solution in the form $y_2(x) = v(x)x^{r_1}$. We first list this function and its derivatives,

$$
\begin{aligned}
y_2(x) &= vx^{r_1}, \\
y_2'(x) &= (xv' + r_1 v)x^{r_1 - 1}, \\
y_2''(x) &= (x^2 v'' + 2r_1 xv' + r_1(r_1 - 1)v)x^{r_1 - 2}.
\end{aligned}
\tag{2.119}
$$

Inserting these forms into the differential equation, we have

$$
\begin{aligned}
0 &= x^2 y'' + (1 - 2r_1)xy' + r_1^2 y \\
&= (xv'' + v')x^{r_1 + 1}.
\end{aligned}
\tag{2.120}
$$

Thus, we need to solve the equation

$$xv'' + v' = 0,$$

or

$$\frac{v''}{v'} = -\frac{1}{x}.$$

Integrating, we have

$$\ln |v'| = -\ln |x| + C,$$

where $A = \pm e^C$ absorbs C and the signs from the absolute values. Exponentiating, we obtain one last differential equation to solve,

$$v' = \frac{A}{x}.$$

Thus,

$$v(x) = A \ln |x| + k.$$

So, we have found that the second linearly independent equation can be written as

$$y_2(x) = x^{r_1} \ln |x|.$$

Therefore, the general solution is found as $y(x) = (c_1 + c_2 \ln |x|)x^r$.

For one root, $r_1 = r_2 = r$, the general solution is of the form

$$y(x) = (c_1 + c_2 \ln |x|)x^r.$$

Example 2.21. *Solve the initial value problem:* $t^2 y'' + 3ty' + y = 0$, *with the initial conditions* $y(1) = 0$, $y'(1) = 1$.

For this example, the characteristic equation takes the form

$$r(r-1) + 3r + 1 = 0,$$

or

$$r^2 + 2r + 1 = 0.$$

There is only one real root, $r = -1$. Therefore, the general solution is

$$y(t) = (c_1 + c_2 \ln|t|) t^{-1}.$$

However, this problem is an initial value problem. At $t = 1$, we know the values of y and y'. Using the general solution, we first have that

$$0 = y(1) = c_1.$$

Thus, we have so far that $y(t) = c_2 \ln|t| t^{-1}$. Now, using the second condition and

$$y'(t) = c_2(1 - \ln|t|) t^{-2},$$

we have

$$1 = y(1) = c_2.$$

Therefore, the solution of the initial value problem is $y(t) = \ln|t| t^{-1}$.

We now turn to the case of complex conjugate roots, $r = \alpha \pm i\beta$. When dealing with the Cauchy–Euler Equations, we have solutions of the form $y(x) = x^{\alpha + i\beta}$. The key to obtaining real solutions is to first rewrite x^y:

$$x^y = e^{\ln x^y} = e^{y \ln x}.$$

Thus, a power can be written as an exponential and the solution can be written as

$$y(x) = x^{\alpha + i\beta} = x^\alpha e^{i\beta \ln x}, \quad x > 0.$$

Recalling that

$$e^{i\beta \ln x} = \cos(\beta \ln|x|) + i \sin(\beta \ln|x|),$$

we can now find two real, linearly independent solutions, $x^\alpha \cos(\beta \ln|x|)$ and $x^\alpha \sin(\beta \ln|x|)$ following the same steps as earlier for the constant coefficient case. This gives the general solution as

$$y(x) = x^\alpha(c_1 \cos(\beta \ln|x|) + c_2 \sin(\beta \ln|x|)).$$

For complex conjugate roots, $r = \alpha \pm i\beta$, the general solution takes the form

$$y(x) = x^\alpha(c_1 \cos(\beta \ln|x|) + c_2 \sin(\beta \ln|x|)).$$

Example 2.22. *Solve:* $x^2 y'' - xy' + 5y = 0$.

The characteristic equation takes the form

$$r(r-1) - r + 5 = 0,$$

or

$$r^2 - 2r + 5 = 0.$$

The roots of this equation are complex, $r_{1,2} = 1 \pm 2i$. Therefore, the general solution is $y(x) = x(c_1 \cos(2 \ln|x|) + c_2 \sin(2 \ln|x|))$.

The three cases are summarized in the table below.

Classification of Roots of the Characteristic Equation for Cauchy–Euler Differential Equations

1. Real, distinct roots r_1, r_2. In this case, the solutions corresponding to each root are linearly independent. Therefore, the general solution is simply $y(x) = c_1 x^{r_1} + c_2 x^{r_2}$.

2. Real, equal roots $r_1 = r_2 = r$. In this case, the solutions corresponding to each root are linearly dependent. To find a second linearly independent solution, one uses the Method of Reduction of Order. This gives the second solution as $x^r \ln |x|$. Therefore, the general solution is found as $y(x) = (c_1 + c_2 \ln |x|) x^r$.

3. Complex conjugate roots $r_1, r_2 = \alpha \pm i\beta$. In this case, the solutions corresponding to each root are linearly independent. These complex exponentials can be rewritten in terms of trigonometric functions. Namely, one has that $x^\alpha \cos(\beta \ln |x|)$ and $x^\alpha \sin(\beta \ln |x|)$ are two linearly independent solutions. Therefore, the general solution becomes $y(x) = x^\alpha (c_1 \cos(\beta \ln |x|) + c_2 \sin(\beta \ln |x|))$.

Nonhomogeneous Cauchy–Euler Equations

We can also solve some nonhomogeneous Cauchy–Euler equations using the Method of Undetermined Coefficients or the Method of Variation of Parameters. We will demonstrate this with a couple of examples.

Example 2.23. *Find the solution of $x^2 y'' - xy' - 3y = 2x^2$.*

First we find the solution of the homogeneous equation. The characteristic equation is $r^2 - 2r - 3 = 0$. So, the roots are $r = -1, 3$ and the solution is $y_h(x) = c_1 x^{-1} + c_2 x^3$.

We next need a particular solution. Let's guess $y_p(x) = Ax^2$. Inserting the guess into the nonhomogeneous differential equation, we have

$$
\begin{aligned}
2x^2 &= x^2 y'' - xy' - 3y = 2x^2 \\
&= 2Ax^2 - 2Ax^2 - 3Ax^2 \\
&= -3Ax^2.
\end{aligned}
\tag{2.121}
$$

So, $A = -2/3$. Therefore, the general solution of the problem is

$$
y(x) = c_1 x^{-1} + c_2 x^3 - \frac{2}{3} x^2.
$$

Example 2.24. *Find the solution of $x^2 y'' - xy' - 3y = 2x^3$.*

In this case, the nonhomogeneous term is a solution of the homogeneous problem, which we solved in the previous example. So, we will need a modification of the method. We have a problem of the form

$$
a x^2 y'' + bxy' + cy = dx^r,
$$

where r is a solution of $ar(r-1) + br + c = 0$. Let's guess a solution of the form $y = Ax^r \ln x$. Then one finds that the differential equation reduces to $Ax^r(2ar - a + b) = dx^r$. [You should verify this for yourself.]

With this in mind, we can now solve the problem at hand. Let $y_p = Ax^3 \ln x$. Inserting into the equation, we obtain $4Ax^3 = 2x^3$, or $A = 1/2$. The general solution of the problem can now be written as

$$y(x) = c_1 x^{-1} + c_2 x^3 + \frac{1}{2} x^3 \ln x.$$

Example 2.25. *Find the solution of $x^2 y'' - xy' - 3y = 2x^3$ using Variation of Parameters.*

As noted in the previous examples, the solution of the homogeneous problem has two linearly independent solutions, $y_1(x) = x^{-1}$ and $y_2(x) = x^3$. Assuming a particular solution of the form $y_p(x) = c_1(x)y_1(x) + c_2(x)y_2(x)$, we need to solve the system (2.88):

$$
\begin{aligned}
c_1'(x)x^{-1} + c_2'(x)x^3 &= 0 \\
-c_1'(x)x^{-2} + 3c_2'(x)x^2 &= \frac{2x^3}{x^2} = 2x.
\end{aligned}
\tag{2.122}
$$

From the first equation of the system we have $c_1'(x) = -x^4 c_2'(x)$. Substituting this into the second equation gives $c_2'(x) = \frac{1}{2x}$. So, $c_2(x) = \frac{1}{2} \ln |x|$ and, therefore, $c_1(x) = \frac{1}{8} x^4$. The particular solution is

$$y_p(x) = c_1(x)y_1(x) + c_2(x)y_2(x) = \frac{1}{8} x^3 + \frac{1}{2} x^3 \ln |x|.$$

Adding this to the homogeneous solution, we obtain the same solution as in the last example using the Method of Undetermined Coefficients. However, because $\frac{1}{8} x^3$ is a solution of the homogeneous problem, it can be absorbed into the first terms, leaving

$$y(x) = c_1 x^{-1} + c_2 x^3 + \frac{1}{2} x^3 \ln x.$$

2.9 Numerical Solutions of ODEs

SO FAR WE HAVE SEEN SOME OF THE STANDARD METHODS for solving first- and second-order differential equations. However, we have had to restrict ourselves to special cases in order to get nice analytical solutions to initial value problems. While these are not the only equations for which we can get exact results, there are many cases in which exact solutions are not possible. In such cases, we have to rely on approximation techniques, including the numerical solution of the equation at hand.

The use of numerical methods to obtain approximate solutions of differential equations and systems of differential equations has been known for some time. However, with the advent of powerful computers and desktop computers, we can now solve many of these problems with relative ease. The simple ideas used to solve first-order differential equations can be extended to the solutions of more complicated systems of partial differential equations, such as the large-scale problems of modeling ocean dynamics, weather systems, and even cosmological problems stemming from general relativity.

2.9.1   Euler's Method

IN THIS SECTION WE WILL LOOK AT THE SIMPLEST METHOD for solving first-order equations, Euler's Method. While it is not the most efficient method, it does provide us with a picture of how one proceeds and can be improved by introducing better techniques, which are typically covered in a numerical analysis text.

Let's consider the class of first-order initial value problems of the form

$$\frac{dy}{dx} = f(x,y), \quad y(x_0) = y_0. \tag{2.123}$$

We are interested in finding the solution $y(x)$ of this equation that passes through the initial point (x_0, y_0) in the xy-plane for values of x in the interval $[a,b]$, where $a = x_0$. We will seek approximations of the solution at N points, labeled x_n for $n = 1, \ldots, N$. For equally spaced points, we have $\Delta x = x_1 - x_0 = x_2 - x_1$, etc. We can write these as

$$x_n = x_0 + n\Delta x.$$

In Figure 2.22 we show three such points on the x-axis.

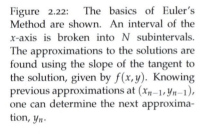

Figure 2.22: The basics of Euler's Method are shown. An interval of the x-axis is broken into N subintervals. The approximations to the solutions are found using the slope of the tangent to the solution, given by $f(x,y)$. Knowing previous approximations at (x_{n-1}, y_{n-1}), one can determine the next approximation, y_n.

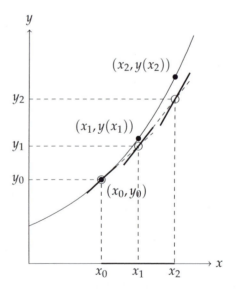

The first step in Euler's Method is to use the initial condition. We represent this as a point on the solution curve, $(x_0, y(x_0)) = (x_0, y_0)$, as shown in Figure 2.22. The next step is to develop a method for obtaining approximations to the solution for the other x_n's.

We first note that the differential equation gives the slope of the tangent line at $(x, y(x))$ of the solution curve as the slope is the derivative, $y'(x)'$

From the differential equation the slope is $f(x, y(x))$. Referring to Figure 2.22, we see the tangent line drawn at (x_0, y_0). We look now at $x = x_1$. The vertical line $x = x_1$ intersects both the solution curve and the tangent line passing through (x_0, y_0). This is shown by a heavy dashed line.

While we do not know the solution at $x = x_1$, we can determine the tangent line and find the intersection point that it makes with the vertical. As seen in the figure, this intersection point is in theory close to the point on the solution curve. So, we will designate y_1 as the approximation of the solution $y(x_1)$. We just need to determine y_1.

The idea is simple. We approximate the derivative in the differential equation by its difference quotient:

$$\frac{dy}{dx} \approx \frac{y_1 - y_0}{x_1 - x_0} = \frac{y_1 - y_0}{\Delta x}. \tag{2.124}$$

Because the slope of the tangent to the curve at (x_0, y_0) is $y'(x_0) = f(x_0, y_0)$, we can write

$$\frac{y_1 - y_0}{\Delta x} \approx f(x_0, y_0). \tag{2.125}$$

Solving this equation for y_1, we obtain

$$y_1 = y_0 + \Delta x f(x_0, y_0). \tag{2.126}$$

This gives y_1 in terms of quantities that we know.

We now proceed to approximate $y(x_2)$. Referring to Figure 2.22, we see that this can be done using the slope of the solution curve at (x_1, y_1). The corresponding tangent line is shown passing through (x_1, y_1) and we can then get the value of y_2 from the intersection of the tangent line with a vertical line, $x = x_2$. Following the previous arguments, we find that

$$y_2 = y_1 + \Delta x f(x_1, y_1). \tag{2.127}$$

Continuing this procedure for all x_n, $n = 1, \ldots N$, we arrive at the following scheme for determining a numerical solution to the initial value problem:

$$\begin{aligned} y_0 &= y(x_0), \\ y_n &= y_{n-1} + \Delta x f(x_{n-1}, y_{n-1}), \quad n = 1, \ldots, N. \end{aligned} \tag{2.128}$$

This is referred to as Euler's Method.

Example 2.26. *Use Euler's Method to solve the initial value problem $\frac{dy}{dx} = x + y$, $y(0) = 1$ and obtain an approximation for $y(1)$.*

First, we will do this by hand. We break up the interval $[0, 1]$, because we want the solution at $x = 1$ and the initial value is at $x = 0$. Let $\Delta x = 0.50$. Then, $x_0 = 0$, $x_1 = 0.5$, and $x_2 = 1.0$. Note that there are $N = \frac{b-a}{\Delta x} = 2$ subintervals and thus three points.

We next carry out Euler's Method systematically by setting up a table for the needed values. Such a table is shown in Table 2.2. Note how the table is set up. There is a column for each x_n and y_n. The first row is the initial condition. We

Table 2.2: Application of Euler's Method
for $y' = x + y$, $y(0) = 1$ and $\Delta x = 0.5$

n	x_n	$y_n = y_{n-1} + \Delta x f(x_{n-1}, y_{n-1} = 0.5x_{n-1} + 1.5y_{n-1}$
0	0	1
1	0.5	$0.5(0) + 1.5(1.0) = 1.5$
2	1.0	$0.5(0.5) + 1.5(1.5) = 2.5$

Table 2.2: Application of Euler's Method for $y' = x + y$, $y(0) = 1$ and $\Delta x = 0.5$

also made use of the function $f(x, y)$ in computing the y_n's from Equation (2.128). This sometimes makes the computation easier. As a result, we find that the desired approximation is given as $y_2 = 2.5$.

Is this a good result? Well, we could make the spatial increments smaller. Let's repeat the procedure for $\Delta x = 0.2$, or $N = 5$. The results are in Table 2.3.

Table 2.3: Application of Euler's Method for $y' = x + y$, $y(0) = 1$ and $\Delta x = 0.2$

n	x_n	$y_n = 0.2x_{n-1} + 1.2y_{n-1}$
0	0	1
1	0.2	$0.2(0) + 1.2(1.0) = 1.2$
2	0.4	$0.2(0.2) + 1.2(1.2) = 1.48$
3	0.6	$0.2(0.4) + 1.2(1.48) = 1.856$
4	0.8	$0.2(0.6) + 1.2(1.856) = 2.3472$
5	1.0	$0.2(0.8) + 1.2(2.3472) = 2.97664$

Now we see that the approximation is $y_1 = 2.97664$. So, it looks like the value is near 3, but we cannot say much more. Decreasing Δx more shows that we are beginning to converge to a solution. We see this in Table 2.4.

Table 2.4: Results of Euler's Method for $y' = x + y$, $y(0) = 1$ and Varying Δx

Δx	$y_N \approx y(1)$
0.5	2.5
0.2	2.97664
0.1	3.187484920
0.01	3.409627659
0.001	3.433847864
0.0001	3.436291854

Of course, these values were not done by hand. The last computation would have taken 1,000 lines in the table, or at least 40 pages! One could use a computer to do this. A simple code in Maple would look like the following:

```
> restart:
> f:=(x,y)->y+x;
> a:=0:  b:=1:  N:=100:  h:=(b-a)/N;
> x[0]:=0: y[0]:=1:
  for i from 1 to N do
  y[i]:=y[i-1]+h*f(x[i-1],y[i-1]):
  x[i]:=x[0]+h*(i):
  od:
  evalf(y[N]);
```

In this case, we could simply use the exact solution. The exact solution is easily found as

$$y(x) = 2e^x - x - 1.$$

(The reader can verify this.) So, the value we are seeking is

$$y(1) = 2e - 2 = 3.4365636\ldots.$$

Thus, even the last numerical solution was off by about 0.00027.

Adding a few extra lines for plotting, we can visually see how well the approximations compare to the exact solution. The Maple code for doing such a plot is given below:

```
> with(plots):
> Data:=[seq([x[i],y[i]],i=0..N)]:
> P1:=pointplot(Data,symbol=DIAMOND):
> Sol:=t->-t-1+2*exp(t);
> P2:=plot(Sol(t),t=a..b,Sol=0..Sol(b)):
> display({P1,P2});
```

We show in Figures 2.23 and 2.24 the results for $N = 10$ and $N = 100$. In Figure 2.23 we can see how quickly the numerical solution diverges from the exact solution. In Figure 2.24 we can see that visually the solutions agree, but we note that from Table 2.4 that for $\Delta x = 0.01$, the solution is still off in the second decimal place with a relative error of about 0.8%.

Why would we use a numerical method when we have the exact solution? Exact solutions can serve as test cases for our methods. We can make sure our code works before applying them to problems whose solution is not known.

There are many other methods for solving first-order equations. One commonly used method is the fourth-order Runge–Kutta Method. This method has smaller errors at each step as compared to Euler's Method. It is well suited for programming and comes built-in in many packages like Maple and MATLAB. Typically, it is set up to handle systems of first-order equations.

In fact, it is well known that nth-order equations can be written as a system of n first-order equations. Consider the simple second-order equation

$$y'' = f(x, y).$$

This is a larger class of equations than the second-order constant coefficient equation. We can turn this into a system of two first-order differential equations by letting $u = y$ and $v = y' = u'$. Then, $v' = y'' = f(x, u)$. So, we have the first-order system

$$
\begin{aligned}
u' &= v, \\
v' &= f(x, u).
\end{aligned}
\tag{2.129}
$$

We will go further into the Runge–Kutta Method in Section 2.9.3 and you can find more about it in a numerical analysis text. We will see that systems of differential equations do arise naturally in physics. Such systems are often coupled equations and lead to interesting behaviors.

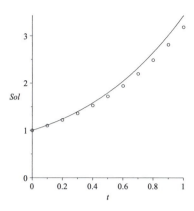

Figure 2.23: A comparison of the results of Euler's Method to the exact solution for $y' = x + y$, $y(0) = 1$ and $N = 10$.

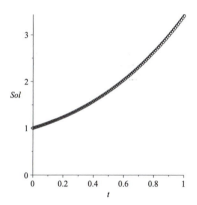

Figure 2.24: A comparison of the results Euler's Method to the exact solution for $y' = x + y$, $y(0) = 1$ and $N = 100$.

2.9.2   Higher-Order Taylor Methods

EULER'S METHOD FOR SOLVING DIFFERENTIAL EQUATIONS is easy to understand but is not efficient in the sense that it is what is called a first-order method. The error at each step, the local truncation error, is of order Δx, for x the independent variable. The accumulation of the local truncation errors results in what is called the global error. In order to generalize Euler's Method, we need to rederive it. Also, because these methods are typically used for initial value problems, we will cast the problem to be solved as

$$\frac{dy}{dt} = f(t, y), \quad y(a) = y_0, \quad t \in [a, b]. \tag{2.130}$$

The first step toward obtaining a numerical approximation to the solution of this problem is to divide the t-interval, $[a, b]$, into N subintervals,

$$t_i = a + ih, \quad i = 0, 1, \dots, N, \quad t_0 = a, \quad t_N = b,$$

where

$$h = \frac{b - a}{N}.$$

We then seek the numerical solutions

$$\tilde{y}_i \approx y(t_i), \quad i = 1, 2, \dots, N,$$

with $\tilde{y}_0 = y(t_0) = y_0$. Figure 2.25 graphically shows how these quantities are related.

Euler's Method can be derived using the Taylor series expansion of the solution $y(t_i + h)$ about $t = t_i$ for $i = 1, 2, \dots, N$. This is given by

$$
\begin{aligned}
y(t_{i+1}) &= y(t_i + h) \\
&= y(t_i) + y'(t_i)h + \frac{h^2}{2} y''(\xi_i), \quad \xi_i \in (t_i, t_{i+1}). \tag{2.131}
\end{aligned}
$$

Here the term $\frac{h^2}{2} y''(\xi_i)$ captures all of the higher order terms and represents the error made using a linear approximation to $y(t_i + h)$.

Dropping the remainder term, noting that $y'(t) = f(t, y)$, and defining the resulting numerical approximations by $\tilde{y}_i \approx y(t_i)$, we have

$$
\begin{aligned}
\tilde{y}_{i+1} &= \tilde{y}_i + hf(t_i, \tilde{y}_i), \quad i = 0, 1, \dots, N - 1, \\
\tilde{y}_0 &= y(a) = y_0. \tag{2.132}
\end{aligned}
$$

This is Euler's Method.

Euler's Method is not used in practice because the error is of order h. However, it is simple enough for understanding the idea of solving differential equations numerically. Also, it is easy to study the numerical error, which we will show next.

The error that results for a single step of the method is called the local truncation error, which is defined by

$$\tau_{i+1}(h) = \frac{y(t_{i+1}) - \tilde{y}_i}{h} - f(t_i, y_i).$$

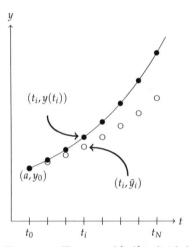

Figure 2.25: The interval $[a, b]$ is divided into N equally spaced subintervals. The exact solution $y(t_i)$ is shown with the numerical solution, $\tilde{y}_i$ with $t_i = a + ih$, $i = 0, 1, \dots, N$.

A simple computation gives

$$\tau_{i+1}(h) = \frac{h}{2}y''(\xi_i), \quad \xi_i \in (t_i, t_{i+1}).$$

Because the local truncation error is of order h, this scheme is said to be of order one. More generally, for a numerical scheme of the form

$$\begin{aligned}
\tilde{y}_{i+1} &= \tilde{y}_i + hF(t_i, \tilde{y}_i), \quad i = 0, 1, \ldots, N-1, \\
\tilde{y}_0 &= y(a) = y_0,
\end{aligned} \tag{2.133}$$

the local truncation error is defined by

$$\tau_{i+1}(h) = \frac{y(t_{i+1}) - \tilde{y}_i}{h} - F(t_i, y_i).$$

The local truncation error.

The accumulation of these errors leads to the global error. In fact, one can show that if f is continuous, satisfies the Lipschitz condition,

$$|f(t, y_2) - f(t, y_1)| \le L|y_2 - y_1|$$

for a particular domain $D \subset R^2$, and

$$|y''(t)| \le M, \quad t \in [a, b],$$

then

$$|y(t_i) - \tilde{y}_i| \le \frac{hM}{2L}\left(e^{L(t_i - a)} - 1\right), \quad i = 0, 1, \ldots, N.$$

Furthermore, if one introduces round-off errors, bounded by δ, in both the initial condition and at each step, the global error is modified as

$$|y(t_i) - \tilde{y}_i| \le \frac{1}{L}\left(\frac{hM}{2} + \frac{\delta}{h}\right)\left(e^{L(t_i - a)} - 1\right) + |\delta_0|e^{L(t_i - a)}, \quad i = 0, 1, \ldots, N.$$

Then for small enough steps h, there is a point when the round-off error will dominate the error. [See Burden and Faires, *Numerical Analysis* for the details.]

Can we improve upon Euler's Method? The natural next step toward finding a better scheme would be to keep more terms in the Taylor series expansion. This leads to Taylor series methods of order n.

Taylor series methods of order n take the form

$$\begin{aligned}
\tilde{y}_{i+1} &= \tilde{y}_i + hT^{(n)}(t_i, \tilde{y}_i), \quad i = 0, 1, \ldots, N-1, \\
\tilde{y}_0 &= y_0,
\end{aligned} \tag{2.134}$$

where we have defined

$$T^{(n)}(t, y) = y'(t) + \frac{h}{2}y''(t) + \cdots + \frac{h^{(n-1)}}{n!}y^{(n)}(t).$$

However, because $y'(t) = f(t, y)$, we can write

$$T^{(n)}(t, y) = f(t, y) + \frac{h}{2}f'(t, y) + \cdots + \frac{h^{(n-1)}}{n!}f^{(n-1)}(t, y).$$

We note that for $n = 1$, we retrieve Euler's Method as a special case. We demonstrate a third-order Taylor's Method in the next example.

Example 2.27. *Apply the third-order Taylor's Method to*

$$\frac{dy}{dt} = t + y, \quad y(0) = 1$$

and obtain an approximation for $y(1)$ for $h = 0.1$.

The third-order Taylor's Method takes the form

$$
\begin{aligned}
\tilde{y}_{i+1} &= \tilde{y}_i + hT^{(3)}(t_i, \tilde{y}_i), \quad i = 0, 1, \ldots, N-1, \\
\tilde{y}_0 &= y_0,
\end{aligned}
\tag{2.135}
$$

where

$$T^{(3)}(t,y) = f(t,y) + \frac{h}{2}f'(t,y) + \frac{h^2}{3!}f''(t,y)$$

and $f(t,y) = t + y(t)$.

In order to set up the scheme, we need the first and second derivative of $f(t,y)$:

$$
\begin{aligned}
f'(t,y) &= \frac{d}{dt}(t+y) \\
&= 1 + y' \\
&= 1 + t + y
\end{aligned}
\tag{2.136}
$$

$$
\begin{aligned}
f''(t,y) &= \frac{d}{dt}(1 + t + y) \\
&= 1 + y' \\
&= 1 + t + y
\end{aligned}
\tag{2.137}
$$

Inserting these expressions into the scheme, we have

$$
\begin{aligned}
\tilde{y}_{i+1} &= \tilde{y}_i + h\left[(t_i + y_i) + \frac{h}{2}(1 + t_i + y_i) + \frac{h^2}{3!}(1 + t_i + y_i)\right], \\
&= \tilde{y}_i + h(t_i + y_i) + h^2(\frac{1}{2} + \frac{h}{6})(1 + t_i + y_i), \\
\tilde{y}_0 &= y_0,
\end{aligned}
\tag{2.138}
$$

for $i = 0, 1, \ldots, N-1$.

In Figure 2.27 we show the results comparing Euler's Method, the third-order Taylor's Method, and the exact solution for $N = 10$. In Table 2.5 we provide the numerical values. The relative error in Euler's method is about 7% and that of the third-order Taylor's Method is about 0.006%. Thus, the third-order Taylor's Method is significantly better than Euler's Method.

In the previous section we provided some Maple code for performing Euler's Method. A similar code in MATLAB looks like the following:

```
a=0;
b=1;
N=10;
h=(b-a)/N;

% Slope function
```

Euler	Taylor	Exact
1.0000	1.0000	1.0000
1.1000	1.1103	1.1103
1.2200	1.2428	1.2428
1.3620	1.3997	1.3997
1.5282	1.5836	1.5836
1.7210	1.7974	1.7974
1.9431	2.0442	2.0442
2.1974	2.3274	2.3275
2.4872	2.6509	2.6511
2.8159	3.0190	3.0192
3.1875	3.4364	3.4366

Table 2.5: Numerical Values for Euler's Method, Third-Order Taylor's Method, and Exact Solution for Solving Example 2.27 with $N = 10$

```
f = inline('t+y','t','y');
sol = inline('2*exp(t)-t-1','t');


% Initial Condition
  t(1)=0;
  y(1)=1;


% Euler's Method
  for i=2:N+1
      y(i)=y(i-1)+h*f(t(i-1),y(i-1));
      t(i)=t(i-1)+h;
  end
```

A simple modification can be made for the third-order Taylor's Method by replacing the Euler's Method part of the preceding code with

```
% Taylor's Method, Order 3
  y(1)=1;
  h3 = h^2*(1/2+h/6);
  for i=2:N+1
      y(i)=y(i-1)+h*f(t(i-1),y(i-1))+h3*(1+t(i-1)+y(i-1));
      t(i)=t(i-1)+h;
  end
```

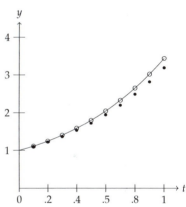

Figure 2.26: Numerical results for Euler's Method (filled circle) and third-order Taylor's Method (open circle) for solving Example 2.27 as compared to exact solution (solid line).

While the accuracy in the previous example seemed sufficient, we have to remember that we only stopped at one unit of time. How can we be confident that the scheme would work as well if we carried out the computation for much longer times. For example, if the time unit were only a second, then one would need 86,400 times longer to predict a day forward. Of course, the scale matters. But often we need to carry out numerical schemes for long times and we hope that the scheme not only converges to a solution, but that it coverges to the solution to the given problem. Also, the previous example was relatively easy to program because we could provide a relatively simple form for $T^{(3)}(t, y)$ with a quick computation of the derivatives of $f(t,y)$. This is not always the case and higher-order Taylor Methods in this form are not typically used. Instead, one can approximate $T^{(n)}(t,y)$ by evaluating the known function $f(t,y)$ at selected values of t and y, leading to Runge–Kutta methods.

2.9.3   Runge–Kutta Methods

As we had seen in the last section, we can use higher-order Taylor Methods to derive numerical schemes for solving

$$\frac{dy}{dt} = f(t,y), \quad y(a) = y_0, \quad t \in [a,b], \tag{2.139}$$

using a scheme of the form

$$
\begin{aligned}
\tilde{y}_{i+1} &= \tilde{y}_i + hT^{(n)}(t_i, \tilde{y}_i), \quad i = 0, 1, \ldots, N-1, \\
\tilde{y}_0 &= y_0,
\end{aligned}
\tag{2.140}
$$

where we have defined

$$T^{(n)}(t,y) = y'(t) + \frac{h}{2}y''(t) + \cdots + \frac{h^{(n-1)}}{n!}y^{(n)}(t).$$

In this section we will find approximations of $T^{(n)}(t,y)$ which avoid the need for computing the derivatives.

For example, we could approximate

$$T^{(2)}(t,y) = f(t,y) + \frac{h}{2}fracdfdt(t,y)$$

by

$$T^{(2)}(t,y) \approx af(t + \alpha, y + \beta)$$

for selected values of a, α, and β. This requires the use of a generalization of Taylor's series to functions of two variables. In particular, for small α and β, we have

$$
\begin{aligned}
af(t + \alpha, y + \beta) &= a\left[f(t,y) + \frac{\partial f}{\partial t}(t,y)\alpha + \frac{\partial f}{\partial y}(t,y)\beta \right. \\
&\left. + \frac{1}{2}\left(\frac{\partial^2 f}{\partial t^2}(t,y)\alpha^2 + 2\frac{\partial^2 f}{\partial t \partial y}(t,y)\alpha\beta + \frac{\partial^2 f}{\partial y^2}(t,y)\beta^2 \right) \right] \\
&+ \text{ higher order terms.} \tag{2.141}
\end{aligned}
$$

Furthermore, we need $\frac{df}{dt}(t,y)$. Because $y = y(t)$, this can be found using a generalization of the Chain Rule from Calculus III:

$$\frac{df}{dt}(t,y) = \frac{\partial f}{\partial t} + \frac{\partial f}{\partial y}\frac{dy}{dt}.$$

Thus,

$$T^{(2)}(t,y) = f(t,y) + \frac{h}{2}\left[\frac{\partial f}{\partial t} + \frac{\partial f}{\partial y}\frac{dy}{dt} \right].$$

Comparing this expression to the linear (Taylor series) approximation of $af(t + \alpha, y + \beta)$, we have

$$T^{(2)} \approx af(t + \alpha, y + \beta)$$

$$f + \frac{h}{2}\frac{\partial f}{\partial t} + \frac{h}{2}f\frac{\partial f}{\partial y} \approx af + a\alpha\frac{\partial f}{\partial t} + \beta\frac{\partial f}{\partial y}. \qquad (2.142)$$

We see that we can choose

$$a = 1, \quad \alpha = \frac{h}{2}, \quad \beta = \frac{h}{2}f.$$

This leads to the numerical scheme

The Midpoint or second-order Runge–Kutta Method.

$$\tilde{y}_{i+1} = \tilde{y}_i + hf\left(t_i + \frac{h}{2}, \tilde{y}_i + \frac{h}{2}f(t_i, \tilde{y}_i)\right), \quad i = 0, 1, \ldots, N-1,$$

$$\tilde{y}_0 = y_0. \qquad (2.143)$$

This Runge–Kutta Scheme is called the Midpoint Method, or second-order Runge–Kutta Method, and it has order 2 if all second-order derivatives of $f(t, y)$ are bounded.

Often, in implementing Runge–Kutta schemes, one computes the arguments separately as shown in the following MATLAB code snippet. (This code snippet could replace the Euler's Method section in the code in the previous section.)

```
% Midpoint Method
  y(1)=1;
  for i=2:N+1
      k1=h/2*f(t(i-1),y(i-1));
      k2=h*f(t(i-1)+h/2,y(i-1)+k1);
      y(i)=y(i-1)+k2;
      t(i)=t(i-1)+h;
  end
```

Example 2.28. *Compare the Midpoint Method with the second-order Taylor's Method for the problem*

$$y' = t^2 + y, \quad y(0) = 1, \quad t \in [0, 1]. \qquad (2.144)$$

The solution to this problem is $y(t) = 3e^t - 2 - 2t - t^2$. In order to implement the Second-Order Taylor's Method, we need

$$T^{(2)} = f(t, y) + \frac{h}{2}f'(t, y)$$

$$= t^2 + y + \frac{h}{2}(2t + t^2 + y). \qquad (2.145)$$

The results of the implementation are shown in Table 2.28.

There are other way to approximate higher-order Taylor polynomials. For example, we can approximate $T^{(3)}(t, y)$ using four parameters by

$$T^{(3)}(t, y) \approx af(t, y) + bf(t + \alpha, y + \beta f(t, y)).$$

Table 2.6: Numerical Values for Second-Order Taylor's Method, Midpoint Method, Exact Solution, and Errors for Solving Example 2.28 with $N = 10$

Exact	Taylor	Error	Midpoint	Error
1.0000	1.0000	0.0000	1.0000	0.0000
1.1055	1.1050	0.0005	1.1053	0.0003
1.2242	1.2231	0.0011	1.2236	0.0006
1.3596	1.3577	0.0019	1.3585	0.0010
1.5155	1.5127	0.0028	1.5139	0.0016
1.6962	1.6923	0.0038	1.6939	0.0023
1.9064	1.9013	0.0051	1.9032	0.0031
2.1513	2.1447	0.0065	2.1471	0.0041
2.4366	2.4284	0.0083	2.4313	0.0053
2.7688	2.7585	0.0103	2.7620	0.0068
3.1548	3.1422	0.0126	3.1463	0.0085

Expanding this approximation and using

$$T^{(3)}(t,y) \approx f(t,y) + \frac{h}{2}\frac{df}{dt}(t,y) + \frac{h^2}{6}\frac{df}{dt}(t,y),$$

we find that we cannot get rid of $O(h^2)$ terms. Thus, the best we can do is derive second-order schemes. In fact, following a procedure similar to the derivation of the Midpoint Method, we find that

$$a + b = 1, \quad , \alpha b = \frac{h}{2}, \beta = \alpha.$$

There are three equations and four unknowns. Therefore there are many second-order methods. Two classic methods are given by the modified Euler method ($a = b = \frac{1}{2}$, $\alpha = \beta = h$) and Huen's method ($a = \frac{1}{4}$, $b = \frac{3}{4}$, $\alpha = \beta = \frac{2}{3}h$).

The fourth-order Runge–Kutta Method.

The fourth-order Runge–Kutta Method, which is most often used, is given by the scheme

$$
\begin{aligned}
\tilde{y}_0 &= y_0, \\
k_1 &= hf(t_i, \tilde{y}_i), \\
k_2 &= hf(t_i + \frac{h}{2}, \tilde{y}_i + \frac{1}{2}k_1), \\
k_3 &= hf(t_i + \frac{h}{2}, \tilde{y}_i + \frac{1}{2}k_2), \\
k_4 &= hf(t_i + h, \tilde{y}_i + k_3), \\
\tilde{y}_{i+1} &= \tilde{y}_i + \frac{1}{6}(k_1 + 2k_2 + 2k_3 + k_4), \quad i = 0, 1, \ldots, N-1. \quad (2.146)
\end{aligned}
$$

Again, we can test this on Example 2.28 with $N = 10$. The MATLAB implementation is given by

```
% Runge-Kutta 4th Order to solve dy/dt = f(t,y), y(a)=y0, on [a,b]
  clear

a=0;
b=1;
```

```
    N=10;
    h=(b-a)/N;

% Slope function
    f = inline('t^2+y','t','y');
    sol = inline('-2-2*t-t^2+3*exp(t)','t');

% Initial Condition
    t(1)=0;
    y(1)=1;

% RK4 Method
    y1(1)=1;
    for i=2:N+1
        k1=h*f(t(i-1),y1(i-1));
        k2=h*f(t(i-1)+h/2,y1(i-1)+k1/2);
        k3=h*f(t(i-1)+h/2,y1(i-1)+k2/2);
        k4=h*f(t(i-1)+h,y1(i-1)+k3);
        y1(i)=y1(i-1)+(k1+2*k2+2*k3+k4)/6;
        t(i)=t(i-1)+h;
    end
```

MATLAB has built-in ODE solvers, such as **ode45** for a Fourth-Order Runge–Kutta Method. Its implementation is given by

```
[t,y]=ode45(f,[0 1],1);
```

In this case, f is given by an inline function such as in the above RK4 code. The time interval is enetered as $[0, 1]$, and the 1 is the initial condition, $y(0) = 1$.

However, **ode45** is not a straightforward RK4 implementation. It is a hybrid method in which a combination of fourth- and fifth-order methods are combined allowing for adaptive methods to handled subintervals of the integration region which need more care. In this case, it implements a fourth-order Runge–Kutta–Fehlberg method. Running this code for the above example actually results in values for $N = 41$ and not $N = 10$. If we wanted to have the routine output numerical solutions at specific times, then one could use the following form

```
tspan=0:h:1;
[t,y]=ode45(f,tspan,1);
```

In Table 2.7 we show the solutions which result for Example 2.28 comparing the RK4 snippet above with **ode45**. As you can see, RK4 is much better than the previous implementation of the second order RK (Midpoint) Method. However, the MATLAB routine is two orders of magnitude better that RK4.

MATLAB has built-in ODE solvers, as do other software packages, like Maple and Mathematica. You should also note that there are currently open source packages, such as Python based NumPy and Matplotlib, or Octave, of which some packages are contained within the Sage Project.

Table 2.7: Numerical Values for Fourth-Order Runge–Kutta Method, rk45, Exact Solution, and Errors for Solving Example 2.28 with $N = 10$

Exact	Taylor	Error	Midpoint	Error
1.0000	1.0000	0.0000	1.0000	0.0000
1.1055	1.1055	4.5894e-08	1.1055	−2.5083e-10
1.2242	1.2242	1.2335e-07	1.2242	−6.0935e-10
1.3596	1.3596	2.3850e-07	1.3596	−1.0954e-09
1.5155	1.5155	3.9843e-07	1.5155	−1.7319e-09
1.6962	1.6962	6.1126e-07	1.6962	−2.5451e-09
1.9064	1.9064	8.8636e-07	1.9064	−3.5651e-09
2.1513	2.1513	1.2345e-06	2.1513	−4.8265e-09
2.4366	2.4366	1.6679e-06	2.4366	−6.3686e-09
2.7688	2.7688	2.2008e-06	2.7688	−8.2366e-09
3.1548	3.1548	2.8492e-06	3.1548	−1.0482e-08

There are many ODE solvers in MATLAB. These are typically useful if RK4 is having difficulty solving particular problems. For the most part, one is fine using RK4, especially as a starting point. For example, there is **ode23**, which is similar to **ode45** but combining a second- and third-order scheme. Applying the results to Example 2.28 we obtain the results in Table 2.8. We compare these to the Second-Order Runge–Kutta Method. The code snippets are shown below.

```
% Second Order RK Method
  y1(1)=1;
  for i=2:N+1
      k1=h*f(t(i-1),y1(i-1));
      k2=h*f(t(i-1)+h/2,y1(i-1)+k1/2);
      y1(i)=y1(i-1)+k2;
      t(i)=t(i-1)+h;
end

tspan=0:h:1;
[t,y]=ode23(f,tspan,1);
```

Table 2.8: Numerical Values for Second-Order Runge–Kutta Method, rk23, Exact Solution, and Errors for Solving Example 2.28 with $N = 10$

Exact	Taylor	Error	Midpoint	Error
1.0000	1.0000	0.0000	1.0000	0.0000
1.1055	1.1053	0.0003	1.1055	2.7409e-06
1.2242	1.2236	0.0006	1.2242	8.7114e-06
1.3596	1.3585	0.0010	1.3596	1.6792e-05
1.5155	1.5139	0.0016	1.5154	2.7361e-05
1.6962	1.6939	0.0023	1.6961	4.0853e-05
1.9064	1.9032	0.0031	1.9063	5.7764e-05
2.1513	2.1471	0.0041	2.1512	7.8665e-05
2.4366	2.4313	0.0053	2.4365	0.0001
2.7688	2.7620	0.0068	2.7687	0.0001
3.1548	3.1463	0.0085	3.1547	0.0002

We have seen several numerical schemes for solving initial value problems. There are other methods, or combinations of methods, that aim to refine the numerical approximations efficiently as if the step size in the current methods were taken to be much smaller. Some methods extrapolate solutions to obtain information outside the solution interval. Others use one scheme to get a guess to the solution while refining, or correcting, this to obtain better solutions as the iteration through time proceeds. Such methods are described in courses in numerical analysis and in the literature. At this point we will apply these methods to several physics problems before continuing with analytical solutions.

2.10 Numerical Applications

IN THIS SECTION WE APPLY VARIOUS NUMERICAL METHODS to several physics problems after setting them up. We first describe how to work with second-order equations, such as the nonlinear pendulum problem. We will see that there is a bit more to numerically solving differential equations than just running standard routines. As we explore these problems, we will introduce other methods and provide some MATLAB code indicating how one might set up the system.

Other problems covered in these applications are various free-fall problems, beginning with a falling body from a large distance from the Earth, to flying soccer balls, and falling raindrops. We will also discuss the numerical solution of the two-body problem and the Friedmann equation as nonterrestrial applications.

2.10.1 Nonlinear Pendulum

NOW WE WILL INVESTIGATE THE USE OF NUMERCIAL METHODS for solving the nonlinear pendulum problem.

Example 2.29. Nonlinear pendulum *Solve*

$$\ddot{\theta} = -\frac{g}{L}\sin\theta, \quad \theta(0) = \theta_0, \quad \omega(0) = 0, \quad t \in [0,8],$$

using Euler's Method. Use the parameter values of m = 0.005 kg, L = 0.500 m, and g = 9.8 m/s^2.

This is a second-order differential equation. As described later, we can write this differential equation as a system of two first-order differential equations,

$$\begin{aligned} \dot{\theta} &= \omega, \\ \dot{\omega} &= -\frac{g}{L}\sin\theta. \end{aligned} \tag{2.147}$$

Defining the vector

$$\Theta(t) = \begin{pmatrix} \theta(t) \\ \omega(t) \end{pmatrix},$$

we can write the first-order system as

$$\frac{d\Theta}{dt} = \mathbf{F}(t, \Theta), \quad \Theta(0) = \begin{pmatrix} \theta_0 \\ 0 \end{pmatrix},$$

where

$$F(t, \Theta) = \begin{pmatrix} \omega(t) \\ -\frac{g}{L}\sin\theta(t) \end{pmatrix}.$$

This allows us to use the the methods we have discussed for this first-order equation for $\Theta(t)$.

For example, Euler's Method for this system becomes

$$\Theta_{i+1} = \Theta_{i+1} + h\mathbf{F}(t_i, \Theta_i)$$

with $\Theta_0 = \Theta(0)$.

We can write this scheme in component form as

$$\begin{pmatrix} \theta_{i+1} \\ \omega_{i+1} \end{pmatrix} = \begin{pmatrix} \theta_i \\ \omega_i \end{pmatrix} + h \begin{pmatrix} \omega_i \\ -\frac{g}{L}\sin\theta_i \end{pmatrix},$$

or

$$\begin{aligned} \theta_{i+1} &= \theta_i + h\omega_i, \\ \omega_{i+1} &= \omega_i - h\frac{g}{L}\sin\theta_i, \end{aligned} \tag{2.148}$$

starting with $\theta_0 = \theta_0$ and $\omega_0 = 0$.

The MATLAB code that can be used to implement this scheme takes the form

```
g=9.8;
L=0.5;
m=0.005;

a=0;
b=8;
N=500;
h=(b-a)/N;

% Initial Condition
  t(1)=0;
  theta(1)=pi/6;
  omega(1)=0;

% Euler's Method
  for i=2:N+1
      omega(i)=omega(i-1)-g/L*h*sin(theta(i-1));
      theta(i)=theta(i-1)+h*omega(i-1);
      t(i)=t(i-1)+h;
  end
```

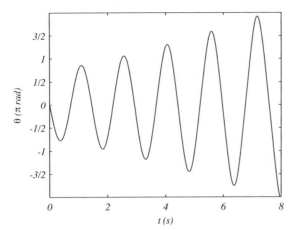

Figure 2.27: Solution for the nonlinear pendulum problem using Euler's Method on $t \in [0,8]$ with $N = 500$.

In Figure 2.27 we plot the solution for a starting position of 30° with N = 500. Notice that the amplitude of oscillation is increasing, contrary to our experience. So, we increase N and see if that helps. In Figure 2.28 we show the results for N = 500, 1000, and 2000 points, or h = 0.016, 0.008, and 0.004, respectively. We note that the amplitude is not increasing as much.

The problem with the solution is that Euler's Method is not an energy conserving method. As conservation of energy is important in physics, we would like to be able to seek problems that conserve energy. Such schemes used to solve oscillatory problems in classical mechanics are called symplectic integrators. A simple example is the Euler–Cromer, or semi-implicit Euler Method. We only need to make a small modification of Euler's Method. Namely, in the second equation of the method, we use the updated value of the dependent variable as computed in the first line.

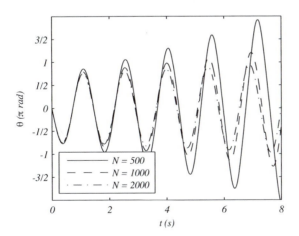

Figure 2.28: Solution for the nonlinear pendulum problem using Euler's Method on $t \in [0,8]$ with $N = 500, 1000, 2000$.

Let's write the Euler scheme as

$$
\begin{aligned}
\omega_{i+1} &= \omega_i - h\frac{g}{L}\sin\theta_i, \\
\theta_{i+1} &= \theta_i + h\omega_i.
\end{aligned}
\tag{2.149}
$$

Then, we replace ω_i in the second line by ω_{i+1} to obtain the new scheme

$$\begin{aligned}
\omega_{i+1} &= \omega_i - h\frac{g}{L}\sin\theta_i, \\
\theta_{i+1} &= \theta_i + h\omega_{i+1}.
\end{aligned} \qquad (2.150)$$

The MATLAB code is easily changed as shown below.

```
g=9.8;
L=0.5;
m=0.005;

a=0;
b=8;
N=500;
h=(b-a)/N;

% Initial Condition
  t(1)=0;
  theta(1)=pi/6;
  omega(1)=0;

% Euler-Cromer Method
  for i=2:N+1
      omega(i)=omega(i-1)-g/L*h*sin(theta(i-1));
      theta(i)=theta(i-1)+h*omega(i);
      t(i)=t(i-1)+h;
  end
```

We then run the new scheme for $N = 500$ and compare this with what we obtained previously. The results are shown in Figure 2.29. We see that the oscillation amplitude seems to be under control. However, the best test would be to investigate if the energy is conserved.

Figure 2.29: Solution for the nonlinear pendulum problem comparing Euler's Method and the Euler-Cromer Method on $t \in [0,8]$ with $N = 500$.

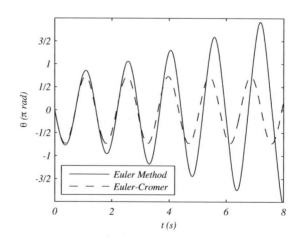

Recall that the total mechanical energy for a pendulum consists of the kinetic and gravitational potential energies:

$$E = \frac{1}{2}mv^2 + mgh.$$

For the pendulum, the tangential velocity is given by $v = L\omega$, and the height of the pendulum mass from the lowest point of the swing is $h = L(1 - \cos\theta)$. Therefore, in terms of the dynamical variables, we have

$$E = \frac{1}{2}mL^2\omega^2 + mgL(1 - \cos\theta).$$

We can compute the energy at each time step in the numerical simulation. In MATLAB it is easy to do using

```
E = 1/2*m*L^2*omega.^2+m*g*L*(1-cos(theta)){;}
```

after implementing the scheme. In other programming environments, one needs to loop through the times steps and compute the energy along the way. In Figure 2.30 we show the results for Euler's Method for $N = 500, 1000, 2000$ and the Euler-Cromer Method for $N = 500$. It is clear that the Euler-Cromer Method does a much better job of maintaining energy conservation.

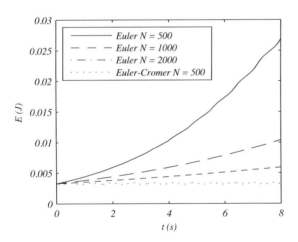

Figure 2.30: Total energy for the nonlinear pendulum problem.

2.10.2   Extreme Sky Diving

ON OCTOBER 14, 2012, FELIX BAUMGARTNER JUMPED from a helium balloon at an altitude of 39,045 m (24.26 mi or 128,100 ft). According to preliminary data from the Red Bull Stratos Mission[3], as of November 6, 2012, Baumgartner experienced free fall until he opened his parachute at 1,585 m after 4 minutes and 20 seconds. Within the first minute he had broken the record set by Joe Kittinger on August 16, 1960. Kittinger jumped from 102,800 feet (31 km) and fell freely for 4 minutes and 36 seconds to an altitude of 18,000 ft (5,500 m). Both set records for their times. Kittinger

[3] The original estimated data was found at the Red Bull Stratos site, http://www.redbullstratos.com/. Some of the data has since been updated. The reader can redo the solution using the updated data.

reached 614 mph (Mach 0.9) and Baumgartner reached 833.9 mph (Mach 1.24). Another record that was broken was that over 8 million watched the event on YouTube, breaking current live stream viewing events at the time.

This much attention also peaked interest in the physics of free fall. Free fall at constant g through a height of h should take a time of

$$t = \sqrt{\frac{2h}{g}} = \sqrt{\frac{2(36,529)}{9.8}} = 86 \text{ s}.$$

Of course, g is not constant. In fact, at an altitude of 39 km, we have

$$g = \frac{GM}{R+h} = \frac{6.67 \times 10^{-11} \text{ N m}^2\text{kg}^2(5.97 \times 10^{24} \text{ kg})}{6375 + 39 \text{ km}} = 9.68 \text{ m/s}^2.$$

So, g is roughly constant.

Next, we need to consider the drag force as one free falls through the atmosphere, $F_D = \frac{1}{2}CA\rho_a v^2$. One needs some values for the parameters in this problem. Let's take $m = 90$ kg, $A = 1.0$ m^2, and $\rho = 1.29$ kg/m^3, $C = 0.42$. Then, a simple model would give

$$m\dot{v} = -mg + \frac{1}{2}CA\rho v^2,$$

or

$$\dot{v} = -g + .0030v^2.$$

The Reynolds number is used several times in this chapter. It is defined as

$$Re = \frac{2rv}{\nu},$$

where ν is the kinematic viscosity. The kinematic viscosity of air at 60^0 F is about 1.47×10^{-5} m^2/s.
[4] http://www.grc.nasa.gov/WWW/k-12/rocket/atmos.html

This gives a terminal velocity of 57.2 m/s, or 128 mph. However, we again have assumed that the drag coefficient and air density are constant. Because the Reynolds number is high, we expect that C is roughly constant. However, the density of the atmosphere is a function of altitude and we need to take this into account.

A simple model for $\rho = \rho(h)$ can be found at the NASA site.[4]. Using their data, we have

$$\rho(h) = \begin{cases} \dfrac{101290(1.000 - 0.2253 \times 10^{-4}h)^{5.256}}{83007 - 1.8696h}, & h < 11000, \\ .3629e^{1.73 - 0.157 \times 10^{-3}h}, & h, < 25000, \\ \dfrac{2488}{(.6551 + 0.1380 \times 10^{-4}h)^{11.388}(40876 + .8614h)}, & h > 25000. \end{cases}$$

$$(2.151)$$

In Figure 2.31, the atmospheric density is shown as a function of altitude.

In order to use the methods for solving first-order equations, we write the system of equations in the form

$$\frac{dh}{dt} = v,$$

$$\frac{dv}{dt} = -\frac{GM}{(R+h)^2} + \frac{1}{5}\rho(h)CAv^2. \qquad (2.152)$$

This is now in the form of a system of first-order differential equations.

Then we define a function to be called and store it as **gravf.m** as shown below.

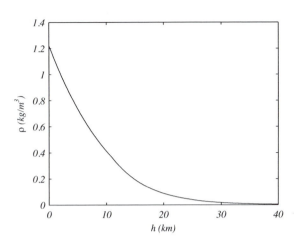

```
function dy=gravf(t,y);

G=6.67E-11;
M=5.97E24;
R=6375000;
m=90;
C=.42;
A=1;

dy(1,1)=y(2);
dy(2,1)=-G*M/(R+y(1)).^2+.5*density2(y(1))*C*A*y(2).^2/m;
```

Now we are ready to call the function in our favorite routine.

```
h0=1000;
tmax=20;
tmin=0;
[t,y]=ode45('dgravf',[tmin tmax],[h0;0]);% Const rho
plot(t,y(:,1),'k--')
```

Here we are simulating free fall from an altitude of 1 km. In Figure 2.32 we compare different models of free fall with g taken as constant or derived from Newton's Law of Gravitation. We also consider constant density or the density dependence on the altitude as given earlier. We chose to keep the drag coefficient constant at $C = 0.42$.

We can see from these plots that the slight variation in the acceleration due to gravity does not have as much an effect as the variation of density with distance.

Now we can push the model to Baumgartner's jump from 39 km. In Figure 2.33 we compare the general model with that with no air resistance, though both taking into account the variation in g. As a body falls through the atmosphere, we see the changing effects of the denser atmosphere on the free fall. For the parameters chosen, we find that it takes 238.8 s, or a little less than 4 minutes to reach the point where Baumgartner opened his parachute. While not exactly the same as the real fall, it is amazingly close.

Figure 2.32: Comparison of different models of free fall from 1 km above the Earth.

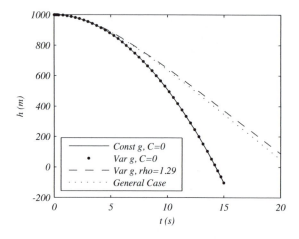

Figure 2.33: Free fall from 39 km at constant g as compared to nonconstant g and nonconstant atmospheric density with drag coefficient $C = .42$.

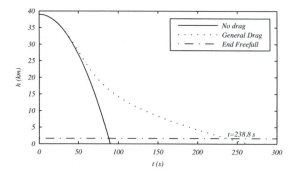

2.10.3   Flight of Sports Balls

ANOTHER INTERESTING PROBLEM IS THE PROJECTILE MOTION of a sports ball. In an introductory physics course, one typically ignores air resistance and the path of the ball is a nice parabolic curve. However, adding air resistance complicates the problem significantly and cannot be solved analytically. Examples in sports are flying soccer balls, golf balls, ping pong balls, baseballs, and other spherical balls.

We will consider a ball moving in the xz-plane spinning about an axis perpendicular to the plane of motion. Such an analysis was reported in Goff and Carré, AJP 77(11) 1020. The typical trajectory of the ball is shown in Figure 2.34. The forces acting on the ball are the drag force $\mathbf{F}_D$, the lift force $\mathbf{F}_L$, and the gravitational force $\mathbf{F}_W$. These are indicated in Figure 2.35. The equation of motion takes the form

$$m\mathbf{a} = \mathbf{F}_W + \mathbf{F}_D + \mathbf{F}_L.$$

Writing out the components, we have

$$ma_x = -F_D \cos\theta - F_L \sin\theta, \qquad (2.153)$$

$$ma_z = -mg - F_D \sin\theta + F_L \cos\theta. \qquad (2.154)$$

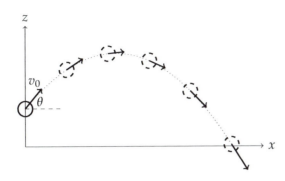

Figure 2.34: Projectile path.

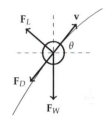

Figure 2.35: Forces acting on ball.

As we have seen before, the magnitude of the damping (drag) force is given by

$$F_D = \frac{1}{2} C_D \rho A v^2.$$

For the case of soccer ball dynamics, Goff and Carré noted that the Reynolds number, $Re = \frac{2rv}{\nu}$, is between 70,000 and 490,000 using a kinematic viscosity of $\nu = 1.54 \times 10^{-5}$ m^2/s and typical speeds of $v = 4.5 - 31$ m/s. Their analysis gives $C_D \approx 0.2$. The parameters used for the ball were $m = 0.424$ kg and cross-sectional area $A = 0.035$ m^2; and the density of air was taken as 1.2 kg/m^3.

The lift force takes a similar form,

$$F_L = \frac{1}{2} C_L \rho A v^2.$$

The sign of C_L indicates if the ball has top spin ($C_L < 0$) or bottom spin *$C_L > 0$). The lift force is just one component of a more general Magnus force, which is the force on a spinning object in a fluid and is perpendicular to the motion. In this example we assume that the spin axis is perpendicular to the plane of motion. Allowing for spinning balls to veer from this plane would mean that we would also need a component of the Magnus force perpendicular to the plane of motion. This would lead to an additional sideways component (in the **k** direction) leading to a third acceleration equation. We will leave that case for the reader.

So far, the problem has been reduced to

$$\frac{dv_x}{dt} = -\frac{\rho A}{2m}(C_D \cos\theta + C_L \sin\theta)v^2, \qquad (2.155)$$

$$\frac{dv_z}{dt} = -g - \frac{\rho A}{2m}(C_D \sin\theta - C_L \cos\theta)v^2, \qquad (2.156)$$

for v_x and v_z the components of the velocity. Also, $v^2 = v_x^2 + v_z^2$. Furthermore, from Figure 2.35, we can write

$$\cos\theta = \frac{v_x}{v}, \quad \sin\theta = \frac{v_z}{v}.$$

The lift coefficient can be related to the spin as

$$C_L = \frac{1}{2 + \frac{v}{v_{spin}}},$$

where $v_{spin} = r\omega$ is the peripheral speed of the ball. Here, R is the ball radius and ω is the angular speed in rad/s. If $v = 20$ m/s, $\omega = 200$ rad/s, and $r = 20$ mm, then $C_L = 0.45$.

So, the equations can be written entirely as a system of differential equations for the velocity components,

$$\frac{dv_x}{dt} = -\alpha(C_D v_x + C_L v_z)(v_x^2 + v_z^2)^{1/2}, \tag{2.157}$$

$$\frac{dv_z}{dt} = -g - \alpha(C_D v_z - C_L v_x)(v_x^2 + v_z^2)^{1/2}, \tag{2.158}$$

where $\alpha = \rho A/2m = 0.0530 \text{ m}^{-1}$.

Such systems of equations can be solved numerically by thinking of this as a vector differential equation,

$$\frac{d\mathbf{v}}{dt} = \mathbf{F}(t, \mathbf{v}),$$

and applying one of the numerical methods for solving first-order equations.

Because we are interested in the trajectory, $z = z(x)$, we would like to determine the parametric form of the path, $(x(t), z(t))$. So, instead of solving two first-order equations for the velocity components, we can rewrite the two second-order differential equations for $x(t)$ and $z(t)$ as four first-order differential equations of the form

$$\frac{d\mathbf{y}}{dt} = \mathbf{F}(t, \mathbf{y}).$$

We first define

$$\mathbf{y} = \begin{bmatrix} y_1(t) \\ y_2(t) \\ y_3(t) \\ y_4(t) \end{bmatrix} = \begin{bmatrix} x(t) \\ z(t) \\ v_x(t) \\ v_z(t) \end{bmatrix}$$

Then, the systems of first-order differential equations becomes

$$\frac{dy_1}{dt} = y_3,$$

$$\frac{dy_2}{dt} = y_4,$$

$$\frac{dy_3}{dt} = -\alpha(C_D v_x + C_L v_z)(v_x^2 + v_z^2)^{1/2},$$

$$\frac{dy_4}{dt} = -g - \alpha(C_D v_z - C_L v_x)(v_x^2 + v_z^2)^{1/2}. \tag{2.159}$$

The system can be placed into a function file that can be called by an ODE solver, such as the MATLAB m-file below.

```
function dy = ballf(t,y)
global g CD CL alpha

dy = zeros(4,1);    % a column vector
v = sqrt(y(3).^2+y(4).^2);   % speed v

dy(1) = y(3);
```

```
dy(2) = y(4);
dy(3) = -alpha*v.*(CD*y(3)+CL*y(4));
dy(4) = alpha*v.*(-CD*y(4)+CL*y(3))-g;
```

Then, the solver can be called using

```
[T,Y] = ode45('ballf',[0 2.5],[x0,z0,v0x,v0z]);
```

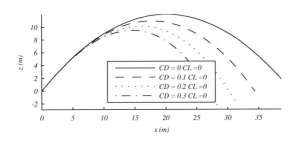

Figure 2.36: Example of soccer ball under the influence of drag.

In Figures 2.36 and 2.37, we indicate what typical solutions would look like for different values of drag and lift coefficients. In the case of nonzero lift coefficients, we indicate positive and negative values leading to flight with top spin, $C_L < 0$, or bottom spin, $C_L > 0$.

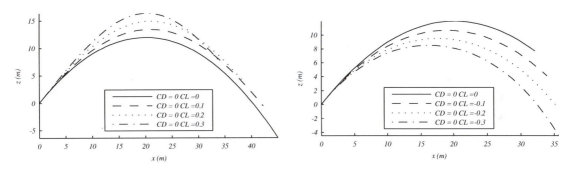

Figure 2.37: Example of soccer ball under the influence of lift with $CL > 0$ and $CL < 0$.

2.10.4 Falling Raindrops

A SIMPLE PROBLEM THAT APPEARS IN MECHANICS is that of a falling raindrop through a mist. The raindrop not only undergoes free fall, but the mass of the drop grows as it interacts with the mist. There have been several papers written on this problem, and it is a nice example to explore using numerical methods. In this section we look at models of a falling raindrop with and without air drag.

First we consider the case in which there is no air drag. A simple model of free fall from Newton's Second Law of Motion is

$$\frac{d(mv)}{dt} = mg.$$

In this discussion we will take downward as positive. Because the mass is not constant, we have

$$m\frac{dv}{dt} = mg - v\frac{dm}{dt}.$$

In order to proceed, we need to specify the rate at which the mass is changing. There are several models one can adapt. We will borrow some of the ideas and in some cases the numerical values from Sokal(2010)[5] and Edwards, Wilder, and Scime (2001).[6] These papers also quote other interesting work on the topic.

[5] A. D. Sokal, The falling raindrop, revisited, *Am. J. Phys.* 78, 643-645, (2010).
[6] B. F. Edwards, J. W. Wilder, and E. E. Scime, Dynamics of Falling Raindrops, *Eur. J. Phys.* 22, 113-118, (2001).

While v and m are functions of time, one can look for a way to eliminate time by assuming the rate of change of mass is an explicit function of m and v alone. For example, Sokal (2010) assumes the form

$$\frac{dm}{dt} = \lambda m^{\sigma} v^{\beta}, \quad \lambda > 0.$$

This contains two commonly assumed models of accretion:

1. $\sigma = 2/3, \beta = 0$. This corresponds to growth of the raindrop proportional to the surface area. Because $m \propto r^3$ and $A \propto r^2$, then $\dot{m} \propto A$ implies that $\dot{m} \propto m^{2/3}$.

2. $\sigma = 2/3, \beta = 1$. In this case, the growth of the raindrop is proportional to the volume swept out along the path. Thus, $\Delta m \propto A(v\Delta t)$, where A is the cross-sectional area and $v\Delta t$ is the distance traveled in time Δt.

In both cases, the limiting value of the acceleration is a constant. It is $g/4$ in the first case and $g/7$ in the second case.

Another approach might be to use the effective radius of the drop, assuming that the raindrop remains close to spherical during the fall. According to Edwards, Wilder, and Scime (2001), raindrops with a Reynolds number greater than 1,000 and radii larger than 1 mm will flatten. Even larger raindrops will break up when the drag force exceeds the surface tension. Therefore, they take 0.1 mm $< r <$ 1 mm and $10 < Re < 1,000$. We will return to a discussion of the drag later.

It might seem more natural to make the radius the dynamic variable, rather than the mass. In this case, we can assume the accretion rate takes the form

$$\frac{dr}{dt} = \gamma r^{\alpha} v^{\beta}, \quad \gamma > 0.$$

Because $m = \frac{4}{3}\pi \rho_d r^3$,

$$\frac{dm}{dt} \sim r^2 \frac{dr}{dt} \sim m^{2/3}\frac{dr}{dt}.$$

Therefore, the two special cases become

1. $\alpha = 0, \beta = 0$. This corresponds to a growth of the raindrop proportional to the surface area.

2. $\alpha = 0, \beta = 1$. In this case, the growth of the raindrop is proportional to the volume swept out along the path.

Here, ρ_d is the density of the raindrop.

We will also need

$$
\begin{aligned}
\frac{v}{m}\frac{dm}{dt} &= \frac{4\pi\rho_d r^2}{\frac{4}{3}\pi\rho_d r^3}v\frac{dr}{dt} \\
&= 3\frac{v}{r}\frac{dr}{dt} \\
&= 3\gamma r^{\alpha-1}v^{\beta+1}.
\end{aligned} \tag{2.160}
$$

Putting this all together, we have a systems of two equations for $v(t)$ and $r(t)$:

$$
\begin{aligned}
\frac{dv}{dt} &= g - 3\gamma r^{\alpha-1}v^{\beta+1}, \\
\frac{dr}{dt} &= \gamma r^\alpha v^\beta.
\end{aligned} \tag{2.161}
$$

Example 2.30. *Determine $v = v(r)$ for the case $\alpha = 0$, $\beta = 0$ and the initial conditions $r(0) = 0.1$ mm and $v(0) = 0$ m/s.*

In this case, Equations (2.161) become

$$
\begin{aligned}
\frac{dv}{dt} &= g - 3\gamma r^{-1}v, \\
\frac{dr}{dt} &= \gamma.
\end{aligned} \tag{2.162}
$$

Noting that

$$
\frac{dv}{dt} = \frac{dv}{dr}\frac{dr}{dt} = \gamma\frac{dv}{dr},
$$

we can convert the problem to one of finding the solution $v(r)$ subject to the equation

$$
\frac{dv}{dr} = \frac{g}{\gamma} - 3\frac{v}{r}
$$

with the initial condition $v(r_0) = 0$ m/s for $r_0 = 0.0001$ m.

Rearranging the differential equation, we find that it is a linear first-order differential equation,

$$
\frac{dv}{dr} + \frac{3}{r}v = \frac{g}{\gamma}.
$$

This equation can be solved using an integrating factor, $\mu = r^3$, obtaining

$$
\frac{d}{dr}(r^3 v) = \frac{g}{\gamma}r^3.
$$

Integrating, we obtain the solution

$$
v(r) = \frac{g}{4\gamma}r\left(1 - \left(\frac{r_0}{r}\right)^4\right).
$$

Note that for large r, $v \sim \frac{g}{4\gamma}r$. Therefore, $\frac{dv}{dt} \sim \frac{g}{4}$.

While this case was easily solved in terms of elementary operations, it is not always easy to generate solutions to Equations (2.161) analytically. Sokal (2010) derived a general solution in terms of incomplete Beta functions, though this does not help visualize the solution. Also, as we will see, adding air drag will lead to a nonintegrable system. So, we turn to numerical solutions.

In MATLAB, we can use the function in **raindropf.m** to capture the system (2.161). Here we put the velocity in y(1) and the radius in y(2).

```
function dy=raindropf(t,y);
global alpha beta gamma g

dy=[g-3*gamma*y(2)^(alpha-1)*y(1)^(beta+1); ...
    gamma*y(2)^alpha*y(1)^beta];
```

We then use the Runge–Kutta solver, **ode45**, to solve the system. An implementation is shown below that calls the function containing the system. The value $\gamma = 2.5 \times 10^{-7}$ is based on empirical results quoted by Edwards, Wilder, and Scime (2001).

Figure 2.38: The plots of position and velocity as a function of time for $\alpha = \beta = 0$.

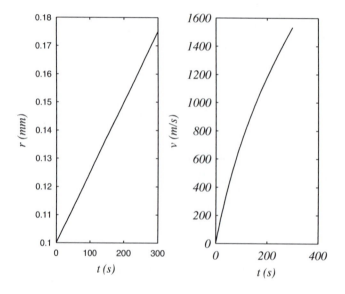

```
clear
global alpha beta gamma g

alpha=0;
beta=0;
gamma=2.5e-07;
g=9.81;

r0=0.0001;
v0=0;
y0=[v0;r0];
tspan=[0 1000];
```

```
[t,y]=ode45(@raindropf,tspan,y0);
plot(1000*y(:,2),y(:,1),'k')
```

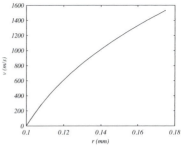

The resulting plots are shown in Figures 2.38 and 2.39. The plot of velocity as a function of position agrees with the exact solution, which we derived in the previous example. We note that these drops do not grow much, but they seem to attain large speeds.

For the second case, $\alpha = 0, \beta = 1$, one can also obtain an exact solution. The result is

$$v(r) = \left[\frac{2g}{7\gamma} r \left(1 - \left(\frac{r_0}{r} \right)^7 \right) \right]^{\frac{1}{2}}.$$

Figure 2.39: The plot the velocity as a function of position for $\alpha = \beta = 0$.

For large r, one can show that $\frac{dv}{dt} \sim \frac{g}{7}$. In Figures 2.40 and 2.41 we see again large velocities, though about a third as fast over the same time interval. However, we also see that the raindrop has significantly grown well past the point it would break up.

In this simple model of a falling raindrop, we have not considered air drag. Earlier in the chapter we discussed the free fall of a body with air resistance and this led to a terminal velocity. Recall that the drag force is given by

$$f_D(v) = -\frac{1}{2} C_D A \rho_a v^2, \tag{2.163}$$

where C_D is the drag coefficient, A is the cross-sectional area, and ρ_a is the air density. Also, we assume that the body is falling downward and downward is positive, so that $f_D(v) < 0$ so as to oppose the motion.

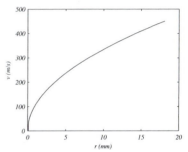

Figure 2.40: The plot the velocity as a function of position for $\alpha = 0, \beta = 1$.

We would like to incorporate this force into our model (2.161). The first equation came from the force law, which now becomes

$$m\frac{dv}{dt} = mg - v\frac{dm}{dt} - \frac{1}{2} C_D A \rho_a v^2,$$

or

$$\frac{dv}{dt} = g - \frac{v}{m}\frac{dm}{dt} - \frac{1}{2m} C_D A \rho_a v^2.$$

The next step is to eliminate the dependence on the mass, m, in favor of the radius, r. The drag force term can be written as

$$\begin{aligned}
\frac{f_D}{m} &= \frac{1}{2m} C_D A \rho_a v^2 \\
&= \frac{1}{2} C_D \frac{\pi r^2}{\frac{4}{3}\pi \rho_d r^3} \rho_a v^2 \\
&= \frac{3}{8} \frac{\rho_a}{\rho_d} C_D \frac{v^2}{r}.
\end{aligned} \tag{2.164}$$

We had already done this for the second term; however, Edwards, Wilder, and Scime (2001) point to experimental data and propose that

$$\frac{dm}{dt} = \pi \rho_m r^2 v,$$

where ρ_m is the mist density. So, the second term leads to

$$\frac{v}{m}\frac{dm}{dt} = \frac{3}{4} \frac{\rho_m}{\rho_d} \frac{v^2}{r}.$$

Figure 2.41: The plots of position and velocity as a function of time for $\alpha = 0, \beta = 1$.

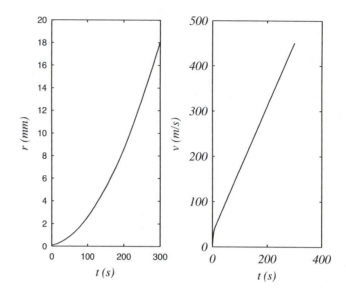

But because $m = \frac{4}{3}\pi\rho_d r^3$,

$$\frac{dm}{dt} = 4\pi\rho_d r^2 \frac{dr}{dt}.$$

So,

$$\frac{dr}{dt} = \frac{\rho_m}{4\rho_d}v.$$

This suggests that their model corresponds to $\alpha = 0$, $\beta = 1$, and $\gamma = \frac{\rho_m}{4\rho_d}$.

Now we can write the modified system

$$\frac{dv}{dt} = g - 3\gamma r^{\alpha-1}v^{\beta+1} - \frac{3}{8}\frac{\rho_a}{\rho_d}C_D\frac{v^2}{r},$$
$$\frac{dr}{dt} = \gamma r^\alpha v^\beta. \tag{2.165}$$

Edwards, Wilder, and Scime (2001) assume that the densities are constant with values $\rho_a = .856$ kg/m^3, $\rho_d = 1.000$ kg/m^3, and $\rho_m = 1.00 \times 10^{-3}$ kg/m^3. However, the drag coefficient is not constant. As described later in Section 2.10.7, there are various models indicating the dependence of C_D on the Reynolds number,

$$Re = \frac{2rv}{\nu},$$

where ν is the kinematic viscosity, which Edwards, Wilder, and Scime (2001) set to $\nu = 2.06 \times 10^{-5}$ m^2/s. For raindrops of the range $r = 0.1$ mm to 1 mm, the Reynolds number is below 1,000. Edwards, Wilder, and Scime (2001) modeled $C_D = 12Re^{-1/2}$. In the plots in Section 2.10.7, we include this model and see that this is a good approximation for these raindrops. In Chapter 10 we discuss least squares curve fitting and using these methods, one can use the models of Putnam (1961) and Schiller-Naumann (1933) to obtain a power law fit similar to that used here.

So, introducing

$$C_D = 12Re^{-1/2} = 12\left(\frac{2rv}{\nu}\right)^{-1/2}$$

and defining

$$\delta = \frac{9}{2^{3/2}}\frac{\rho_a}{\rho_d}\nu^{1/2},$$

we can write the system of equations (2.165) as

$$\frac{dv}{dt} = g - 3\gamma\frac{v^2}{r} - \delta\left(\frac{v}{r}\right)^{\frac{3}{2}},$$

$$\frac{dr}{dt} = \gamma v. \tag{2.166}$$

Now we can modify the MATLAB code for the raindrop by adding the extra term to the first equation; setting $\alpha = 0$, $\beta = 1$; and using $\delta = 0.0124$ and $\gamma = 2.5 \times 10^{-7}$ from Edwards, Wilder, and Scime (2001).

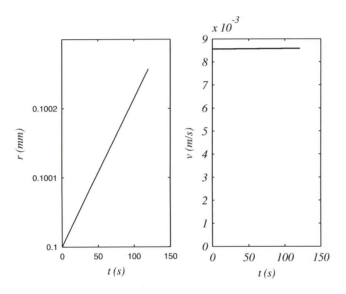

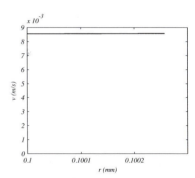

Figure 2.42: The plots of position and velocity as a function of time with air drag included.

Figure 2.43: The plot the velocity as a function of position with air drag included.

In Figures 2.42 and 2.43 we see different behaviors as compared to the previous models. It appears that the velocity quickly reaches a terminal velocity and the radius continues to grow linearly in time, though at a slow rate.

We might be able to understand this behavior. Terminal, or constant v, would occur when

$$g - 3\gamma\frac{v^2}{r} - \delta\left(\frac{v}{r}\right)^{\frac{3}{2}} = 0.$$

Looking at these terms, one finds that the second term is significantly smaller than the other terms and thus

$$\delta\left(\frac{v}{r}\right)^{\frac{3}{2}} \approx g,$$

or

$$\frac{v}{r} \approx \left(\frac{g}{\delta}\right)^{2/3} \approx 85.54 \text{ s}^{-1}.$$

This agrees with the numerical data, which gives the slope of the v versus r plot as 85.5236 s^{-1}.

2.10.5   Two-Body Problem

A STANDARD PROBLEM IN CLASSICAL DYNAMICS is the study of the motion of several bodies under the influence of Newton's Law of Gravitation. The so-called n-body problem is not solvable. However, the two body problem is. Such problems can model the motion of a planet around the sun, the moon around the Earth, or a satellite around the Earth. Further interesting and more realistic problems would involve perturbations of these orbits due to additional bodies. For example, one can study problems such as the influence of large planets on the asteroid belt. Because there are no analytic solutions to these problems, we have to resort to finding numerical solutions. We will look at the two-body problem because we can compare the numerical methods to the exact solutions.

We consider two masses, m_1 and m_2, located at positions, $\mathbf{r}_1$ and $\mathbf{r}_2$, respectively, as shown in Figure 2.44. Newton's Law of Gravitation for the force between two masses separated by position vector $\mathbf{r}$ is given by

$$\mathbf{F} = -\frac{Gm_1m_2}{r^2}\frac{\mathbf{r}}{r}.$$

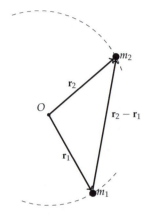

Figure 2.44: Two masses interact under Newton's Law of Gravitation.

Each mass experiences this force due to the other mass. This gives the system of equations

$$m_1\ddot{\mathbf{r}}_1 = -\frac{Gm_1m_2}{|\mathbf{r}_2 - \mathbf{r}_1|^3}(\mathbf{r}_1 - \mathbf{r}_2), \tag{2.167}$$

$$m_2\ddot{\mathbf{r}}_2 = -\frac{Gm_1m_2}{|\mathbf{r}_2 - \mathbf{r}_1|^3}(\mathbf{r}_2 - \mathbf{r}_1). \tag{2.168}$$

Now we seek to set up this system so that we can find numerical solutions for the positions of the masses. From the conservation of angular momentum, we know that the motion takes place in a plane. [Note: The solution of the Kepler Problem is discussed in Chapter 9.] We will choose the orbital plane to be the xy-plane. We define $r_{12} = |\mathbf{r}_2 - \mathbf{r}_1|$, and $(x_i, y_i) = \mathbf{r}_i$, $i = 1, 2$. Furthermore, we write the two second-order equations as four first-order equations. So, defining the velocity components as $(u_i, v_i) = \mathbf{v}_i$, the system of equations can be written in the form

$$\frac{d}{dt}\begin{pmatrix} x_1 \\ y_1 \\ x_2 \\ y_2 \\ u_1 \\ v_1 \\ u_2 \\ v_2 \end{pmatrix} = \begin{pmatrix} u_1 \\ v_1 \\ u_2 \\ v_2 \\ -\alpha m_2(x_1 - x_2) \\ -\alpha m_2(y_1 - y_2) \\ -\alpha m_1(x_2 - x_1) \\ -\alpha m_1(y_2 - y_1). \end{pmatrix}, \tag{2.169}$$

where $\alpha = \dfrac{G}{r_{12}^3}$.

This system can be encoded in MATLAB as indicated in the function **twobody**:

```
function dz = twobody(t,z)
dz = zeros(8,1);
G = 1;
m1 = .1;
m2 = 2;
r=((z(1) - z(3)).^2 + (z(2) - z(4)).^2).^(3/2);
alpha=G/r;
dz(1) = z(5);
dz(2) = z(6);
dz(3) = z(7);
dz(4) = z(8);
dz(5) = alpha*m2*(z(3) - z(1));
dz(6) = alpha*m2*(z(4) - z(2));
dz(7) = alpha*m1*(z(1) - z(3));
dz(8) = alpha*m1*(z(2) - z(4));
```

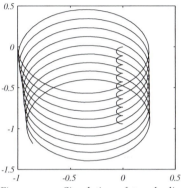

Figure 2.45: Simulation of two bodies under gravitational attraction.

In the above code we picked some seemingly nonphysical numbers for G and the masses. Calling **ode45** with a set of initial conditions,

```
[t,z] = ode45('twobody',[0 20], [-1 0 0 0 0 -1 0 0]);
plot(z(:,1),z(:,2),'k',z(:,3),z(:,4),'k');
```

we obtain the plot shown in Figure 2.45. We see each mass moves along what looks like elliptical helices, with the smaller body tracing out a larger orbit.

In the case of a very large body, most of the motion will be due to the smaller body. So, it might be better to plot the relative motion of the small body with respect to the larger body. Actually, an analysis of the two-body problem shows that the center of mass

$$\mathbf{R} = \frac{m_1\mathbf{r}_1 + m_2\mathbf{r}_2}{m_1 + m_2}$$

satisfies $\ddot{\mathbf{R}} = 0$. Therefore, the system moves with a constant velocity.

The relative position of the masses is defined through the variable $\mathbf{r} = \mathbf{r}_1 - \mathbf{r}_2$. Dividing the masses from the left hand side of Equations (2.168) and subtracting, we have the motion of m_1 about m_2

$$\ddot{\mathbf{r}} = -G(m_1 + m_2)\frac{\mathbf{r}}{r^3},$$

where $r = |\mathbf{r}| = |\mathbf{r}_1 - \mathbf{r}_2|$. Note that $\mathbf{r} \times \ddot{\mathbf{r}} = 0$. Integrating, this gives $\mathbf{r} \times \dot{\mathbf{r}} =$ constant. This is just a statement of the conservation of angular momentum.

The orbiting body will remain in a plane and, therefore, we can take the z-axis perpendicular to $\mathbf{r} \times \dot{\mathbf{r}}$, the position as $\mathbf{r} = (x(t), y(t))$, and the velocity as $\dot{\mathbf{r}} = (u(t), v(t))$. Then, the equations of motion can be written as four first-order equations:

$$\dot{x} = u$$
$$\dot{y} = v$$

$$\dot{u} = -\mu \frac{x}{r^3}$$

$$\dot{v} = -\mu \frac{y}{r^3}, \tag{2.170}$$

where $\mu = G(m_1 + m_2)$ and $r = \sqrt{x^2 + y^2}$.

While we have established a system of equations that can be integrated, we should note a few results from the study of the Kepler Problem in classical dynamics which we review in Section 9.1.6. Kepler's Laws of Planetary Motion state:

1. All planets travel in ellipses.

   The polar equation for the path is given by

   $$r = \frac{a(1 - e^2)}{1 + e \cos \phi},$$

   where e is the eccentricity and a is the length of the semimajor axis. For $0 \leq e < 1$, the orbit is an ellipse.

2. A planet sweeps out equal areas in equal times.

3. The square of the period of the orbit is proportional to the cube of the semimajor axis. In particular, one can show that

   $$T^2 = \frac{4\pi^2}{\mu} a^3.$$

By an appropriate choice of units, we can make $\mu = G(m_1 + m_2)$ a reasonable number. For the Earth-Sun system,

$$\mu = 6.67 \times 10^{-11} \text{m}^3\text{kg}^{-1}\text{s}^{-2}(1.99 \times 10^{30} + 5.97 \times 10^{24})\text{kg}$$
$$= 1.33 \times 10^{20} \text{m}^3\text{s}^{-1}.$$

That is a large number and can cause problems in the numerics. However, if one uses astronomical scales, such as putting lengths in astronomical units, 1 AU $= 1.50 \times 10^8$ km, and time in years, then

$$\mu = \frac{4\pi^2}{T^2} a^3 = 4\pi^2$$

in units of AU^3/yr^2.

Setting $\phi = 0$, the location of the perigee is given by

$$r = \frac{a(1 - e^2)}{1 + e} = a(1 - e),$$

or

$$\mathbf{r} = (a(1 - e), 0).$$

At this point the velocity is given by

$$\dot{\mathbf{r}} = \left(0, \sqrt{\frac{\mu}{a} \frac{1 + e}{1 - e}}\right).$$

Knowing the position and velocity at $\phi = 0$, we can set the initial conditions for a bound orbit. The MATLAB code based on the above analysis is given below and the solution can be seen in Figure 2.46.

```
e=0.9;
tspan=[0 100];
z0=[1-e;0;0;sqrt((1+e)/(1-e))];
[t,z] = ode45('twobodyf',tspan, z0);
plot(z(:,1),z(:,2),'k');
axis equal

function dz = twobodyf(t,z)
dz = zeros(4,1);
GM = 1;


r=(z(1).^2 + z(2).^2).^(3/2);
alpha=GM/r;
dz(1) = z(3);
dz(2) = z(4);
dz(3) = -alpha*z(1);
dz(4) = -alpha*z(2);
```

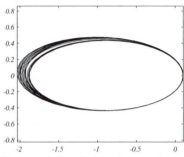

Figure 2.46: Simulation of one-body orbiting a larger body under gravitational attraction.

While it is clear that the mass is following an elliptical orbit, we see that it will only do so for a finite period of time, partly because the Runge–Kutta code does not conserve energy and it does not conserve the angular momentum. The conservation of energy is found (up to a factor of m_1) as

$$\frac{1}{2}(\dot{x}^2 + \dot{y}^2) - \frac{\mu}{t} = -\frac{\mu}{2a}.$$

Similarly, the conservation of (specific) angular momentum is given by

$$\mathbf{r} \times \mathbf{v} = (x\dot{y} - y\dot{x})\mathbf{k} = \sqrt{\mu a(1 - e^2)}\mathbf{k}.$$

As was the case with the nonlinear pendulum example, we saw that an implicit Euler Method, or Cromer's Method, was better at conserving energy. So, we compare the Euler's Method version with the Implicit-Euler Method. In general, we seek to solve the system

$$\dot{\mathbf{r}} = \mathbf{F}(\mathbf{r}, \mathbf{v}),$$
$$\dot{\mathbf{v}} = \mathbf{G}(\mathbf{r}, \mathbf{v}). \qquad (2.171)$$

As seen earlier, Euler's Method is given by

$$\mathbf{v}_n = \mathbf{v}_{n-1} + \Delta t * \mathbf{G}(t_{n-1}, \mathbf{x}_{n-1}),$$
$$\mathbf{r}_n = \mathbf{r}_{n-1} + \Delta t * \mathbf{F}(t_{n-1}, \mathbf{v}_{n-1}). \qquad (2.172)$$

For the two-body problem, we can write out the Euler Method steps using $\mathbf{v} = (u, v)$, $\mathbf{r} = (x, y)$, $\mathbf{F} = (u, v)$, and $\mathbf{G} = -\frac{\mu}{r^3}(x, y)$. The MATLAB code would use the loop

Euler's Method for the two-body problem.

```
for i=2:N+1
    alpha=mu/(x(i-1).^2 + y(i-1).^2).^(3/2);
    u(i)=u(i-1)-h*alpha*x(i-1);
```

```
        v(i)=v(i-1)-h*alpha*y(i-1);
        x(i)=x(i-1)+h*u(i-1);
        y(i)=y(i-1)+h*v(i-1);
        t(i)=t(i-1)+h;
    end
```

Note that more compact forms can be used but they are not readily adaptable to other packages or programming languages.

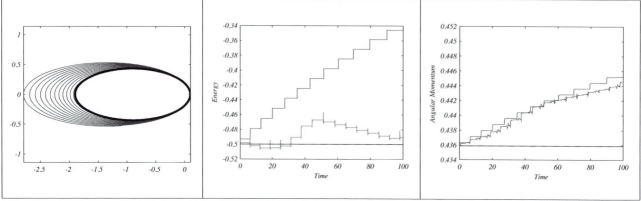

Table 2.9: Results for using Euler's Method for $N = 4,000,000$ and $t \in [0, 100]$. The parameters are $\mu = 1$, $e = 0,9$, and $a = 1$.

In Figure 2.9 we show the results along with the energy and angular momentum plots for $N = 4,000,000$ and $t \in [0, 100]$ for the case of $\mu = 1$, $e = 0,9$, and $a = 1$. The orbit based on the exact solution is in the center of the figure on the left. The energy and angular momentum as a function of time are shown along with the similar plots obtained using **ode45**. In neither case are these two quantities conserved.

```
    for i=2:N+1
        alpha=mu/(x(i-1).^2 + y(i-1).^2).^(3/2);
        u(i)=u(i-1)-h*alpha*x(i-1);
        v(i)=v(i-1)-h*alpha*y(i-1);
        x(i)=x(i-1)+h*u(i);
        y(i)=y(i-1)+h*v(i);
        t(i)=t(i-1)+h;
    end
```

Implicit-Euler Method for the two-body problem.

The Implicit-Euler Method is a slight modification of the Euler Method and has a better chance of handling the conserved quantities as the Implicit-Euler Method is one of many symplectic integrators. The modification uses the new value of the velocities in the updating of the position. Thus, we have

$$\mathbf{v}_n = \mathbf{v}_{n-1} + \Delta t * \mathbf{G}(t_{n-1}, \mathbf{x}_{n-1}),$$
$$\mathbf{r}_n = \mathbf{r}_{n-1} + \Delta t * \mathbf{F}(t_{n-1}, \mathbf{v}_n). \tag{2.173}$$

It is a simple matter to update the MATLAB code. In Figure 2.10 we show the results along with the energy and angular momentum plots for

$N = 200,000$ and $t \in [0, 100]$ for the case of $\mu = 1$, $e = 0, 9$, and $a = 1$. The orbit based on the exact solution coincides with the orbit as seen in the left figure. The energy and angular momentum as functions of time appear to be conserved. The energy fluctuates about -0.5 and the angular momentum remains constant. Again, the **ode45** results are shown in comparison. The number of time steps has been decreased from the Euler Method by a factor of 20.

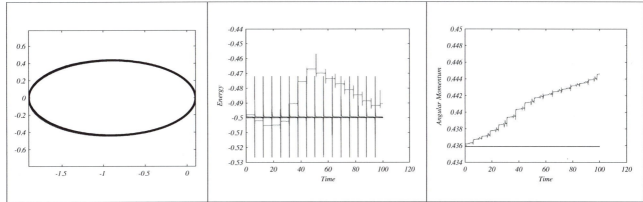

Table 2.10: Results for using the Implicit-Euler method for $N = 200,000$ and $t \in [0, 100]$. The parameters are $\mu = 1$, $e = 0, 9$, and $a = 1$.

The Euler and Implicit Euler are first-order methods. We can attempt a faster and more accurate process which is also a symplectic method. As a final example, we introduce the Velocity Verlet Method for solving

$$\ddot{\mathbf{r}} = \mathbf{a}(\mathbf{r}(t)).$$

The derivation is based on a simple Taylor expansion:

$$\mathbf{r}(t + \Delta t) = \mathbf{r}(t) + \mathbf{v}(t)\Delta t + \frac{1}{2}\mathbf{a}(\mathbf{t})\Delta t^2 + \cdots.$$

Replace Δt with $-\Delta t$ to obtain

$$\mathbf{r}(t - \Delta t) = \mathbf{r}(t) - \mathbf{v}(t)\Delta t + \frac{1}{2}\mathbf{a}(\mathbf{t})\Delta t^2 - \cdots.$$

Now, adding these expressions leads to some cancellations,

$$\mathbf{r}(t + \Delta t) = 2\mathbf{r}(t) - \mathbf{r}(t - \Delta t) + \mathbf{a}(\mathbf{t})\Delta t^2 + O(\Delta t^4).$$

Writing this in a more useful form, we have

$$\mathbf{r}_{n+1} = 2\mathbf{r}_n - \mathbf{r}_{n-1} + \mathbf{a}(\mathbf{r}_n)\Delta t^2.$$

Thus, we can find $\mathbf{r}_{n+1}$ from the previous two values without knowing the velocity. This method is called the Verlet, or Störmer-Verlet Method.

It is useful to know the velocity so that we can check energy conservation and angular momentum conservation. The Verlet Method can be rewritten in an equivalent form known as the Velocity Verlet Method. We use

$$\mathbf{r}(t) - \mathbf{r}(t - \Delta t) \approx \mathbf{v}(t)\Delta t - \frac{1}{2}\mathbf{a}\Delta t^2$$

Loup Verlet (1931–) is a physicist who works on molecular dynamics and Fredrik Carl Mülertz Störmer (1874–1957) was a mathematician and physicist who modeled the motion of charged particles in his studies of the aurora borealis.

in the Stömer-Verlet Method and write

$$\mathbf{r}_n = \mathbf{r}_{n-1} + \mathbf{v}_{n-1}u + \frac{h^2}{2}\mathbf{a}(\mathbf{r}_{n-1}),$$

$$\mathbf{v}_{n-1/2} = \mathbf{v}_{n-1} + \frac{h}{2}\mathbf{a}(\mathbf{r}_{n-1}),$$

$$\mathbf{a}_n = \mathbf{a}(\mathbf{r}_n),$$

$$\mathbf{v}_n = \mathbf{v}_{n-1/2} + \frac{h}{2}\mathbf{a}_n, \tag{2.174}$$

Störmer-Verlet Method for the two-body problem.

where $h = \Delta t$. For the current problem, $\mathbf{a}(\mathbf{r}_n) = -\frac{\mu}{r_n^2}\mathbf{r}_n$.
The MATLAB snippet is given as

```
for i=2:N+1
    alpha=mu/(x(i-1).^2 + y(i-1).^2).^(3/2);
    x(i)=x(i-1)+h*u(i-1)-h^2/2*alpha*x(i-1);
    y(i)=y(i-1)+h*v(i-1)-h^2/2*alpha*y(i-1);
    u(i)=u(i-1)-h/2*alpha*x(i-1);
    v(i)=v(i-1)-h/2*alpha*y(i-1);
    alpha=mu/(x(i).^2 + y(i).^2).^(3/2);
    u(i)=u(i)-h/2*alpha*x(i);
    v(i)=v(i)-h/2*alpha*y(i);
    t(i)=t(i-1)+h;
end
```

The results using the Velocity Verlet Method are shown in Figure 2.11. For only 50,000 steps, we have much better results for the conservation laws and the orbit appears stable.

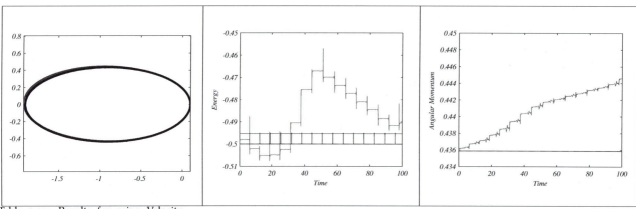

Table 2.11: Results for using Velocity Verlet Method for $N = 50,000$ and $t \in [0, 100]$. The parameters are $\mu = 1$, $e = 0, 9$, and $a = 1$.

2.10.6 Expanding Universe

ONE OF THE REMARKABLE STORIES of the twentieth century was the development of both the theory and the experimental data leading to our current understanding of the large-scale structure of the universe. In 1916, Albert

Einstein (1879–1955) published his general theory of relativity. It is a geometric theory of gravitation that relates the curvature of spacetime to its energy and momentum content. This relationship is embodied in the Einstein field equations, which are written compactly as

$$G_{\mu\nu} + \Lambda g_{\mu\nu} = \frac{8\pi G}{c^4} T_{\mu\nu}.$$

The left side contains the curvature of spacetime as determined by the metric $g_{\mu\nu}$. The Einstein tensor, $G_{\mu\nu} = R_{\mu\nu} - \frac{1}{2} R g_{\mu\nu}$, is determined from the curvature tensor $R_{\mu\nu}$ and the scalar curvature R. These in turn are obtained from the metric tensor. Λ is the famous cosmological constant, which Einstein originally introduced to maintain a static universe, that has since taken on a different role. The right-hand side of Einstein's equation involves the familiar gravitational constant, the speed of light, and the stress-energy tensor $T_{\mu\nu}$.

In 1917, Einstein applied general relativity to cosmology. However, it was Alexander Alexandrovich Friedmann (1888–1925) who was the first to provide solutions to Einstein's equation based on the assumptions of homogeneity and isotropy and leading to the expansion of the universe. Unfortunately, Friedmann died in 1925 of typhoid.

Georges Lemaître (1894–1966) had actually predicted the expansion of the universe in 1927 and proposed what later became known as the Big Bang Theory.

In 1929, Edwin Hubble (1889–1953) showed that the radial velocities of galaxies are proportional to their distance, resulting in what is now called Hubble's Law. Hubble's Law takes the form

$$v = H_0 r,$$

where H_0 is the Hubble constant and indicates that the universe is expanding. The current values of the Hubble constant are (70 ± 7) km s^{-1} Mpc^{-1}, and some recent WMAP results[7] indicate it could be (71.0 ± 2.5) km s^{-1} Mpc^{-1}.

[7] These strange units are in common usage. Mpc stands for 1 megaparsec = 3.086×10^{22} m and 1 km s^{-1} Mpc^{-1} = 3.24×10^{-20} s^{-1}. The recent value was reported at the NASA website on March 25, 2013, `http://map.gsfc.nasa.gov/universe/bb_tests_exp.html`

In this section we are interested in Friedmann's Equation, which is the simple differential equation

$$\left(\frac{\dot{a}}{a}\right)^2 = \frac{8\pi G}{3c^2} \varepsilon(t) - \frac{\kappa c^2}{R_0^2} + \frac{\Lambda}{3}.$$

Here, $a(t)$ is the scale factor of the universe, which is taken to be one at present time; $\varepsilon(t)$ is the energy density; R_0 is the radius of curvature; and κ is the curvature constant, ($\kappa = +1$ for positively curved space, $\kappa = 0$ for flat space, $\kappa = -1$ for negatively curved space.) The cosmological constant Λ is now added to account for dark energy. The idea is that if we know the right side of Friedmann's equation, then we can say something about the future size of the unverse. This is a simple differential equation that comes from applying Einstein's Equation to an isotropic, homogenous, and curved spacetime. Einstein's equation actually gives us a little more than this equation, but we will only focus on the (first) Friedmann equation. The reader can read more in books on cosmology, such as B. Ryden's *Introduction to Cosmology*.

Friedmann's Equation can be written in a simpler form by taking into account the different contributions to the energy density. For $\Lambda = 0$ and zero curvature, one has

$$\left(\frac{\dot{a}}{a}\right)^2 = \frac{8\pi G}{3c^2}\varepsilon(t).$$

We define the Hubble parameter as $H(t) = \dot{a}/a$. At the current time t_0, $H(t_0) = H_0$, Hubble's constant, and we take $a(t_0) = 1$. The energy density in this case is called the critical density,

$$\varepsilon_c(t) = \frac{3c^2}{8\pi G}H(t)^2.$$

It is typical to introduce the density parameter, $\Omega = \varepsilon/\varepsilon_c$. Then, the Friedmann Equation can be written as

$$1 - \Omega = -\frac{\kappa c^2}{R_0^2 a(t)^2 H(t)^2}.$$

Evaluating this expression at the current time then

$$1 - \Omega_0 = -\frac{\kappa c^2}{R_0^2 H_0^2};$$

and therefore,

$$1 - \Omega = -\frac{H_0^2(1 - \Omega_0)}{a^2 H^2}.$$

Solving for H^2, we have the differential equation

$$\left(\frac{\dot{a}}{a}\right)^2 = H_0^2\left[\Omega(t) + \frac{1 - \Omega_0}{a^2}\right],$$

where Ω takes into account the contributions to the energy density of the universe. These contributions are due to nonrelativistic matter density, contributions due to photons and neutrinos, and the cosmological constant, which might represent dark energy. This is discussed in Ryden (2003). In particular, Ω is a function of $a(t)$ for certain models. So, we write

$$\Omega = \frac{\Omega_{r,0}}{a^4} + \frac{\Omega_{m,0}}{a^3} + \Omega_{\Lambda,0},$$

where current estimates (Ryden (2003)) are $\Omega_{r,0} = 8.4 \times 10^{-5}$, $\Omega_{m,0} = 0.3$, $\Omega_{\Lambda,0} \approx 0.7$. In general, we require

$$\Omega_{r,0} + \Omega_{m,0} + \Omega_{\Lambda,0} = \Omega_0.$$

So, in later examples, we will take this relationship into account.

The compact form of Friedmann's Equation.

Therefore, the Friedmann Equation can be written as

$$\left(\frac{\dot{a}}{a}\right)^2 = H_0^2\left[\frac{\Omega_{r,0}}{a^4} + \frac{\Omega_{m,0}}{a^3} + \Omega_{\Lambda,0} + \frac{1 - \Omega_0}{a^2}\right]. \tag{2.175}$$

Taking the square root of this expression, we obtain a first-order equation for the scale factor,

$$\dot{a} = \pm H_0\sqrt{\frac{\Omega_{r,0}}{a^2} + \frac{\Omega_{m,0}}{a} + \Omega_{\Lambda,0}a^2 + 1 - \Omega_0}.$$

The appropriate sign will be used when the scale factor is increasing or decreasing.

For special universes, by restricting the contributions to Ω, one can get analytic solutions. But, in general one has to solve this equation numerically. We will leave most of these cases to the reader or for homework problems and will consider some simple examples here.

Example 2.31. *Determine $a(t)$ for a flat universe with nonrelativistic matter only. (This is called an Einstein-de Sitter universe.)*

In this case, we have $\Omega_{r,0} = 0$, $\Omega_{\Lambda,0} = 0$, and $\Omega_0 = 1$. Because $\Omega_{r,0} + \Omega_{m,0} + \Omega_{\Lambda,0} = \Omega_0$, $\Omega_{m,0} = 1$ and the Friedman Equation takes the form

$$\dot{a} = H_0 \sqrt{\frac{1}{a}}.$$

This is a simple separable first-order equation. Thus,

$$H_0 \, dt = \sqrt{a} \, da.$$

Integrating, we have

$$H_0 t = \frac{2}{3} a^{3/2} + C.$$

Taking $a(0) = 0$, we have

$$a(t) = \left(\frac{t}{\frac{2}{3} H_0} \right)^{2/3}.$$

Because $a(t_0) = 1$, we find

$$t_0 = \frac{2}{3 H_0}.$$

This would give the age of the universe in this model as roughly $t_0 = 9.3$ Gyr.

Example 2.32. *Determine $a(t)$ for a curved universe with nonrelativistic matter only.*

We will consider $\Omega_0 > 1$. In this case, the Friedman equation takes the form

$$\dot{a} = \pm H_0 \sqrt{\frac{\Omega_0}{a} + (1 - \Omega_0)}.$$

Note that there is an extremum a_{max} that occurs for $\dot{a} = 0$. This occurs for

$$a = a_{max} \equiv \frac{\Omega_0}{\Omega_0 - 1}.$$

Analytic solutions are possible for this problem in parametric form. Note that we can write the differential equation in the form

$$\begin{aligned}
\dot{a} &= \pm H_0 \sqrt{\frac{\Omega_0}{a} + (1 - \Omega_0)} \\
&= \pm H_0 \sqrt{\frac{\Omega_0}{a}} \sqrt{1 + \frac{a(1 - \Omega_0)}{\Omega_0}} \\
&= \pm H_0 \sqrt{\frac{\Omega_0}{a}} \sqrt{1 - \frac{a}{a_{max}}}.
\end{aligned} \tag{2.176}$$

A separation of variables gives

$$H_0 \sqrt{\Omega_0}\, dt = \pm \frac{\sqrt{a}}{\sqrt{1 - \frac{a}{a_{max}}}}\, da.$$

This form suggests a trigonometric substitution,

$$\frac{a}{a_{max}} = \sin^2 \theta$$

with $da = 2a_{max} \sin \theta \cos \theta\, d\theta$. Thus, the integration becomes

$$H_0 \sqrt{\Omega_0}\, t = \pm \int \frac{\sqrt{a_{max} \sin^2 \theta}}{\sqrt{\cos^2 \theta}}\, 2a_{max} \sin \theta \cos \theta\, d\theta.$$

In proceeding, we should be careful. Recall that for real numbers $\sqrt{x^2} = |x|$. In order to remove the roots of squares, we need to consider the quadrant θ is in. Because $a = 0$ at $t = 0$, and it will vanish again for $\theta = \pi$, we will assume $0 \leq \theta \leq \pi$. For this range, $\sin \theta \geq 0$. However, $\cos \theta$ is not of one sign for this domain. In fact, a reaches it maximum at $\theta = \pi/2$. So, $\dot{a} > 0$. This corresponds to the upper sign in front of the integral. For $\theta > \pi/2$, $\dot{a} < 0$ and thus we need the lower sign and $\sqrt{\cos^2 \theta} = -\cos \theta$ for that part of the domain. Thus, it is safe to simplify the square roots and we obtain

$$\begin{aligned}
H_0 \sqrt{\Omega_0}\, t &= 2a_{max}^{3/2} \int \sin^2 \theta, d\theta \\
&= a_{max}^{3/2} \int (1 - \cos 2\theta)\, d\theta \\
&= a_{max}^{3/2} \left(\theta - \frac{1}{2} \sin 2\theta \right)
\end{aligned} \tag{2.177}$$

for $t = 0$ at $\theta = 0$.

We have arrived at a parametric solution to the example,

$$\begin{aligned}
a &= = a_{max} \sin^2 \theta, \\
t &= \frac{a_{max}^{3/2}}{H_0 \sqrt{\Omega_0}} \left(\theta - \frac{1}{2} \sin 2\theta \right),
\end{aligned} \tag{2.178}$$

for $0 \leq \theta \leq \pi$. Letting, $\phi = 2\theta$, this solution can be written as

$$\begin{aligned}
a &= \frac{1}{2} a_{max}(1 - \cos \phi), \\
t &= \frac{a_{max}^{3/2}}{2H_0 \sqrt{\Omega_0}} (\phi - \sin \phi),
\end{aligned} \tag{2.179}$$

for $0 \leq \phi \leq 2\pi$. As we will see in Chapter 10, the curve described by these equations is a cycloid.

A similar computation can be performed for $\Omega_0 < 1$. This will be left as a homework exercise. The answer takes the form

$$\begin{aligned}
a &= \frac{\Omega_0}{2(1 - \Omega_0)} (\cosh \eta - 1), \\
t &= \frac{\Omega_0}{2H_0(1 - \Omega_0)^{3/2}} (\sinh \eta - \eta),
\end{aligned} \tag{2.180}$$

for $\eta \geq 0$.

Example 2.33. *Determine the numerical solution of Friedmann's Equation for a curved universe with nonrelativistic matter only.*

Because Friedmann's Equation is a differential equation, we can use our favorite solver to obtain a solution. Not all universe types are amenable to obtaining an analytic solution as the last example. We can create a function in MATLAB for use in **ode45**:

```
function da=cosmosf(t,a)
global Omega
f=Omega./a+1-Omega;
da=sqrt(f);
end
```

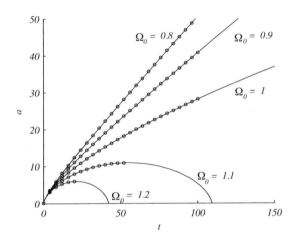

Figure 2.47: Numerical solution (circles) of the Friedmann equation superimposed on the analytic solutions for a matter plus curvature ($\Omega_0 \neq 1$) or no curvature ($\Omega_0 = 1$) universe.

We can then solve the Friedmann Equation and compare the solutions to the analytic forms in the previous two examples. The code for doing this is given below:

```
clear
global Omega

for Omega=0.8:.1:1.2;
    if Omega<1
        amax=50;
        tmax=100;
    elseif Omega==1
        amax=50;
        tmax=100;
    else
        amax=Omega/(Omega-1);
        tmax=Omega/(Omega-1)^1.5/2*pi;
    end

    tspan=0:4:tmax;
    a0=.1;
```

```
[t,a]=ode45(@cosmosf,tspan,a0);
plot(t,a,'ok')
hold on

if Omega<1
    eta=0:.1:4;
    a3 = Omega/(1-Omega)/2*(cosh(eta)-1);
    t3 = Omega/(1-Omega)^1.5/2*(sinh(eta)-eta);
    plot(t3,a3,'k')
    axis([0,max(t3),0,max(a3)])
    xlabel('t')
    ylabel('a')
elseif Omega==1
    t3=0:.1:1.5*tmax;
    a3=(3*t3/2).^(2/3);
    plot(t3,a3,'k')
else
    phi=0:.1:2*pi;
    a3 = Omega/(Omega-1)/2*(1-cos(phi));
    t3 = Omega/(Omega-1)^1.5/2*(phi-sin(phi));
    plot(t3,a3,'k')
end
end
hold off
axis([0,150,0,50])
xlabel('t')
ylabel('a')
```

In Figure 2.47 we show the results. For $\Omega_0 > 1$, the solutions lie on the first half of the cycloid solution. The other solutions indicate that the universe continues to expand, leading to what is called the Big Chill. The analytic solutions to the $\Omega_0 > 1$ cases eventually collapse to $a = 0$ in finite time. These final states are what Stephen Hawking calls the Big Crunch.

The numerical solutions for $\Omega_0 > 1$ run into difficulty because the radicand in the square root is negative. However, this corresponds to when $\dot{a} < 0$. So, we have to modify the code by estimating the maximum on the curve and run the numerical algorithm with new initial conditions and using the fact that $\dot{a} < 0$ in the function **cosmosf** by setting da=-sqrt(f). The modified code is shown below, and the resulting numerical solutions are shown in Figure 2.48.

```
tspan=0:4:tmax;
a0=.1;
[t,a]=ode45(@cosmosf,tspan,a0);
plot(t,a,'ok','MarkerSize',2)
hold on
if Omega>1
    tspan=tmax+.0001:4:2*tmax;
```

```
        a0=amax-.0001;
        [t2,a2]=ode45(@cosmosf2,tspan,a0);
        plot(t2,a2,'ok','MarkerSize',2)
    end
```

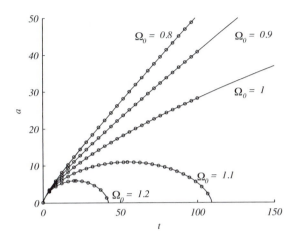

Figure 2.48: Modified numerical solution (circles) of the Friedmann Equation superimposed on the analytic solutions for a matter plus curvature ($\Omega_0 \neq 1$) or no curvature ($\Omega_0 = 1$) universe with the extension past the maximum value of a when $\Omega_0 > 1$.

2.10.7   Coefficient of Drag

WE HAVE SEEN THAT AIR DRAG can play a role in interesting physics problems in differential equations. This also is an important concept in fluid flow and fluid mechanics when looking at flows around obstacles, or when the obstacle is moving with respect to the background fluid. The simplest such object is a sphere, such as a baseball, soccer ball, golf ball, or ideal spherical raindrop. The resistive force is characterized by the dimensionless drag coefficient

$$C_D = \frac{F_D}{\frac{1}{2}\rho U^2 L^2},$$

where L and U are the characteristic length and speed of the object moving through the fluid, respectively.

There has been much attention focused on relating the drag coefficient to the Reynolds number. The Reynolds number is given by

$$Re = \frac{\rho L U}{\eta} = \frac{LV}{\nu},$$

The Reynolds number, Re, is named after Osborne Reynolds (1842–1912) who first determined it in 1883.

where η is the viscosity and $\nu = \frac{\eta}{\rho}$ is the kinematic viscosity. It is a measure of the size of the kinematic to viscous forces in the problem at hand. There are different ranges of fluid behavior, depending on the order of the Reynolds number. These range from laminar flow ($Re < 1,000$) to turbulent flow ($Re > 2.5 \times 10^5$). There are a range of other types of flows, such as creeping flow ($Re << 1$) and transitional flows, which are a mix of laminar and turbulent flow.

For low Reynolds number, the inertial forces are small compared to the viscous forces, leading to the Stokes drag force $C_D = 24Re^{-1}$. This result can be determined analytically. Similarly, for large Reynolds number, the drag coefficient is a constant. This is the Newtonian regime. Somewhere in between, the form of the drag coefficient is found through empirical studies. There have been many empirical expressions developed and all are within a few percent of the data in the range of applicability. Some of the commonly used expressions are given below.

Models that are useful for $Re < 10^3$:

$$C_2 = \frac{24}{Re} + \frac{4}{Re^{1/3}}, \qquad \text{Putnam (1961),} \qquad (2.181)$$

$$C_3 = \frac{24}{Re}\left(1 + 0.15Re^{0.687}\right), \quad \text{Schiller-Naumann (1933),} \quad (2.182)$$

$$C_4 = 12Re^{-.5}; \qquad \text{Edwards et al. (2000),} \qquad (2.183)$$

Models that are useful for $Re < 2 \times 10^5$ are the White (1991) and Clift-Gavin (1970), respectively:

$$C_1 = \frac{24}{Re} + \frac{6.}{1 + \sqrt{Re}} + 0.4, \qquad (2.184)$$

$$C_5 = \frac{24}{Re}\left(1 + 0.15Re^{0.687}\right) + \frac{.42}{1 + 42500/Re^{1.16}}. \qquad (2.185)$$

A more recent model was proposed by Morrison (2010) for $Re < 10^6$:

$$C_6 = \frac{24}{Re} + \frac{2.6\left(\frac{Re}{5.0}\right)}{1 + \left(\frac{Re}{5.0}\right)^{1.52}} + \frac{.411\left(\frac{Re}{263000}\right)^{-7.94}}{1 + \left(\frac{Re}{263000}\right)^{-8.00}} + \frac{Re^{0.80}}{461000}. \qquad (2.186)$$

Plots for these models are shown in Figures 2.49 and 2.50. In Figure 2.49, we see that the models differ significantly for large Reynolds numbers.

Figure 2.49: Drag coefficient as a function of Reynolds number for spheres.

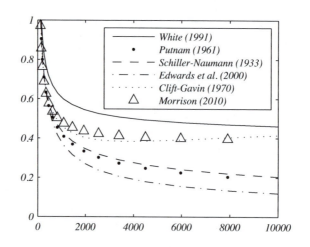

Figure 2.50 shows a log-log plot of the drag coefficient as a function of Reynolds number. In Example 10.15, we show a power law fit for Reynolds number less than 1,000, confirming the model used by Edwards, Wilder, and Scime (2001) as described in the raindrop problem.

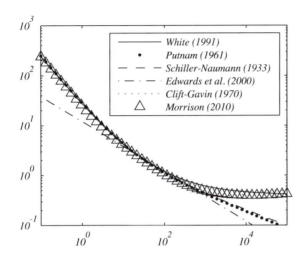

Figure 2.50: Log-log plot of the drag coefficient as a function of Reynolds number for spheres.

2.11   Linear Systems

2.11.1   Coupled Oscillators

IN THE PREVIOUS SECTION WE SAW that the numerical solution of second-order equations, or higher, can be cast into systems of first-order equations. Such systems are typically coupled in the sense that the solution of at least one of the equations in the system depends on knowing one of the other solutions in the system. In many physical systems, this coupling takes place naturally. We will introduce a simple model in this section to illustrate the coupling of simple oscillators.

There are many problems in physics that result in systems of equations. This is because the most basic law of physics is given by Newton's Second Law, which states that if a body experiences a net force, it will accelerate. Thus,

$$\sum \mathbf{F} = m\mathbf{a}.$$

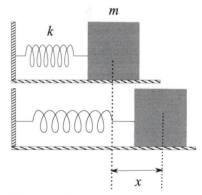

Figure 2.51: Spring-mass system.

Because $\mathbf{a} = \ddot{\mathbf{x}}$ we have a system of second-order differential equations in general for three-dimensional problems, or one second-order differential equation for one-dimensional problems for a single mass.

We have already seen the simple problem of a mass on a spring as shown in Figure 2.4. Recall that the net force in this case is the restoring force of the spring given by Hooke's Law,

$$F_s = -kx,$$

where $k > 0$ is the spring constant and x is the elongation of the spring. When the spring constant is positive, the spring force is negative, and when the spring constant is negative, the spring force is positive. The equation for simple harmonic motion for the mass-spring system was found to be given by

$$m\ddot{x} + kx = 0.$$

This second-order equation can be written as a system of two first-order equations in terms of the unknown position and velocity. We first set $y = \dot{x}$. Noting that $\ddot{x} = \dot{y}$, we rewrite the second-order equation in terms of x and $\dot{y}$. Thus, we have

$$\dot{x} = y$$
$$\dot{y} = -\frac{k}{m}x. \qquad (2.187)$$

One can look at more complicated spring-mass systems. Consider two blocks attached with two springs as in Figure 2.52. In this case, we apply Newton's Second Law for each block. We will designate the elongations of each spring from equilibrium as x_1 and x_2. These are shown in Figure 2.52.

For mass m_1, the forces acting on it are due to each spring. The first spring with spring constant k_1 provides a force on m_1 of $-k_1 x_1$. The second spring is stretched, or compressed, based upon the relative locations of the two masses. So, the second spring will exert a force on m_1 of $k_2(x_2 - x_1)$.

Figure 2.52: System of two masses and two springs.

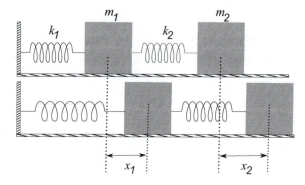

Similarly, the only force acting directly on mass m_2 is provided by the restoring force from spring 2. So, that force is given by $-k_2(x_2 - x_1)$. The reader should think about the signs in each case.

Putting this all together, we apply Newton's Second Law to both masses. We obtain the two equations

$$m_1\ddot{x}_1 = -k_1 x_1 + k_2(x_2 - x_1),$$
$$m_2\ddot{x}_2 = -k_2(x_2 - x_1). \qquad (2.188)$$

Thus, we see that we have a coupled system of two second-order differential equations. Each equation depends on the unknowns x_1 and x_2.

One can rewrite this system of two second-order equations as a system of four first order equations by letting $x_3 = \dot{x}_1$ and $x_4 = \dot{x}_2$. This leads to the system

$$
\begin{aligned}
\dot{x}_1 &= x_3 \\
\dot{x}_2 &= x_4 \\
\dot{x}_3 &= -\frac{k_1}{m_1}x_1 + \frac{k_2}{m_1}(x_2 - x_1) \\
\dot{x}_4 &= -\frac{k_2}{m_2}(x_2 - x_1).
\end{aligned}
\tag{2.189}
$$

As we will see in the next chapter, this system can be written more compactly in matrix form:

$$
\frac{d}{dt}\begin{pmatrix} x_1 \\ x_2 \\ x_3 \\ x_4 \end{pmatrix} = \begin{pmatrix} 0 & 0 & 1 & 0 \\ 0 & 0 & 0 & 1 \\ -\frac{k_1+k_2}{m_1} & \frac{k_2}{m_1} & 0 & 0 \\ \frac{k_2}{m_2} & -\frac{k_2}{m_2} & 0 & 0 \end{pmatrix}\begin{pmatrix} x_1 \\ x_2 \\ x_3 \\ x_4 \end{pmatrix}
\tag{2.190}
$$

We can solve this system of first-order equations using matrix methods. However, we will first need to recall a few things from linear algebra. This will be done in the next chapter. For now, we will return to simpler systems and explore the behavior of typical solutions in planar systems.

2.11.2 Planar Systems

WE NOW CONSIDER EXAMPLES of solving a coupled system of first-order differential equations in the plane. We will focus on the theory of linear systems with constant coefficients. Understanding these simple systems will help in the study of nonlinear systems, which contain much more interesting behaviors, such as the onset of chaos. In the next chapter we will return to these systems and describe a matrix approach to obtaining the solutions.

A general form for first-order systems in the plane is given by a system of two equations for unknowns $x(t)$ and $y(t)$:

$$
\begin{aligned}
x'(t) &= P(x,y,t), \\
y'(t) &= Q(x,y,t).
\end{aligned}
\tag{2.191}
$$

An autonomous system is one in which there is no explicit time dependence:

Autonomous systems.

$$
\begin{aligned}
x'(t) &= P(x,y) \\
y'(t) &= Q(x,y).
\end{aligned}
\tag{2.192}
$$

Otherwise the system is called nonautonomous.

A *linear system* takes the form

$$
\begin{aligned}
x' &= a(t)x + b(t)y + e(t), \\
y' &= c(t)x + d(t)y + f(t).
\end{aligned}
\tag{2.193}
$$

A homogeneous linear system results when $e(t) = 0$ and $f(t) = 0$.

A linear, constant coefficient system of first-order differential equations is given by

$$
\begin{aligned}
x' &= ax + by + e, \\
y' &= cx + dy + f.
\end{aligned}
\tag{2.194}
$$

We will focus on linear, homogeneous systems of constant coefficient, first-order differential equations:

A linear, homogeneous system of constant coefficient, first-order differential equations in the plane.

$$
\boxed{
\begin{aligned}
x' &= ax + by, \\
y' &= cx + dy.
\end{aligned}
}
\tag{2.195}
$$

As we will see later, such systems can result by a simple translation of the unknown functions. These equations are said to be coupled if either $b \neq 0$ or $c \neq 0$.

We begin by noting that the system (2.195) can be rewritten as a second-order constant coefficient linear differential equation, which we already know how to solve. We differentiate the first equation in system (2.195) and systematically replace occurrences of y and y', as we also know from the first equation that $y = \frac{1}{b}(x' - ax)$. Thus, we have

$$
\begin{aligned}
x'' &= ax' + by' \\
&= ax' + b(cx + dy) \\
&= ax' + bcx + d(x' - ax).
\end{aligned}
\tag{2.196}
$$

Rewriting the last line, we have

$$
\boxed{x'' - (a + d)x' + (ad - bc)x = 0.}
\tag{2.197}
$$

This is a linear, homogeneous, constant coefficient ordinary differential equation. We know that we can solve this by first looking at the roots of the characteristic equation

$$
r^2 - (a + d)r + ad - bc = 0
\tag{2.198}
$$

and writing the appropriate general solution for $x(t)$. Then we can find $y(t)$ using Equation (2.195):

$$
y = \frac{1}{b}(x' - ax).
$$

We now demonstrate this for a specific example.

Example 2.34. *Consider the system of differential equations*

$$
\begin{aligned}
x' &= -x + 6y, \\
y' &= x - 2y.
\end{aligned}
\tag{2.199}
$$

Carrying out the above outlined steps, we have that $x'' + 3x' - 4x = 0$. This can be shown as follows:

$$
\begin{aligned}
x'' &= -x' + 6y' \\
&= -x' + 6(x - 2y) \\
&= -x' + 6x - 12\left(\frac{x' + x}{6}\right) \\
&= -3x' + 4x
\end{aligned}
\tag{2.200}
$$

The resulting differential equation has a characteristic equation of $r^2 + 3r - 4 = 0$. The roots of this equation are $r = 1, -4$. Therefore, $x(t) = c_1 e^t + c_2 e^{-4t}$. But we still need $y(t)$. From the first equation of the system we have

$$
y(t) = \frac{1}{6}(x' + x) = \frac{1}{6}(2c_1 e^t - 3c_2 e^{-4t}).
$$

Thus, the solution to the system is

$$
\begin{aligned}
x(t) &= c_1 e^t + c_2 e^{-4t}, \\
y(t) &= \tfrac{1}{3}c_1 e^t - \tfrac{1}{2}c_2 e^{-4t}.
\end{aligned}
\tag{2.201}
$$

Sometimes one needs initial conditions. For these systems, we would specify conditions like $x(0) = x_0$ and $y(0) = y_0$. These would allow the determination of the arbitrary constants as before.

Solving systems with initial conditions.

Example 2.35. *Solve*

$$
\begin{aligned}
x' &= -x + 6y, \\
y' &= x - 2y,
\end{aligned}
\tag{2.202}
$$

given $x(0) = 2$, $y(0) = 0$.

We already have the general solution of this system in Equations (2.201). Inserting the initial conditions, we have

$$
\begin{aligned}
2 &= c_1 + c_2, \\
0 &= \tfrac{1}{3}c_1 - \tfrac{1}{2}c_2.
\end{aligned}
\tag{2.203}
$$

Solving for c_1 and c_2 gives $c_1 = 6/5$ and $c_2 = 4/5$. Therefore, the solution of the initial value problem is

$$
\begin{aligned}
x(t) &= \tfrac{2}{5}\left(3e^t + 2e^{-4t}\right), \\
y(t) &= \tfrac{2}{5}\left(e^t - e^{-4t}\right).
\end{aligned}
\tag{2.204}
$$

2.11.3 Equilibrium Solutions and Nearby Behaviors

IN STUDYING SYSTEMS OF DIFFERENTIAL EQUATIONS, it is often useful to study the behavior of solutions without obtaining an algebraic form for the solution. This is done by exploring equilibrium solutions and solutions nearby equilibrium solutions. Such techniques will be seen to be useful later in studying nonlinear systems.

We begin this section by studying equilibrium solutions of system (2.194). For equilibrium solutions, the system does not change in time. Therefore, equilibrium solutions satisfy the equations $x' = 0$ and $y' = 0$. Of course, this can only happen for constant solutions. Let x_0 and y_0 be the (constant) equilibrium solutions. Then, x_0 and y_0 must satisfy the system

Equilibrium solutions.

$$
\begin{aligned}
0 &= ax_0 + by_0 + e, \\
0 &= cx_0 + dy_0 + f.
\end{aligned}
$$
(2.205)

This is a linear system of nonhomogeneous algebraic equations. One only has a unique solution when the determinant of the system is not zero, that is, $ad - bc \neq 0$. Using Cramer's (determinant) Rule (3.61) for solving such systems, we have

$$
x_0 = -\frac{\begin{vmatrix} e & b \\ f & d \end{vmatrix}}{\begin{vmatrix} a & b \\ c & d \end{vmatrix}}, \qquad y_0 = -\frac{\begin{vmatrix} a & e \\ c & f \end{vmatrix}}{\begin{vmatrix} a & b \\ c & d \end{vmatrix}}.
$$
(2.206)

If the system is homogeneous, $e = f = 0$, then we have that the origin is the equilibrium solution; that is, $(x_0, y_0) = (0, 0)$. Often we will have this case because one can always make a change in coordinates from (x, y) to (u, v) by $u = x - x_0$ and $v = y - y_0$. Then, $u_0 = v_0 = 0$.

Next we are interested in the behavior of solutions near the equilibrium solutions. Later this behavior will be useful in analyzing more complicated nonlinear systems. We will look at some simple systems that are readily solved.

Example 2.36. Stable Node (Sink)
Consider the system

$$
\begin{aligned}
x' &= -2x \\
y' &= -y.
\end{aligned}
$$
(2.207)

This is a simple uncoupled system. Each equation is simply solved to give

$$
x(t) = c_1 e^{-2t} \text{ and } y(t) = c_2 e^{-t}.
$$

In this case we see that all solutions tend toward the equilibrium point, $(0, 0)$. This will be called a stable node, or a sink.

Before looking at other types of solutions, we will explore the stable node in the above example. There are several methods of looking at the behavior of solutions. We can look at solution plots of the dependent versus the independent variables, or we can look in the xy-plane at the parametric curves $(x(t), y(t))$.

Solution Plots: One can plot each solution as a function of t given a set of initial conditions. Examples are shown in Figure 2.53 for several initial conditions. Note that the solutions decay for large t. Special cases result for various initial conditions. Note that for $t = 0$, $x(0) = c_1$ and $y(0) = c_2$. (Of course, one can provide initial conditions at any $t = t_0$. It is generally easier to pick $t = 0$ in our general explanations.) If we pick an initial condition with $c_1 = 0$, then $x(t) = 0$ for all t. One obtains similar results when setting $y(0) = 0$.

Phase Portrait: There are other types of plots that can provide additional information about the solutions even if we cannot find the exact solutions as we can for these simple examples. In particular, one can consider the solutions $x(t)$ and $y(t)$ as the coordinates along a parameterized path, or curve, in the plane: $\mathbf{r} = (x(t), y(t))$ Such curves are called trajectories or orbits. The xy-plane is called the phase plane, and a collection of such orbits gives a phase portrait for the family of solutions of the given system.

One method for determining the equations of the orbits in the phase plane is to eliminate the parameter t between the known solutions to get a relationship between x and y. Because the solutions are known for the previous example, we can do this, as the solutions are known. In particular, we have

$$x = c_1 e^{-2t} = c_1 \left(\frac{y}{c_2} \right)^2 \equiv A y^2.$$

Another way to obtain information about the orbits comes from noting that the slopes of the orbits in the xy-plane are given by dy/dx. For autonomous systems, we can write this slope just in terms of x and y. This leads to a first-order differential equation, which possibly could be solved analytically or numerically.

First we will obtain the orbits for Example 2.36 by solving the corresponding slope equation. Recall that for trajectories defined parametrically by $x = x(t)$ and $y = y(t)$, we have from the Chain Rule for $y = y(x(t))$ that

$$\frac{dy}{dt} = \frac{dy}{dx} \frac{dx}{dt}.$$

Therefore,

$$\frac{dy}{dx} = \frac{\frac{dy}{dt}}{\frac{dx}{dt}}. \tag{2.208}$$

For the system in Equations (2.207) we use Equation (2.208) to obtain the equation for the slope at a point on the orbit:

$$\frac{dy}{dx} = \frac{y}{2x}.$$

The general solution of this first-order differential equation is found using separation of variables as $x = Ay^2$ for A an arbitrary constant. Plots of these solutions in the phase plane are given in Figure 2.54. [Note that this is the same form for the orbits that we had obtained above by eliminating t from the solution of the system.]

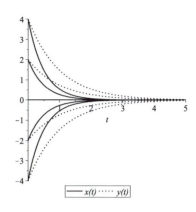

Figure 2.53: Plots of solutions of Example 2.36 for several initial conditions.

The slope of a parametric curve.

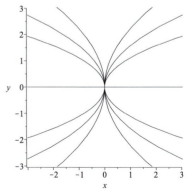

Figure 2.54: Orbits for Example 2.36.

Once one has solutions to differential equations, we often are interested in the long time behavior of the solutions. Given a particular initial condition (x_0, y_0), how does the solution behave as time increases? For orbits near an equilibrium solution, do the solutions tend toward, or away from, the equilibrium point? The answer is obvious when one has the exact solutions $x(t)$ and $y(t)$. However, this is not always the case.

Let's consider the above example for initial conditions in the first quadrant of the phase plane. For a point in the first quadrant, we have that

$$dx/dt = -2x < 0,$$

meaning that as $t \to \infty$, $x(t)$ get more negative. Similarly,

$$dy/dt = -y < 0,$$

indicating that $y(t)$ is also getting smaller for this problem. Thus, these orbits tend toward the origin as $t \to \infty$. This qualitative information was obtained without relying on the known solutions to the problem.

Direction Fields: Another way to determine the behavior of the solutions of the system of differential equations is to draw the direction field. A direction field is a vector field in which one plots arrows in the direction of tangents to the orbits at selected points in the plane. This is done because the slopes of the tangent lines are given by dy/dx. For the general system (2.195), the slope is

$$\frac{dy}{dx} = \frac{cx + dy}{ax + by}.$$

This is a first-order differential equation that can be solved as we show in the following examples.

Example 2.37. *Draw the direction field for Example 2.36.*

We can use software to draw direction fields. However, one can sketch these fields by hand. We have that the slope of the tangent at this point is given by

$$\frac{dy}{dx} = \frac{-y}{-2x} = \frac{y}{2x}.$$

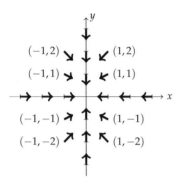

Figure 2.55: Sketch of tangent vectors using Example 2.36.

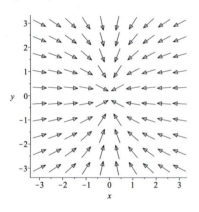

Figure 2.56: Direction field for Example 2.36.

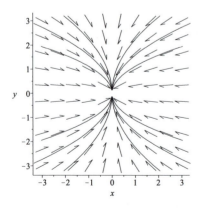

Figure 2.57: Phase portrait for Example 2.36. This is a stable node, or sink

For each point in the plane, one draws a piece of tangent line with this slope. In Figure 2.55 we show a few of these. For $(x, y) = (1, 1)$ the slope is $dy/dx = 1/2$. So, we draw an arrow with slope $1/2$ at this point. From system (2.207), we have that x' and y' are both negative at this point. Therefore, the vector points down and to the left.

We can do this for several points, as shown in Figure 2.55. Sometimes one can quickly sketch vectors with the same slope. For this example, when $y = 0$, the slope is zero and when $x = 0$, the slope is infinite. So, several vectors can be provided. Such vectors are tangent to curves known as isoclines in which $\frac{dy}{dx} =$ constant.

It is often difficult to provide an accurate sketch of a direction field. Computer software can be used to provide a better rendition. For Example 2.36, the direction field is shown in Figure 2.56. Looking at this direction field, one can begin to "see" the orbits by following the tangent vectors.

Of course, one can superimpose the orbits on the direction field. This is shown in Figure 2.57. Are these the patterns you saw in Figure 2.56?

In this example, we see all orbits "flow" toward the origin, or equilibrium point. Again, this is an example of what is called a *stable node* or a *sink*. (Imagine what happens to the water in a sink when the drain is unplugged.)

Example 2.38. Saddle *Consider the system*

$$\begin{aligned} x' &= -x \\ y' &= y. \end{aligned} \qquad (2.209)$$

This is another uncoupled system. The solutions are again simply obtained by integration. We have that $x(t) = c_1 e^{-t}$ and $y(t) = c_2 e^t$. Here we have that x decays as t gets large and y increases as t gets large. In particular, if one picks initial conditions with $c_2 = 0$, then orbits follow the x-axis toward the origin. For initial points with $c_1 = 0$, orbits originating on the y-axis will flow away from the origin. Of course, in these cases the origin is an equilibrium point and once at equilibrium, one remains there.

In fact, there is only one line on which to pick initial conditions such that the orbit leads toward the equilibrium point. No matter how small c_2 is, sooner, or later, the exponential growth term will dominate the solution. One can see this behavior in Figure 2.58.

Similar to the first example, we can look at plots of solutions orbits in the phase plane. These are given by Figures 2.58 and 2.59. The orbits can be obtained from the system as

$$\frac{dy}{dx} = \frac{dy/dt}{dx/dt} = -\frac{y}{x}.$$

The solution is $y = \frac{A}{x}$. For different values of $A \neq 0$, we obtain a family of hyperbolae. These are the same curves one might obtain for the level curves of a surface known as a saddle surface, $z = xy$. Thus, this type of equilibrium point is classified as a saddle point. From the phase portrait we can verify that there are many orbits that lead away from the origin (equilibrium point), but there is one line of initial conditions that leads to the origin and that is the x-axis. In this case, the line of initial conditions is given by the x-axis.

Example 2.39. Unstable Node (Source) *This example is similar to Example 2.36. The solutions are obtained by replacing t with $-t$. The solutions, orbits, and direction fields can be seen in Figures 2.60 and 2.61. This is once again a node, but all orbits lead away from the equilibrium point. It is called an unstable node or a source.*

$$\begin{aligned} x' &= 2x \\ y' &= y. \end{aligned} \qquad (2.210)$$

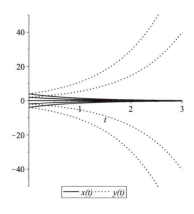

Figure 2.58: Plots of solutions of Example 2.38 for several initial conditions.

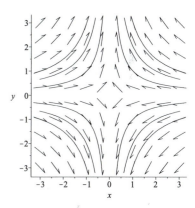

Figure 2.59: Phase portrait for Example 2.38. This is a saddle.

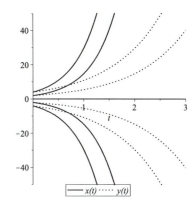

Figure 2.60: Plots of solutions of Example 2.39 for several initial conditions.

Example 2.40. Center *This system is a simple, coupled system. Neither equation can be solved without some information about the other unknown function. However, we can differentiate the first equation and use the second equation to obtain*

$$
\begin{aligned}
x' &= y \\
y' &= -x. \\
x'' + x &= 0.
\end{aligned}
\tag{2.211}
$$

We recognize this equation as one that appears in the study of simple harmonic motion. The solutions are pure sinusoidal oscillations:

$$
x(t) = c_1 \cos t + c_2 \sin t, \quad y(t) = -c_1 \sin t + c_2 \cos t.
$$

In the phase plane, the trajectories can be determined either by looking at the direction field or solving the first-order equation

$$
\frac{dy}{dx} = -\frac{x}{y}.
$$

Performing a separation of variables and integrating, we find that

$$
x^2 + y^2 = C.
$$

Thus, we have a family of circles for $C > 0$. (Can you prove this using the general solution?) Looking at the results graphically in Figures 2.62 and 2.63 confirms this result. This type of point is called a center.

Example 2.41. Focus (Spiral) *In this example, we will see an additional set of behaviors of equilibrium points in planar systems. We have added one term, αx, to the system in Example 2.40. We will consider the effects for two specific values of the parameter: $\alpha = 0.1, -0.2$. The resulting behaviors are shown in Figures 2.64 through 2.67. We see orbits that look like spirals. These orbits are stable and unstable spirals (or foci, the plural of focus.)*

$$
\begin{aligned}
x' &= \alpha x + y \\
y' &= -x.
\end{aligned}
\tag{2.212}
$$

We can understand these behaviors by once again relating the system of first-order differential equations to a second-order differential equation. Using the usual method for obtaining a second-order equation from a system, we find that $x(t)$ satisfies the differential equation

$$
x'' - \alpha x' + x = 0.
$$

We recall from our first course that this is a form of damped simple harmonic motion. The characteristic equation is $r^2 - \alpha r + 1 = 0$. The solution of this quadratic equation is

$$
r = \frac{\alpha \pm \sqrt{\alpha^2 - 4}}{2}.
$$

There are five special cases to consider as shown in the classification below.

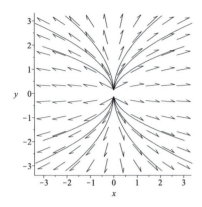

Figure 2.61: Phase portrait for Example 2.39, an unstable node or source.

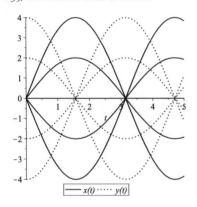

Figure 2.62: Plots of solutions of Example 2.40 for several initial conditions.

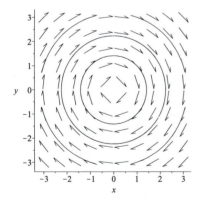

Figure 2.63: Phase portrait for Example 2.40, a center.

Classification of Solutions of $x'' - \alpha x' + x = 0$

1. $\alpha = -2$. *There is one real solution. This case is called critical damping since the solution $r = -1$ leads to exponential decay. The solution is $x(t) = (c_1 + c_2 t)e^{-t}$.*

2. $\alpha < -2$. *There are two real, negative solutions, $r = -\mu, -\nu$, $\mu, \nu > 0$. The solution is $x(t) = c_1 e^{-\mu t} + c_2 e^{-\nu t}$. In this case we have what is called overdamped motion. There are no oscillations*

3. $-2 < \alpha < 0$. *There are two complex conjugate solutions $r = \alpha/2 \pm i\beta$ with real part less than zero and $\beta = \frac{\sqrt{4-\alpha^2}}{2}$. The solution is $x(t) = (c_1 \cos \beta t + c_2 \sin \beta t)e^{\alpha t/2}$. Since $\alpha < 0$, this consists of a decaying exponential times oscillations. This is often called an underdamped oscillation.*

4. $\alpha = 0$. *This leads to simple harmonic motion.*

5. $0 < \alpha < 2$. *This is similar to the underdamped case, except $\alpha > 0$. The solutions are growing oscillations.*

6. $\alpha = 2$. *There is one real solution. The solution is $x(t) = (c_1 + c_2 t)e^t$. It leads to unbounded growth in time.*

7. *For $\alpha > 2$. There are two real, positive solutions $r = \mu, \nu > 0$. The solution is $x(t) = c_1 e^{\mu t} + c_2 e^{\nu t}$, which grows in time.*

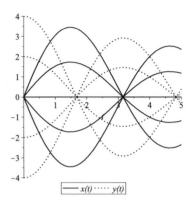

Figure 2.64: Plots of solutions of Example 2.41 for several initial conditions, $\alpha = -0.2$.

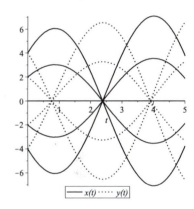

Figure 2.65: Plots of solutions of Example 2.41 for several initial conditions, $\alpha = 0.1$.

For $\alpha < 0$ the solutions are losing energy, so the solutions can oscillate with a diminishing amplitude. (See Figure 2.64.) For $\alpha > 0$, there is a growth in the amplitude, which is not typical. (See Figure 2.65.) Of course, there can be overdamped motion if the magnitude of α is too large.

Example 2.42. Degenerate Node *For this example, we will write out the solutions. It is a coupled system for which only the second equation is coupled.*

$$
\begin{aligned}
x' &= -x \\
y' &= -2x - y.
\end{aligned}
\tag{2.213}
$$

There are two possible approaches:

a. We could solve the first equation to find $x(t) = c_1 e^{-t}$. Inserting this solution into the second equation, we have

$$y' + y = -2c_1 e^{-t}.$$

This is a relatively simple linear first-order equation for $y = y(t)$. The integrating factor is $\mu = e^t$. The solution is found as $y(t) = (c_2 - 2c_1 t)e^{-t}$.

b. Another method would be to proceed to rewrite this as a second-order equation. Computing x'' does not get us very far. So, we look at

$$
\begin{aligned}
y'' &= -2x' - y' \\
&= 2x - y' \\
&= -2y' - y.
\end{aligned}
\tag{2.214}
$$

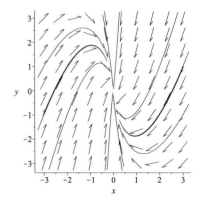

Figure 2.66: Phase portrait for Example 2.41 with $\alpha = 0.1$. This is an unstable focus, or spiral.

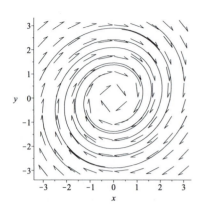

Figure 2.67: Phase portrait for Example 2.41 with $\alpha = -0.2$. This is a stable focus, or spiral.

Therefore, y satisfies

$$y'' + 2y' + y = 0.$$

The characteristic equation has one real root, $r = -1$. So, we write

$$y(t) = (k_1 + k_2 t)e^{-t}.$$

This is a stable degenerate node. Combining this with the solution $x(t) = c_1 e^{-t}$, we can show that $y(t) = (c_2 - 2c_1 t)e^{-t}$ as before.

In Figure 2.68 we see several orbits in this system. It differs from the stable node shown in Figure 2.54 in that there is only one direction along which the orbits approach the origin instead of two. If one picks $c_1 = 0$, then $x(t) = 0$ and $y(t) = c_2 e^{-t}$. This leads to orbits running along the y-axis as seen in the figure.

Example 2.43. A Line of Equilibria, Zero Root

$$
\begin{aligned}
x' &= 2x - y \\
y' &= -2x + y.
\end{aligned}
\tag{2.215}
$$

In this last example, we have a coupled set of equations. We rewrite it as a second-order differential equation:

$$
\begin{aligned}
x'' &= 2x' - y' \\
&= 2x' - (-2x + y) \\
&= 2x' + 2x + (x' - 2x) = 3x'.
\end{aligned}
\tag{2.216}
$$

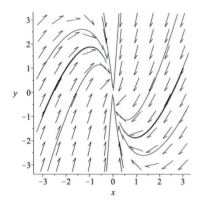

Figure 2.68: Phase portrait for Example 2.44. This is a degenerate node.

So, the second-order equation is

$$x'' - 3x' = 0,$$

and the characteristic equation is $0 = r(r - 3)$. This gives the general solution as

$$x(t) = c_1 + c_2 e^{3t}$$

and thus

$$y = 2x - x' = 2(c_1 + c_2 e^3 t) - (3c_2 e^{3t}) = 2c_1 - c_2 e^{3t}.$$

In Figure 2.69 we show the direction field. The constant slope field seen in this example is confirmed by a simple computation:

$$\frac{dy}{dx} = \frac{-2x + y}{2x - y} = -1.$$

Furthermore, looking at initial conditions with $y = 2x$, we have at $t = 0$,

$$2c_1 - c_2 = 2(c_1 + c_2) \quad \Rightarrow \quad c_2 = 0.$$

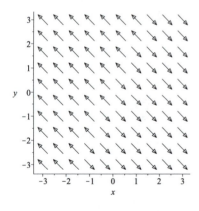

Figure 2.69: Plots of direction field of Example 2.43.

Therefore, points on this line remain on this line forever, $(x, y) = (c_1, 2c_1)$. This line of fixed points is called a line of equilibria.

2.11.4 *Polar Representation of Spirals*

IN THE EXAMPLES WITH A CENTER OR A SPIRAL, one might be able to write the solutions in polar coordinates. Recall that a point in the plane can be described by either Cartesian (x, y) or polar (r, θ) coordinates. Given the polar form, one can find the Cartesian components using

$$x = r \cos \theta \text{ and } y = r \sin \theta.$$

Given the Cartesian coordinates, one can find the polar coordinates using

$$r^2 = x^2 + y^2 \text{ and } \tan \theta = \frac{y}{x}. \tag{2.217}$$

Because x and y are functions of t, then naturally we can think of r and θ as functions of t. Converting a system of equations in the plane for x' and y' to polar form requires knowing r' and θ'. So, we first find expressions for r' and θ' in terms of x' and y'.

Differentiating the first equation in (2.217) gives

$$rr' = xx' + yy'.$$

Inserting the expressions for x' and y' from system 2.195, we have

$$rr' = x(ax + by) + y(cx + dy).$$

In some cases this may be written entirely in terms of r's. Similarly, we have that

$$\theta' = \frac{xy' - yx'}{r^2},$$

which the reader can prove for homework.

In summary, when converting first-order equations from rectangular to polar form, one needs the relations below.

Derivatives of Polar Variables

$$
\begin{aligned}
r' &= \frac{xx' + yy'}{r}, \\
\theta' &= \frac{xy' - yx'}{r^2}.
\end{aligned}
\tag{2.218}
$$

Example 2.44. *Rewrite the following system in polar form and solve the resulting system.*

$$
\begin{aligned}
x' &= ax + by, \\
y' &= -bx + ay.
\end{aligned}
\tag{2.219}
$$

We first compute r' and θ':

$$rr' = xx' + yy' = x(ax + by) + y(-bx + ay) = ar^2.$$

$$r^2\theta' = xy' - yx' = x(-bx + ay) - y(ax + by) = -br^2.$$

This leads to a simpler system:

$$\begin{aligned} r' &= ar \\ \theta' &= -b. \end{aligned} \qquad (2.220)$$

This system is uncoupled. The second equation in this system indicates that we traverse the orbit at a constant rate in the clockwise direction. Solving these equations, we have that $r(t) = r_0 e^{at}$, $\theta(t) = \theta_0 - bt$. Eliminating t between these solutions, we finally find the polar equation of the orbits:

$$r = r_0 e^{-a(\theta - \theta_0)t/b}.$$

If you graph this for $a \neq 0$, you will get stable or unstable spirals.

Example 2.45. *Consider the specific system*

$$\begin{aligned} x' &= -y + x \\ y' &= x + y. \end{aligned} \qquad (2.221)$$

In order to convert this system into polar form, we compute

$$rr' = xx' + yy' = x(-y + x) + y(x + y) = r^2,$$

$$r^2\theta' = -xy' - yx' = x(x + y) - y(-y + x) = r^2.$$

This leads to the simpler system

$$\begin{aligned} r' &= r, \\ \theta' &= 1. \end{aligned} \qquad (2.222)$$

Solving these equations yields

$$r(t) = r_0 e^t, \quad \theta(t) = t + \theta_0.$$

Eliminating t from this solution gives the orbits in the phase plane, $r(\theta) = r_0 e^{\theta - \theta_0}$.

A more complicated example arises for a nonlinear system of differential equations. Consider the following example.

Example 2.46.

$$\begin{aligned} x' &= -y + x(1 - x^2 - y^2), \\ y' &= x + y(1 - x^2 - y^2). \end{aligned} \qquad (2.223)$$

Transforming to polar coordinates, one can show that in order to convert this system into polar form, we compute

$$r' = r(1 - r^2), \quad \theta' = 1.$$

This uncoupled system can be solved and this is left for Problem 2.33.

Problems

1. Find all the solutions of the first-order differential equations. When an initial condition is given, find the particular solution satisfying that condition.

a. $\dfrac{dy}{dx} = \dfrac{e^x}{2y}$.

b. $\dfrac{dy}{dt} = y^2(1 + t^2),\ y(0) = 1$.

c. $\dfrac{dy}{dx} = \dfrac{\sqrt{1 - y^2}}{x}$.

d. $xy' = y(1 - 2y),\quad y(1) = 2$.

e. $y' - (\sin x)y = \sin x$.

f. $xy' - 2y = x^2,\, y(1) = 1$.

g. $\dfrac{ds}{dt} + 2s = st^2,\quad s(0) = 1$.

h. $x' - 2x = te^{2t}$.

i. $\dfrac{dy}{dx} + y = \sin x,\, y(0) = 0$.

j. $\dfrac{dy}{dx} - \dfrac{3}{x}y = x^3,\, y(1) = 4$.

2. Find all the solutions of the second-order differential equations. When an initial condition is given, find the particular solution satisfying that condition.

a. $y'' - 9y' + 20y = 0$.

b. $y'' - 3y' + 4y = 0,\quad y(0) = 0,\quad y'(0) = 1$.

c. $x^2y'' + 5xy' + 4y = 0,\quad x > 0$.

d. $x^2y'' - 2xy' + 3y = 0,\quad x > 0$.

3. Consider the differential equation

$$\frac{dy}{dx} = \frac{x}{y} - \frac{x}{1 + y}.$$

a. Find the 1-parameter family of solutions (general solution) of this equation.

b. Find the solution of this equation satisfying the initial condition $y(0) = 1$. Is this a member of the 1-parameter family?

4. The initial value problem

$$\frac{dy}{dx} = \frac{y^2 + xy}{x^2},\quad y(1) = 1$$

does not fall into the class of problems considered in this chapter. However, if one substitutes $y(x) = xz(x)$ into the differential equation, one obtains an equation for $z(x)$ that can be solved. Use this substitution to solve the initial value problem for $y(x)$.

5. Consider the nonhomogeneous differential equation $x'' - 3x' + 2x = 6e^{3t}$.

    a. Find the general solution of the homogenous equation.

    b. Find a particular solution using the Method of Undetermined Co-efficients by guessing $x_p(t) = Ae^{3t}$.

    c. Use your answers in the previous parts to write the general solution for this problem.

6. Find the general solution of the given equation by the method given.

    a. $y'' - 3y' + 2y = 10$. Method of Undetermined Coefficients.

    b. $y'' + y' = 3x^2$. Variation of Parameters.

7. Find the general solution of each differential equation. When an initial condition is given, find the particular solution satisfying that condition.

    a. $y'' - 3y' + 2y = 20e^{-2x}$, $y(0) = 0$, $y'(0) = 6$.

    b. $y'' + y = 2\sin 3x$.

    c. $y'' + y = 1 + 2\cos x$.

    d. $x^2 y'' - 2xy' + 2y = 3x^2 - x$, $x > 0$.

8. Verify that the given function is a solution and use Reduction of Order to find a second linearly independent solution.

    a. $x^2 y'' - 2xy' - 4y = 0$, $y_1(x) = x^4$.

    b. $xy'' - y' + 4x^3 y = 0$, $y_1(x) = \sin(x^2)$.

9. Use the Method of Variation of Parameters to determine the general solution for the following problems.

    a. $y'' + y = \tan x$.

    b. $y'' - 4y' + 4y = 6xe^{2x}$.

10. Instead of assuming that $c_1' y_1 + c_2' y_2 = 0$ in the derivation of the solution using Variation of Parameters, assume that $c_1' y_1 + c_2' y_2 = h(x)$ for an arbitrary function $h(x)$ and show that one gets the same particular solution.

11. Find the solution of each initial value problem using the appropriate initial value Green's function.

    a. $y'' - 3y' + 2y = 20e^{-2x}$, $y(0) = 0$, $y'(0) = 6$.

    b. $y'' + y = 2\sin 3x$, $y(0) = 5$, $y'(0) = 0$.

    c. $y'' + y = 1 + 2\cos x$, $y(0) = 2$, $y'(0) = 0$.

    d. $x^2 y'' - 2xy' + 2y = 3x^2 - x$, $y(1) = \pi$, $y'(1) = 0$.

12. Use the initial value Green's function for $x'' + x = f(t)$, $x(0) = 4$, $x'(0) = 0$, to solve the following problems.

    a. $x'' + x = 5t^2$.

b. $x'' + x = 2 \tan t$.

13. For the problem $y'' - k^2 y = f(x)$, $y(0) = 0$, $y'(0) = 1$,

    a. Find the initial value Green's function.

    b. Use the Green's function to solve $y'' - y = e^{-x}$.

    c. Use the Green's function to solve $y'' - 4y = e^{2x}$.

14. Find and use the initial value Green's function to solve

$$x^2 y'' + 3xy' - 15y = x^4 e^x, y(1) = 1, y'(1) = 0.$$

15. A ball is thrown upward with an initial velocity of 49 m/s from 539 m high. How high does the ball get, and how long does in take before it hits the ground? [Use results from the simple free fall problem, $y'' = -g$.]

16. Consider the case of free fall with a damping force proportional to the velocity, $f_D = \pm kv$ with $k = 0.1$ kg/s.

    a. Using the correct sign, consider a 50-kg mass falling from rest at a height of 100 m. Find the velocity as a function of time. Does the mass reach terminal velocity?

    b. Let the mass be thrown upward from the ground with an initial speed of 50 m/s. Find the velocity as a function of time as it travels upward and then falls to the ground. How high does the mass get? What is its speed when it returns to the ground?

17. A piece of a satellite falls to the ground from a height of 10,000 m. Ignoring air resistance, find the height as a function of time. [Hint: For free fall from large distances,

$$\ddot{h} = -\frac{GM}{(R+h)^2}.$$

Multiplying both sides by $\dot{h}$, show that

$$\frac{d}{dt}\left(\frac{1}{2}\dot{h}^2\right) = \frac{d}{dt}\left(\frac{GM}{R+h}\right).$$

Integrate and solve for $\dot{h}$. Further integrating gives $h(t)$.]

18. The problem of growth and decay is stated as follows: The rate of change of a quantity is proportional to the quantity. The differential equation for such a problem is

$$\frac{dy}{dt} = \pm ky.$$

The solution of this growth and decay problem is $y(t) = y_0 e^{\pm kt}$. Use this solution to answer the following questions if 40 percent of a radioactive substance disappears in 100 years.

    a. What is the half-life of the substance?

b. After how many years will 90 percent be gone?

19. A spring fixed at its upper end is stretched 6 inches by a 10-pound weight attached at its lower end. The spring-mass system is suspended in a viscous medium so that the system is subjected to a damping force of $5\frac{dx}{dt}$ lbs. Describe the motion of the system if the weight is drawn down an additional 4 inches and released. What would happen if you changed the coefficient "5" to "4"? [You may need to consult your introductory physics text.]

20. Consider an LRC circuit with $L = 1.00$ H, $R = 1.00 \times 10^2$ Ω, $C = 1.00 \times 10^{-4}$ F, and $V = 1.00 \times 10^3$ V. Suppose that no charge is present and no current is flowing at time $t = 0$ when a battery of voltage V is inserted. Find the current and the charge on the capacitor as functions of time. Describe how the system behaves over time.

21. Consider the problem of forced oscillations as described in Section 2.7.2.

  a. Derive the general solution in Equation (2.77).

  b. Plot the solutions in Equation (2.77) for the following cases: Let $c_1 = 0.5$, $c_2 = 0$, $F_0 = 1.0$ N, and $m = 1.0$ kg for $t \in [0, 100]$.

  i. $\omega_0 = 2.0$ rad/s, $\omega = 0.1$ rad/s.

  ii. $\omega_0 = 2.0$ rad/s, $\omega = 0.5$ rad/s.

  iii. $\omega_0 = 2.0$ rad/s, $\omega = 1.5$ rad/s.

  iv. $\omega_0 = 2.0$ rad/s, $\omega = 2.2$ rad/s.

  v. $\omega_0 = 1.0$ rad/s, $\omega = 1.2$ rad/s.

  vi. $\omega_0 = 1.5$ rad/s, $\omega = 1.5$ rad/s.

  c. Derive the form in Equation (2.78).

  d. Confirm that the solution in Equation (2.78) is the same as the solution in Equation (2.77) for $F_0 = 2.0$ N, $m = 10.0$ kg, $\omega_0 = 1.5$ rad/s, and $\omega = 1.25$ rad/s, by plotting both solutions for $t \in [0, 100]$.

22. A certain model of the motion of a light plastic ball tossed into the air is given by

$$mx'' + cx' + mg = 0, \quad x(0) = 0, \quad x'(0) = v_0.$$

Here m is the mass of the ball, $g = 9.8$ m/s^2 is the acceleration due to gravity, and c is a measure of the damping. Because there is no x term, we can write this as a first order equation for the velocity $v(t) = x'(t)$:

$$mv' + cv + mg = 0.$$

  a. Find the general solution for the velocity $v(t)$ of the linear first-order differential equation above.

  b. Use the solution of part a to find the general solution for the position $x(t)$.

c. Find an expression to determine how long it takes for the ball to reach its maximum height?

d. Assume that $c/m = 5\,\text{s}^{-1}$. For $v_0 = 5, 10, 15, 20$ m/s, plot the solution, $x(t)$, versus the time.

e. From your plots and the expression in part c, determine the rise time. Do these answers agree?

f. What can you say about the time it takes for the ball to fall as compared to the rise time?

23. Use i) Euler's Method and ii) the Midpoint Method to determine the given value of y for the following problems:

a. $\dfrac{dy}{dx} = 2y$, $y(0) = 2$. Find $y(1)$ with $h = 0.1$.

b. $\dfrac{dy}{dx} = x - y$, $y(0) = 1$. Find $y(2)$ with $h = 0.2$.

c. $\dfrac{dy}{dx} = x\sqrt{1 - y^2}$, $y(1) = 0$. Find $y(2)$ with $h = 0.2$.

24. Numerically solve the nonlinear pendulum problem using the Euler-Cromer Method for a pendulum with length $L = 0.5$ m using initial angles of $\theta_0 = 10^\circ$, and $\theta_0 = 70^\circ$. In each case, run the routines long enough and with an appropriate h such that you can determine the period in each case. Compare your results with the linear pendulum period.

25. For the Baumgartner sky dive we had obtained the results for his position as a function of time. There are other questions that could be asked:

a. Find the velocity as a function of time for the model developed in the text.

b. Find the velocity as a function of altitude for the model developed in the text.

c. What maximum speed is obtained in the model? At what time and position?

d. Does the model indicate that terminal velocity was reached?

e. What speed is predicted for the point at which the parachute opened?

f. How do these numbers compare with reported data?

26. Consider the flight of a golf ball with mass 46 g and a diameter of 42.7 mm. Assume it is projected at 30° with a speed of 36 m/s and no spin.

a. Ignoring air resistance, analytically find the path of the ball and determine the range, maximum height, and time of flight for it to land at the height that the ball had started.

b. Now consider a drag force $f_D = \frac{1}{2}C_D\rho\pi r^2 v^2$, with $C_D = 0.42$ and $\rho = 1.21$ kg/m^3. Determine the range, maximum height, and time of flight for the ball to land at the height that it had started.

c. Plot the Reynolds number as a function of time. [Take the kinematic viscosity of air, $\nu = 1.47 \times 10^{-5}$.]

d. Based on the plot in part c, create a model to incorporate the change in Reynolds number and repeat part b. Compare the results from parts a, b, and d.

27. Consider the flight of a tennis ball with mass 57 g and a diameter of 66.0 mm. Assume the ball is served 6.40 meters from the net at a speed of 50.0 m/s down the center line from a height of 2.8 m. It needs to just clear the net (0.914 m).

a. Ignoring air resistance and spin, analytically find the path of the ball assuming it just clears the net. Determine the angle to clear the net and the time of flight.

b. Find the angle to clear the net assuming the tennis ball is given a topspin with $w = 50$ rad/s.

c. Repeat part b assuming the tennis ball is given a bottom spin with $\omega = 50$ rad/s.

d. Repeat parts a, b, and c with a drag force, taking $C_D = 0.55$.

28. In Example 2.32, $a(t)$ was determined for a curved universe with non-relativistic matter for $\Omega_0 > 1$. Derive the parametric equations for $\Omega_0 < 1$,

$$
\begin{aligned}
a &= \frac{\Omega_0}{2(1 - \Omega_0)} (\cosh \eta - 1), \\
t &= \frac{\Omega_0}{2H_0(1 - \Omega_0)^{3/2}} (\sinh \eta - \eta),
\end{aligned} \tag{2.224}
$$

for $\eta \geq 0$.

29. Find numerical solutions for other models of the universe.

a. A flat universe with nonrelativistic matter only with $\Omega_{m,0} = 1$.

b. A curved universe with radiation only with curvature of different types.

c. A flat universe with nonrelativistic matter and radiation with several values of $\Omega_{m,0}$ and $\Omega_{r,0} + \Omega_{m,0} = 1$.

d. Look up the current values of $\Omega_{r,0}$, $\Omega_{m,0}$, and $\Omega_{\Lambda,0}$ and κ. Use these values to predict future values of $a(t)$.

e. Investigate other types of universes of your choice, but different from the previous problems and examples.

30. Consider the system

$$
\begin{aligned}
x' &= -4x - y, \\
y' &= x - 2y.
\end{aligned}
$$

a. Determine the second-order differential equation satisfied by $x(t)$.

b. Solve the differential equation for $x(t)$.

    c. Using this solution, find $y(t)$.

    d. Verify your solutions for $x(t)$ and $y(t)$.

    e. Find a particular solution to the system given the initial conditions $x(0) = 1$ and $y(0) = 0$.

31. Consider the following systems. Determine the families of orbits for each system and sketch several orbits in the phase plane and classify them by their type (stable node, etc.).

    a.

$$
\begin{aligned}
x' &= 3x, \\
y' &= -2y.
\end{aligned}
$$

    b.

$$
\begin{aligned}
x' &= -y, \\
y' &= -5x.
\end{aligned}
$$

    c.

$$
\begin{aligned}
x' &= 2y, \\
y' &= -3x.
\end{aligned}
$$

    d.

$$
\begin{aligned}
x' &= x - y, \\
y' &= y.
\end{aligned}
$$

    e.

$$
\begin{aligned}
x' &= 2x + 3y, \\
y' &= -3x + 2y.
\end{aligned}
$$

32. Use the transformations relating polar and Cartesian coordinates to prove that

$$
\frac{d\theta}{dt} = \frac{1}{r^2}\left[x\frac{dy}{dt} - y\frac{dx}{dt}\right].
$$

33. Consider the system of equations in Example 2.46.

    a. Derive the polar form of the system.

    b. Solve the radial equation, $r' = r(1 - r^2)$, for the initial values $r(0) = 0, 0.5, 1.0, 2.0$.

    c. Based upon these solutions, plot and describe the behavior of all solutions to the original system in Cartesian coordinates.

3
Linear Algebra

"Physics is much too hard for physicists."—David Hilbert (1862–1943)

AS THE READER IS AWARE BY NOW, calculus has its roots in physics and has become a very useful tool for modeling the physical world. Another very important area of mathematics is linear algebra. Physics students who have taken a course in linear algebra in a mathematics department might not come away with this perception. It is not until students take more advanced classes in physics that they begin to realize that a good grounding in linear algebra can lead to a better understanding of the behavior of physical systems.

In this chapter we will introduce some of the basics of linear algebra for finite dimensional vector spaces, and we will reinforce these concepts through generalizations in later chapters to infinite dimensional vector spaces. In keeping with the theme of the text, we will apply these ideas to the coupled systems introduced in the previous chapter. Such systems lead to linear and nonlinear oscillations in dynamical systems.

Linear algebra is the backbone of most of applied mathematics and underlies many areas of physics, such as quantum mechanics.

3.1 Finite Dimensional Vector Spaces

MUCH OF THE DISCUSSION AND TERMINOLOGY that we will use comes from the theory of vector spaces. Until now you may only have dealt with finite dimensional vector spaces. Even then, you might only be comfortable with two and three dimensions. We will review a little of what we know about finite dimensional spaces so that we can introduce more general function spaces later.

The notion of a vector space is a generalization of three-dimensional vectors and operations on them. In three dimensions, we have objects called vectors,[1] which are represented by arrows of a specific length and pointing in a given direction. To each vector, we can associate a point in a three-dimensional Cartesian system. We just attach the tail of the vector $\mathbf{v}$ to the origin and the head lands at the point (x, y, z).[2] We then use unit vectors $\mathbf{i}$, $\mathbf{j}$, and $\mathbf{k}$ along the coordinate axes to write

$$\mathbf{v} = x\mathbf{i} + y\mathbf{j} + z\mathbf{k}.$$

[1] In introductory physics, one defines a vector as any quantity that has both magnitude and direction.

[2] In multivariate calculus, one concentrates on the component form of vectors. These representations are easily generalized, as we will see.

Having defined vectors, we then learned how to add vectors and multiply vectors by numbers, or scalars. We learned that there were two types of multiplication of vectors. We could multiply them to get a scalar or a vector. This led to dot products and cross products, respectively. The dot product is useful for determining the length of a vector, the angle between two vectors, if the vectors are perpendicular, or projections of one vector onto another. The cross product is used to produce orthogonal vectors, areas of parallelograms, and volumes of parallelepipeds.

In physics you first learned about vector products when you defined work, $W = \mathbf{F} \cdot \mathbf{r}$. Cross products were useful in describing things like torque, $\tau = \mathbf{r} \times \mathbf{F}$, or the force on a moving charge in a magnetic field, $\mathbf{F} = q\mathbf{v} \times \mathbf{B}$. We will return to these more complicated vector operations later when we need them.

Properties and definition of vector spaces.

The properties three-dimensional vectors are generalized to spaces of more than three dimensions in linear algebra courses. The properties roughly outlined above need to be preserved. So, we will start with a space of vectors and the operations of addition and scalar multiplication. We will need a set of scalars, which generally come from some field. However, in our applications, the field will either be the set of real numbers or the set of complex numbers.

A vector space V over a field F is a set that is closed under addition and scalar multiplication and satisfies the following conditions:

A field is a set together with two operations, usually addition and multiplication, such that we have

- Closure under addition and multiplication
- Associativity of addition and multiplication
- Commutativity of addition and multiplication
- Additive and multiplicative identity
- Additive and multiplicative inverses
- Distributivity of multiplication over addition

For any $\mathbf{u}, \mathbf{v}, \mathbf{w} \in V$ and $a, b \in F$,

1. $\mathbf{u} + \mathbf{v} = \mathbf{v} + \mathbf{u}$.
2. $(\mathbf{u} + \mathbf{v}) + \mathbf{w} = \mathbf{u} + (\mathbf{v} + \mathbf{w})$.
3. There exists a $\mathbf{0}$ such that $\mathbf{0} + \mathbf{v} = \mathbf{v}$.
4. There exists an additive inverse, $-\mathbf{v}$, such that $\mathbf{v} + (-\mathbf{v}) = \mathbf{0}$.

   There are also several distributive properties:
5. $a(b\mathbf{v}) = (ab)\mathbf{v}$.
6. $(a + b)\mathbf{v} = a\mathbf{v} + b\mathbf{v}$.
7. $a(\mathbf{u} + \mathbf{v}) = a\mathbf{u} + a\mathbf{v}$.
8. There exists a multiplicative identity, 1, such that $1(\mathbf{v}) = \mathbf{v}$.

For now, we will restrict our examples to two and three dimensions, and the field will consist of the set of real numbers.

Basis vectors.

In three dimensions, the unit vectors $\mathbf{i}$, $\mathbf{j}$, and $\mathbf{k}$ play an important role. Any vector in the three-dimensional space can be written as a linear combination of these vectors,

$$\mathbf{v} = x\mathbf{i} + y\mathbf{j} + z\mathbf{k}.$$

In fact, given any three non-coplanar vectors, $\{\mathbf{a}_1, \mathbf{a}_2, \mathbf{a}_3\}$, all vectors can be written as a linear combination of those vectors,

$$\mathbf{v} = c_1\mathbf{a}_1 + c_2\mathbf{a}_2 + c_3\mathbf{a}_3.$$

Such vectors are said to span the space and are called a basis for the space.

We can generalize these ideas. In an n-dimensional vector space, any vector in the space can be represented as the sum over n linearly independent vectors (the equivalent of non-coplanar vectors). Such a linearly independent set of vectors $\{\mathbf{v}_j\}_{j=1}^n$ satisfies the condition

$$\sum_{j=1}^n c_j \mathbf{v}_j = \mathbf{0} \quad \Leftrightarrow \quad c_j = 0.$$

Note that we will often use summation notation instead of writing out all of the terms in the sum. Also, the symbol $\Leftrightarrow$ means "if and only if," or "is equivalent to." Each side of the symbol implies the other side.

Now we can define a basis for an n-dimensional vector space. We begin with the standard basis in an n-dimensional vector space. It is a generalization of the standard basis in three dimensions ($\mathbf{i}$, $\mathbf{j}$, and $\mathbf{k}$).

We define the standard basis with the notation

$$\mathbf{e}_k = (0,\ldots,0,\underbrace{1}_{k\text{th space}},0,\ldots,0), \quad k = 1,\ldots,n. \tag{3.1}$$

We can expand any $\mathbf{v} \in V$ as

$$\mathbf{v} = \sum_{k=1}^n v_k \mathbf{e}_k, \tag{3.2}$$

where the v_k's are called the components of the vector in this basis. Sometimes we will write $\mathbf{v}$ as an n-tuple $(v_1, v_2, \ldots, v_n)$. This is similar to the ambiguous use of (x, y, z) to denote both vectors and points in the three-dimensional space.

The only other thing we will need at this point is to generalize the dot product. Recall that there are two forms for the dot product in three dimensions. First, one has that

$$\mathbf{u} \cdot \mathbf{v} = uv \cos\theta, \tag{3.3}$$

where u and v denote the length of the vectors. The other form is the component form:

$$\mathbf{u} \cdot \mathbf{v} = u_1 v_1 + u_2 v_2 + u_3 v_3 = \sum_{k=1}^3 u_k v_k. \tag{3.4}$$

Of course, this form is easier to generalize. So, we define the scalar product between two n-dimensional vectors as

$$\langle \mathbf{u}, \mathbf{v} \rangle = \sum_{k=1}^n u_k v_k. \tag{3.5}$$

Actually, there are a number of notations that are used in other texts. One can write the scalar product as $(\mathbf{u}, \mathbf{v})$ or even in the Dirac bra-ket notation[3] $\langle \mathbf{u}|\mathbf{v}\rangle$.

We note that the (real) scalar product satisfies some simple properties. For vectors $\mathbf{v}$, $\mathbf{w}$ and real scalar α, we have

n-dimensional vector spaces.

Linearly independent vectors.

The standard basis vectors, $\mathbf{e}_k$ are a natural generalization of $\mathbf{i}$, $\mathbf{j}$ and $\mathbf{k}$.

For more general vector spaces, the term *inner product* is used to generalize the notions of dot and scalar products, as we will see below.

[3] The bra-ket notation was introduced by Paul Adrien Maurice Dirac (1902–1984) in order to facilitate computations of inner products in quantum mechanics. In the notation $\langle \mathbf{u}|\mathbf{v}\rangle$, $\langle \mathbf{u}|$ is the bra and $|\mathbf{v}\rangle$ is the ket. The kets live in a vector space and represented by column vectors with respect to a given basis. The bras live in the dual vector space and are represented by row vectors. The correspondence between bra and kets is $|\mathbf{v}\rangle = \overline{|\mathbf{v}\rangle}^T$. One can operate on kets, $A|\mathbf{v}\rangle$, and make sense out of operations like $\langle \mathbf{u}|A|\mathbf{v}\rangle$, which are used to obtain expectation values associated with the operator. Finally, the outer product, $|\mathbf{v}\rangle\langle \mathbf{v}|$ is used to perform vector space projections.

1. $\langle \mathbf{v}, \mathbf{v} \rangle \geq 0$ and $\langle \mathbf{v}, \mathbf{v} \rangle = 0$ if and only if $\mathbf{v} = \mathbf{0}$.

2. $\langle \mathbf{v}, \mathbf{w} \rangle = \langle \mathbf{w}, \mathbf{v} \rangle$.

3. $\langle \alpha \mathbf{v}, \mathbf{w} \rangle = \alpha \langle \mathbf{v}, \mathbf{w} \rangle$.

While it does not always make sense to talk about angles between general vectors in higher-dimensional vector spaces, there is one concept that is useful. It is that of orthogonality, which in three dimensions is another way of saying the vectors are perpendicular to each other. So, we also say that vectors $\mathbf{u}$ and $\mathbf{v}$ are orthogonal if and only if $\langle \mathbf{u}, \mathbf{v} \rangle = 0$. If $\{\mathbf{a}_k\}_{k=1}^{n}$, is a set of basis vectors such that

$$\langle \mathbf{a}_j, \mathbf{a}_k \rangle = 0, \quad k \neq j,$$

Orthogonal basis vectors.

then it is called an *orthogonal basis*.

If in addition each basis vector is a unit vector, then one has an orthonormal basis. This generalization of the unit basis can be expressed more compactly. We will denote such a basis of unit vectors by $\mathbf{e}_j$ for $j = 1 \ldots n$. Then,

$$\langle \mathbf{e}_j, \mathbf{e}_k \rangle = \delta_{jk}, \tag{3.6}$$

where we have introduced the Kronecker delta (named after Leopold Kronecker (1823–1891))

$$\delta_{jk} \equiv \begin{cases} 0, & j \neq k \\ 1, & j = k \end{cases} \tag{3.7}$$

Normalization of vectors.

The process of making vectors have unit length is called normalization. This is simply done by dividing by the length of the vector. Recall that the length of a vector, $\mathbf{v}$, is obtained as $v = \sqrt{\mathbf{v} \cdot \mathbf{v}}$. So, if we want to find a unit vector in the direction of $\mathbf{v}$, then we simply normalize it as

$$\hat{\mathbf{v}} = \frac{\mathbf{v}}{v}.$$

Notice that we used a hat to indicate that we have a unit vector. Furthermore, if $\{\mathbf{a}_j\}_{j=1}^{n}$ is a set of orthogonal basis vectors, then

$$\hat{\mathbf{a}}_j = \frac{\mathbf{a}_j}{\sqrt{\langle \mathbf{a}_j, \mathbf{a}_j \rangle}}, \quad j = 1 \ldots n.$$

Example 3.1. *Find the angle between the vectors* $\mathbf{u} = (-2, 1, 3)$ *and* $\mathbf{v} = (1, 0, 2)$. *We need the lengths of each vector,*

$$u = \sqrt{(-2)^2 + 1^2 + 3^2} = \sqrt{14},$$

$$v = \sqrt{1^2 + 0^2 + 2^2} = \sqrt{5}.$$

We also need the scalar product of these vectors:

$$\mathbf{u} \cdot \mathbf{v} = -2 + 6 = 4.$$

This gives

$$\cos \theta = \frac{\mathbf{u} \cdot \mathbf{v}}{uv} = \frac{4}{\sqrt{5}\sqrt{14}}.$$

So, $\theta = 61.4°$.

Example 3.2. *Normalize the vector* $\mathbf{v} = 2\mathbf{i} + \mathbf{j} - 2\mathbf{k}$.

The length of the vector is $v = \sqrt{2^2 + 1^2 + (-2)^2} = \sqrt{9} = 3$. *So, the unit vector in the direction of* $\mathbf{v}$ *is* $\hat{\mathbf{v}} = \frac{2}{3}\mathbf{i} + \frac{1}{3}\mathbf{j} - \frac{2}{3}\mathbf{k}$.

Let $\{\mathbf{a}_k\}_{k=1}^{n}$ be a set of orthogonal basis vectors for vector space V. We know that any vector $\mathbf{v}$ can be represented in terms of this basis, $\mathbf{v} = \sum_{k=1}^{n} v_k \mathbf{a}_k$. If we know the basis and vector, can we find the components v_k? The answer is yes. We can use the scalar product of $\mathbf{v}$ with each basis element $\mathbf{a}_j$. Using the properties of the scalar product, we have for $j = 1, \ldots, n$

$$
\begin{aligned}
\langle \mathbf{a}_j, \mathbf{v} \rangle &= \langle \mathbf{a}_j, \sum_{k=1}^{n} v_k \mathbf{a}_k \rangle \\
&= \sum_{k=1}^{n} v_k \langle \mathbf{a}_j, \mathbf{a}_k \rangle.
\end{aligned}
\tag{3.8}
$$

Because we know the basis elements, we can easily compute the numbers

$$
A_{jk} \equiv \langle \mathbf{a}_j, \mathbf{a}_k \rangle
$$

and

$$
b_j \equiv \langle \mathbf{a}_j, \mathbf{v} \rangle.
$$

Therefore, the system (3.8) for the v_k's is a linear algebraic system, that takes the form

$$
b_j = \sum_{k=1}^{n} A_{jk} v_k.
\tag{3.9}
$$

We can write this set of equations in a more compact form. The set of numbers A_{jk}, $j, k = 1, \ldots n$, are the elements of an $n \times n$ matrix A with A_{jk} being an element in the jth row and kth column. We write such matrices with the n^2 entries A_{ij} as

$$
A = \begin{pmatrix}
A_{11} & A_{12} & \cdots & A_{1n} \\
A_{21} & A_{22} & \cdots & A_{2n} \\
\vdots & \vdots & \ddots & \vdots \\
A_{n1} & A_{n2} & \cdots & A_{nn}
\end{pmatrix}.
\tag{3.10}
$$

Also, v_j and b_j can be written as column vectors, $\mathbf{v}$ and $\mathbf{b}$, respectively. Thus, system (3.8) can be written in matrix form as

$$
A\mathbf{v} = \mathbf{b}.
$$

However, if the basis is orthogonal, then the matrix $A_{jk} \equiv \langle \mathbf{a}_j, \mathbf{a}_k \rangle$ is diagonal,

$$
A = \begin{pmatrix}
A_{11} & 0 & \cdots & \cdots & 0 \\
0 & A_{22} & \ddots & \ddots & 0 \\
\vdots & 0 & \ddots & \ddots & \vdots \\
\vdots & \ddots & \ddots & \ddots & 0 \\
0 & 0 & \cdots & 0 & A_{nn}
\end{pmatrix}.
\tag{3.11}
$$

and the system is easily solvable. Recall that two vectors are orthogonal if and only if

$$\langle \mathbf{a}_i, \mathbf{a}_j \rangle = 0, \quad i \neq j. \tag{3.12}$$

Thus, in this case we have that

$$\langle \mathbf{a}_j, \mathbf{v} \rangle = v_j \langle \mathbf{a}_j, \mathbf{a}_j \rangle, \quad j = 1, \ldots, n, \tag{3.13}$$

or

$$v_j = \frac{\langle \mathbf{a}_j, \mathbf{v} \rangle}{\langle \mathbf{a}_j, \mathbf{a}_j \rangle}. \tag{3.14}$$

In fact, if the basis is orthonormal, that is, the basis consists of an orthogonal set of unit vectors, then A is the identity matrix and the solution takes on a simpler form:

$$v_j = \langle \mathbf{a}_j, \mathbf{v} \rangle. \tag{3.15}$$

Example 3.3. *Consider the set of vectors* $\mathbf{a}_1 = \mathbf{i} + \mathbf{j}$ *and* $\mathbf{a}_2 = \mathbf{i} - 2\mathbf{j}$.

1. *Determine the matrix elements* $A_{jk} = \langle \mathbf{a}_j, \mathbf{a}_k \rangle$.

2. *Is this an orthogonal basis?*

3. *Expand the vector* $\mathbf{v} = 2\mathbf{i} + 3\mathbf{j}$ *in the basis* $\{\mathbf{a}_1, \mathbf{a}_2\}$.

First, we compute the matrix elements of A:

$$
\begin{aligned}
A_{11} &= \langle \mathbf{a}_1, \mathbf{a}_1 \rangle = 2, \\
A_{12} &= \langle \mathbf{a}_1, \mathbf{a}_2 \rangle = -1, \\
A_{21} &= \langle \mathbf{a}_2, \mathbf{a}_1 \rangle = -1, \\
A_{22} &= \langle \mathbf{a}_2, \mathbf{a}_2 \rangle = 5,
\end{aligned} \tag{3.16}
$$

So,

$$A = \begin{pmatrix} 2 & -1 \\ -1 & 5 \end{pmatrix}.$$

Because $A_{12} = A_{21} \neq 0$, *the vectors are not orthogonal. However, they are linearly independent. Obviously, if* $c_1 = c_2 = 0$, *then the linear combination* $c_1 \mathbf{a}_1 + c_2 \mathbf{a}_2 = \mathbf{0}$. *Conversely, we assume that* $c_1 \mathbf{a}_1 + c_2 \mathbf{a}_2 = \mathbf{0}$ *and solve for the coefficients. Inserting the given vectors, we have*

$$
\begin{aligned}
\mathbf{0} &= c_1 (\mathbf{i} + \mathbf{j}) + c_2 (\mathbf{i} - 2\mathbf{j}) \\
&= (c_1 + c_2)\mathbf{i} + (c_1 - 2c_2)\mathbf{j}.
\end{aligned} \tag{3.17}
$$

This implies that

$$
\begin{aligned}
c_1 + c_2 &= 0, \\
c_1 - 2c_2 &= 0.
\end{aligned} \tag{3.18}
$$

Solving this system, one has $c_1 = 0$, $c_2 = 0$. *Therefore, the two vectors are linearly independent.*

In order to determine the components of **v** *with respect to the new basis, we need to set up the system (3.8) and solve for the v_k's. We have first,*

$$\mathbf{b} = \begin{pmatrix} \langle \mathbf{a}_1, \mathbf{v} \rangle \\ \langle \mathbf{a}_2, \mathbf{v} \rangle \end{pmatrix}$$
$$= \begin{pmatrix} \langle \mathbf{i} + \mathbf{j}, 2\mathbf{i} + 3\mathbf{j} \rangle \\ \langle \mathbf{i} - 2\mathbf{j}, 2\mathbf{i} + 3\mathbf{j} \rangle \end{pmatrix}$$
$$= \begin{pmatrix} 5 \\ -4 \end{pmatrix}. \tag{3.19}$$

So, now we have to solve the system $A\mathbf{v} = \mathbf{b}$ for **v**:

$$\begin{pmatrix} 2 & -1 \\ -1 & 5 \end{pmatrix} \begin{pmatrix} v_1 \\ v_2 \end{pmatrix} = \begin{pmatrix} 5 \\ -4 \end{pmatrix}. \tag{3.20}$$

We can solve this with matrix methods, $\mathbf{v} = A^{-1}\mathbf{b}$, *or rewrite it as a system of two equations and two unknowns as*

$$\begin{aligned} 2v_1 - v_2 &= 5, \\ -v_1 + 5v_2 &= -4. \end{aligned} \tag{3.21}$$

The solution of this set of algebraic equations is $v_1 = \frac{7}{3}$, $v_2 = -\frac{1}{3}$. Thus, $\mathbf{v} = \frac{7}{3}\mathbf{a}_1 - \frac{1}{3}\mathbf{a}_2$. *We will return later to using matrix methods to solve such systems.*

3.2 Linear Transformations

A MAIN THEME IN LINEAR ALGEBRA is to study linear transformations between vector spaces. These come in many forms and there are an abundance of applications in physics. For example, the transformation between the spacetime coordinates of observers moving in inertial frames in the theory of special relativity constitute such a transformation. Other, less exotic linear transformations are rotations. It is not only the transformations, but what the transformations do not affect, or leave invariant, that have shaped some of modern physics.

What is a linear transformation? It is a function between two vector spaces V and W over the same field of scalars F, $f : V \rightarrow W$, satisfying: For any two vectors $\mathbf{v}, \mathbf{w}$ in V and scalar α in F, then

a. $f(\mathbf{v} + \mathbf{w}) = f(\mathbf{v}) + f(\mathbf{w})$, and

b. $f(\alpha \mathbf{v}) = \alpha f(\mathbf{v})$.

As simple as the definition seems, there are some interesting consequences:

a. For any scalars α and β, in F, $f(\alpha \mathbf{v} + \beta \mathbf{w}) = \alpha f(\mathbf{v}) + \beta f(\mathbf{w})$.

b. $f(\mathbf{0}) = \mathbf{0}$.

c. In one dimension, $f(\mathbf{v}) = f(1 \cdot v) = v f(1)$.

d. In three dimensions, $f(x\mathbf{i} + y\mathbf{j} + z\mathbf{k}) = xf(\mathbf{i}) + yf(\mathbf{j}) + zf(\mathbf{k})$.

Item a. is equivalent to the two properties of a linear transformation and can be used instead, as often is the case. Item b. says that linear transformations take the zero vector (in V) to the zero vector (in W).

Items c. and d. indicate that if one wants to know how a linear transformation acts on vectors, one need only know how the transformation acts on the basis vectors. This turns out to play an important role in applications and leads to what are called matrix representations of linear transformations. This is the connection between linear algebra and the study of matrices, as we will see later in the chapter. However, we will first look at an important set of transformation in physics—rotations.

3.2.1   Active and Passive Rotations

A SIMPLE EXAMPLE OFTEN ENCOUNTERED IN PHYSICS COURSES is the rotation by a fixed angle. This is the description of points in space using two different coordinate bases, one just a rotation of the other by some angle. We begin with the position vector $\mathbf{r}$ as described with respect to a set of axes in the standard orientation, as shown in Figure 3.1. Also displayed in this figure are the unit vectors $\mathbf{i}$ and $\mathbf{j}$. In order to find the coordinates, (x, y), of the tip of the arrow, one needs only to draw perpendiculars to the axes and read off the coordinates.

We now consider a second set of axes at an angle θ to the old set of axes. Such a system is shown in Figure 3.2. We will designate these axes as x' and y'. Note that the basis vectors are different in this system and are denoted $\mathbf{i}'$ and $\mathbf{j}'$. Projections of $\mathbf{r}$ to the axes are shown.

We would like to compare the coordinates in both systems shown in Figures 3.1 and 3.2. We see that the primed coordinates are not the same as the unprimed ones. In terms of the standard basis, we have

$$\mathbf{r} = x\mathbf{i} + y\mathbf{j},$$

and in terms of the rotated system,

$$\mathbf{r} = x'\mathbf{i}' + y'\mathbf{j}'.$$

The goal is to find a relation between the components (x, y) and (x', y').

In order to derive this transformation we will make use of polar coordinates. In Figure 3.1 we see that the vector makes an angle of ϕ with respect to the positive x-axis. The components (x, y) of the vector can be determined from this angle and the magnitude of $\mathbf{r}$ as

$$
\begin{aligned}
x &= r\cos\phi, \\
y &= r\sin\phi.
\end{aligned}
\tag{3.22}
$$

In Figure 3.3, the two systems are superimposed on each other. The polar form for the primed system is given by

$$x' = r\cos(\phi - \theta)$$

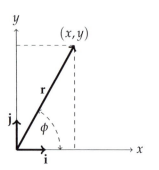

Figure 3.1: The position vector $\mathbf{r}$ in a standard coordinate system.

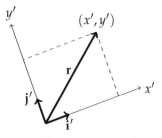

Figure 3.2: Vector $\mathbf{r}$ in a rotated coordinate system.

$$y' = r\sin(\phi - \theta). \tag{3.23}$$

We can use this form to find a relationship between the two systems. Namely, we use the addition formula for trigonometric functions to obtain

$$
\begin{aligned}
x' &= r\cos\phi\cos\theta + v\sin\phi\sin\theta, \\
y' &= r\sin\phi\cos\theta - v\cos\phi\sin\theta.
\end{aligned}
\tag{3.24}
$$

Noting that these expressions involve products of r with $\cos\phi$ and $\sin\phi$, we can use the polar form for x and y to write

$$
\begin{aligned}
x' &= x\cos\theta + y\sin\theta, \\
y' &= -x\sin\theta + y\cos\theta.
\end{aligned}
\tag{3.25}
$$

This is an example of a transformation between two coordinate systems. It is called a rotation by θ. We will designate this transformation by

$$(x', y') = \hat{R}_\theta(x, y).$$

It is referred to as a passive transformation because it does not affect the vector. [Note: We will use the hat for passive rotations.]

Instead of rotating coordinate axes, we could rotate the vector $\mathbf{r}$, keeping the axes fixed. This is called an active rotation and is shown in Figure 3.4. We could derive a transformation for how the coordinates of the vector change under this active rotation. Denoting the new vector as $\mathbf{r}'$ with new coordinates (x'', y''), it is easy to show

$$
\begin{aligned}
x'' &= x\cos\theta - y\sin\theta, \\
y'' &= x\sin\theta + y\cos\theta.
\end{aligned}
\tag{3.26}
$$

We will define this transformation as R_θ without a hat and write it compactly as

$$(x'', y'') = R_\theta(x, y).$$

Note that the active and passive rotations are related. We replace θ by $-\theta$ in Equations (3.25) to obtain Equations (3.26). Therefore,

$$R_\theta(x, y) = \hat{R}_{-\theta}(x, y).$$

3.2.2 Rotation Matrices

IN THIS SECTION WE WILL REWRITE the rotation transformations in a more compact matrix form. We begin with the rotation transformation as applied to the axes in Equations (3.25). We write vectors like $\mathbf{v}$ as a column matrix

$$\mathbf{v} = \begin{pmatrix} x \\ y \end{pmatrix}.$$

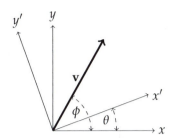

Figure 3.3: Comparison of the coordinate systems.

Passive rotation.

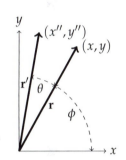

Figure 3.4: This is an active rotation of vector $\mathbf{r}$ into $\mathbf{r}'$.

We can also write the trigonometric functions in a 2×2 matrix form as

$$\hat{R}_\theta = \begin{pmatrix} \cos\theta & \sin\theta \\ -\sin\theta & \cos\theta \end{pmatrix}.$$

Then, the transformation takes the form

$$\begin{pmatrix} x' \\ y' \end{pmatrix} = \begin{pmatrix} \cos\theta & \sin\theta \\ -\sin\theta & \cos\theta \end{pmatrix} \begin{pmatrix} x \\ y \end{pmatrix}. \tag{3.27}$$

This can be written in the more compact form

$$\mathbf{v}' = \hat{R}_\theta \mathbf{v}.$$

2D rotation matrix for the passive rotation of coordinates.

In using the matrix form of the transformation, we have employed the definition of matrix multiplication. Namely, we have multiplied a 2×2 matrix times a 2×1 matrix. (Note that an $n \times m$ matrix has n rows and m columns.) The multiplication proceeds by selecting the ith row of the first matrix and the jth column of the second matrix. Multiply corresponding elements of each and add them. Then, place the result into the ijth entry of the product matrix. This operation can only be performed if the number of columns of the first matrix is the same as the number of columns of the second matrix.

Introduction to matrix multiplication.

Example 3.4. *As an example, we multiply a 3×2 matrix times a 2×2 matrix to obtain a 3×2 matrix. The details are shown below, and the diagram in Figure 3.4 indicates visually how the entries are selected.*

$$\begin{pmatrix} 1 & 2 \\ 5 & -1 \\ 3 & 2 \end{pmatrix} \begin{pmatrix} 3 & 2 \\ 1 & 4 \end{pmatrix} = \begin{pmatrix} 1(3)+2(1) & 1(2)+2(4) \\ 5(3)+(-1)(1) & 5(2)+(-1)(4) \\ 3(3)+2(1) & 3(2)+2(4) \end{pmatrix}$$

$$= \begin{pmatrix} 5 & 10 \\ 14 & 6 \\ 11 & 14 \end{pmatrix} \tag{3.28}$$

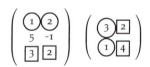

Figure 3.5: Diagram for multiplying matrices. In order to get the 5 in the final matrix, take the elements of the first row and multiply them, respectively, with the elements of the first column of the second matrix. These are all given by circles. The rectangles are used to obtain the 14 in the lower right corner of the result.

Example 3.5. *Show that the matrix product in Equation (3.27) produces the equations in (3.25).*

In Equation (3.27), consider the first row $(\cos\theta, \sin\theta)$ of $\hat{R}_\theta$ and the column $\begin{pmatrix} x \\ y \end{pmatrix}$. Multiplying these, we obtain $x\cos\theta + y\sin\theta$. Performing the same operations for the second row, we have

$$\begin{pmatrix} x' \\ y' \end{pmatrix} = \begin{pmatrix} \cos\theta & \sin\theta \\ -\sin\theta & \cos\theta \end{pmatrix} \begin{pmatrix} x \\ y \end{pmatrix}$$

$$= \begin{pmatrix} x\cos\theta + y\sin\theta \\ -x\sin\theta + y\cos\theta \end{pmatrix}. \tag{3.29}$$

Therefore, $x' = x\cos\theta + y\sin\theta$ and $y' = -x\sin\theta + y\cos\theta$, as was to be shown.

In the previous section we also introduced active rotations. These were rotations of vectors keeping the coordinate system fixed. In general, we start with a vector $\mathbf{v}$ and rotate it by θ to get a new vector $\mathbf{u}$. That transformation can be written as

$$\mathbf{u} = R_\theta \mathbf{v}, \qquad (3.30)$$

where

$$R_\theta = \begin{pmatrix} \cos\theta & -\sin\theta \\ \sin\theta & \cos\theta \end{pmatrix}.$$

2D rotation matrix for the active rotation of vectors.

Now consider a rotation by $-\theta$. Due to the symmetry properties of the sines and cosines, we have

$$R_{-\theta} = \begin{pmatrix} \cos\theta & \sin\theta \\ -\sin\theta & \cos\theta \end{pmatrix}.$$

Matrix transpose.

We see that if the 12 and 21 elements of this matrix are interchanged, we recover R_θ. This is an example of what is called the transpose of R_θ. Given a matrix A, its transpose A^T is the matrix obtained by interchanging the rows and columns of A. Formally, let A_{ij} be the elements of A. Then

$$A_{ij}^T = A_{ji}.$$

Also, the transpose of a row matrix is a column matrix, and vice versa.

It is also the case that the matrices R_θ and $R_{-\theta}$ are inverses of each other. This makes intuitive sense. We first rotate the vector by θ as $\mathbf{u} = R_\theta \mathbf{v}$ and then rotate $\mathbf{u}$ by $-\theta$ obtaining $\mathbf{w} = R_{-\theta}\mathbf{u}$. Thus, the "composition" of these two transformations leads to

$$\mathbf{w} = R_{-\theta}\mathbf{u} = R_{-\theta}(R_\theta \mathbf{v}). \qquad (3.31)$$

We can view this as a net transformation from $\mathbf{v}$ to $\mathbf{w}$ given by

$$\mathbf{w} = (R_{-\theta}R_\theta)\mathbf{v},$$

where the transformation matrix for the composition is given by $R_{-\theta}R_\theta$. Actually, if you think about it, we should end up with the original vector. We can compute the resulting matrix by carrying out the multiplication. We obtain

$$R_{-\theta}R_\theta = \begin{pmatrix} \cos\theta & \sin\theta \\ -\sin\theta & \cos\theta \end{pmatrix} \begin{pmatrix} \cos\theta & -\sin\theta \\ \sin\theta & \cos\theta \end{pmatrix} = \begin{pmatrix} 1 & 0 \\ 0 & 1 \end{pmatrix}. \qquad (3.32)$$

The matrix

$$I \equiv \begin{pmatrix} 1 & 0 \\ 0 & 1 \end{pmatrix}$$

is the 2×2 identity matrix. So, we have shown that the product of these two matrices yields the identity, $R_{-\theta}R_\theta = I$. This is like the multiplication of real numbers. If $ab = 1$, then a and b are multiplicative inverses of each other. So, we see that R_θ and $R_{-\theta}$ are inverses of each other as well. In fact, we have determined that

$$R_{-\theta} = R_\theta^{-1} = R_\theta^T,$$

(3.33)

where the T designates the transpose.

We note that matrices satisfying the relation $A^T = A^{-1}$, or $AA^T = I$, are called orthogonal matrices. These special matrices get their name from the fact that requiring $AA^T = I$ means that the columns of A have to be an orthonormal set of vectors. We can see this for R_θ. The columns are the vectors $\mathbf{x}_1 = (\cos\theta, \sin\theta)$ and $\mathbf{x}_2 = (-\sin\theta, \cos\theta)$. It is easy to see that $\langle \mathbf{x}_1, \mathbf{x}_2 \rangle = 0$.

Orthogonal matrices satisfy $A^T = A^{-1}$.

We can easily extend this discussion to three dimensions. Rotations in the xy-plane can be viewed as rotations about the z-axis. Rotating a vector about the z-axis by angle α will leave the z-component fixed. This can be represented by the rotation matrix

$$R_z(\alpha) = \begin{pmatrix} \cos\alpha & -\sin\alpha & 0 \\ \sin\alpha & \cos\alpha & 0 \\ 0 & 0 & 1 \end{pmatrix}.$$

(3.34)

We can also rotate vectors about the other axes, so that we would have two other rotation matrices:

$$R_y(\beta) = \begin{pmatrix} \cos\beta & 0 & \sin\beta \\ 0 & 1 & 0 \\ -\sin\beta & 0 & \cos\beta \end{pmatrix},$$

(3.35)

$$R_x(\gamma) = \begin{pmatrix} 1 & 0 & 0 \\ 0 & \cos\gamma & -\sin\gamma \\ 0 & \sin\gamma & \cos\gamma \end{pmatrix}.$$

(3.36)

As before, passive rotations of the coordinate axes are obtained by replacing the angles above by their negatives; for example, $\hat{R}_z(\alpha) = R_z(-\alpha)$ as shown in Figure 3.6.

3.2.3 Matrix Representations

In classical dynamics, one describes a general rotation in terms of the so-called Euler angles. These are the angles (ϕ, θ, ψ) such that the combined rotation $\hat{R}_z(\psi)\hat{R}_x(\theta)\hat{R}_z(\phi)$ rotates the initial coordinate system into a new one.

LINEAR TRANSFORMATIONS SUCH AS ROTATIONS can be represented by matrices as we saw in the previous section. Such matrix representations often become the core of a linear algebra class to the extent that one might lose sight of their meaning. We will introduce matrix representations in this section and then discuss various matrix operations in the next section. One application of matrices is in the study of coupled systems of differential equations, which we will cover later in the chapter.

First, we will generalize what we have seen with the simple example of rotation to other linear transformations. We begin with a vector $\mathbf{v}$ in an n-dimensional vector space. We can consider a transformation L that takes $\mathbf{v}$ into a new vector $\mathbf{u}$ as

$$\mathbf{u} = L(\mathbf{v}).$$

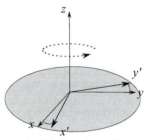

Figure 3.6: This is a three-dimensional rotation of coordinates about the z-axis. It would be represented by $\hat{R}_z(\alpha)$.

We will restrict the discussion to linear transformations between two n-dimensional vector spaces. Recall that a linear transformation satisfies the following linearity property:

$$L(\alpha\mathbf{a} + \beta\mathbf{b}) = \alpha L(\mathbf{a}) + \beta L(\mathbf{b}) \tag{3.37}$$

for any vectors $\mathbf{a}$ and $\mathbf{b}$ and scalars α and β.

Such linear transformations can be represented by matrices. Take any vector $\mathbf{v}$. It can be represented in terms of a basis. Let's use the standard basis $\{\mathbf{e}_i\}$, $i = 1, \ldots n$. Then, we have

$$\mathbf{v} = \sum_{i=1}^{n} v_i\mathbf{e}_i.$$

Now consider the effect of the transformation L on $\mathbf{v}$, using the following linearity property:

$$L(\mathbf{v}) = L\left(\sum_{i=1}^{n} v_i\mathbf{e}_i\right) = \sum_{i=1}^{n} v_i L(\mathbf{e}_i). \tag{3.38}$$

Thus, we see that determining how L acts on $\mathbf{v}$ requires that we know how L acts on the basis vectors. Namely, we need $L(\mathbf{e}_i)$. Because $\mathbf{e}_i$ is a vector, this produces another vector in the space. But the resulting vector can also be expanded in the basis. Let's assume that the resulting vector takes the form

$$L(\mathbf{e}_i) = \sum_{j=1}^{n} L_{ji}\mathbf{e}_j, \tag{3.39}$$

where L_{ji} is the jth component of $L(\mathbf{e}_i)$ for each $i = 1, \ldots, n$. The matrix of L_{ji}'s is called the matrix representation of the operator L. We are using L to represent both the operator and the matrix representation and knowing which one is being used will depend upon the context.

The matrix representation depends on the basis used. We used the standard basis above. However, we could have started with a different basis, such as dictated by another coordinate system. Using a different basis leads to a different matrix representation.

Example 3.6. *Find the matrix representation of the transformation*

$$L(u, v) = (3u - v, v + u) = (x, y).$$

This is the linear transformation in the form $\mathbf{Lv} = \mathbf{u}$ for $\mathbf{v} = (u, v)$ and $\mathbf{u} = (x, y)$. The matrix representation for this transformation is found by considering how L acts on the basis vectors. We have $L(1, 0) = (3, 1)$ and $L(0, 1) = (-1, 1)$. Thus, the representation is given as

$$L = \begin{pmatrix} 3 & -1 \\ 1 & 1 \end{pmatrix}.$$

Example 3.7. *Construct the matrix representation of a 90^0 rotation about the z-axis that takes unit vector $\mathbf{i}$ into unit vector $\mathbf{j}$.*

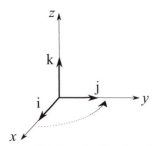

Figure 3.7: Rotation of unit vectors about the z-axis.

In order to find the matrix representation, we need to know how the rotation affects unit vectors. We can determine these from Figure 3.7. As noted in the problem statement, the transformation rotates $\mathbf{i}$ into $\mathbf{j}$. From the figure we see that $\mathbf{j}$ rotates into $-\mathbf{i}$ and $\mathbf{k}$ stays the same. Writing this information in component form, we have

$$
\begin{aligned}
L(1,0,0) &= (0,1,0) \\
L(0,1,0) &= (-1,0,0) \\
L(0,0,1) &= (0,0,1).
\end{aligned}
\tag{3.40}
$$

Therefore, taking these vectors as columns we can write down the matrix representation as

$$
L = \begin{pmatrix} 0 & -1 & 0 \\ 1 & 0 & 0 \\ 0 & 0 & 1 \end{pmatrix}.
\tag{3.41}
$$

Note that this agrees with the matrix $R_z(90^\circ)$.

Example 3.8. *Construct the matrix representation of a 90° rotation counterclockwise about the y-axis.*

Figure 3.8: Rotation of unit vectors about the y-axis. (a) Standard orientation of axes. (b) Orientation with y-axis pointing upward.

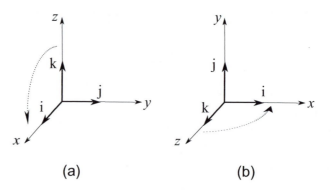

(a)   (b)

This problem is similar to the previous example. Just like that example, we are verifying the location of the negative sign in the rotation matrix R_y. The rotation we seek in shown in Figure 3.8(a). In order to understand that this is the appropriate rotation, we can reorient the axes into the more familiar picture as in Figure 3.7. This is shown in Figure 3.8(b).

We are ready to construct the transformation matrix. From the figures we see that the transformation rotates $\mathbf{k}$ into $\mathbf{i}$, $\mathbf{i}$ into $-\mathbf{k}$ and this time $\mathbf{j}$ remains fixed. In component form, we have

$$
\begin{aligned}
L(1,0,0) &= (0,0,-1), \\
L(0,1,0) &= (0,1,0), \\
L(0,0,1) &= (1,0,0).
\end{aligned}
\tag{3.42}
$$

Therefore, taking these vectors as columns, we can write the matrix representation as

$$L = \begin{pmatrix} 0 & 0 & 1 \\ 0 & 1 & 0 \\ -1 & 0 & 0 \end{pmatrix}. \tag{3.43}$$

This agrees with the matrix $R_y(90^o)$.

Now that we know how L acts on basis vectors, what does this say about how L acts on other vectors in the space? Inserting Expression (3.39) into Equation (3.38), we find

$$\begin{aligned} L(\mathbf{v}) &= \sum_{i=1}^{n} v_i L(\mathbf{e}_i) \\ &= \sum_{i=1}^{n} v_i \left(\sum_{j=1}^{n} L_{ji}\mathbf{e}_j \right) \\ &= \sum_{j=1}^{n} \left(\sum_{i=1}^{n} L_{ji}v_i \right) \mathbf{e}_j. \end{aligned} \tag{3.44}$$

Because $L(\mathbf{v}) = \mathbf{u}$, we see that the jth component of $\mathbf{u}$ can be written as

$$u_j = \sum_{i=1}^{n} L_{ji}v_i, \quad j = 1\dots n. \tag{3.45}$$

This equation can be written in matrix form as

$$\mathbf{u} = L\mathbf{v},$$

where L now takes the role of a matrix. It is similar to the multiplication of the rotation matrix times a vector, as seen in the previous section. We will mostly work with matrix representations from here on.

Example 3.9. *For the transformation $L(u,v) = (3u - v, v + u) = (x,y)$ in Example 3.6, what does $\mathbf{v} = 5\mathbf{i} + 3\mathbf{j}$ get mapped into?*

We know the matrix representation from the previous example, so we have

$$\mathbf{u} = \begin{pmatrix} 3 & -1 \\ 1 & 1 \end{pmatrix} \begin{pmatrix} 5 \\ 3 \end{pmatrix} = \begin{pmatrix} 12 \\ 2 \end{pmatrix}.$$

Next, we can compose transformations like we had done with the two rotation matrices. Let A be a transformation, $\mathbf{u} = A(\mathbf{v})$, taking vector $\mathbf{v}$ in an n-dimensional space V to $\mathbf{u}$ in an m-dimensional space U. Let B be a transformation, $\mathbf{w} = A(\mathbf{u})$, taking vector $\mathbf{u}$ in an m-dimensional space U to $\mathbf{w}$ in an ℓ-dimensional space W. (Thus, $\mathbf{v} \to \mathbf{u} \to \mathbf{w}$.) Then the composition of these transformations is given by

$$\mathbf{w} = B(\mathbf{u}) = B(A\mathbf{v}).$$

This can be viewed as a transformation from $\mathbf{v}$ to $\mathbf{w}$ as

$$\mathbf{w} = BA(\mathbf{v}),$$

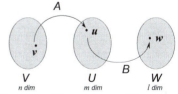

Figure 3.9: The spaces used for composition of transformations.

where the matrix representation of BA is given by the product of the matrix representations of A and B.

To see this, we look at the ijth element of the matrix representation of BA. We first note that the transformation from $\mathbf{v}$ to $\mathbf{w}$ is given by

$$w_i = \sum_{j=1}^{n} (BA)_{ij} v_j, \quad i = 1, \ldots, \ell. \tag{3.46}$$

However, if we use the successive transformations, we have

$$
\begin{aligned}
w_i &= \sum_{k=1}^{m} B_{ik} u_k, \quad i = 1, \ldots, \ell, \\
&= \sum_{k=1}^{m} B_{ik} \left(\sum_{j=1}^{n} A_{kj} v_j \right) \\
&= \sum_{j=1}^{n} \left(\sum_{k=1}^{m} B_{ik} A_{kj} \right) v_j.
\end{aligned} \tag{3.47}
$$

We have two expressions for w_i as sums over v_j. So, the coefficients must be equal. This leads to the following result:

$$\boxed{(BA)_{ij} = \sum_{k=1}^{n} B_{ik} A_{kj}, \quad i = 1, \ldots, \ell, j = 1, \ldots, n. \tag{3.48}}$$

Expression for matrix multiplication of an $\ell \times m$ matrix with an $m \times n$ matrix.

Thus, we have found the component form of matrix multiplication, which resulted from the composition of two linear transformations. This agrees with our earlier example of matrix multiplication: The ij-th component of the product is obtained by multiplying elements in the ith row of B and the jth column of A and summing. The resulting matrix is an $\ell \times n$ matrix.

We pause to note the dimensions of the matrices. Transformation A maps n-dimensional vectors into m-dimensional vectors. The matrix representation is given by an $m \times n$ matrix. Similarly, transformation B is represented by an $\ell \times m$ matrix. Multiplying these matrices give the matrix representation of the composition, BA. Thus, we multiply an $\ell \times m$ matrix times an $m \times n$ matrix. This is only possible if the inner dimensions of the matrices, m, agree. In most applications we will be restricted to mapping spaces of the same dimension

Example 3.10. *Consider the rotation in two dimensions of the axes by an angle θ. Now apply the scaling transformation[4]*

[4] This scaling transformation will rescale x-components by a and y-components by b. If either is negative, it will also provide an additional reflection.

$$L_s = \begin{pmatrix} a & 0 \\ 0 & b \end{pmatrix}.$$

What is the matrix representation of this combination of transformations? The result is a simple product of the matrix representations (in reverse order of application):

$$L_s \hat{R}_\theta = \begin{pmatrix} a & 0 \\ 0 & b \end{pmatrix} \begin{pmatrix} \cos\theta & \sin\theta \\ -\sin\theta & \cos\theta \end{pmatrix} = \begin{pmatrix} a\cos\theta & a\sin\theta \\ -b\sin\theta & b\cos\theta \end{pmatrix}.$$

Example 3.11. *Find the matrix representation of the transformation in which one first rescales coordinates using L_s followed by a rotation by θ. Is this the same as the previous transformation in reverse?*

A simple computation yields

$$\hat{R}_\theta L_s = \begin{pmatrix} \cos\theta & \sin\theta \\ -\sin\theta & \cos\theta \end{pmatrix} \begin{pmatrix} a & 0 \\ 0 & b \end{pmatrix} = \begin{pmatrix} a\cos\theta & b\sin\theta \\ -a\sin\theta & b\cos\theta \end{pmatrix}.$$

Comparing this with the result of the last example, we see that

$$L_s\hat{R}_\theta - \hat{R}_\theta L_s = \begin{pmatrix} 0 & (a-b)\sin\theta \\ (a-b)\sin\theta & 0 \end{pmatrix}.$$

So, in general, $L_s\hat{R}_\theta - \hat{R}_\theta L_s \neq 0$. The difference vanishes when either $a = b$, $\theta = 0$, or $\theta = \pi$.

In the last example we saw a property of matrices, which is important in physics. In general, for $n \times n$ matrices A and B, $AB \neq BA$. Such matrices are said to be noncommutative.

Another way to state this is to define the commutator

A and B are noncommutative if and only if $[A, B] \neq 0$.

$$[A, B] \equiv AB - BA \tag{3.49}$$

and write that A and B are noncommutative if and only if $[A, B] \neq 0$. Also, we say that A and B are commutative if and only if $[A, B] = 0$. These definitions also extend to transformations and operators. From the above example, we can write

$$[L_s, \hat{R}_\theta] = [(a-b)\sin\theta]\, I.$$

3.2.4 Matrix Inverses and Determinants

MANY PROBLEMS IN PHYSICS CAN BE POSED as finding the solution of and operator equation, $Ly = f$. If we can find a matrix representation of this equation, such as $Ax = b$, then a formal solution would be $x = A^{-1}b$, where A^{-1} is the inverse of matrix A. Of course, we would need to know if this inverse exists and how to compute it. In this section we will discuss matrix inverses and other related properties of $n \times n$ matrices, such as the transpose and determinant.

An operation that we have seen earlier is the transpose of a matrix. The transpose of a matrix is a new matrix in which the rows and columns are interchanged. If we write an $n \times m$ matrix A in standard form as

Matrix transpose.

$$A = \begin{pmatrix} a_{11} & a_{12} & \cdots & a_{1m} \\ a_{21} & a_{22} & \cdots & a_{2m} \\ \vdots & \vdots & \ddots & \vdots \\ a_{n1} & a_{n2} & \cdots & a_{nm} \end{pmatrix}, \tag{3.50}$$

then the transpose is defined as

$$A^T = \begin{pmatrix} a_{11} & a_{21} & \cdots & a_{1n} \\ a_{12} & a_{22} & \cdots & a_{2n} \\ \vdots & \vdots & \ddots & \vdots \\ a_{1m} & a_{2m} & \cdots & a_{mn} \end{pmatrix}. \tag{3.51}$$

In index form, we have

$$(A^T)_{ij} = A_{ji}, \quad i, j = 1, \ldots, n.$$

As we had seen in the previous section, a matrix satisfying

$$A^T = A^{-1}, \quad \text{or} \quad AA^T = A^T A = I,$$

is called an orthogonal matrix. One also can show that

$$(AB)^T = B^T A^T.$$

Trace of a matrix.

Finally, the trace of a square matrix is the sum of its diagonal elements:

$$\text{Tr}(A) = a_{11} + a_{22} + \ldots + a_{nn} = \sum_{i=1}^{n} a_{ii}.$$

We can show that for two square matrices

$$\text{Tr}(AB) = \text{Tr}(BA).$$

Identity matrix.

There are special matrices, and the simplest of these is the $n \times n$ identity matrix, I. The identity is defined as that matrix satisfying

$$IA = AI = A \tag{3.52}$$

for any $n \times n$ matrix A. The $n \times n$ identity matrix takes the form

$$I = \begin{pmatrix} 1 & 0 & \cdots & 0 \\ 0 & 1 & \cdots & 0 \\ \vdots & \vdots & \ddots & \vdots \\ 0 & 0 & \vdots & 1 \end{pmatrix}. \tag{3.53}$$

Kronecker delta, δ_{ij}.

A component form is given by the Kronecker delta. Namely, we have that

$$I_{ij} = \delta_{ij} \equiv \begin{cases} 0, & i \neq j \\ 1, & i = j \end{cases}. \tag{3.54}$$

The main goal of this section is to find the inverse of a matrix. The inverse of A is the matrix A^{-1} that satisfies the equation

$$AA^{-1} = A^{-1}A = I. \tag{3.55}$$

There is a systematic method for determining the inverse in terms of cofactors, which we describe a little later. However, the inverse of a 2×2 matrix is easily obtained without learning about cofactors.

Let

$$A = \begin{pmatrix} a & b \\ c & d \end{pmatrix}.$$

Now consider the matrix

$$B = \begin{pmatrix} d & -b \\ -c & a \end{pmatrix}.$$

Multiplying these matrices, we find that

$$AB = \begin{pmatrix} a & b \\ c & d \end{pmatrix}\begin{pmatrix} d & -b \\ -c & a \end{pmatrix} = \begin{pmatrix} ad - bc & 0 \\ 0 & ad - bc \end{pmatrix}.$$

This is not quite the identity, but it is a multiple of the identity. Dividing by $ad - bc$, we obtain the inverse matrix:

$$A^{-1} = \frac{1}{ad - bc}\begin{pmatrix} d & -b \\ -c & a \end{pmatrix}.$$

We leave it to the reader to show that $A^{-1}A = I$.

Inverse of a 2×2 matrix.

The factor $ad - bc$ is the difference in the products of the diagonal and off-diagonal elements of matrix A. This factor is called the *determinant* of A. It is denoted as $\det(A)$, det A or $|A|$. Thus, we define

$$\det(A) = \begin{vmatrix} a & b \\ c & d \end{vmatrix} = ad - bc. \tag{3.56}$$

For higher-dimensional matrices, one can write the definition of the determinant. We will for now just indicate the process for 3×3 matrices. We write matrix A as

Determinant of a 3×3 matrix.

$$A = \begin{pmatrix} a_{11} & a_{12} & a_{13} \\ a_{21} & a_{22} & a_{23} \\ a_{31} & a_{32} & a_{33} \end{pmatrix}. \tag{3.57}$$

The determinant of A can be computed in terms of simpler 2×2 determinants. We define

$$\det A = \begin{vmatrix} a_{11} & a_{12} & a_{13} \\ a_{21} & a_{22} & a_{23} \\ a_{31} & a_{32} & a_{33} \end{vmatrix}$$

$$= a_{11}\begin{vmatrix} a_{22} & a_{23} \\ a_{32} & a_{33} \end{vmatrix} - a_{12}\begin{vmatrix} a_{21} & a_{23} \\ a_{31} & a_{33} \end{vmatrix} + a_{13}\begin{vmatrix} a_{21} & a_{22} \\ a_{31} & a_{32} \end{vmatrix}. \tag{3.58}$$

There are many other properties of determinants. For example, if two rows, or columns, of a matrix are multiples of each other, then det $A = 0$. If one multiplies one row, or column, of a matrix by a constant k, then the determinant of the matrix is multiplied by k. The determinant of a product of $n \times n$ matrices is the product of the determinants, $|AB| = |A||B|$.

Singular matrices, det $A = 0$.

If det $A = 0$, A is called a *singular* matrix. Otherwise, it is called *nonsingular*. If a matrix is nonsingular, then the inverse exists. From our example for a general 2×2 system, the inverse exists if $ad - bc \neq 0$.

Computing the inverses of larger matrices is a little more complicated. One first constructs the matrix of cofactors. The ij-th cofactor is obtained by computing the determinant of the matrix, α_{ij} resulting from eliminating the ith row and jth column of A and multiplying by either $+1$ or -1. Thus,

The matrix of cofactors.

$$C_{ij} = (-1)^{i+j}\det(\alpha_{ij}).$$

The inverse matrix is obtained by dividing the transpose of the matrix of cofactors by the determinant of A. This gives the elements of A^{-1} as

$$\left(A^{-1}\right)_{ij} = \frac{C_{ji}}{\det A}.$$

This process is best shown by example.

Example 3.12. *Find the inverse of the matrix*

$$A = \begin{pmatrix} 1 & 2 & -1 \\ 0 & 3 & 2 \\ 1 & -2 & 1 \end{pmatrix}.$$

The determinant of this matrix is easily found as

$$\det A = \begin{vmatrix} 1 & 2 & -1 \\ 0 & 3 & 2 \\ 1 & -2 & 1 \end{vmatrix} = 1 \begin{vmatrix} 3 & 2 \\ -2 & 1 \end{vmatrix} + 1 \begin{vmatrix} 2 & -1 \\ 3 & 2 \end{vmatrix} = 14.$$

Next, we need to construct the matrix of cofactors. Let's look at $C_{11} = (-1)^2\det \alpha_{11}$. We compute α_{11} by writing A without the first row and first column,

$$\alpha_{11} = \begin{pmatrix} 3 & 2 \\ -2 & 1 \end{pmatrix}.$$

Therefore,

$$C_{11} = + \begin{vmatrix} 3 & 2 \\ -2 & 1 \end{vmatrix}.$$

Similarly, we can get $C_{12} = (-1)^{(1+2)}\det \alpha_{12}$, or

$$C_{12} = - \begin{vmatrix} 0 & 2 \\ 1 & 1 \end{vmatrix}.$$

Continuing with the rest of the entries, we find the cofactor matrix:

$$C_{ij} = \begin{pmatrix} +\begin{vmatrix} 3 & 2 \\ -2 & 1 \end{vmatrix} & -\begin{vmatrix} 0 & 2 \\ 1 & 1 \end{vmatrix} & +\begin{vmatrix} 0 & 3 \\ 1 & -2 \end{vmatrix} \\[2mm] -\begin{vmatrix} 2 & -1 \\ -2 & 1 \end{vmatrix} & +\begin{vmatrix} 1 & -1 \\ 1 & 1 \end{vmatrix} & -\begin{vmatrix} 1 & 2 \\ 1 & -2 \end{vmatrix} \\[2mm] +\begin{vmatrix} 2 & -1 \\ 3 & 2 \end{vmatrix} & -\begin{vmatrix} 1 & -1 \\ 0 & 2 \end{vmatrix} & +\begin{vmatrix} 1 & 2 \\ 0 & 3 \end{vmatrix} \end{pmatrix}.$$

Evaluating the 2×2 determinants, we obtain

$$C_{ij} = \begin{pmatrix} 7 & -2 & -3 \\ 0 & 2 & 4 \\ 7 & -2 & 3 \end{pmatrix}.$$

Finally, we compute the inverse using $C^T / \det A$, to obtain

$$\begin{aligned} A^{-1} &= \frac{1}{14} \begin{pmatrix} 7 & -2 & -3 \\ 0 & 2 & 4 \\ 7 & -2 & 3 \end{pmatrix}^T \\ &= \frac{1}{14} \begin{pmatrix} 7 & 0 & 7 \\ -2 & 2 & -2 \\ -3 & 4 & 3 \end{pmatrix} \\ &= \begin{pmatrix} \frac{1}{2} & 0 & \frac{1}{2} \\ -\frac{1}{7} & \frac{1}{7} & -\frac{1}{7} \\ -\frac{3}{14} & \frac{2}{7} & \frac{3}{14} \end{pmatrix}. \end{aligned} \tag{3.59}$$

The reader can verify that this is the inverse by evaluating AA^{-1} and $A^{-1}A$.

3.2.5 Cramer's Rule

A STANDARD APPLICATION OF DETERMINANTS is the solution of a system of linear algebraic equations using Cramer's Rule. As an example, we consider a simple system of two equations and two unknowns, x and y, in the form

> Cramer's Rule for solving algebraic systems of equations.

$$\begin{aligned} ax + by &= e, \\ cx + dy &= f. \end{aligned} \tag{3.60}$$

The standard way to solve this is to eliminate one of the variables. (Just imagine dealing with a bigger system!). So, we can eliminate the x's. Multiply the first equation by c and the second equation by a and subtract. We then get

$$(bc - ad)y = (ec - fa).$$

If $bc - ad \neq 0$, then we can solve to y, getting

$$y = \frac{ec - fa}{bc - ad}.$$

Similarly, we find

$$x = \frac{ed - bf}{ad - bc}.$$

We note the denominators can be replaced with the determinant of the matrix of coefficients,

$$\begin{pmatrix} a & b \\ c & d \end{pmatrix}.$$

In fact, we can also replace each numerator with a determinant. Thus, the solutions may be written as

$$
x = \frac{\begin{vmatrix} e & b \\ f & d \end{vmatrix}}{\begin{vmatrix} a & b \\ c & d \end{vmatrix}}
$$

$$
y = \frac{\begin{vmatrix} a & e \\ c & f \end{vmatrix}}{\begin{vmatrix} a & b \\ c & d \end{vmatrix}}. \tag{3.61}
$$

Cramer's Rule is named after Swiss mathematician Gabriel Cramer (1704–1752), who published the rule in a book he wrote on algebraic curves in 1750.

This form for writing the solutions is an example of what is called is Cramer's Rule. Note that each variable is determined by placing a determinant with e and f placed in the column of the coefficient matrix corresponding to the order of the variable in the equation. The denominator is the determinant of the coefficient matrix. This construction is easily extended to larger systems of equations.

Cramer's Rule can be extended to higher-dimensional systems. As an example, we now solve a system of three equations and three unknowns.

Example 3.13. *Solve the system of equations*

$$
\begin{aligned}
x + 2y - z &= 1, \\
3y + 2z &= 2, \\
x - 2y + z &= 0.
\end{aligned} \tag{3.62}
$$

First, one writes the system in the form $L\mathbf{x} = \mathbf{b}$, *where* L *is the coefficient matrix*

$$
L = \begin{pmatrix} 1 & 2 & -1 \\ 0 & 3 & 2 \\ 1 & -2 & 1 \end{pmatrix}
$$

and

$$
\mathbf{b} = \begin{pmatrix} 1 \\ 2 \\ 0 \end{pmatrix}.
$$

The solution is generally given by $\mathbf{x} = L^{-1}\mathbf{b}$ *if* L^{-1} *exists. Because* $\det L = 14 \neq 0$, L *is nonsingular and its inverse exists.*

The solution of this system of three equations and three unknowns can now be found using Cramer's Rule. Essentially, this is done by construction. Consider the first component of $\mathbf{x} = L^{-1}\mathbf{b}$,

$$
\begin{aligned}
x &= \left(L^{-1}\mathbf{b} \right)_1 \\
&= \sum_{k=1}^{3} \frac{C_{1k}^T b_k}{\det L}
\end{aligned}
$$

$$= \frac{1}{L}(C_{11}b_1 + C_{21}b_2 + C_{31}b_3)$$

$$= \frac{1}{L}\left(b_1 \begin{vmatrix} 3 & 2 \\ -2 & 1 \end{vmatrix} - b_2 \begin{vmatrix} 2 & -1 \\ -2 & 1 \end{vmatrix} + b_3 \begin{vmatrix} 2 & -1 \\ 3 & 2 \end{vmatrix}\right)$$

$$= \frac{\begin{vmatrix} b_1 & 2 & -1 \\ b_2 & 3 & 2 \\ b_3 & -2 & 1 \end{vmatrix}}{L}. \tag{3.63}$$

Note that this answer has the same form as we had seen for Cramer's Rule for two equations and two unknowns. Inserting the components of **b**, *we have found*

$$x = \frac{\begin{vmatrix} 1 & 2 & -1 \\ 2 & 3 & 2 \\ 0 & -2 & 1 \end{vmatrix}}{\det L} = \frac{7}{14} = \frac{1}{2}, \tag{3.64}$$

Similarly, we can compute the other components using Cramer's Rule for three equations and three unknowns without the derivation we had used to get x. The results are below.

$$y = \frac{\begin{vmatrix} 1 & 1 & -1 \\ 0 & 2 & 2 \\ 1 & 0 & 1 \end{vmatrix}}{\det L} = \frac{6}{14} = \frac{3}{7}, \tag{3.65}$$

$$z = \frac{\begin{vmatrix} 1 & 2 & 1 \\ 0 & 3 & 2 \\ 1 & -2 & 0 \end{vmatrix}}{\det L} = \frac{5}{14}. \tag{3.66}$$

We end this section by summarizing the rule for the existence of solutions of systems of algebraic equations, $L\mathbf{x} = \mathbf{b}$.

1. If $\det L \neq 0$, then there exists a unique solution, $\mathbf{x} = L^{-1}\mathbf{b}$. In particular, if $\mathbf{b} = \mathbf{0}$, the system is homogeneous and only has the trivial solution, $\mathbf{x} = \mathbf{0}$.

2. If $\det L = 0$, then the system does not have a unique solution. Either there is no solution, or an infinite number of solutions. For example, consider that the system

$$\begin{aligned} 2x + y &= 5, \\ 4x + 2y &= 2, \end{aligned} \tag{3.67}$$

has no solutions, while

$$\begin{aligned} 2x + y &= 0, \\ 4x + 2y &= 0, \end{aligned} \tag{3.68}$$

has an infinite number of solutions ($y = -2x$).

In both cases, the assumptions hold because we have that $L = \begin{pmatrix} 2 & 1 \\ 4 & 2 \end{pmatrix}$ and thus $\det L = 0$.

3.3 Eigenvalue Problems

3.3.1 An Introduction to Coupled Systems

RECALL THAT ONE OF THE REASONS we have seemingly digressed into topics in linear algebra and matrices is to solve a coupled system of differential equations. The simplest example is a system of linear differential equations of the form

$$\begin{aligned} \frac{dx}{dt} &= ax + by, \\ \frac{dy}{dt} &= cx + dy. \end{aligned} \tag{3.69}$$

We note that this system is coupled. We cannot solve either equation without knowing either $x(t)$ or $y(t)$. A much easier problem would be to solve an uncoupled system like

Uncoupled system.

$$\begin{aligned} \frac{dx}{dt} &= \lambda_1 x, \\ \frac{dy}{dt} &= \lambda_2 y. \end{aligned} \tag{3.70}$$

The solutions are quickly found to be

$$\begin{aligned} x(t) &= c_1 e^{\lambda_1 t}, \\ y(t) &= c_2 e^{\lambda_2 t}. \end{aligned} \tag{3.71}$$

Here, c_1 and c_2 are two arbitrary constants.

We can determine particular solutions of the system by specifying $x(t_0) = x_0$ and $y(t_0) = y_0$ at some time t_0. Thus,

$$\begin{aligned} x(t) &= x_0 e^{\lambda_1 t}, \\ y(t) &= y_0 e^{\lambda_2 t}. \end{aligned} \tag{3.72}$$

Wouldn't it be nice if we could transform the original system into one that is not coupled? Let's write these systems more compactly. We write the coupled system as

$$\frac{d}{dt}\mathbf{x} = A\mathbf{x}$$

and the uncoupled system as

$$\frac{d}{dt}\mathbf{y} = \Lambda\mathbf{y},$$

where

$$\Lambda = \begin{pmatrix} \lambda_1 & 0 \\ 0 & \lambda_2 \end{pmatrix}$$

is a diagonal matrix.

Now, we seek a transformation between $\mathbf{x}$ and $\mathbf{y}$ that will transform the coupled system into the uncoupled system. Define the sought transformation as

$$\mathbf{x} = S\mathbf{y}. \tag{3.73}$$

Inserting this transformation into the coupled system, we have

$$\frac{d}{dt}S\mathbf{y} = AS\mathbf{y}.$$

Because S is a constant matrix,

$$\frac{d}{dt}\mathbf{y} = AS\mathbf{y}. \tag{3.74}$$

If S is invertible, then $\mathbf{y} = S^{-1}\mathbf{x}$. So, we can multiply both sides of Equation (3.74) by S^{-1} to obtain

$$\frac{d}{dt}\mathbf{y} = S^{-1}AS\mathbf{y}.$$

Because we are seeking an uncoupled system of the form

$$\frac{d}{dt}\mathbf{y} = \Lambda\mathbf{y},$$

we will require that

$$S^{-1}AS = \Lambda. \tag{3.75}$$

$S^{-1}AS$ is called a similarity transformation.

The expression $S^{-1}AS$ is called a similarity transformation of matrix A. So, in order to uncouple the system, we seek a similarity transformation that results in a diagonal matrix. This process is called the diagonalization of matrix A.

We do not know S, nor do we know Λ. Multiplying Equation (3.75) by S^{-1}, we have

$$AS = S\Lambda.$$

We can solve this equation if S is real symmetric, that is, $S^T = S$. [In the case of complex matrices, we need the matrix to be Hermitian, $\bar{S}^T = S$, where the bar denotes complex conjugation. Further discussion of diagonalization is left for the end of the chapter.]

We first show that $S\Lambda = \Lambda S$. We look at the ij-th component of $S\Lambda$ and rearrange the terms in the matrix product making use of the assumption that S is real symmetric.

$$\begin{aligned}
(S\Lambda)_{ij} &= \sum_{k=1}^{n} S_{ik}\Lambda_{kj} \\
&= \sum_{k=1}^{n} S_{ik}\lambda_j I_{kj} \\
&= \sum_{k=1}^{n} \lambda_j I_{jk} S_{ki}^T \\
&= \sum_{k=1}^{n} \Lambda_{jk} S_{ki} \\
&= (\Lambda S)_{ij} \tag{3.76}
\end{aligned}$$

This result leads us to the fact that S satisfies the equation

Matrix Eigenvalue Problem: Seek non-trivial eigenvectors $\mathbf{v}$ and eigenvalues λ such that $A\mathbf{v} = \lambda\mathbf{v}$.

$$AS = \Lambda S.$$

In particular, the columns of S (denoted $\mathbf{v}$) satisfy equations of the form

$$A\mathbf{v} = \lambda\mathbf{v} \tag{3.77}$$

for each λ on the diagonal of Λ. This is an equation for vectors $\mathbf{v}$ and numbers λ given matrix A. This is called an eigenvalue problem. The vectors are called eigenvectors and the numbers, λ, are called eigenvalues. In principle, we can solve the eigenvalue problem and this will lead us to solutions of the uncoupled system of differential equations.

3.3.2   *Eigenvalue Problems*

IN THE PREVIOUS SECTION WE WERE INTRODUCED to eigenvalue problems as a way to obtain a solution to a coupled system of linear differential equations. Eigenvalue problems appear in many contexts in physical applications. In this section we will summarize the method of solution of eigenvalue problems

We seek *nontrivial solutions* to the eigenvalue problem

$$A\mathbf{v} = \lambda\mathbf{v}. \tag{3.78}$$

We note that $\mathbf{v} = \mathbf{0}$ is an obvious solution. Furthermore, it does not lead to anything useful. So, it is a trivial solution. Typically, we are given the matrix A and have to determine the eigenvalues, λ, and the associated eigenvectors, $\mathbf{v}$, satisfying the above eigenvalue problem.

In order to demonstrate how this works, we will seek eigenvalues and eigenfunctions for the case where A is the 2×2 matrix $\begin{pmatrix} a & b \\ c & d \end{pmatrix}$. The eigenvalue problem Equation (3.78) takes the form

$$\begin{pmatrix} a & b \\ c & d \end{pmatrix} \begin{pmatrix} v_1 \\ v_2 \end{pmatrix} = \lambda \begin{pmatrix} v_1 \\ v_2 \end{pmatrix}.$$

Multiplying the matrices, we obtain the homogeneous algebraic system

$$(a - \lambda)v_1 + bv_2 = 0,$$
$$cv_1 + (d - \lambda)v_2 = 0. \tag{3.79}$$

The solution of such a system would be unique if the determinant of the system is not zero. However, this would give the trivial solution $v_1 = 0$, $v_2 = 0$.

To get a nontrivial solution, we need to force the determinant to be zero:

$$0 = \begin{vmatrix} a - \lambda & b \\ c & d - \lambda \end{vmatrix} = (a - \lambda)(d - \lambda) - bc.$$

This is a quadratic equation for the eigenvalues that would lead to nontrivial solutions. This is called the eigenvalue equation. Expanding the right-hand side of the equation, we find

$$\lambda^2 - (a + d)\lambda + ad - bc = 0.$$

This is the same equation as the characteristic equation for the general constant coefficient differential equation considered in the previous chapter. Thus, the eigenvalues correspond to the solutions of the characteristic equation for the system.

Once we find the eigenvalues, then we solve the homogeneous system for v_1 in terms of v_2, or vice versa. There are possibly an infinite number solutions to the algebraic system with representing parallel vectors in the plane. So, we need only pick one representative eigenvector as the solution for each eigenvalue.

The method for solving eigenvalue problems consists of just a few simple steps. We list these steps as follows:

Solving Eigenvalue Problems

---

a) Write the coefficient matrix;

b) Find the eigenvalues from the equation $\det(A - \lambda I) = 0$; and,

c) Solve the linear system $(A - \lambda I)\mathbf{v} = 0$ for each λ.

The following example shows how these steps are carried out in practice. We will then solve other eigenvalue problems in the next section as we show how matrix methods can be used to solve linear systems of differential equations.

Example 3.14. *Determine the eigenvalues and eigenvectors for*

$$A = \begin{pmatrix} 1 & -2 \\ -3 & 2 \end{pmatrix}$$

Let $\mathbf{v} = \begin{pmatrix} v_1 \\ v_2 \end{pmatrix}$. *Then the eigenvalue problem can be written as the system*

$$(1 - \lambda)v_1 - 2v_2 = 0,$$
$$-3v_1 + (2 - \lambda)v_2 = 0. \tag{3.80}$$

This is a homogeneous system and has nontrivial solution and therefore the eigenvalue equation must hold:

$$\det(A - \lambda I) = 0. \tag{3.81}$$

We write out this condition for the example at hand. We have that

$$\begin{vmatrix} 1 - \lambda & -2 \\ -3 & 2 - \lambda \end{vmatrix} = 0.$$

This will always be the starting point in solving eigenvalue problems. Note that the matrix is A with λ's subtracted from the diagonal elements.

Computing the determinant, we have

$$(1 - \lambda)(2 - \lambda) - 6 = 0,$$

or

$$\lambda^2 - 3\lambda - 4 = 0.$$

This quadratic can be factored:

$$(\lambda - 4)(\lambda + 1) = 0.$$

So, the eigenvalues are $\lambda = 4, -1$. The second step is to find the eigenvectors. We have to do this for each eigenvalue. We first insert $\lambda = 4$ into the system:

$$-3v_1 - 2v_2 = 0,$$
$$-3v_1 - 2v_2 = 0. \tag{3.82}$$

Note that these equations are the same. So, we have one equation in two unknowns. We will not get a unique solution. This is typical of eigenvalue problems. We can pick anything we want for v_2 and then determine v_1. For example, $v_2 = 1$ gives $v_1 = -2/3$. A nicer solution would be $v_2 = 3$ and $v_1 = -2$. These vectors are different, but they point in the same direction in the $v_1 v_2$ plane.

For $\lambda = -1$, the system becomes

$$2v_1 - 2v_2 = 0,$$
$$-3v_1 + 3v_2 = 0. \tag{3.83}$$

While these equations do not at first look the same, we can divide out the constants and see that, once again, we get the same equation,

$$v_1 = v_2.$$

Picking $v_2 = 1$, we get $v_1 = 1$.

In summary, the solution to the eigenvalue problem has two eigenvalues, each with a different eigenvector:

$$\lambda = 4, \quad \mathbf{v} = \begin{pmatrix} -2 \\ 3 \end{pmatrix},$$

$$\lambda = -1, \quad \mathbf{v} = \begin{pmatrix} 1 \\ 1 \end{pmatrix}.$$

3.3.3   Rotations of Conics

EIGENVALUE PROBLEMS SHOW UP IN APPLICATIONS other than the solution of differential equations. We will see applications of this later in the text. For now, we are content to deal with problems that can be cast into matrix form. One example is the transformation of a simple system through rotation into a more complicated appearing system simply due to the choice of coordinate system. In this section we will explore this through the study of the rotation of conics.

You may have seen the general form for the equation of a conic in Cartesian coordinates in your calculus class. It is given by

$$Ax^2 + 2Bxy + Cy^2 + Ex + Fy = D. \tag{3.84}$$

This equation describes a variety of conics (ellipses, hyperbolae and parabolae), depending on the constants. The E and F terms result from a translation of the origin, and the B term is the result of a rotation of the coordinate system. We leave it to the reader to show that coordinate translations can be made to eliminate the linear terms. So, we will set $E = F = 0$ in our discussion and only consider quadratic equations of the form

$$Ax^2 + 2Bxy + Cy^2 = D.$$

If $B = 0$, then the resulting equation could be an equation for the standard ellipse or hyperbola with center at the origin. In the case of an ellipse, the semimajor and semiminor axes lie along the coordinate axes. However, rotating the ellipse introduces a nonzero B term, as we will see.

This conic equation can be written in matrix form. We note that the product

$$\begin{pmatrix} x & y \end{pmatrix} \begin{pmatrix} A & B \\ B & C \end{pmatrix} \begin{pmatrix} x \\ y \end{pmatrix} = Ax^2 + 2Bxy + Cy^2$$

gives the quadratic form in the conic equation. This equation can be written compactly as

$$\mathbf{x}^T Q \mathbf{x} = D, \tag{3.85}$$

where Q is the matrix of coefficients A, B, and C, and $\mathbf{x}$ is the column vector with components x and y.

We want to determine the transformation that puts this conic into a coordinate system in which there is no B term. Our goal is to obtain an equation of the form

$$A'x'^2 + C'y'^2 = D'$$

in the new coordinates $\mathbf{y}^T = (x', y')$. The matrix form of this equation is given as

$$\mathbf{y}^T \Lambda \mathbf{y} = D', \quad \Lambda = \begin{pmatrix} A' & 0 \\ 0 & C' \end{pmatrix}. \tag{3.86}$$

Let

$$\mathbf{x} = R\mathbf{y},$$

where R is a rotation matrix. Inserting this transformation into Equation (3.85), we find that

$$\begin{aligned} D &= \mathbf{x}^T Q \mathbf{x} \\ &= (R\mathbf{y})^T Q R\mathbf{y} \\ &= \mathbf{y}^T (R^T Q R)\mathbf{y}. \end{aligned} \tag{3.87}$$

Comparing this result to Equation (3.86), we have

$$\Lambda = R^T Q R, \quad D' = D. \tag{3.88}$$

Recalling that the rotation matrix is an orthogonal matrix, $R^T = R^{-1}$, we have that Λ is related to D by a similarity transformation,

$$\Lambda = R^{-1} Q R. \tag{3.89}$$

It is easy to see how the terms $Ex + Fy$ correspond to translations of conics. Consider the simple example $x^2 + y^2 + 2x - 6y = 0$. By completing the squares in both x and y, this equation can be written as $(x+1)^2 + (y-3)^2 = 10$. Now you can recognize that this is a circle whose center has been translated from the origin to $(-1, 3)$.

Thus, the problem reduces to that of diagonalization of Q. Because Q is a real and symmetric matrix, Q is diagonalizable and the eigenvalues are real. The eigenvalues of Q will become the constants in the rotated equation and the eigenvectors, as we will see, will give the directions of the principal axes (the semimajor and semiminor axes of ellipses). We will show this in an example.

Example 3.15. *Determine the principal axes of the ellipse given by*

$$13x^2 - 10xy + 13y^2 - 72 = 0.$$

From Figure 3.10 we see that the graph of this equation is an ellipse. However, we might not know this without plotting it. (Actually, we will see that there are conditions on the coefficients that do allow us to determine the conic.) If the equation were in standard form, we could identify its general shape. So, we will use the method outlined above to find the coordinate system in which the ellipse appears in standard form.

The coefficient matrix for this equation is given by

$$Q = \begin{pmatrix} 13 & -5 \\ -5 & 13 \end{pmatrix}. \tag{3.90}$$

We seek a solution to the eigenvalue problem: $Q\mathbf{v} = \lambda\mathbf{v}$. Recall that the first step is to get the eigenvalue equation from $\det(Q - \lambda I) = 0$. For this problem we have

$$\begin{vmatrix} 13 - \lambda & -5 \\ -5 & 13 - \lambda \end{vmatrix} = 0, \tag{3.91}$$

or

$$(13 - \lambda)^2 - 25 = 0.$$

Thus, the equation in the new system is

$$8x'^2 + 18y'^2 = 72.$$

Dividing out the 72 puts this into the standard form

$$\frac{x'^2}{9} + \frac{y'^2}{4} = 1.$$

Now we can identify the ellipse in the new system. We show the two ellipses in Figure 3.11. We note that the given ellipse is the new one rotated by some angle, which we still need to determine. We can find this angle if we know the eigenvectors corresponding to each eigenvalue.

Eigenvalue 1: $\lambda = 8$

We insert this eigenvalue into the equation $(Q - \lambda I)\mathbf{v} = 0$. The system for the unknown eigenvector is

$$\begin{pmatrix} 13 - 8 & -5 \\ -5 & 13 - 8 \end{pmatrix} \begin{pmatrix} v_1 \\ v_2 \end{pmatrix} = 0. \tag{3.92}$$

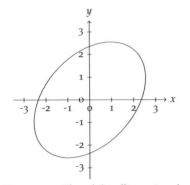

Figure 3.10: Plot of the ellipse given by $13x^2 - 10xy + 13y^2 - 72 = 0$.

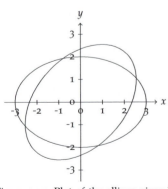

Figure 3.11: Plot of the ellipse given by $13x^2 - 10xy + 13y^2 - 72 = 0$ and the ellipse $\frac{x'^2}{9} + \frac{y'^2}{4} = 1$ showing that the first ellipse is a rotated version of the second ellipse.

The first equation is

$$5v_1 - 5v_2 = 0, \tag{3.93}$$

or $v_1 = v_2$. Thus, we can choose our eigenvector to be

$$\begin{pmatrix} v_1 \\ v_2 \end{pmatrix} = \begin{pmatrix} 1 \\ 1 \end{pmatrix}.$$

Eigenvalue 2: $\lambda = 18$

In the same way, we insert the eigenvalue into the equation $(Q - \lambda I)\mathbf{v} = 0$ and obtain

$$\begin{pmatrix} 13 - 18 & -5 \\ -5 & 13 - 18 \end{pmatrix} \begin{pmatrix} v_1 \\ v_2 \end{pmatrix} = 0. \tag{3.94}$$

The first equation is

$$-5v_1 - 5v_2 = 0, \tag{3.95}$$

or $v_1 = -v_2$. Thus, we can choose our eigenvector to be

$$\begin{pmatrix} v_1 \\ v_2 \end{pmatrix} = \begin{pmatrix} -1 \\ 1 \end{pmatrix}.$$

In Figure 3.12 we superimpose the eigenvectors on the original ellipse. We see that the eigenvectors point in directions along the semimajor and semiminor axes and indicate the angle of rotation. Eigenvector one is at a 45° angle. Thus, the ellipse is a rotated version of one in standard position. Or, we could define new axes that are at 45° to the standard axes and then the ellipse would take the standard form in the new coordinate system.

A general rotation of any conic can be performed. Consider the equation

$$Ax^2 + 2Bxy + Cy^2 = D. \tag{3.96}$$

We would like to find a rotation that puts it in the form

$$\lambda_1 x'^2 + \lambda_2 y'^2 = D. \tag{3.97}$$

We use the rotation matrix

$$\hat{R}_\theta = \begin{pmatrix} \cos\theta & \sin\theta \\ -\sin\theta & \cos\theta \end{pmatrix},$$

which gives a passive rotation of coordinates, $\mathbf{x}' = \hat{R}_\theta \mathbf{x}$. Inverting the transformation, we have $\mathbf{x} = \hat{R}_\theta^{-1}\mathbf{x}' = R_\theta \mathbf{x}'$.

The general equation can be written in matrix form:

$$\mathbf{x}^T Q \mathbf{x} = D, \tag{3.98}$$

where Q is the usual matrix of coefficients A, B, and C. Transforming this equation gives

$$\mathbf{x}'^T R_\theta^{-1} Q R_\theta \mathbf{x}' = D. \tag{3.99}$$

The resulting equation is of the form

$$A'x'^2 + 2B'x'y' + C'y'^2 = D, \tag{3.100}$$

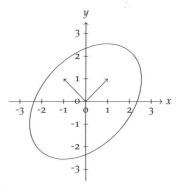

Figure 3.12: Plot of the ellipse given by $13x^2 - 10xy + 13y^2 - 72 = 0$ and the eigenvectors. Note that they are along the semimajor and semiminor axes and indicate the angle of rotation.

where

$$A' = A\cos^2\theta + 2B\sin\theta\cos\theta + C\sin^2\theta, \tag{3.101}$$
$$B' = (C - A)\sin\theta\cos\theta + B(2\cos\theta^2 - 1), \tag{3.102}$$
$$C' = A\sin^2\theta - 2B\sin\theta\cos\theta + C\cos^2\theta. \tag{3.103}$$

(We only need B' for this discussion).

If we want the nonrotated form, then we seek an angle θ such that $B' = 0$. Noting that $2\sin\theta\cos\theta = \sin 2\theta$ and $2\cos\theta^2 - 1 = \cos 2\theta$, this gives

$$\tan(2\theta) = \frac{A - C}{B}. \tag{3.104}$$

Example 3.16. *So, in the previous example, with $A = C = 13$ and $B = -5$, we have $\tan(2\theta) = \infty$. Thus, $2\theta = \pi/2$, or $\theta = \pi/4$.*

Using $\theta = \pi/4$, we find that $A' = 8$, $B' = 0$, and $C' = 18$. This is consistent with the answers we obtained by diagonalizing the coefficient matrix.

Finally, we had noted that knowing the coefficients in the general quadratic is enough to determine the type of conic represented without doing any plotting. This is based on the fact that the determinant of the coefficient matrix is invariant under rotation. We see this from the equation for diagonalization

$$\begin{aligned}
\det(\Lambda) &= \det(R_\theta^{-1} Q R_\theta) \\
&= \det(R_\theta^{-1})\det(Q)\det(R_\theta) \\
&= \det(R_\theta^{-1} R_\theta)\det(Q) \\
&= \det(Q).
\end{aligned} \tag{3.105}$$

Therefore, we have

$$\lambda_1 \lambda_2 = AC - B^2.$$

Looking at Equation (3.97), we have three cases:

1. Ellipse $\lambda_1\lambda_2 > 0$ or $B^2 - AC < 0$.

2. Hyperbola $\lambda_1\lambda_2 < 0$ or $B^2 - AC > 0$.

3. Parabola $\lambda_1\lambda_2 = 0$ or $B^2 - AC = 0$. and one eigenvalue is nonzero. Otherwise, the equation degenerates to a linear equation.

Example 3.17. *For the previous example, $13x^2 - 10xy + 13y^2 - 72 = 0$. In this case, $B^2 - AC = (-5)^2 - 13 \cdot 13 < 0$. Therefore, it is the equation of an ellipse.*

Example 3.18. *Consider the hyperbola $xy = 6$. We can see that this is a rotated hyperbola by plotting $y = 6/x$. Also, we note that $B^2 - AC = (0.5)^2 - 0 > 0$ indicates that it is a hyperbola. A plot is shown in Figure 3.13. Determine the rotation angle needed to transform the hyperbola to new coordinates so that its equation will be in standard form.*

The coefficient matrix for this equation is given by

$$A = \begin{pmatrix} 0 & -0.5 \\ 0.5 & 0 \end{pmatrix}. \tag{3.106}$$

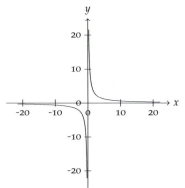

Figure 3.13: Plot of the hyperbola given by $xy = 6$.

The eigenvalue equation is

$$\begin{vmatrix} -\lambda & -0.5 \\ -0.5 & -\lambda \end{vmatrix} = 0. \qquad (3.107)$$

Thus,

$$\lambda^2 - 0.25 = 0,$$

or $\lambda = \pm 0.5$.

Once again, $\tan(2\theta) = \infty$, *so the new system is at* 45^0 *to the old. The equation in new coordinates is* $0.5x^2 + (-0.5)y^2 = 6$, *or* $x^2 - y^2 = 12$. *A plot is shown in Figure 3.14.*

3.4   *Matrix Formulation of Planar Systems*

IN THE PREVIOUS CHAPTER WE INVESTIGATED LINEAR SYSTEMS of differential equations in the plane. In this section we will recast first-order linear systems in matrix form. This will lead to a better understanding of first-order systems and allow for extensions to higher dimensions. In particular, we will see that the solutions obtained for planar systems in later chapters are intimately connected to the underlying eigenvalue problems.

We start with the usual homogeneous system in Equation (2.195),

$$\begin{aligned} x' &= ax + by \\ y' &= cx + dy. \end{aligned} \qquad (3.108)$$

We can write this system in matrix form. Let the unknowns be represented by the vector

$$\mathbf{x}(t) = \begin{pmatrix} x(t) \\ y(t) \end{pmatrix}.$$

Differentiating $\mathbf{x}$ and putting the result in matrix form, we have

$$\mathbf{x}' = \begin{pmatrix} x' \\ y' \end{pmatrix} = \begin{pmatrix} ax + by \\ cx + dy \end{pmatrix} = \begin{pmatrix} a & b \\ c & d \end{pmatrix} \begin{pmatrix} x \\ y \end{pmatrix} \equiv A\mathbf{x}.$$

Here we have introduced the coefficient matrix A.

This is a first-order vector differential equation,

$$\mathbf{x}' = A\mathbf{x}.$$

Formally, we can write the solution as

$$\mathbf{x} = \mathbf{x}_0 e^{At}.$$

Differentiating this solution with respect to t, one can verify that this is a solution. However, we need to know what it means to exponentiate a matrix in order for this simple solution to be useful.[5]

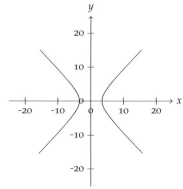

Figure 3.14: Plot of the rotated hyperbola given by $x^2 - y^2 = 12$.

[5] The exponential of a matrix is defined using the Maclaurin series expansion of the exponential,

$$e^A = \sum_{k=0}^{\infty} = I + A + \frac{A^2}{2!} + \frac{A^3}{3!} + \cdots.$$

$$(3.109)$$

In order to find a constructive approach for solving a system using matrix methods, we turn to the specific problem in (2.202),

$$x' = -x + 6y,$$

$$y' = x - 2y.$$

We obtained the solution to this system as

$$x(t) = c_1 e^t + c_2 e^{-4t},$$

$$y(t) = \frac{1}{3} c_1 e^t - \frac{1}{2} c_2 e^{-4t}. \tag{3.110}$$

This solution can be rewritten using matrix operations. Namely, we first write the solution in vector form:

$$
\begin{aligned}
\mathbf{x} &= \begin{pmatrix} x(t) \\ y(t) \end{pmatrix} \\
&= \begin{pmatrix} c_1 e^t + c_2 e^{-4t} \\ \frac{1}{3} c_1 e^t - \frac{1}{2} c_2 e^{-4t} \end{pmatrix} \\
&= \begin{pmatrix} c_1 e^t \\ \frac{1}{3} c_1 e^t \end{pmatrix} + \begin{pmatrix} c_2 e^{-4t} \\ -\frac{1}{2} c_2 e^{-4t} \end{pmatrix} \\
&= c_1 \begin{pmatrix} 1 \\ \frac{1}{3} \end{pmatrix} e^t + c_2 \begin{pmatrix} 1 \\ -\frac{1}{2} \end{pmatrix} e^{-4t}. \tag{3.111}
\end{aligned}
$$

We see that the solution is a linear combination of vectors of the form

$$\mathbf{x} = \mathbf{v} e^{\lambda t}$$

with $\mathbf{v}$ a constant vector and λ a constant number. Similar to how we began to find solutions to second-order constant coefficient differential equations, we guess a solution, $\mathbf{x} = \mathbf{v} e^{\lambda t}$, and insert the guess into the differential equation, $\mathbf{x}' = A\mathbf{x}$. This gives

$$\lambda \mathbf{v} e^{\lambda t} = A \mathbf{v} e^{\lambda t}.$$

For this equation to be true for all t, we find that

$$A\mathbf{v} = \lambda \mathbf{v}. \tag{3.112}$$

What we have shown is that solutions to the system take the form $\mathbf{v} e^{\lambda t}$, where λ is an eigenvalue of matrix A and $\mathbf{v}$ is the associated eigenvector. Thus, we need to be able to solve eigenvalue problems before constructing solutions. Even though for the example A is a 2×2 matrix, we can easily generalize these observations to systems of n first-order differential equations. We will confine our discussion for now to planar systems.

3.4.1   Solving Constant Coefficient Systems in 2D

BEFORE PROCEEDING TO EXAMPLES, we first indicate the types of solutions that could result from the solution of a homogeneous, constant coefficient system of first-order differential equations.

We begin with the linear system of differential equations in matrix form:

$$\frac{d\mathbf{x}}{dt} = \begin{pmatrix} a & b \\ c & d \end{pmatrix} \mathbf{x} = A\mathbf{x}. \qquad (3.113)$$

The type of behavior depends upon the eigenvalues of matrix A. The procedure is to determine the eigenvalues and eigenvectors and use them to construct the general solution.

If we have an initial condition, $\mathbf{x}(t_0) = \mathbf{x}_0$, we can determine the two arbitrary constants in the general solution in order to obtain the particular solution. Thus, if $\mathbf{x}_1(t)$ and $\mathbf{x}_2(t)$ are two linearly independent solutions, then the general solution is given as

$$\begin{aligned} \mathbf{x}(t) &= c_1 \mathbf{x}_1(t) + c_2 \mathbf{x}_2(t) \\ &= c_1 \mathbf{v}_1 e^{\lambda_1 t} + c_2 \mathbf{v}_2 e^{\lambda_2 t}. \end{aligned} \qquad (3.114)$$

Setting $t = 0$, we get two linear equations for c_1 and c_2:

$$\begin{aligned} \mathbf{x}_0 &= c_1 \mathbf{x}_1(0) + c_2 \mathbf{x}_2(0) \\ \begin{pmatrix} x_0 \\ y_0 \end{pmatrix} &= c_1 \mathbf{v}_1 + c_2 \mathbf{v}_2. \end{aligned} \qquad (3.115)$$

The major work is in finding the linearly independent solutions. This depends upon the different types of eigenvalues that one obtains from solving the eigenvalue equation, $\det(A - \lambda I) = 0$. The nature of these roots indicate the form of the general solution. For two-dimensional systems we have to solve a 2×2 eigenvalue problem. The eigenvalue equation will be a quadratic equation and there are three cases that will need to be investigated:

Case I. The eigenvalues are real and distinct.

Case II. There is one real eigenvalue.

Case III. The eigenvalues are complex conjugates.

This parallels the classification of second-order constant coefficient equations in the previous chapter. A summary of the results is given at the end of this section and examples are worked out in the next section.

The construction of the general solution in Case I is straightforward. We find the eigenvectors for each real eigenvalue and form the solution as

$$\mathbf{x}(t) = c_1 e^{\lambda_1 t} \mathbf{v}_1 + c_2 e^{\lambda_2 t} \mathbf{v}_2.$$

However, the other two cases need a little more explanation.

For Case II, there is only one eigenvalue. However, if we recall the second case for constant coefficient equations in the previous chapter, the general solution took the form

$$x(t) = (c_1 + c_2 t) e^{\lambda t}.$$

This suggests that the general solution in the matrix formulation should involve a term of the form $\mathbf{v}te^{\lambda t}$.

Let's guess that the solution is given by

$$\mathbf{x} = te^{\lambda t}\mathbf{v}_1 + e^{\lambda t}\mathbf{v}_2.$$

Inserting this guess into the system $\mathbf{x}' = A\mathbf{x}$ yields

$$(te^{\lambda t}\mathbf{v}_1 + e^{\lambda t}\mathbf{v}_2)' = A\left[te^{\lambda t}\mathbf{v}_1 + e^{\lambda t}\mathbf{v}_2\right].$$

Expanding both sides of the equation,

$$e^{\lambda t}\mathbf{v}_1 + \lambda te^{\lambda t}\mathbf{v}_1 + \lambda e^{\lambda t}\mathbf{v}_2 = \lambda te^{\lambda t}\mathbf{v}_1 + e^{\lambda t}A\mathbf{v}_2,$$

and simplifying, we find

$$e^{\lambda t}\left(\mathbf{v}_1 + \lambda\mathbf{v}_2\right) = e^{\lambda t}A\mathbf{v}_2.$$

Noting this should be true for all t, then we arrive at a modification of the eigenvalue problem,

$$\mathbf{v}_1 + \lambda\mathbf{v}_2 = A\mathbf{v}_2, \tag{3.116}$$

or

$$(A - \lambda I)\mathbf{v}_2 = \mathbf{v}_1.$$

We know everything in the equation except for $\mathbf{v}_2$. So, we just solve Equation (3.116) for $\mathbf{v}_2$ and then form the second linearly independent solution

$$\mathbf{x} = te^{\lambda t}\mathbf{v}_1 + e^{\lambda t}\mathbf{v}_2.$$

Let's turn to Case III. Note that because the original system of equations does not have any complex coefficients, then we would expect real solutions. So, we look at the real and imaginary parts of the complex solution, $\mathbf{y}(t) = \text{Re}(\mathbf{y}(t)) + i\,\text{Im}(\mathbf{y}(t))$. We have that this complex solution satisfies the equation

$$\frac{d}{dt}\left[\text{Re}(\mathbf{y}(t)) + i\,\text{Im}(\mathbf{y}(t))\right] = A[\text{Re}(\mathbf{y}(t)) + i\,\text{Im}(\mathbf{y}(t))].$$

Differentiating the sum and splitting the real and imaginary parts of the equation gives

$$\frac{d}{dt}\text{Re}(\mathbf{y}(t)) + i\frac{d}{dt}\text{Im}(\mathbf{y}(t)) = A[\text{Re}(\mathbf{y}(t))] + iA[\text{Im}(\mathbf{y}(t))].$$

Setting the real and imaginary parts equal, we have

$$\frac{d}{dt}\text{Re}(\mathbf{y}(t)) = A[\text{Re}(\mathbf{y}(t))]$$

and

$$\frac{d}{dt}\text{Im}(\mathbf{y}(t)) = A[\text{Im}(\mathbf{y}(t))].$$

Therefore, the real and imaginary parts of the solution are each solutions of the system. So, the general solution can be written as a linear combination of these expressions,

$$\mathbf{x}(t) = c_1 \operatorname{Re}(\mathbf{y}(t)) + c_2 \operatorname{Im}(\mathbf{y}(t)),$$

where

$$\mathbf{y}(t) = e^{\alpha t}(\cos \beta t + i \sin \beta t)\mathbf{v}.$$

Classification of the Solutions for Two Linear First-Order Differential Equations

Case I: Two real, distinct roots.

Solve the eigenvalue problem $A\mathbf{v} = \lambda\mathbf{v}$ for each eigenvalue obtaining two eigenvectors $\mathbf{v}_1, \mathbf{v}_2$. Then write the general solution as a linear combination $\mathbf{x}(t) = c_1 e^{\lambda_1 t}\mathbf{v}_1 + c_2 e^{\lambda_2 t}\mathbf{v}_2$.

Case II: One repeated root.

Solve the eigenvalue problem $A\mathbf{v} = \lambda\mathbf{v}$ for one eigenvalue λ, obtaining the first eigenvector $\mathbf{v}_1$. One then needs a second linearly independent solution. This is obtained by solving the nonhomogeneous problem $A\mathbf{v}_2 - \lambda\mathbf{v}_2 = \mathbf{v}_1$ for $\mathbf{v}_2$.

The general solution is then given by $\mathbf{x}(t) = c_1 e^{\lambda t}\mathbf{v}_1 + c_2 e^{\lambda t}(\mathbf{v}_2 + t\mathbf{v}_1)$.

Case III: Two complex conjugate roots.

Solve the eigenvalue problem $A\mathbf{x} = \lambda\mathbf{x}$ for one eigenvalue, $\lambda = \alpha + i\beta$, obtaining one eigenvector $\mathbf{v}$. Note that this eigenvector may have complex entries. Thus, one can write the vector

$$\mathbf{y}(t) = e^{\lambda t}\mathbf{v} = e^{\alpha t}(\cos \beta t + i \sin \beta t)\mathbf{v}.$$

Now construct two linearly independent solutions to the problem using the real and imaginary parts of $\mathbf{y}(t)$:

$$\mathbf{y}_1(t) = \operatorname{Re}(\mathbf{y}(t)) \text{ and } \mathbf{y}_2(t) = \operatorname{Im}(\mathbf{y}(t)).$$

Then the general solution can be written as $\mathbf{x}(t) = c_1\mathbf{y}_1(t) + c_2\mathbf{y}_2(t)$.

3.4.2   *Examples of the Matrix Method*

WE WILL GIVE SEVERAL EXAMPLES OF TYPICAL SYSTEMS of differential equations for the three cases mentioned in the previous section. In each example, we specify the coefficient matrix A for the system of equations

$$\mathbf{x}' = A\mathbf{x}.$$

Then we work through what is needed to get the general solution. There are three basic steps in the process:

1. Find the eigenvalues.

2. For each eigenvalue, find the eigenvector.

3. Construct the general solution using the eigenvalues and eigenvectors.

Example 3.19. *Solve*

$$
\begin{aligned}
x' &= 4x + 2y, \\
y' &= 3x + 3y.
\end{aligned}
$$
(3.117)

The coefficient matrix is $A = \begin{pmatrix} 4 & 2 \\ 3 & 3 \end{pmatrix}$.

Eigenvalues: We first determine the eigenvalues.

$$
0 = \begin{vmatrix} 4 - \lambda & 2 \\ 3 & 3 - \lambda \end{vmatrix}.
$$
(3.118)

Therefore,

$$
\begin{aligned}
0 &= (4 - \lambda)(3 - \lambda) - 6 \\
0 &= \lambda^2 - 7\lambda + 6 \\
0 &= (\lambda - 1)(\lambda - 6)
\end{aligned}
$$
(3.119)

The eigenvalues are found as $\lambda = 1, 6$. *This is an example of Case I.*

Eigenvectors: Next we determine the eigenvectors associated with each of these eigenvalues. We have to solve the system $A\mathbf{v} = \lambda\mathbf{v}$ *in each case.*

Case $\lambda = 1$.

The eigenvalue problem in this case is

$$
\begin{pmatrix} 4 & 2 \\ 3 & 3 \end{pmatrix} \begin{pmatrix} v_1 \\ v_2 \end{pmatrix} = \begin{pmatrix} v_1 \\ v_2 \end{pmatrix}.
$$

Rewriting, we have

$$
\begin{pmatrix} 3 & 2 \\ 3 & 2 \end{pmatrix} \begin{pmatrix} v_1 \\ v_2 \end{pmatrix} = \begin{pmatrix} 0 \\ 0 \end{pmatrix}.
$$

Both equations give $3v_1 + 2v_2 = 0$. *One possible solution of this equation yields an eigenvector of*

$$
\begin{pmatrix} v_1 \\ v_2 \end{pmatrix} = \begin{pmatrix} 2 \\ -3 \end{pmatrix}.
$$

Case $\lambda = 6$.

The eigenvalue problem is

$$
\begin{pmatrix} 4 & 2 \\ 3 & 3 \end{pmatrix} \begin{pmatrix} v_1 \\ v_2 \end{pmatrix} = 6 \begin{pmatrix} v_1 \\ v_2 \end{pmatrix}.
$$

Therefore,

$$\begin{pmatrix} -2 & 2 \\ 3 & -3 \end{pmatrix}\begin{pmatrix} v_1 \\ v_2 \end{pmatrix} = \begin{pmatrix} 0 \\ 0 \end{pmatrix}.$$

For this case we need to solve $-2v_1 + 2v_2 = 0.$ *This yields*

$$\begin{pmatrix} v_1 \\ v_2 \end{pmatrix} = \begin{pmatrix} 1 \\ 1 \end{pmatrix}.$$

General Solution: *We can now construct the general solution.*

$$\begin{aligned} \mathbf{x}(t) &= c_1 e^{\lambda_1 t}\mathbf{v}_1 + c_2 e^{\lambda_2 t}\mathbf{v}_2 \\ &= c_1 e^t \begin{pmatrix} 2 \\ -3 \end{pmatrix} + c_2 e^{6t}\begin{pmatrix} 1 \\ 1 \end{pmatrix} \\ &= \begin{pmatrix} 2c_1 e^t + c_2 e^{6t} \\ -3c_1 e^t + c_2 e^{6t} \end{pmatrix}. \end{aligned} \tag{3.120}$$

The individual solutions for $x(t)$ *and* $y(t)$ *are then*

$$x(t) = 2c_1 e^t + c_2 e^{6t}, \quad y(t) = -3c_1 e^t + c_2 e^{6t}.$$

Example 3.20. *Solve*

$$\begin{aligned} x' &= 3x - 5y, \\ y' &= x - y. \end{aligned} \tag{3.121}$$

The coefficient matrix is $A = \begin{pmatrix} 3 & -5 \\ 1 & -1 \end{pmatrix}.$

    Eigenvalues: *Again, one solves the eigenvalue equation.*

$$0 = \begin{vmatrix} 3-\lambda & -5 \\ 1 & -1-\lambda \end{vmatrix}. \tag{3.122}$$

Therefore,

$$\begin{aligned} 0 &= (3-\lambda)(-1-\lambda) + 5, \\ 0 &= \lambda^2 - 2\lambda + 2, \\ \lambda &= 1 \pm i. \end{aligned} \tag{3.123}$$

    The eigenvalues are then $\lambda = 1+i, 1-i.$ *This is an example of Case III.*

    Eigenvectors: *In order to find the general solution, we need only find the eigenvector associated with* $1+i.$ *As before, we write the eigenvalue problem and then form the homogeneous system that we have to solve.*

$$\begin{pmatrix} 3 & -5 \\ 1 & -1 \end{pmatrix}\begin{pmatrix} v_1 \\ v_2 \end{pmatrix} = (1+i)\begin{pmatrix} v_1 \\ v_2 \end{pmatrix},$$

$$\begin{pmatrix} 2-i & -5 \\ 1 & -2-i \end{pmatrix}\begin{pmatrix} v_1 \\ v_2 \end{pmatrix} = \begin{pmatrix} 0 \\ 0 \end{pmatrix}.$$

We need to solve $(2 - i)v_1 - 5v_2 = 0$. Thus,

$$\begin{pmatrix} v_1 \\ v_2 \end{pmatrix} = \begin{pmatrix} 2 + i \\ 1 \end{pmatrix}. \tag{3.124}$$

Complex Solution: *In order to get the two real linearly independent solutions, we need to compute the real and imaginary parts of* $\mathbf{v}e^{\lambda t}$. *We do this using Euler's formula, multiply the terms, and then collect the real and imaginary parts.*

$$\begin{aligned}
e^{\lambda t} \begin{pmatrix} 2 + i \\ 1 \end{pmatrix} &= e^{(1+i)t} \begin{pmatrix} 2 + i \\ 1 \end{pmatrix} \\
&= e^t (\cos t + i \sin t) \begin{pmatrix} 2 + i \\ 1 \end{pmatrix} \\
&= e^t \begin{pmatrix} (2 + i)(\cos t + i \sin t) \\ \cos t + i \sin t \end{pmatrix} \\
&= e^t \begin{pmatrix} (2\cos t - \sin t) + i(\cos t + 2\sin t) \\ \cos t + i \sin t \end{pmatrix} \\
&= e^t \begin{pmatrix} 2\cos t - \sin t \\ \cos t \end{pmatrix} + i e^t \begin{pmatrix} \cos t + 2\sin t \\ \sin t \end{pmatrix}.
\end{aligned}$$

General Solution: *Now we can construct the general solution as a linear combination of the real and imaginary parts:*

$$\begin{aligned}
\mathbf{x}(t) &= c_1 e^t \begin{pmatrix} 2\cos t - \sin t \\ \cos t \end{pmatrix} + c_2 e^t \begin{pmatrix} \cos t + 2\sin t \\ \sin t \end{pmatrix} \\
&= e^t \begin{pmatrix} c_1(2\cos t - \sin t) + c_2(\cos t + 2\sin t) \\ c_1 \cos t + c_2 \sin t \end{pmatrix}. \tag{3.125}
\end{aligned}$$

Note: This can be rewritten as

$$\mathbf{x}(t) = e^t \cos t \begin{pmatrix} 2c_1 + c_2 \\ c_1 \end{pmatrix} + e^t \sin t \begin{pmatrix} 2c_2 - c_1 \\ c_2 \end{pmatrix}.$$

Therefore, we have found that

$$x(t) = [(2c_1 + c_2)\cos t + (2c_2 - c_1)\sin t] \, e^t,$$

$$y(t) = [c_1 \cos t + c_2 \sin t] \, e^t.$$

Example 3.21. *Solve*

$$\begin{aligned}
x' &= 7x - y, \\
y' &= 9x + y. \tag{3.126}
\end{aligned}$$

The coefficient matrix is $A = \begin{pmatrix} 7 & -1 \\ 9 & 1 \end{pmatrix}$.

Eigenvalues: *The eigenvalue equation is given by*

$$0 = \begin{vmatrix} 7 - \lambda & -1 \\ 9 & 1 - \lambda \end{vmatrix}. \tag{3.127}$$

Therefore,

$$
\begin{aligned}
0 &= (7 - \lambda)(1 - \lambda) + 9 \\
0 &= \lambda^2 - 8\lambda + 16 \\
0 &= (\lambda - 4)^2. \tag{3.128}
\end{aligned}
$$

There is only one real eigenvalue, $\lambda = 4$. This is an example of Case II.

Eigenvectors: *In this case, we first solve for $\mathbf{v}_1$ and then get the second linearly independent vector.*

$$
\begin{aligned}
\begin{pmatrix} 7 & -1 \\ 9 & 1 \end{pmatrix} \begin{pmatrix} v_1 \\ v_2 \end{pmatrix} &= 4 \begin{pmatrix} v_1 \\ v_2 \end{pmatrix}, \\
\begin{pmatrix} 3 & -1 \\ 9 & -3 \end{pmatrix} \begin{pmatrix} v_1 \\ v_2 \end{pmatrix} &= \begin{pmatrix} 0 \\ 0 \end{pmatrix}. \tag{3.129}
\end{aligned}
$$

Therefore, we have $3v_1 - v_2 = 0$, which has a solution

$$\begin{pmatrix} v_1 \\ v_2 \end{pmatrix} = \begin{pmatrix} 1 \\ 3 \end{pmatrix}.$$

Second Linearly Independent Solution:

Now we need the second solution. So, we solve $A\mathbf{v}_2 - \lambda\mathbf{v}_2 = \mathbf{v}_1$. Setting up the problem, we have to solve the equation

$$
\begin{aligned}
\begin{pmatrix} 7 & -1 \\ 9 & 1 \end{pmatrix} \begin{pmatrix} u_1 \\ u_2 \end{pmatrix} - 4 \begin{pmatrix} u_1 \\ u_2 \end{pmatrix} &= \begin{pmatrix} 1 \\ 3 \end{pmatrix}, \\
\begin{pmatrix} 3 & -1 \\ 9 & -3 \end{pmatrix} \begin{pmatrix} u_1 \\ u_2 \end{pmatrix} &= \begin{pmatrix} 1 \\ 3 \end{pmatrix}. \tag{3.130}
\end{aligned}
$$

Expanding the matrix product, we obtain the system of equations

$$
\begin{aligned}
3u_1 - u_2 &= 1, \\
9u_1 - 3u_2 &= 3. \tag{3.131}
\end{aligned}
$$

A solution of this system is $\begin{pmatrix} u_1 \\ u_2 \end{pmatrix} = \begin{pmatrix} 1 \\ 2 \end{pmatrix}.$

General Solution: *We now construct the general solution by combining the two solutions, $\mathbf{v}_1$ and $\mathbf{v}_2$,*

$$
\begin{aligned}
\mathbf{y}(t) &= c_1 e^{\lambda t} \mathbf{v}_1 + c_2 e^{\lambda t} (\mathbf{v}_2 + t\mathbf{v}_1) \\
&= c_1 e^{4t} \begin{pmatrix} 1 \\ 3 \end{pmatrix} + c_2 e^{4t} \left[\begin{pmatrix} 1 \\ 2 \end{pmatrix} + t \begin{pmatrix} 1 \\ 3 \end{pmatrix} \right] \\
&= e^{4t} \begin{pmatrix} c_1 + c_2(1 + t) \\ 3c_1 + c_2(2 + 3t) \end{pmatrix}. \tag{3.132}
\end{aligned}
$$

Thus, the solutions for $x(t)$ and $y(t)$ are given as

$$x(t) = (c_1 + c_2(1 + t))e^{4t}, \quad y(t) = (3c_1 + c_2(2 + 3t))e^{4t}.$$

3.4.3 Planar Systems—Summary

THE READER SHOULD HAVE NOTED BY NOW that there is a connection between the behavior of the solutions of planar systems obtained in Chapter 2 and the eigenvalues found from the coefficient matrices in the previous examples. Here we summarize some of these cases.

Table 3.1: List of typical behaviors in planar systems.

Type	Eigenvalues	Stability
Node	Real λ, same signs	$\lambda > 0$, stable
Saddle	Real λ opposite signs	Mostly unstable
Center	λ pure imaginary	—
Focus/spiral	Complex λ, $\text{Re}(\lambda) \neq 0$	$\text{Re}(\lambda > 0)$, stable
Degenerate node	Repeated roots	$\lambda > 0$, stable
Line of equilibria	One zero eigenvalue	$\lambda > 0$, stable

The connection, as we have seen, is that the characteristic equation for the associated second-order differential equation is the same as the eigenvalue equation of the coefficient matrix for the linear system. However, one should be a little careful in cases in which the coefficient matrix is not diagonalizable. In Table 3.2 are three examples of systems with repeated roots. The reader should look at these systems and look at the commonalities and differences in these systems and their solutions. In these cases, one has unstable nodes, although they are degenerate in that there is only one accessible eigenvector.

Table 3.2: Three examples of systems with a repeated root of $\lambda = 2$.

System 1	System 2	System 3
$\mathbf{x'} = \begin{pmatrix} 2 & 0 \\ 0 & 2 \end{pmatrix} \mathbf{x}$	$\mathbf{x'} = \begin{pmatrix} 0 & 1 \\ -4 & 4 \end{pmatrix} \mathbf{x}$	$\mathbf{x'} = \begin{pmatrix} 2 & 1 \\ 0 & 2 \end{pmatrix} \mathbf{x}$

3.5 Applications

IN THIS SECTION WE WILL DESCRIBE SOME SIMPLE APPLICATIONS leading to systems of differential equations that can be solved using the methods in this chapter. These systems are left for homework problems and then as the start of further explorations for student projects.

3.5.1 Mass-Spring Systems

THE FIRST EXAMPLES THAT WE HAD SEEN involved masses on springs. Recall that for a simple mass on a spring, we studied simple harmonic motion, which is governed by the equation

$$m\ddot{x} + kx = 0.$$

This second-order equation can be written as two first-order equations

$$\dot{x} = y$$
$$\dot{y} = -\frac{k}{m}x, \tag{3.133}$$

or

$$\dot{x} = y$$
$$\dot{y} = -\omega^2 x, \tag{3.134}$$

where $\omega^2 = \frac{k}{m}$. The coefficient matrix for this system is

$$A = \begin{pmatrix} 0 & 1 \\ -\omega^2 & 0 \end{pmatrix}.$$

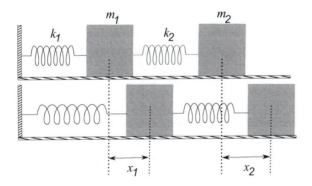

Figure 3.15: System of two masses and two springs.

We also looked at the system of two masses and two springs as shown in Figure 3.15. The equations governing the motion of the masses is

$$
\begin{aligned}
m_1\ddot{x}_1 &= -k_1 x_1 + k_2(x_2 - x_1), \\
m_2\ddot{x}_2 &= -k_2(x_2 - x_1).
\end{aligned}
\tag{3.135}
$$

We can rewrite this system as four first-order equations

$$
\begin{aligned}
\dot{x}_1 &= x_3, \\
\dot{x}_2 &= x_4, \\
\dot{x}_3 &= -\frac{k_1}{m_1}x_1 + \frac{k_2}{m_1}(x_2 - x_1), \\
\dot{x}_4 &= -\frac{k_2}{m_2}(x_2 - x_1).
\end{aligned}
\tag{3.136}
$$

The coefficient matrix for this system is

$$A = \begin{pmatrix} 0 & 0 & 1 & 0 \\ 0 & 0 & 0 & 1 \\ -\frac{k_1+k_2}{m_1} & \frac{k_2}{m_1} & 0 & 0 \\ \frac{k_2}{m_2} & -\frac{k_2}{m_2} & 0 & 0 \end{pmatrix}.$$

Writing the spring-block system as a second order-vector system.

We can study this system for specific values of the constants using the methods covered in the previous sections.

Actually, one can also put the system (3.135) in the matrix form

$$\begin{pmatrix} m_1 & 0 \\ 0 & m_2 \end{pmatrix} \begin{pmatrix} \ddot{x}_1 \\ \ddot{x}_2 \end{pmatrix} = \begin{pmatrix} -(k_1+k_2) & k_2 \\ k_2 & -k_2 \end{pmatrix} \begin{pmatrix} x_1 \\ x_2 \end{pmatrix}. \tag{3.137}$$

This system can then be written compactly as

$$M\ddot{\mathbf{x}} = -K\mathbf{x}, \tag{3.138}$$

where

$$M = \begin{pmatrix} m_1 & 0 \\ 0 & m_2 \end{pmatrix}, \quad K = \begin{pmatrix} k_1+k_2 & -k_2 \\ -k_2 & k_2 \end{pmatrix}.$$

This system can be solved by guessing a form for the solution. We could guess

$$\mathbf{x} = \mathbf{a}e^{i\omega t}$$

or

$$\mathbf{x} = \begin{pmatrix} a_1 \cos(\omega t - \delta_1) \\ a_2 \cos(\omega t - \delta_2) \end{pmatrix}$$

where δ_i are phase shifts determined from initial conditions.

Inserting $\mathbf{x} = \mathbf{a}e^{i\omega t}$ into the system gives

$$(K - \omega^2 M)\mathbf{a} = \mathbf{0}.$$

This is a homogeneous system. It is a generalized eigenvalue problem for eigenvalues ω^2 and eigenvectors $\mathbf{a}$. We solve this in a similar way to the standard matrix eigenvalue problems. The eigenvalue equation is found as

$$\det(K - \omega^2 M) = 0.$$

Once the eigenvalues are found, then one determines the eigenvectors and constructs the solution.

Example 3.22. *Let $m_1 = m_2 = m$ and $k_1 = k_2 = k$. Then we have to solve the system*

$$\omega^2 \begin{pmatrix} m & 0 \\ 0 & m \end{pmatrix} \begin{pmatrix} a_1 \\ a_2 \end{pmatrix} = \begin{pmatrix} 2k & -k \\ -k & k \end{pmatrix} \begin{pmatrix} a_1 \\ a_2 \end{pmatrix}.$$

The eigenvalue equation is given by

$$\begin{aligned} 0 &= \begin{vmatrix} 2k - m\omega^2 & -k \\ -k & k - m\omega^2 \end{vmatrix} \\ &= (2k - m\omega^2)(k - m\omega^2) - k^2 \\ &= m^2\omega^4 - 3km\omega^2 + k^2. \end{aligned} \tag{3.139}$$

Solving this quadratic equation for ω^2, we have

$$\omega^2 = \frac{3 \pm 1}{2} \frac{k}{m}.$$

For positive values of ω, one can show that

$$\omega = \frac{1}{2}\left(\pm 1 + \sqrt{5}\right)\sqrt{\frac{k}{m}}.$$

The eigenvectors can be found for each eigenvalue by solving the homogeneous system

$$\begin{pmatrix} 2k - m\omega^2 & -k \\ -k & k - m\omega^2 \end{pmatrix}\begin{pmatrix} a_1 \\ a_2 \end{pmatrix} = 0.$$

The eigenvectors are given by

$$\mathbf{a_1} = \begin{pmatrix} -\frac{\sqrt{5}+1}{2} \\ 1 \end{pmatrix}, \quad \mathbf{a_2} = \begin{pmatrix} \frac{\sqrt{5}-1}{2} \\ 1 \end{pmatrix}.$$

We are now ready to construct the real solutions to the problem. Similar to solving two first-order systems with complex roots, we take the real and imaginary parts and take a linear combination of the solutions. In this problem there are four terms, giving the solution in the form

$$\mathbf{x}(t) = c_1\mathbf{a_1}\cos\omega_1 t + c_2\mathbf{a_1}\sin\omega_1 t + c_3\mathbf{a_2}\cos\omega_2 t + c_4\mathbf{a_2}\sin\omega_2 t,$$

where the ω's are the eigenvalues and the $\mathbf{a}$'s are the corresponding eigenvectors. The constants are determined from the initial conditions, $\mathbf{x}(0) = \mathbf{x_0}$ and $\dot{\mathbf{x}}(0) = \mathbf{v_0}$.

3.5.2 Circuits

IN THE CHAPTER 2 WE INVESTIGATED SIMPLE SERIES LRC CIRCUITS. More complicated circuits are possible by looking at parallel connections, or other combinations, of resistors, capacitors and inductors. This results in several equations for each loop in the circuit, leading to larger systems of differential equations. An example of another circuit setup is shown in Figure 3.16. This is not a problem that can be covered in the first-year physics course.

There are two loops, indicated in Figure 3.17 as traversed clockwise. For each loop we need to apply Kirchoff's Loop Rule. There are three oriented currents, labeled I_i, $i = 1, 2, 3$. Corresponding to each current is a changing charge, q_i such that

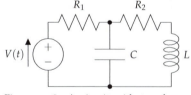

Figure 3.16: A circuit with two loops containing several different circuit elements.

$$I_i = \frac{dq_i}{dt}, \quad i = 1, 2, 3.$$

We have for loop 1

$$I_1 R_1 + \frac{q_2}{C} = V(t) \tag{3.140}$$

and for loop 2

$$I_3 R_2 + L\frac{dI_3}{dt} = \frac{q_2}{C}. \tag{3.141}$$

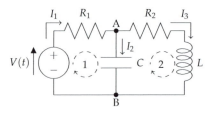

Figure 3.17: The previous parallel circuit with the directions indicated for traversing the loops in Kirchoff's Laws.

There are three unknown functions for the charge. Once we know the charge functions, differentiation will yield the three currents. However, we only have two equations. We need a third equation. This equation is found from Kirchoff's Point (Junction) Rule.

Consider the points A and B in Figure 3.17. Any charge (current) entering these junctions must be the same as the total charge (current) leaving the junctions. For point A we have

$$I_1 = I_2 + I_3, \tag{3.142}$$

or

$$\dot{q}_1 = \dot{q}_2 + \dot{q}_3. \tag{3.143}$$

Equations (3.140), (3.141), and (3.143) form a coupled system of differential equations for this problem. There are both first- and second-order derivatives involved. We can write the whole system in terms of charges as

$$R_1 \dot{q}_1 + \frac{q_2}{C} = V(t),$$
$$R_2 \dot{q}_3 + L \ddot{q}_3 = \frac{q_2}{C},$$
$$\dot{q}_1 = \dot{q}_2 + \dot{q}_3. \tag{3.144}$$

The question is whether, or not, we can write this as a system of first-order differential equations. Because there is only one second-order derivative, we can introduce the new variable $q_4 = \dot{q}_3$. The first equation can be solved for $\dot{q}_1$. The third equation can be solved for $\dot{q}_2$ with appropriate substitutions for the other terms. $\dot{q}_3$ is obtained from the definition of q_4 and the second equation can be solved for $\ddot{q}_3$ and substitutions made to obtain the system

$$\dot{q}_1 = \frac{V}{R_1} - \frac{q_2}{R_1 C},$$
$$\dot{q}_2 = \frac{V}{R_1} - \frac{q_2}{R_1 C} - q_4,$$
$$\dot{q}_3 = q_4$$
$$\dot{q}_4 = \frac{q_2}{LC} - \frac{R_2}{L} q_4.$$

So, we have a nonhomogeneous first-order system of differential equations.

3.5.3 Mixture Problems

THERE ARE MANY TYPES OF MIXTURE PROBLEMS. Such problems are standard in a first course on differential equations as examples of first-order differential equations. Typically, these examples consist of a tank of brine, water containing a specific amount of salt with pure water entering and the mixture leaving, or the flow of a pollutant into, or out of, a lake.

In general, one has a rate of flow of some concentration of mixture entering a region and a mixture leaving the region. The goal is to determine how much stuff is in the region at a given time. This is governed by the equation

Rate of Change of Substance $=$ Rate In $-$ Rate Out

This can be generalized to the case of two interconnected tanks. We provide some examples.

Example 3.23. Single Tank Problem

A 50-gal tank of pure water has a brine mixture with concentration of 2 lb/gal entering at the rate of 5 gal/min. [See Figure 3.18.] At the same time, the well-mixed contents drain out at the rate of 5 gal/min. Find the amount of salt in the tank at time t. In all such problems, one assumes that the solution is well mixed at each instant of time.

Let $x(t)$ be the amount of salt at time t. Then the rate at which the salt in the tank increases is due to the amount of salt entering the tank less that leaving the tank. To figure out these rates, one notes that dx/dt has units of pounds per minute (lb./min). The amount of salt entering per minute is given by the product of the entering concentration times the rate at which the brine enters. This gives the correct units:

$$\left(2\frac{lb}{gal}\right)\left(5\frac{gal}{min}\right) = 10\frac{lb}{min}.$$

Similarly, one can determine the rate out as

$$\left(\frac{x\ lb}{50\ gal}\right)\left(5\frac{gal}{min}\right) = \frac{x}{10}\frac{lb}{min}.$$

Thus, we have

$$\frac{dx}{dt} = 10 - \frac{x}{10}, \quad x(0) = 0.$$

This equation is easily solved using the methods for first-order equations. The general solution is given by

$$x(t) = 100 + Ae^{-t/10}.$$

Figure 3.18: A typical mixing problem.

Using the initial condition, one finds the particular solution

$$x(t) = 100(1 - e^{-t/10}).$$

Often one is interested in the long-term behavior of a system. In this case, we have that $\lim_{t\to\infty} x(t) = 100$ lb. This makes sense because 2 lb./gal enter during this time to eventually leave the entire 50 gal with this concentration. Thus,

$$50\,gal \times 2\frac{lb}{gal} = 100\,lb.$$

Example 3.24. Double Tank Problem

One has two tanks connected together, labeled tank X and tank Y, as shown in Figure 3.19.

Figure 3.19: The two tank problem.

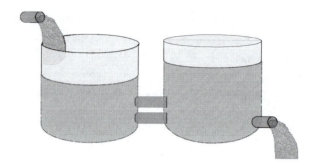

Let tank X initially have 100 gal of brine made with 100 lb. of salt. Tank Y initially has 100 gal of pure water. Pure water is pumped into tank X at a rate of 2.0 gal/min. Some of the mixture of brine and pure water flows into tank Y at 3 gal/min. To keep the tank levels the same, 1 gal of the Y mixture flows back into tank X at a rate of 1 gal/min and 2.0 gal/min drains out. Find the amount of salt at any given time in the tanks. What happens over a long period of time?

In this problem we set up two equations. Let $x(t)$ be the amount of salt in tank X and $y(t)$ the amount of salt in tank Y. Again, we carefully look at the rates into and out of each tank in order to set up the system of differential equations. We obtain the system

$$\frac{dx}{dt} = \frac{y}{100} - \frac{3x}{100}$$
$$\frac{dy}{dt} = \frac{3x}{100} - \frac{3y}{100}. \tag{3.145}$$

This is a linear, homogenous constant coefficient system of two first-order equations, which we know how to solve. The matrix form of the system is given by

$$\dot{\mathbf{x}} = \begin{pmatrix} -\frac{3}{100} & \frac{1}{100} \\ \frac{3}{100} & -\frac{3}{100} \end{pmatrix} \mathbf{x}, \quad \mathbf{x}(0) = \begin{pmatrix} 100 \\ 0 \end{pmatrix}.$$

The eigenvalues for the problem are given by $\lambda = -3 \pm \sqrt{3}$ and the eigenvectors are

$$\begin{pmatrix} 1 \\ \pm\sqrt{3} \end{pmatrix}.$$

Because the eigenvalues are real and distinct, the general solution is easily written down:

$$\mathbf{x}(t) = c_1 \begin{pmatrix} 1 \\ \sqrt{3} \end{pmatrix} e^{(-3+\sqrt{3})t} + c_2 \begin{pmatrix} 1 \\ -\sqrt{3} \end{pmatrix} e^{(-3-\sqrt{3})t}.$$

Finally, we need to satisfy the initial conditions. So,

$$\mathbf{x}(0) = c_1 \begin{pmatrix} 1 \\ \sqrt{3} \end{pmatrix} + c_2 \begin{pmatrix} 1 \\ -\sqrt{3} \end{pmatrix} = \begin{pmatrix} 100 \\ 0 \end{pmatrix},$$

or

$$c_1 + c_2 = 100, \quad (c_1 - c_2)\sqrt{3} = 0.$$

So, $c_2 = c_1 = 50$. The final solution is

$$\mathbf{x}(t) = 50 \left(\begin{pmatrix} 1 \\ \sqrt{3} \end{pmatrix} e^{(-3+\sqrt{3})t} + \begin{pmatrix} 1 \\ -\sqrt{3} \end{pmatrix} e^{(-3-\sqrt{3})t} \right),$$

or

$$
\begin{aligned}
x(t) &= 50 \left(e^{(-3+\sqrt{3})t} + e^{(-3-\sqrt{3})t} \right) \\
y(t) &= 50\sqrt{3} \left(e^{(-3+\sqrt{3})t} - e^{(-3-\sqrt{3})t} \right).
\end{aligned}
\tag{3.146}
$$

3.5.4   Chemical Kinetics

THERE ARE MANY PROBLEMS IN THE CHEMISTRY of chemical reactions that lead to systems of differential equations. The simplest reaction is when a chemical A turns into chemical B. This happens at a certain rate, $k > 0$. This reaction can be represented by the chemical formula

$$A \xrightarrow{\ k\ } B$$

In this case, we have that the rates of change of the concentrations of A, $[A]$, and B, $[B]$, are given by

The chemical reactions used in these examples are first order reactions. Second-order reactions have rates proportional to the square of the concentration.

$$
\begin{aligned}
\frac{d[A]}{dt} &= -k[A] \\
\frac{d[B]}{dt} &= k[A]
\end{aligned}
\tag{3.147}
$$

Think about this as it is a key to understanding the next reactions.

A more complicated reaction is given by

$$A \xrightarrow{\ k_1\ } B \xrightarrow{\ k_2\ } C$$

Here there are three concentrations and two rates of change. The system of equations governing the reaction is

$$
\begin{aligned}
\frac{d[A]}{dt} &= -k_1[A], \\
\frac{d[B]}{dt} &= k_1[A] - k_2[B], \\
\frac{d[C]}{dt} &= k_2[B].
\end{aligned}
\tag{3.148}
$$

The more complicated rate of change is when [B] increases from [A] changing to [B] and decrease when [B] changes to [C]. Thus, there are two terms in the rate of change equation for concentration [B].

One can further consider reactions in which a reverse reaction is possible. Thus, a further generalization occurs for the reaction

$$A \underset{k_1}{\overset{k_3}{\rightleftarrows}} B \xrightarrow{\ k_2\ } C$$

The reverse reaction rates contribute to the reaction equations for [A] and [B]. The resulting system of equations is

$$
\begin{aligned}
\frac{d[A]}{dt} &= -k_1[A] + k_3[B], \\
\frac{d[B]}{dt} &= k_1[A] - k_2[B] - k_3[B], \\
\frac{d[C]}{dt} &= k_2[B].
\end{aligned}
\tag{3.149}
$$

Nonlinear chemical reactions will be discussed in the next chapter.

3.5.5 Predator Prey Models

ANOTHER COMMON POPULATION MODEL is that describing the coexistence of species. For example, we could consider a population of rabbits and foxes. Left to themselves, rabbits would tend to multiply, and thus

$$
\frac{dR}{dt} = aR,
$$

with $a > 0$. In such a model, the rabbit population would grow exponentially. Similarly, a population of foxes would decay without the rabbits to feed on. So, we have that

$$
\frac{dF}{dt} = -bF
$$

for $b > 0$.

Now, if we put these populations together on a deserted island, they would interact. The more foxes, the more the rabbit population would decrease. However, the more rabbits, the foxes would have plenty to eat and the population would thrive. Thus, we could model the competing populations as

$$
\begin{aligned}
\frac{dR}{dt} &= aR - cF, \\
\frac{dF}{dt} &= -bF + dR,
\end{aligned}
\tag{3.150}
$$

where all of the constants are positive numbers. Studying this coupled system would lead to a study of the dynamics of these populations. The nonlinear version of this system, the Lotka–Volterra Model, will be discussed in the next chapter.

3.5.6 Love Affairs

THE NEXT APPLICATION IS ONE THAT WAS INTRODUCED in 1988 by Strogatz as a cute system involving relationships.[6] One considers what happens to the affections that two people have for each other over time. Let R denote the affection that Romeo has for Juliet and J be the affection that Juliet

[6] Steven H. Strogatz introduced this problem as an interesting example of systems of differential equations in *Mathematics Magazine*, Vol. 61, No. 1 (Feb. 1988) p 35. He also describes it in his book *Nonlinear Dynamics and Chaos* (1994).

has for Romeo. Positive values indicate love and negative values indicate dislike.

One possible model is given by

$$\frac{dR}{dt} = bJ$$

$$\frac{dJ}{dt} = cR \tag{3.151}$$

with $b > 0$ and $c < 0$. In this case, Romeo loves Juliet the more she likes him. But Juliet backs away when she finds his love for her increasing.

A typical system relating the combined changes in affection can be modeled as

$$\frac{dR}{dt} = aR + bJ$$

$$\frac{dJ}{dt} = cR + dJ. \tag{3.152}$$

Several scenarios are possible for various choices of the constants. For example, if $a > 0$ and $b > 0$, Romeo gets more and more excited by Juliet's love for him. If $c > 0$ and $d < 0$, Juliet is being cautious about her relationship with Romeo. For specific values of the parameters and initial conditions, one can explore this match of an overly zealous lover with a cautious lover.

3.5.7 Epidemics

ANOTHER INTERESTING AREA OF APPLICATION of differential equations is in predicting the spread of disease. Typically, one has a population of susceptible people or animals. Several infected individuals are introduced into the population and one is interested in how the infection spreads and if the number of infected people drastically increases or dies off. Such models are typically nonlinear, and we will look at what is called the SIR model in the next chapter. In this section we will model a simple linear model.

Let us divide the population into three classes. First, we let $S(t)$ represent the healthy people, who are susceptible to infection. Let $I(t)$ be the number of infected people. Of these infected people, some will die from the infection and others could recover. We will consider the case that initially there is one infected person and the rest, say N, are healthy. Can we predict how many deaths have occurred by time t?

We model this problem using the compartmental analysis we had seen for mixing problems. The total rate of change of any population would be due to those entering the group less those leaving the group. For example, the number of healthy people decreases due to infection and can increase when some of the infected group recovers. Let's assume that (a) the rate of infection is proportional to the number of healthy people, aS, and (b) the number who recover is proportional to the number of infected people, rI. Thus, the rate of change of healthy people is found as

$$\frac{dS}{dt} = -aS + rI.$$

Let the number of deaths be $D(t)$. Then, the death rate could be taken to be proportional to the number of infected people. So,

$$\frac{dD}{dt} = dI.$$

Finally, the rate of change of infected people is due to healthy people getting infected and the infected people who either recover or die. Using the corresponding terms in the other equations, we can write the rate of change of infected people as

$$\frac{dI}{dt} = aS - rI - dI.$$

This linear system of differential equations can be written in matrix form:

$$\frac{d}{dt} \begin{pmatrix} S \\ I \\ D \end{pmatrix} = \begin{pmatrix} -a & r & 0 \\ a & -d-r & 0 \\ 0 & d & 0 \end{pmatrix} \begin{pmatrix} S \\ I \\ D \end{pmatrix}. \qquad (3.153)$$

The reader can find the solutions of this system and determine if this is a realistic model.

3.6   Appendix: Diagonalization and Linear Systems

AS WE HAVE SEEN, THE MATRIX FORMULATION FOR LINEAR SYSTEMS can be powerful, especially for n differential equations involving n unknown functions. Our ability to proceed toward solutions depended on the solution of eigenvalue problems. However, in the case of repeated eigenvalues, we saw some additional complications. This all depends deeply on the background linear algebra. Namely, we relied on being able to diagonalize the given coefficient matrix. In this section we will discuss the limitations of diagonalization and introduce the Jordan canonical form.

We begin with the notion of similarity. Matrix A is similar to matrix B if and only if there exists a nonsingular matrix P such that

$$B = P^{-1}AP. \qquad (3.154)$$

Recall that a nonsingular matrix has a nonzero determinant and is invertible.

We note that the similarity relation is an equivalence relation. Namely, it satisfies the following:

1. A is similar to itself.

2. If A is similar to B, then B is similar to A.

3. If A is similar to B and B is similar to C, then A is similar to C.

Similar matrices have the same eigenvalues.

Also, if A is similar to B, then they have the same eigenvalues. This follows from a simple computation of the eigenvalue equation. Namely,

$$
\begin{aligned}
0 &= \det(B - \lambda I) \\
&= \det(P^{-1}AP - \lambda P^{-1}IP) \\
&= \det(P)^{-1}\det(A - \lambda I)\det(P) \\
&= \det(A - \lambda I). \quad\quad\quad (3.155)
\end{aligned}
$$

Therefore, $\det(A - \lambda I) = 0$, and λ is an eigenvalue of both A and B.

An $n \times n$ matrix A is diagonalizable if and only if A is similar to a diagonal matrix D; that is, there exists a nonsingular matrix P such that

$$
D = P^{-1}AP. \quad\quad\quad (3.156)
$$

One of the most important theorems in linear algebra is the Spectral Theorem. This theorem tells us when a matrix can be diagonalized. In fact, it goes beyond matrices to the diagonalization of linear operators. We learn in linear algebra that linear operators can be represented by matrices once we pick a particular representation basis. Diagonalization is simplest for finite dimensional vector spaces and requires some generalization for infinite dimensional vectors spaces. Examples of operators to which the spectral theorem applies are self-adjoint operators (more generally normal operators on Hilbert spaces). We will explore some of these ideas later in the course. The spectral theorem provides a canonical decomposition, called the spectral decomposition, or eigendecomposition, of the underlying vector space on which it acts.

The next theorem tells us how to diagonalize a matrix:

Theorem 3.1. *Let A be an $n \times n$ matrix. Then A is diagonalizable if and only if A has n linearly independent eigenvectors. If so, then*

$$
D = P^{-1}AP.
$$

If $\{v_1, \ldots, v_n\}$ are the eigenvectors of A and $\{\lambda_1, \ldots, \lambda_n\}$ are the corresponding eigenvalues, then v_j is the jth column of P and $D_{jj} = \lambda_j$.

A simpler determination results by noting

Theorem 3.2. *Let A be an $n \times n$ matrix with n real and distinct eigenvalues. Then A is diagonalizable.*

Therefore, we need only look at the eigenvalues and determine diagonalizability. In fact, one also has from linear algebra the following result.

Theorem 3.3. *Let A be an $n \times n$ real symmetric matrix. Then A is diagonalizable.*

Recall that a symmetric matrix is one whose transpose is the same as the matrix, or $A_{ij} = A_{ji}$.

Example 3.25. *Consider the matrix*

$$
A = \begin{pmatrix} 1 & 2 & 2 \\ 2 & 3 & 0 \\ 2 & 0 & 3 \end{pmatrix}.
$$

This is a real symmetric matrix. The characteristic polynomial is found to be

$$\det(A - \lambda I) = -(\lambda - 5)(\lambda - 3)(\lambda + 1) = 0.$$

As before, we can determine the corresponding eigenvectors (for $\lambda = -1, 3, 5$, respectively) as

$$\begin{pmatrix} -2 \\ 1 \\ 1 \end{pmatrix}, \quad \begin{pmatrix} 0 \\ -1 \\ 1 \end{pmatrix}, \quad \begin{pmatrix} 1 \\ 1 \\ 1 \end{pmatrix}.$$

We can use these to construct the diagonalizing matrix P. Namely, we have

$$\begin{aligned} P^{-1}AP &= \begin{pmatrix} -2 & 0 & 1 \\ 1 & -1 & 1 \\ 1 & 1 & 1 \end{pmatrix}^{-1} \begin{pmatrix} 1 & 2 & 2 \\ 2 & 3 & 0 \\ 2 & 0 & 3 \end{pmatrix} \begin{pmatrix} -2 & 0 & 1 \\ 1 & -1 & 1 \\ 1 & 1 & 1 \end{pmatrix} \\ &= \begin{pmatrix} -1 & 0 & 0 \\ 0 & 3 & 0 \\ 0 & 0 & 5 \end{pmatrix}. \end{aligned} \tag{3.157}$$

It is not always possible to diagonalize a given square matrix. This is because some matrices do not have enough linearly independent vectors, or we have repeated eigenvalues. However, we have the following theorem:

Theorem 3.4. *Every $n \times n$ matrix A is similar to a matrix of the form*

$$J = diag[J_1, J_2, \dots, J_n],$$

where

$$J_i = \begin{pmatrix} \lambda_i & 1 & 0 & \cdots & 0 \\ 0 & \lambda_i & 1 & \cdots & 0 \\ \vdots & \ddots & \ddots & \ddots & \vdots \\ 0 & \cdots & 0 & \lambda_i & 1 \\ 0 & 0 & \cdots & 0 & \lambda_i \end{pmatrix}. \tag{3.158}$$

We will not go into the details of how one finds this **Jordan Canonical Form** or proving the theorem. In practice, you can use a computer algebra system to determine this and the similarity matrix. However, we would still need to know how to use it to solve some systems of differential equations. In the next example we indicate how to solve a system with repeated eigenvalues.

Example 3.26. *Let's consider a simple system with the 3×3 Jordan block:*

$$A = \begin{pmatrix} 2 & 1 & 0 \\ 0 & 2 & 1 \\ 0 & 0 & 2 \end{pmatrix}.$$

The corresponding system of coupled first-order differential equations takes the form

$$\begin{aligned} \frac{dx_1}{dt} &= 2x_1 + x_2, \\ \frac{dx_2}{dt} &= 2x_2 + x_3, \\ \frac{dx_3}{dt} &= 2x_3. \end{aligned} \tag{3.159}$$

The last equation is simple to solve, giving $x_3(t) = c_3 e^{2t}$. Inserting into the second equation, you have

$$\frac{dx_2}{dt} = 2x_2 + c_3 e^{2t}.$$

Using the integrating factor, e^{-2t}, one can solve this equation to get $x_2(t) = (c_2 + c_3 t)e^{2t}$. Similarly, one can solve the first equation to obtain $x_1(t) = (c_1 + c_2 t + \frac{1}{2}c_3 t^2)e^{2t}$.

This should remind you of a problem we had solved earlier leading to the generalized eigenvalue problem in Expression (3.116). This suggests that there is a more general theory when there are multiple eigenvalues and relating to Jordan canonical forms.

Let's write the solution we just obtained in vector form. We have

$$\mathbf{x}(t) = \left[c_1 \begin{pmatrix} 1 \\ 0 \\ 0 \end{pmatrix} + c_2 \begin{pmatrix} t \\ 1 \\ 0 \end{pmatrix} + c_3 \begin{pmatrix} \frac{1}{2}t^2 \\ t \\ 1 \end{pmatrix} \right] e^{2t}. \tag{3.160}$$

It looks like this solution is a linear combination of three linearly independent solutions:

$$\begin{aligned} \mathbf{x}_1 &= \mathbf{v}_1 e^{2\lambda t}, \\ \mathbf{x}_2 &= (t\mathbf{v}_1 + \mathbf{v}_2)e^{\lambda t}, \\ \mathbf{x}_3 &= (\frac{1}{2}t^2 \mathbf{v}_1 + t\mathbf{v}_2 + \mathbf{v}_3)e^{\lambda t}, \end{aligned} \tag{3.161}$$

where $\lambda = 2$ and the vectors satisfy the equations

$$\begin{aligned} (A - \lambda I)\mathbf{v}_1 &= 0, \\ (A - \lambda I)\mathbf{v}_2 &= \mathbf{v}_1, \\ (A - \lambda I)\mathbf{v}_3 &= \mathbf{v}_2, \end{aligned} \tag{3.162}$$

and

$$\begin{aligned} (A - \lambda I)\mathbf{v}_1 &= 0, \\ (A - \lambda I)^2 \mathbf{v}_2 &= 0, \\ (A - \lambda I)^3 \mathbf{v}_3 &= 0. \end{aligned} \tag{3.163}$$

It is easy to generalize this result to build linearly independent solutions corresponding to multiple roots (eigenvalues) of the characteristic equation.

Problems

1. Express the vector $\mathbf{v} = (1,2,3)$ as a linear combination of the vectors $\mathbf{a}_1 = (1,1,1)$, $\mathbf{a}_2 = (1,0,-2)$, and $\mathbf{a}_3 = (2,1,0)$.

2. A symmetric matrix is one for which the transpose of the matrix is the same as the original matrix, $A^T = A$. An antisymmetric matrix is one that satisfies $A^T = -A$.

a. Show that the diagonal elements of an $n \times n$ antisymmetric matrix are all zero.

b. Show that a general 3×3 antisymmetric matrix has three independent off-diagonal elements.

c. How many independent elements does a general 3×3 symmetric matrix have?

d. How many independent elements does a general $n \times n$ symmetric matrix have?

e. How many independent elements does a general $n \times n$ antisymmetric matrix have?

3. Consider the matrix representations for two-dimensional rotations of vectors by angles α and β, denoted by R_α and R_β, respectively.

a. Find R_α^{-1} and R_α^T. How do they relate?

b. Prove that $R_{\alpha+\beta} = R_\alpha R_\beta = R_\beta R_\alpha$.

4. Consider the matrix

$$A = \begin{pmatrix} \frac{1}{2} & \frac{1}{\sqrt{2}} & \frac{1}{2} \\ -\frac{1}{\sqrt{2}} & 0 & \frac{1}{\sqrt{2}} \\ \frac{1}{2} & -\frac{1}{\sqrt{2}} & \frac{1}{2} \end{pmatrix}.$$

a. Verify that this is a rotation matrix.

b. Find the angle and axis of rotation.

c. Determine the corresponding similarity transformation using the results from part b.

5. Consider the matrix

$$A = \begin{pmatrix} -0.8124 & -0.5536 & -0.1830 \\ -0.3000 & 0.6660 & -0.6830 \\ 0.5000 & -0.5000 & -0.7071 \end{pmatrix}.$$

This matrix represents the active rotation through three Euler angles. Determine the possible angles of rotation leading to this matrix.

6. Consider the three-dimensional Euler rotation matrix $\hat{R}(\phi, \theta, \psi) = \hat{R}_z(\psi)\hat{R}_x(\theta)\hat{R}_z(\phi)$.

a. Find the elements of $\hat{R}(\phi, \theta, \psi)$.

b. Compute $Tr(\hat{R}(\phi, \theta, \psi)$.

c. Show that $\hat{R}^{-1}(\phi, \theta, \psi) = \hat{R}^T(\phi, \theta, \psi)$.

d. Show that $\hat{R}^{-1}(\phi, \theta, \psi) = \hat{R}(-\psi, -\theta, -\phi)$.

7. The Pauli spin matrices in quantum mechanics are given by the following matrices: $\sigma_1 = \begin{pmatrix} 0 & 1 \\ 1 & 0 \end{pmatrix}$, $\sigma_2 = \begin{pmatrix} 0 & -i \\ i & 0 \end{pmatrix}$, and $\sigma_3 = \begin{pmatrix} 1 & 0 \\ 0 & -1 \end{pmatrix}$. Show that

a. $\sigma_1^2 = \sigma_2^2 = \sigma_3^2 = I$.

b. $\{\sigma_i, \sigma_j\} \equiv \sigma_i\sigma_j + \sigma_j\sigma_i = 2\delta_{ij}I$, for $i, j = 1, 2, 3$ and I the 2×2 identity matrix. $\{,\}$ is the anti-commutation operation.

c. $[\sigma_1, \sigma_2] \equiv \sigma_1\sigma_2 - \sigma_2\sigma_1 = 2i\sigma_3$, and similarly for the other pairs. $[,]$ is the commutation operation.

d. Show that an arbitrary 2×2 matrix M can be written as a linear combination of Pauli matrices, $M = a_0 I + \sum_{j=1}^{3} a_j\sigma_j$, where the a_j's are complex numbers.

8. Use Cramer's Rule to solve the system:

$$
\begin{aligned}
2x - 5z &= 7 \\
x - 2y &= 1 \\
3x - 5y - z &= 4.
\end{aligned}
\tag{3.164}
$$

9. Find the eigenvalue(s) and eigenvector(s) for the following:

a. $\begin{pmatrix} 4 & 2 \\ 3 & 3 \end{pmatrix}$

b. $\begin{pmatrix} 3 & -5 \\ 1 & -1 \end{pmatrix}$

c. $\begin{pmatrix} 4 & 1 \\ 0 & 4 \end{pmatrix}$

d. $\begin{pmatrix} 1 & -1 & 4 \\ 3 & 2 & -1 \\ 2 & 1 & -1 \end{pmatrix}$

10. For the matrices in the previous problem, compute the determinants and find the inverses, if they exist.

11. Consider the conic $5x^2 - 4xy + 2y^2 = 30$.

a. Write the left side in matrix form.

b. Diagonalize the coefficient matrix, finding the eigenvalues and eigenvectors.

c. Construct the rotation matrix from the information in part b. What is the angle of rotation needed to bring the conic into standard form?

d. What is the conic?

12. In Equation (3.109), the exponential of a matrix was defined.

a. Let

$$
A = \begin{pmatrix} 2 & 0 \\ 0 & 0 \end{pmatrix}.
$$

Compute e^A.

b. Give a definition of $\cos A$ and compute $\cos \begin{pmatrix} 1 & 0 \\ 0 & 2 \end{pmatrix}$ in simplest form.

c. Using the definition of e^A, prove $e^{PAP^{-1}} = Pe^A P^{-1}$ for general A.

13. Prove the following for matrices A, B, and C.

    a. $(AB)C = A(BC)$.

    b. $(AB)^T = B^T A^T$

    c. $tr(A)$ is invariant under similarity transformations.

    d. If A and B are orthogonal, then AB is orthogonal.

14. Consider the following systems. For each system, determine the coefficient matrix. When possible, solve the eigenvalue problem for each matrix and use the eigenvalues and eigenvectors to provide solutions to the given systems. Finally, in the common cases that you investigated in Problem 2.31, make comparisons with your previous answers, such as what type of eigenvalues correspond to stable nodes.

    a.

$$\begin{aligned} x' &= 3x - y, \\ y' &= 2x - 2y. \end{aligned}$$

    b.

$$\begin{aligned} x' &= -y, \\ y' &= -5x. \end{aligned}$$

    c.

$$\begin{aligned} x' &= x - y, \\ y' &= y. \end{aligned}$$

    d.

$$\begin{aligned} x' &= 2x + 3y, \\ y' &= -3x + 2y. \end{aligned}$$

    e.

$$\begin{aligned} x' &= -4x - y, \\ y' &= x - 2y. \end{aligned}$$

    f.

$$\begin{aligned} x' &= x - y, \\ y' &= x + y. \end{aligned}$$

15. In Example 3.22 we investigated a coupled mass-spring system as a pair of second-order differential equations.

a. In that problem, we used $\sqrt{\frac{3\pm\sqrt{5}}{2}} = \frac{\sqrt{5}\pm1}{2}$. Prove this result.

b. Rewrite the system as a system of four first-order equations.

c. Find the eigenvalues and eigenfunctions for the system of equations in part b to arrive at the solution found in Example 3.22.

d. Let $k = 5.00$ N/m and $m = 0.250$ kg. Assume that the masses are initially at rest and plot the positions as a function of time if initially i) $x_1(0) = x_2(0) = 10.0$ cm and i) $x_1(0) = -x_2(0) = 10.0$ cm. Describe the resulting motion.

16. Add a third spring connected to mass 2 in the coupled system shown in Figure 3.15 to a wall on the far right. Assume that the masses are the same and the springs are the same.

a. Model this system with a set of first-order differential equations.

b. If the masses are all 2.0 kg and the spring constants are all 10.0 N/m, then find the general solution for the system.

c. Move mass 1 to the left (of equilibrium) 10.0 cm and mass 2 to the right 5.0 cm. Let them go. find the solution and plot it as a function of time. Where is each mass at 5.0 seconds?

d. Model this initial value problem with a set of two second-order differential equations setup the system in the form $M\ddot{x} = Kx$ and solve using the values in part b.

17. Consider the series circuit in Figure 2.7 with $L = 1.00$ H, $R = 1.00 \times 10^2$ Ω, $C = 1.00 \times 10^{-4}$ F, and $V_0 = 1.00 \times 10^3$ V.

a. Set up the problem as a system of two first-order differential equations for the charge and the current.

b. Suppose that no charge is present and no current is flowing at time $t = 0$ when V_0 is applied. Find the current and the charge on the capacitor as functions of time.

c. Plot your solutions and describe how the system behaves over time.

18. Consider the series circuit in Figure 3.16 with $L = 1.00$ H, $R_1 = R_2 = 1.00 \times 10^2$ Ω, $C = 1.00 \times 10^{-4}$ F, and $V_0 = 1.00 \times 10^3$ V.

a. Set up the problem as a system of first order differential equations for the charges and the currents in each loop.

b. Suppose that no charge is present at time $t = 0$ when V_0 is applied. Find the current and the charge on the capacitor as functions of time.

c. Plot your solutions and describe how the system behaves over time.

19. Initially, a 200-gallon tank is filled with pure water. At time $t = 0$, a salt concentration with 3 pounds of salt per gallon is added to the container at the rate of 4 gallons per minute, and the well-stirred mixture is drained from the container at the same rate.

   a. Find the number of pounds of salt in the container as a function of time.

   b. How many minutes does it take for the concentration to reach 2 pounds per gallon?

   c. What does the concentration in the container approach for large values of time? Does this agree with your intuition?

   d. Assuming that the tank holds much more than 200 gallons, and everything is the same except that the mixture is drained at 3 gallons per minute, what would the answers to parts a and b become?

20. You make 2 quarts of salsa for a party. The recipe calls for 5 teaspoons of lime juice per quart, but you had accidentally put in 5 tablespoons per quart. You decide to feed your guests the salsa anyway. Assume that the guests take a quarter cup of salsa per minute and that you replace what was taken with chopped tomatoes and onions without any lime juice. [1 quart = 4 cups and 1 Tb = 3 tsp.]

   a. Write the differential equation and initial condition for the amount of lime juice as a function of time in this mixture-type problem.

   b. Solve this initial value problem.

   c. How long will it take to get the salsa back to the recipe's suggested concentration?

21. Consider the chemical reaction leading to the system in (3.149). Let the rate constants be $k_1 = 0.20$ m/s, $k_2 = 0.05$ m/s, and $k_3 = 0.10$ m/s. What do the eigenvalues of the coefficient matrix say about the behavior of the system? Find the solution of the system assuming $[A](0) = A_0 = 1.0$ μmol, $[B](0) = 0$, and $[C](0) = 0$. Plot the solutions for $t = 0.0$ to 50.0 ms and describe what is happening over this time.

22. Consider the epidemic model leading to the system in Expression (3.153). Choose the constants as $a = 2.0$ days^{-1}, $d = 3.0$ days^{-1}, and $r = 1.0$ day^{-1}. What are the eigenvalues of the coefficient matrix? Find the solution of the system assuming an initial population of $1,000$ and one infected individual. Plot the solutions for $t = 0.0$ to 5.0 days and describe what is happening over this time. Is this model realistic?

23. Find and classify any equilibrium points in the Romeo and Juliet problem for the following cases. Solve the systems and describe their affections as a function of time.

   a. $a = 0, b = 2, c = -1, d = 0, R(0) = 1, J(0) = 1$.

   b. $a = 0, b = 2, c = 1, d = 0, R(0) = 1, J(0) = 1$.

   c. $a = -1, b = 2, c = -1, d = 0, R(0) = 1, J(0) = 1$.

4
Nonlinear Dynamics

"The scientist does not study nature because it is useful; he studies it because he delights in it, and he delights in it because it is beautiful."—Jules Henri Poincaré (1854–1912)

4.1 Introduction

SOME OF THE MOST INTERESTING PHENOMENA in the world are modeled by nonlinear systems. These systems can be modeled by differential equations when time is considered a continuous variable or difference equations when time is treated in discrete steps. Applications involving differential equations can be found in many physical systems such as planetary systems, weather prediction, electrical circuits, and kinetics. Even in some simple dynamical systems, a combination of damping and a driving force can lead to chaotic behavior. Namely, small changes in initial conditions could lead to very different outcomes. In this chapter we will explore a few different nonlinear systems and introduce some of the tools needed to investigate them. These tools are based on some of the material in Chapters 2 and 3 for linear systems of differential equations.

Nonlinear differential equations are either integrable but difficult to solve, or they are not integrable and can only be solved numerically. We will see that we can sometimes approximate the solutions of nonlinear systems with linear systems in small regions of phase space and determine the qualitative behavior of the system without knowledge of the exact solution.

Nonlinear problems occur naturally. We will see problems from many of the same fields we explored in Section 3.5. We will concentrate mainly on continuous dynamical systems. We will begin with a simple population model and look at the behavior of equilibrium solutions of first-order autonomous differential equations. We will then look at nonlinear systems in the plane, such as the nonlinear pendulum and other nonlinear oscillations. We will conclude by discussing a few other interesting physical examples, stressing some of the key ideas of nonlinear dynamics.

4.2   Logistic Equation

IN THIS SECTION WE WILL EXPLORE a simple nonlinear population model. Typically, we want to model the growth of a given population, $y(t)$, and the differential equation governing the growth behavior of this population is developed in a manner similar to that used previously for mixing problems. Namely, we note that the rate of change of the population is given by an equation of the form

$$\frac{dy}{dt} = \text{Rate In} \ - \ \text{Rate Out}.$$

The *Rate In* could be due to the number of births per unit time and the *Rate Out* by the number of deaths per unit time. While there are other potential contributions to these rates, we will consider the birth and death rates in the simplest examples.

A simple population model can be obtained if one assumes that these rates are linear in the population. Thus, we assume that the

$$\text{Rate In} \ = \ by, \text{ and the } \ \text{Rate Out} \ = \ my.$$

Here we have denoted the birth rate as b and the mortality rate as m. This gives the rate of change of population as

$$\frac{dy}{dt} = by - my. \tag{4.1}$$

Generally, these rates could depend on the time. In the case that they are both constant rates, we can define $k = b - m$ and obtain the familiar exponential model of population growth:

$$\frac{dy}{dt} = ky.$$

This is easily solved and one obtains exponential growth ($k > 0$) or decay ($k < 0$). This Malthusian growth model was named after Thomas Robert Malthus (1766–1834), a clergyman who used this model to warn of the impending doom of the human race if its reproductive practices continued.

When populations get large enough, there is competition for resources, such as space and food, which can lead to a higher mortality rate. Thus, the mortality rate may be a function of the population size, $m = m(y)$. The simplest model would be a linear dependence, $m = \tilde{m} + cy$. Then, the previous exponential model takes the form

$$\frac{dy}{dt} = ky - cy^2, \tag{4.2}$$

where $k = b - \tilde{m}$. This is known as the *logistic model* of population growth. Typically, c is small and the added nonlinear term does not really kick in until the population gets large enough.

Malthusian population growth.

The logistic model was first published in 1838 by Pierre François Verhulst (1804-1849) in the form

$$\frac{dN}{dt} = rN\left(1 - \frac{N}{K}\right),$$

where N is the population at time t, r is the growth rate, and K is what is called the carrying capacity. Note that in our model, $r = k = Kc$.

Example 4.1. *Show that Equation (4.2) can be written in the form*

$$z' = kz(1 - z)$$

which has only one parameter.

We carry this out be rescaling the population, $y(t) = \alpha z(t)$, where α is to be determined. Inserting this transformation, we have

$$
\begin{aligned}
y' &= ky - cy^2 \\
\alpha z' &= \alpha kz - c\alpha^2 z^2,
\end{aligned}
$$

or

$$z' = kz \left(1 - \alpha \frac{c}{k} z\right).$$

Thus, we obtain the result, $z' = kz(1 - z)$, if we pick $\alpha = \frac{k}{c}$.

Before we obtain the exact solution, it is instructive to study the qualitative behavior of the solutions without actually writing down any explicit solutions. Such methods are useful for more difficult nonlinear equations, as we will see later in this chapter.

We will demonstrate this analysis with a simple logistic equation example. We will first look for constant solutions, called equilibrium solutions, satisfying $y'(t) = 0$. Then we will look at the behavior of solutions near the equilibrium solutions, or fixed points, and determine the stability of the equilibrium solutions. In the next section we will extend these ideas to other first-order differential equations.

Example 4.2. *Find and classify the equilibrium solutions of the logistic equation,*

$$\frac{dy}{dt} = y - y^2. \tag{4.3}$$

First, we determine the equilibrium, or constant, solutions given by $y' = 0$. For this case, we have $y - y^2 = 0$. So, the equilibrium solutions are $y = 0$ and $y = 1$.

These solutions divide the ty-plane into three regions, $y < 0$, $0 < y < 1$, and $y > 1$. Solutions that originate in one of these regions at $t = t_0$ will remain in that region for all $t > t_0$ as solutions of this differential equation cannot intersect.

Next, we determine the behavior of solutions in the three regions. Noting that $y'(t)$ gives the slope of any solution in the plane, then we find that the solutions are monotonic in each region. Namely, in regions where $y'(t) > 0$, we have monotonically increasing functions and in regions where $y'(t) < 0$, we have monotonically decreasing functions. We determine the sign of $y'(t)$ from the right-hand side of the differential equation.

For example, in this problem $y - y^2 > 0$ only for the middle region and $y - y^2 < 0$ for the other two regions. Thus, the slope is positive in the middle region, giving a rising solution as shown in Figure 4.1. Note that this solution does not cross the equilibrium solutions. Similar statements can be made about the solutions in the other regions.

We further note that the solutions on either side of the equilibrium solution $y = 1$ tend to approach this equilibrium solution for large values of t. In fact, no matter

Note: If two solutions of the differential equation intersect, then they have common values y_1 at time t_1. Using this information, we could set up an initial value problem for which the initial condition is $y(t_1) = y_1$. Because the two different solutions intersect at this point in the phase plane, we would have an initial value problem with two different solutions. This would violate the uniqueness theorem for initial value problems.

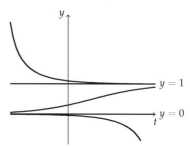

Figure 4.1: Representative solution behavior for $y' = y - y^2$.

Stable and unstable equilibria.

how far these solutions are from $y = 1$, as long as $y(t) > 0$, the solutions will eventually approach this equilibrium solution as $t \to \infty$. We then say that the equilibrium solution, $y = 1$, is a stable equilibrium.

Similarly, we note that the solutions on either side of the equilibrium solution $y = 0$ tend away from $y = 0$ for large values of t. No matter how close a solution is to $y = 0$ at some given time, eventually these solutions will diverge as $t \to \infty$. We say that such equilibrium solutions are unstable equilibria.

Figure 4.2: Representative solution behavior and the phase line for $y' = y - y^2$.

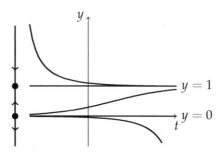

Phase lines.

If we are only interested in the behavior of the equilibrium solutions, we could just display a phase line. In Figure 4.2 we place a vertical line to the right of the ty-plane plot. On this line we first place dots at the corresponding equilibrium solutions and label the solutions. These points divide the phase line into three intervals.

In each interval we then place arrows pointing upward or downward, indicating solutions with positive or negative slopes, respectively. For example, for the interval $y > 1$, there is a downward-pointing arrow indicating that the slope is negative in that region.

Looking at the resulting phase line, we can determine if a given equilibrium is stable (arrows pointing towards the point) or unstable (arrows pointing away from the point). In Figure 4.3 we draw the final phase line by itself. We see that $y = 1$ is a stable equilibrium point and $y = 0$ is an unstable equilibrium point.

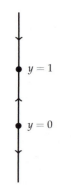

Figure 4.3: Phase line for $y' = y - y^2$.

4.2.1 Riccati Equation

The Riccati Equation is named after the Italian mathematician Jacopo Francesco Riccati (1676–1754). When $a(t) = 0$, the equation becomes a Bernoulli Equation.

WE HAVE SEEN THAT ONE DOES NOT NEED an explicit solution of the logistic equation (4.2) in order to study the behavior of its solutions. However, the logistic equation is an example of a nonlinear first-order equation that is solvable. It is also an example of a general Riccati equation, a first-order differential equation quadratic in the unknown function.

The general form of the *Riccati Equation* is

$$\frac{dy}{dt} = a(t) + b(t)y + c(t)y^2. \tag{4.4}$$

As long as $c(t) \neq 0$, this equation can be reduced to a second-order linear differential equation through the transformation

$$y(t) = -\frac{1}{c(t)}\frac{x'(t)}{x(t)}.$$

We will demonstrate the use of this transformation in obtaining the solution of the logistic equation.

Example 4.3. *Solve the logistic equation*

$$\frac{dy}{dt} = ky - cy^2 \tag{4.5}$$

using the transformation

$$y = \frac{1}{c}\frac{x'}{x}.$$

Differentiating this transformation with respect to t, we obtain

$$
\begin{aligned}
\frac{dy}{dt} &= \frac{1}{c}\left[\frac{x''}{x} - \left(\frac{x'}{x}\right)^2\right] \\
&= \frac{1}{c}\left[\frac{x''}{x} - (cy)^2\right] \\
&= \frac{1}{c}\frac{x''}{x} - cy^2. \tag{4.6}
\end{aligned}
$$

Inserting this result into the logistic equation (4.5), we have

$$\frac{1}{c}\frac{x''}{x} - cy^2 = k\frac{1}{c}\left(\frac{x'}{x}\right) - cy^2.$$

Simplifying, we see that the logistic equation has been reduced to a second-order linear, differential equation,

$$x'' = kx'.$$

This equation is readily solved. One integration gives

$$x'(t) = Be^{kt}.$$

A second integration gives

$$x(t) = A + Be^{kt},$$

where A and B are two arbitrary constants.

Inserting this result into the Riccati transformation, we obtain

$$y(t) = \frac{1}{c}\frac{x'}{x} = \frac{kBe^{kt}}{c(A + Be^{kt})}.$$

It appears that we have two arbitrary constants. However, we started out with a first-order differential equation and so we expect only one arbitrary constant. We can resolve this dilemma by dividing[1] the numerator and denominator by Be^{kt} and defining $C = \frac{A}{B}$. Then, we have the solution

$$y(t) = \frac{k/c}{1 + Ce^{-kt}}, \tag{4.7}$$

showing that there really is only one arbitrary constant in the solution.

[1] This general solution holds for $B \neq 0$. If $B = 0$, then we have $x(t) = A$ and, thus, $y(t)$ is the constant equilibrium solution.

Plots of the solution Expressiom (4.7) of the logistic equation for different initial conditions gives the solutions seen in the previous section. In particular, setting all of the constants to unity, we have the sigmoid function,

$$y(t) = \frac{1}{1 + e^{-t}}.$$

This is the signature S-shaped curve of the logistic model as shown in Figure 4.4. We should note that this is not the only way to obtain the solution to the logistic equation, although this approach has provided us with an introduction to Riccati Equations. A more direct approach would be to use separation of variables on the logistic equation, which is Problem 1.

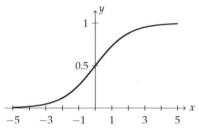

Figure 4.4: Plot of the sigmoid function.

4.3   Autonomous First-Order Equations

IN THIS SECTION WE WILL STUDY THE STABILITY of nonlinear first-order autonomous equations. We will then extend this study in the next section to looking at families of first-order equations that are connected through a parameter.

Recall that a first-order autonomous equation is given in the form

$$\frac{dy}{dt} = f(y). \tag{4.8}$$

We will assume that f and $\frac{\partial f}{\partial y}$ are continuous functions of y, so that we know that solutions of initial value problems exist and are unique.

A solution $y(t)$ of Equation (4.8) is called an *equilibrium solution*, or a *fixed point* solution, if it is a constant solution satisfying $y'(t) = 0$. Such solutions are the roots of the right-hand side of the differential equation $f(y) = 0$.

Example 4.4. *Find the equilibrium solutions of $y' = 1 - y^2$.*

The equilibrium solutions are the roots of $f(y) = 1 - y^2 = 0$. The equilibria are found to be $y = \pm 1$.

Once we have determined the equilibrium solutions, we would like to classify them. Are they stable or unstable? As we had seen previously, we are interested in the behavior of solutions near the equilibria. This classification can be determined using a linearization of the given equation. This will provide an analytic criteria to establish the stability of equilibrium solutions without geometrically drawing the phase lines as we had done previously.

Linearization of first-order equations.

Let y^* be an equilibrium solution of Equation (4.8). Then, any solution can be written in the form

$$y(t) = y^* + \xi(t),$$

where $\xi(t)$ measures how far the solution is from the equilibrium at any given time.

Inserting this form into Equation (4.8), we have

$$\frac{d\xi}{dt} = f(y^* + \xi).$$

We now consider small $\xi(t)$ in order to study solutions near the equilibrium solution. For such solutions, we can expand $f(y)$ about the equilibrium solution,

$$f(y^* + \xi) = f(y^*) + f'(y^*)\xi + \frac{1}{2!}f''(y^*)\xi^2 + \cdots.$$

Since y^* is an equilibrium solution, $f(y^*) = 0$, the first term in the Taylor series vanishes. If the first derivative does not vanish, then for solutions close to equilibrium, we can neglect higher order terms in the expansion. Then, $\xi(t)$ approximately satisfies the differential equation

$$\frac{d\xi}{dt} = f'(y^*)\xi. \tag{4.9}$$

This is called a linearization of the original nonlinear equation about the equilibrium point. This equation has exponential solutions for $f'(y^*) \neq 0$,

$$\xi(t) = \xi_0 e^{f'(y^*)t}.$$

Now we see how the stability criteria arise. If $f'(y^*) > 0$, $\xi(t)$ grows in time. Therefore, nearby solutions stray from the equilibrium solution for large times. On the other hand, if $f'(y^*) < 0$, $\xi(t)$ decays in time and nearby solutions approach the equilibrium solution for large t. Thus, we have the results:

The stability criteria for equilibrium solutions of a first-order differential equation.

$$\boxed{\begin{aligned} f'(y^*) < 0, \quad & y^* \text{ is stable} \\ f'(y^*) > 0, \quad & y^* \text{ is unstable} \end{aligned}} \tag{4.10}$$

Example 4.5. *Determine the stability of the equilibrium solutions of $y' = 1 - y^2$.*

In the last example we found the equilibrium solutions, $y^* = \pm 1$. The stability criteria require computing

$$f'(y^*) = -2y^*.$$

For this problem, we have $f'(\pm 1) = \mp 2$. Therefore, $y^* = 1$ is a stable equilibrium and $y^* = -1$ is an unstable equilibrium.

Example 4.6. *Find and classify the equilibria for the logistic equation $y' = y - y^2$.*

We have already investigated this problem using phase lines. There are two equilibria: $y = 0$ and $y = 1$.

We next apply the stability criteria. Noting that $f'(y) = 1 - 2y$, the first equilibrium solution gives $f'(0) = 1$. So, $y = 0$ is an unstable equilibrium. Because $f'(1) = -1 < 0$, we see that $y = 1$ is a stable equilibrium. These results are the same as we determined earlier using phase lines.

4.4   Bifurcations for First-Order Equations

WE NOW CONSIDER FAMILIES of first-order autonomous differential equations of the form

$$\frac{dy}{dt} = f(y; \mu).$$

Bifurcations and bifurcation points.

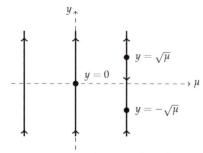

Figure 4.5: Phase lines for $y' = y^2 - \mu$. On the right, $\mu > 0$; and on the left, $\mu < 0$.

Here, μ is a parameter that we can change and then observe the resulting behaviors of the solutions of the differential equation. When a small change in the parameter leads to changes in the behavior of the solution, then the system is said to undergo a *bifurcation*. The value of the parameter μ at which the bifurcation occurs is called a *bifurcation point*.

We will consider several generic examples, leading to special classes of bifurcations of first-order autonomous differential equations. We will study the stability of equilibrium solutions using both phase lines and the stability criteria developed in the previous section

Example 4.7. $y' = y^2 - \mu$.

First note that equilibrium solutions occur for $y^2 = \mu$. In this problem, there are three cases to consider.

1. $\mu > 0$.

   In this case, there are two real solutions of $y^2 = \mu$, $y = \pm\sqrt{\mu}$. Note that $y^2 - \mu < 0$ for $|y| < \sqrt{\mu}$. So, we have the right phase line in Figure 4.5.

2. $\mu = 0$.

   There is only one equilibrium point at $y = 0$. The equation becomes $y' = y^2$. It is obvious that the right side of this equation is never negative. So, the phase line, which is shown as the middle line in Figure 4.5, has upward-pointing arrows.

3. $\mu < 0$.

   In this case, there are no equilibrium solutions. Because $y^2 - \mu > 0$, the slopes for all solutions are positive, as indicated by the last phase line in Figure 4.5.

We can also confirm the behaviors of the equilibrium points by noting that $f'(y) = 2y$. Then, $f'(\pm\sqrt{\mu}) = \pm 2\sqrt{\mu}$ for $\mu \geq 0$. Therefore, the equilibria $y = +\sqrt{\mu}$ are unstable equilibria for $\mu > 0$. Similarly, the equilibria $y = -\sqrt{\mu}$ are stable equilibria for $\mu > 0$.

Figure 4.6: (a) The typical phase lines for $y' = y^2 - \mu$. (b) Bifurcation diagram for $y' = y^2 - \mu$. This is an example of a saddle-node bifurcation.

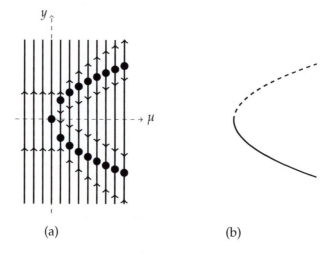

(a)   (b)

We can combine these results for the phase lines into one diagram known as a bifurcation diagram. We will plot the equilibrium solutions and their phase lines $y = \pm\sqrt{\mu}$ in the μy-plane. We begin by lining up the phase lines for various μ's. These are shown on the left side of Figure 4.6. Note that the pattern of equilibrium points lies on the parabolic curve $y^2 = \mu$. The upper branch of this curve is a collection of unstable equilibria and the bottom is a stable branch. So, we can dispose of the phase lines and just keep the equilibria. However, we will draw the unstable branch as a dashed line and the stable branch as a solid line.

The bifurcation diagram is displayed on the right side of Figure 4.6. This type of bifurcation is called a saddle-node bifurcation. The point $\mu = 0$ at which the behavior changes is the bifurcation point. As μ changes from negative to positive values, the system goes from having no equilibria to having one stable and one unstable equilibrium point.

Example 4.8. $y' = y^2 - \mu y$.

Writing this equation in factored form, $y' = y(y - \mu)$, we see that there are two equilibrium points: $y = 0$ and $y = \mu$. The behavior of the solutions depends upon the sign of $y' = y(y - \mu)$. This leads to four cases with the indicated signs of the derivative. The regions indicating the signs of y' are shown in Figure 4.7.

1. $y > 0, y - \mu > 0 \Rightarrow y' > 0$.
2. $y < 0, y - \mu > 0 \Rightarrow y' < 0$.
3. $y > 0, y - \mu < 0 \Rightarrow y' < 0$.
4. $y < 0, y - \mu < 0 \Rightarrow y' > 0$.

The corresponding phase lines and superimposed bifurcation diagram are shown in Figure 4.8. The bifurcation diagram is on the right side of Figure 4.8, and this type of bifurcation is called a transcritical bifurcation.

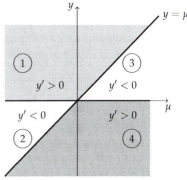

Figure 4.7: The regions indicating the different signs of the derivative for $y' = y^2 - \mu y$.

Figure 4.8: (a) Collection of phase lines for $y' = y^2 - \mu y$. (b) Bifurcation diagram for $y' = y^2 - \mu y$. This is an example of a transcritical bifurcation.

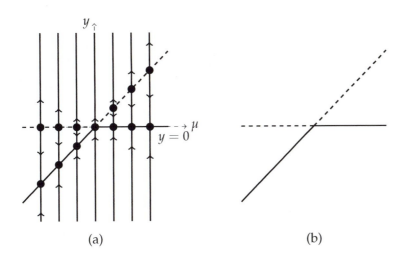

(a)   (b)

Again, the stability can be determined from the derivative $f'(y) = 2y - \mu$ evaluated at $y = 0, \mu$. From $f'(0) = -\mu$, we see that $y = 0$ is stable for $\mu > 0$ and unstable for $\mu < 0$. Similarly, $f'(\mu) = \mu$ implies that $y = \mu$ is unstable for $\mu > 0$ and stable for $\mu < 0$. These results are consistent with the phase line plots.

Example 4.9. $y' = y^3 - \mu y$.

For this last example, we find from $y^3 - \mu y = y(y^2 - \mu) = 0$ that there are two cases.

1. $\mu < 0$. In this case, there is only one equilibrium point at $y = 0$. For positive values of y, we have that $y' > 0$ and for negative values of y, we have that $y' < 0$. Therefore, this is an unstable equilibrium point.

2. $\mu > 0$. Here we have three equilibria: $y = 0, \pm\sqrt{\mu}$. A careful investigation shows that $y = 0$ is a stable equilibrium point and that the other two equilibria are unstable.

Figure 4.9: (a) The phase lines for $y' = y^3 - \mu y$. The left one corresponds to $\mu < 0$ and the right phase line is for $\mu > 0$. (b)Bifurcation diagram for $y' = y^3 - \mu y$. This is an example of a pitchfork bifurcation.

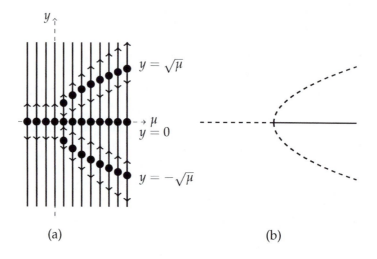

(a)   (b)

In Figure 4.9 we show the phase lines for these two cases. The corresponding bifurcation diagram is then sketched on the right side of Figure 4.9. For obvious reasons, this has been labeled a pitchfork bifurcation.

Because $f'(y) = 3y^2 - \mu$, the stability analysis gives that $f'(0) = -\mu$. So, $y = 0$ is stable for $\mu > 0$ and unstable for $\mu < 0$. For $\mu > 0$, we have that $f'(\pm\sqrt{\mu}) = 2\mu$. Therefore, $y = \pm\sqrt{\mu}$, $\mu > 0$, is unstable. Thus, we have a subcritical pitchfork bifurcation.

When two of the prongs of the pitchfork are unstable branches, the bifurcation is called a subcritical pitchfork bifurcation. When two prongs are stable branches, the bifurcation is a supercritical pitchfork bifurcation.

4.5   Nonlinear Pendulum

IN THIS SECTION WE RETURN TO THE NONLINEAR PENDULUM as an example of periodic motion in a nonlinear system. Oscillations are important in many areas of physics. We have already seen the motion of a mass on a spring, leading to simple, damped, and forced harmonic motions. Later we will explore these effects on a simple nonlinear system. In this section we will investigate the nonlinear pendulum equation (2.36) and determine its period of oscillation.

Recall that the derivation of the pendulum equation was based upon a simple point mass m hanging on a string of length L from some support as shown in Figure 4.10. One pulls the mass back to some starting angle, θ_0, and releases it. The goal is to find the angular position as a function of time, $\theta(t)$.

In Chapter 2 we derived the nonlinear pendulum equation,

$$L\ddot{\theta} + g\sin\theta = 0. \tag{4.11}$$

There are several variations of Equation (4.11) that we have used in this text. The first one is the linear pendulum, which was obtained using a small angle approximation:

$$L\ddot{\theta} + g\theta = 0. \tag{4.12}$$

We also made the system more realistic by adding damping and forcing. A variety of these oscillation problems are summarized in the table below.

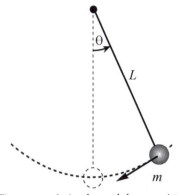

Figure 4.10: A simple pendulum consists of a point mass m attached to a string of length L. It is released from an angle θ_0.

Equations for Pendulum Motion

1. Nonlinear pendulum: $L\ddot{\theta} + g\sin\theta = 0$.

2. Damped nonlinear pendulum: $L\ddot{\theta} + b\dot{\theta} + g\sin\theta = 0$.

3. Linear pendulum: $L\ddot{\theta} + g\theta = 0$.

4. Damped linear pendulum: $L\ddot{\theta} + b\dot{\theta} + g\theta = 0$.

5. Forced damped nonlinear pendulum: $L\ddot{\theta} + b\dot{\theta} + g\sin\theta = F\cos\omega t$.

6. Forced damped linear pendulum: $L\ddot{\theta} + b\dot{\theta} + g\theta = F\cos\omega t$.

4.5.1 Period of the Nonlinear Pendulum

RECALL THAT THE PERIOD OF THE SIMPLE PENDULUM is given by

$$T = \frac{2\pi}{\omega} = 2\pi\sqrt{\frac{L}{g}} \tag{4.13}$$

for

$$\omega \equiv \sqrt{\frac{g}{L}}. \tag{4.14}$$

This was based upon the solving the linear pendulum equation (4.12). This equation was derived assuming a small angle approximation. How good is this approximation? What is meant by a *small* angle?

We recall that the Taylor series approximation of $\sin\theta$ about $\theta = 0$:

$$\sin\theta = \theta - \frac{\theta^3}{3!} + \frac{\theta^5}{5!} + \dots. \tag{4.15}$$

One can obtain a bound on the error when truncating this series to one term after taking a numerical analysis course. But we can just simply plot the relative error, which is defined as

$$\text{Relative Error} = \left|\frac{\sin\theta - \theta}{\sin\theta}\right|.$$

Relative error in $\sin\theta$ approximation.

A plot of the relative error is given in Figure 4.11. Thus, for $\theta \approx 0.4$ radians (or, 23°), we have that the relative error is about 2.6 percent.

We would like to do better than this. So, we now turn to the nonlinear pendulum equation (4.11) in the simpler form

$$\ddot{\theta} + \omega^2 \sin\theta = 0. \qquad (4.16)$$

Figure 4.11: The relative error in percent when approximating $\sin\theta$ by θ.

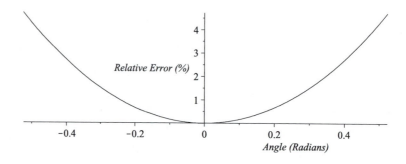

Solution of nonlinear pendulum equation.

We next employ a technique that is useful for equations of the form

$$\ddot{\theta} + F(\theta) = 0$$

when it is easy to integrate the function $F(\theta)$. Namely, we note that

$$\frac{d}{dt}\left[\frac{1}{2}\dot{\theta}^2 + \int^{\theta(t)} F(\phi)\,d\phi\right] = (\ddot{\theta} + F(\theta))\dot{\theta}.$$

For the nonlinear pendulum problem, we multiply Equation (4.16) by $\dot{\theta}$,

$$\ddot{\theta}\dot{\theta} + \omega^2 \sin\theta\dot{\theta} = 0$$

and note that the left side of this equation is a perfect derivative. Thus,

$$\frac{d}{dt}\left[\frac{1}{2}\dot{\theta}^2 - \omega^2 \cos\theta\right] = 0.$$

Therefore, the quantity in the brackets is a constant. So, we can write

$$\frac{1}{2}\dot{\theta}^2 - \omega^2 \cos\theta = c. \qquad (4.17)$$

Solving for $\dot{\theta}$, we obtain

$$\frac{d\theta}{dt} = \sqrt{2(c + \omega^2 \cos\theta)}.$$

This equation is a separable first-order equation, and we can rearrange and integrate the terms to find that

$$t = \int dt = \int \frac{d\theta}{\sqrt{2(c + \omega^2 \cos\theta)}}.$$

Of course, we need to be able to do the integral. When one finds a solution in this implicit form, one says that the problem has been solved by quadratures. Namely, the solution is given in terms of some integral.

In fact, the above integral can be transformed into what is known as an elliptic integral of the first kind. We will rewrite this result and then use it to obtain an approximation to the period of oscillation of the nonlinear pendulum, leading to corrections to the linear result found earlier.

We will first rewrite the constant found in Equation (4.17). This requires a little physics. The swinging of a mass on a string, assuming no energy loss at the pivot point, is a conservative process. Namely, the total mechanical energy is conserved. Thus, the total of the kinetic and gravitational potential energies is a constant. The kinetic energy of the mass on the string is given as

$$T = \frac{1}{2}mv^2 = \frac{1}{2}mL^2\dot{\theta}^2.$$

The potential energy is the gravitational potential energy. If we set the potential energy to zero at the bottom of the swing, then the potential energy is $U = mgh$, where h is the height that the mass is from the bottom of the swing. A little trigonometry gives that $h = L(1 - \cos\theta)$. So,

$$U = mgL(1 - \cos\theta).$$

So, the total mechanical energy is

$$E = \frac{1}{2}mL^2\dot{\theta}^2 + mgL(1 - \cos\theta). \tag{4.18}$$

Total mechanical energy for the nonlinear pendulum.

We note that a little rearranging shows that we can relate this result to Equation (4.17). Dividing by m and L^2 and using the definition of $\omega^2 = g/L$, we have

$$\frac{1}{2}\dot{\theta}^2 - \omega^2\cos\theta = \frac{1}{mL^2}E - \omega^2.$$

Therefore, we have determined the integration constant in terms of the total mechanical energy,

$$c = \frac{1}{mL^2}E - \omega^2.$$

We can use Equation (4.18) to get a value for the total energy. At the top of the swing, the mass is not moving, if only for a moment. Thus, the kinetic energy is zero and the total mechanical energy is pure potential energy. Letting θ_0 denote the angle at the highest angular position, we have that

$$E = mgL(1 - \cos\theta_0) = mL^2\omega^2(1 - \cos\theta_0).$$

Therefore, we have found that

$$\frac{1}{2}\dot{\theta}^2 - \omega^2\cos\theta = -\omega^2\cos\theta_0. \tag{4.19}$$

We can solve for $\dot{\theta}$ and integrate the differential equation to obtain

$$t = \int dt = \int \frac{d\theta}{\omega\sqrt{2(\cos\theta - \cos\theta_0)}}.$$

Using the half angle formula,

$$\sin^2\frac{\theta}{2} = \frac{1}{2}(1 - \cos\theta),$$

we can rewrite the argument in the radical as

$$\cos\theta - \cos\theta_0 = 2\left[\sin^2\frac{\theta_0}{2} - \sin^2\frac{\theta}{2}\right].$$

Noting that a motion from $\theta = 0$ to $\theta = \theta_0$ is a quarter of a cycle, we have that

$$T = \frac{2}{\omega}\int_0^{\theta_0}\frac{d\theta}{\sqrt{\sin^2\frac{\theta_0}{2} - \sin^2\frac{\theta}{2}}}. \tag{4.20}$$

[2] Elliptic integrals were first studied by Leonhard Euler and Giulio Carlo de' Toschi di Fagnano (1682–1766) , who studied the lengths of curves such as the ellipse and the lemniscate,

$$(x^2 + y^2)^2 = x^2 - y^2.$$

This result can now be transformed into an elliptic integral.[2] We define

$$z = \frac{\sin\frac{\theta}{2}}{\sin\frac{\theta_0}{2}}$$

and

$$k = \sin\frac{\theta_0}{2}.$$

Then, Equation (4.20) becomes

$$T = \frac{4}{\omega}\int_0^1\frac{dz}{\sqrt{(1-z^2)(1-k^2z^2)}}. \tag{4.21}$$

This is done by noting that $dz = \frac{1}{2k}\cos\frac{\theta}{2}\,d\theta = \frac{1}{2k}(1-k^2z^2)^{1/2}\,d\theta$ and that $\sin^2\frac{\theta_0}{2} - \sin^2\frac{\theta}{2} = k^2(1-z^2)$. The integral in this result is called the complete elliptic integral of the first kind.

The complete elliptic integral of the first kind.

We note that the incomplete elliptic integral of the first kind is defined as

The incomplete elliptic integral of the first kind.

$$F(\phi, k) \equiv \int_0^\phi\frac{d\theta}{\sqrt{1-k^2\sin^2\theta}} = \int_0^{\sin\phi}\frac{dz}{\sqrt{(1-z^2)(1-k^2z^2)}}.$$

Then, the complete elliptic integral of the first kind is given by $K(k) = F(\frac{\pi}{2}, k)$, or

$$K(k) = \int_0^{\pi/2}\frac{d\theta}{\sqrt{1-k^2\sin^2\theta}} = \int_0^1\frac{dz}{\sqrt{(1-z^2)(1-k^2z^2)}}.$$

Therefore, the period of the nonlinear pendulum is given by

$$T = \frac{4}{\omega}K\left(\sin\frac{\theta_0}{2}\right). \tag{4.22}$$

There are table of values for elliptic integrals. However, one can use a computer algebra system to compute values of such integrals. We will look for small angle approximations.

For small angles ($\theta_0 \ll \frac{\pi}{2}$), we have that k is small. So, we can develop a series expansion for the period, T, for small k. This is simply done using the binomial expansion

$$(1-k^2z^2)^{-1/2} = 1 + \frac{1}{2}k^2z^2 + \frac{3}{8}k^2z^4 + O((kz)^6).$$

Inserting this expansion into the integrand for the complete elliptic integral and integrating term by term, we find that

$$T = 2\pi\sqrt{\frac{L}{g}}\left[1 + \frac{1}{4}k^2 + \frac{9}{64}k^4 + \dots\right]. \tag{4.23}$$

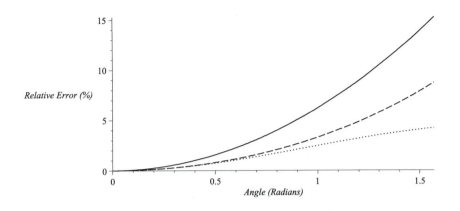

Figure 4.12: The relative error in percent when approximating the exact period of a nonlinear pendulum with one (solid), two (dashed), or three (dotted) terms in Equation (4.23).

The first term of the expansion gives the well-known period of the simple pendulum for small angles. The next terms in the expression give further corrections to the linear result and are useful for larger amplitudes of oscillation. In Figure 4.12 we show the relative errors incurred when keeping the k^2 (quadratic) and k^4 (quartic) terms as compared to the exact value of the period.

4.6   Stability of Fixed Points in Nonlinear Systems

WE NEXT INVESTIGATE THE STABILITY OF THE EQUILIBRIUM SOLUTIONS of the nonlinear pendulum. Along the way we will develop some basic methods for studying the stability of equilibria in nonlinear systems in general.

There are two simple systems that we will consider: the damped linear pendulum,

$$x'' + bx' + \omega^2 x = 0,$$

and the the damped nonlinear pendulum,

$$x'' + bx' + \omega^2 \sin x = 0.$$

These are second-order differential equations and can be cast as a system of two first-order differential equations using the methods of Chapter 2.

The linear equation can be written as

$$
\begin{aligned}
x' &= y, \\
y' &= -by - \omega^2 x.
\end{aligned}
\tag{4.24}
$$

This system has only one equilibrium solution: $x = 0, y = 0$.

The damped nonlinear pendulum takes the form

$$
\begin{aligned}
x' &= y, \\
y' &= -by - \omega^2 \sin x.
\end{aligned}
\tag{4.25}
$$

This system also has the equilibrium solution $x = 0$, $y = 0$. However, there are actually an infinite number of solutions. The equilibria are determined from

$$y = 0 \text{ and } -by - \omega^2 \sin x = 0. \tag{4.26}$$

These equations imply that $y = 0$ and $\sin x = 0$. There are an infinite number of solutions to the latter equation: $x = n\pi, n = 0, \pm1, \pm2, \ldots$. So, this system has an infinite number of equilibria, $(n\pi, 0), n = 0, \pm1, \pm2, \ldots$.

The next step is to determine the stability of the equilibrium solutions for these systems. This can be accomplished just as we did for first-order equations. To do this, we need a more general theory for nonlinear systems. So, we will develop the needed machinery.

We begin with the n-dimensional system

$$\mathbf{x}' = \mathbf{f}(\mathbf{x}), \quad \mathbf{x} \in R^n. \tag{4.27}$$

Here, $\mathbf{f} : R^n \to R^n$ is a mapping from R^n to R^n. We define the equilibrium solutions, or fixed points, of this system as the points $\mathbf{x}^*$ satisfying $\mathbf{f}(\mathbf{x}^*) = \mathbf{0}$.

The stability in the neighborhood of equilibria will now be determined. We are interested in what happens to solutions of the system with initial conditions starting near a fixed point. We will represent a general point in the plane, which is near the fixed point, in the form $\mathbf{x} = \mathbf{x}^* + \boldsymbol{\xi}$. We note that the length of $\boldsymbol{\xi}$ gives an indication of how close we are to the fixed point. So, we consider that initially, $|\boldsymbol{\xi}| \ll 1$.

As the system evolves, $\boldsymbol{\xi}$ will change. The change of $\boldsymbol{\xi}$ in time is, in turn, governed by a system of equations. We can approximate this evolution as follows. First, we note that

$$\mathbf{x}' = \boldsymbol{\xi}'.$$

Next, we have that

$$\mathbf{f}(\mathbf{x}) = \mathbf{f}(\mathbf{x}^* + \boldsymbol{\xi}).$$

We can expand the right side about the fixed point using a multidimensional version of Taylor's Theorem. Thus, we have that

$$\mathbf{f}(\mathbf{x}^* + \boldsymbol{\xi}) = \mathbf{f}(\mathbf{x}^*) + D\mathbf{f}(\mathbf{x}^*)\boldsymbol{\xi} + O(|\boldsymbol{\xi}|^2).$$

Here, $D\mathbf{f}(\mathbf{x})$ is the *Jacobian matrix*, defined as

$$D\mathbf{f}(\mathbf{x}^*) \equiv \begin{pmatrix} \frac{\partial f_1}{\partial x_1} & \frac{\partial f_1}{\partial x_2} & \cdots & \frac{\partial f_1}{\partial x_n} \\ \frac{\partial f_2}{\partial x_1} & \ddots & \ddots & \vdots \\ \vdots & \ddots & \ddots & \vdots \\ \frac{\partial f_n}{\partial x_1} & \cdots & \cdots & \frac{\partial f_n}{\partial x_n} \end{pmatrix}.$$

Noting that $\mathbf{f}(\mathbf{x}^*) = \mathbf{0}$, we then have that system (4.27) becomes

$$\boldsymbol{\xi}' \approx D\mathbf{f}(\mathbf{x}^*)\boldsymbol{\xi}. \tag{4.28}$$

It is this equation that describes the behavior of the system near the fixed point. As with first-order equations, we say that system (4.27) has been linearized or that Equation (4.28) is the linearization of system (4.27).

Linear stability analysis of systems.

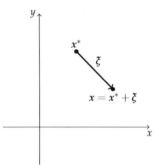

Figure 4.13: A general point in the plane, which is near the fixed point, in the form $\mathbf{x} = \mathbf{x}^* + \boldsymbol{\xi}$,

The Jacobian matrix.

Linearization of the system $\mathbf{x}' = \mathbf{f}(\mathbf{x})$.

The stability of the equilibrium point of the nonlinear system is now reduced to analyzing the behavior of the linearized system given by Equation (4.28). We can use the methods from the two previous chapters to investigate the eigenvalues of the Jacobian matrix evaluated at each equilibrium point. We will demonstrate this procedure with several examples.

Example 4.10. *Determine the equilibrium points and their stability for the system*

$$
\begin{aligned}
x' &= -2x - 3xy, \\
y' &= 3y - y^2.
\end{aligned} \tag{4.29}
$$

We first determine the fixed points. Setting the right-hand side equal to zero and factoring, we have

$$
\begin{aligned}
-x(2+3y) &= 0, \\
y(3-y) &= 0.
\end{aligned} \tag{4.30}
$$

From the second equation, we see that either $y = 0$ or $y = 3$. The first equation then gives $x = 0$ in either case. So, there are two fixed points: $(0,0)$ and $(0,3)$.

Next, we linearize the system of differential equations about each fixed point. First, we note that the Jacobian matrix is given by

$$
\mathbf{Df}(x,y) = \begin{pmatrix} -2 - 3y & -3x \\ 0 & 3 - 2y \end{pmatrix}. \tag{4.31}
$$

1. *Case I Equilibrium point $(0,0)$.*

   In this case, we find that

   $$
   \mathbf{Df}(0,0) = \begin{pmatrix} -2 & 0 \\ 0 & 3 \end{pmatrix}. \tag{4.32}
   $$

   Therefore, the linearized equation becomes

   $$
   \xi' = \begin{pmatrix} -2 & 0 \\ 0 & 3 \end{pmatrix} \xi. \tag{4.33}
   $$

   This is equivalently written out as the system

   $$
   \begin{aligned}
   \xi_1' &= -2\xi_1, \\
   \xi_2' &= 3\xi_2.
   \end{aligned} \tag{4.34}
   $$

   This is the linearized system about the origin. Note the similarity with the original system.

   We should emphasize that the linearized equations are constant coefficient equations and we can use matrix methods to determine the nature of the equilibrium point. The eigenvalues of this system are obviously $\lambda = -2, 3$. Therefore, we have that the origin is a saddle point.

2. *Case II Equilibrium point $(0,3)$.*

Again we evaluate the Jacobian matrix at the equilibrium point and look at its eigenvalues to determine the type of fixed point. The Jacobian matrix for this case becomes

$$D\mathbf{f}(0,3) = \begin{pmatrix} -11 & 0 \\ 0 & -3 \end{pmatrix}. \qquad (4.35)$$

The eigenvalues are $\lambda = -11, -3$. So, this fixed point is a stable node.

This analysis has given us a saddle and a stable node. We know what the behavior is like near each fixed point, but we have to resort to other means to say anything about the behavior far from these points. The phase portrait for this system is given in Figure 4.14. You should be able to locate the saddle point and the node in the figure. Notice how solutions behave in regions far from these points.

Figure 4.14: Phase plane for the system $x' = -2x - 3xy$, $y' = 3y - y^2$.

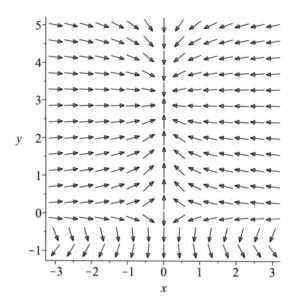

We can expect to be able to perform a linearization under general conditions. These are given in the *Hartman-Großman Theorem*:

Theorem 4.1. *A continuous map exists between the linear and nonlinear systems when $D\mathbf{f}(\mathbf{x}^*)$ does not have any eigenvalues with zero real part.*

Generally, there are several types of behavior that one can see in nonlinear systems. One can see sinks or sources, hyperbolic (saddle) points, elliptic points (centers) or foci. We have defined some of these for planar systems. In general, if at least two eigenvalues have real parts with opposite signs, then the fixed point is a *hyperbolic point*. If the real part of a nonzero eigenvalue is zero, then we have a center, or *elliptic point*.

For linear systems in the plane, this classification was done in Chapter 3. The Jacobian matrix evaluated at the equilibrium points is simply a 2×2 matrix.

$$J = \begin{pmatrix} a & b \\ c & d \end{pmatrix}. \qquad (4.36)$$

Here we are using $J = D\mathbf{f}(\mathbf{x}^*)$.

The eigenvalue equation is given by

$$\lambda^2 - (a + d)\lambda + (ad - bc) = 0.$$

However, $a + d$ is the trace, $\mathrm{tr}(J)$ and $\det(J) = ad - bc$. Therefore, we can write the eigenvalue equation as

$$\lambda^2 - \mathrm{tr}(J)\lambda + \det(J) = 0.$$

The solution of this equation is found using the quadratic formula,

$$\lambda = \frac{1}{2}\left[-\mathrm{tr}(J) \pm \sqrt{\mathrm{tr}^2(J) - 4\det(J)} \right].$$

We had seen in Chapters 2 and 3 that equilibrium points in planar systems can be classified as nodes, saddles, centers, or spirals (foci). The type of behavior can be determined from solutions of the eigenvalue equation. Because the nature of the eigenvalues depends on the trace and determinant of the Jacobian matrix at the equilibrium point, we can relate the types of equilibria to points in the det-tr plane. This is shown in Figure 4.15.

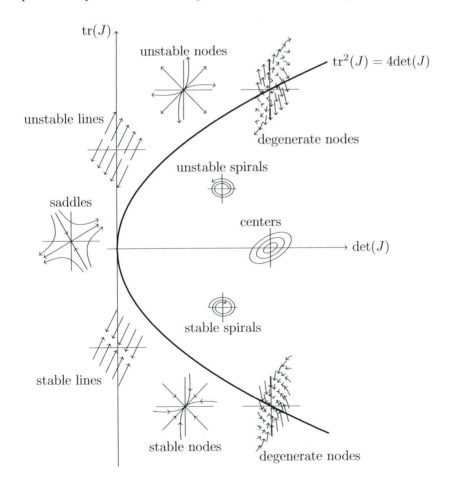

Figure 4.15: Diagram indicating the behavior of equilibrium points in the det-tr plane. The parabolic curve

$$\mathrm{tr}^2(J) = 4\det(J)$$

indicates where the discriminant vanishes.

In Figure 4.15, the parabola $\mathrm{tr}^2(J) = 4\det(J)$ divides the det-tr plane. Points on this curve give a vanishing discriminant in the computation of the eigenvalues. In these cases, one finds repeated roots, or eigenvalues. Along this curve one can find stable and unstable degenerate nodes. Also along this line are stable and unstable proper nodes, called star nodes. These arise from systems of the form $x' = ax$, $y' = ay$.

In the case that $\det(J) < 0$, we have that the discriminant

$$\Delta \equiv \text{tr}^2(J) - 4\det(J)$$

is positive. Not only that, $\Delta > \text{tr}^2(J)$. Thus, we obtain two real and distinct eigenvalues with opposite signs. These lead to saddle points.

In the case that $\det(J) > 0$, we can have either $\Delta > 0$ or $\Delta < 0$. The discriminant is negative for points inside the parabolic curve. It is in this region that one finds centers and spirals, corresponding to complex eigenvalues. When $\text{tr}(J) > 0$, there are unstable spirals. There are stable spirals when $\text{tr}(J) < 0$. For the case that $\text{tr}(J) = 0$, the eigenvalues are pure imaginary, giving centers.

There are several other types of behavior depicted in the figure, but we will now turn to studying a few of examples.

Example 4.11. *Find and classify all of the equilibrium solutions of the nonlinear system*

$$
\begin{aligned}
x' &= 2x - y + 2xy + 3(x^2 - y^2), \\
y' &= x - 3y + xy - 3(x^2 - y^2).
\end{aligned}
\tag{4.37}
$$

In Figure 4.16, we show the direction field for this system. Try to locate and classify the equilibrium points visually. After the stability analysis, you should return to this figure and determine if you identified the equilibrium points correctly.

Figure 4.16: Phase plane for the system
$$x' = 2x - y + 2xy + 3(x^2 - y^2),$$
$$y' = x - 3y + xy - 3(x^2 - y^2).$$

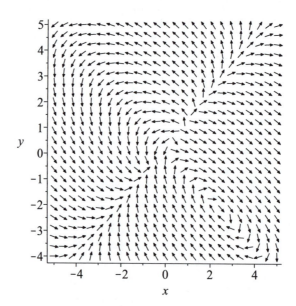

We will first determine the equilibrium points. Setting the right-hand side of each differential equation to zero, we have

$$2x - y + 2xy + 3(x^2 - y^2) = 0,$$

$$x - 3y + xy - 3(x^2 - y^2) \;=\; 0. \qquad\qquad (4.38)$$

This system of algebraic equations can be solved exactly. Adding the equations, we have

$$3x - 4y + 3xy = 0.$$

Solving for x,

$$x = \frac{4y}{3(1+y)},$$

and substituting the result for x into the first algebraic equation, we find an equation for y:

$$\frac{y(1-y)(9y^2 + 22y + 5)}{3(1+y)^2} = 0.$$

The solutions to this equation are

$$y = 0, 1, -\frac{11}{9} \pm \frac{2}{9}\sqrt{19}.$$

The corresponding values for x are

$$x = 0, \frac{2}{3}, 1 \mp \frac{\sqrt{19}}{3}.$$

   Now that we have located the equilibria, we can classify them. The Jacobian matrix is given by

$$D\mathbf{f}(x,y) = \begin{pmatrix} 6x + 2y + 2 & 2x - 6y - 1 \\ -6x + y + 1 & x + 6y - 3 \end{pmatrix}. \qquad (4.39)$$

Now we evaluate the Jacobian at each equilibrium point and find the eigenvalues.

1. *Case I. Equilibrium point $(0,0)$.*

   In this case, we find that

   $$D\mathbf{f}(0,0) = \begin{pmatrix} -2 & -1 \\ 1 & -3 \end{pmatrix}. \qquad\qquad (4.40)$$

   The eigenvalues of this matrix are $\lambda = -\frac{1}{2} \pm \frac{\sqrt{21}}{2}$. Therefore, the origin is a saddle point.

2. *Case II. Equilibrium point $\left(\frac{2}{3}, 1\right)$.*

   Again we evaluate the Jacobian matrix at the equilibrium point and look at its eigenvalues to determine the type of fixed point. The Jacobian matrix for this case becomes

   $$D\mathbf{f}\left(\frac{2}{3}, 1\right) = \begin{pmatrix} 8 & -\frac{17}{3} \\ -2 & \frac{11}{3} \end{pmatrix}. \qquad (4.41)$$

   The eigenvalues are $\lambda = \frac{35}{6} \pm \frac{\sqrt{577}}{6} \approx 9.84, 1.83$. This fixed point is an unstable node.

3. *Case III. Equilibrium point* $(1 \mp \frac{\sqrt{19}}{3}, -\frac{11}{9} \pm \frac{2}{9}\sqrt{19})$.

The Jacobian matrix for this case becomes

$$D\mathbf{f}\left(1 \mp \frac{\sqrt{19}}{3}, -\frac{11}{9} \pm \frac{2}{9}\sqrt{19}\right) = \begin{pmatrix} \frac{50}{9} \mp \frac{14}{9}\sqrt{19} & \frac{25}{3} \mp 2\sqrt{19} \\ -\frac{56}{9} \pm \frac{20}{9}\sqrt{19} & -\frac{28}{3} \pm \sqrt{19} \end{pmatrix}.$$
(4.42)

There are two equilibrium points under this case. The first is given by

$$(1 - \frac{\sqrt{19}}{3}, -\frac{11}{9} + \frac{2}{9}\sqrt{19}) \approx (0.453, -0.254).$$

The eigenvalues for this point are

$$\lambda = -\frac{17}{9} - \frac{5}{18}\sqrt{19} \pm \frac{1}{18}\sqrt{3868\sqrt{19} - 16153}.$$

These are approximately -4.58 and -1.62 So, this equilibrium point is a stable node.

The other equilibrium is $(1 + \frac{\sqrt{19}}{3}, -\frac{11}{9} - \frac{2}{9}\sqrt{19}) \approx (2.45, -2.19)$. The corresponding eigenvalues are complex with negative real parts,

$$\lambda = -\frac{17}{9} + \frac{5}{18}\sqrt{19} \pm \frac{i}{18}\sqrt{16153 + 3868\sqrt{19}},$$

or $\lambda \approx -0.678 \pm 10.1i$. This point is a stable spiral.

Plots of the phase plane are given in Figures 4.14 and 4.16. The reader can look at the direction field and verify these results for the behavior of equilibrium solutions. A zoomed-in view is shown in Figure 4.17 with several orbits indicated.

Figure 4.17: A closer look at the phase plane for the system

$$x' = 2x - y + 2xy + 3(x^2 - y^2),$$

$$y' = x - 3y + xy - 3(x^2 - y^2)$$

with a few trajectories shown.

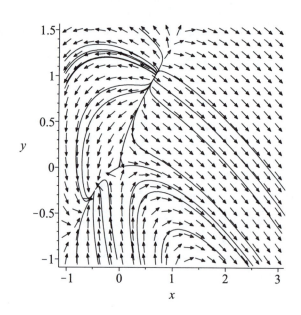

Example 4.12. Damped Nonlinear Pendulum Equilibria

We are now ready to establish the behavior of the fixed points of the damped nonlinear pendulum system in Equation (4.25). Recall that the system for the *damped nonlinear pendulum was given by*

$$x' = y,$$
$$y' = -by - \omega^2 \sin x. \tag{4.43}$$

For a damped system, we will need $b > 0$. We had found that there are an infinite number of equilibrium points at $(n\pi, 0)$, $n = 0, \pm1, \pm2, \ldots$

The Jacobian matrix for this systems is

$$D\mathbf{f}(x, y) = \begin{pmatrix} 0 & 1 \\ -\omega^2 \cos x & -b \end{pmatrix}. \tag{4.44}$$

Evaluating this matrix at the fixed points, we find that

$$D\mathbf{f}(n\pi, 0) = \begin{pmatrix} 0 & 1 \\ -\omega^2 \cos n\pi & -b \end{pmatrix} = \begin{pmatrix} 0 & 1 \\ (-1)^{n+1}\omega^2 & -b \end{pmatrix}. \tag{4.45}$$

The eigenvalue equation is given by

$$\lambda^2 + b\lambda + (-1)^n \omega^2 = 0.$$

There are two cases to consider: n even and n odd. For the first case, we find the eigenvalues

$$\lambda = \frac{-b \pm \sqrt{b^2 - 4\omega^2}}{2}.$$

For $b^2 < 4\omega^2$, we have two complex conjugate roots with a negative real part. Thus, we have stable foci for even n values. If there is no damping, then we obtain centers ($\lambda = \pm i\omega$).

In the second case, n odd, we find

$$\lambda = \frac{-b \pm \sqrt{b^2 + 4\omega^2}}{2}.$$

Because $b^2 + 4\omega^2 > b^2$, these roots will be real with opposite signs. Thus, we have hyperbolic points, or saddles. If there is no damping, the eigenvalues reduce to $\lambda = \pm \omega$.

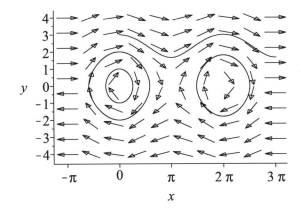

Figure 4.18: Phase plane for the undamped nonlinear pendulum. Solution curves are shown for initial conditions $(x_0, y_0) = (0, 3), (0, 2), (0, 1), (5, 1)$.

In Figure (4.18) we show the phase plane for the undamped nonlinear pendulum with $\omega = 1.25$. We see that we have a mixture of centers and saddles. There are orbits for which there is periodic motion. In the case that $\theta = \pi$, we have an inverted pendulum. This is an unstable position and this is reflected in the presence of saddle points, especially if the pendulum is constructed using a massless rod.

There are also unbounded orbits, going through all possible angles. These correspond to the mass spinning around the pivot in one direction forever due to initially having large enough energies.

We have indicated in the figure solution curves with the initial conditions $(x_0, y_0) = (0,3), (0,2), (0,1), (5,1)$. These show the various types of motions that we have described.

Figure 4.19: Phase plane for the damped nonlinear pendulum. Solution curves are shown for initial conditions $(x_0, y_0) = (0,3), (0,2), (0,1), (5,1)$.

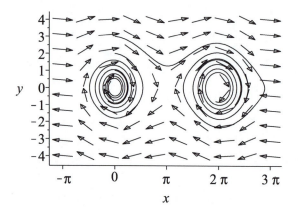

When there is damping, we see that we can have a variety of other behaviors, as seen in Figure 4.19. In this example, we have set $b = 0.08$ and $\omega = 1.25$. We see that energy loss results in the mass settling around one of the stable fixed points. This leads to an understanding as to why there are an infinite number of equilibria, even though physically the mass traces out a bound set of Cartesian points. We have indicated in Figure 4.19 solution curves with the initial conditions $(x_0, y_0) = (0,3), (0,2), (0,1), (5,1)$.

In Figure 4.20, we show a region of the phase plane that corresponds to oscillations about $x = 0$. For small angles, the pendulum oscillates following somewhat elliptical orbits. As the angles get larger, due to greater initial energies, these orbits begin to change from ellipses to other periodic orbits. There is a limiting orbit, beyond which one has unbounded motion. The limiting orbit connects the saddle points on either side of the center. The curve is called a separatrix and being that these trajectories connect two saddles, they are often referred to as heteroclinic orbits.

In Figures 4.21 and 4.21, we show more orbits, including both bound and unbound motion beyond the interval $x \in [-\pi, \pi]$. For both plots we have chosen $\omega = 5$ and the same set of initial conditions, $x(0) = \pi k/10$, $k = -20, \dots, 20$. for $y(0) = 0, \pm 10$. The time interval is taken for $t \in [-3, 3]$. The only difference is that in the damped case, we have $b = 0.5$. In these plots one can see what happens to the heteroclinic orbits and nearby unbounded orbits under damping.

Heteroclinc orbits and separatrices.

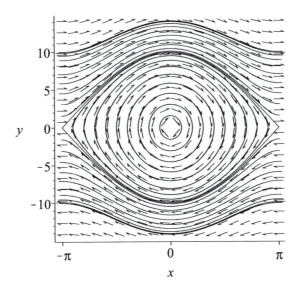

Figure 4.20: Several orbits in the phase plane for the undamped nonlinear pendulum with $\omega = 5.0$. The orbits surround a center at $(0,0)$. At the edges there are saddle points, $(\pm\pi, 0)$.

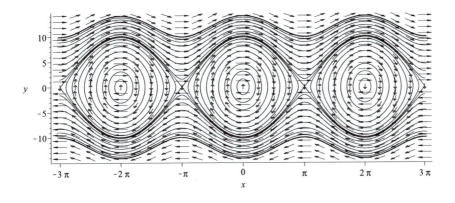

Figure 4.21: Several orbits in the phase plane for the undamped nonlinear pendulum with $\omega = 5.0$.

Before leaving this problem, we should note that the orbits in the phase plane for the undamped nonlinear pendulum can be obtained graphically. Recall from Equation (4.18) that the total mechanical energy for the nonlinear pendulum is

$$E = \frac{1}{2}mL^2\dot{\theta}^2 + mgL(1 - \cos\theta).$$

From this equation we obtained Equation (4.19),

$$\frac{1}{2}\dot{\theta}^2 - \omega^2\cos\theta = -\omega^2\cos\theta_0.$$

Letting $y = \dot{\theta}$, $x = \theta$, and defining $z = -\omega^2\cos\theta_0$, this equation can be written as

$$\frac{1}{2}y^2 - \omega^2\cos x = z. \tag{4.46}$$

For each energy (z), this gives a constant energy curve. Plotting the family of energy curves, we obtain the phase portrait shown in Figure 4.23.

Figure 4.22: Several orbits in the phase plane for the damped nonlinear pendulum with $\omega = 5.0$ and $b = 0.5$.

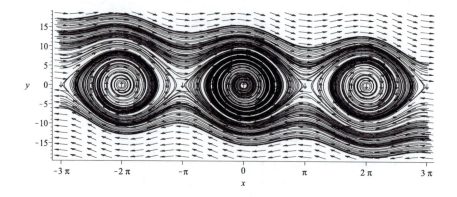

Figure 4.23: A family of energy curves in the phase plane for $\frac{1}{2}\dot{\theta}^2 - \omega^2 \cos\theta = z$. Here we took $\omega = 1.0$ and $z \in [-5, 15]$.

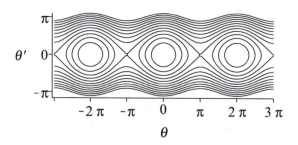

4.7 Nonlinear Population Models

WE HAVE ALREADY ENCOUNTERED SEVERAL MODELS of population dynamics in this and previous chapters. Of course, one could dream up several other examples. While such models might seem far from applications in physics, it turns out that these models lead to systems od differential equations which also appear in physical systems such as the coupling of waves in lasers, in plasma physics, and in chemical reactions.

Two well-known nonlinear population models are the predator-prey and competing species models. In the predator-prey model, one typically has one species, the predator, feeding on the other, the prey. We will look at the standard Lotka–Volterra model in this section. The competing species model looks similar, except there are a few sign changes, as one species is not feeding on the other. Also, we can build logistic terms into our model. We will save this latter type of model for the homework.

The Lotka–Volterra Model is named after Alfred James Lotka (1880–1949) and Vito Volterra (1860–1940).

The Lotka–Volterra Model of population dynamics.

The Lotka–Volterra Model   takes the form

$$
\begin{aligned}
\dot{x} &= ax - bxy, \\
\dot{y} &= -dy + cxy,
\end{aligned}
\qquad (4.47)
$$

where a, b, c, and d are positive constants. In this model, we can think of x as the population of rabbits (prey) and y is the population of foxes (predators). Choosing all constants to be positive, we can describe the terms.

- *ax*: When left alone, the rabbit population will grow. Thus, *a* is the natural growth rate without predators.

- *−dy*: When there are no rabbits, the fox population should decay. Thus, the coefficient needs to be negative.

- *−bxy*: We add a nonlinear term corresponding to the depletion of the rabbits when the foxes are around.

- *cxy*: The more rabbits there are, the more food for the foxes. So, we add a nonlinear term giving rise to an increase in the fox population.

Example 4.13. *Determine the equilibrium points and their stability for the Lotka–Volterra system.*

The analysis of the Lotka–Volterra Model begins with determining the fixed points. So, we have from Equation (4.47)

$$
\begin{aligned}
x(a - by) &= 0, \\
y(-d + cx) &= 0.
\end{aligned}
\tag{4.48}
$$

Therefore, the origin, $(0,0)$, and $(\frac{d}{c}, \frac{a}{b})$ are the fixed points.

Next, we determine their stability, by linearization about the fixed points. We can use the Jacobian matrix, or we could just expand the right-hand side of each equation in (4.47) about the equilibrium points as shown in he next example. The Jacobian matrix for this system is

$$
Df(x,y) = \begin{pmatrix} a - by & -bx \\ cy & -d + cx \end{pmatrix}.
$$

Evaluating at each fixed point, we have

$$
Df(0,0) = \begin{pmatrix} a & 0 \\ 0 & -d \end{pmatrix},
\tag{4.49}
$$

$$
Df\left(\frac{d}{c}, \frac{a}{b}\right) = \begin{pmatrix} 0 & -\frac{bd}{c} \\ \frac{ac}{b} & 0 \end{pmatrix}.
\tag{4.50}
$$

The eigenvalues of Expression (4.49) are $\lambda = a, -d$. So, the origin is a saddle point.

The eigenvalues of Expression (4.50) satisfy $\lambda^2 + ad = 0$. So, the other point is a center. In Figure 4.24, we show a sample direction field for the Lotka–Volterra system.

Another way to carry out the linearization of the system of differential equations is to expand the equations about the fixed points. For fixed points (x^*, y^*), we let

$$
(x,y) = (x^* + u, y^* + v).
$$

Inserting this translation of the origin into the equations of the system, and dropping nonlinear terms in *u* and *v*, results in the linearized system. This method is equivalent to analyzing the Jacobian matrix for each fixed point.

Direct linearization of a system is carried out by introducing $x = x^* + \xi$, or $(x,y) = (x^* + u, y^* + v)$ into the system and dropping nonlinear terms in *u* and *v*.

Figure 4.24: Phase plane for the Lotka–Volterra system given by $\dot{x} = x - 0.2xy$, $\dot{y} = -y + 0.2xy$. Solution curves are shown for initial conditions $(x_0, y_0) = (8, 3), (1, 5)$.

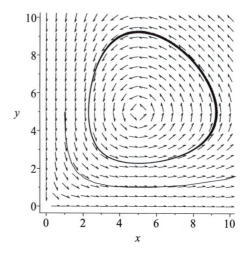

Example 4.14. *Expand the Lotka–Volterra system about the equilibrium points.*

For the origin $(0, 0)$, the linearization about the origin amounts to simply dropping the nonlinear terms. In this case, we have

$$
\begin{aligned}
\dot{u} &= au, \\
\dot{v} &= -dv.
\end{aligned}
\tag{4.51}
$$

The coefficient matrix for this system is the same as $Df(0, 0)$.

For the second fixed point, we let

$$
(x, y) = \left(\frac{d}{c} + u, \frac{a}{b} + v \right).
$$

Inserting this transformation into the system gives

$$
\begin{aligned}
\dot{u} &= a\left(\frac{d}{c} + u \right) - b\left(\frac{d}{c} + u \right)\left(\frac{a}{b} + v \right), \\
\dot{v} &= -d\left(\frac{a}{b} + v \right) + c\left(\frac{d}{c} + u \right)\left(\frac{a}{b} + v \right).
\end{aligned}
\tag{4.52}
$$

Expanding, we obtain

$$
\begin{aligned}
\dot{u} &= \frac{ad}{c} + au - b\left(\frac{ad}{bc} + \frac{d}{c}v + \frac{a}{b}u + uv \right), \\
\dot{v} &= -\frac{ad}{b} - dv + c\left(\frac{ad}{bc} + \frac{d}{c}v + \frac{a}{b}u + uv \right).
\end{aligned}
\tag{4.53}
$$

In both equations, the constant terms cancel and linearization is simply getting rid of the uv terms. This leaves the linearized system

$$
\begin{aligned}
\dot{u} &= au - b\left(\frac{d}{c}v + \frac{a}{b}u \right), \\
\dot{v} &= -dv + c\left(+\frac{d}{c}v + \frac{a}{b}u \right),
\end{aligned}
\tag{4.54}
$$

or

$$
\begin{aligned}
\dot{u} &= -\frac{bd}{c}v, \\
\dot{v} &= \frac{ac}{b}u.
\end{aligned}
\tag{4.55}
$$

The coefficient matrix for this linearized system is the same as $Df\left(\frac{d}{c},\frac{a}{b}\right)$. In fact, for nearby orbits, they are almost circular orbits. From this linearized system, we have $\ddot{u} + adu = 0$.

We can take $u = A\cos(\sqrt{ad}t + \phi)$, where A and ϕ can be determined from the initial conditions. Then,

$$
\begin{aligned}
v &= -\frac{c}{bd}\dot{u} \\
&= \frac{c}{bd}A\sqrt{ad}\sin(\sqrt{ad}t + \phi) \\
&= \frac{c}{b}\sqrt{\frac{a}{d}}A\sin(\sqrt{ad}t + \phi).
\end{aligned}
$$

(4.56)

Therefore, the solutions near the center are given by

$$(x,y) = \left(\frac{d}{c} + A\cos(\sqrt{ad}t + \phi), \frac{a}{b} + \frac{c}{b}\sqrt{\frac{a}{d}}A\sin(\sqrt{ad}t + \phi)\right).$$

For $a = d = 1$, $b = c = 0.2$, and initial values of $(x_0, y_0) = (5.5, 5)$, these solutions become

$$x(t) = 5.0 + 0.5\cos t, \quad y(t) = 5.0 + 0.5\sin t.$$

Plots of these solutions are shown in Figure 4.25.

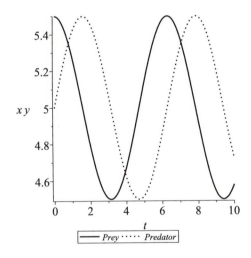

Figure 4.25: The linearized solutions of Lotka–Volterra system $\dot{x} = x - 0.2xy$, $\dot{y} = -y + 0.2xy$ for the initial conditions $(x_0, y_0) = (5.5, 5)$.

It is also possible to find a first integral of the Lotka–Volterra system whose level curves give the phase portrait of the system. As done in Chapter 2, we can write

$$
\begin{aligned}
\frac{dy}{dx} &= \frac{\dot{y}}{\dot{x}} \\
&= \frac{-dy + cxy}{ax - bxy} \\
&= \frac{y(-d + cx)}{x(a - by)}.
\end{aligned}
$$

(4.57)

A function $F(x, y)$ is said to be homogeneous of degree k if $F(tx, ty) = t^k F(x, y)$.

The function on the right-hand side is a homogenous function of degree 1.

The first integral of the Lotka–Volterra system.

Figure 4.26: Phase plane for the Lotka–Volterra system given by $\dot{x} = x - 0.2xy$, $\dot{y} = -y + 0.2xy$ based upon the first integral of the system.

This is an equation is now a separable differential equation. The solution of this differential equation is given in implicit form as

$$a \ln y + d \ln x - cx - by = C,$$

where C is an arbitrary constant. This expression is known as the first integral of the Lotka–Volterra system. These level curves are shown in Figure 4.26.

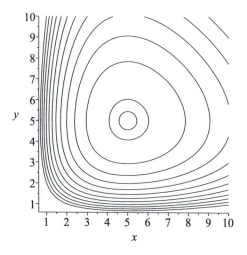

4.8 Limit Cycles

So far we have just been concerned with equilibrium solutions and their behavior. However, asymptotically stable fixed points are not the only attractors. There are other types of solutions, known as limit cycles, toward which a solution may tend. In this section we will look at some examples of these periodic solutions.

Such solutions are common in nature. Rayleigh investigated the problem

$$x'' + c\left(\frac{1}{3}(x')^2 - 1\right)x' + x = 0 \tag{4.58}$$

in the study of the vibrations of a violin string. Balthasar van der Pol (1889–1959) studied an electrical circuit, modeling this behavior. Others have looked into biological systems, such as neural systems, chemical reactions, such as Michaelis-Menten kinetics, and other chemical systems leading to chemical oscillations. One of the most important models in the historical study of dynamical systems is that of planetary motion and investigating the stability of planetary orbits. As is well known, these orbits are periodic.

Limit cycles are isolated periodic solutions toward which neighboring states might tend when stable. A key example exhibiting a limit cycle is given in the next example.

Example 4.15. *Find the limit cycle in the system*

$$x' = \mu x - y - x(x^2 + y^2),$$
$$y' = x + \mu y - y(x^2 + y^2). \tag{4.59}$$

It is clear that the origin is a fixed point. The Jacobian matrix is given as

$$Df(0,0) = \begin{pmatrix} \mu & -1 \\ 1 & \mu \end{pmatrix}. \tag{4.60}$$

The eigenvalues are found to be $\lambda = \mu \pm i$. For $\mu = 0$, we have a center. For $\mu < 0$, we have a stable spiral; and for $\mu > 0$, we have an unstable spiral. However, this spiral does not wander off to infinity. We see in Figure 4.27 that the equilibrium point is a spiral. However, in Figure 4.28 it is clear that the solution does not spiral out to infinity. It is bounded by a circle.

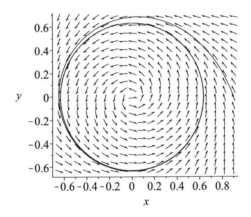

Figure 4.27: Phase plane for system (4.59) with $\mu = 0.4$.

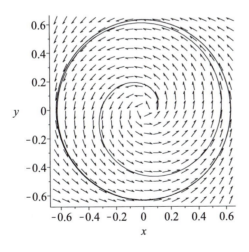

Figure 4.28: Phase plane for system (4.59) with $\mu = 0.4$, showing that the inner spiral is bounded by a limit cycle.

One can actually find the radius of this circle. This requires rewriting the system in polar form. Recall from Chapter 2 that we can change derivatives of Cartesian coordinates to derivatives of polar coordinates using the following relations:

$$rr' = xx' + yy', \tag{4.61}$$
$$r^2\theta' = xy' - yx'. \tag{4.62}$$

Inserting the system (4.59) into these expressions, we have

$$rr' = \mu r^2 - r^4, \qquad r^2 \theta' = r^2.$$

This leads to the system

$$\begin{aligned} r' &= \mu r - r^3, \\ \theta' &= 1. \end{aligned} \qquad (4.63)$$

Of course, for a circle the radius is constant: $r = const$. Therefore, in order to find the limit cycle, we need to look at the equilibrium solutions of Equation (4.63). This amounts to finding the constant solutions of $\mu r - r^3 = 0$. The equilibrium solutions are $r = 0, \pm\sqrt{\mu}$. The limit cycle corresponds to the positive radius solution, $r = \sqrt{\mu}$.

In Figures 4.27 and 4.28, we take $\mu = 0.4$. In this case, we expect a circle with $r = \sqrt{0.4} \approx 0.63$. From the θ equation, we have that $\theta' > 0$. This means that we follow the limit cycle in a counterclockwise direction as time increases.

Limit cycles are not always circles. In Figures 4.29 and 4.30, we show the behavior of the Rayleigh system (4.58) for $c = 0.4$ and $c = 2.0$. In this case, we see that solutions tend toward a noncircular limit cycle in a clockwise direction.

Figure 4.29: Phase plane for the Rayleigh system (4.58) with $c = 0.4$.

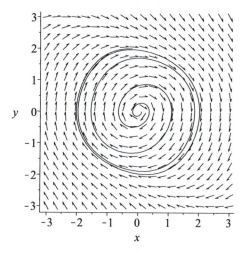

A slight change of the Rayleigh system leads to the van der Pol equation:

$$x'' + c(x^2 - 1)x' + x = 0 \qquad (4.64)$$

The van der Pol system.

The limit cycle for $c = 2.0$ is shown in Figure 4.31.

Can one determine ahead of time if a given nonlinear system will have a limit cycle? In order to answer this question, we will introduce some definitions.

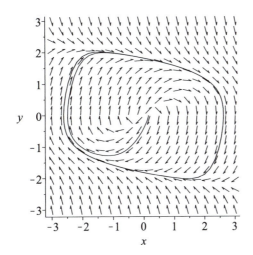

Figure 4.30: Phase plane for the Rayleigh system (4.58) with $c = 2.0$.

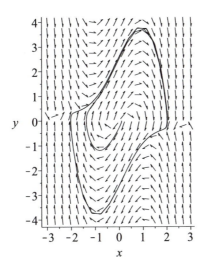

Figure 4.31: Phase plane for the van der Pol system (4.64) with $c = 2.0$.

We first describe different trajectories and families of trajectories. A *flow* on R^2 is a function ϕ that satisfies the following:

1. $\phi(\mathbf{x}, t)$ is continuous in both arguments.

2. $\phi(\mathbf{x}, 0) = \mathbf{x}$ for all $\mathbf{x} \in R^2$.

3. $\phi(\phi(\mathbf{x}, t_1), t_2) = \phi(\mathbf{x}, t_1 + t_2)$.

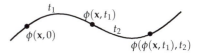

Figure 4.32: A sketch depicting the idea of trajectory, or orbit, passing through $\mathbf{x}$.

The *orbit*, or *trajectory*, through $\mathbf{x}$ is defined as $\gamma = \{\phi(\mathbf{x}, t) | t \in I\}$. In Figure 4.32 we demonstrate these properties. For $t = 0$, $\phi(\mathbf{x}, 0) = \mathbf{x}$. Increasing t, one follows the trajectory until one reaches the point $\phi(\mathbf{x}, t_1)$. Continuing t_2 further, one is then at $\phi(\phi(\mathbf{x}, t_1), t_2)$. By the third property, this is the same as going from $\mathbf{x}$ to $\phi(\mathbf{x}, t_1 + t_2)$ for $t = t_1 + t_2$.

Orbits and trajectories.

Having defined the orbits, we need to define the asymptotic behavior of the orbit for both positive and negative large times. We define the *positive semiorbit* through $\mathbf{x}$ as $\gamma^+ = \{\phi(\mathbf{x}, t) | t > 0\}$. The *negative semiorbit* through $\mathbf{x}$ is defined as $\gamma^- = \{\phi(\mathbf{x}, t) | t < 0\}$. Thus, we have $\gamma = \gamma^+ \cup \gamma^-$.

Limit sets and limit points.

The *positive limit set*, or *ω-limit set*, of point **x** is defined as

$$\Lambda^+ = \{\mathbf{y}|\ \text{there exists a sequence of } t_n \to \infty \text{ such that } \phi(\mathbf{x}, t_n) \to \mathbf{y}\}.$$

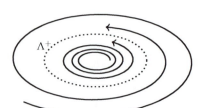

Figure 4.33: A sketch depicting an *ω*-limit set. Note that the orbits tend toward the set as *t* increases.

The **y**'s are referred to as *ω-limit points*. This is shown in Figure 4.33.

Similarly, we define the *negative limit set*, or the *α-limit set*, of point **x** is defined as

$$\Lambda^- = \{\mathbf{y}|\ \text{there exists a sequences of } t_n \to -\infty \text{ such that } \phi(\mathbf{x}, t_n) \to \mathbf{y}\}$$

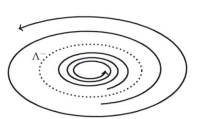

Figure 4.34: A sketch depicting an *α*-limit set. Note that the orbits tend away from the set as *t* increases.

and the corresponding **y**'s are *α-limit points*. This is shown in Figure 4.34.

There are several types of orbits that a system might possess. A *cycle* or *periodic orbit* is any closed orbit that is not an equilibrium point. A periodic orbit is stable if for every neighborhood of the orbit such that all nearby orbits stay inside the neighborhood. Otherwise, it is unstable. The orbit is asymptotically stable if all nearby orbits converge to the periodic orbit.

A limit cycle is a cycle that is the *α* or *ω*-limit set of some trajectory other than the limit cycle. A limit cycle Γ is stable if $\Lambda^+ = \Gamma$ for all **x** in some neighborhood of Γ. A limit cycle Γ is unstable if $\Lambda^- = \Gamma$ for all **x** in some neighborhood of Γ. Finally, a limit cycle is semistable if it is attracting on one side and repelling on the other side. In the previous examples, we saw limit cycles that were stable. Figures 4.33 and 4.34 depict stable and unstable limit cycles, respectively.

We now state a theorem that describes the type of orbits we might find in our system.

Cycles and periodic orbits.

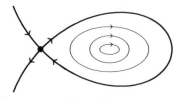

Figure 4.35: A heteroclinic orbit connecting two critical points.

Figure 4.36: A homoclinic orbit returning to the point it left.

> **Theorem 4.2. Poincaré–Bendixon Theorem** *Let* γ^+ *be contained in a bounded region in which there are finitely many critical points. Then* Λ^+ *is either*
>
> 1. *A single critical point;*
>
> 2. *A single closed orbit;*
>
> 3. *A set of critical points joined by heteroclinic orbits.*
>   *[Compare Figures 4.35 and 4.36.]*

We are interested in determining when limit cycles may, or may not, exist. A consequence of the Poincaré–Bendixon Theorem is given by the following corollary.

Corollary Let D be a bounded closed set containing no critical points and suppose that $\gamma^+ \subset D$. Then there exists a limit cycle contained in D.

More specific criteria allow us to determine if there is a limit cycle in a given region. These are given by Dulac's Criteria and Bendixon's Criteria.

Dulac's Criteria Consider the autonomous planar system

$$x' = f(x,y), \quad y' = g(x,y)$$

and a continuously differentiable function ψ defined on an annular region D contained in some open set. If

$$\frac{\partial}{\partial x}(\psi f) + \frac{\partial}{\partial y}(\psi g)$$

does not change sign in D, then there is at most one limit cycle contained entirely in D.

Bendixon's Criteria Consider the autonomous planar system

$$x' = f(x,y), \quad y' = g(x,y)$$

defined on a simply connected domain D such that

$$\frac{\partial}{\partial x}(\psi f) + \frac{\partial}{\partial y}(\psi g) \neq 0$$

in D. Then, there are no limit cycles entirely in D.

Proof. These are easily proved using Green's Theorem in the Plane (see Equation 9.61). We prove Bendixon's Criteria. Let $\mathbf{f} = (f, g)$. Assume that Γ is a closed orbit lying in D. Let S be the interior of Γ. Then

$$
\begin{aligned}
\int_S \nabla \cdot \mathbf{f} \, dx dy &= \oint_\Gamma (f \, dy - g \, dx) \\
&= \int_0^T (f \dot{y} - g \dot{x}) dt \\
&= \int_0^T (fg - gf) dt = 0.
\end{aligned}
\tag{4.65}
$$

So, if $\nabla \cdot \mathbf{f}$ is not identically zero and does not change sign in S, then from the continuity of $\nabla \cdot \mathbf{f}$ in S, we have that the right side above is either positive or negative. Thus, we have a contradiction and there is no closed orbit lying in D. $\qquad\square$

Example 4.16. *Consider the earlier example in (4.59) with $\mu = 1$.*

$$
\begin{aligned}
x' &= x - y - x(x^2 + y^2) \\
y' &= x + y - y(x^2 + y^2).
\end{aligned}
\tag{4.66}
$$

We already know that a limit cycle exists at $x^2 + y^2 = 1$. A simple computation gives that

$$\nabla \cdot \mathbf{f} = 2 - 4x^2 - 4y^2.$$

For an arbitrary annulus $a < x^2 + y^2 < b$, we have

$$2 - 4b < \nabla \cdot \mathbf{f} < 2 - 4a.$$

For $a = 3/4$ and $b = 5/4$, $-3 < \nabla \cdot \mathbf{f} < -1$. Thus, $\nabla \cdot \mathbf{f} < 0$ in the annulus $3/4 < x^2 + y^2 < 5/4$. Therefore, by Dulac's Criteria, there is at most one limit cycle in this annulus.

Example 4.17. *Consider the system*

$$
\begin{aligned}
x' &= y \\
y' &= -ax - by + cx^2 + dy^2.
\end{aligned}
\tag{4.67}
$$

Let $\psi(x, y) = e^{-2dx}$. *Then,*

$$
\frac{\partial}{\partial x}(\psi y) + \frac{\partial}{\partial y}(\psi(-ax - by + cx^2 + dy^2)) = -be^{-2dx} \neq 0.
$$

We conclude by Bendixon's Criteria that there are no limit cycles for this system.

4.9   Nonautonomous Nonlinear Systems

IN THIS SECTION WE DISCUSS NONAUTONOMOUS SYSTEMS. Recall that an autonomous system is one in which there is no explicit time dependence. A simple example is the forced nonlinear pendulum given by the nonhomogeneous equation

$$
\ddot{x} + \omega^2 \sin x = f(t).
\tag{4.68}
$$

We can set this up as a system of two first-order equations:

$$
\begin{aligned}
\dot{x} &= y, \\
\dot{y} &= -\omega^2 \sin x + f(t).
\end{aligned}
\tag{4.69}
$$

This system is not in a form for which we could use the earlier methods. Namely, it is a nonautonomous system. However, we introduce a new variable $z(t) = t$ and turn it into an autonomous system in one more dimension. The new system takes the form

$$
\begin{aligned}
\dot{x} &= y, \\
\dot{y} &= -\omega^2 \sin x + f(z), \\
\dot{z} &= 1.
\end{aligned}
\tag{4.70}
$$

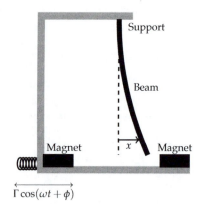

Figure 4.37: One model of the Duffing Equation describes a periodically forced beam that interacts with two magnets.

The system is now a three dimensional autonomous, possibly nonlinear, system and can be explored using methods from Chapters 2 and 3.

A more interesting model is provided by the Duffing Equation. This equation, named after Georg Wilhelm Christian Caspar Duffing (1861–1944), models hard spring and soft spring oscillations. It also models a periodically forced beam as shown in Figure 4.37. It is of interest because it is a simple system that exhibits chaotic dynamics and will motivate us toward using new visualization methods for nonautonomous systems.

The most general form of Duffing's equation is given by the damped, forced system

$$
\ddot{x} + k\dot{x} + (\beta x^3 \pm \omega_0^2 x) = \Gamma \cos(\omega t + \phi).
\tag{4.71}
$$

This equation models hard spring, $(\beta > 0)$, and soft spring, $(\beta < 0)$, oscillations. However, we will use the following simpler version of the Duffing Equation:

$$\ddot{x} + k\dot{x} + x^3 - x = \Gamma \cos \omega t. \tag{4.72}$$

An equation of this form can be obtained by setting $\phi = 0$ and rescaling x and t in the original equation. We will explore the behavior of the system as we vary the remaining parameters. In Figures 4.38 through 4.41, we show some typical solution plots superimposed on the direction field.

The undamped, unforced Duffing Equation.

We start with the undamped $(k = 0)$ and unforced $(\Gamma = 0)$ Duffing Equation,

$$\ddot{x} + x^3 - x = 0.$$

We can write this second order equation as the autonomous system

$$\begin{aligned} \dot{x} &= y \\ \dot{y} &= x(1 - x^2). \end{aligned} \tag{4.73}$$

We see that there are three equilibrium points at $(0,0)$ and $(\pm 1, 0)$. In Figure 4.38 we plot several orbits for $k = 0$, and $\Gamma = 0$. We see that the three equilibrium points consist of two centers and a saddle.

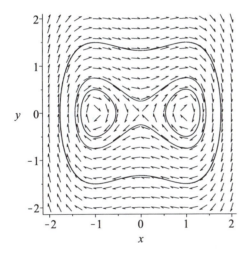

Figure 4.38: Phase plane for the undamped, unforced Duffing Equation $(k = 0, \Gamma = 0)$.

The unforced Duffing Equation.

We now turn on the damping. The system becomes

$$\begin{aligned} \dot{x} &= y \\ \dot{y} &= -ky + x(1 - x^2). \end{aligned} \tag{4.74}$$

In Figures 4.39 and 4.40 we show what happens when $k = 0.1$. These plots are reminiscent of the plots for the nonlinear pendulum; however, there are fewer equilibria. Note that the centers become stable spirals for $k > 0$.

The damped, forced Duffing Equation.

Next we turn on the forcing to obtain a damped, forced Duffing Equation. The system is now nonautonomous.

$$\begin{aligned} \dot{x} &= y \\ \dot{y} &= x(1 - x^2) + \Gamma \cos \omega t. \end{aligned} \tag{4.75}$$

In Figure 4.41 we only show one orbit with $k = 0.1$, $\Gamma = 0.5$, and $\omega = 1.25$. The solution intersects itself and looks a bit messy. We can imagine what we would get if we added any more orbits. For completeness, we show in Figure 4.42 an example with four different orbits.

Figure 4.39: Phase plane for the unforced Duffing Equation with $k = 0.1$ and $\Gamma = 0$.

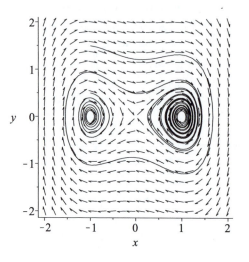

Figure 4.40: Display of two orbits for the unforced Duffing Equation with $k = 0.1$ and $\Gamma = 0$.

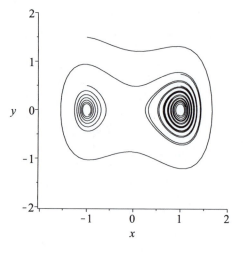

In cases for which one has periodic orbits such as the Duffing Equation, Poincaré introduced the notion of *surfaces of section*. One embeds the orbit in a higher-dimensional space so that there are no self-intersections, like we saw in Figures 4.41 and 4.42. In Figure 4.44, we show an example where a simple orbit is shown as it periodically pierces a given surface.

In order to simplify the resulting pictures, one only plots the points at which the orbit pierces the surface as sketched in Figure 4.43. In practice, there is a natural frequency, such as ω in the forced Duffing Equation. Then one plots points at times that are multiples of the period $T = \frac{2\pi}{\omega}$. In Figure 4.45 we show what the plot for one orbit would look like for the damped, unforced Duffing Equation.

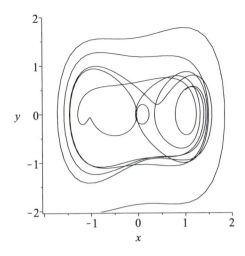

Figure 4.41: Phase plane for the Duffing Equation with $k = 0.1$, $\Gamma = 0.5$, and $\omega = 1.25$. In this case, we show only one orbit that was generated from the initial condition ($x_0 = 1.0$, $y_0 = 0.5$).

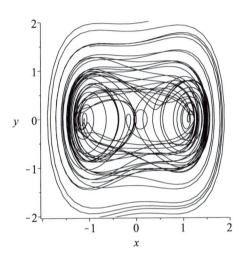

Figure 4.42: Phase plane for the Duffing Equation with $k = 0.1$, $\Gamma = 0.5$, and $\omega = 1.25$. In this case, four initial conditions were used to generate four orbits.

The more interesting case is when there is forcing and damping. In this case, the surface of section plot is given in Figure 4.46. While this is not as busy as the solution plot in Figure 4.41, it still provides some interesting behavior. What one finds is what is called a strange attractor. Plotting many orbits, we find that after a long time, all the orbits are attracted to a small region in the plane, much like a stable node attracts nearby orbits. However, this set consists of more than one point. Also, the flow on the attractor is chaotic in nature. Thus, points wander in an irregular way throughout the attractor. This is one of the interesting topics in chaos theory, and this whole theory of dynamical systems has only been touched in this text, leaving the reader to wander off into further depth into this fascinating field.

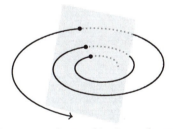

Figure 4.43: As an orbit pierces the surface of section, one plots the point of intersection in that plane to produce the surface of section plot.

The surface of section plots at the end of the previous section were obtained using code from S. Lynch's book *Dynamical Systems with Applications Using Maple*. For reference, the plots in Figures 4.38 and 4.39 were generated in Maple using the following commands:

```
> with(DEtools):
```

Figure 4.44: Poincaré's surface of section. One notes each time the orbit pierces the surface.

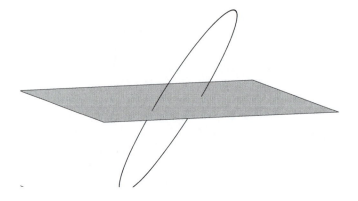

Figure 4.45: Poincaré's surface of section plot for the damped, unforced Duffing Equation.

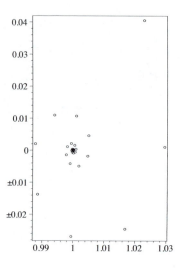

Figure 4.46: Poincaré's surface of section plot for the damped, forced Duffing Equation. This leads to what is known as a strange attractor.

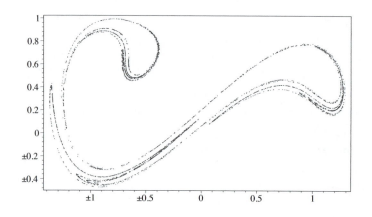

```
> Gamma:=0.5:omega:=1.25:k:=0.1:
> DEplot([diff(x(t),t)=y(t), diff(y(t),t)=x(t)-k*y(t)-(x(t))^3
  + Gamma*cos(omega*t)], [x(t),y(t)],t=0..500,[[x(0)=1,y(0)=0.5],
  [x(0)=-1,y(0)=0.5], [x(0)=1,y(0)=0.75], [x(0)=-1,y(0)=1.5]],
  x=-2..2,y=-2..2, stepsize=0.1, linecolor=blue, thickness=1,
  color=black);
```

4.10   *Exact Solutions Using Elliptic Functions*

THE SOLUTION OF THE NONLINEAR PENDULUM EQUATION led to the introduction of elliptic integrals. The incomplete elliptic integral of the first kind is defined as

$$F(\phi,k) \equiv \int_0^\phi \frac{d\theta}{\sqrt{1-k^2\sin^2\theta}} = \int_0^{\sin\phi} \frac{dz}{\sqrt{(1-z^2)(1-k^2z^2)}}. \qquad (4.76)$$

The complete integral of the first kind is given by $K(k) = F(\frac{\pi}{2},k)$, or

$$K(k) = \int_0^{\pi/2} \frac{d\theta}{\sqrt{1-k^2\sin^2\theta}} = \int_0^1 \frac{dz}{\sqrt{(1-z^2)(1-k^2z^2)}}.$$

Recall that a first integration of the nonlinear pendulum equation from Equation (4.18),

$$\left(\frac{d\theta}{dt}\right)^2 - \omega^2\cos\theta = -\omega^2\cos\theta_0.$$

or

$$\left(\frac{d\theta}{dt}\right)^2 = 2\omega^2\left[\sin^2\frac{\theta}{2} - \sin^2\frac{\theta_0}{2}\right].$$

Letting

$$kz = \sin\frac{\theta}{2} \text{ and } k = \sin\frac{\theta_0}{2},$$

the differential equation becomes

$$\frac{dz}{d\tau} = \pm\omega\sqrt{1-z^2}\sqrt{1-k^2z^2}.$$

Applying separation of variables, we find

$$\pm\omega(t-t_0) = \frac{1}{\omega}\int_1^z \frac{dz}{\sqrt{1-z^2}\sqrt{1-k^2z^2}} \qquad (4.77)$$

$$= \int_0^1 \frac{dz}{\sqrt{1-z^2}\sqrt{1-k^2z^2}} - \int_0^z \frac{dz}{\sqrt{1-z^2}\sqrt{1-k^2z^2}} \qquad (4.78)$$

$$= K(k) - F(\sin^{-1}(k^{-1}\sin\theta),k). \qquad (4.79)$$

The solution, $\theta(t)$, is then found by solving for z and using $kz = \sin\frac{\theta}{2}$ to solve for θ. This requires that we know how to invert the elliptic integral, $F(z,k)$.

The inverse of Equation (4.76) is defined as $\phi = F^{-1}(u,k) = \text{am}(u,k)$, where $u = \sin\phi$. The function $\text{am}(u,k)$ is called the Jacobi amplitude function and k is the elliptic modulus. [In some references and software like

MATLAB packages, $m = k^2$ is used as the parameter.] Three of the Jacobi elliptic functions, shown in Figure 4.47, can be defined in terms of the amplitude function by

$$\mathrm{sn}(u,k) = \sin \mathrm{am}(u,k) = \sin \phi,$$

$$\mathrm{cn}(u,k) = \cos \mathrm{am}(u,k) = \cos \phi,$$

Jacobi elliptic functions.

and the delta amplitude

$$\mathrm{dn}(u,k) = \sqrt{1 - k^2 \sin^2 \phi}.$$

They are related through the identities

$$\mathrm{cn}^2(u,k) + \mathrm{sn}^2(u,k) = 1, \tag{4.80}$$

$$\mathrm{dn}^2(u,k) + k^2 \mathrm{sn}^2(u,k) = 1. \tag{4.81}$$

Figure 4.47: Plots of the Jacobi elliptic functions $\mathrm{sn}(u,k)$, $\mathrm{cn}(u,k)$, and $\mathrm{dn}(u,k)$ for $m = k^2 = 0.5$. Here, $K(k) = 1.8541$.

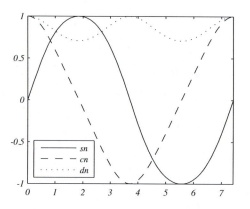

Also, we see that these functions are periodic. The period is given in terms of the complete elliptic integral of the first kind, $K(k)$; namely,

The elliptic functions can be extended to the complex plane. In this case, the functions are doubly periodic. However, we will not need to consider this in the current text.

$$\mathrm{sn}(u + K, k) = \frac{\mathrm{cn}\, u}{\mathrm{dn}\, u}, \quad \mathrm{sn}(u + 2K, k) = -\,\mathrm{sn}\, u,$$

$$\mathrm{cn}(u + K, k) = -\sqrt{1 - k^2}\,\frac{\mathrm{sn}\, u}{\mathrm{dn}\, u}, \quad \mathrm{dn}(u + 2K, k) = -\,\mathrm{cn}\, u,$$

$$\mathrm{dn}(u + K, k) = \frac{\sqrt{1 - k^2}}{\mathrm{dn}\, u}, \quad \mathrm{dn}(u + 2K, k) = \mathrm{dn}\, u.$$

Therefore, dn and cn have a period of $4K$, and dn has a period of $2K$.

Special values found in Figure 4.47 are seen as

$$\mathrm{sn}(K, k) = 1,$$

$$\mathrm{cn}(K, k) = 0,$$

$$\mathrm{dn}(K, k) = \sqrt{1 - k^2} = k',$$

where k' is called the complementary modulus.

Important to this section are the derivatives of these elliptic functions,

$$\frac{\partial}{\partial u} \operatorname{sn}(u,k) = \operatorname{cn}(u,k) \operatorname{dn}(u,k),$$

$$\frac{\partial}{\partial u} \operatorname{cn}(u,k) = -\operatorname{sn}(u,k) \operatorname{dn}(u,k),$$

$$\frac{\partial}{\partial u} \operatorname{dn}(u,k) = -k^2 \operatorname{sn}(u,k) \operatorname{cn}(u,k),$$

and the amplitude function

$$\frac{\partial}{\partial u} \operatorname{am}(u,k) = \operatorname{dn}(u,k).$$

Sometimes the Jacobi elliptic functions are displayed without reference to the elliptic modulus, such as $\operatorname{sn}(u) = \operatorname{sn}(u,k)$. When k is understood, we can do the same.

Example 4.18. *Show that* $\operatorname{sn}(u)$ *satisfies the differential equation*

$$y'' + (1 + k^2)y = 2k^2 y^3.$$

From the above derivatives, we have that

$$\begin{aligned}
\frac{d^2}{du^2} \operatorname{sn}(u) &= \frac{d}{du}(\operatorname{cn}(u)\operatorname{dn}(u)) \\
&= -\operatorname{sn}(u)\operatorname{dn}^2(u) - k^2 \operatorname{sn}(u)\operatorname{cn}^2(u).
\end{aligned} \qquad (4.82)$$

Letting $y(u) = \operatorname{sn}(u)$ *and using the identities (4.80) through (4.81), we have that*

$$y'' = -y(1 - k^2 y^2) - k^2 y(1 - y^2) = -(1 + k^2)y + 2k^2 y^3.$$

This is the desired result.

Example 4.19. *Show that* $\theta(t) = 2\sin^{-1}(k\operatorname{sn}t)$ *is a solution of the equation* $\ddot{\theta} + \sin\theta = 0$.

Differentiating $\theta(t) = 2\sin^{-1}(k\operatorname{sn}t)$, *we have*

$$\begin{aligned}
\frac{d^2}{dt^2}\left(2\sin^{-1}(k\operatorname{sn}t)\right) &= \frac{d}{dt}\left(2\frac{k\operatorname{cn}t\operatorname{dn}t}{\sqrt{1 - k^2\operatorname{sn}^2 t}}\right) \\
&= \frac{d}{dt}(2k\operatorname{cn}t) \\
&= -2k\operatorname{sn}t\operatorname{dn}t.
\end{aligned} \qquad (4.83)$$

However, we can evaluate $\sin\theta$ *for a range of* θ. *Thus, we have*

$$\begin{aligned}
\sin\theta &= \sin(2\sin^{-1}(k\operatorname{sn}t)) \\
&= 2\sin(\sin^{-1}(k\operatorname{sn}t))\cos(\sin^{-1}(k\operatorname{sn}t)) \\
&= 2k\operatorname{sn}t\sqrt{1 - k^2\operatorname{sn}^2 t} \\
&= 2k\operatorname{sn}t\operatorname{dn}t.
\end{aligned} \qquad (4.84)$$

Comparing these results, we have shown that $\ddot{\theta} + \sin\theta = 0$.

Figure 4.48: Comparison of exact solutions of the linear and nonlinear pendulum problems for $L = 1.0$ m and $\theta_0 = 10^\circ$.

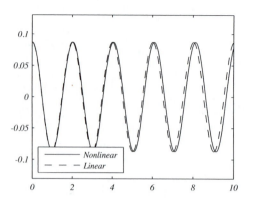

The solution to the last example can be used to obtain the exact solution to the nonlinear pendulum problem, $\ddot{\theta} + \omega^2 \sin\theta = 0$, $\theta(0) = \theta_0$, $\dot{\theta}(0) = 0$. The general solution is given by $\theta(t) = 2\sin^{-1}(k\,\mathrm{sn}(\omega t + \phi))$, where ϕ must be determined from the initial conditions. We note that

$$
\begin{aligned}
\frac{d\,\mathrm{sn}(u+K)}{du} &= \mathrm{cn}(u+K)\,\mathrm{dn}(u+K) \\
&= \left(-\sqrt{1-k^2}\,\frac{\mathrm{sn}\,u}{\mathrm{dn}\,u}\right)\left(\frac{\sqrt{1-k^2}}{\mathrm{dn}\,u}\right) \\
&= -(1-k^2)\frac{\mathrm{sn}\,u}{\mathrm{dn}^2\,u}.
\end{aligned}
\tag{4.85}
$$

Evaluating at $u = 0$, we have $\mathrm{sn}'(K) = 0$.

Therefore, if we pick $\phi = K$, then $\dot{\theta}(0) = 0$ and the solution is

$$\theta(t) = 2\sin^{-1}(k\,\mathrm{sn}(\omega t + K)).$$

Furthermore, the other initial value is found to be

$$\theta(0) = 2\sin^{-1}(k\,\mathrm{sn}\,K) = 2\sin^{-1}k.$$

Thus, $k = \sin\frac{\theta_0}{2}$, as we had seen in the earlier derivation of the elliptic integral solution. The solution is given as

$$\theta(t) = 2\sin^{-1}\left(\sin\frac{\theta_0}{2}\,\mathrm{sn}(\omega t + K)\right).$$

In Figures 4.48 and 4.49 we show comparisons of the exact solutions of the linear and nonlinear pendulum problems for $L = 1.0$ m and initial angles $\theta_0 = 10^\circ$ and $\theta_0 = 50^\circ$.

Problems

1. Solve the general logistic problem,

$$\frac{dy}{dt} = ky - cy^2, \quad y(0) = y_0 \tag{4.86}$$

using separation of variables.

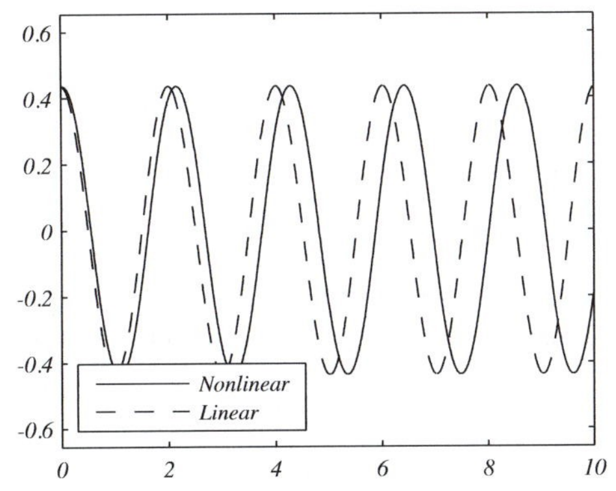

Figure 4.49: Comparison of the exact solutions of the linear and nonlinear pendulum problems for $L = 1.0$ m and $\theta_0 = 50^\circ$.

2. Find the equilibrium solutions and determine their stability for the following systems. For each case, draw representative solutions and phase lines.

    a. $y' = y^2 - 6y - 16$.

    b. $y' = \cos y$.

    c. $y' = y(y - 2)(y + 3)$.

    d. $y' = y^2(y + 1)(y - 4)$.

3. For $y' = y - y^2$, find the general solution corresponding to $y(0) = y_0$. Provide specific solutions for the following initial conditions and sketch them: a. $y(0) = 0.25$, b. $y(0) = 1.5$, and c. $y(0) = -0.5$.

4. For each problem, determine equilibrium points, bifurcation points, and construct a bifurcation diagram. Discuss the different behaviors in each system.

    a. $y' = y - \mu y^2$

    b. $y' = y(\mu - y)(\mu - 2y)$

    c. $x' = \mu - x^3$

    d. $x' = x - \frac{\mu x}{1+x^2}$

5. Consider the family of differential equations $x' = x^3 + \delta x^2 - \mu x$.

    a. Sketch a bifurcation diagram in the $x\mu$-plane for $\delta = 0$.

    b. Sketch a bifurcation diagram in the $x\mu$-plane for $\delta > 0$.

Hint: Pick a few values of δ and μ in order to get a feel for how this system behaves.

6. System (4.63) can be solved exactly. Integrate the r-equation using separation of variables. For initial conditions (a) $r(0) = 0.25$, $\theta(0) = 0$, and (b) $r(0) = 1.5$, $\theta(0) = 0$, and $\mu = 1.0$, find and plot the solutions in the xy-plane, showing the approach to a limit cycle.

7. Consider the system

$$
\begin{aligned}
x' &= -y + x\left[\mu - x^2 - y^2\right], \\
y' &= x + y\left[\mu - x^2 - y^2\right].
\end{aligned}
$$

Rewrite this system in polar form. Look at the behavior of the r equation and construct a bifurcation diagram in μr space. What might this diagram look like in the three-dimensional μxy space? (Think about the symmetry in this problem.) This leads to what is called a *Hopf bifurcation*.

8. Find the fixed points of the following systems. Linearize the system about each fixed point and determine the nature and stability in the neighborhood of each fixed point, when possible. Verify your findings by plotting phase portraits using a computer.

   a.

$$
\begin{aligned}
x' &= x(100 - x - 2y), \\
y' &= y(150 - x - 6y).
\end{aligned}
$$

   b.

$$
\begin{aligned}
x' &= x + x^3, \\
y' &= y + y^3.
\end{aligned}
$$

   c.

$$
\begin{aligned}
x' &= x - x^2 + xy, \\
y' &= 2y - xy - 6y^2.
\end{aligned}
$$

   d.

$$
\begin{aligned}
x' &= -2xy, \\
y' &= -x + y + xy - y^3.
\end{aligned}
$$

9. Plot phase portraits for the Lienard system

$$
\begin{aligned}
x' &= y - \mu(x^3 - x), \\
y' &= -x,
\end{aligned}
$$

for a small and a not so small value of μ. Describe what happens as one varies μ.

10. Consider the period of a nonlinear pendulum. Let the length be $L = 1.0$ m and $g = 9.8$ m/s^2. Sketch T versus the initial angle θ_0, and compare the linear and nonlinear values for the period. For what angles can you use the linear approximation confidently?

11. Another population model is one in which species compete for resources, such as a limited food supply. Such a model is given by

$$x' = ax - bx^2 - cxy,$$
$$y' = dy - ey^2 - fxy.$$

In this case, assume that all constants are positive.

    a. Describe the effects/purpose of each terms.

    b. Find the fixed points of the model.

    c. Linearize the system about each fixed point and determine the stability.

    d. From the above, describe the types of solution behavior you might expect, in terms of the model.

12. Consider a model of a food chain of three species. Assume that each population on its own can be modeled by logistic growth. Let the species be labeled by $x(t)$, $y(t)$, and $z(t)$. Assume that population x is at the bottom of the chain. That population will be depleted by population y. Population y is sustained by x's, but eaten by z's. A simple, but scaled, model for this system can be given by the system

$$\begin{aligned} x' &= x(1-x) - xy, \\ y' &= y(1-y) + xy - yz, \\ z' &= z(1-z) + yz. \end{aligned}$$

    a. Find the equilibrium points of the system.

    b. Find the Jacobian matrix for the system and evaluate it at the equilibrium points.

    c. Find the eigenvalues and eigenvectors.

    d. Describe the solution behavior near each equilibrium point.

    e. Which of these equilibria are important in the study of the population model and describe the interactions of the species in the neighborhood of these point(s).

13. Derive the first integral of the Lotka–Volterra system, $a \ln y + d \ln x - cx - by = C$.

14. Show that the system $x' = x - y - x^3$, $y' = x + y - y^3$, has at least one limit cycle by picking an appropriate $\psi(x,y)$ in Dulac's Criteria.

15. The Lorenz Model is a simple model for atmospheric convection developed by Edward Lorenz in 1963. The system is given by three equations:

$$\begin{aligned} \frac{dx}{dt} &= \sigma(y - x), \\ \frac{dy}{dt} &= x(\rho - z) - y, \\ \frac{dz}{dt} &= xy - \beta z. \end{aligned}$$

a. Find the equilibrium points of the system.

b. Find the Jacobian matrix for the system and evaluate it at the equilibrium points.

c. Determine any bifurcation points and describe what happens near bifurcation point(s). Consider $\sigma = 10$, $\beta = 8/3$, and vary ρ.

d. This system is known to exhibit chaotic behavior. Lorenz found a so-called strange attractor for parameter values $\sigma = 10$, $\beta = 8/3$, and $\rho = 28$. Using a computer, locate this strange attractor.

16. The Michaelis-Menten kinetics reaction is given by

$$E + S \underset{k_1}{\overset{k_3}{\rightleftarrows}} ES \xrightarrow{k_2} E + P$$

The resulting system of equations for the chemical concentrations is

$$\begin{aligned}
\frac{d[S]}{dt} &= -k_1[E][S] + k_3[ES], \\
\frac{d[E]}{dt} &= -k_1[E][S] + (k_2 + k_2)[ES], \\
\frac{d[ES]}{dt} &= k_1[E][S] - (k_2 + k_2)[ES], \\
\frac{d[P]}{dt} &= k_3[ES].
\end{aligned} \tag{4.87}$$

In chemical kinetics, one seeks to determine the rate of product formation ($v = d[P]/dt = k_3[ES]$). Assuming that $[ES]$ is a constant, find v as a function of $[S]$ and the total enzyme concentration $[E_T] = [E] + [ES]$. As a nonlinear dynamical system, what are the equilibrium points?

17. In Equation (3.153), we saw a linear version of an epidemic model. The commonly used nonlinear SIR model is given by

$$\begin{aligned}
\frac{dS}{dt} &= -\beta SI, \\
\frac{dI}{dt} &= \beta SI - \gamma I, \\
\frac{dR}{dt} &= \gamma I,
\end{aligned} \tag{4.88}$$

where S is the number of susceptible individuals, I is the number of infected individuals, and R is the number who have been removed from the other groups, either by recovering or dying.

a. Let $N = S + I + R$ be the total population. Prove that $N = $ constant. Thus, one need only solve the first two equations and find $R = N - S - I$ afterward.

b. Find and classify the equilibria. Describe the equilibria in terms of the population behavior.

c. Let $\beta = 0.05$ and $\gamma = 0.2$. Assume that in a population of 100 there is one infected person. Numerically solve the system of equations for $S(t)$ and $I(t)$ and describe the solution being careful to determine the units of population and the constants.

d. The equations can be modified by adding constant birth and death rates. Assuming these rates are the same, one would have a new system.

$$\frac{dS}{dt} = -\beta SI + \mu(N - S)$$
$$\frac{dI}{dt} = \beta SI - \gamma I - \mu I$$
$$\frac{dR}{dt} = -\gamma I - \mu R. \tag{4.89}$$

How does this affect any equilibrium solutions?

e. Again, let $\beta = 0.05$ and $\gamma = 0.2$. Let $\mu = 0.1$ For a population of 100 with one infected person, numerically solve the system of equations for $S(t)$ and $I(t)$ and describe the solution being careful to determine the units of populations and the constants.

18. An undamped, unforced Duffing Equation, $\ddot{x} + \omega^2 x + \epsilon x^3 = 0$, can be solved exactly in terms of elliptic functions. Using the results of Example 4.18, determine the solution of this equation and determine if there are any restrictions on the parameters.

19. Differentiate the Fourier sine series by term in Problem 18. Show that the result is not the derivative of $f(x) = x$.

20. Evaluate the following in terms of elliptic integrals, and compute the values to four decimal places.

a. $\int_0^{\pi/4} \frac{d\theta}{\sqrt{1 - \frac{1}{2}\sin^2\theta}}$.

b. $\int_0^{\pi/2} \frac{d\theta}{\sqrt{1 - \frac{1}{4}\sin^2\theta}}$.

c. $\int_0^2 \frac{dx}{\sqrt{(9-x^2)(4-x^2)}}$.

d. $\int_0^{\pi/2} \frac{d\theta}{\sqrt{\cos\theta}}$.

e. $\int_1^\infty \frac{dx}{\sqrt{x^4-1}}$.

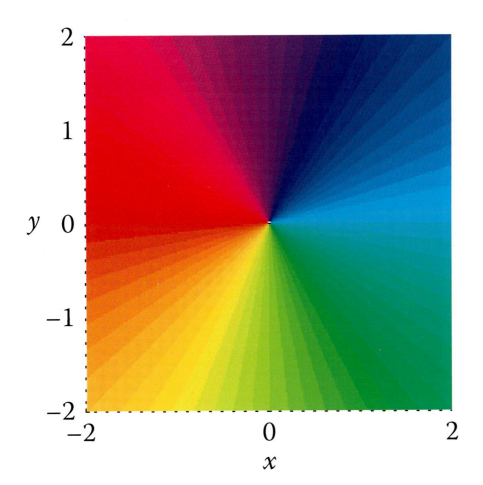

Figure 7.9: Domain coloring of the complex z-plane assigning colors to $\arg(z)$.

Figure 7.10: Domain coloring for $f(z) = z^2$. The top figure shows the phase coloring. The bottom figure shows the colored surface with height $|f(z)|$.

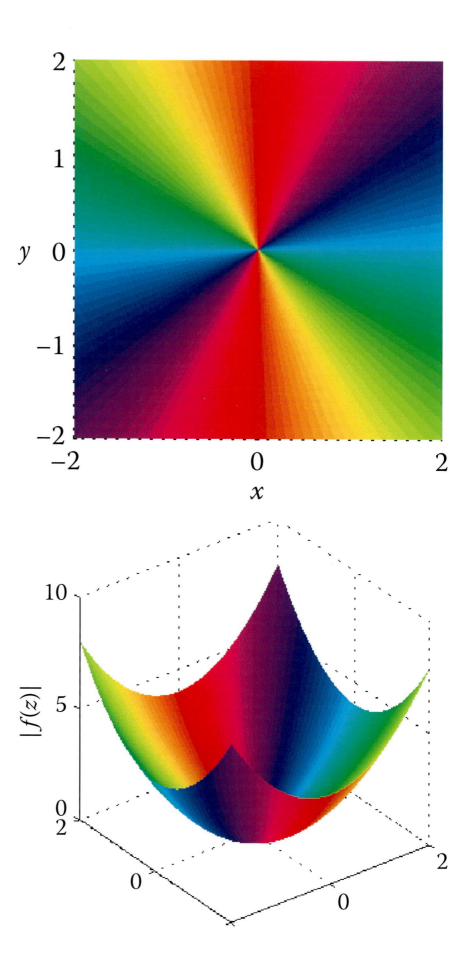

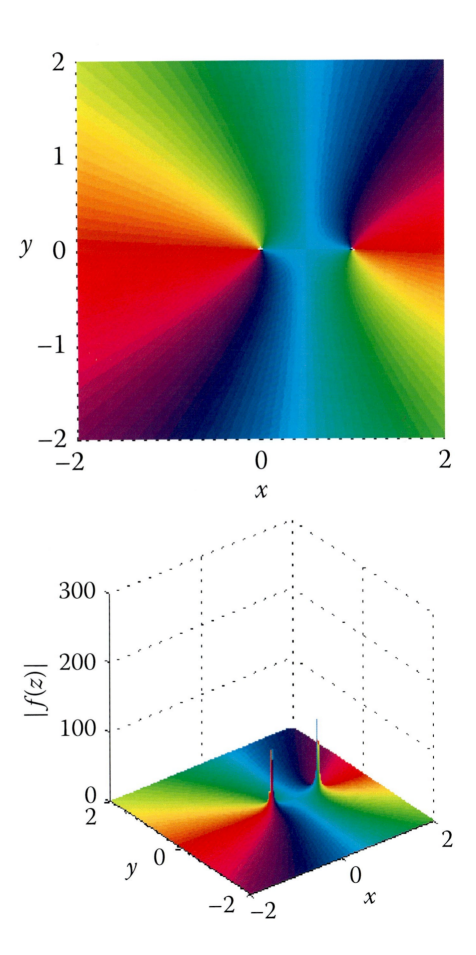

Figure 7.11: Domain coloring for $f(z) = 1/z(1 - z)$. The top figure shows the phase coloring. The bottom figure shows the colored surface with height $|f(z)|$.

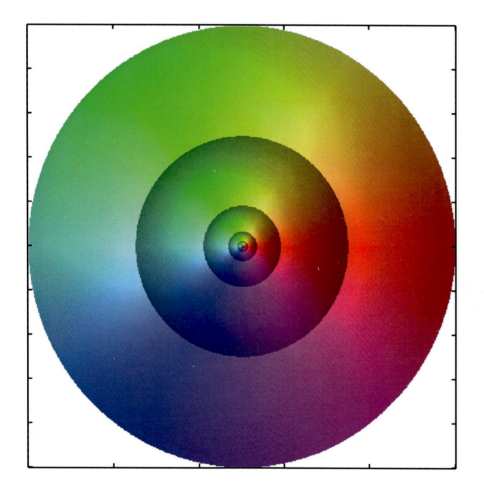

Figure 7.12: Domain coloring for the function $f(z) = z$ showing a coloring for $\arg(z)$ and brightness based on $|f(z)|$.

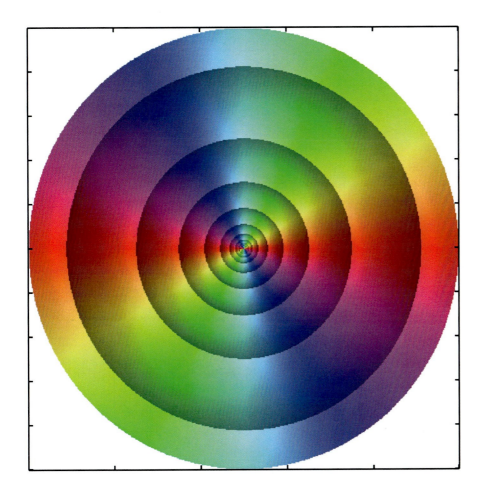

Figure 7.13: Domain coloring for the function $f(z) = z^2$.

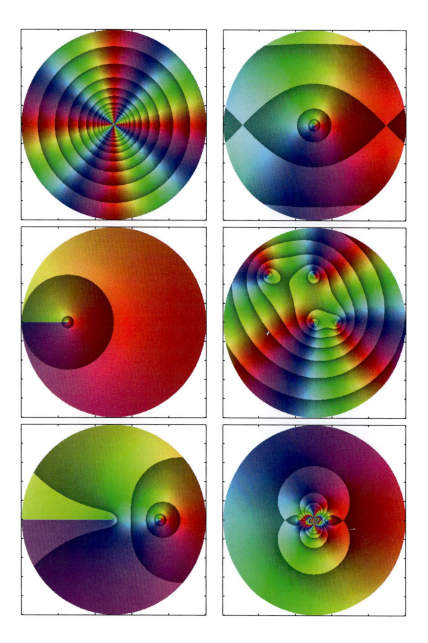

Figure 7.14: Domain coloring for several functions. On the top row, the domain coloring is shown for $f(z) = z^4$ and $f(z) = \sin z$. On the second row, plots for $f(z) = \sqrt{1+z}$ and $f(z) = \frac{1}{z(1/2-z)(z-i)(z-i+1)}$ are shown. In the last row, domain colorings for $f(z) = \ln z$ and $f(z) = \sin(1/z)$ are shown.

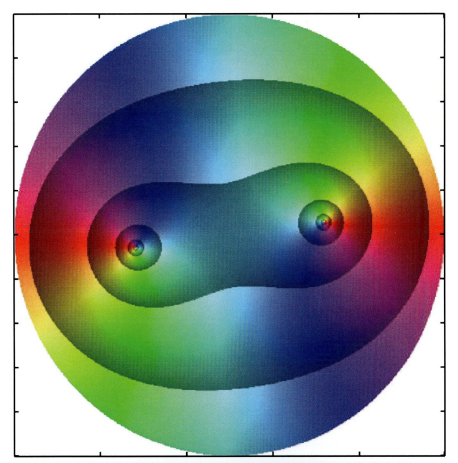

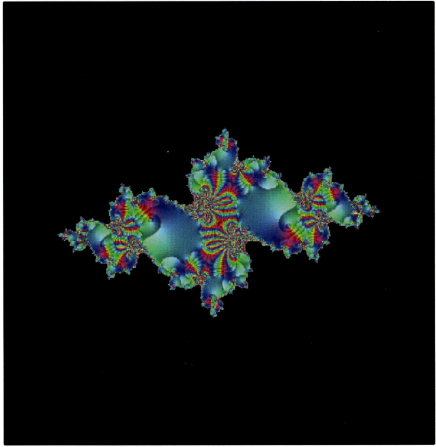

Figure 7.15: Domain coloring for $f(z) = z^2 - 0.75 - 0.2i$. The top figure shows the phase coloring. On the bottom is the plot for $f^{20}(z)$.

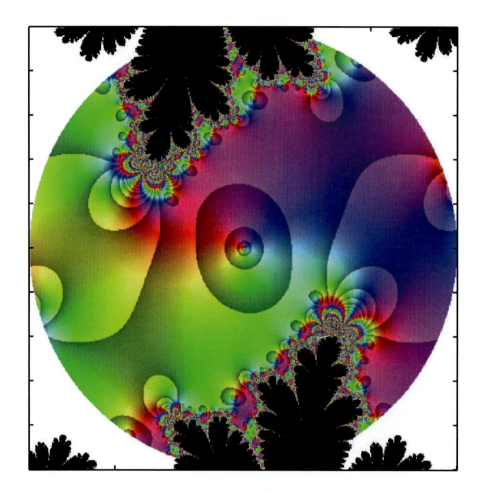

Figure 7.16: Domain coloring for six iterations of $f(z) = (1 - i/2) \sin x$.

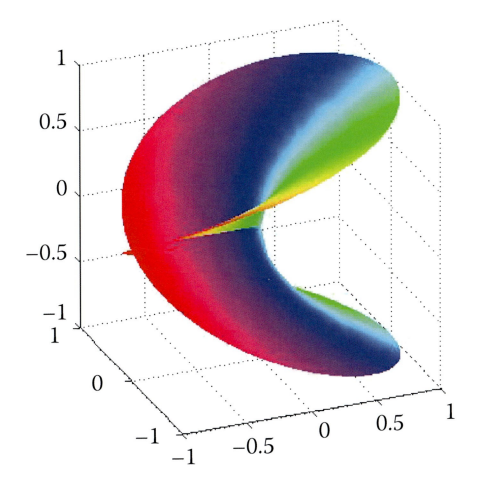

Figure 7.50: Riemann surface for $f(z) = z^{1/2}$.

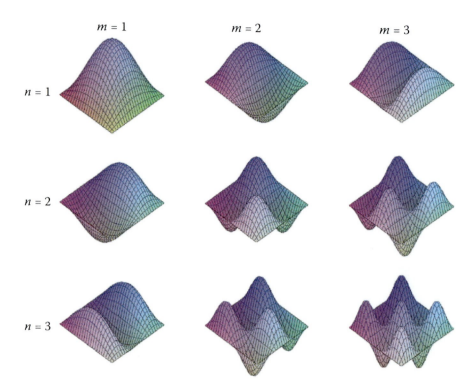

$m = 1$　　$m = 2$　　$m = 3$

$n = 1$

$n = 2$

$n = 3$

Table 11.1: A Three-Dimensional View of the Vibrating Rectangular Membrane for the Lowest Modes. Compare these images with the nodal lines in Figure 11.3.

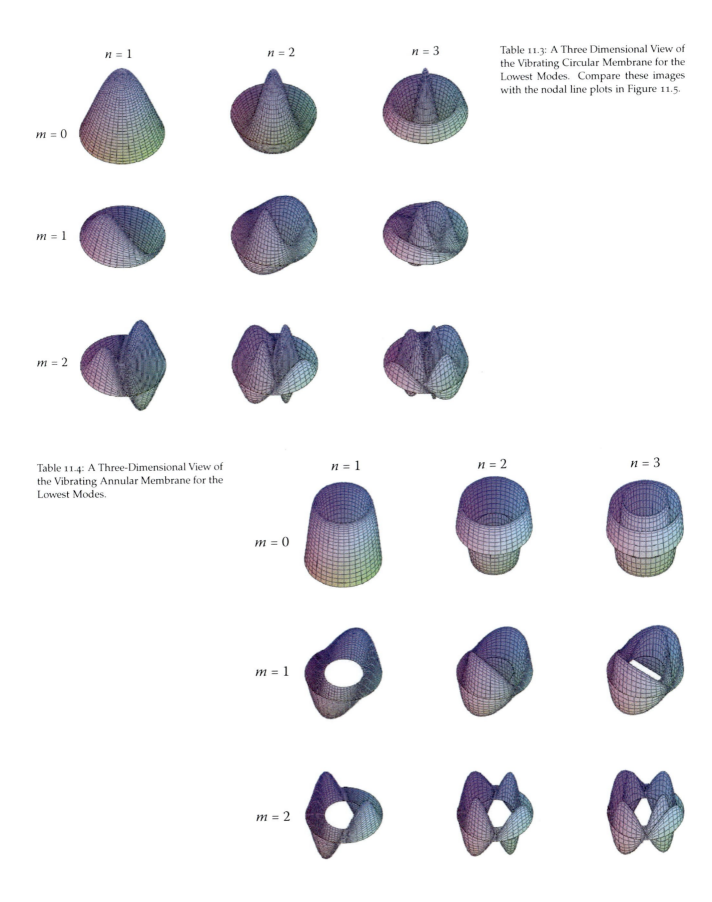

$n = 1$ $n = 2$ $n = 3$

$m = 0$

$m = 1$

$m = 2$

Table 11.3: A Three Dimensional View of the Vibrating Circular Membrane for the Lowest Modes. Compare these images with the nodal line plots in Figure 11.5.

Table 11.4: A Three-Dimensional View of the Vibrating Annular Membrane for the Lowest Modes.

$n = 1$ $n = 2$ $n = 3$

$m = 0$

$m = 1$

$m = 2$

Figure 11.14: Temperature evolution for a 13″ × 9″ × 2″ cake shown as vertical slices at the indicated length in feet.

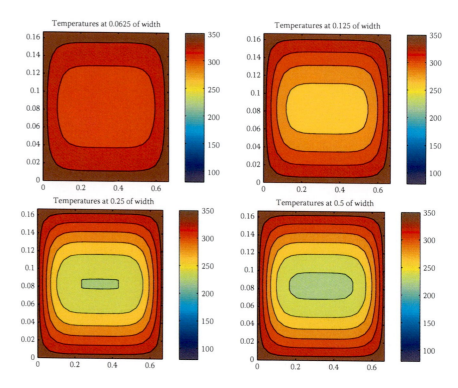

Figure 11.17: Temperature evolution for a standard 9″ cake shown as vertical slices through the center.

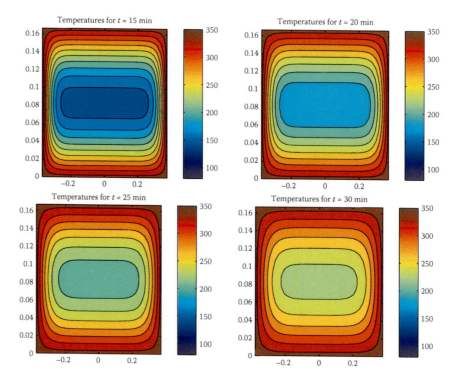

Table 11.6: The First Few Spherical Harmonics, $|Y_{\ell m}(\theta, \phi)|^2$.

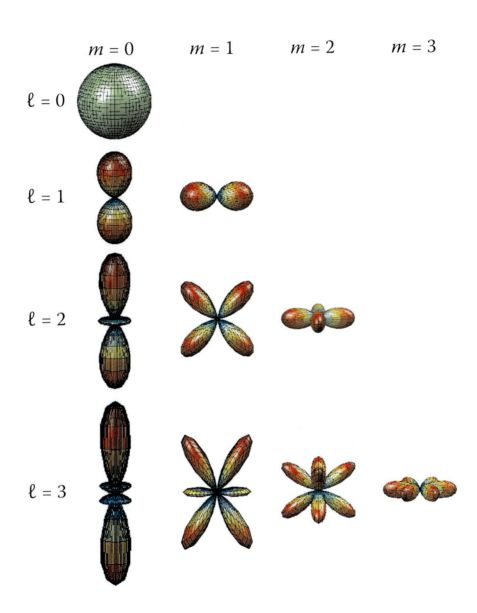

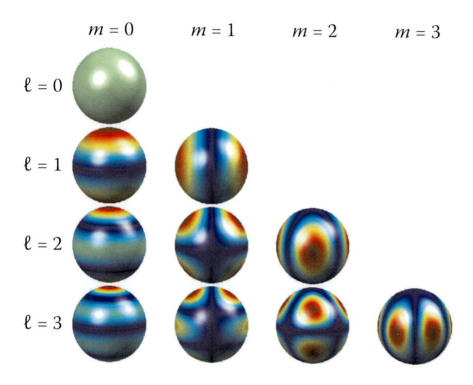

Table 11.7: Spherical Harmonic Contours for $|Y_{\ell m}(\theta, \phi)|^2$.

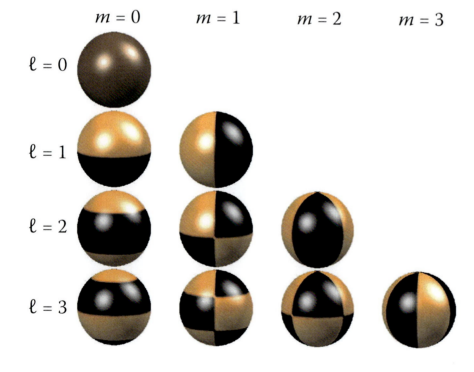

Table 11.8: In these figures we show the nodal curves of $|Y_{\ell m}(\theta, \phi)|^2$. Along the first column ($m = 0$) are the zonal harmonics seen as ℓ horizontal circles. Along the top diagonal ($m = \ell$) are the sectional harmonics. These look like orange sections formed from m vertical circles. The remaining harmonics are tesseral harmonics. They look like a checkerboard pattern formed from intersections of $\ell - m$ horizontal circles and m vertical circles.

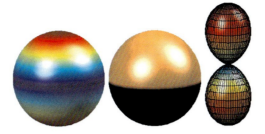

Figure 11.21: Zonal harmonics, $\ell = 1$, $m = 0$.

Figure 11.22: Zonal harmonics, $\ell = 2$, $m = 0$.

Figure 11.23: Sectoral harmonics, $\ell = 2$, $m = 2$.

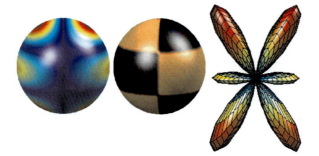

Figure 11.24: Tesseral harmonics, $\ell = 3$, $m = 1$.

Figure 11.25: Sectoral harmonics, $\ell = 3$, $m = 3$.

Figure 11.26: Tesseral harmonics, $\ell = 4$, $m = 3$.

5
Harmonics of Vibrating Strings

"What I am going to tell you about is what we teach our physics students in the third or fourth year of graduate school . . . It is my task to convince you not to turn away because you don't understand it. You see my physics students don't understand it . . . That is because I don't understand it. Nobody does."—Richard Feynman (1918–1988)

5.1 Harmonics and Vibrations

UNTIL NOW WE HAVE STUDIED OSCILLATIONS in several physical systems. These led to ordinary differential equations describing the time evolution of the systems and required the solution of initial value problems. In this chapter we will extend our study to include oscillations in space. The typical example is the vibrating string. When one plucks a violin or guitar string, the string vibrates, exhibiting a variety of sounds. These are enhanced by the violin case, but we will only focus on the simpler vibrations of the string.

We will consider the one-dimensional wave motion in the string. Physically, the speed of these waves depends on the tension in the string and its mass density. The frequencies we hear are then related to the string shape, or the allowed wavelengths across the string. We will be interested in the harmonics, or pure sinusoidal waves, of the vibrating string and how a general wave on the string can be represented as a sum over such harmonics. This will take us into the field of spectral, or Fourier, analysis.

Such systems are governed by partial differential equations. The vibrations of a string are governed by the one-dimensional wave equation. Another simple partial differential equation is the heat, or diffusion, equation. This equation governs heat flow. We will consider the flow of heat through a one-dimensional rod. The solution of the heat equation also involves the use of Fourier analysis. However, in this case there are no oscillations in time.

There are many applications that are studied using spectral analysis. At the root of these studies is the belief that continuous waveforms are comprised of a number of harmonics. Such ideas stretch back to the Pythagoreans study of the vibrations of strings, which led to their program of a world

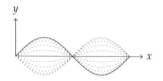

Figure 5.1: Plot of the second harmonic of a vibrating string at different times.

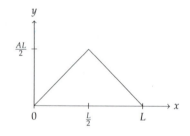

Figure 5.2: Plot of an initial condition for a plucked string.

The one-dimensional version of the heat equation is a partial differential equation for $u(x, t)$ of the form

$$\frac{\partial u}{\partial t} = k\frac{\partial^2 u}{\partial x^2}.$$

Solutions satisfying boundary conditions $u(0, t) = 0$ and $u(L, t) = 0$, are of the form

$$u(x, t) = \sum_{n=0}^{\infty} b_n \sin\frac{n\pi x}{L} e^{-n^2\pi^2 t/L^2}.$$

In this case, setting $u(x, 0) = f(x)$, one has to satisfy the condition

$$f(x) = \sum_{n=0}^{\infty} b_n \sin\frac{n\pi x}{L}.$$

This is another example leading to an infinite series of trigonometric functions.

of harmony. This idea was carried further by Johannes Kepler (1571–1630) in his harmony of the spheres approach to planetary orbits. In the 1700s, others worked on the superposition theory for vibrating waves on a stretched spring, starting with the wave equation and leading to the superposition of right and left traveling waves. This work was carried out by people such as John Wallis (1616–1703), Brook Taylor (1685–1731), and Jean le Rond d'Alembert (1717–1783).

In 1742, d'Alembert solved the wave equation

$$c^2\frac{\partial^2 y}{\partial x^2} - \frac{\partial^2 y}{\partial t^2} = 0,$$

where y is the string height and c is the wave speed. However, this solution led him and others, like Leonhard Euler (1707–1783) and Daniel Bernoulli (1700–1782), to investigate what "functions" could be the solutions of this equation. In fact, this led to a more rigorous approach to the study of analysis by first coming to grips with the concept of a function. For example, in 1749, Euler sought the solution for a plucked string in which case the initial condition $y(x, 0) = h(x)$ has a discontinuous derivative! (We will see how this led to important questions in analysis.)

In 1753, Daniel Bernoulli viewed the solutions as a superposition of simple vibrations, or harmonics. Such superpositions amounted to looking at solutions of the form

$$y(x, t) = \sum_k a_k \sin\frac{k\pi x}{L}\cos\frac{k\pi ct}{L},$$

where the string extends over the interval $[0, L]$ with fixed ends at $x = 0$ and $x = L$.

However, the initial profile for such superpositions is given by

$$y(x, 0) = \sum_k a_k \sin\frac{k\pi x}{L}.$$

It was determined that many functions could not be represented by a finite number of harmonics, even for the simply plucked string in Figure 5.2 given by an initial condition of the form

$$y(x, 0) = \begin{cases} Ax, & 0 \leq x \leq L/2, \\ A(L - x), & L/2 \leq x \leq L. \end{cases}$$

Thus, the solution consists generally of an infinite series of trigonometric functions.

Such series expansions were also of importance in Joseph Fourier's (1768–1830) solution of the heat equation. The use of Fourier expansions has become an important tool in the solution of linear partial differential equations, such as the wave equation and the heat equation. More generally, using a technique called the Method of Separation of Variables, allowed higher-dimensional problems to be reduced to one-dimensional boundary value problems. However, these studies led to very important questions, which in turn opened the doors to whole fields of analysis. Some of the problems raised were

1. What functions can be represented as the sum of trigonometric functions?

2. How can a function with discontinuous derivatives be represented by a sum of smooth functions, such as the above sums of trigonometric functions?

3. Do such infinite sums of trigonometric functions actually converge to the functions they represent?

There are many other systems for which it makes sense to interpret the solutions as sums of sinusoids of particular frequencies. For example, we can consider ocean waves. Ocean waves are affected by the gravitational pull of the moon and the sun and numerous other forces. These lead to the tides, which in turn have their own periods of motion. In an analysis of wave heights, one can separate out the tidal components by making use of Fourier analysis.

5.2   Boundary Value Problems

UNTIL THIS POINT WE HAVE SOLVED INITIAL VALUE PROBLEMS. For an initial value problem, one has to solve a differential equation subject to conditions on the unknown function and its derivatives at one value of the independent variable. For example, for $x = x(t)$, we could have the initial value problem

$$x'' + x = 2, \quad x(0) = 1, \quad x'(0) = 0. \tag{5.1}$$

Typically, initial value problems involve time-dependent functions and boundary value problems are spatial. So, with an initial value problem, one knows how a system evolves in terms of the differential equation and the state of the system at some fixed time. Then one seeks to determine the state of the system at a later time.

For boundary values problems, one knows how each point responds to its neighbors, but there are conditions that must be satisfied at the endpoints. An example would be a horizontal beam supported at the ends, like a bridge. The shape of the beam under the influence of gravity, or other forces, would lead to a differential equation and the boundary conditions at the beam ends would affect the solution of the problem. There are also a variety of other types of boundary conditions. In the case of a beam, one end could be fixed and the other end could be free to move. We will explore the effects of different boundary conditions in our discussions and examples. But, we will first solve a simple boundary value problem which is a slight modification of the above problem.

Example 5.1. *Solve the boundary value problem, $x'' + x = 2, \quad x(0) = 1, \quad x(1) = 0$.*

Note that the conditions at $t = 0$ and $t = 1$ make this a boundary value problem because the conditions are given at two different points. As with initial value *problems, we need to find the general solution and then apply any conditions that we may have. This is a nonhomogeneous differential equation, so the solution is a sum of a solution of the homogeneous equation and a particular solution of the nonhomogeneous equation, $x(t) = x_h(t) + x_p(t)$. The solution of $x'' + x = 0$ is easily found as*

$$x_h(t) = c_1 \cos t + c_2 \sin t.$$

The particular solution is found using the Method of Undetermined Coefficients,

$$x_p(t) = 2.$$

Thus, the general solution is

$$x(t) = 2 + c_1 \cos t + c_2 \sin t.$$

We now apply the boundary conditions and see if there are values of c_1 and c_2 that yield a solution to this boundary value problem. The first condition, $x(0) = 0$, gives

$$0 = 2 + c_1.$$

Thus, $c_1 = -2$. Using this value for c_1, the second condition, $x(1) = 1$, gives

$$0 = 2 - 2\cos 1 + c_2 \sin 1.$$

This yields

$$c_2 = \frac{2(\cos 1 - 1)}{\sin 1}.$$

We have found that there is a solution to the boundary value problem and it is given by

$$x(t) = 2\left(1 - \cos t \frac{(\cos 1 - 1)}{\sin 1} \sin t\right).$$

Boundary value problems arise in many physical systems, just as the initial value problems we have seen earlier. We will see in the next sections that boundary value problems for ordinary differential equations often appear in the solutions of partial differential equations. However, there is no guarantee that we will have unique solutions of our boundary value problems as we found in the example above.

5.3 Partial Differential Equations

IN THIS SECTION WE WILL INTRODUCE several generic partial differential equations and see how such equations lead naturally to the study of boundary value problems for ordinary differential equations.

For ordinary differential equations, the unknown functions are functions of a single variable, for example $y = y(x)$. Partial differential equations are equations involving an unknown function of several variables, such as $u = u(x,t)$, $u = u(x,y)$, and $u = u(x,y,z,t)$, and its (partial) derivatives. We will often use the standard notations for partial derivatives, $u_x = \frac{\partial u}{\partial x}$, $u_{xx} = \frac{\partial^2 u}{\partial x^2}$, etc.

There are a few standard equations that one encounters. These can be studied in one to three dimensions, and all are linear differential equations. A list is provided in Table 5.1. Here we have introduced the Laplacian operator, $\nabla^2 u = u_{xx} + u_{yy} + u_{zz}$. Depending on the types of boundary conditions imposed and on the geometry of the system (rectangular, cylindrical, spherical, etc.), one encounters many interesting boundary value problems.

Name	2 Vars	3D
Heat Equation	$u_t = ku_{xx}$	$u_t = k\nabla^2 u$
Wave Equation	$u_{tt} = c^2 u_{xx}$	$u_{tt} = c^2 \nabla^2 u$
Laplace's Equation	$u_{xx} + u_{yy} = 0$	$\nabla^2 u = 0$
Poisson's Equation	$u_{xx} + u_{yy} = F(x,y)$	$\nabla^2 u = F(x,y,z)$
Schrödinger's Equation	$iu_t = u_{xx} + F(x,t)u$	$iu_t = \nabla^2 u + F(x,y,z,t)u$

Table 5.1: List of Generic Partial Differential Equations

Let's look at the heat equation in one dimension. This could describe the heat conduction in a thin insulated rod of length L. It could also describe the diffusion of pollutant in a long narrow stream, or the flow of traffic down a road. In problems involving diffusion processes, one instead calls this equation the diffusion equation. [See the derivation in Section 5.14.2.]

A typical initial-boundary value problem for the heat equation would be that initially one has a temperature distribution $u(x,0) = f(x)$. Placing the bar in an ice bath and assuming the heat flow is only through the ends of the bar, one has the boundary conditions $u(0,t) = 0$ and $u(L,t) = 0$. Of course, we are dealing with Celsius temperatures and we assume there is plenty of ice to keep that temperature fixed at each end for all time. So, the problem one would need to solve is given as [IC = initial condition(s) and BC = boundary conditions.]

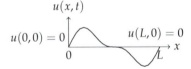

Figure 5.3: One-dimensional string of length L.

1D Heat Equation

$$
\begin{aligned}
\text{PDE} \quad & u_t = ku_{xx}, & 0 < t, \quad 0 \le x \le L, \\
\text{IC} \quad & u(x,0) = f(x), & 0 < x < L, \\
\text{BC} \quad & u(0,t) = 0, & t > 0, \\
& u(L,t) = 0, & t > 0,
\end{aligned}
\tag{5.2}
$$

Here, k is the heat conduction constant and is determined using properties of the bar.

Another problem that will come up in later discussions is that of the vibrating string. A string of length L is stretched out horizontally with both ends fixed such as a violin string. Let $u(x,t)$ be the vertical displacement of the string at position x and time t. The motion of the string is governed by the one-dimensional wave equation. [See the derivation in Section 5.14.1.] The string might be plucked, giving the string an initial profile, $u(x,0) = f(x)$, and possibly each point on the string has an initial velocity $u_t(x,0) = g(x)$. The initial-boundary value problem for this problem is given below.

	1D Wave Equation		
PDE	$u_{tt} = c^2 u_{xx}$	$0 < t, \quad 0 \le x \le L$	
IC	$u(x,0) = f(x)$	$0 < x < L$	
	$u_t(x,0) = g(x)$	$0 < x < L$	(5.3)
BC	$u(0,t) = 0$	$t > 0$	
	$u(L,t) = 0$	$t > 0$	

In this problem, c is the wave speed in the string. It depends on the mass per unit length of the string, μ, and the tension, τ, placed on the string.

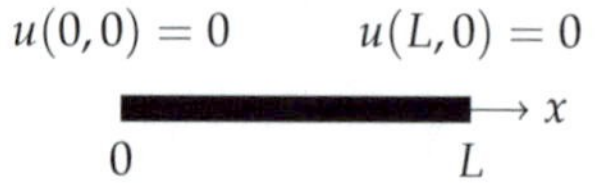

$$u(0,0) = 0 \qquad u(L,0) = 0$$

Figure 5.4: One-dimensional heated rod of length L.

In the next two sections we describe how to go about solving these equations. We will find that in order to accommodate the initial conditions, we will need to introduce Fourier series before we can complete the problems.

5.4 1D Heat Equation

SOLVING MANY OF THE LINEAR PARTIAL DIFFERENTIAL EQUATIONS presented in the last section can be reduced to solving ordinary differential equations. We will demonstrate this by solving the initial-boundary value problem for the heat equation as given in Equation (5.2). We will employ a method typically used in studying linear partial differential equations, called the Method of Separation of Variables. We want to solve the heat equation,

$$u_t = k u_{xx}, \quad 0 < t, \quad 0 \le x \le L,$$

subject to the boundary conditions

$$u(0,t) = 0, u(L,t) = 0, \quad t > 0,$$

and the initial condition

$$u(x,0) = f(x), \quad 0 < x < L.$$

Solution of the 1D heat equation using the method of separation of variables.

We begin by assuming that u can be written as a product of single variable functions of each independent variable,

$$u(x,t) = X(x)T(t).$$

Substituting this guess into the heat equation, we find that

$$XT' = kX''T.$$

The prime denotes differentiation with respect to the independent variable and we will suppress the independent variable in the following unless needed for emphasis.

Dividing both sides of this result by k and $u = XT$ yields

$$\frac{1}{k}\frac{T'}{T} = \frac{X''}{X}.$$

We have separated the functions of time on one side and space on the other side. The constant k could be on either side of this expression, but we moved it to make later computations simpler.

The only way that a function of t equals a function of x is if the functions are constant functions. Therefore, we set each function equal to a constant, λ : [For example, if $Ae^{ct} = ax^2 + b$ is possible for any x or t, then this is only possible if $a = 0$, $c = 0$, and $b = A$.]

$$\underbrace{\frac{1}{k}\frac{T'}{T}}_{\text{function of } t} = \underbrace{\frac{X''}{X}}_{\text{function of } x} = \underbrace{\lambda.}_{\text{constant}}$$

This leads to two equations:

$$T' = k\lambda T, \tag{5.4}$$

$$X'' = \lambda X. \tag{5.5}$$

These are ordinary differential equations. The general solutions to these constant coefficient equations are readily found as

$$T(t) = Ae^{k\lambda t}, \tag{5.6}$$

$$X(x) = c_1 e^{\sqrt{\lambda}x} + c_2 e^{-\sqrt{\lambda}x}. \tag{5.7}$$

We need to be a little careful at this point. The aim is to force the final solutions to satisfy both the boundary conditions and initial conditions. Also, we should note that λ is arbitrary and may be positive, zero, or negative. We first look at how the boundary conditions on $u(x,t)$ lead to conditions on $X(x)$.

The first boundary condition is $u(0,t) = 0$. This implies that

$$X(0)T(t) = 0, \quad \text{for all } t.$$

The only way that this is true is if $X(0) = 0$. Similarly, $u(L,t) = 0$ for all t implies that $X(L) = 0$. So, we have to solve the boundary value problem

$$X'' - \lambda X = 0, \quad X(0) = 0 = X(L). \tag{5.8}$$

An obvious solution is $X \equiv 0$. However, this implies that $u(x,t) = 0$, which is not an interesting solution. We call such solutions, $X \equiv 0$, trivial solutions and will seek nontrivial solution for these problems.

There are three cases to consider, depending on the sign of λ.

Case I. $\lambda > 0$

In this case, we have the exponential solutions

$$X(x) = c_1 e^{\sqrt{\lambda}x} + c_2 e^{-\sqrt{\lambda}x}. \tag{5.9}$$

For $X(0) = 0$, we have

$$0 = c_1 + c_2.$$

We will take $c_2 = -c_1$. Then,

$$X(x) = c_1(e^{\sqrt{\lambda}x} - e^{-\sqrt{\lambda}x}) = 2c_1 \sinh \sqrt{\lambda}x.$$

Applying the second condition, $X(L) = 0$ yields

$$c_1 \sinh \sqrt{\lambda}L = 0.$$

This will be true only if $c_1 = 0$, because $\lambda > 0$. Thus, the only solution in this case is the trivial solution, $X(x) = 0$.

Case II. $\lambda = 0$

For this case, it is easier to set λ to zero in the differential equation. So, $X'' = 0$. Integrating twice, one finds

$$X(x) = c_1 x + c_2.$$

Setting $x = 0$, we have $c_2 = 0$, leaving $X(x) = c_1 x$. Setting $x = L$, we find $c_1 L = 0$. So, $c_1 = 0$ and we are once again left with a trivial solution.

Case III. $\lambda < 0$

In this case, it would be simpler to write $\lambda = -\mu^2$. Then the differential equation is

$$X'' + \mu^2 X = 0.$$

The general solution is

$$X(x) = c_1 \cos \mu x + c_2 \sin \mu x.$$

At $x = 0$, we get $0 = c_1$. This leaves $X(x) = c_2 \sin \mu x$.
At $x = L$, we find

$$0 = c_2 \sin \mu L.$$

So, either $c_2 = 0$ or $\sin \mu L = 0$. $c_2 = 0$ leads to a trivial solution again. But there are cases when the sine is zero, namely,

$$\mu L = n\pi, \quad n = 1, 2, \ldots.$$

Note that $n = 0$ is not included because this leads to a trivial solution. Also, negative values of n are redundant, as the sine function is an odd function.

In summary, we can find solutions to the boundary value problem (5.8) for particular values of λ. The solutions are

$$X_n(x) = \sin \frac{n\pi x}{L}, \quad n = 1, 2, 3, \ldots,$$

for

$$\lambda_n = -\mu_n^2 = -\left(\frac{n\pi}{L}\right)^2, \quad n = 1, 2, 3, \ldots.$$

We should note that the boundary value problem in Equation (5.8) is an eigenvalue problem. We can recast the differential equation as

$$LX = \lambda X,$$

where

$$L = D^2 = \frac{d^2}{dx^2}$$

is a linear differential operator. The solutions, $X_n(x)$, are called eigenfunctions and the λ_n's are the eigenvalues. We will elaborate more on this characterization later in the next chapter.

We have found the product solutions of the heat equation (5.2) satisfying Product solutions. the boundary conditions. These are

$$u_n(x,t) = e^{k\lambda_n t} \sin\frac{n\pi x}{L}, \quad n = 1, 2, 3, \ldots. \tag{5.10}$$

However, these do not necessarily satisfy the initial condition $u(x,0) = f(x)$. What we do get is

$$u_n(x,0) = \sin\frac{n\pi x}{L}, \quad n = 1, 2, 3, \ldots.$$

So, if the initial condition is in one of these forms, we can pick out the right value for n and we are done.

For other initial conditions, we have to do more work. Note: Because General solution. the heat equation is linear, the linear combination of the product solutions is also a solution of the heat equation. The general solution satisfying the given boundary conditions is given as

$$u(x,t) = \sum_{n=1}^{\infty} b_n e^{k\lambda_n t} \sin\frac{n\pi x}{L}. \tag{5.11}$$

The coefficients in the general solution are determined using the initial condition. Namely, setting $t = 0$ in the general solution, we have

$$f(x) = u(x,0) = \sum_{n=1}^{\infty} b_n \sin\frac{n\pi x}{L}.$$

So, if we know $f(x)$, can we find the coefficients, b_n? If we can, then we will have the solution to the full initial-boundary value problem.

The expression for $f(x)$ is a Fourier sine series. We will need to digress into the study of Fourier series in order to see how one can find the Fourier series coefficients given $f(x)$. Before proceeding, we will show that this process is not uncommon by applying the Method of Separation of Variables to the wave equation in the next section.

5.5   1D Wave Equation

IN THIS SECTION WE WILL APPLY the Method of Separation of Variables to the one-dimensional wave equation, given by

$$\frac{\partial^2 u}{\partial^2 t} = c^2 \frac{\partial^2 u}{\partial^2 x}, \quad t > 0, \quad 0 \le x \text{l} L, \tag{5.12}$$

subject to the boundary conditions

$$u(0,t) = 0, u(L,t) = 0, \quad t > 0,$$

and the initial conditions

$$u(x,0) = f(x), u_t(x,0) = g(x), \quad 0 < x < L.$$

This problem applies to the propagation of waves on a string of length L with both ends fixed so that they do not move; $u(x,t)$ represents the vertical displacement of the string over time. The derivation of the wave equation assumes that the vertical displacement is small and the string is uniform. The constant c is the wave speed, given by

$$c = \sqrt{\frac{\tau}{\mu}},$$

where τ is the tension in the string and μ is the mass per unit length. We can understand this in terms of string instruments. The tension can be adjusted to produce different tones and the makeup of the string (nylon or steel, thick or thin) also has an effect. In some cases, the mass density is changed simply by using thicker strings. Thus, the thicker strings in a piano produce lower frequency notes.

The u_{tt} term gives the acceleration of a piece of the string. The u_{xx} is the concavity, of the string. Thus, for a positive concavity, the string is curved upward near the point of interest. Thus, neighboring points tend to pull upward toward the equilibrium position. If the concavity is negative, it would cause a negative acceleration.

Solution of the 1D wave equation using the Method of Separation of Variables.

The solution of this problem is easily found using separation of variables. We let $u(x,t) = X(x)T(t)$. Then we find

$$XT'' = c^2 X''T,$$

which can be rewritten as

$$\frac{1}{c^2}\frac{T''}{T} = \frac{X''}{X}.$$

Again, we have separated the functions of time on one side and space on the other side. Therefore, we set each function equal to a constant, λ.

$$\underbrace{\frac{1}{c^2}\frac{T''}{T}}_{\text{function of } t} = \underbrace{\frac{X''}{X}}_{\text{function of } x} = \underbrace{\lambda.}_{\text{constant}}$$

This leads to two equations:

$$T'' = c^2 \lambda T, \tag{5.13}$$

$$X'' = \lambda X. \tag{5.14}$$

As before, we have the boundary conditions on $X(x)$:

$$X(0) = 0, \quad \text{and} \quad X(L) = 0,$$

giving the solutions, as shown in Figure 5.5,

$$X_n(x) = \sin\frac{n\pi x}{L}, \quad \lambda_n = -\left(\frac{n\pi}{L}\right)^2.$$

The main difference from the solution of the heat equation is the form of the time function. Namely, from Equation (5.13) we have to solve

$$T'' + \left(\frac{n\pi c}{L}\right)^2 T = 0. \tag{5.15}$$

This equation takes a familiar form. We let

$$\omega_n = \frac{n\pi c}{L};$$

then we have

$$T'' + \omega_n^2 T = 0.$$

This is the differential equation for simple harmonic motion and ω_n is the angular frequency. The solutions are easily found as

$$T(t) = A_n \cos\omega_n t + B_n \sin\omega_n t. \tag{5.16}$$

Therefore, we have found that the product solutions of the wave equation take the forms $\sin\frac{n\pi x}{L}\cos\omega_n t$ and $\sin\frac{n\pi x}{L}\sin\omega_n t$. The general solution, a superposition of all product solutions, is given by

$$u(x,t) = \sum_{n=1}^{\infty}\left[A_n\cos\frac{n\pi ct}{L} + B_n\sin\frac{n\pi ct}{L}\right]\sin\frac{n\pi x}{L}. \tag{5.17}$$

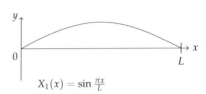

$X_1(x) = \sin\frac{\pi x}{L}$

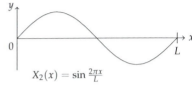

$X_2(x) = \sin\frac{2\pi x}{L}$

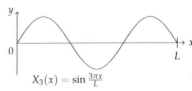

$X_3(x) = \sin\frac{3\pi x}{L}$

Figure 5.5: The first three harmonics of the vibrating string.

General solution.

This solution satisfies the wave equation and the boundary conditions. We still need to satisfy the initial conditions. Note that there are two initial conditions, as the wave equation is second order in time.

First, we have $u(x,0) = f(x)$. Thus,

$$f(x) = u(x,0) = \sum_{n=1}^{\infty} A_n\sin\frac{n\pi x}{L}. \tag{5.18}$$

In order to obtain the condition on the initial velocity, $u_t(x,0) = g(x)$, we need to differentiate the general solution with respect to t:

$$u_t(x,t) = \sum_{n=1}^{\infty}\frac{n\pi c}{L}\left[-A_n\sin\frac{n\pi ct}{L} + B_n\cos\frac{n\pi ct}{L}\right]\sin\frac{n\pi x}{L}. \tag{5.19}$$

Then, we have from the initial velocity

$$g(x) = u_t(x,0) = \sum_{n=1}^{\infty}\frac{n\pi c}{L}B_n\sin\frac{n\pi x}{L}. \tag{5.20}$$

So, applying the two initial conditions, we have found that $f(x)$ and $g(x)$ are represented as Fourier sine series. In order to complete the problem, we need to determine the coefficients A_n and B_n for $n = 1,2,3,\ldots$. Once we have these, we have the complete solution to the wave equation. We had seen similar results for the heat equation. In the next section we will find out how to determine these Fourier coefficients for such series of sinusoidal functions.

5.6   Introduction to Fourier Series

WE WILL NOW TURN TO THE STUDY of trigonometric series. You have seen that functions have series representations as expansions in powers of x, or $x - a$, in the form of Maclaurin and Taylor series. Recall that the Taylor series expansion is given by

$$f(x) = \sum_{n=0}^{\infty} c_n (x - a)^n,$$

where the expansion coefficients are determined as

$$c_n = \frac{f^{(n)}(a)}{n!}.$$

From the study of the heat equation and wave equation, we have found that there are infinite series expansions over other functions, such as sine functions. We now turn to such expansions and in the next chapter we will find out that expansions over special sets of functions are not uncommon in physics. But first we turn to Fourier trigonometric series.

We will begin with the study of the Fourier trigonometric series expansion

$$f(x) = \frac{a_0}{2} + \sum_{n=1}^{\infty} a_n \cos \frac{n\pi x}{L} + b_n \sin \frac{n\pi x}{L}.$$

We will find expressions useful for determining the Fourier coefficients $\{a_n, b_n\}$ given a function $f(x)$ defined on $[-L, L]$. We will also see if the resulting infinite series reproduces $f(x)$. However, we first begin with some basic ideas involving simple sums of sinusoidal functions.

There is a natural appearance of such sums over sinusoidal functions in music. A pure note can be represented as

$$y(t) = A \sin(2\pi f t),$$

where A is the amplitude, f is the frequency in Hertz (Hz), and t is time in seconds. The amplitude is related to the volume of the sound. The larger the amplitude, the louder the sound. In Figure 5.6 we show plots of two such tones with $f = 2$ Hz in the top plot and $f = 5$ Hz in the bottom one.

In these plots you should notice the difference due to the amplitudes and the frequencies. You can easily reproduce these plots and others in your favorite plotting utility.

As an aside, you should be cautious when plotting functions, or sampling data. The plots you get might not be what you expect, even for a simple sine function. In Figure 5.7 we show four plots of the function $y(t) = 2 \sin(4\pi t)$. In the top left, you see a proper rendering of this function. However, if you use a different number of points to plot this function, the results may be surprising. In this example, we show what happens if you use $N = 200, 100, 101$ points instead of the 201 points used in the first plot. Such disparities are not only possible when plotting functions, but are also present when collecting data. Typically, when you sample a set of data, you only gather a finite

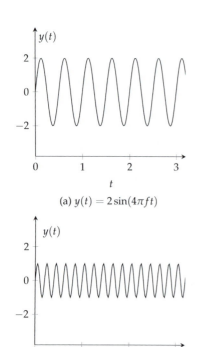

(a) $y(t) = 2 \sin(4\pi f t)$

(b) $y(t) = \sin(10\pi f t)$

Figure 5.6: Plots of $y(t) = A \sin(2\pi f t)$ on $[0, 5]$ for $f = 2$ Hz and $f = 5$ Hz.

amount of information at a fixed rate. This could happen when getting data on ocean wave heights, digitizing music, and other audio to put on your computer, or any other process when you attempt to analyze a continuous signal.

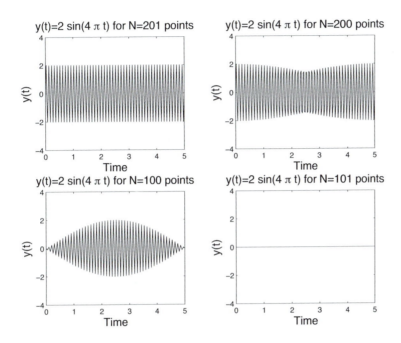

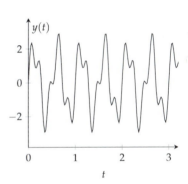

(a) Sum of signals with frequencies $f = 2$ Hz and $f = 5$ Hz.

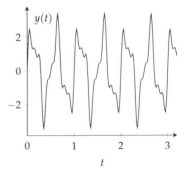

(b) Sum of signals with frequencies $f = 2$ Hz, $f = 5$ Hz, and $f = 8$ Hz.

Figure 5.8: Superposition of several sinusoids.

Figure 5.7: Problems can occur while plotting. Here we plot the function $y(t) = 2 \sin 4\pi t$ using $N = 201, 200, 100, 101$ points.

Next, we consider what happens when we add several pure tones. After all, most of the sounds that we hear are, in fact, a combination of pure tones with different amplitudes and frequencies. In Figure 5.8 we see what happens when we add several sinusoids. Note that as one adds more and more tones with different characteristics, the resulting signal gets more complicated. However, we still have a function of time. In this chapter we will ask, "Given a function $f(t)$, can we find a set of sinusoidal functions whose sum converges to $f(t)$?"

Looking at the superpositions in Figure 5.8, we see that the sums yield functions that appear to be periodic. This is not unexpected. We recall that a periodic function is one in which the function values repeat over the domain of the function. The length of the smallest part of the domain that repeats is called the period. We can define this more precisely: A function is said to be periodic with period T if $f(t + T) = f(t)$ for all t, and the smallest such positive number T is called the period.

For example, we consider the functions used in Figure 5.8. We began with $y(t) = 2\sin(4\pi t)$. Recall from your first studies of trigonometric functions that one can determine the period by dividing the coefficient of t into 2π to get the period. In this case, we have

$$T = \frac{2\pi}{4\pi} = \frac{1}{2}.$$

Looking at the top plot in Figure 5.6, we can verify this result. (You can count the full number of cycles in the graph and divide this into the total time to get a more accurate value of the period.)

In general, if $y(t) = A\sin(2\pi f t)$, the period is found as

$$T = \frac{2\pi}{2\pi f} = \frac{1}{f}.$$

Of course, this result makes sense, as the unit of frequency, the Hertz, is also defined as s^{-1}, or cycles per second.

Returning to Figure 5.8, the functions $y(t) = 2\sin(4\pi t)$, $y(t) = \sin(10\pi t)$, and $y(t) = 0.5\sin(16\pi t)$ have periods of 0.5 s, 0.2 s, and 0.125 s, respectively. Each superposition in Figure 5.8 retains a period that is the least common multiple of the periods of the signals added. For both plots, this is 1.0 s $= 2(0.5)$ s $= 5(.2)$ s $= 8(.125)$ s.

Our goal will be to start with a function and then determine the amplitudes of the simple sinusoids needed to sum to that function. We will see that this might involve an infinite number of such terms. Thus, we will be studying an infinite series of sinusoidal functions.

Second, we will find that using just sine functions will not be enough either. This is because we can add sinusoidal functions that do not necessarily peak at the same time. We will consider two signals that originate at different times. This is similar to when your music teacher would make sections of the class sing a song like "Row, Row, Row Your Boat" starting at slightly different times.

We can easily add shifted sine functions. In Figure 5.9, we show the functions $y(t) = 2\sin(4\pi t)$ and $y(t) = 2\sin(4\pi t + 7\pi/8)$ and their sum. Note that this shifted sine function can be written as $y(t) = 2\sin(4\pi(t + 7/32))$. Thus, this corresponds to a time shift of $-7/32$.

So, we should account for shifted sine functions in the general sum. Of course, we would then need to determine the unknown time shift as well as the amplitudes of the sinusoidal functions that make up the signal, $f(t)$. While this is one approach that some researchers use to analyze signals, there is a more common approach. This results from another reworking of the shifted function.

Consider the general shifted function

$$y(t) = A\sin(2\pi f t + \phi). \tag{5.21}$$

Note that $2\pi f t + \phi$ is called the phase of the sine function and ϕ is called the phase shift. We can use the trigonometric identity (1.17) for the sine of the sum of two angles[1] to obtain

$$\begin{aligned} y(t) &= A\sin(2\pi f t + \phi) \\ &= A\sin(\phi)\cos(2\pi f t) + A\cos(\phi)\sin(2\pi f t). \end{aligned} \tag{5.22}$$

Defining $a = A\sin(\phi)$ and $b = A\cos(\phi)$, we can rewrite this as

$$y(t) = a\cos(2\pi f t) + b\sin(2\pi f t).$$

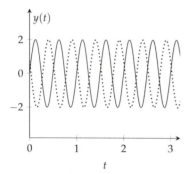

(a) Plot of each function.

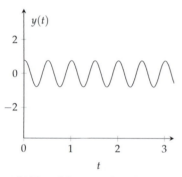

(b) Plot of the sum of the functions.

Figure 5.9: Plot of the functions $y(t) = 2\sin(4\pi t)$ and $y(t) = 2\sin(4\pi t + 7\pi/8)$ and their sum.

We should note that the form in the lower plot of Figure 5.9 looks like a simple sinusoidal function for a reason. Let

$$y_1(t) = 2\sin(4\pi t),$$

$$y_2(t) = 2\sin(4\pi t + 7\pi/8).$$

Then,

$$\begin{aligned} y_1 + y_2 &= 2\sin(4\pi t + 7\pi/8) + 2\sin(4\pi t) \\ &= 2[\sin(4\pi t + 7\pi/8) + \sin(4\pi t)] \\ &= 4\cos\frac{7\pi}{16}\sin\left(4\pi t + \frac{7\pi}{16}\right). \end{aligned}$$

[1] Recall the identities (1.17) and (1.18)

$$\sin(x+y) = \sin x \cos y + \sin y \cos x,$$
$$\cos(x+y) = \cos x \cos y - \sin x \sin y.$$

Thus, we see that the signal in Equation (5.21) is a sum of sine and cosine functions with the same frequency and different amplitudes. If we can find a and b, then we can easily determine A and ϕ:

$$A = \sqrt{a^2 + b^2}, \quad \tan \phi = \frac{b}{a}.$$

We are now in a position to state our goal:

Goal–Fourier Analysis

Given a signal $f(t)$, we would like to determine its frequency content by finding out what combinations of sines and cosines of varying frequencies and amplitudes will sum to the given function. This is called Fourier Analysis.

5.7 Fourier Trigonometric Series

AS WE HAVE SEEN IN THE PREVIOUS SECTION, we are interested in finding representations of functions in terms of sines and cosines. Given a function $f(x)$, we seek a representation in the form

$$f(x) \sim \frac{a_0}{2} + \sum_{n=1}^{\infty} \left[a_n \cos nx + b_n \sin nx \right]. \tag{5.23}$$

Notice that we have opted to drop the references to the time-frequency form of the phase. This will lead to a simpler discussion for now, and one can always make the transformation $nx = 2\pi f_n t$ when applying these ideas to applications.

The series representation in Equation (5.23) is called a Fourier trigonometric series. We will simply refer to this as a Fourier series for now. The set of constants $a_0, a_n, b_n, n = 1, 2, \ldots$ are called the Fourier coefficients. The constant term is chosen in this form to make later computations simpler, though some other authors choose to write the constant term as a_0. Our goal is to find the Fourier series representation given $f(x)$. Having found the Fourier series representation, we will be interested in determining when the Fourier series converges and to what function it converges.

From our discussion in the previous section, we see that The Fourier series is periodic. The periods of $\cos nx$ and $\sin nx$ are $\frac{2\pi}{n}$. Thus, the largest period, $T = 2\pi$, comes from the $n = 1$ terms and the Fourier series has period 2π. This means that the series should be able to represent functions that are periodic of period 2π.

While this appears restrictive, we could also consider functions that are defined over one period. In Figure 5.10 we show a function defined on $[0, 2\pi]$. In the same figure, we show its periodic extension. These are just copies of the original function shifted by the period and glued together. The extension can now be represented by a Fourier series and restricting the Fourier series to $[0, 2\pi]$ will give a representation of the original function.

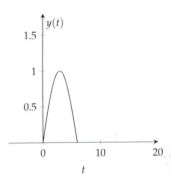

(a) Plot of function $f(t)$.

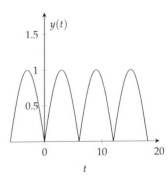

(b) Periodic extension of $f(t)$.

Figure 5.10: Plot of the function $f(t)$ defined on $[0, 2\pi]$ and its periodic extension.

Therefore, we will first consider Fourier series representations of functions defined on this interval. Note that we could just as easily consider functions defined on $[-\pi, \pi]$ or any interval of length 2π. We will consider more general intervals later in the chapter.

Fourier Coefficients

Theorem 5.1. *The Fourier series representation of $f(x)$ defined on $[0, 2\pi]$, when it exists, is given by Equation (5.23) with Fourier coefficients*

$$a_n = \frac{1}{\pi} \int_0^{2\pi} f(x) \cos nx \, dx, \quad n = 0, 1, 2, \ldots,$$

$$b_n = \frac{1}{\pi} \int_0^{2\pi} f(x) \sin nx \, dx, \quad n = 1, 2, \ldots. \tag{5.24}$$

These expressions for the Fourier coefficients are obtained by considering special integrations of the Fourier series. We will now derive the a_n integrals in Equation (5.24).

We begin with the computation of a_0. Integrating the Fourier series term by term in Equation (5.23), we have

$$\int_0^{2\pi} f(x) \, dx = \int_0^{2\pi} \frac{a_0}{2} \, dx + \int_0^{2\pi} \sum_{n=1}^{\infty} [a_n \cos nx + b_n \sin nx] \, dx. \tag{5.25}$$

Evaluating the integral of an infinite series by integrating term by term depends on the convergence properties of the series.

We will assume that we can integrate the infinite sum term by term. Then we will need to compute

$$\int_0^{2\pi} \frac{a_0}{2} \, dx = \frac{a_0}{2}(2\pi) = \pi a_0,$$

$$\int_0^{2\pi} \cos nx \, dx = \left[\frac{\sin nx}{n} \right]_0^{2\pi} = 0,$$

$$\int_0^{2\pi} \sin nx \, dx = \left[\frac{-\cos nx}{n} \right]_0^{2\pi} = 0. \tag{5.26}$$

From these results we see that only one term in the integrated sum does not vanish, leaving

$$\int_0^{2\pi} f(x) \, dx = \pi a_0.$$

This confirms the value for a_0.[2]

[2] Note that $\frac{a_0}{2}$ is the average of $f(x)$ over the interval $[0, 2\pi]$. Recall from the first semester of calculus, that the average of a function defined on $[a, b]$ is given by

$$f_{\text{ave}} = \frac{1}{b-a} \int_a^b f(x) \, dx.$$

For $f(x)$ defined on $[0, 2\pi]$, we have

$$f_{\text{ave}} = \frac{1}{2\pi} \int_0^{2\pi} f(x) \, dx = \frac{a_0}{2}.$$

Next, we will find the expression for a_n. We multiply the Fourier series in Equation (5.23) by $\cos mx$ for some positive integer m. This is like multiplying by $\cos 2x$, $\cos 5x$, etc. We are multiplying by all possible $\cos mx$ functions for different integers m all at the same time. We will see that this will allow us to solve for the a_n's.

We find the integrated sum of the series times $\cos mx$ is given by

$$\int_0^{2\pi} f(x) \cos mx \, dx = \int_0^{2\pi} \frac{a_0}{2} \cos mx \, dx$$

$$+ \int_0^{2\pi} \sum_{n=1}^{\infty} [a_n \cos nx + b_n \sin nx] \cos mx \, dx.$$

$$\tag{5.27}$$

Integrating term by term, the right side becomes

$$\int_0^{2\pi} f(x) \cos mx \, dx = \frac{a_0}{2} \int_0^{2\pi} \cos mx \, dx$$

$$+ \sum_{n=1}^{\infty} \left[a_n \int_0^{2\pi} \cos nx \cos mx \, dx + b_n \int_0^{2\pi} \sin nx \cos mx \, dx \right].$$

(5.28)

We have already established that $\int_0^{2\pi} \cos mx \, dx = 0$, which implies that the first term vanishes.

Next we need to compute integrals of products of sines and cosines. This requires that we make use of some of the trigonometric identities listed in Chapter 1. For quick reference, we list these here.

Useful Trigonometric Identities		
$\sin(x \pm y)$ = $\sin x \cos y \pm \sin y \cos x$		(5.29)
$\cos(x \pm y)$ = $\cos x \cos y \mp \sin x \sin y$		(5.30)
$\sin^2 x$ = $\frac{1}{2}(1 - \cos 2x)$		(5.31)
$\cos^2 x$ = $\frac{1}{2}(1 + \cos 2x)$		(5.32)
$\sin x \sin y$ = $\frac{1}{2}(\cos(x - y) - \cos(x + y))$		(5.33)
$\cos x \cos y$ = $\frac{1}{2}(\cos(x + y) + \cos(x - y))$		(5.34)
$\sin x \cos y$ = $\frac{1}{2}(\sin(x + y) + \sin(x - y))$		(5.35)

We first want to evaluate $\int_0^{2\pi} \cos nx \cos mx \, dx$. We do this using the product identity (5.34). We have

$$\int_0^{2\pi} \cos nx \cos mx \, dx = \frac{1}{2} \int_0^{2\pi} [\cos(m + n)x + \cos(m - n)x] \, dx$$

$$= \frac{1}{2} \left[\frac{\sin(m + n)x}{m + n} + \frac{\sin(m - n)x}{m - n} \right]_0^{2\pi}$$

$$= 0.$$

(5.36)

There is one caveat when doing such integrals. What if one of the denominators $m \pm n$ vanishes? For this problem, $m + n \neq 0$, as both m and n are positive integers. However, it is possible for $m = n$. This means that the vanishing of the integral can only happen when $m \neq n$. So, what can we do about the $m = n$ case? One way is to start from scratch with our integration. (Another way is to compute the limit as n approaches m in our result and use L'Hopital's Rule. Try it!)

For $n = m$, we have to compute $\int_0^{2\pi} \cos^2 mx \, dx$. This can also be handled using a trigonometric identity. Using the half angle formula, Equation (5.32), with $\theta = mx$, we find

$$\int_0^{2\pi} \cos^2 mx \, dx = \frac{1}{2} \int_0^{2\pi} (1 + \cos 2mx) \, dx$$

$$= \frac{1}{2} \left[x + \frac{1}{2m} \sin 2mx \right]_0^{2\pi}$$

$$= \frac{1}{2} (2\pi) = \pi. \tag{5.37}$$

To summarize, we have shown that

$$\int_0^{2\pi} \cos nx \cos mx \, dx = \begin{cases} 0, & m \neq n, \\ \pi, & m = n. \end{cases} \tag{5.38}$$

This holds true for $m, n = 0, 1, \ldots$. [Why did we include $m, n = 0$?] When we have such a set of functions, they are said to be an orthogonal set over the integration interval. A set of (real) functions $\{\phi_n(x)\}$ is said to be orthogonal on $[a, b]$ if $\int_a^b \phi_n(x)\phi_m(x) \, dx = 0$ when $n \neq m$. Furthermore, if we also have that $\int_a^b \phi_n^2(x) \, dx = 1$, these functions are called orthonormal.

Definition of an orthogonal set of functions and orthonormal functions.

The set of functions $\{\cos nx\}_{n=0}^\infty$ are orthogonal on $[0, 2\pi]$. Actually, they are orthogonal on any interval of length 2π. We can make them orthonormal by dividing each function by $\sqrt{\pi}$, as indicated by Equation (5.37). This is sometimes referred to as normalization of the set of functions.

The notion of orthogonality is actually a generalization of the orthogonality of vectors in finite dimensional vector spaces. The integral $\int_a^b f(x)f(x) \, dx$ is the generalization of the dot product, and is called the scalar product of $f(x)$ and $g(x)$, which are thought of as vectors in an infinite dimensional vector space spanned by a set of orthogonal functions. We will return to these ideas in the next chapter.

Returning to the integrals in Equation (5.28), we still have to evaluate $\int_0^{2\pi} \sin nx \cos mx \, dx$. We can use the trigonometric identity involving products of sines and cosines, Equation (5.35). Setting $A = nx$ and $B = mx$, we find that

$$\int_0^{2\pi} \sin nx \cos mx \, dx = \frac{1}{2} \int_0^{2\pi} [\sin(n + m)x + \sin(n - m)x] \, dx$$

$$= \frac{1}{2} \left[\frac{-\cos(n + m)x}{n + m} + \frac{-\cos(n - m)x}{n - m} \right]_0^{2\pi}$$

$$= (-1 + 1) + (-1 + 1) = 0. \tag{5.39}$$

So,

$$\int_0^{2\pi} \sin nx \cos mx \, dx = 0. \tag{5.40}$$

For these integrals we should also be careful about setting $n = m$. In this special case, we have the integrals

$$\int_0^{2\pi} \sin mx \cos mx \, dx = \frac{1}{2} \int_0^{2\pi} \sin 2mx \, dx = \frac{1}{2} \left[\frac{-\cos 2mx}{2m} \right]_0^{2\pi} = 0.$$

Finally, we can finish evaluating the expression in Equation (5.28). We have determined that all but one integral vanishes. In that case, $n = m$. This leaves us with

$$\int_0^{2\pi} f(x) \cos mx \, dx = a_m \pi.$$

Solving for a_m gives

$$a_m = \frac{1}{\pi} \int_0^{2\pi} f(x) \cos mx \, dx.$$

Because this is true for all $m = 1, 2, \ldots$, we have proven this part of the theorem. The only part left is finding the b_n's This will be left as an exercise for the reader.

We now consider examples of finding Fourier coefficients for given functions. In all of these cases, we define $f(x)$ on $[0, 2\pi]$.

Example 5.2. $f(x) = 3 \cos 2x$, $x \in [0, 2\pi]$.

We first compute the integrals for the Fourier coefficients:

$$
\begin{aligned}
a_0 &= \frac{1}{\pi} \int_0^{2\pi} 3 \cos 2x \, dx = 0, \\
a_n &= \frac{1}{\pi} \int_0^{2\pi} 3 \cos 2x \cos nx \, dx = 0, \quad n \neq 2, \\
a_2 &= \frac{1}{\pi} \int_0^{2\pi} 3 \cos^2 2x \, dx = 3, \\
b_n &= \frac{1}{\pi} \int_0^{2\pi} 3 \cos 2x \sin nx \, dx = 0, \forall n.
\end{aligned}
$$

$$(5.41)$$

The integrals for a_0, $a_n, n \neq 2$, and b_n are the result of orthogonality. For a_2, the integral can be computed as follows:

$$
\begin{aligned}
a_2 &= \frac{1}{\pi} \int_0^{2\pi} 3 \cos^2 2x \, dx \\
&= \frac{3}{2\pi} \int_0^{2\pi} [1 + \cos 4x] \, dx \\
&= \frac{3}{2\pi} \left[x + \underbrace{\frac{1}{4} \sin 4x}_{\text{This term vanishes!}} \right]_0^{2\pi} = 3.
\end{aligned}
$$

$$(5.42)$$

Therefore, we have that the only nonvanishing coefficient is $a_2 = 3$. So there is one term and $f(x) = 3 \cos 2x$.

Well, we should have known the answer to the previous example before doing all those integrals. If we have a function expressed simply in terms of sums of simple sines and cosines, then it should be easy to write the Fourier coefficients without much work. This is seen by writing out the Fourier series,

$$
\begin{aligned}
f(x) &\sim \frac{a_0}{2} + \sum_{n=1}^{\infty} [a_n \cos nx + b_n \sin nx] . \\
&= \frac{a_0}{2} + a_1 \cos x + b_1 \sin x + + a_2 \cos 2x + b_2 \sin 2x + \ldots. \quad (5.43)
\end{aligned}
$$

For the last problem, $f(x) = 3 \cos 2x$. Comparing this to the expanded Fourier series, one can immediately read off the Fourier coefficients without doing any integration. In the next example we, emphasize this point.

Example 5.3. $f(x) = \sin^2 x$, $x \in [0, 2\pi]$.

We could determine the Fourier coefficients by integrating as in the previous example. However, it is easier to use trigonometric identities. We know that

$$\sin^2 x = \frac{1}{2}(1 - \cos 2x) = \frac{1}{2} - \frac{1}{2}\cos 2x.$$

There are no sine terms, so $b_n = 0$, $n = 1, 2, \ldots$. There is a constant term, implying $a_0/2 = 1/2$. So, $a_0 = 1$. There is a $\cos 2x$ term, corresponding to $n = 2$, so $a_2 = -\frac{1}{2}$. That leaves $a_n = 0$ for $n \neq 0, 2$. So, $a_0 = 1$, $a_2 = -\frac{1}{2}$, and all other Fourier coefficients vanish.

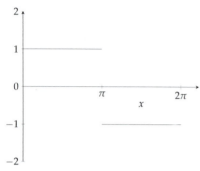

Figure 5.11: Plot of discontinuous function in Example 5.4.

Example 5.4. $f(x) = \begin{cases} 1, & 0 < x < \pi, \\ -1, & \pi < x < 2\pi, \end{cases}$.

This example will take a little more work. We cannot bypass evaluating any integrals this time. As seen in Figure 5.11, this function is discontinuous. So, we will break up any integration into two integrals, one over $[0, \pi]$ and the other over $[\pi, 2\pi]$.

$$
\begin{aligned}
a_0 &= \frac{1}{\pi}\int_0^{2\pi} f(x)\,dx \\
&= \frac{1}{\pi}\int_0^{\pi} dx + \frac{1}{\pi}\int_\pi^{2\pi}(-1)\,dx \\
&= \frac{1}{\pi}(\pi) + \frac{1}{\pi}(-2\pi + \pi) = 0.
\end{aligned}
\tag{5.44}
$$

$$
\begin{aligned}
a_n &= \frac{1}{\pi}\int_0^{2\pi} f(x)\cos nx\,dx \\
&= \frac{1}{\pi}\left[\int_0^{\pi}\cos nx\,dx - \int_\pi^{2\pi}\cos nx\,dx\right] \\
&= \frac{1}{\pi}\left[\left(\frac{1}{n}\sin nx\right)_0^{\pi} - \left(\frac{1}{n}\sin nx\right)_\pi^{2\pi}\right] \\
&= 0.
\end{aligned}
\tag{5.45}
$$

$$
\begin{aligned}
b_n &= \frac{1}{\pi}\int_0^{2\pi} f(x)\sin nx\,dx \\
&= \frac{1}{\pi}\left[\int_0^{\pi}\sin nx\,dx - \int_\pi^{2\pi}\sin nx\,dx\right] \\
&= \frac{1}{\pi}\left[\left(-\frac{1}{n}\cos nx\right)_0^{\pi} + \left(\frac{1}{n}\cos nx\right)_\pi^{2\pi}\right] \\
&= \frac{1}{\pi}\left[-\frac{1}{n}\cos n\pi + \frac{1}{n} + \frac{1}{n} - \frac{1}{n}\cos n\pi\right] \\
&= \frac{2}{n\pi}(1 - \cos n\pi).
\end{aligned}
\tag{5.46}
$$

Often we see expressions involving $\cos n\pi = (-1)^n$ and $1 \pm \cos n\pi = 1 \pm (-1)^n$. This is an example showing how to re-index series containing $\cos n\pi$.

We have found the Fourier coefficients for this function. Before inserting them into the Fourier series (5.23), we note that $\cos n\pi = (-1)^n$. Therefore,

$$1 - \cos n\pi = \begin{cases} 0, & n \text{ even}, \\ 2, & n \text{ odd}. \end{cases} \tag{5.47}$$

So, half of the b_n's are zero. While we could write the Fourier series representation as

$$f(x) \sim \frac{4}{\pi} \sum_{\substack{n=1 \\ n \text{ odd}}}^{\infty} \frac{1}{n} \sin nx,$$

we could let $n = 2k - 1$ in order to capture the odd numbers only. The answer can be written as

$$f(x) = \frac{4}{\pi} \sum_{k=1}^{\infty} \frac{\sin(2k-1)x}{2k-1},$$

Having determined the Fourier representation of a given function, we would like to know if the infinite series can be summed; that is, does the series converge? Does it converge to $f(x)$? We will discuss this question later in the chapter after we generalize the Fourier series to intervals other than for $x \in [0, 2\pi]$.

5.8   Fourier Series over Other Intervals

IN MANY APPLICATIONS WE ARE INTERESTED in determining Fourier series representations of functions defined on intervals other than $[0, 2\pi]$. In this section, we will determine the form of the series expansion and the Fourier coefficients in these cases.

The most general type of interval is given as $[a, b]$. However, this often is too general. More common intervals are of the form $[-\pi, \pi]$, $[0, L]$, or $[-L/2, L/2]$. The simplest generalization is to the interval $[0, L]$. Such intervals arise often in applications. For example, for the problem of a one-dimensional string of length L, we set up the axes with the left end at $x = 0$ and the right end at $x = L$. Similarly for the temperature distribution along a one-dimensional rod of length L, we set the interval to $x \in [0, 2\pi]$. Such problems naturally lead to the study of Fourier series on intervals of length L. We will see later that symmetric intervals, $[-a, a]$, are also useful.

Given an interval $[0, L]$, we could apply a transformation to an interval of length 2π by simply rescaling the interval. Then we could apply this transformation to the Fourier series representation to obtain an equivalent one useful for functions defined on $[0, L]$.

We define $x \in [0, 2\pi]$ and $t \in [0, L]$. A linear transformation relating these intervals is simply $x = \frac{2\pi t}{L}$ as shown in Figure 5.12. So, $t = 0$ maps to $x = 0$ and $t = L$ maps to $x = 2\pi$. Furthermore, this transformation maps $f(x)$ to a new function $g(t) = f(x(t))$, which is defined on $[0, L]$. We will determine the Fourier series representation of this function using the representation for $f(x)$ from the previous section.

Recall the form of the Fourier representation for $f(x)$ in Equation (5.23):

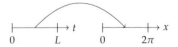

Figure 5.12: A sketch of the transformation between intervals $x \in [0, 2\pi]$ and $t \in [0, L]$.

$$f(x) \sim \frac{a_0}{2} + \sum_{n=1}^{\infty} \left[a_n \cos nx + b_n \sin nx \right]. \tag{5.48}$$

Inserting the transformation relating x and t, we have

$$g(t) \sim \frac{a_0}{2} + \sum_{n=1}^{\infty} \left[a_n \cos \frac{2n\pi t}{L} + b_n \sin \frac{2n\pi t}{L} \right]. \tag{5.49}$$

This gives the form of the series expansion for $g(t)$ with $t \in [0, L]$. But we still need to determine the Fourier coefficients.

Recall that

$$a_n = \frac{1}{\pi} \int_0^{2\pi} f(x) \cos nx \, dx.$$

We need to make a substitution in the integral of $x = \frac{2\pi t}{L}$. We also will need to transform the differential $dx = \frac{2\pi}{L} dt$. Thus, the resulting form for the Fourier coefficients is

$$a_n = \frac{2}{L} \int_0^{L} g(t) \cos \frac{2n\pi t}{L} \, dt. \tag{5.50}$$

Similarly, we find that

$$b_n = \frac{2}{L} \int_0^{L} g(t) \sin \frac{2n\pi t}{L} \, dt. \tag{5.51}$$

We note first that when $L = 2\pi$, we get back the series representation that we first studied. Also, the period of $\cos \frac{2n\pi t}{L}$ is L/n, which means that the representation for $g(t)$ has a period of L corresponding to $n = 1$.

At the end of this section, we present the derivation of the Fourier series representation for a general interval for the interested reader. In Table 5.2 we summarize some commonly used Fourier series representations.

At this point we need to remind the reader about the integration of even and odd functions on symmetric intervals.

We first recall that $f(x)$ is an even function if $f(-x) = f(x)$ for all x. One can recognize even functions as they are symmetric with respect to the y-axis as shown in Figure 5.13.

If one integrates an even function over a symmetric interval, then one has that

$$\int_{-a}^{a} f(x) \, dx = 2 \int_0^{a} f(x) \, dx. \tag{5.52}$$

One can prove this by splitting off the integration over negative values of x, using the substitution $x = -y$, and employing the evenness of $f(x)$. Thus,

$$\begin{aligned}
\int_{-a}^{a} f(x) \, dx &= \int_{-a}^{0} f(x) \, dx + \int_0^{a} f(x) \, dx \\
&= -\int_{a}^{0} f(-y) \, dy + \int_0^{a} f(x) \, dx \\
&= \int_0^{a} f(y) \, dy + \int_0^{a} f(x) \, dx \\
&= 2 \int_0^{a} f(x) \, dx. \tag{5.53}
\end{aligned}$$

Integration of even and odd functions over symmetric intervals, $[-a, a]$.

Even Functions.

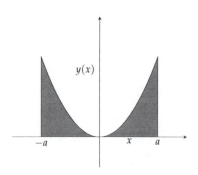

Figure 5.13: Area under an even function on a symmetric interval, $[-a, a]$.

Fourier Series on $[0, L]$

$$f(x) \sim \frac{a_0}{2} + \sum_{n=1}^{\infty} \left[a_n \cos \frac{2n\pi x}{L} + b_n \sin \frac{2n\pi x}{L} \right]. \tag{5.54}$$

$$a_n = \frac{2}{L} \int_0^L f(x) \cos \frac{2n\pi x}{L} \, dx. \quad n = 0, 1, 2, \ldots,$$

$$b_n = \frac{2}{L} \int_0^L f(x) \sin \frac{2n\pi x}{L} \, dx. \quad n = 1, 2, \ldots. \tag{5.55}$$

Fourier Series on $\left[-\frac{L}{2}, \frac{L}{2}\right]$

$$f(x) \sim \frac{a_0}{2} + \sum_{n=1}^{\infty} \left[a_n \cos \frac{2n\pi x}{L} + b_n \sin \frac{2n\pi x}{L} \right]. \tag{5.56}$$

$$a_n = \frac{2}{L} \int_{-\frac{L}{2}}^{\frac{L}{2}} f(x) \cos \frac{2n\pi x}{L} \, dx. \quad n = 0, 1, 2, \ldots,$$

$$b_n = \frac{2}{L} \int_{-\frac{L}{2}}^{\frac{L}{2}} f(x) \sin \frac{2n\pi x}{L} \, dx. \quad n = 1, 2, \ldots. \tag{5.57}$$

Fourier Series on $[-\pi, \pi]$

$$f(x) \sim \frac{a_0}{2} + \sum_{n=1}^{\infty} \left[a_n \cos nx + b_n \sin nx \right]. \tag{5.58}$$

$$a_n = \frac{1}{\pi} \int_{-\pi}^{\pi} f(x) \cos nx \, dx. \quad n = 0, 1, 2, \ldots,$$

$$b_n = \frac{1}{\pi} \int_{-\pi}^{\pi} f(x) \sin nx \, dx. \quad n = 1, 2, \ldots. \tag{5.59}$$

Table 5.2: Special Fourier Series Representations on Different Intervals

Odd Functions.

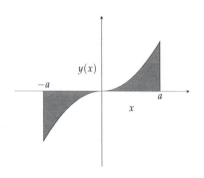

Figure 5.14: Area under an odd function on a symmetric interval, $[-a, a]$.

This can be visually verified by looking at Figure 5.13.

A similar computation could be done for odd functions. $f(x)$ is an odd function if $f(-x) = -f(x)$ for all x. The graphs of such functions are symmetric with respect to the origin, as shown in Figure 5.14. If one integrates an odd function over a symmetric interval, then one has that

$$\int_{-a}^{a} f(x)\, dx = 0. \tag{5.60}$$

Example 5.5. *Let $f(x) = |x|$ on $[-\pi, \pi]$ We compute the coefficients, beginning as usual with a_0. We have, using the fact that $|x|$ is an even function,*

$$
\begin{aligned}
a_0 &= \frac{1}{\pi} \int_{-\pi}^{\pi} |x|\, dx \\
&= \frac{2}{\pi} \int_{0}^{\pi} x\, dx = \pi \tag{5.61}
\end{aligned}
$$

We continue with the computation of the general Fourier coefficients for $f(x) = |x|$ on $[-\pi, \pi]$. We have

$$a_n = \frac{1}{\pi} \int_{-\pi}^{\pi} |x| \cos nx\, dx = \frac{2}{\pi} \int_{0}^{\pi} x \cos nx\, dx. \tag{5.62}$$

Here we have made use of the fact that $|x| \cos nx$ is an even function.

In order to compute the resulting integral, we need to use integration by parts,

$$\int_{a}^{b} u\, dv = uv \Big|_{a}^{b} - \int_{a}^{b} v\, du,$$

by letting $u = x$ and $dv = \cos nx\, dx$. Thus, $du = dx$ and $v = \int dv = \frac{1}{n} \sin nx$. Continuing with the computation, we have

$$
\begin{aligned}
a_n &= \frac{2}{\pi} \int_{0}^{\pi} x \cos nx\, dx. \\
&= \frac{2}{\pi} \left[\frac{1}{n} x \sin nx \Big|_{0}^{\pi} - \frac{1}{n} \int_{0}^{\pi} \sin nx\, dx \right] \\
&= -\frac{2}{n\pi} \left[-\frac{1}{n} \cos nx \right]_{0}^{\pi} \\
&= -\frac{2}{\pi n^2} (1 - (-1)^n). \tag{5.63}
\end{aligned}
$$

Here we have used the fact that $\cos n\pi = (-1)^n$ for any integer n. This leads to a factor $(1 - (-1)^n)$. This factor can be simplified as

$$1 - (-1)^n = \begin{cases} 2, & n \text{ odd}, \\ 0, & n \text{ even} \end{cases}. \tag{5.64}$$

So, $a_n = 0$ for n even and $a_n = -\frac{4}{\pi n^2}$ for n odd.

Computing the b_n's is simpler. We note that we have to integrate $|x| \sin nx$ from $x = -\pi$ to π. The integrand is an odd function and this is a symmetric interval. So, the result is that $b_n = 0$ for all n.

Putting this all together, the Fourier series representation of $f(x) = |x|$ on $[-\pi, \pi]$ is given as

$$f(x) \sim \frac{\pi}{2} - \frac{4}{\pi} \sum_{\substack{n=1 \\ n\, odd}}^{\infty} \frac{\cos nx}{n^2}. \tag{5.65}$$

While this is correct, we can rewrite the sum over only odd n by re-indexing. We let $n = 2k - 1$ for $k = 1, 2, 3, \ldots$. Then we only get the odd integers. The series can then be written as

$$f(x) \sim \frac{\pi}{2} - \frac{4}{\pi} \sum_{k=1}^{\infty} \frac{\cos(2k-1)x}{(2k-1)^2}. \tag{5.66}$$

Throughout our discussion we have referred to such results as Fourier representations. We have not looked at the convergence of these series. Here is an example of an infinite series of functions. What does this series sum to? We show in Figure 5.15 the first few partial sums. They appear to be converging to $f(x) = |x|$ fairly quickly.

Even though $f(x)$ was defined on $[-\pi, \pi]$, we can still evaluate the Fourier series at values of x outside this interval. In Figure 5.16, we see that the representation agrees with $f(x)$ on the interval $[-\pi, \pi]$. Outside this interval, we have a periodic extension of $f(x)$ with period 2π.

Another example is the Fourier series representation of $f(x) = x$ on $[-\pi, \pi]$ as left for Problem 7. This is determined to be

$$f(x) \sim 2 \sum_{n=1}^{\infty} \frac{(-1)^{n+1}}{n} \sin nx. \tag{5.67}$$

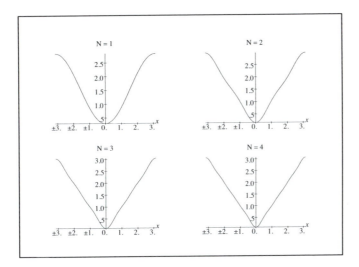

Figure 5.15: Plot of the first partial sums of the Fourier series representation for $f(x) = |x|$.

As seen in Figure 5.17, we again obtain the periodic extension of the function. In this case, we needed many more terms. Also, the vertical parts of the first plot are nonexistent. In the second plot, we only plot the points and not the typical connected points that most software packages plot as the default style.

Figure 5.16: Plot of the first 10 terms
of the Fourier series representation for
$f(x) = |x|$ on the interval $[-2\pi, 4\pi]$.

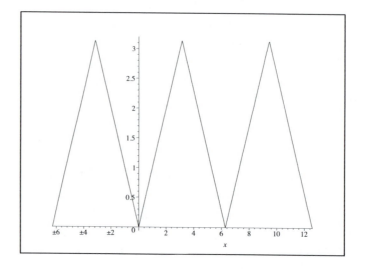

Figure 5.17: Plot of the first 10 terms
and 200 terms of the Fourier series representation for $f(x) = x$ on the interval
$[-2\pi, 4\pi]$.

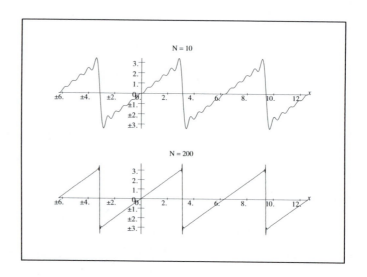

Example 5.6. *It is interesting to note that one can use Fourier series to obtain sums of some infinite series. For example, in the previous example, we found that*

$$x \sim 2 \sum_{n=1}^{\infty} \frac{(-1)^{n+1}}{n} \sin nx.$$

Now, what if we chose $x = \frac{\pi}{2}$? Then, we have

$$\frac{\pi}{2} = 2 \sum_{n=1}^{\infty} \frac{(-1)^{n+1}}{n} \sin \frac{n\pi}{2} = 2 \left[1 - \frac{1}{3} + \frac{1}{5} - \frac{1}{7} + \cdots \right].$$

This gives a well-known expression for π:

$$\pi = 4 \left[1 - \frac{1}{3} + \frac{1}{5} - \frac{1}{7} + \cdots \right].$$

5.8.1   *Fourier Series on $[a, b]$*

A FOURIER SERIES REPRESENTATION is also possible for a general interval, $t \in [a, b]$. As before, we just need to transform this interval to $[0, 2\pi]$. Let

$$x = 2\pi \frac{t - a}{b - a}.$$

This section can be skipped on first reading. It is here for completeness and the end result, Theorem 5.2, provides the result of the section.

Inserting this into the Fourier series (5.23) representation for $f(x)$, we obtain

$$g(t) \sim \frac{a_0}{2} + \sum_{n=1}^{\infty} \left[a_n \cos \frac{2n\pi(t-a)}{b-a} + b_n \sin \frac{2n\pi(t-a)}{b-a} \right]. \tag{5.68}$$

Well, this expansion is ugly. It is not like the previous example, where the transformation was straightforward. If one were to apply the theory to applications, it might seem to make sense to just shift the data so that $a = 0$ and be done with any complicated expressions. However, some students enjoy the challenge of developing such generalized expressions. So, let's see what is involved.

First, we apply the addition identities for trigonometric functions and rearrange the terms.

$$
\begin{aligned}
g(t) \quad \sim \quad & \frac{a_0}{2} + \sum_{n=1}^{\infty} \left[a_n \cos \frac{2n\pi(t-a)}{b-a} + b_n \sin \frac{2n\pi(t-a)}{b-a} \right] \\
= \quad & \frac{a_0}{2} + \sum_{n=1}^{\infty} \left[a_n \left(\cos \frac{2n\pi t}{b-a} \cos \frac{2n\pi a}{b-a} + \sin \frac{2n\pi t}{b-a} \sin \frac{2n\pi a}{b-a} \right) \right. \\
& \left. + b_n \left(\sin \frac{2n\pi t}{b-a} \cos \frac{2n\pi a}{b-a} - \cos \frac{2n\pi t}{b-a} \sin \frac{2n\pi a}{b-a} \right) \right] \\
= \quad & \frac{a_0}{2} + \sum_{n=1}^{\infty} \left[\cos \frac{2n\pi t}{b-a} \left(a_n \cos \frac{2n\pi a}{b-a} - b_n \sin \frac{2n\pi a}{b-a} \right) \right. \\
& \left. + \sin \frac{2n\pi t}{b-a} \left(a_n \sin \frac{2n\pi a}{b-a} + b_n \cos \frac{2n\pi a}{b-a} \right) \right]. \tag{5.69}
\end{aligned}
$$

Defining $A_0 = a_0$ and

$$A_n \equiv a_n \cos \frac{2n\pi a}{b-a} - b_n \sin \frac{2n\pi a}{b-a}$$

$$B_n \equiv a_n \sin \frac{2n\pi a}{b-a} + b_n \cos \frac{2n\pi a}{b-a}, \tag{5.70}$$

we arrive at the more desirable form for the Fourier series representation of a function defined on the interval $[a, b]$.

$$g(t) \sim \frac{A_0}{2} + \sum_{n=1}^{\infty} \left[A_n \cos \frac{2n\pi t}{b-a} + B_n \sin \frac{2n\pi t}{b-a} \right]. \tag{5.71}$$

We next need to find expressions for the Fourier coefficients. We insert the known expressions for a_n and b_n and rearrange. First, we note that under the transformation $x = 2\pi \frac{t-a}{b-a}$, we have

$$
\begin{aligned}
a_n &= \frac{1}{\pi} \int_0^{2\pi} f(x) \cos nx\, dx \\
&= \frac{2}{b-a} \int_a^b g(t) \cos \frac{2n\pi(t-a)}{b-a}\, dt, \tag{5.72}
\end{aligned}
$$

and

$$
\begin{aligned}
b_n &= \frac{1}{\pi} \int_0^{2\pi} f(x) \cos nx\, dx \\
&= \frac{2}{b-a} \int_a^b g(t) \sin \frac{2n\pi(t-a)}{b-a}\, dt. \tag{5.73}
\end{aligned}
$$

Then, inserting these integrals in A_n, combining integrals, and making use of the addition formula for the cosine of the sum of two angles, we obtain

$$
\begin{aligned}
A_n &\equiv a_n \cos \frac{2n\pi a}{b-a} - b_n \sin \frac{2n\pi a}{b-a} \\
&= \frac{2}{b-a} \int_a^b g(t) \left[\cos \frac{2n\pi(t-a)}{b-a} \cos \frac{2n\pi a}{b-a} - \sin \frac{2n\pi(t-a)}{b-a} \sin \frac{2n\pi a}{b-a} \right] dt \\
&= \frac{2}{b-a} \int_a^b g(t) \cos \frac{2n\pi t}{b-a}\, dt. \tag{5.74}
\end{aligned}
$$

A similar computation gives

$$B_n = \frac{2}{b-a} \int_a^b g(t) \sin \frac{2n\pi t}{b-a}\, dt. \tag{5.75}$$

Summarizing, we have shown that:

Theorem 5.2. *The Fourier series representation of $f(x)$ defined on $[a, b]$ when it exists, is given by*

$$f(x) \sim \frac{a_0}{2} + \sum_{n=1}^{\infty} \left[a_n \cos \frac{2n\pi x}{b-a} + b_n \sin \frac{2n\pi x}{b-a} \right]. \tag{5.76}$$

with Fourier coefficients

$$
\begin{aligned}
a_n &= \frac{2}{b-a} \int_a^b f(x) \cos \frac{2n\pi x}{b-a}\, dx. \quad n = 0, 1, 2, \dots, \\
b_n &= \frac{2}{b-a} \int_a^b f(x) \sin \frac{2n\pi x}{b-a}\, dx. \quad n = 1, 2, \dots. \tag{5.77}
\end{aligned}
$$

5.9   Sine and Cosine Series

IN THE TWO PREVIOUS EXAMPLES ($f(x) = |x|$ and $f(x) = x$ on $[-\pi, \pi]$), we have seen Fourier series representations that contain only sine or cosine terms. As we know, the sine functions are odd functions and thus sum to odd functions. Similarly, cosine functions sum to even functions. Such occurrences happen often in practice. Fourier representations involving just sines are called sine series, and those involving just cosines (and the constant term) are called cosine series.

Another interesting result, based upon these examples, is that the original functions, $|x|$ and x, agree on the interval $[0, \pi]$. Note from Figures 5.15 through 5.17 that their Fourier series representations do as well. Thus, more than one series can be used to represent functions defined on finite intervals. All they need to do is agree with the function over that particular interval. Sometimes one of these series is more useful because it has additional properties needed in the given application.

We have made the following observations from the previous examples:

1. There are several trigonometric series representations for a function defined on a finite interval.

2. Odd functions on a symmetric interval are represented by sine series, and even functions on a symmetric interval are represented by cosine series.

These two observations are related and are the subject of this section. We begin by defining a function $f(x)$ on interval $[0, L]$. We have seen that the Fourier series representation of this function appears to converge to a periodic extension of the function.

In Figure 5.18, we show a function defined on $[0, 1]$. To the right is its periodic extension to the whole real axis. This representation has a period of $L = 1$. The bottom left plot is obtained by first reflecting f about the y-axis to make it an even function and then graphing the periodic extension of this new function. Its period will be $2L = 2$. Finally, in the last plot, we flip the function about each axis and graph the periodic extension of the new odd function. It will also have a period of $2L = 2$.

In general, we obtain three different periodic representations. In order to distinguish these, we will refer to them simply as the periodic, even, and odd extensions. Now, starting with $f(x)$ defined on $[0, L]$, we would like to determine the Fourier series representations leading to these extensions. [For easy reference, the results are summarized in Table 5.3]

We have already seen from Table 5.2 that the periodic extension of $f(x)$, defined on $[0, L]$, is obtained through the Fourier series representation

$$f(x) \sim \frac{a_0}{2} + \sum_{n=1}^{\infty} \left[a_n \cos \frac{2n\pi x}{L} + b_n \sin \frac{2n\pi x}{L} \right], \tag{5.78}$$

Figure 5.18: This is a sketch of a function and its various extensions. The original function $f(x)$ is defined on $[0,1]$ and graphed in the upper left corner. To its right is the periodic extension, obtained by adding replicas. The two lower plots are obtained by first making the original function even or odd and then creating the periodic extensions of the new function.

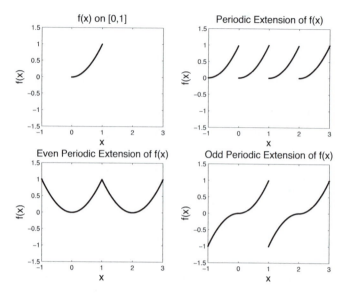

where

$$a_n = \frac{2}{L}\int_0^L f(x)\cos\frac{2n\pi x}{L}\,dx. \quad n = 0,1,2,\ldots,$$

$$b_n = \frac{2}{L}\int_0^L f(x)\sin\frac{2n\pi x}{L}\,dx. \quad n = 1,2,\ldots. \tag{5.79}$$

Even periodic extension.

Given $f(x)$ defined on $[0,L]$, the even periodic extension is obtained by simply computing the Fourier series representation for the even function

$$f_e(x) \equiv \begin{cases} f(x), & 0 < x < L, \\ f(-x) & -L < x < 0. \end{cases} \tag{5.80}$$

Because $f_e(x)$ is an even function on a symmetric interval $[-L, L]$, we expect that the resulting Fourier series will not contain sine terms. Therefore, the series expansion will be given by [Use the general case in Equation (5.76) with $a = -L$ and $b = L$.]:

$$f_e(x) \sim \frac{a_0}{2} + \sum_{n=1}^{\infty} a_n \cos\frac{n\pi x}{L}. \tag{5.87}$$

with Fourier coefficients

$$a_n = \frac{1}{L}\int_{-L}^{L} f_e(x)\cos\frac{n\pi x}{L}\,dx. \quad n = 0,1,2,\ldots. \tag{5.88}$$

However, we can simplify this by noting that the integrand is even, and the interval of integration can be replaced by $[0, L]$. On this interval $f_e(x) = f(x)$. So, we have the Cosine Series Representation of $f(x)$ for $x \in [0, L]$ is given as

Fourier Cosine Series.

$$f(x) \sim \frac{a_0}{2} + \sum_{n=1}^{\infty} a_n \cos\frac{n\pi x}{L}. \tag{5.89}$$

where

$$a_n = \frac{2}{L}\int_0^L f(x)\cos\frac{n\pi x}{L}\,dx. \quad n = 0,1,2,\ldots. \tag{5.90}$$

Fourier Series on $[0, L]$

$$f(x) \sim \frac{a_0}{2} + \sum_{n=1}^{\infty} \left[a_n \cos \frac{2n\pi x}{L} + b_n \sin \frac{2n\pi x}{L} \right]. \tag{5.81}$$

$$\begin{aligned} a_n &= \frac{2}{L} \int_0^L f(x) \cos \frac{2n\pi x}{L} \, dx. \quad n = 0, 1, 2, \ldots, \\ b_n &= \frac{2}{L} \int_0^L f(x) \sin \frac{2n\pi x}{L} \, dx. \quad n = 1, 2, \ldots. \end{aligned} \tag{5.82}$$

Fourier Cosine Series on $[0, L]$

$$f(x) \sim a_0/2 + \sum_{n=1}^{\infty} a_n \cos \frac{n\pi x}{L}. \tag{5.83}$$

where

$$a_n = \frac{2}{L} \int_0^L f(x) \cos \frac{n\pi x}{L} \, dx. \quad n = 0, 1, 2, \ldots. \tag{5.84}$$

Fourier Sine Series on $[0, L]$

$$f(x) \sim \sum_{n=1}^{\infty} b_n \sin \frac{n\pi x}{L}. \tag{5.85}$$

where

$$b_n = \frac{2}{L} \int_0^L f(x) \sin \frac{n\pi x}{L} \, dx. \quad n = 1, 2, \ldots. \tag{5.86}$$

Table 5.3: Fourier Cosine and Sine Series Representations on $[0, L]$

Odd periodic extension.

Similarly, given $f(x)$ defined on $[0, L]$, the odd periodic extension is obtained by simply computing the Fourier series representation for the odd function

$$f_o(x) \equiv \begin{cases} f(x), & 0 < x < L, \\ -f(-x) & -L < x < 0. \end{cases} \tag{5.91}$$

The resulting series expansion leads to defining the Sine Series Representation of $f(x)$ for $x \in [0, L]$ as

Fourier Sine Series Representation.

$$f(x) \sim \sum_{n=1}^{\infty} b_n \sin \frac{n\pi x}{L}, \tag{5.92}$$

where

$$b_n = \frac{2}{L} \int_0^L f(x) \sin \frac{n\pi x}{L} \, dx. \quad n = 1, 2, \dots. \tag{5.93}$$

Example 5.7. *In Figure 5.18, we actually provided plots of the various extensions of the function $f(x) = x^2$ for $x \in [0, 1]$. Let's determine the representations of the periodic, even, and odd extensions of this function.*

For a change, we will use a CAS (Computer Algebra System) package to do the integrals. In this case, we can use Maple. A general code for doing this for the periodic extension is shown in Table 5.4.

Example 5.8. Periodic Extension – Trigonometric Fourier Series *Using the code in Table 5.4, we have that $a_0 = \frac{2}{3}$, $a_n = \frac{1}{n^2\pi^2}$, and $b_n = -\frac{1}{n\pi}$. Thus, the resulting series is given as*

$$f(x) \sim \frac{1}{3} + \sum_{n=1}^{\infty} \left[\frac{1}{n^2\pi^2} \cos 2n\pi x - \frac{1}{n\pi} \sin 2n\pi x \right].$$

In Figure 5.19, we see the sum of the first 50 terms of this series. Generally, we see that the series seems to be converging to the periodic extension of f. There appear to be some problems with the convergence around integer values of x. We will later see that this is because of the discontinuities in the periodic extension and the resulting overshoot is referred to as the Gibbs phenomenon, which is discussed in the final section of this chapter.

Figure 5.19: The periodic extension of $f(x) = x^2$ on $[0, 1]$.

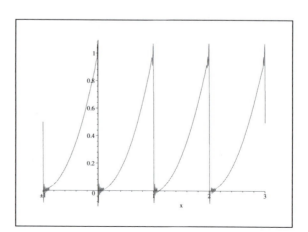

```
> restart:
> L:=1:
> f:=x^2:
> assume(n,integer):
> a0:=2/L*int(f,x=0..L);
```

$$a0 := 2/3$$

```
> an:=2/L*int(f*cos(2*n*Pi*x/L),x=0..L);
```

$$an := \frac{1}{n{\sim}^2 \ Pi^2}$$

```
> bn:=2/L*int(f*sin(2*n*Pi*x/L),x=0..L);
```

$$bn := - \frac{1}{n{\sim} \ Pi}$$

```
> F:=a0/2+sum((1/(k*Pi)^2)*cos(2*k*Pi*x/L)
    -1/(k*Pi)*sin(2*k*Pi*x/L),k=1..50):
> plot(F,x=-1..3,title='Periodic Extension',
    titlefont=[TIMES,ROMAN,14],font=[TIMES,ROMAN,14]);
```

Table 5.4: Maple code for computing Fourier coefficients and plotting partial sums of the Fourier series.

Example 5.9. Even Periodic Extension – Cosine Series

In this case we compute $a_0 = \frac{2}{3}$ and $a_n = \frac{4(-1)^n}{n^2 \pi^2}$. Therefore, we have

$$f(x) \sim \frac{1}{3} + \frac{4}{\pi^2} \sum_{n=1}^{\infty} \frac{(-1)^n}{n^2} \cos n\pi x.$$

In Figure 5.20, we see the sum of the first 50 terms of this series. In this case, the convergence seems to be much better than in the periodic extension case. We also see that it is converging to the even extension.

Figure 5.20: The even periodic extension of $f(x) = x^2$ on $[0,1]$.

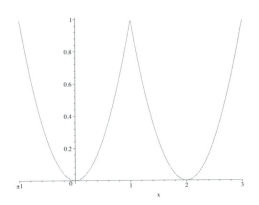

Example 5.10. Odd Periodic Extension – Sine Series

Finally, we look at the sine series for this function. We find that

$$b_n = -\frac{2}{n^3\pi^3}(n^2\pi^2(-1)^n - 2(-1)^n + 2).$$

Therefore,

$$f(x) \sim -\frac{2}{\pi^3}\sum_{n=1}^{\infty}\frac{1}{n^3}(n^2\pi^2(-1)^n - 2(-1)^n + 2)\sin n\pi x.$$

Once again we see discontinuities in the extension as seen in Figure 5.21. However, we have verified that our sine series appears to be converging to the odd extension as we first sketched in Figure 5.18.

Figure 5.21: The odd periodic extension of $f(x) = x^2$ on $[0,1]$.

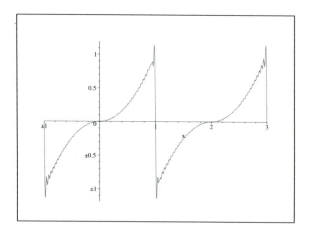

5.10   Solution of the Heat Equation

We started this chapter seeking solutions of initial-boundary value problems involving the heat equation and the wave equation. In particular, we found the general solution for the problem of heat flow in a one-dimensional rod of length L with fixed zero temperature ends. The problem was given by

$$
\begin{array}{llll}
\text{PDE} & u_t = ku_{xx}, & 0 < t, & 0 \le x \le L, \\
\text{IC} & u(x,0) = f(x), & & 0 < x < L, \\
\text{BC} & u(0,t) = 0, & & t > 0, \\
& u(L,t) = 0, & & t > 0.
\end{array}
\tag{5.94}
$$

We found the solution using separation of variables. This resulted in a sum over various product solutions:

$$u(x,t) = \sum_{n=1}^{\infty} b_n e^{k\lambda_n t}\sin\frac{n\pi x}{L},$$

where

$$\lambda_n = -\left(\frac{n\pi}{L}\right)^2.$$

This equation satisfies the boundary conditions. However, we were only able to state initial condition using this solution. Namely,

$$f(x) = u(x,0) = \sum_{n=1}^{\infty} b_n \sin \frac{n\pi x}{L}.$$

We were left with having to determine the constants b_n. Once we know them, we have the solution.

Now we can get the Fourier coefficients when we are given the initial condition, $f(x)$. They are given by

$$b_n = \frac{2}{L} \int_0^L f(x) \sin \frac{n\pi x}{L}\, dx, \quad n = 1, 2, \ldots.$$

We consider a couple of examples with different initial conditions.

Example 5.11. *Consider the solution of the heat equation with $f(x) = \sin x$ and $L = \pi$.*

In this case, the solution takes the form

$$u(x,t) = \sum_{n=1}^{\infty} b_n e^{k\lambda_n t} \sin nx.$$

However, the initial condition takes the form of the first term in the expansion; that is, the $n = 1$ term. So, we need not carry out the integral because we can immediately write $b_1 = 1$ and $b_n = 0$, $n = 2, 3, \ldots$. Therefore, the solution consists of just one term,

$$u(x,t) = e^{-kt} \sin x.$$

In Figure 5.22, we see that how this solution behaves for $k = 1$ and $t \in [0,1]$.

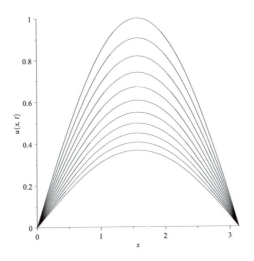

Figure 5.22: The evolution of the initial condition $f(x) = \sin x$ for $L = \pi$ and $k = 1$.

Example 5.12. *Consider solutions of the heat equation with $f(x) = x(1 - x)$ and $L = 1$.*

This example requires a bit more work. The solution takes the form

$$u(x,t) = \sum_{n=1}^{\infty} b_n e^{-n^2 \pi^2 kt} \sin n\pi x,$$

where

$$b_n = 2 \int_0^1 f(x) \sin n\pi x \, dx.$$

This integral is easily computed using integration by parts:

$$
\begin{aligned}
b_n &= 2 \int_0^1 x(1-x) \sin n\pi x \, dx \\
&= \left[2x(1-x) \left(-\frac{1}{n\pi} \cos n\pi x \right) \right]_0^1 + \frac{2}{n\pi} \int_0^1 (1-2x) \cos n\pi x \, dx \\
&= -\frac{2}{n^2 \pi^2} \left\{ [(1-2x) \sin n\pi x]_0^1 + 2 \int_0^1 \sin n\pi x \, dx \right\} \\
&= \frac{4}{n^3 \pi^3} [\cos n\pi x]_0^1 \\
&= \frac{4}{n^3 \pi^3} (\cos n\pi - 1) \\
&= \begin{cases} 0, & n \text{ even} \\ -\frac{8}{n^3 \pi^3}, & n \text{ odd} \end{cases}.
\end{aligned}
$$

(5.95)

So, we have that the solution can be written as

$$u(x,t) = \frac{8}{\pi^3} \sum_{\ell=1}^{\infty} \frac{1}{(2\ell-1)^3} e^{-(2\ell-1)^2 \pi^2 kt} \sin(2\ell-1)\pi x.$$

In Figure 5.23, we see that how this solution behaves for $k = 1$ and $t \in [0,1]$. Twenty terms were used. We see that this solution diffuses much faster than in the previous example. Most of the terms damp out quickly as the solution asymptotically approaches the first term.

Figure 5.23: The evolution of the initial condition $f(x) = x(1-x)$ for $L = 1$ and $k = 1$.

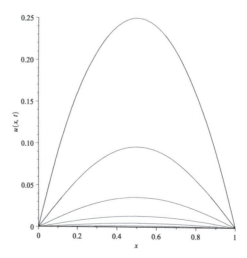

5.11 Finite Length Strings

WE NOW RETURN TO THE PHYSICAL EXAMPLE of wave propagation in a string. We found that the general solution can be represented as a sum over product solutions. We will restrict our discussion to the special case that the initial velocity is zero and the original profile is given by $u(x,0) = f(x)$. The solution is then

$$u(x,t) = \sum_{n=1}^{\infty} A_n \sin \frac{n\pi x}{L} \cos \frac{n\pi c t}{L} \tag{5.96}$$

satisfying

$$f(x) = \sum_{n=1}^{\infty} A_n \sin \frac{n\pi x}{L}. \tag{5.97}$$

We have seen that the Fourier sine series coefficients are given by

$$A_n = \frac{2}{L} \int_0^L f(x) \sin \frac{n\pi x}{L} \, dx. \tag{5.98}$$

We can rewrite this solution in a more compact form. First, we define the wave numbers,

$$k_n = \frac{n\pi}{L}, \quad n = 1, 2, \ldots,$$

and the angular frequencies,

$$\omega_n = c k_n = \frac{n\pi c}{L}.$$

Then, the product solutions take the form

$$\sin k_n x \cos \omega_n t.$$

Using trigonometric identities, these products can be written as

$$\sin k_n x \cos \omega_n t = \frac{1}{2} \left[\sin(k_n x + \omega_n t) + \sin(k_n x - \omega_n t) \right].$$

Inserting this expression in the solution, we have

$$u(x,t) = \frac{1}{2} \sum_{n=1}^{\infty} A_n \left[\sin(k_n x + \omega_n t) + \sin(k_n x - \omega_n t) \right]. \tag{5.99}$$

Because $\omega_n = c k_n$, we can put this into a more suggestive form:

$$u(x,t) = \frac{1}{2} \left[\sum_{n=1}^{\infty} A_n \sin k_n(x + ct) + \sum_{n=1}^{\infty} A_n \sin k_n(x - ct) \right]. \tag{5.100}$$

We see that each sum is simply the sine series for $f(x)$ but evaluated at either $x + ct$ or $x - ct$. Thus, the solution takes the form

The solution of the wave equation can be written as the sum of right and left traveling waves.

$$u(x,t) = \frac{1}{2} \left[f(x + ct) + f(x - ct) \right]. \tag{5.101}$$

If $t = 0$, then we have $u(x,0) = \frac{1}{2}[f(x) + f(x)] = f(x)$. So, the solution satisfies the initial condition. At $t = 1$, the sum has a term $f(x - c)$.

Recall from your mathematics classes that this is simply a shifted version of $f(x)$. Namely, it is shifted to the right. For general times, the function is shifted by ct to the right. For larger values of t, this shift is further to the right. The function (wave) shifts to the right with velocity c. Similarly, $f(x + ct)$ is a wave traveling to the left with velocity $-c$.

Thus, the waves on the string consist of waves traveling to the right and to the left. However, the story does not stop here. We have a problem when needing to shift $f(x)$ across the boundaries. The original problem only defines $f(x)$ on $[0, L]$. If we are not careful, we would think that the function leaves the interval, leaving nothing left inside. However, we have to recall that our sine series representation for $f(x)$ has a period of $2L$. So, before we apply this shifting, we need to account for its periodicity. In fact, being a sine series, we really have the odd periodic extension of $f(x)$ being shifted. The details of such analysis would take us too far from our current goal. However, we can illustrate this with a few figures.

We begin by plucking a string of length L. This can be represented by the function

$$f(x) = \begin{cases} \frac{x}{a} & 0 \le x \le a \\ \frac{L-x}{L-a} & a \le x \le L, \end{cases} \tag{5.102}$$

where the string is pulled up one unit at $x = a$. This is shown in Figure 5.24.

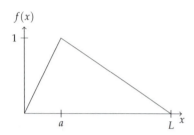

Figure 5.24: The initial profile for a string of length 1 plucked at $x = a$.

Next, we create an odd function by extending the function to a period of $2L$. This is shown in Figure 5.25.

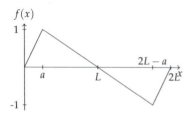

Figure 5.25: Odd extension about the right end of a plucked string.

Finally, we construct the periodic extension of this to the entire line. In Figure 5.26 we show in the lower part of the figure copies of the periodic extension, one moving to the right and the other moving to the left. (Actually, the copies are $\frac{1}{2}f(x \pm ct)$.) The top plot is the sum of these solutions. The physical string lies in the interval $[0,1]$. Of course, this is better seen when the solution is animated.

The time evolution for this plucked string is shown for several times in Figure 5.27. This results in a wave that appears to reflect from the ends as time increases.

The relation between the angular frequency and the wave number, $\omega = ck$, is called a dispersion relation. In this case, ω depends on k linearly. If one knows the dispersion relation, then one can find the wave speed as $c = \frac{\omega}{k}$. In this case, all the harmonics travel at the same speed. In cases where they do not, we have nonlinear dispersion, which we will discuss later.

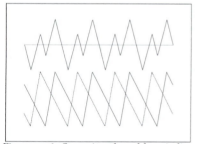

Figure 5.26: Summing the odd periodic extensions. The lower plot shows copies of the periodic extension, one moving to the right and the other moving to the left. The upper plot is the sum.

5.12 Gibbs Phenomenon

WE HAVE SEEN THE GIBBS PHENOMENON when there is a jump discontinuity in the periodic extension of a function, whether the function originally had a discontinuity or developed one due to a mismatch in the values of the endpoints. This can be seen in Figures 5.17, 5.19, and 5.21. The Fourier series has a difficult time converging at the point of discontinuity, and these graphs of the Fourier series show a distinct overshoot that does not go away.

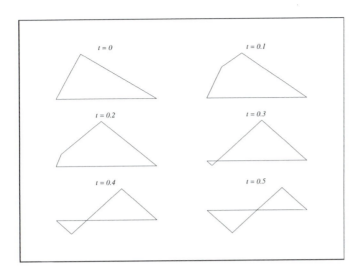

Figure 5.27: This figure shows the plucked string at six successive times.

[3] The Gibbs phenomenon was named after Josiah Willard Gibbs (1839–1903) even though it was discovered earlier by the Englishman Henry Wilbraham (1825–1883). Wilbraham published a soon-forgotten paper about the effect in 1848. In 1889 Albert Abraham Michelson (1852–1931), an American physicist, observed an overshoot in his mechanical graphing machine. Shortly afterwards, J. Willard Gibbs published papers describing this phenomenon, which was later to be called the Gibbs phenomenon. Gibbs was a mathematical physicist and chemist and is considered the father of physical chemistry.

This is called the Gibbs phenomenon,[3] and the amount of overshoot can be computed.

In one of our first examples, Example 5.4, we found the Fourier series representation of the piecewise defined function

$$f(x) = \begin{cases} 1, & 0 < x < \pi, \\ -1, & \pi < x < 2\pi, \end{cases}$$

to be

$$f(x) \sim \frac{4}{\pi} \sum_{k=1}^{\infty} \frac{\sin(2k-1)x}{2k-1}.$$

Figure 5.28: The Fourier series representation of a step function on $[-\pi, \pi]$ for $N = 10$.

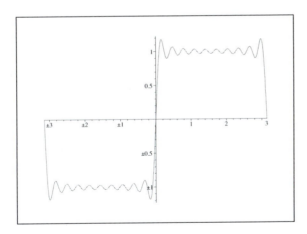

In Figure 5.28, we display the sum of the first 10 terms. Note the wiggles, overshoots, and undershoots. These are seen more when we plot the representation for $x \in [-3\pi, 3\pi]$, as shown in Figure 5.29.

We note that the overshoots and undershoots occur at discontinuities in the periodic extension of $f(x)$. These occur whenever $f(x)$ has a disconti-

Figure 5.29: The Fourier series representation of a step function on $[-\pi, \pi]$ for $N = 10$ plotted on $[-3\pi, 3\pi]$ displaying the periodicity.

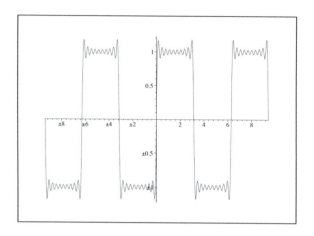

nuity or if the values of $f(x)$ at the endpoints of the domain do not agree.

Figure 5.30: The Fourier series representation of a step function on $[-\pi, \pi]$ for $N = 20$.

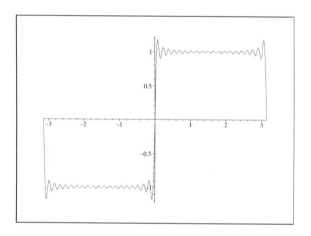

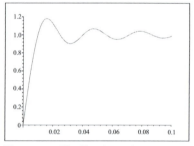

Figure 5.31: The Fourier series representation of a step function on $[-\pi, \pi]$ for $N = 100$.

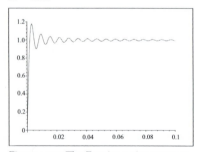

Figure 5.32: The Fourier series representation of a step function on $[-\pi, \pi]$ for $N = 500$.

One might expect that we only need to add more terms. In Figure 5.30, we show the sum for twenty terms. Note the sum appears to converge better for points far from the discontinuities. But, the overshoots and undershoots are still present. Figures 5.31 and 5.32 show magnified plots of the overshoot at $x = 0$ for $N = 100$ and $N = 500$, respectively. We see that the overshoot persists. The peak is at about the same height, but its location seems to be getting closer to the origin. We will show how one can estimate the size of the overshoot.

We can study the Gibbs phenomenon by looking at the partial sums of general Fourier trigonometric series for functions $f(x)$ defined on the interval $[-L, L]$. Writing out the partial sums, inserting the Fourier coefficients, and rearranging, we have

$$S_N(x) = a_0 + \sum_{n=1}^{N} \left[a_n \cos \frac{n\pi x}{L} + b_n \sin \frac{n\pi x}{L} \right]$$

$$
= \frac{1}{2L} \int_{-L}^{L} f(y)\, dy + \sum_{n=1}^{N} \left[\left(\frac{1}{L} \int_{-L}^{L} f(y) \cos \frac{n\pi y}{L}\, dy \right) \cos \frac{n\pi x}{L} \right.
$$

$$
\left. + \left(\frac{1}{L} \int_{-L}^{L} f(y) \sin \frac{n\pi y}{L}\, dy. \right) \sin \frac{n\pi x}{L} \right]
$$

$$
= \frac{1}{L} \int_{-L}^{L} \left\{ \frac{1}{2} \right.
$$

$$
\left. + \sum_{n=1}^{N} \left(\cos \frac{n\pi y}{L} \cos \frac{n\pi x}{L} + \sin \frac{n\pi y}{L} \sin \frac{n\pi x}{L} \right) \right\} f(y)\, dy
$$

$$
= \frac{1}{L} \int_{-L}^{L} \left\{ \frac{1}{2} + \sum_{n=1}^{N} \cos \frac{n\pi(y-x)}{L} \right\} f(y)\, dy
$$

$$
\equiv \frac{1}{L} \int_{-L}^{L} D_N(y-x) f(y)\, dy
$$

We have defined

$$
D_N(x) = \frac{1}{2} + \sum_{n=1}^{N} \cos \frac{n\pi x}{L},
$$

which is called the N-th Dirichlet kernel.

We now prove

Lemma 5.1. *The N-th Dirichlet kernel is given by*

$$
D_N(x) = \begin{cases} \dfrac{\sin\left((N+\frac{1}{2})\frac{\pi x}{L}\right)}{2 \sin \frac{\pi x}{2L}}, & \sin \frac{\pi x}{2L} \neq 0, \\ N + \frac{1}{2}, & \sin \frac{\pi x}{2L} = 0. \end{cases}
$$

Proof. Let $\theta = \frac{\pi x}{L}$ and multiply $D_N(x)$ by $2 \sin \frac{\theta}{2}$ to obtain

$$
\begin{aligned}
2 \sin \frac{\theta}{2} D_N(x) &= 2 \sin \frac{\theta}{2} \left[\frac{1}{2} + \cos \theta + \cdots + \cos N\theta \right] \\
&= \sin \frac{\theta}{2} + 2 \cos \theta \sin \frac{\theta}{2} + 2 \cos 2\theta \sin \frac{\theta}{2} + \cdots + 2 \cos N\theta \sin \frac{\theta}{2} \\
&= \sin \frac{\theta}{2} + \left(\sin \frac{3\theta}{2} - \sin \frac{\theta}{2} \right) + \left(\sin \frac{5\theta}{2} - \sin \frac{3\theta}{2} \right) + \cdots \\
&\quad + \left[\sin \left(N + \frac{1}{2} \right)\theta - \sin \left(N - \frac{1}{2} \right)\theta \right] \\
&= \sin \left(N + \frac{1}{2} \right)\theta. \quad\quad\quad (5.103)
\end{aligned}
$$

Thus,

$$
2 \sin \frac{\theta}{2} D_N(x) = \sin \left(N + \frac{1}{2} \right)\theta.
$$

If $\sin \frac{\theta}{2} \neq 0$, then

$$
D_N(x) = \frac{\sin \left(N + \frac{1}{2} \right)\theta}{2 \sin \frac{\theta}{2}}, \qquad \theta = \frac{\pi x}{L}.
$$

If $\sin\frac{\theta}{2} = 0$, then one needs to apply L'Hopital's Rule as $\theta \to 2m\pi$:

$$
\begin{aligned}
\lim_{\theta \to 2m\pi} \frac{\sin\left(N+\frac{1}{2}\right)\theta}{2\sin\frac{\theta}{2}} &= \lim_{\theta \to 2m\pi} \frac{\left(N+\frac{1}{2}\right)\cos\left(N+\frac{1}{2}\right)\theta}{\cos\frac{\theta}{2}} \\
&= \frac{\left(N+\frac{1}{2}\right)\cos\left(2m\pi N + m\pi\right)}{\cos m\pi} \\
&= \frac{\left(N+\frac{1}{2}\right)(\cos 2m\pi N \cos m\pi - \sin 2m\pi N \sin m\pi)}{\cos m\pi} \\
&= N + \frac{1}{2}.
\end{aligned}
\tag{5.104}
$$

□

We further note that $D_N(x)$ is periodic with period $2L$ and is an even function.

So far, we have found that the Nth partial sum is given by

$$
S_N(x) = \frac{1}{L} \int_{-L}^{L} D_N(y-x) f(y)\, dy.
\tag{5.105}
$$

Making the substitution $\xi = y - x$, we have

$$
\begin{aligned}
S_N(x) &= \frac{1}{L} \int_{-L-x}^{L-x} D_N(\xi) f(\xi + x)\, d\xi \\
&= \frac{1}{L} \int_{-L}^{L} D_N(\xi) f(\xi + x)\, d\xi.
\end{aligned}
\tag{5.106}
$$

In the second integral, we have made use of the fact that $f(x)$ and $D_N(x)$ are periodic with period $2L$ and shifted the interval back to $[-L, L]$.

We now write the integral as the sum of two integrals over positive and negative values of ξ and use the fact that $D_N(x)$ is an even function. Then,

$$
\begin{aligned}
S_N(x) &= \frac{1}{L} \int_{-L}^{0} D_N(\xi) f(\xi + x)\, d\xi + \frac{1}{L} \int_{0}^{L} D_N(\xi) f(\xi + x)\, d\xi \\
&= \frac{1}{L} \int_{0}^{L} [f(x - \xi) + f(\xi + x)]\, D_N(\xi)\, d\xi.
\end{aligned}
\tag{5.107}
$$

We can use this result to study the Gibbs phenomenon whenever it occurs. In particular, we will only concentrate on the earlier example. For this case, we have

$$
S_N(x) = \frac{1}{\pi} \int_{0}^{\pi} [f(x - \xi) + f(\xi + x)]\, D_N(\xi)\, d\xi
\tag{5.108}
$$

for

$$
D_N(x) = \frac{1}{2} + \sum_{n=1}^{N} \cos nx.
$$

Also, one can show that

$$
f(x - \xi) + f(\xi + x) = \begin{cases} 2, & 0 \le \xi < x, \\ 0, & x \le \xi < \pi - x, \\ -2, & \pi - x \le \xi < \pi. \end{cases}
$$

Thus, we have

$$
\begin{aligned}
S_N(x) &= \frac{2}{\pi} \int_0^x D_N(\zeta)\, d\zeta - \frac{2}{\pi} \int_{\pi-x}^{\pi} D_N(\zeta)\, d\zeta \\
&= \frac{2}{\pi} \int_0^x D_N(z)\, dz + \frac{2}{\pi} \int_0^x D_N(\pi - z)\, dz. \qquad (5.109)
\end{aligned}
$$

Here we made the substitution $z = \pi - \zeta$ in the second integral.

The Dirichlet kernel for $L = \pi$ is given by

$$
D_N(x) = \frac{\sin(N + \frac{1}{2})x}{2 \sin \frac{x}{2}}.
$$

For N large, we have $N + \frac{1}{2} \approx N$; and for small x, we have $\sin \frac{x}{2} \approx \frac{x}{2}$. So, under these assumptions,

$$
D_N(x) \approx \frac{\sin Nx}{x}.
$$

Therefore,

$$
S_N(x) \to \frac{2}{\pi} \int_0^x \frac{\sin N\zeta}{\zeta}\, d\zeta \qquad \text{for large } N \text{ and small } x.
$$

If we want to determine the locations of the minima and maxima, where the undershoot and overshoot occur, then we apply the first derivative test for extrema to $S_N(x)$. Thus,

$$
\frac{d}{dx} S_N(x) = \frac{2}{\pi} \frac{\sin Nx}{x} = 0.
$$

The extrema occur for $Nx = m\pi$, $m = \pm 1, \pm 2, \ldots$. One can show that there is a maximum at $x = \pi/N$ and a minimum for $x = 2\pi/N$. The value for the overshoot can be computed as

$$
\begin{aligned}
S_N(\pi/N) &= \frac{2}{\pi} \int_0^{\pi/N} \frac{\sin N\zeta}{\zeta}\, d\zeta \\
&= \frac{2}{\pi} \int_0^{\pi} \frac{\sin t}{t}\, dt \\
&= \frac{2}{\pi} \mathrm{Si}(\pi) \\
&= 1.178979744\ldots. \qquad (5.110)
\end{aligned}
$$

Note that this value is independent of N and is given in terms of the sine integral,

$$
\mathrm{Si}(x) \equiv \int_0^x \frac{\sin t}{t}\, dt.
$$

5.13 Green's Functions for 1D Partial Differential Equations

IN SECTION 2.7.4 WE ENCOUNTERED THE INITIAL VALUE GREEN'S FUNCTION for initial value problems for ordinary differential equations. In that case, we were able to express the solution of the differential equation $L[y] = f$ in the form

$$y(t) = \int G(t, \tau) f(\tau) \, d\tau,$$

where the Green's function $G(t, \tau)$ was used to handle the nonhomogeneous term in the differential equation. In a similar spirit, we can introduce Green's functions of different types to handle nonhomogeneous terms, nonhomogeneous boundary conditions, or nonhomogeneous initial conditions. Occasionally, we will stop and rearrange the solutions of different problems and recast the solution and identify the Green's function for the problem.

In this section we will rewrite the solutions of the heat equation and wave equation on a finite interval to obtain an initial value Green's function. Assuming homogeneous boundary conditions and a homogeneous differential operator, we can write the solution of the heat equation in the form

$$u(x, t) = \int_0^L G(x, \xi; t, t_0) f(\xi) \, d\xi,$$

where $u(x, t_0) = f(x)$, and the solution of the wave equation as

$$u(x, t) = \int_0^L G_c(x, \xi, t, t_0) f(\xi) \, d\xi + \int_0^L G_s(x, \xi, t, t_0) g(\xi) \, d\xi.$$

where $u(x, t_0) = f(x)$ and $u_t(x, t_0) = g(x)$. The functions $G(x, \xi; t, t_0)$, $G(x, \xi; t, t_0)$, and $G(x, \xi; t, t_0)$ are initial value Green's functions, and we will need to explore some more methods before we can discuss the properties of these functions. [For example, see Section 6.7.]

We will now turn to showing that for the solutions of the one-dimensional heat and wave equations with fixed, homogeneous boundary conditions, we can construct the particular Green's functions.

5.13.1   Heat Equation

IN SECTION 5.10 WE OBTAINED THE SOLUTION to the one-dimensional heat equation on a finite interval satisfying homogeneous Dirichlet conditions,

$$
\begin{aligned}
u_t &= k u_{xx}, & 0 < t, \quad 0 \le x \le L, \\
u(x, 0) &= f(x), & 0 < x < L, \\
u(0, t) &= 0, & t > 0, \\
u(L, t) &= 0, & t > 0.
\end{aligned}
\tag{5.111}
$$

The solution we found was the Fourier sine series

$$u(x, t) = \sum_{n=1}^{\infty} b_n e^{\lambda_n k t} \sin \frac{n \pi x}{L},$$

where

$$\lambda_n = - \left(\frac{n \pi}{L} \right)^2,$$

and the Fourier sine coefficients are given in terms of the initial temperature distribution,

$$b_n = \frac{2}{L} \int_0^L f(x) \sin \frac{n\pi x}{L} \, dx, \quad n = 1, 2, \ldots.$$

Inserting the coefficients b_n into the solution, we have

$$u(x,t) = \sum_{n=1}^{\infty} \left(\frac{2}{L} \int_0^L f(\xi) \sin \frac{n\pi \xi}{L} \, d\xi \right) e^{\lambda_n k t} \sin \frac{n\pi x}{L}.$$

Interchanging the sum and integration, we obtain

$$u(x,t) = \int_0^L \left(\frac{2}{L} \sum_{n=1}^{\infty} \sin \frac{n\pi x}{L} \sin \frac{n\pi \xi}{L} e^{\lambda_n k t} \right) f(\xi) \, d\xi.$$

This solution is of the form

$$u(x,t) = \int_0^L G(x,\xi;t,0) f(\xi) \, d\xi.$$

Here, the function $G(x,\xi;t,0)$ is the initial value Green's function for the heat equation in the form

$$G(x,\xi;t,0) = \frac{2}{L} \sum_{n=1}^{\infty} \sin \frac{n\pi x}{L} \sin \frac{n\pi \xi}{L} e^{\lambda_n k t},$$

which involves a sum over eigenfunctions of the spatial eigenvalue problem, $X_n(x) = \sin \frac{n\pi x}{L}$.

5.13.2   Wave Equation

RECALL THAT THE SOLUTION OF THE ONE-DIMENSIONAL WAVE EQUATION (5.3),

$$\begin{aligned}
u_{tt} &= c^2 u_{xx}, \quad 0 < t, \quad 0 \le x \le L, \\
u(0,t) &= 0, u(L,0) = 0, \quad t > 0, \\
u(x,0) &= f(x), u_t(x,0) = g(x), \quad 0 < x < L,
\end{aligned} \tag{5.112}$$

was found as

$$u(x,t) = \sum_{n=1}^{\infty} \left[A_n \cos \frac{n\pi c t}{L} + B_n \sin \frac{n\pi c t}{L} \right] \sin \frac{n\pi x}{L}.$$

The Fourier coefficients were determined from the initial conditions,

$$\begin{aligned}
f(x) &= \sum_{n=1}^{\infty} A_n \sin \frac{n\pi x}{L}, \\
g(x) &= \sum_{n=1}^{\infty} \frac{n\pi c}{L} B_n \sin \frac{n\pi x}{L},
\end{aligned} \tag{5.113}$$

as

$$
A_n = \frac{2}{L} \int_0^L f(\xi) \sin \frac{n\pi\xi}{L} \, d\xi,
$$

$$
B_n = \frac{L}{n\pi c} \frac{2}{L} \int_0^L f(\xi) \sin \frac{n\pi\xi}{L} \, d\xi. \tag{5.114}
$$

Inserting these coefficients into the solution and interchanging integration with summation, we have

$$
\begin{aligned}
u(x,t) &= \int_0^\infty \left[\frac{2}{L} \sum_{n=1}^\infty \sin \frac{n\pi x}{L} \sin \frac{n\pi\xi}{L} \cos \frac{n\pi ct}{L} \right] f(\xi) \, d\xi \\
&\quad + \int_0^\infty \left[\frac{2}{L} \sum_{n=1}^\infty \sin \frac{n\pi x}{L} \sin \frac{n\pi\xi}{L} \frac{\sin \frac{n\pi ct}{L}}{n\pi c/L} \right] g(\xi) \, d\xi \\
&= \int_0^L G_c(x,\xi,t,0) f(\xi) \, d\xi + \int_0^L G_s(x,\xi,t,0) g(\xi) \, d\xi. \tag{5.115}
\end{aligned}
$$

In this case, we have defined two Green's functions,

$$
\begin{aligned}
G_c(x,\xi,t,0) &= \frac{2}{L} \sum_{n=1}^\infty \sin \frac{n\pi x}{L} \sin \frac{n\pi\xi}{L} \cos \frac{n\pi ct}{L}, \\
G_s(x,\xi,t,0) &= \frac{2}{L} \sum_{n=1}^\infty \sin \frac{n\pi x}{L} \sin \frac{n\pi\xi}{L} \frac{\sin \frac{n\pi ct}{L}}{n\pi c/L}. \tag{5.116}
\end{aligned}
$$

The first, G_c, provides the response to the initial profile and the second, G_s, to the initial velocity.

5.14   Derivation of Generic 1D Equations

5.14.1   Derivation of Wave Equation for String

THE WAVE EQUATION FOR A ONE-DIMENSIONAL STRING is derived based upon simply looking at Newton's Second Law of Motion for a piece of the string plus a few simple assumptions, such as small-amplitude oscillations and constant density.

We begin with $\mathbf{F} = m\mathbf{a}$. The mass of a piece of string of length ds is $m = \rho(x)ds$. From Figure 5.33 an incremental length f the string is given by

$$
\Delta s^2 = \Delta x^2 + \Delta u^2.
$$

The piece of string undergoes an acceleration of $a = \frac{\partial^2 u}{\partial t^2}$.

We will assume that the main force acting on the string is that of tension. Let $T(x,t)$ be the magnitude of the tension acting on the left end of the piece of string. Then, on the right end, the tension is $T(x + \Delta x, t)$. At these points, the tension makes an angle to the horizontal of $\theta(x,t)$ and $\theta(x + \Delta x, t)$, respectively.

The wave equation is derived from $\mathbf{F} = m\mathbf{a}$.

Assuming that there is no horizontal acceleration, the x-component in the Second Law, $m\mathbf{a} = \mathbf{F}$, for the string element is given by

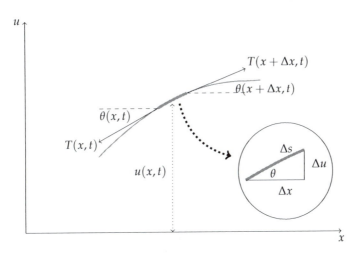

$$0 = T(x + \Delta x, t) \cos \theta(x + \Delta x, t) - T(x, t) \cos \theta(x, t).$$

The vertical component is given by

$$\rho(x)\Delta s \frac{\partial^2 u}{\partial t^2} = T(x + \Delta x, t) \sin \theta(x + \Delta x, t) - T(x, t) \sin \theta(x, t)$$

The length of the piece of string can be written in terms of Δx,

$$\Delta s = \sqrt{\Delta x^2 + \Delta u^2} = \sqrt{1 + \left(\frac{\Delta u}{\Delta x}\right)^2}\, \Delta x.$$

and the right-hand sides of the component equation can be expanded about $\Delta x = 0$ to obtain

$$T(x + \Delta x, t) \cos \theta(x + \Delta x, t) - T(x, t) \cos \theta(x, t) \approx \frac{\partial(T \cos \theta)}{\partial x}(x, t)\Delta x,$$

$$T(x + \Delta x, t) \sin \theta(x + \Delta x, t) - T(x, t) \sin \theta(x, t) \approx \frac{\partial(T \sin \theta)}{\partial x}(x, t)\Delta x.$$

Furthermore, we note that

$$\tan \theta = \lim_{\Delta x \to 0} \frac{\Delta u}{\Delta x} = \frac{\partial u}{\partial x}.$$

Now we can divide these component equations by Δx and let $\Delta x \to 0$. This gives the approximations

$$0 = \frac{T(x + \Delta x, t) \cos \theta(x + \Delta x, t) - T(x, t) \cos \theta(x, t)}{\Delta x}$$

$$\approx \frac{\partial(T \cos \theta)}{\partial x}(x, t)$$

$$\rho(x)\frac{\partial^2 u}{\partial t^2}\frac{\delta s}{\delta s} = \frac{T(x + \Delta x, t) \sin \theta(x + \Delta x, t) - T(x, t) \sin \theta(x, t)}{\Delta x}$$

$$\rho(x)\frac{\partial^2 u}{\partial t^2}\sqrt{1 + \left(\frac{\partial u}{\partial x}\right)^2} \approx \frac{\partial(T \sin \theta)}{\partial x}(x, t). \tag{5.117}$$

We will assume a small angle approximation, giving

$$\sin \theta \approx \tan \theta = \frac{\partial u}{\partial x},$$

$\cos\theta \approx 1$, and

$$\sqrt{1 + \left(\frac{\partial u}{\partial x}\right)^2} \approx 1.$$

Then, the horizontal component becomes

$$\frac{\partial T(x,t)}{\partial x} = 0.$$

Therefore, the magnitude of the tension $T(x,t) = T(t)$ is, at most, time dependent.

The vertical component equation is now

$$\rho(x)\frac{\partial^2 u}{\partial t^2} = T(t)\frac{\partial}{\partial x}\left(\frac{\partial u}{\partial x}\right) = T(t)\frac{\partial^2 u}{\partial x^2}.$$

Assuming that ρ and T are constant and defining

$$c^2 = \frac{T}{\rho},$$

we obtain the one-dimensional wave equation,

$$\frac{\partial^2 u}{\partial t^2} = c^2\frac{\partial^2 u}{\partial x^2}.$$

5.14.2   Derivation of 1D Heat Equation

CONSIDER A ONE-DIMENSIONAL ROD of length L as shown in Figure 5.34. It is heated and allowed to sit. The heat equation is the governing equation that allows us to determine the temperature of the rod at a later time.

We begin with some simple thermodynamics. Recall that to raise the temperature of a mass m by ΔT takes thermal energy given by

$$Q = mc\Delta T,$$

assuming the mass does not go through a phase transition. Here, c is the specific heat capacity of the substance. So, we will begin with the heat content of the rod as

$$Q = mcT(x,t)$$

and assume that m and c are constant.

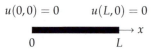

$u(0,0) = 0 \qquad u(L,0) = 0$

Figure 5.34:   One-dimensional heated rod of length L.

We will also need Fourier's Law of heat transfer or heat conduction . This law simply states that heat energy flows from warmer to cooler regions and is written in terms of the heat energy flux, $\phi(x,t)$. The heat energy flux, or flux density, gives the rate of energy flow per area. Thus, the amount of heat energy flowing over the left end of the region of cross-section A in time Δt is given by $\phi(x,t)\Delta t A$. The units of $\phi(x,t)$ are then $J/s/m^2 = W/m^2$.

Fourier's Law of heat conduction states that the flux density is proportional to the gradient of the temperature,

$$\phi = -K\frac{\partial T}{\partial x}.$$

Here, K is the thermal conductivity, and the negative sign takes into account the direction of flow from higher to lower temperatures.

Now we make use of the conservation of energy. Consider a small section of the rod of width Δx as shown in Figure 5.35. The rate of change of the energy through this section is due to energy flow through the ends; namely,

$$\text{Rate of change of heat energy} = \text{Heat In} - \text{Heat Out}.$$

The energy content of the small segment of the rod is given by

$$\Delta Q = (\rho A \Delta x) c T(x, t + \Delta t) - (\rho A \Delta x) c T(x, t).$$

The flow rates across the boundaries are given by the flux:

$$(\rho A \Delta x) c T(x, t + \Delta t) - (\rho A \Delta x) c T(x, t) = [\phi(x, t) - \phi(x + \Delta x, t)] \Delta t A.$$

Dividing by Δx and Δt and letting $\Delta x, \Delta t \to 0$, we obtain

$$\frac{\partial T}{\partial t} = -\frac{1}{c\rho} \frac{\partial \phi}{\partial x}.$$

Using Fourier's Law of heat conduction,

$$\frac{\partial T}{\partial t} = \frac{1}{c\rho} \frac{\partial}{\partial x} \left(K \frac{\partial T}{\partial x} \right).$$

Assuming that K, c, and ρ are constant, we have the one-dimensional heat equation as used in the text:

$$\frac{\partial T}{\partial t} = k \frac{\partial^2 T}{\partial x^2},$$

where $k = \frac{k}{c\rho}$.

Problems

1. Solve the following boundary value problems directly, when possible.

    a. $x'' + x = 2$, $x(0) = 0$, $x'(1) = 0$.

    b. $y'' + 2y' - 8y = 0$, $y(0) = 1$, $y(1) = 0$.

    c. $y'' + y = 0$, $y(0) = 1$, $y(\pi) = 0$.

2. Find product solutions, $u(x, t) = b(t)\phi(x)$, to the heat equation satisfying the boundary conditions $u_x(0, t) = 0$ and $u(L, t) = 0$. Use these solutions to find a general solution of the heat equation satisfying these boundary conditions.

3. Find product solutions, $u(x,t) = b(t)\phi(x)$, to the wave equation satisfying the boundary conditions $u(0,t) = 0$ and $u_x(1,t) = 0$. Use these solutions to find a general solution of the heat equation satisfying these boundary conditions.

4. Consider the following boundary value problems. Determine the eigenvalues λ and eigenfunctions $y(x)$ for each problem.

    a. $y'' + \lambda y = 0$, $y(0) = 0$, $y'(1) = 0$.

    b. $y'' - \lambda y = 0$, $y(-\pi) = 0$, $y'(\pi) = 0$.

    c. $x^2 y'' + xy' + \lambda y = 0$, $y(1) = 0$, $y(2) = 0$.

    d. $(x^2 y')' + \lambda y = 0$, $y(1) = 0$, $y'(e) = 0$.

In problem d, you will not get exact eigenvalues. Show that you obtain a transcendental equation for the eigenvalues in the form $\tan z = 2z$. Find the first three eigenvalues numerically.

5. Consider the boundary value problem for the deflection of a horizontal beam fixed at one end,

$$\frac{d^4 y}{dx^4} = C, \quad y(0) = 0, \quad y'(0) = 0, \quad y''(L) = 0, \quad y'''(L) = 0.$$

Solve this problem assuming that C is a constant.

6. Write $y(t) = 3\cos 2t - 4\sin 2t$ in the form $y(t) = A\cos(2\pi f t + \phi)$.

7. Derive the coefficients b_n in Equation (5.24).

8. Let $f(x)$ be defined for $x \in [-L, L]$. Parseval's identity is given by

$$\frac{1}{L} \int_{-L}^{L} f^2(x)\, dx = \frac{a_0^2}{2} + \sum_{n=1}^{\infty} a_n^2 + b_n^2.$$

Assuming the the Fourier series of $f(x)$ converges uniformly in $(-L, L)$, prove Parseval's identity by multiplying the Fourier series representation by $f(x)$ and integrating from $x = -L$ to $x = L$. [In Section 8.6.3 we will encounter Parseval's equality for Fourier transforms, which is a continuous version of this identity.]

9. Consider the square wave function

$$f(x) = \begin{cases} 1, & 0 < x < \pi, \\ -1, & \pi < x < 2\pi. \end{cases}$$

    a. Find the Fourier series representation of this function and plot the first 50 terms.

    b. Apply Parseval's identity in Problem 8 to the result in part a.

    c. Use the result of part b to show $\frac{\pi^2}{8} = \sum_{n=1}^{\infty} \frac{1}{(2n-1)^2}$.

10. For the following sets of functions: (i) show that each is orthogonal on the given interval, and (ii) determine the corresponding orthonormal set. [See page 288.]

    a. $\{\sin 2nx\}$, $n = 1, 2, 3, \ldots$, $0 \le x \le \pi$.

b. $\{\cos n\pi x\}$, $n = 0, 1, 2, \ldots$, $0 \leq x \leq 2$.

c. $\{\sin \frac{n\pi x}{L}\}$, $n = 1, 2, 3, \ldots$, $x \in [-L, L]$.

11. Consider $f(x) = 4 \sin^3 2x$.

    a. Derive the trigonometric identity giving $\sin^3 \theta$ in terms of $\sin \theta$ and $\sin 3\theta$ using DeMoivre's Formula.

    b. Find the Fourier series of $f(x) = 4 \sin^3 2x$ on $[0, 2\pi]$ without computing any integrals.

12. Find the Fourier series of the following:

    a. $f(x) = x$, $x \in [0, 2\pi]$.

    b. $f(x) = \frac{x^2}{4}$, $|x| < \pi$.

    c. $f(x) = \begin{cases} \frac{\pi}{2}, & 0 < x < \pi, \\ -\frac{\pi}{2}, & \pi < x < 2\pi. \end{cases}$

13. Find the Fourier series of each function $f(x)$ of period 2π. For each series, plot the Nth partial sum,

$$S_N = \frac{a_0}{2} + \sum_{n=1}^{N} [a_n \cos nx + b_n \sin nx],$$

for $N = 5, 10, 50$ and describe the convergence (Is it fast? What is it converging to?, etc.) [Some simple Maple code for computing partial sums is shown in the notes.]

    a. $f(x) = x$, $|x| < \pi$.

    b. $f(x) = |x|$, $|x| < \pi$.

    c. $f(x) = \begin{cases} 0, & -\pi < x < 0, \\ 1, & 0 < x < \pi. \end{cases}$

14. Find the Fourier series of $f(x) = x$ on the given interval. Plot the Nth partial sums and describe what you see.

    a. $0 < x < 2$.

    b. $-2 < x < 2$.

    c. $1 < x < 2$.

15. The result in Problem 12b, above gives a Fourier series representation of $\frac{x^2}{4}$. By picking the right value for x and a little arrangement of the series, show that [See Example 5.6.]

    a.

$$\frac{\pi^2}{6} = 1 + \frac{1}{2^2} + \frac{1}{3^2} + \frac{1}{4^2} + \cdots.$$

    b.

$$\frac{\pi^2}{8} = 1 + \frac{1}{3^2} + \frac{1}{5^2} + \frac{1}{7^2} + \cdots.$$

    Hint: Consider how the series in part a. can be used to do this.

16. Sketch (by hand) the graphs of each of the following functions over four periods. Then sketch the extensions of each of the functions as both an even and odd periodic function. Determine the corresponding Fourier sine and cosine series, and verify the convergence to the desired function using Maple.

a. $f(x) = x^2, 0 < x < 1$.

b. $f(x) = x(2 - x), 0 < x < 2$.

c. $f(x) = \begin{cases} 0, & 0 < x < 1, \\ 1, & 1 < x < 2. \end{cases}$

d. $f(x) = \begin{cases} \pi, & 0 < x < \pi, \\ 2\pi - x, & \pi < x < 2\pi. \end{cases}$

17. Consider the function $f(x) = x, -\pi < x < \pi$.

a. Show that $x = 2\sum_{n=1}^{\infty}(-1)^{n+1}\frac{\sin nx}{n}$.

b. Integrate the series in part a and show that $x^2 = \frac{\pi^2}{3} - 4\sum_{n=1}^{\infty}(-1)^{n+1}\frac{\cos nx}{n^2}$.

c. Find the Fourier cosine series of $f(x) = x^2$ on $(0, \pi)$ and compare the result in part b.

18. Consider the function $f(x) = x, 0 < x < 2$.

a. Find the Fourier sine series representation of this function and plot the first 50 terms.

b. Find the Fourier cosine series representation of this function and plot the first 50 terms.

c. Extend and apply Parseval's identity in Problem 8 to the result in part b.

d. Use the result of part c, to find the sum $\sum_{n=1}^{\infty}\frac{1}{n^4}$.

19. Differentiate the Fourier sine series term by term in Problem 18. Show that the result is not true. Why not?

20. Rewrite the solution to Problem 2 and identify the initial value Green's function.

21. Rewrite the solution to Problem 3 and identify the initial value Green's functions.

6

Non-Sinusoidal Harmonics and Special Functions

"To the pure geometer the radius of curvature is an incidental characteristic—like the grin of the Cheshire cat. To the physicist it is an indispensable characteristic. It would be going too far to say that to the physicist the cat is merely incidental to the grin. Physics is concerned with interrelatedness such as the interrelatedness of cats and grins. In this case the "cat without a grin" and the "grin without a cat" are equally set aside as purely mathematical phantasies."
—Sir Arthur Stanley Eddington (1882–1944)

IN THIS CHAPTER WE PROVIDE A GLIMPSE into generalized Fourier series in which the normal modes of oscillation are not sinusoidal. For vibrating strings, we saw that the harmonics were sinusoidal basis functions for a large, infinite dimensional function space. Now we will extend these ideas to non-sinusoidal harmonics and explore the underlying structure behind these ideas. In particular, we will explore Legendre polynomials and Bessel functions, which will later arise in problems having cylindrical or spherical symmetry.

The background for the study of generalized Fourier series is that of function spaces. We begin by exploring the general context in which one finds oneself when discussing Fourier series and (later) Fourier transforms. We can view the sine and cosine functions in the Fourier trigonometric series representations as basis vectors in an infinite dimensional function space. A given function in that space may then be represented as a linear combination over this infinite basis. With this in mind, we might wonder

- Do we have enough basis vectors for the function space?

- Are the infinite series expansions convergent?

- What functions can be represented by such expansions?

In the context of the boundary value problems that typically appear in physics, one is led to the study of boundary value problems in the form of Sturm–Liouville eigenvalue problems. These lead to an appropriate set of basis vectors for the function space under consideration. We will touch a little on these ideas, leaving some of the deeper results for more advanced courses in mathematics. For now, we will turn to function spaces and explore some typical basis functions, many which originated from the study of physical problems. The common basis functions are often referred to as special functions in physics. Examples are the classical orthogonal polynomials (Legendre, Hermite, Laguerre, Tchebychef) and Bessel functions. But first we will introduce function spaces.

6.1   Function Spaces

EARLIER WE STUDIED FINITE DIMENSIONAL VECTOR SPACES. Given a set of basis vectors, $\{\mathbf{a}_k\}_{k=1}^n$, in vector space V, we showed that we can expand any vector $\mathbf{v} \in \mathbf{V}$ in terms of this basis, $\mathbf{v} = \sum_{k=1}^n v_k \mathbf{a}_k$. We then spent some time looking at the simple case of extracting the components v_k of the vector. The keys to doing this simply were to have a scalar product and an orthogonal basis set. These are also the key ingredients that we will need in the infinite dimensional case. In fact, we already did this when we studied Fourier series.

Recall when we found Fourier trigonometric series representations of functions, we started with a function (vector) that we wanted to expand in a set of trigonometric functions (basis) and we sought the Fourier coefficients (components). In this section we will extend our notions from finite dimensional spaces to infinite dimensional spaces, and we will develop the needed background in which to think about more general Fourier series expansions. This conceptual framework is very important in other areas in mathematics (such as ordinary and partial differential equations) and physics (such as quantum mechanics and electrodynamics).

We note that the above determination of vector components for finite dimensional spaces is precisely what we did to compute the Fourier coefficients using trigonometric bases. Reading further, you will see how this works.

We will consider various infinite dimensional function spaces. Functions in these spaces would differ by their properties. For example, we could consider the space of continuous functions on [0,1], the space of differentiably continuous functions, or the set of functions integrable from a to b. As you will see, there are many types of function spaces . In order to view these spaces as vector spaces, we must be able to add functions and multiply them by scalars in such as way that they satisfy the definition of a vector space as defined in Chapter 3.

We will also need a scalar product defined on this space of functions. There are several types of scalar products, or inner products, that we can define. An inner product $\langle , \rangle$ on a real vector space V is a mapping from $V \times V$ into R such that for $u, v, w \in V$ and $\alpha \in R$, one has

1. $\langle v, v \rangle \geq 0$ and $\langle v, v \rangle = 0$ iff $v = 0$.

2. $\langle v, w \rangle = \langle w, v \rangle$.

3. $\langle \alpha v, w \rangle = \alpha \langle v, w \rangle$.

4. $\langle u + v, w \rangle = \langle u, w \rangle + \langle v, w \rangle$.

A real vector space equipped with the above inner product leads to what is called a real inner product space. For complex inner product spaces, the above properties hold, with the third property replaced with $\langle v, w \rangle = \overline{\langle w, v \rangle}$.

For the time being, we will only deal with real valued functions and, thus we will need an inner product appropriate for such spaces. One such definition is the following. Let $f(x)$ and $g(x)$ be functions defined on $[a, b]$ and introduce the weight function $\sigma(x) > 0$. Then, we define the inner product, if the integral exists, as

$$\langle f, g \rangle = \int_a^b f(x) g(x) \sigma(x) \, dx. \tag{6.1}$$

Spaces in which $\langle f, f \rangle < \infty$ under this inner product are called the space of square integrable functions on (a, b) under weight σ and are denoted as $L_\sigma^2(a, b)$. In what follows, we will assume for simplicity that $\sigma(x) = 1$. This is possible to do using a change of variables.

The space of square integrable functions.

Now that we have function spaces equipped with an inner product, we seek a basis for the space. For an n-dimensional space, we need n basis vectors. For an infinite dimensional space, how many will we need? How do we know when we have enough? We will provide some answers to these questions later.

Let's assume that we have a basis of functions $\{\phi_n(x)\}_{n=1}^\infty$. Given a function $f(x)$, how can we go about finding the components of f in this basis? In other words, let

$$f(x) = \sum_{n=1}^\infty c_n \phi_n(x).$$

How do we find the c_n's? Does this remind you of Fourier series expansions? Does it remind you of the problem we had earlier for finite dimensional spaces? [You may want to review the discussion at the end of Section 3.1 as you read the next derivation.]

Formally, we take the inner product of f with each ϕ_j and use the properties of the inner product to find

$$\begin{aligned} \langle \phi_j, f \rangle &= \left\langle \phi_j, \sum_{n=1}^\infty c_n \phi_n \right\rangle \\ &= \sum_{n=1}^\infty c_n \langle \phi_j, \phi_n \rangle. \end{aligned} \tag{6.2}$$

If the basis is an orthogonal basis, then we have

$$\langle \phi_j, \phi_n \rangle = N_j \delta_{jn}, \tag{6.3}$$

where δ_{jn} is the Kronecker delta. Recall from Chapter 3 that the Kronecker delta is defined as

$$\delta_{ij} = \begin{cases} 0, & i \neq j \\ 1, & i = j. \end{cases} \tag{6.4}$$

Continuing with the derivation, we have

$$\begin{aligned} \langle \phi_j, f \rangle &= \sum_{n=1}^\infty c_n \langle \phi_j, \phi_n \rangle \\ &= \sum_{n=1}^\infty c_n N_j \delta_{jn}. \end{aligned} \tag{6.5}$$

For the generalized Fourier series expansion $f(x) = \sum_{n=1}^\infty c_n \phi_n(x)$, we have determined the generalized Fourier coefficients to be $c_j = \langle \phi_j, f \rangle / \langle \phi_j, \phi_j \rangle$.

Expanding the sum, we see that the Kronecker delta picks out one nonzero term:

$$
\begin{aligned}
\langle \phi_j, f \rangle &= c_1 N_j \delta_{j1} + c_2 N_j \delta_{j2} + \ldots + c_j N_j \delta_{jj} + \ldots \\
&= c_j N_j.
\end{aligned}
\tag{6.6}
$$

So, the expansion coefficients are

$$
c_j = \frac{\langle \phi_j, f \rangle}{N_j} = \frac{\langle \phi_j, f \rangle}{\langle \phi_j, \phi_j \rangle} \quad j = 1, 2, \ldots.
$$

We summarize this important result:

Generalized Basis Expansion

Let $f(x)$ be represented by an expansion over a basis of orthogonal functions, $\{\phi_n(x)\}_{n=1}^{\infty}$,

$$
f(x) = \sum_{n=1}^{\infty} c_n \phi_n(x).
$$

Then, the expansion coefficients are formally determined as

$$
c_n = \frac{\langle \phi_n, f \rangle}{\langle \phi_n, \phi_n \rangle}.
$$

This will be referred to as the general Fourier series expansion and the c_j's are called the Fourier coefficients. Technically, equality only holds when the infinite series converges to the given function on the interval of interest.

Example 6.1. *Find the coefficients of the Fourier sine series expansion of $f(x)$, given by*

$$
f(x) = \sum_{n=1}^{\infty} b_n \sin nx, \quad x \in [-\pi, \pi].
$$

In the previous chapter we established that the set of functions $\phi_n(x) = \sin nx$ for $n = 1, 2, \ldots$ is orthogonal on the interval $[-\pi, \pi]$. Recall that, using trigonometric identities, we have for $n \neq m$

$$
\langle \phi_n, \phi_m \rangle = \int_{-\pi}^{\pi} \sin nx \sin mx \, dx = \pi \delta_{nm}.
\tag{6.7}
$$

Therefore, the set $\phi_n(x) = \sin nx$ for $n = 1, 2, \ldots$ is an orthogonal set of functions on the interval $[-\pi, \pi]$.

We determine the expansion coefficients using

$$
b_n = \frac{\langle \phi_n, f \rangle}{N_n} = \frac{\langle \phi_n, f \rangle}{\langle \phi_n, \phi_n \rangle} = \frac{1}{\pi} \int_{-\pi}^{\pi} f(x) \sin nx \, dx.
$$

Does this result look familiar?

Just as with vectors in three dimensions, we can normalize these basis functions to arrive at an orthonormal basis. This is simply done by dividing by the length of *the vector. Recall that the length of a vector is obtained as $v = \sqrt{\mathbf{v} \cdot \mathbf{v}}$. In the same way, we define the norm of a function by*

$$\|f\| = \sqrt{\langle f, f \rangle}.$$

Note that there are many types of norms, but this induced norm will be sufficient.[1]

For this example, the norms of the basis functions are $\|\phi_n\| = \sqrt{\pi}$. Defining $\psi_n(x) = \frac{1}{\sqrt{\pi}}\phi_n(x)$, we can normalize the ϕ_n's and have obtained an orthonormal basis of functions on $[-\pi, \pi]$.

We can also use the normalized basis to determine the expansion coefficients. In this case we have

$$b_n = \frac{\langle \psi_n, f \rangle}{N_n} = \langle \psi_n, f \rangle = \frac{1}{\pi} \int_{-\pi}^{\pi} f(x) \sin nx\, dx.$$

6.2 Classical Orthogonal Polynomials

THERE ARE OTHER BASIS FUNCTIONS that can be used to develop series representations of functions. In this section we introduce the classical orthogonal polynomials. We begin by noting that the sequence of functions $\{1, x, x^2, \ldots\}$ is a basis of linearly independent functions. In fact, by the Stone–Weierstraß Approximation Theorem,[2] this set is a basis of $L_\sigma^2(a, b)$, the space of square integrable functions over the interval $[a, b]$ relative to weight $\sigma(x)$. However, we will show that the sequence of functions $\{1, x, x^2, \ldots\}$ does not provide an orthogonal basis for these spaces. We will then proceed to find an appropriate orthogonal basis of functions.

We are familiar with being able to expand functions over the basis $\{1, x, x^2, \ldots\}$, as these expansions are just Maclaurin series representations of the functions about $x = 0$,

$$f(x) \sim \sum_{n=0}^{\infty} c_n x^n.$$

However, this basis is not an orthogonal set of basis functions. One can easily see this by integrating the product of two even, or two odd, basis functions with $\sigma(x) = 1$ and $(a, b) = (-1, 1)$. For example,

$$\int_{-1}^{1} x^0 x^2\, dx = \frac{2}{3}.$$

Because we have found that orthogonal bases have been useful in determining the coefficients for expansions of given functions, we might ask, "Given a set of linearly independent basis vectors, can one find an orthogonal basis of the given space?" The answer is yes. We recall from introductory linear algebra, which mostly covers finite dimensional vector spaces, that there is a method for carrying out this so-called Gram–Schmidt Orthogonalization Process. We will review this process for finite dimensional vectors and then generalize to function spaces.

[1] The norm defined here is the natural, or induced, norm on the inner product space. Norms are a generalization of the concept of lengths of vectors. Denoting $\|\mathbf{v}\|$ the norm of $\mathbf{v}$, it needs to satisfy the properties

1. $\|\mathbf{v}\| \geq 0$. $\|\mathbf{v}\| = 0$ if and only if $\mathbf{v} = \mathbf{0}$.

2. $\|\alpha \mathbf{v}\| = |\alpha| \|\mathbf{v}\|$.

3. $\|\mathbf{u} + \mathbf{v}\| \leq \|\mathbf{u}\| + \|\mathbf{v}\|$.

Examples of common norms are

1. Euclidean norm:

$$\|\mathbf{v}\| = \sqrt{v_1^2 + \cdots + v_n^2}.$$

2. Taxicab norm:

$$\|\mathbf{v}\| = |v_1| + \cdots + |v_n|.$$

3. L^p norm:

$$\|f\| = \left(\int [f(x)]^p\, dx \right)^{\frac{1}{p}}.$$

[2] **Stone–Weierstraß Approximation Theorem** Suppose f is a continuous function defined on the interval $[a, b]$. For every $\epsilon > 0$, there exists a polynomial function $P(x)$ such that for all $x \in [a, b]$, we have $|f(x) - P(x)| < \epsilon$. Therefore, every continuous function defined on $[a, b]$ can be uniformly approximated as closely as we wish by a polynomial function.

The Gram–Schmidt Orthogonalization Process.

Let's assume that we have three vectors that span the usual three-dimensional space, $\mathbf{R}^3$, given by $\mathbf{a}_1$, $\mathbf{a}_2$, and $\mathbf{a}_3$ and shown in Figure 6.1. We seek an orthogonal basis $\mathbf{e}_1$, $\mathbf{e}_2$, and $\mathbf{e}_3$, beginning one vector at a time.

First we take one of the original basis vectors, say $\mathbf{a}_1$, and define

$$\mathbf{e}_1 = \mathbf{a}_1.$$

It is sometimes useful to normalize these basis vectors, denoting such a normalized vector with a "hat":

$$\hat{\mathbf{e}}_1 = \frac{\mathbf{e}_1}{e_1},$$

where $e_1 = \sqrt{\mathbf{e}_1 \cdot \mathbf{e}_1}$.

Next, we want to determine an $\mathbf{e}_2$ that is orthogonal to $\mathbf{e}_1$. We take another element of the original basis, $\mathbf{a}_2$. In Figure 6.2 we show the orientation of the vectors. Note that the desired orthogonal vector is $\mathbf{e}_2$. We can now write $\mathbf{a}_2$ as the sum of $\mathbf{e}_2$ and the projection of $\mathbf{a}_2$ on $\mathbf{e}_1$. Denoting this projection by $\mathbf{pr}_1\mathbf{a}_2$, we then have

$$\mathbf{e}_2 = \mathbf{a}_2 - \mathbf{pr}_1\mathbf{a}_2. \tag{6.8}$$

Recall the projection of one vector onto another from your vector calculus class.

$$\mathbf{pr}_1\mathbf{a}_2 = \frac{\mathbf{a}_2 \cdot \mathbf{e}_1}{e_1^2}\mathbf{e}_1. \tag{6.9}$$

This is easily proven by writing the projection as a vector of length $a_2 \cos\theta$ in direction $\hat{\mathbf{e}}_1$, where θ is the angle between $\mathbf{e}_1$ and $\mathbf{a}_2$. Using the definition of the dot product, $\mathbf{a} \cdot \mathbf{b} = ab\cos\theta$, the projection formula follows.

Combining Equations (6.8) and (6.9), we find that

$$\mathbf{e}_2 = \mathbf{a}_2 - \frac{\mathbf{a}_2 \cdot \mathbf{e}_1}{e_1^2}\mathbf{e}_1. \tag{6.10}$$

It is a simple matter to verify that $\mathbf{e}_2$ is orthogonal to $\mathbf{e}_1$:

$$\begin{aligned}
\mathbf{e}_2 \cdot \mathbf{e}_1 &= \mathbf{a}_2 \cdot \mathbf{e}_1 - \frac{\mathbf{a}_2 \cdot \mathbf{e}_1}{e_1^2}\mathbf{e}_1 \cdot \mathbf{e}_1 \\
&= \mathbf{a}_2 \cdot \mathbf{e}_1 - \mathbf{a}_2 \cdot \mathbf{e}_1 = 0. \tag{6.11}
\end{aligned}$$

Next, we seek a third vector $\mathbf{e}_3$ that is orthogonal to both $\mathbf{e}_1$ and $\mathbf{e}_2$. Pictorially, we can write the given vector $\mathbf{a}_3$ as a combination of vector projections along $\mathbf{e}_1$ and $\mathbf{e}_2$ with the new vector. This is shown in Figure 6.3. Thus, we can see that

$$\mathbf{e}_3 = \mathbf{a}_3 - \frac{\mathbf{a}_3 \cdot \mathbf{e}_1}{e_1^2}\mathbf{e}_1 - \frac{\mathbf{a}_3 \cdot \mathbf{e}_2}{e_2^2}\mathbf{e}_2. \tag{6.12}$$

Again, it is a simple matter to compute the scalar products with $\mathbf{e}_1$ and $\mathbf{e}_2$ to verify orthogonality.

We can easily generalize this procedure to the N-dimensional case. Let $\mathbf{a}_n$, $n = 1, ..., N$ be a set of linearly independent vectors in $\mathbf{R}^N$. Then, an orthogonal basis can be found by setting $\mathbf{e}_1 = \mathbf{a}_1$ and defining

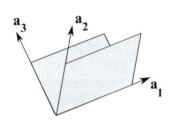

Figure 6.1: The basis $\mathbf{a}_1$, $\mathbf{a}_2$, and $\mathbf{a}_3$, of $\mathbf{R}^3$.

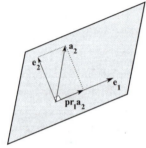

Figure 6.2: A plot of the vectors $\mathbf{e}_1$, $\mathbf{a}_2$, and $\mathbf{e}_2$ needed to find the projection of $\mathbf{a}_2$ on $\mathbf{e}_1$.

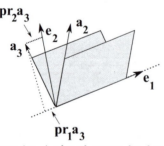

Figure 6.3: A plot of vectors for determining $\mathbf{e}_3$.

$$\mathbf{e}_n = \mathbf{a}_n - \sum_{j=1}^{n-1} \frac{\mathbf{a}_n \cdot \mathbf{e}_j}{e_j^2} \mathbf{e}_j, \quad n = 2, 3, \dots, N. \tag{6.13}$$

Now we can generalize this idea to (real) function spaces. Let $f_n(x)$, $n \in N_0 = \{0, 1, 2, \dots\}$, be a linearly independent sequence of continuous functions defined for $x \in [a, b]$. Then, an orthogonal basis of functions, $\phi_n(x)$, $n \in N_0$ can be found and is given by

$$\phi_0(x) = f_0(x)$$

and

$$\phi_n(x) = f_n(x) - \sum_{j=0}^{n-1} \frac{\langle f_n, \phi_j \rangle}{\|\phi_j\|^2} \phi_j(x), \quad n = 1, 2, \dots. \tag{6.14}$$

Here we are using inner products relative to weight $\sigma(x)$,

$$\langle f, g \rangle = \int_a^b f(x) g(x) \sigma(x) \, dx. \tag{6.15}$$

Note the similarity between the orthogonal basis in Equation (6.14) and the expression for the finite dimensional case in Equation (6.13).

Example 6.2. *Apply the Gram–Schmidt Orthogonalization Process to the set* $f_n(x) = x^n$, $n \in N_0$, *when* $x \in (-1, 1)$ *and* $\sigma(x) = 1$.
  First, we have $\phi_0(x) = f_0(x) = 1$. *Note that*

$$\int_{-1}^1 \phi_0^2(x) \, dx = 2.$$

We could use this result to fix the normalization of the new basis, but we will hold off doing that for now.
  Now we compute the second basis element:

$$\begin{aligned}
\phi_1(x) &= f_1(x) - \frac{\langle f_1, \phi_0 \rangle}{\|\phi_0\|^2} \phi_0(x) \\
&= x - \frac{\langle x, 1 \rangle}{\|1\|^2} 1 = x, \tag{6.16}
\end{aligned}$$

as $\langle x, 1 \rangle$ *is the integral of an odd function over a symmetric interval.*
  For $\phi_2(x)$, *we have*

$$\begin{aligned}
\phi_2(x) &= f_2(x) - \frac{\langle f_2, \phi_0 \rangle}{\|\phi_0\|^2} \phi_0(x) - \frac{\langle f_2, \phi_1 \rangle}{\|\phi_1\|^2} \phi_1(x) \\
&= x^2 - \frac{\langle x^2, 1 \rangle}{\|1\|^2} 1 - \frac{\langle x^2, x \rangle}{\|x\|^2} x \\
&= x^2 - \frac{\int_{-1}^1 x^2 \, dx}{\int_{-1}^1 dx} \\
&= x^2 - \frac{1}{3}. \tag{6.17}
\end{aligned}$$

So far, we have the orthogonal set $\{1, x, x^2 - \frac{1}{3}\}$. If one chooses to normalize these by forcing $\phi_n(1) = 1$, then one obtains the classical Legendre polynomials, $P_n(x)$. Thus,

$$P_2(x) = \frac{1}{2}(3x^2 - 1).$$

Note that this normalization is different from the usual one. In fact, we see that $P_2(x)$ does not have a unit norm,

$$\|P_2\|^2 = \int_{-1}^{1} P_2^2(x)\, dx = \frac{2}{5}.$$

[3] Adrien-Marie Legendre (1752–1833) was a French mathematician who made many contributions to analysis and algebra.

The set of Legendre[3] polynomials is just one set of classical orthogonal polynomials that can be obtained in this way. Many of these special functions had originally appeared as solutions of important boundary value problems in physics. They all have similar properties and we will just elaborate some of these for the Legendre functions in the next section. Others in this group are shown in Table 6.1.

Table 6.1: Common Classical Orthogonal Polynomials with the Interval and Weight Function Used to Define Them

Polynomial	Symbol	Interval	$\sigma(x)$
Hermite	$H_n(x)$	$(-\infty, \infty)$	e^{-x^2}
Laguerre	$L_n^\alpha(x)$	$[0, \infty)$	e^{-x}
Legendre	$P_n(x)$	$(-1,1)$	1
Gegenbauer	$C_n^\lambda(x)$	$(-1,1)$	$(1 - x^2)^{\lambda - 1/2}$
Tchebychef of the 1st kind	$T_n(x)$	$(-1,1)$	$(1 - x^2)^{-1/2}$
Tchebychef of the 2nd kind	$U_n(x)$	$(-1,1)$	$(1 - x^2)^{-1/2}$
Jacobi	$P_n^{(\nu,\mu)}(x)$	$(-1,1)$	$(1 - x)^\nu (1 - x)^\mu$

6.3 Fourier–Legendre Series

IN THE PREVIOUS CHAPTER WE SAW how useful Fourier series expansions were for solving the heat and wave equations. In Chapter 11 we will investigate partial differential equations in higher dimensions and find that problems with spherical symmetry may lead to the series representations in terms of a basis of Legendre polynomials. For example, we could consider the steady-state temperature distribution inside a hemispherical igloo, which takes the form

$$\phi(r, \theta) = \sum_{n=0}^{\infty} A_n r^n P_n(\cos\theta)$$

in spherical coordinates. Evaluating this function at the surface $r = a$ as $\phi(a, \theta) = f(\theta)$, leads to a Fourier–Legendre series expansion of function f:

$$f(\theta) = \sum_{n=0}^{\infty} c_n P_n(\cos\theta),$$

where $c_n = A_n a^n$.

In this section we would like to explore Fourier-Legendre series expansions of functions $f(x)$ defined on $(-1, 1)$:

$$f(x) \sim \sum_{n=0}^{\infty} c_n P_n(x). \tag{6.18}$$

As with Fourier trigonometric series, we can determine the expansion coefficients by multiplying both sides of Equation (6.18) by $P_m(x)$ and integrating for $x \in [-1, 1]$. Orthogonality gives the usual form for the generalized Fourier coefficients,

$$c_n = \frac{\langle f, P_n \rangle}{\|P_n\|^2}, n = 0, 1, \ldots.$$

We will later show that

$$\|P_n\|^2 = \frac{2}{2n+1}.$$

Therefore, the Fourier-Legendre coefficients are

$$c_n = \frac{2n+1}{2} \int_{-1}^{1} f(x) P_n(x) \, dx. \tag{6.19}$$

6.3.1 Properties of Legendre Polynomials

The Rodrigues Formula.

WE CAN DO EXAMPLES OF FOURIER-LEGENDRE EXPANSIONS given just a few facts about Legendre polynomials. The first property that the Legendre polynomials have is the Rodrigues Formula:

$$P_n(x) = \frac{1}{2^n n!} \frac{d^n}{dx^n} (x^2 - 1)^n, \quad n \in N_0. \tag{6.20}$$

From the Rodrigues formula, one can show that $P_n(x)$ is an nth degree polynomial. Also, for n odd, the polynomial is an odd function and for n even, the polynomial is an even function.

Example 6.3. *Determine* $P_2(x)$ *from the Rodrigues Formula:*

$$\begin{aligned} P_2(x) &= \frac{1}{2^2 2!} \frac{d^2}{dx^2} (x^2 - 1)^2 \\ &= \frac{1}{8} \frac{d^2}{dx^2} (x^4 - 2x^2 + 1) \\ &= \frac{1}{8} \frac{d}{dx} (4x^3 - 4x) \\ &= \frac{1}{8} (12x^2 - 4) \\ &= \frac{1}{2} (3x^2 - 1). \end{aligned} \tag{6.21}$$

Note that we get the same result as we found in the previous section using orthogonalization.

The first several Legendre polynomials are given in Table 6.2. In Figure 6.4, we show plots of these Legendre polynomials.

The Three-Term Recursion Formula.

Table 6.2: Tabular computation of the Legendre polynomials using the Rodrigues Formula.

n	$(x^2 - 1)^n$	$\frac{d^n}{dx^n}(x^2 - 1)^n$	$\frac{1}{2^n n!}$	$P_n(x)$
0	1	1	1	1
1	$x^2 - 1$	$2x$	$\frac{1}{2}$	x
2	$x^4 - 2x^2 + 1$	$12x^2 - 4$	$\frac{1}{8}$	$\frac{1}{2}(3x^2 - 1)$
3	$x^6 - 3x^4 + 3x^2 - 1$	$120x^3 - 72x$	$\frac{1}{48}$	$\frac{1}{2}(5x^3 - 3x)$

Figure 6.4: Plots of the Legendre polynomials $P_2(x)$, $P_3(x)$, $P_4(x)$, and $P_5(x)$.

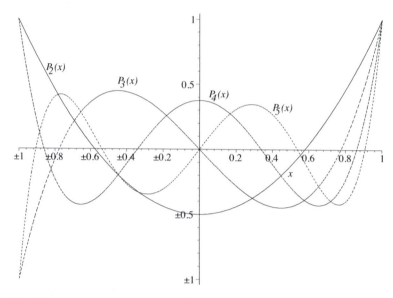

All of the classical orthogonal polynomials satisfy a three-term recursion formula (or, recurrence relation or formula). In the case of the Legendre polynomials, we have

$$(n + 1)P_{n+1}(x) = (2n + 1)xP_n(x) - nP_{n-1}(x), \quad n = 1, 2, \ldots. \quad (6.22)$$

This can also be rewritten by replacing n with $n - 1$ as

$$(2n - 1)xP_{n-1}(x) = nP_n(x) + (n - 1)P_{n-2}(x), \quad n = 1, 2, \ldots. \quad (6.23)$$

Example 6.4. *Use the recursion formula to find $P_2(x)$ and $P_3(x)$, given that $P_0(x) = 1$ and $P_1(x) = x$.*

We first begin by inserting $n = 1$ into Equation (6.22):

$$2P_2(x) = 3xP_1(x) - P_0(x) = 3x^2 - 1.$$

So, $P_2(x) = \frac{1}{2}(3x^2 - 1)$.

For $n = 2$, we have

$$\begin{aligned}
3P_3(x) &= 5xP_2(x) - 2P_1(x) \\
&= \frac{5}{2}x(3x^2 - 1) - 2x \\
&= \frac{1}{2}(15x^3 - 9x). \quad (6.24)
\end{aligned}$$

The first proof of the three-term recursion formula is based upon the nature of the Legendre polynomials as an orthogonal basis, while the second proof is derived using generating functions.

This gives $P_3(x) = \frac{1}{2}(5x^3 - 3x)$. These expressions agree with the earlier results.

We will prove the three-term recursion formula in two ways. First, we use the orthogonality properties of Legendre polynomials and the following lemma.

Lemma 6.1. *The leading coefficient of x^n in $P_n(x)$ is $\frac{1}{2^n n!} \frac{(2n)!}{n!}$.*

Proof. We can prove this using the Rodrigues Formula. First, we focus on the leading coefficient of $(x^2 - 1)^n$, which is x^{2n}. The first derivative of x^{2n} is $2nx^{2n-1}$. The second derivative is $2n(2n-1)x^{2n-2}$. The jth derivative is

$$\frac{d^j x^{2n}}{dx^j} = [2n(2n-1)\dots(2n-j+1)]x^{2n-j}.$$

Thus, the nth derivative is given by

$$\frac{d^n x^{2n}}{dx^n} = [2n(2n-1)\dots(n+1)]x^n.$$

This proves that $P_n(x)$ has degree n. The leading coefficient of $P_n(x)$ can now be written as

$$\frac{[2n(2n-1)\dots(n+1)]}{2^n n!} = \frac{[2n(2n-1)\dots(n+1)]}{2^n n!} \frac{n(n-1)\dots 1}{n(n-1)\dots 1}$$

$$= \frac{1}{2^n n!} \frac{(2n)!}{n!}. \tag{6.25}$$

$\square$

Theorem 6.1. *Legendre polynomials satisfy the three-term recursion formula*

$$(2n-1)xP_{n-1}(x) = nP_n(x) + (n-1)P_{n-2}(x), \quad n = 1, 2, \dots. \tag{6.26}$$

Proof. In order to prove the three-term recursion formula, we consider the expression $(2n-1)xP_{n-1}(x) - nP_n(x)$. While each term is a polynomial of degree n, the leading order terms cancel. We need only look at the coefficient of the leading order term first expression. It is

$$\frac{2n-1}{2^{n-1}(n-1)!} \frac{(2n-2)!}{(n-1)!} = \frac{1}{2^{n-1}(n-1)!} \frac{(2n-1)!}{(n-1)!} = \frac{(2n-1)!}{2^{n-1}[(n-1)!]^2}.$$

The coefficient of the leading term for $nP_n(x)$ can be written as

$$n\frac{1}{2^n n!} \frac{(2n)!}{n!} = n\left(\frac{2n}{2n^2}\right)\left(\frac{1}{2^{n-1}(n-1)!}\right)\frac{(2n-1)!}{(n-1)!} \frac{(2n-1)!}{2^{n-1}[(n-1)!]^2}.$$

It is easy to see that the leading order terms in the expression $(2n-1)xP_{n-1}(x) - nP_n(x)$ cancel.

The next terms will be of degree $n-2$. This is because the P_n's are either even or odd functions, thus only containing even, or odd, powers of x. We conclude that

$$(2n-1)xP_{n-1}(x) - nP_n(x) = \text{ polynomial of degree } n - 2.$$

Therefore, because the Legendre polynomials form a basis, we can write this polynomial as a linear combination of Legendre polynomials:

$$(2n-1)xP_{n-1}(x) - nP_n(x) = c_0 P_0(x) + c_1 P_1(x) + \dots + c_{n-2}P_{n-2}(x). \tag{6.27}$$

Multiplying Equation (6.27) by $P_m(x)$ for $m = 0, 1, \ldots, n - 3$, integrating from -1 to 1, and using orthogonality, we obtain

$$0 = c_m \|P_m\|^2, \quad m = 0, 1, \ldots, n - 3.$$

[Note: $\int_{-1}^{1} x^k P_n(x)\, dx = 0$ for $k \leq n - 1$. Thus, $\int_{-1}^{1} x P_{n-1}(x) P_m(x)\, dx = 0$ for $m \leq n - 3$.]

Thus, all these c_m's are zero, leaving Equation (6.27) as

$$(2n - 1)x P_{n-1}(x) - n P_n(x) = c_{n-2} P_{n-2}(x).$$

The final coefficient can be found using the normalization condition, $P_n(1) = 1$. Thus, $c_{n-2} = (2n - 1) - n = n - 1$. □

6.3.2   Generating Functions: Generating Function for Legendre Polynomials

A SECOND PROOF OF THE THREE-TERM RECURSION FORMULA can be obtained from the generating function of the Legendre polynomials. Many special functions have such generating functions. In this case, it is given by

$$g(x, t) = \frac{1}{\sqrt{1 - 2xt + t^2}} = \sum_{n=0}^{\infty} P_n(x) t^n, \quad |x| \leq 1, |t| < 1. \tag{6.28}$$

This generating function occurs often in applications. In particular, it arises in potential theory, such as electromagnetic or gravitational potentials. These potential functions are $\frac{1}{r}$ type functions.

Figure 6.5: The position vectors used to describe the tidal force on the Earth due to the moon.

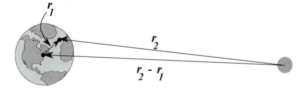

For example, the gravitational potential between the Earth and the moon is proportional to the reciprocal of the magnitude of the difference between their positions relative to some coordinate system. An even better example would be to place the origin at the center of the Earth and consider the forces on the non-pointlike Earth due to the moon. Consider a piece of the Earth at position $\mathbf{r}_1$ and the moon at position $\mathbf{r}_2$ as shown in Figure 6.5. The tidal potential Φ is proportional to

$$\Phi \propto \frac{1}{|\mathbf{r}_2 - \mathbf{r}_1|} = \frac{1}{\sqrt{(\mathbf{r}_2 - \mathbf{r}_1) \cdot (\mathbf{r}_2 - \mathbf{r}_1)}} = \frac{1}{\sqrt{r_1^2 - 2r_1 r_2 \cos\theta + r_2^2}},$$

where θ is the angle between $\mathbf{r}_1$ and $\mathbf{r}_2$.

Typically, one of the position vectors is much larger than the other. Let's assume that $r_1 \ll r_2$. Then, one can write

$$\Phi \propto \frac{1}{\sqrt{r_1^2 - 2r_1 r_2 \cos\theta + r_2^2}} = \frac{1}{r_2} \frac{1}{\sqrt{1 - 2\frac{r_1}{r_2}\cos\theta + \left(\frac{r_1}{r_2}\right)^2}}.$$

Now, define $x = \cos\theta$ and $t = \frac{r_1}{r_2}$. We then have that the tidal potential is proportional to the generating function for the Legendre polynomials! So, we can write the tidal potential as

$$\Phi \propto \frac{1}{r_2} \sum_{n=0}^{\infty} P_n(\cos\theta) \left(\frac{r_1}{r_2} \right)^n.$$

The first term in the expansion, $\frac{1}{r_2}$, is the gravitational potential that gives the usual force between the Earth and the moon. [Recall that the gravitational potential for mass m at distance r from M is given by $\Phi = -\frac{GMm}{r}$, and that the force is the gradient of the potential, $\mathbf{F} = -\nabla\Phi \propto \nabla\left(\frac{1}{r}\right)$.] The next terms will give expressions for the tidal effects.

Now that we have some idea as to where this generating function might have originated, we can proceed to use it. First of all, the generating function can be used to obtain special values of the Legendre polynomials.

Example 6.5. *Evaluate $P_n(0)$ using the generating function. $P_n(0)$ is found by considering $g(0, t)$. Setting $x = 0$ in Equation (6.28), we have*

$$
\begin{aligned}
g(0, t) &= \frac{1}{\sqrt{1 + t^2}} \\
&= \sum_{n=0}^{\infty} P_n(0) t^n \\
&= P_0(0) + P_1(0)t + P_2(0)t^2 + P_3(0)t^3 + \dots. \quad (6.29)
\end{aligned}
$$

We can use the binomial expansion to find the final answer. Namely, we have

$$\frac{1}{\sqrt{1 + t^2}} = 1 - \frac{1}{2}t^2 + \frac{3}{8}t^4 + \dots.$$

Comparing these expansions, we have the $P_n(0) = 0$ for n odd and for even integers one can show (see Problem 12) that[4]

$$P_{2n}(0) = (-1)^n \frac{(2n-1)!!}{(2n)!!}, \quad (6.30)$$

where $n!!$ is the double factorial,

$$
n!! = \begin{cases} n(n-2)\dots(3)1, & n > 0, \text{odd}, \\ n(n-2)\dots(4)2, & n > 0, \text{even}, \\ 1 & n = 0, -1. \end{cases}
$$

Example 6.6. *Evaluate $P_n(-1)$. This is a simpler problem. In this case, we have*

$$g(-1, t) = \frac{1}{\sqrt{1 + 2t + t^2}} = \frac{1}{1 + t} = 1 - t + t^2 - t^3 + \dots.$$

Therefore, $P_n(-1) = (-1)^n$.

Example 6.7. *Prove the three-term recursion formula,*

$$(k+1)P_{k+1}(x) - (2k+1)xP_k(x) + kP_{k-1}(x) = 0, \quad k = 1, 2, \dots,$$

[4] This example can be finished by first proving that

$$(2n)!! = 2^n n!$$

and

$$(2n-1)!! = \frac{(2n)!}{(2n)!!} = \frac{(2n)!}{2^n n!}.$$

Proof of the three-term recursion formula using the generating function.

using the generating function.

We can also use the generating function to find recurrence relations. To prove the three-term recursion (6.22) that we introduced above, we need only differentiate the generating function with respect to t in Equation (6.28) and rearrange the result. First note that

$$\frac{\partial g}{\partial t} = \frac{x - t}{(1 - 2xt + t^2)^{3/2}} = \frac{x - t}{1 - 2xt + t^2} g(x, t).$$

Combining this with

$$\frac{\partial g}{\partial t} = \sum_{n=0}^{\infty} n P_n(x) t^{n-1},$$

we have

$$(x - t) g(x, t) = (1 - 2xt + t^2) \sum_{n=0}^{\infty} n P_n(x) t^{n-1}.$$

Inserting the series expression for $g(x, t)$ and distributing the sum on the right side, we obtain

$$(x - t) \sum_{n=0}^{\infty} P_n(x) t^n = \sum_{n=0}^{\infty} n P_n(x) t^{n-1} - \sum_{n=0}^{\infty} 2nx P_n(x) t^n + \sum_{n=0}^{\infty} n P_n(x) t^{n+1}.$$

Multiplying out the $x - t$ factor and rearranging leads to three separate sums:

$$\sum_{n=0}^{\infty} n P_n(x) t^{n-1} - \sum_{n=0}^{\infty} (2n + 1) x P_n(x) t^n + \sum_{n=0}^{\infty} (n + 1) P_n(x) t^{n+1} = 0. \quad (6.31)$$

Each term contains powers of t that we would like to combine into a single sum. This is done by reindexing. For the first sum, we could use the new index $k = n - 1$. Then the first sum can be written

$$\sum_{n=0}^{\infty} n P_n(x) t^{n-1} = \sum_{k=-1}^{\infty} (k + 1) P_{k+1}(x) t^k.$$

Using different indices is just another way of writing out the terms. Note that

$$\sum_{n=0}^{\infty} n P_n(x) t^{n-1} = 0 + P_1(x) + 2P_2(x)t + 3P_3(x)t^2 + \ldots$$

and

$$\sum_{k=-1}^{\infty} (k + 1) P_{k+1}(x) t^k = 0 + P_1(x) + 2P_2(x)t + 3P_3(x)t^2 + \ldots$$

actually give the same sum. The indices are sometimes referred to as dummy indices because they do not show up in the expanded expression and can be replaced with another letter.

If we want to do so, we could now replace all the k's with n's. However, we will leave the k's in the first term and now reindex the next sums in Equation (6.31). The second sum just needs the replacement $n = k$, and the last sum we reindex using $k = n + 1$. Therefore, Equation (6.31) becomes

$$\sum_{k=-1}^{\infty} (k + 1) P_{k+1}(x) t^k - \sum_{k=0}^{\infty} (2k + 1) x P_k(x) t^k + \sum_{k=1}^{\infty} k P_{k-1}(x) t^k = 0. \quad (6.32)$$

We can now combine all the terms, noting that the $k = -1$ term is automatically zero and the $k = 0$ terms give

$$P_1(x) - xP_0(x) = 0. \tag{6.33}$$

Of course, we know this already. So that leaves the $k > 0$ terms:

$$\sum_{k=1}^{\infty} \left[(k+1)P_{k+1}(x) - (2k+1)xP_k(x) + kP_{k-1}(x)\right] t^k = 0. \tag{6.34}$$

Because this is true for all t, the coefficients of the t^k's are zero, or

$$(k+1)P_{k+1}(x) - (2k+1)xP_k(x) + kP_{k-1}(x) = 0, \quad k = 1, 2, \ldots.$$

While this is the standard form for the three-term recurrence relation, the earlier form is obtained by setting $k = n - 1$.

There are other recursion relations that we list in the box below. Equation (6.35) was derived using the generating function. Differentiating it with respect to x, we find Equation (6.36). Equation (6.37) can be proven using the generating function by differentiating $g(x, t)$ with respect to x and rearranging the resulting infinite series just as in this last manipulation. This will be left as Problem 4. Combining this result with Equation (6.35), we can derive Equations (6.38) and (6.39). Adding and subtracting these equations yields Equations (6.40) and (6.41).

Recursion Formulae for Legendre Polynomials for $n = 1, 2, \ldots$.

$$(n+1)P_{n+1}(x) = (2n+1)xP_n(x) - nP_{n-1}(x) \tag{6.35}$$

$$(n+1)P'_{n+1}(x) = (2n+1)[P_n(x) + xP'_n(x)] - nP'_{n-1}(x)$$
$$\tag{6.36}$$

$$P_n(x) = P'_{n+1}(x) - 2xP'_n(x) + P'_{n-1}(x) \tag{6.37}$$

$$P'_{n-1}(x) = xP'_n(x) - nP_n(x) \tag{6.38}$$

$$P'_{n+1}(x) = xP'_n(x) + (n+1)P_n(x) \tag{6.39}$$

$$P'_{n+1}(x) + P'_{n-1}(x) = 2xP'_n(x) + P_n(x). \tag{6.40}$$

$$P'_{n+1}(x) - P'_{n-1}(x) = (2n+1)P_n(x). \tag{6.41}$$

$$(x^2 - 1)P'_n(x) = nxP_n(x) - nP_{n-1}(x) \tag{6.42}$$

Finally, Equation (6.42) can be obtained using Equations (6.38) and (6.39). Just multiply Equation (6.38) by x,

$$x^2 P'_n(x) - nxP_n(x) = xP'_{n-1}(x).$$

Now use Equation (6.39), but first replace n with $n - 1$ to eliminate the $xP'_{n-1}(x)$ term:

$$x^2 P'_n(x) - nxP_n(x) = P'_n(x) - nP_{n-1}(x).$$

Rearranging gives the Equation (6.42).

Example 6.8. *Use the generating function to prove*

$$\|P_n\|^2 = \int_{-1}^{1} P_n^2(x)\, dx = \frac{2}{2n+1}.$$

The normalization constant.

Another use of the generating function is to obtain the normalization constant. This can be done by first squaring the generating function in order to get the products $P_n(x)P_m(x)$, and then integrating over x.

Squaring the generating function must be done with care, as we need to make proper use of the dummy summation index. So, we first write

$$
\begin{aligned}
\frac{1}{1 - 2xt + t^2} &= \left[\sum_{n=0}^{\infty} P_n(x) t^n \right]^2 \\
&= \sum_{n=0}^{\infty} \sum_{m=0}^{\infty} P_n(x) P_m(x) t^{n+m}.
\end{aligned}
\tag{6.43}
$$

Integrating from $x = -1$ to $x = 1$ and using the orthogonality of the Legendre polynomials, we have

$$
\begin{aligned}
\int_{-1}^{1} \frac{dx}{1 - 2xt + t^2} &= \sum_{n=0}^{\infty} \sum_{m=0}^{\infty} t^{n+m} \int_{-1}^{1} P_n(x) P_m(x)\, dx \\
&= \sum_{n=0}^{\infty} t^{2n} \int_{-1}^{1} P_n^2(x)\, dx.
\end{aligned}
\tag{6.44}
$$

[5] You will need the integral

$$\int \frac{dx}{a + bx} = \frac{1}{b} \ln(a + bx) + C.$$

However, one can show that[5]

$$\int_{-1}^{1} \frac{dx}{1 - 2xt + t^2} = \frac{1}{t} \ln\left(\frac{1+t}{1-t} \right).$$

[6] You will need the series expansion

$$
\begin{aligned}
\ln(1 + x) &= \sum_{n=1}^{\infty} (-1)^{n+1} \frac{x^n}{n} \\
&= x - \frac{x^2}{2} + \frac{x^3}{3} - \cdots.
\end{aligned}
$$

Expanding this expression about $t = 0$, we obtain[6]

$$\frac{1}{t} \ln\left(\frac{1+t}{1-t} \right) = \sum_{n=0}^{\infty} \frac{2}{2n+1} t^{2n}.$$

Comparing this result with Equation (6.44), we find that

$$\|P_n\|^2 = \int_{-1}^{1} P_n^2(x)\, dx = \frac{2}{2n+1}.
\tag{6.45}$$

6.3.3 Differential Equation for Legendre Polynomials

THE LEGENDRE POLYNOMIALS SATISFY a second-order linear differential equation. This differential equation occurs naturally in the solution of initial-boundary value problems in three dimensions that possess some spherical symmetry. We will see this in Chapter 11. There are two approaches we could take in showing that the Legendre polynomials satisfy a particular differential equation. Either we can write the equations and attempt to solve it, or we could use the above properties to obtain the equation. For now, we will seek the differential equation satisfied by $P_n(x)$ using the above recursion relations.

We begin by differentiating Equation (6.42) and using Equation (6.38) to simplify:

$$\frac{d}{dx}\left((x^2 - 1)P_n'(x)\right) = nP_n(x) + nxP_n'(x) - nP_{n-1}'(x)$$
$$= nP_n(x) + n^2 P_n(x)$$
$$= n(n+1)P_n(x). \tag{6.46}$$

Therefore, Legendre polynomials, or Legendre functions of the first kind, are solutions of the differential equation

$$(1 - x^2)y'' - 2xy' + n(n+1)y = 0.$$

As this is a linear second-order differential equation, we expect two linearly independent solutions. The second solution, called the Legendre function of the second kind, is given by $Q_n(x)$ and is not well behaved at $x = \pm 1$. For example,

$$Q_0(x) = \frac{1}{2}\ln\frac{1+x}{1-x}.$$

We will not need these for physically interesting examples in this book.

A generalization of the Legendre equation is given by $(1 - x^2)y'' - 2xy' + \left[n(n+1) - \frac{m^2}{1-x^2}\right]y = 0$. Solutions to this equation, $P_n^m(x)$ and $Q_n^m(x)$, are called the associated Legendre functions of the first and second kind.

6.3.4 Fourier–Legendre Series

WITH THESE PROPERTIES OF LEGENDRE FUNCTIONS, we are now prepared to compute the expansion coefficients for the Fourier-Legendre series representation of a given function.

Example 6.9. *Expand $f(x) = x^3$ in a Fourier-Legendre series.*
  We simply need to compute

$$c_n = \frac{2n+1}{2}\int_{-1}^{1} x^3 P_n(x)\,dx. \tag{6.47}$$

We first note that

$$\int_{-1}^{1} x^m P_n(x)\,dx = 0 \quad \text{for } m > n.$$

As a result, we have that $c_n = 0$ for $n > 3$. We could just compute $\int_{-1}^{1} x^3 P_m(x)\,dx$ for $m = 0, 1, 2, \ldots$ outright by looking up Legendre polynomials. But note that x^3 is an odd function. So, $c_0 = 0$ and $c_2 = 0$.

  This leaves us with only two coefficients to compute. We refer to Table 6.2 and find that

$$c_1 = \frac{3}{2}\int_{-1}^{1} x^4\,dx = \frac{3}{5}$$

$$c_3 = \frac{7}{2}\int_{-1}^{1} x^3\left[\frac{1}{2}(5x^3 - 3x)\right]dx = \frac{2}{5}.$$

  Thus,

$$x^3 = \frac{3}{5}P_1(x) + \frac{2}{5}P_3(x).$$

Of course, this is simple to check using Table 6.2:

$$\frac{3}{5}P_1(x) + \frac{2}{5}P_3(x) = \frac{3}{5}x + \frac{2}{5}\left[\frac{1}{2}(5x^3 - 3x)\right] = x^3.$$

We could have obtained this result without doing any integration. Write x^3 as a linear combination of $P_1(x)$ and $P_3(x)$:

$$\begin{aligned} x^3 &= c_1 x + \frac{1}{2}c_2(5x^3 - 3x) \\ &= (c_1 - \frac{3}{2}c_2)x + \frac{5}{2}c_2 x^3. \end{aligned} \tag{6.48}$$

Equating coefficients of like terms, we have that $c_2 = \frac{2}{5}$ and $c_1 = \frac{3}{2}c_2 = \frac{3}{5}$.

[7] Oliver Heaviside (1850–1925) was an English mathematician, physicist, and engineer who used complex analysis to study circuits and was a co-founder of vector analysis. The Heaviside function is also called the step function.

Example 6.10. *Expand the Heaviside[7] function in a Fourier-Legendre series.*
   The Heaviside function is defined as

$$H(x) = \begin{cases} 1, & x > 0, \\ 0, & x < 0. \end{cases} \tag{6.49}$$

In this case, we cannot find the expansion coefficients without some integration. We have to compute

$$\begin{aligned} c_n &= \frac{2n+1}{2}\int_{-1}^{1} f(x)P_n(x)\,dx \\ &= \frac{2n+1}{2}\int_{0}^{1} P_n(x)\,dx. \end{aligned} \tag{6.50}$$

We can make use of identity (6.41),

$$P_{n+1}'(x) - P_{n-1}'(x) = (2n+1)P_n(x), \quad n > 0. \tag{6.51}$$

We have for $n > 0$

$$c_n = \frac{1}{2}\int_0^1 [P_{n+1}'(x) - P_{n-1}'(x)]\,dx = \frac{1}{2}[P_{n-1}(0) - P_{n+1}(0)].$$

For $n = 0$, we have

$$c_0 = \frac{1}{2}\int_0^1 dx = \frac{1}{2}.$$

This leads to the expansion

$$f(x) \sim \frac{1}{2} + \frac{1}{2}\sum_{n=1}^{\infty}[P_{n-1}(0) - P_{n+1}(0)]P_n(x).$$

We still need to evaluate the Fourier-Legendre coefficients

$$c_n = \frac{1}{2}[P_{n-1}(0) - P_{n+1}(0)].$$

Because $P_n(0) = 0$ for n odd, the c_n's vanish for n even. Letting $n = 2k - 1$, we re-index the sum, obtaining

$$f(x) \sim \frac{1}{2} + \frac{1}{2}\sum_{k=1}^{\infty}[P_{2k-2}(0) - P_{2k}(0)]P_{2k-1}(x).$$

We can compute the nonzero Fourier coefficients, $c_{2k-1} = \frac{1}{2}[P_{2k-2}(0) - P_{2k}(0)]$,
using a result from Problem 12:

$$P_{2k}(0) = (-1)^k \frac{(2k-1)!!}{(2k)!!}. \tag{6.52}$$

Namely, we have

$$
\begin{aligned}
c_{2k-1} &= \frac{1}{2}[P_{2k-2}(0) - P_{2k}(0)] \\
&= \frac{1}{2}\left[(-1)^{k-1}\frac{(2k-3)!!}{(2k-2)!!} - (-1)^k\frac{(2k-1)!!}{(2k)!!}\right] \\
&= -\frac{1}{2}(-1)^k\frac{(2k-3)!!}{(2k-2)!!}\left[1 + \frac{2k-1}{2k}\right] \\
&= -\frac{1}{2}(-1)^k\frac{(2k-3)!!}{(2k-2)!!}\frac{4k-1}{2k}. \tag{6.53}
\end{aligned}
$$

Thus, the Fourier-Legendre series expansion for the Heaviside function is given by

$$f(x) \sim \frac{1}{2} - \frac{1}{2}\sum_{n=1}^{\infty}(-1)^n\frac{(2n-3)!!}{(2n-2)!!}\frac{4n-1}{2n}P_{2n-1}(x). \tag{6.54}$$

The sum of the first 21 terms of this series are shown in Figure 6.6. We note the slow convergence to the Heaviside function. Also, we see that the Gibbs phenomenon is present due to the jump discontinuity at $x = 0$. [See Section 5.12.]

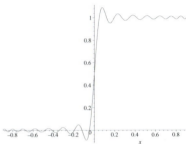

Figure 6.6: Sum of first 21 terms for Fourier-Legendre series expansion of Heaviside function.

6.4 Gamma Function

A FUNCTION THAT OFTEN OCCURS IN THE STUDY OF SPECIAL FUNCTIONS is the Gamma function. We will need the Gamma function in the next section on Fourier–Bessel series.

For $x > 0$, we define the Gamma function as

$$\Gamma(x) = \int_0^{\infty} t^{x-1}e^{-t}\,dt, \quad x > 0. \tag{6.55}$$

The name and symbol for the Gamma function were first given by Legendre in 1811. However, the search for a generalization of the factorial extends back to the 1720s when Euler provided the first representation of the factorial as an infinite product, later to be modified by others like Gauß, Weierstraß, and Legendre.

The Gamma function is a generalization of the factorial function and a plot is shown in Figure 6.7. In fact, we have

$$\Gamma(1) = 1$$

and

$$\Gamma(x+1) = x\Gamma(x).$$

The reader can prove this identity by simply performing an integration by parts. (See Problem 7.) In particular, for integers $n \in Z^+$, we then have

$$\Gamma(n+1) = n\Gamma(n) = n(n-1)\Gamma(n-2) = n(n-1)\cdots 2\Gamma(1) = n!.$$

We can also define the Gamma function for negative, non-integer values of x. We first note that by iteration on $n \in Z^+$, we have

$$\Gamma(x+n) = (x+n-1)\cdots(x+1)x\Gamma(x), \quad x+n > 0.$$

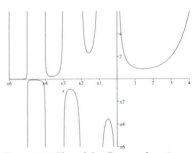

Figure 6.7: Plot of the Gamma function.

Solving for $\Gamma(x)$, we then find

$$\Gamma(x) = \frac{\Gamma(x+n)}{(x+n-1)\cdots(x+1)x}, \quad -n < x < 0.$$

Note that the Gamma function is undefined at zero and the negative integers.

Example 6.11. *We now prove that*

$$\Gamma\left(\frac{1}{2}\right) = \sqrt{\pi}.$$

This is done by direct computation of the integral:

$$\Gamma\left(\frac{1}{2}\right) = \int_0^\infty t^{-\frac{1}{2}} e^{-t}\, dt.$$

Letting $t = z^2$, we have

$$\Gamma\left(\frac{1}{2}\right) = 2\int_0^\infty e^{-z^2}\, dz.$$

Due to the symmetry of the integrand, we obtain the classic integral

$$\Gamma\left(\frac{1}{2}\right) = \int_{-\infty}^\infty e^{-z^2}\, dz,$$

[8] In Example 8.5 we show the more general result:

$$\int_{-\infty}^\infty e^{-\beta y^2}\, dy = \sqrt{\frac{\pi}{\beta}}.$$

which can be performed using a standard trick.[8] Consider the integral

$$I = \int_{-\infty}^\infty e^{-x^2}\, dx.$$

Then,

$$I^2 = \int_{-\infty}^\infty e^{-x^2}\, dx \int_{-\infty}^\infty e^{-y^2}\, dy.$$

Note that we changed the integration variable. This will allow us to write this product of integrals as a double integral:

$$I^2 = \int_{-\infty}^\infty \int_{-\infty}^\infty e^{-(x^2+y^2)}\, dxdy.$$

This is an integral over the entire xy-plane. We can transform this Cartesian integration to an integration over polar coordinates. The integral becomes

$$I^2 = \int_0^{2\pi} \int_0^\infty e^{-r^2}\, rdrd\theta.$$

This is simple to integrate and we have $I^2 = \pi$. So, the final result is found by taking the square root of both sides:

$$\Gamma\left(\frac{1}{2}\right) = I = \sqrt{\pi}.$$

In Problem 12, the reader will prove the identity

$$\Gamma\left(n + \frac{1}{2}\right) = \frac{(2n-1)!!}{2^n}\sqrt{\pi}.$$

Another useful relation, which we only state, is

$$\Gamma(x)\Gamma(1-x) = \frac{\pi}{\sin \pi x}.$$

The are many other important relations, including infinite products, that we will not need at this point. The reader is encouraged to read about these elsewhere. In the meantime, we move on to the discussion of another important special function in physics and mathematics.

6.5   Fourier–Bessel Series

BESSEL FUNCTIONS ARISE IN MANY PROBLEMS in physics possessing cylindrical symmetry, such as the vibrations of circular drumheads and the radial modes in optical fibers. They also provide us with another orthogonal set of basis functions.

The first occurrence of Bessel functions (zeroeth order) was in the work of Daniel Bernoulli on heavy chains (1738). More general Bessel functions were studied by Leonhard Euler in 1781 and in his study of the vibrating membrane in 1764. Joseph Fourier found them in the study of heat conduction in solid cylinders and Siméon Poisson (1781–1840) in heat conduction of spheres (1823).

Bessel functions have a long history and were named after Friedrich Wilhelm Bessel (1784–1846).

The history of Bessel functions did not just originate in the study of the wave and heat equations. These solutions originally came up in the study of the Kepler problem, describing planetary motion. According to G. N. Watson in his *Treatise on Bessel Functions*, the formulation and solution of Kepler's Problem was discovered by Joseph-Louis Lagrange (1736–1813), in 1770. Namely, the problem was to express the radial coordinate and what is called the eccentric anomaly, E, as functions of time. Lagrange found expressions for the coefficients in the expansions of r and E in trigonometric functions of time. However, he only computed the first few coefficients. In 1816, Friedrich Wilhelm Bessel (1784–1846) had shown that the coefficients in the expansion for r could be given an integral representation. In 1824, he presented a thorough study of these functions, which are now called Bessel functions.

You might have seen Bessel functions in a course on differential equations as solutions of the differential equation

$$x^2 y'' + x y' + (x^2 - p^2)y = 0. \tag{6.56}$$

Solutions to this equation are obtained in the form of series expansions. Namely, one seeks solutions of the form

$$y(x) = \sum_{j=0}^{\infty} a_j x^{j+n}$$

by determining the form the coefficients must take. We will leave this for a homework exercise and simply report the results.

One solution of the differential equation is the *Bessel function of the first kind of order p*, given as

$$y(x) = J_p(x) = \sum_{n=0}^{\infty} \frac{(-1)^n}{\Gamma(n+1)\Gamma(n+p+1)} \left(\frac{x}{2}\right)^{2n+p}. \qquad (6.57)$$

Figure 6.8: Plots of the Bessel functions $J_0(x)$, $J_1(x)$, $J_2(x)$, and $J_3(x)$.

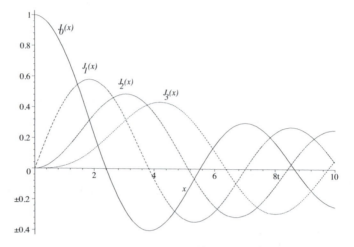

In Figure 6.8, we display the first few Bessel functions of the first kind of integer order. Note that these functions can be described as decaying oscillatory functions.

A second linearly independent solution is obtained for p not an integer as $J_{-p}(x)$. However, for p an integer, the $\Gamma(n+p+1)$ factor leads to evaluations of the Gamma function at zero, or negative integers, when p is negative. Thus, the above series is not defined in these cases.

Another method for obtaining a second linearly independent solution is through a linear combination of $J_p(x)$ and $J_{-p}(x)$ as

$$N_p(x) = Y_p(x) = \frac{\cos \pi p J_p(x) - J_{-p}(x)}{\sin \pi p}. \qquad (6.58)$$

These functions are called the Neumann functions, or Bessel functions of the second kind of order p.

In Figure 6.9, we display the first few Bessel functions of the second kind of integer order. Note that these functions are also decaying oscillatory functions. However, they are singular at $x = 0$.

In many applications, one desires bounded solutions at $x = 0$. These functions do not satisfy this boundary condition. For example, we will later study one standard problem: to describe the oscillations of a circular drumhead. For this problem, one solves the two-dimensional wave equation using separation of variables in cylindrical coordinates. The radial equation leads to a Bessel equation. The Bessel function solutions describe the radial part of the solution, and one does not expect a singular solution at the center of the drum. The amplitude of the oscillation must remain finite. Thus, only Bessel functions of the first kind can be used.

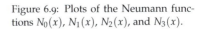

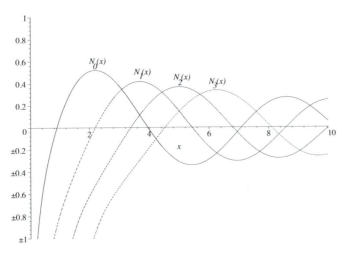

Bessel functions satisfy a variety of properties, which we will only list at this time for Bessel functions of the first kind. The reader will have the opportunity to prove these for homework.

Derivative Identities These identities follow directly from the manipulation of the series solution.

$$\frac{d}{dx}\left[x^p J_p(x)\right] = x^p J_{p-1}(x). \tag{6.59}$$

$$\frac{d}{dx}\left[x^{-p} J_p(x)\right] = -x^{-p} J_{p+1}(x). \tag{6.60}$$

Recursion Formulae The next identities follow from adding, or subtracting, the derivative identities.

$$J_{p-1}(x) + J_{p+1}(x) = \frac{2p}{x} J_p(x). \tag{6.61}$$

$$J_{p-1}(x) - J_{p+1}(x) = 2J_p'(x). \tag{6.62}$$

Orthogonality As we will see in the next chapter, one can recast the Bessel equation into an eigenvalue problem whose solutions form an orthogonal basis of functions on $L_x^2(0,a)$. Using Sturm–Liouville Theory, one can show that

$$\int_0^a x J_p(j_{pn}\frac{x}{a}) J_p(j_{pm}\frac{x}{a})\, dx = \frac{a^2}{2}\left[J_{p+1}(j_{pn})\right]^2 \delta_{n,m}, \tag{6.63}$$

where j_{pn} is the nth root of $J_p(x)$, $J_p(j_{pn}) = 0$, $n = 1, 2, \ldots$. A list of some of these roots is provided in Table 6.3.

Generating Function

$$e^{x(t-\frac{1}{t})/2} = \sum_{n=-\infty}^{\infty} J_n(x)t^n, \quad x > 0, t \neq 0. \tag{6.64}$$

Table 6.3: The Zeros of Bessel Functions, $J_m(j_{mn}) = 0$

n	$m = 0$	$m = 1$	$m = 2$	$m = 3$	$m = 4$	$m = 5$
1	2.405	3.832	5.136	6.380	7.588	8.771
2	5.520	7.016	8.417	9.761	11.065	12.339
3	8.654	10.173	11.620	13.015	14.373	15.700
4	11.792	13.324	14.796	16.223	17.616	18.980
5	14.931	16.471	17.960	19.409	20.827	22.218
6	18.071	19.616	21.117	22.583	24.019	25.430
7	21.212	22.760	24.270	25.748	27.199	28.627
8	24.352	25.904	27.421	28.908	30.371	31.812
9	27.493	29.047	30.569	32.065	33.537	34.989

Integral Representation

$$J_n(x) = \frac{1}{\pi} \int_0^\pi \cos(x \sin\theta - n\theta)\, d\theta, \quad x > 0, n \in \mathbb{Z}. \tag{6.65}$$

Fourier–Bessel Series

Because the Bessel functions are an orthogonal set of functions of a Sturm–Liouville problem, we can expand square integrable functions on this basis. In fact, the Sturm–Liouville problem is given in the form

$$x^2 y'' + xy' + (\lambda x^2 - p^2)y = 0, \quad x \in [0, a], \tag{6.66}$$

satisfying the boundary conditions: $y(x)$ is bounded at $x = 0$ and $y(a) = 0$. The solutions are then of the form $J_p(\sqrt{\lambda}x)$, as can be shown by making the substitution $t = \sqrt{\lambda}x$ in the differential equation. Namely, we let $y(x) = u(t)$ and note that

$$\frac{dy}{dx} = \frac{dt}{dx}\frac{du}{dt} = \sqrt{\lambda}\frac{du}{dt}.$$

Then,

$$t^2 u'' + tu' + (t^2 - p^2)u = 0,$$

In the study of boundary value problems in differential equations, Sturm–Liouville problems are a bountiful source of basis functions for the space of square integrable functions, as will be seen in the next section.

which has a solution $u(t) = J_p(t)$.

Using Sturm–Liouville theory, one can show that $J_p(j_{pn}\frac{x}{a})$ is a basis of eigenfunctions and the resulting *Fourier–Bessel series expansion* of $f(x)$ defined on $x \in [0, a]$ is

$$f(x) = \sum_{n=1}^\infty c_n J_p(j_{pn}\frac{x}{a}), \tag{6.67}$$

where the Fourier–Bessel coefficients are found using the orthogonality relation as

$$c_n = \frac{2}{a^2 \left[J_{p+1}(j_{pn})\right]^2} \int_0^a x f(x) J_p(j_{pn}\frac{x}{a})\, dx. \tag{6.68}$$

Example 6.12. *Expand $f(x) = 1$ for $0 < x < 1$ in a Fourier–Bessel series of the form*

$$f(x) = \sum_{n=1}^\infty c_n J_0(j_{0n}x)$$

We need only compute the Fourier–Bessel coefficients in Equation (6.68):

$$c_n = \frac{2}{[J_1(j_{0n})]^2} \int_0^1 x J_0(j_{0n}x)\, dx.\qquad(6.69)$$

From the identity

$$\frac{d}{dx}\left[x^p J_p(x)\right] = x^p J_{p-1}(x),\qquad(6.70)$$

we have

$$
\begin{aligned}
\int_0^1 x J_0(j_{0n}x)\, dx &= \frac{1}{j_{0n}^2}\int_0^{j_{0n}} y J_0(y)\, dy \\
&= \frac{1}{j_{0n}^2}\int_0^{j_{0n}} \frac{d}{dy}\left[y J_1(y)\right] dy \\
&= \frac{1}{j_{0n}^2}\left[y J_1(y)\right]_0^{j_{0n}} \\
&= \frac{1}{j_{0n}} J_1(j_{0n}).\qquad(6.71)
\end{aligned}
$$

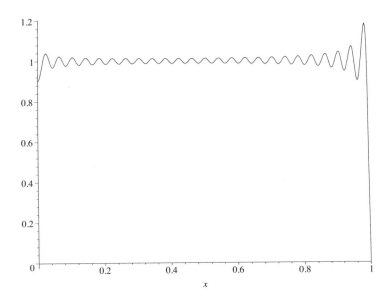

Figure 6.10: Plot of the first 50 terms of the Fourier–Bessel series in Equation (6.72) for $f(x) = 1$ on $0 < x < 1$.

As a result, the desired Fourier–Bessel expansion is given as

$$1 = 2\sum_{n=1}^\infty \frac{J_0(j_{0n}x)}{j_{0n}J_1(j_{0n})},\quad 0 < x < 1.\qquad(6.72)$$

In Figure 6.10, we show the partial sum for the first 50 terms of this series. Note once again the slow convergence due to the Gibbs phenomenon.

6.6 Sturm–Liouville Eigenvalue Problems

WE HAVE SEEN THAT TRIGONOMETRIC FUNCTIONS and special functions are the solutions of differential equations. These solutions give orthogonal sets of functions that can be used to represent functions in generalized Fourier series expansions. At the same time, we would like to generalize the techniques we had first used to solve the heat equation in order to solve more general initial-boundary value problems. Namely, we use separation of variables to separate the given partial differential equation into a set of ordinary differential equations. A subset of those equations provides us with a set of boundary value problems whose eigenfunctions are useful in representing solutions of the partial differential equation. Hopefully, those solutions will form a useful basis in some function space.

A class of problems to which our previous examples belong are the Sturm–Liouville eigenvalue problems. These problems involve self-adjoint (differential) operators, which play an important role in the spectral theory of linear operators and the existence of the eigenfunctions needed to solve the interesting physics problems described by the above initial-boundary value problems. In this section we will introduce the Sturm–Liouville eigenvalue problem as a general class of boundary value problems containing the Legendre and Bessel equations and supplying the theory needed to solve a variety of problems.

6.6.1   Sturm–Liouville Operators

IN PHYSICS, MANY PROBLEMS ARISE IN THE FORM of boundary value problems involving second-order ordinary differential equations. For example, we will explore the wave equation and the heat equation in three dimensions. Separating out the time dependence leads to a three-dimensional boundary value problem in both cases. Further separation of variables leads to a set of boundary value problems involving second-order ordinary differential equations.

In general, we might obtain equations of the form

$$a_2(x)y'' + a_1(x)y' + a_0(x)y = f(x) \tag{6.73}$$

subject to boundary conditions. We can write such an equation in operator form by defining the differential operator

$$L = a_2(x)D^2 + a_1(x)D + a_0(x),$$

where $D = d/dx$. Then, Equation (6.73) takes the form

$$Ly = f.$$

Recall that we had solved such nonhomogeneous differential equations in Chapter 2. In this section we will show that these equations can be solved using eigenfunction expansions. Namely, we seek solutions to the eigenvalue problem

$$L\phi = \lambda\phi$$

with homogeneous boundary conditions on ϕ and then seek a solution of the nonhomogeneous problem, $Ly = f$, as an expansion over these eigenfunctions. Formally, we let

$$y(x) = \sum_{n=1}^{\infty} c_n \phi_n(x).$$

However, we are not guaranteed a nice set of eigenfunctions. We need an appropriate set to form a basis in the function space. Also, it would be nice to have orthogonality so that we can easily solve for the expansion coefficients.

It turns out that any linear second-order differential operator can be turned into an operator that possesses just the right properties (self-adjointedness) to carry out this procedure. The resulting operator is referred to as a Sturm–Liouville operator. We will highlight some of the properties of these operators and see how they are used in applications.

We define the Sturm–Liouville operator as

The Sturm–Liouville operator.

$$\mathcal{L} = \frac{d}{dx} p(x) \frac{d}{dx} + q(x). \tag{6.74}$$

The Sturm–Liouville eigenvalue problem is given by the differential equation

The Sturm–Liouville eigenvalue problem.

$$\mathcal{L}y = -\lambda \sigma(x) y,$$

or

$$\frac{d}{dx} \left(p(x) \frac{dy}{dx} \right) + q(x)y + \lambda \sigma(x) y = 0, \tag{6.75}$$

for $x \in (a, b)$, $y = y(x)$, plus boundary conditions. The functions $p(x)$, $p'(x)$, $q(x)$, and $\sigma(x)$ are assumed to be continuous on (a, b) and $p(x) > 0$, $\sigma(x) > 0$ on $[a, b]$. If the interval is finite and these assumptions on the coefficients are true on $[a, b]$, then the problem is said to be a regular Sturm–Liouville problem. Otherwise, it is called a singular Sturm–Liouville problem.

We also need to impose the set of homogeneous boundary conditions

Types of boundary conditions.

$$\alpha_1 y(a) + \beta_1 y'(a) = 0,$$
$$\alpha_2 y(b) + \beta_2 y'(b) = 0. \tag{6.76}$$

The α's and β's are constants. For different values, one has special types of boundary conditions. For $\beta_i = 0$, we have what are called Dirichlet boundary conditions. Namely, $y(a) = 0$ and $y(b) = 0$. For $\alpha_i = 0$, we have Neumann boundary conditions. In this case, $y'(a) = 0$ and $y'(b) = 0$. In terms of the heat equation example, Dirichlet conditions correspond to maintaining a fixed temperature at the ends of the rod. The Neumann boundary conditions would correspond to no heat flow across the ends, or insulating conditions, as there would be no temperature gradient at those points. The more general boundary conditions allow for partially insulated boundaries.

Dirichlet boundary conditions—the solution takes fixed values on the boundary. These are named after Gustav Lejeune Dirichlet (1805–1859).

Neumann boundary conditions—the derivative of the solution takes fixed values on the boundary. These are named after Carl Neumann (1832–1925).

Another type of boundary condition that is often encountered is the periodic boundary condition. Consider the heated rod that has been bent to

Differential   equations   of   Sturm–
Liouville form.

form a circle. Then the two end points are physically the same. So, we would expect that the temperature and the temperature gradient should agree at those points. For this case, we write $y(a) = y(b)$ and $y'(a) = y'(b)$. Boundary value problems using these conditions must be handled differently than the above homogeneous conditions. These conditions lead to different types of eigenfunctions and eigenvalues.

As previously mentioned, equations of the form in Equation (6.73) occur often. We now show that any second-order linear operator can be put into the form of the Sturm–Liouville operator. In particular, Equation (6.73) can be put into the form

$$\frac{d}{dx}\left(p(x)\frac{dy}{dx}\right) + q(x)y = F(x). \tag{6.77}$$

The proof of this is straightforward as we soon show. Let's first consider Equation (6.73) for the case that $a_1(x) = a_2'(x)$. Then we can write the equation in a form in which the first two terms combine:

$$\begin{aligned} f(x) &= a_2(x)y'' + a_1(x)y' + a_0(x)y \\ &= (a_2(x)y')' + a_0(x)y. \end{aligned} \tag{6.78}$$

The resulting equation is now in Sturm–Liouville form. We just identify $p(x) = a_2(x)$ and $q(x) = a_0(x)$.

Not all second-order differential equations are as simple to convert. Consider the differential equation

$$x^2y'' + xy' + 2y = 0.$$

In this case, $a_2(x) = x^2$ and $a_2'(x) = 2x \neq a_1(x)$. So, this does not fall into this case. However, we can change the operator in this equation, $x^2D + xD$, to a Sturm–Liouville operator, $Dp(x)D$ for a $p(x)$ that depends on the coefficients x^2 and x.

In the Sturm–Liouville operator, the derivative terms are gathered together into one perfect derivative, $Dp(x)D$. This is similar to what we saw in the Chapter 2 when we solved linear first-order equations. In that case, we sought an integrating factor. We can do the same thing here. We seek a multiplicative function $\mu(x)$ that we can multiply through Equation (6.73) so that it can be written in Sturm–Liouville form.

We first divide out the $a_2(x)$, giving

$$y'' + \frac{a_1(x)}{a_2(x)}y' + \frac{a_0(x)}{a_2(x)}y = \frac{f(x)}{a_2(x)}.$$

Next, we multiply this differential equation by μ,

$$\mu(x)y'' + \mu(x)\frac{a_1(x)}{a_2(x)}y' + \mu(x)\frac{a_0(x)}{a_2(x)}y = \mu(x)\frac{f(x)}{a_2(x)}.$$

The first two terms can now be combined into an exact derivative $(\mu y')'$ if the second coefficient is $\mu'(x)$. Therefore, $\mu(x)$ satisfies a first-order, separable differential equation:

$$\frac{d\mu}{dx} = \mu(x)\frac{a_1(x)}{a_2(x)}.$$

This is formally solved to give the sought-after integrating factor

$$\mu(x) = e^{\int \frac{a_1(x)}{a_2(x)} dx}.$$

Thus, the original equation can be multiplied by the factor

$$\frac{\mu(x)}{a_2(x)} = \frac{1}{a_2(x)} e^{\int \frac{a_1(x)}{a_2(x)} dx}$$

to turn it into Sturm–Liouville form.

In summary,

Equation (6.73),

$$a_2(x)y'' + a_1(x)y' + a_0(x)y = f(x), \qquad (6.79)$$

can be put into the Sturm–Liouville form

$$\frac{d}{dx}\left(p(x)\frac{dy}{dx}\right) + q(x)y = F(x), \qquad (6.80)$$

where

$$
\begin{aligned}
p(x) &= e^{\int \frac{a_1(x)}{a_2(x)} dx}, \\
q(x) &= p(x)\frac{a_0(x)}{a_2(x)}, \\
F(x) &= p(x)\frac{f(x)}{a_2(x)}. \qquad (6.81)
\end{aligned}
$$

Conversion of a linear second-order differential equation to Sturm–Liouville form.

Example 6.13. *Convert $x^2y'' + xy' + 2y = 0$ to Sturm–Liouville form.*
We can multiply this equation by

$$\frac{\mu(x)}{a_2(x)} = \frac{1}{x^2} e^{\int \frac{dx}{x}} = \frac{1}{x}$$

to put the equation in Sturm–Liouville form:

$$
\begin{aligned}
0 &= xy'' + y' + \frac{2}{x}y, \\
&= (xy')' + \frac{2}{x}y. \qquad (6.82)
\end{aligned}
$$

6.6.2 Properties of Sturm–Liouville Eigenvalue Problems

THERE ARE SEVERAL PROPERTIES THAT CAN BE PROVEN for the (regular) Sturm–Liouville eigenvalue problem in Equation (6.75). However, we will not prove them all here. We will merely list some of the important facts and focus on a few of the properties.

Real, countable eigenvalues.

1. The eigenvalues are real, countable, ordered, and there is a smallest eigenvalue. Thus, we can write them as $\lambda_1 < \lambda_2 < \dots$. However, there is no largest eigenvalue and $n \to \infty$, $\lambda_n \to \infty$.

Oscillatory eigenfunctions.

2. For each eigenvalue λ_n, there exists an eigenfunction ϕ_n with $n-1$ zeros on (a, b).

3. Eigenfunctions corresponding to different eigenvalues are orthogonal with respect to the weight function, $\sigma(x)$. Defining the inner product of $f(x)$ and $g(x)$ as

$$\langle f, g \rangle >= \int_a^b f(x) g(x) \sigma(x)\, dx, \qquad (6.83)$$

Orthogonality of eigenfunctions.

the orthogonality of the eigenfunctions can be written in the form

$$\langle \phi_n, \phi_m \rangle = \langle \phi_n, \phi_n \rangle \delta_{nm}, \quad n, m = 1, 2, \ldots. \qquad (6.84)$$

4. The set of eigenfunctions is complete; that is, any piecewise smooth function can be represented by a generalized Fourier series expansion of the eigenfunctions

$$f(x) \sim \sum_{n=1}^{\infty} c_n \phi_n(x),$$

where

$$c_n = \frac{\langle f, \phi_n \rangle}{\langle \phi_n, \phi_n \rangle}.$$

Complete basis of eigenfunctions.

Actually, one needs $f(x) \in L_\sigma^2(a, b)$, the set of square integrable functions over $[a, b]$ with weight function $\sigma(x)$. By square integrable, we mean that $\langle f, f \rangle < \infty$. One can show that such a space is isomorphic to a Hilbert space, a complete inner product space. Hilbert spaces play a special role in quantum mechanics.

5. The eigenvalues satisfy the Rayleigh quotient

The Rayleigh quotient is named after Lord Rayleigh, John William Strutt, 3rd Baron Raleigh (1842–1919).

$$\lambda_n = \frac{-p\phi_n \frac{d\phi_n}{dx}\Big|_a^b + \int_a^b \left[p \left(\frac{d\phi_n}{dx} \right)^2 - q\phi_n^2 \right] dx}{\langle \phi_n, \phi_n \rangle}.$$

This is verified by multiplying the eigenvalue problem

$$\mathcal{L}\phi_n = -\lambda_n \sigma(x) \phi_n$$

by ϕ_n and integrating. Solving this result for λ_n, we obtain the Rayleigh quotient. The Rayleigh quotient is useful for getting estimates of eigenvalues and proving some of the other properties.

Example 6.14. *Verify some of these properties for the eigenvalue problem*

$$y'' = -\lambda y, \quad y(0) = y(\pi) = 0.$$

This is a problem we have seen many times. The eigenfunctions for this eigenvalue problem are $\phi_n(x) = \sin nx$, with eigenvalues $\lambda_n = n^2$ for $n = 1, 2, \ldots$. These satisfy the properties listed above.

First of all, the eigenvalues are real, countable, and ordered, $1 < 4 < 9 < \ldots$. There is no largest eigenvalue and there is a first one.

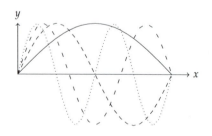

Figure 6.11: Plot of the eigenfunctions $\phi_n(x) = \sin nx$ for $n = 1, 2, 3, 4$.

The eigenfunctions corresponding to each eigenvalue have $n - 1$ zeros on $(0, \pi)$. This is demonstrated for several eigenfunctions in Figure 6.11.

We also know that the set $\{\sin nx\}_{n=1}^{\infty}$ is an orthogonal set of basis functions of length

$$\|\phi_n\| = \sqrt{\frac{\pi}{2}}.$$

Thus, the Rayleigh quotient can be computed using $p(x) = 1$, $q(x) = 0$, and the eigenfunctions. It is given by

$$
\begin{aligned}
R &= \frac{-\phi_n \phi_n' \Big|_0^{\pi} + \int_0^{\pi} (\phi_n')^2 \, dx}{\frac{\pi}{2}} \\
&= \frac{2}{\pi} \int_0^{\pi} \left(-n^2 \cos nx\right)^2 dx = n^2.
\end{aligned}
\tag{6.85}
$$

Therefore, knowing the eigenfunction, the Rayleigh quotient returns the eigenvalues as expected.

Example 6.15. *We seek the eigenfunctions of the operator found in Example 6.13. Namely, we want to solve the eigenvalue problem*

$$\mathcal{L}y = (xy')' + \frac{2}{x}y = -\lambda \sigma y \tag{6.86}$$

subject to a set of homogeneous boundary conditions. Let's use the boundary conditions

$$y'(1) = 0, \quad y'(2) = 0.$$

[Note that we do not know $\sigma(x)$ yet, but will choose an appropriate function to obtain solutions.]

Expanding the derivative, we have

$$xy'' + y' + \frac{2}{x}y = -\lambda \sigma y.$$

Multiply through by x to obtain

$$x^2 y'' + xy' + (2 + \lambda x \sigma) y = 0.$$

Notice that if we choose $\sigma(x) = x^{-1}$, then this equation can be made a Cauchy-Euler type equation. Thus, we have

$$x^2 y'' + xy' + (\lambda + 2) y = 0.$$

The characteristic equation is

$$r^2 + \lambda + 2 = 0.$$

For oscillatory solutions, we need $\lambda + 2 > 0$. Thus, the general solution is

$$y(x) = c_1 \cos(\sqrt{\lambda + 2} \ln |x|) + c_2 \sin(\sqrt{\lambda + 2} \ln |x|). \tag{6.87}$$

Next we apply the boundary conditions. $y'(1) = 0$ forces $c_2 = 0$. This leaves

$$y(x) = c_1 \cos(\sqrt{\lambda + 2} \ln x).$$

The second condition, $y'(2) = 0$, yields

$$\sin(\sqrt{\lambda + 2} \ln 2) = 0.$$

This will give nontrivial solutions when

$$\sqrt{\lambda + 2} \ln 2 = n\pi, \quad n = 0, 1, 2, 3 \dots.$$

In summary, the eigenfunctions for this eigenvalue problem are

$$y_n(x) = \cos\left(\frac{n\pi}{\ln 2} \ln x\right), \quad 1 \le x \le 2,$$

and the eigenvalues are $\lambda_n = \left(\frac{n\pi}{\ln 2}\right)^2 - 2$ for $n = 0, 1, 2, \dots$.

Note: We include the $n = 0$ case because $y(x) = $ constant is a solution of the $\lambda = -2$ case. More specifically, in this case the characteristic equation reduces to $r^2 = 0$. Thus, the general solution of this Cauchy-Euler equation is

$$y(x) = c_1 + c_2 \ln|x|.$$

Setting $y'(1) = 0$, forces $c_2 = 0$. $y'(2)$ automatically vanishes, leaving the solution in this case as $y(x) = c_1$.

We note that some of the properties listed in the beginning of the section hold for this example. The eigenvalues are seen to be real, countable, and ordered. There is a least one, $\lambda_0 = -2$. Next, one can find the zeros of each eigenfunction on $[1,2]$. Then the argument of the cosine, $\frac{n\pi}{\ln 2} \ln x$, takes values 0 to $n\pi$ for $x \in [1, 2]$. The cosine function has $n - 1$ roots on this interval.

Orthogonality can be checked as well. We set up the integral and use the substitution $y = \pi \ln x / \ln 2$. This gives

$$
\begin{aligned}
\langle y_n, y_m \rangle &= \int_1^2 \cos\left(\frac{n\pi}{\ln 2} \ln x\right) \cos\left(\frac{m\pi}{\ln 2} \ln x\right) \frac{dx}{x} \\
&= \frac{\ln 2}{\pi} \int_0^\pi \cos ny \cos my \, dy \\
&= \frac{\ln 2}{2} \delta_{n,m}.
\end{aligned}
\tag{6.88}
$$

6.6.3   Adjoint Operators

IN THE STUDY OF THE SPECTRAL THEORY OF MATRICES, one learns about the adjoint of the matrix, $A^\dagger$, and the role that self-adjoint, or Hermitian, matrices play in diagonalization. Also, one needs the concept of adjoint to discuss the existence of solutions to the matrix problem $\mathbf{y} = A\mathbf{x}$. In the same spirit, one is interested in the existence of solutions of the operator equation $Lu = f$ and solutions of the corresponding eigenvalue problem. The study of linear operators on a Hilbert space is a generalization of what the reader had seen in a linear algebra course.

Just as one can find a basis of eigenvectors and diagonalize Hermitian, or self-adjoint, matrices (or, real symmetric in the case of real matrices), we will see that the Sturm–Liouville operator is self-adjoint. In this section we will define the domain of an operator and introduce the notion of adjoint operators. In the last section we discuss the role the adjoint plays in the existence of solutions to the operator equation $Lu = f$.

We begin by defining the adjoint of an operator. The adjoint, $L^\dagger$, of operator L satisfies

$$\langle u, Lv \rangle = \langle L^\dagger u, v \rangle$$

for all v in the domain of L and u in the domain of $L^\dagger$. Here, the domain of a differential operator L is the set of all $u \in L^2_\sigma(a, b)$ satisfying a given set of homogeneous boundary conditions. This is best understood through example.

Example 6.16. *Find the adjoint of $L = a_2(x)D^2 + a_1(x)D + a_0(x)$ for $D = d/dx$.*

In order to find the adjoint, we place the operator inside an integral. Consider the inner product

$$\langle u, Lv \rangle = \int_a^b u(a_2 v'' + a_1 v' + a_0 v)\, dx.$$

We have to move the operator L from v and determine what operator is acting on u in order to formally preserve the inner product. For a simple operator like $L = \frac{d}{dx}$, this is easily done using integration by parts. For the given operator, we will need to apply several integrations by parts to the individual terms. We consider each derivative term in the integrand separately.

For the $a_1 v'$ term, we integrate by parts to find

$$\int_a^b u(x)a_1(x)v'(x)\, dx = a_1(x)u(x)v(x)\Big|_a^b - \int_a^b (u(x)a_1(x))'v(x)\, dx. \quad (6.89)$$

Now we consider the $a_2 v''$ term. In this case, it will take two integrations by parts:

$$
\begin{aligned}
\int_a^b u(x)a_2(x)v''(x)\, dx &= a_2(x)u(x)v'(x)\Big|_a^b - \int_a^b (u(x)a_2(x))'v(x)'\, dx \\
&= \left[a_2(x)u(x)v'(x) - (a_2(x)u(x))'v(x) \right]\Big|_a^b \\
&\quad + \int_a^b (u(x)a_2(x))''v(x)\, dx. \quad (6.90)
\end{aligned}
$$

Combining these results, we obtain

$$
\begin{aligned}
\langle u, Lv \rangle &= \int_a^b u(a_2 v'' + a_1 v' + a_0 v)\, dx \\
&= \left[a_1(x)u(x)v(x) + a_2(x)u(x)v'(x) - (a_2(x)u(x))'v(x) \right]\Big|_a^b \\
&\quad + \int_a^b \left[(a_2 u)'' - (a_1 u)' + a_0 u \right] v\, dx. \quad (6.91)
\end{aligned}
$$

Inserting the boundary conditions for v, one has to determine boundary conditions for u such that

$$\left[a_1(x)u(x)v(x) + a_2(x)u(x)v'(x) - (a_2(x)u(x))'v(x) \right]\Big|_a^b = 0.$$

This leaves

$$\langle u, Lv \rangle = \int_a^b \left[(a_2 u)'' - (a_1 u)' + a_0 u \right] v \, dx \equiv \langle L^\dagger u, v \rangle.$$

Therefore,

$$L^\dagger = \frac{d^2}{dx^2} a_2(x) - \frac{d}{dx} a_1(x) + a_0(x). \tag{6.92}$$

When $L^\dagger = L$, the operator is called formally self-adjoint. When the domain of L is the same as the domain of $L^\dagger$, the term self-adjoint is used. As the domain is important in establishing self-adjointness, we need to do a complete example in which the domain of the adjoint is found.

Example 6.17. *Determine $L^\dagger$ and its domain for operator $Lu = \frac{du}{dx}$, where u satisfies the boundary conditions $u(0) = 2u(1)$ on $[0, 1]$.*

We need to find the adjoint operator satisfying $\langle v, Lu \rangle = \langle L^\dagger v, u \rangle$. Therefore, we rewrite the integral

$$\langle v, Lu \rangle \ge= \int_0^1 v \frac{du}{dx} \, dx = uv \Big|_0^1 - \int_0^1 u \frac{dv}{dx} \, dx = \langle L^\dagger v, u \rangle.$$

From this we have the adjoint problem consisting of an adjoint operator and the associated boundary condition (or, domain of $L^\dagger$):

1. $L^\dagger = -\frac{d}{dx}$.

2. $uv \Big|_0^1 = 0 \Rightarrow 0 = u(1)[v(1) - 2v(0)] \Rightarrow v(1) = 2v(0)$.

6.6.4   Lagrange's and Green's Identities

BEFORE TURNING TO THE PROOFS that the eigenvalues of a Sturm–Liouville problem are real and the associated eigenfunctions orthogonal, we first need to introduce two important identities. For the Sturm–Liouville operator

$$\mathcal{L} = \frac{d}{dx} \left(p \frac{d}{dx} \right) + q,$$

we have the two identities:

Lagrange's identity:	$u\mathcal{L}v - v\mathcal{L}u$	$= [p(uv' - vu')]'.$
Green's identity:	$\int_a^b (u\mathcal{L}v - v\mathcal{L}u) \, dx$	$= [p(uv' - vu')]\Big|_a^b.$

The proof of Lagrange's identity follows by a simple manipulations of the operator:

$$\begin{aligned}
u\mathcal{L}v - v\mathcal{L}u &= u \left[\frac{d}{dx} \left(p \frac{dv}{dx} \right) + qv \right] - v \left[\frac{d}{dx} \left(p \frac{du}{dx} \right) + qu \right] \\
&= u \frac{d}{dx} \left(p \frac{dv}{dx} \right) - v \frac{d}{dx} \left(p \frac{du}{dx} \right) \\
&= u \frac{d}{dx} \left(p \frac{dv}{dx} \right) + p \frac{du}{dx} \frac{dv}{dx} - v \frac{d}{dx} \left(p \frac{du}{dx} \right) - p \frac{du}{dx} \frac{dv}{dx} \\
&= \frac{d}{dx} \left[pu \frac{dv}{dx} - pv \frac{du}{dx} \right]. \tag{6.93}
\end{aligned}$$

Green's identity is simply proven by integrating Lagrange's identity.

6.6.5   *Orthogonality and Reality*

WE ARE NOW READY TO PROVE that the eigenvalues of a Sturm–Liouville problem are real and the corresponding eigenfunctions are orthogonal. These are easily established using Green's identity, which in turn is a statement about the Sturm–Liouville operator being self-adjoint.

Example 6.18. *The eigenvalues of the Sturm–Liouville problem in Equation (6.75) are real.*

Let $\phi_n(x)$ be a solution of the eigenvalue problem associated with λ_n:

$$\mathcal{L}\phi_n = -\lambda_n \sigma \phi_n.$$

We want to show that $\overline{\lambda}_n = \lambda_n$, where the bar means complex conjugate. So, we also consider the complex conjugate of this equation

$$\mathcal{L}\overline{\phi}_n = -\overline{\lambda}_n \sigma \overline{\phi}_n.$$

Now, multiply the first equation by $\overline{\phi}_n$, the second equation by ϕ_n, and then subtract the results. We obtain

$$\overline{\phi}_n \mathcal{L}\phi_n - \phi_n \mathcal{L}\overline{\phi}_n = (\overline{\lambda}_n - \lambda_n)\sigma \phi_n \overline{\phi}_n.$$

Integrating both sides of this equation, we have

$$\int_a^b \left(\overline{\phi}_n \mathcal{L}\phi_n - \phi_n \mathcal{L}\overline{\phi}_n \right) dx = (\overline{\lambda}_n - \lambda_n) \int_a^b \sigma \phi_n \overline{\phi}_n \, dx.$$

We apply Green's identity to the left-hand side to find

$$[p(\overline{\phi}_n \phi'_n - \phi_n \overline{\phi}'_n)]|_a^b = (\overline{\lambda}_n - \lambda_n) \int_a^b \sigma \phi_n \overline{\phi}_n \, dx.$$

Using the homogeneous boundary conditions (6.76) for a self-adjoint operator, the left side vanishes. This leaves

$$0 = (\overline{\lambda}_n - \lambda_n) \int_a^b \sigma \|\phi_n\|^2 \, dx.$$

The integral is nonnegative, so we must have $\overline{\lambda}_n = \lambda_n$. Therefore, the eigenvalues are real.

Example 6.19. *The eigenfunctions corresponding to different eigenvalues of the Sturm–Liouville problem in Equation (6.75) are orthogonal.*

This is proven similar to the last example. Let $\phi_n(x)$ be a solution of the eigenvalue problem associated with λ_n,

$$\mathcal{L}\phi_n = -\lambda_n \sigma \phi_n,$$

and let $\phi_m(x)$ be a solution of the eigenvalue problem associated with $\lambda_m \neq \lambda_n$,

$$\mathcal{L}\phi_m = -\lambda_m \sigma \phi_m,$$

Now multiply the first equation by ϕ_m and the second equation by ϕ_n. Subtracting these results, we obtain

$$\phi_m \mathcal{L} \phi_n - \phi_n \mathcal{L} \phi_m = (\lambda_m - \lambda_n)\sigma \phi_n \phi_m.$$

Integrating both sides of the equation, using Green's identity, and using the homogeneous boundary conditions, we obtain

$$0 = (\lambda_m - \lambda_n) \int_a^b \sigma \phi_n \phi_m \, dx.$$

Because the eigenvalues are distinct, we can divide by $\lambda_m - \lambda_n$, leaving the desired result:

$$\int_a^b \sigma \phi_n \phi_m \, dx = 0.$$

Therefore, the eigenfunctions are orthogonal with respect to the weight function $\sigma(x)$.

6.6.6 Rayleigh Quotient

THE RAYLEIGH QUOTIENT IS USEFUL for obtaining estimates of eigenvalues and proving some of the other properties associated with Sturm–Liouville eigenvalue problems. The Rayleigh quotient is general and finds applications for both matrix eigenvalue problems as well as self-adjoint operators. For a Hermitian matrix M, the Rayleigh quotient is given by

$$R(\mathbf{v}) = \frac{\langle \mathbf{v}, M\mathbf{v} \rangle}{\langle \mathbf{v}, \mathbf{v} \rangle}.$$

One can show that the critical values of the Rayleigh quotient, as a function of $\mathbf{v}$, are the eigenvectors of M, and the values of R at these critical values are the corresponding eigenvectors. In particular, minimizing $R(\mathbf{v}$ over the vector space will give the lowest eigenvalue. This leads to the Rayleigh-Ritz method for computing the lowest eigenvalues when the eigenvectors are not known.

This definition can easily be extended to Sturm–Liouville operators,

$$R(\phi_n) = \frac{\langle \phi_n \mathcal{L} \phi_n \rangle}{\langle \phi_n, \phi_n \rangle}.$$

We begin by multiplying the eigenvalue problem

$$\mathcal{L}\phi_n = -\lambda_n \sigma(x) \phi_n$$

by ϕ_n and integrating. This gives

$$\int_a^b \left[\phi_n \frac{d}{dx} \left(p \frac{d\phi_n}{dx} \right) + q\phi_n^2 \right] dx = -\lambda_n \int_a^b \phi_n^2 \sigma \, dx.$$

One can solve the last equation for λ to find

$$\lambda_n = \frac{-\int_a^b \left[\phi_n \frac{d}{dx} \left(p \frac{d\phi_n}{dx} \right) + q\phi_n^2 \right] dx}{\int_a^b \phi_n^2 \sigma \, dx} = R(\phi_n).$$

It appears that we have solved for the eigenvalues and have not needed the machinery we developed in Chapter 4 for studying boundary value problems. However, we really cannot evaluate this expression when we do not know the eigenfunctions $\phi_n(x)$ yet. Nevertheless, we will see what we can determine from the Rayleigh quotient.

One can rewrite this result by performing an integration by parts on the first term in the numerator. Namely, pick $u = \phi_n$ and $dv = \frac{d}{dx}\left(p\frac{d\phi_n}{dx}\right)dx$ for the standard integration by parts formula. Then, we have

$$\int_a^b \phi_n \frac{d}{dx}\left(p\frac{d\phi_n}{dx}\right) dx = p\phi_n\frac{d\phi_n}{dx}\Big|_a^b - \int_a^b \left[p\left(\frac{d\phi_n}{dx}\right)^2 - q\phi_n^2\right] dx.$$

Inserting the new formula into the expression for λ leads to the Rayleigh quotient

$$\lambda_n = \frac{-p\phi_n\frac{d\phi_n}{dx}\Big|_a^b + \int_a^b \left[p\left(\frac{d\phi_n}{dx}\right)^2 - q\phi_n^2\right] dx}{\int_a^b \phi_n^2 \sigma\, dx}. \tag{6.94}$$

In many applications, the sign of the eigenvalue is important. As we have seen in the solution of the heat equation, $T' + k\lambda T = 0$. Because we expect the heat energy to diffuse, the solutions should decay in time. Thus, we would expect $\lambda > 0$. In studying the wave equation, one expects vibrations, and these are only possible with the correct sign of the eigenvalue (positive again). Thus, in order to have nonnegative eigenvalues, we see from Equation (6.94) that

    a. $q(x) \leq 0$, and

    b. $-p\phi_n\frac{d\phi_n}{dx}\Big|_a^b \geq 0$.

Furthermore, if λ is a zero eigenvalue, then $q(x) \equiv 0$ and $\alpha_1 = \alpha_2 = 0$ in the homogeneous boundary conditions. This can be seen by setting the numerator equal to zero. Then, $q(x) = 0$ and $\phi_n'(x) = 0$. The second of these conditions inserted into the boundary conditions forces the restriction on the type of boundary conditions.

One of the properties of Sturm–Liouville eigenvalue problems with homogeneous boundary conditions is that the eigenvalues are ordered, $\lambda_1 < \lambda_2 < \dots$. Thus, there is a smallest eigenvalue. It turns out that for any continuous function $y(x)$,

$$\lambda_1 = \min_{y(x)} \frac{-py\frac{dy}{dx}\Big|_a^b + \int_a^b \left[p\left(\frac{dy}{dx}\right)^2 - qy^2\right] dx}{\int_a^b y^2 \sigma\, dx}, \tag{6.95}$$

and this minimum is obtained when $y(x) = \phi_1(x)$. This result can be used to get estimates of the minimum eigenvalue using trial functions that are continuous and satisfy the boundary conditions, but do not necessarily satisfy the differential equation.

Example 6.20. *We have already solved the eigenvalue problem $\phi'' + \lambda\phi = 0$, $\phi(0) = 0$, $\phi(1) = 0$. In this case, the lowest eigenvalue is $\lambda_1 = \pi^2$. We can pick a nice function satisfying the boundary conditions, say $y(x) = x - x^2$. Inserting this into Equation (6.95), we find*

$$\lambda_1 \leq \frac{\int_0^1 (1 - 2x)^2 \, dx}{\int_0^1 (x - x^2)^2 \, dx} = 10.$$

Indeed, $10 \geq \pi^2$.

6.6.7   Eigenfunction Expansion Method

IN THIS SECTION WE SOLVE THE NONHOMOGENEOUS PROBLEM $\mathcal{L}y = f$ using expansions over the basis of Sturm–Liouville eigenfunctions. We have seen that Sturm–Liouville eigenvalue problems have the requisite set of orthogonal eigenfunctions. In this section we will apply the eigenfunction expansion method to solve a particular nonhomogeneous boundary value problem.

Recall that one starts with a nonhomogeneous differential equation

$$\mathcal{L}y = f,$$

where $y(x)$ is to satisfy given homogeneous boundary conditions. The method makes use of the eigenfunctions satisfying the eigenvalue problem

$$\mathcal{L}\phi_n = -\lambda_n \sigma \phi_n$$

subject to the given boundary conditions. Then one assumes that $y(x)$ can be written as an expansion in the eigenfunctions,

$$y(x) = \sum_{n=1}^{\infty} c_n \phi_n(x),$$

and inserts the expansion into the nonhomogeneous equation. This gives

$$f(x) = \mathcal{L}\left(\sum_{n=1}^{\infty} c_n \phi_n(x)\right) = -\sum_{n=1}^{\infty} c_n \lambda_n \sigma(x)\phi_n(x).$$

The expansion coefficients are then found by making use of the orthogonality of the eigenfunctions. Namely, we multiply the last equation by $\phi_m(x)$ and integrate. We obtain

$$\int_a^b f(x)\phi_m(x)\,dx = -\sum_{n=1}^{\infty} c_n \lambda_n \int_a^b \phi_n(x)\phi_m(x)\sigma(x)\,dx.$$

Orthogonality yields

$$\int_a^b f(x)\phi_m(x)\,dx = -c_m \lambda_m \int_a^b \phi_m^2(x)\sigma(x)\,dx.$$

Solving for c_m, we have

$$c_m = -\frac{\int_a^b f(x)\phi_m(x)\,dx}{\lambda_m \int_a^b \phi_m^2(x)\sigma(x)\,dx}.$$

Example 6.21. *As an example, we consider the solution of the boundary value problem*

$$(xy')' + \frac{y}{x} = \frac{1}{x}, \quad x \in [1, e], \tag{6.96}$$

$$y(1) = 0 = y(e). \tag{6.97}$$

This equation is already in self-adjoint form. So, we know that the associated Sturm–Liouville eigenvalue problem has an orthogonal set of eigenfunctions. We first determine this set. Namely, we need to solve

$$(x\phi')' + \frac{\phi}{x} = -\lambda\sigma\phi, \quad \phi(1) = 0 = \phi(e). \tag{6.98}$$

Rearranging the terms and multiplying by x, we have that

$$x^2\phi'' + x\phi' + (1 + \lambda\sigma x)\phi = 0.$$

This is almost an equation of Cauchy-Euler type. Picking the weight function $\sigma(x) = \frac{1}{x}$, we have

$$x^2\phi'' + x\phi' + (1 + \lambda)\phi = 0.$$

This is easily solved. The characteristic equation is

$$r^2 + (1 + \lambda) = 0.$$

One obtains nontrivial solutions of the eigenvalue problem satisfying the boundary conditions when $\lambda > -1$. The solutions are

$$\phi_n(x) = A\sin(n\pi \ln x), \quad n = 1, 2, \ldots,$$

where $\lambda_n = n^2\pi^2 - 1$.

It is often useful to normalize the eigenfunctions. This means that one chooses A so that the norm of each eigenfunction is one. Thus, we have

$$\begin{aligned}
1 &= \int_1^e \phi_n(x)^2 \sigma(x)\,dx \\
&= A^2 \int_1^e \sin(n\pi \ln x)\frac{1}{x}\,dx \\
&= A^2 \int_0^1 \sin(n\pi y)\,dy = \frac{1}{2}A^2. \tag{6.99}
\end{aligned}$$

Thus, $A = \sqrt{2}$. Several of these eigenfunctions are shown in Figure 6.12.

We now turn towards solving the nonhomogeneous problem, $\mathcal{L}y = \frac{1}{x}$. We first expand the unknown solution in terms of the eigenfunctions,

$$y(x) = \sum_{n=1}^{\infty} c_n \sqrt{2}\sin(n\pi \ln x).$$

Inserting this solution into the differential equation, we have

$$\frac{1}{x} = \mathcal{L}y = -\sum_{n=1}^{\infty} c_n \lambda_n \sqrt{2}\sin(n\pi \ln x)\frac{1}{x}.$$

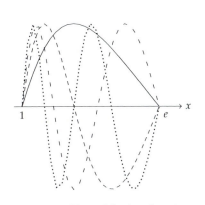

Figure 6.12: Plots of the first five eigenfunctions, $y(x) = \sqrt{2}\sin(n\pi \ln x)$.

Next, we make use of orthogonality. Multiplying both sides by $\phi_m(x) = \sqrt{2}\sin(m\pi \ln x)$ and integrating gives

$$\lambda_m c_m = \int_1^e \sqrt{2}\sin(m\pi \ln x)\frac{1}{x}\,dx = \frac{\sqrt{2}}{m\pi}[(-1)^m - 1].$$

Solving for c_m, we have

$$c_m = \frac{\sqrt{2}}{m\pi}\frac{[(-1)^m - 1]}{m^2\pi^2 - 1}.$$

Finally, we insert these coefficients into the expansion for $y(x)$. The solution is then

$$y(x) = \sum_{n=1}^{\infty} \frac{2}{n\pi}\frac{[(-1)^n - 1]}{n^2\pi^2 - 1}\sin(n\pi \ln(x)).$$

We plot this solution in Figure 6.13.

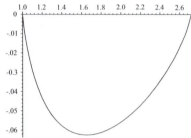

Figure 6.13: Plot of the solution in Example 6.21.

6.7 Nonhomogeneous Boundary Value Problems: Green's Functions

6.7.1 Boundary Value Green's Function

WE SOLVED NONHOMOGENEOUS INITIAL VALUE PROBLEMS in Section 2.7.4 using a Green's function. In this section we will extend this method to the solution of nonhomogeneous boundary value problems using a boundary value Green's function. Recall that the goal is to solve the nonhomogeneous differential equation

$$L[y] = f, \quad a \le x \le b,$$

where L is a differential operator and $y(x)$ satisfies boundary conditions at $x = a$ and $x = b$.. The solution is formally given by

$$y = L^{-1}[f].$$

The inverse of a differential operator is an integral operator, which we seek to write in the form

$$y(x) = \int_a^b G(x, \xi)f(\xi)\,d\xi.$$

The function $G(x, \xi)$ is referred to as the kernel of the integral operator and is called the *Green's function*.

We will consider boundary value problems in Sturm–Liouville form,

$$\frac{d}{dx}\left(p(x)\frac{dy(x)}{dx}\right) + q(x)y(x) = f(x), \quad a < x < b, \tag{6.100}$$

with fixed values of $y(x)$ at the boundary, $y(a) = 0$ and $y(b) = 0$. However, the general theory works for other forms of homogeneous boundary conditions.

We seek the Green's function by first solving the nonhomogeneous differential equation using the Method of Variation of Parameters. Recall this method from Section 2.7.3. We assume a particular solution of the form

$$y_p(x) = c_1(x)y_1(x) + c_2(x)y_2(x),$$

which is formed from two linearly independent solutions of the homogeneous problem, $y_i(x)$, $i = 1, 2$. We had found that the coefficient functions satisfy the equations

$$
\begin{aligned}
c_1'(x)y_1(x) + c_2'(x)y_2(x) &= 0, \\
c_1'(x)y_1'(x) + c_2'(x)y_2'(x) &= \frac{f(x)}{p(x)}.
\end{aligned}
\tag{6.101}
$$

Solving this system, we obtain

$$
c_1'(x) = -\frac{fy_2}{pW(y_1, y_2)},
$$

$$
c_1'(x) = \frac{fy_1}{pW(y_1, y_2)},
$$

where $W(y_1, y_2) = y_1 y_2' - y_1' y_2$ is the Wronskian. Integrating these forms and inserting the results back into the particular solution, we find

$$
y(x) = y_2(x) \int_{x_1}^{x} \frac{f(\xi)y_1(\xi)}{p(\xi)W(\xi)} \, d\xi - y_1(x) \int_{x_0}^{x} \frac{f(\xi)y_2(\xi)}{p(\xi)W(\xi)} \, d\xi,
$$

where x_0 and x_1 are to be determined using the boundary values. In particular, we will seek x_0 and x_1 so that the solution to the boundary value problem can be written as a single integral involving a Green's function. Note that we can absorb the solution to the homogeneous problem, $y_h(x)$, into the integrals with an appropriate choice of limits on the integrals.

We now look to satisfy the conditions $y(a) = 0$ and $y(b) = 0$. First we use solutions of the homogeneous differential equation that satisfy $y_1(a) = 0$, $y_2(b) = 0$ and $y_1(b) \neq 0$, $y_2(a) \neq 0$. Evaluating $y(x)$ at $x = 0$, we have

$$
\begin{aligned}
y(a) &= y_2(a) \int_{x_1}^{a} \frac{f(\xi)y_1(\xi)}{p(\xi)W(\xi)} \, d\xi - y_1(a) \int_{x_0}^{a} \frac{f(\xi)y_2(\xi)}{p(\xi)W(\xi)} \, d\xi \\
&= y_2(a) \int_{x_1}^{a} \frac{f(\xi)y_1(\xi)}{p(\xi)W(\xi)} \, d\xi.
\end{aligned}
\tag{6.102}
$$

We can satisfy the condition at $x = a$ if we choose $x_1 = a$.

Similarly, at $x = b$ we find that

$$
\begin{aligned}
y(b) &= y_2(b) \int_{x_1}^{b} \frac{f(\xi)y_1(\xi)}{p(\xi)W(\xi)} \, d\xi - y_1(b) \int_{x_0}^{b} \frac{f(\xi)y_2(\xi)}{p(\xi)W(\xi)} \, d\xi \\
&= -y_1(b) \int_{x_0}^{b} \frac{f(\xi)y_2(\xi)}{p(\xi)W(\xi)} \, d\xi.
\end{aligned}
\tag{6.103}
$$

This expression vanishes for $x_0 = b$.

So, we have found that the solution takes the form

The general solution of the boundary value problem.

$$
y(x) = y_2(x) \int_{a}^{x} \frac{f(\xi)y_1(\xi)}{p(\xi)W(\xi)} \, d\xi - y_1(x) \int_{b}^{x} \frac{f(\xi)y_2(\xi)}{p(\xi)W(\xi)} \, d\xi.
\tag{6.104}
$$

This solution can be written in a compact form just like we did for the initial value problem in Section 2.7.4. We seek a Green's function so that the solution can be written as a single integral. We can move the functions of

x under the integral. Also, because $a < x < b$, we can flip the limits in the second integral. This gives

$$y(x) = \int_a^x \frac{f(\xi)y_1(\xi)y_2(x)}{p(\xi)W(\xi)} \, d\xi + \int_x^b \frac{f(\xi)y_1(x)y_2(\xi)}{p(\xi)W(\xi)} \, d\xi. \tag{6.105}$$

This result can now be written in compact form:

Boundary Value Green's Function

The solution of the boundary value problem

$$\frac{d}{dx}\left(p(x)\frac{dy(x)}{dx}\right) + q(x)y(x) = f(x), \quad a < x < b,$$

$$y(a) = 0, \quad y(b) = 0, \tag{6.106}$$

takes the form

$$y(x) = \int_a^b G(x,\xi)f(\xi)\, d\xi, \tag{6.107}$$

where the Green's function is the piecewise defined function

$$G(x,\xi) = \begin{cases} \dfrac{y_1(\xi)y_2(x)}{pW}, & a \le \xi \le x, \\ \dfrac{y_1(x)y_2(\xi)}{pW}, & x \le \xi \le b, \end{cases} \tag{6.108}$$

where $y_1(x)$ and $y_2(x)$ are solutions of the homogeneous problem satisfying $y_1(a) = 0, y_2(b) = 0$ and $y_1(b) \ne 0, y_2(a) \ne 0$.

The Green's function satisfies several properties, which we will explore further in the next section. For example, the Green's function satisfies the boundary conditions at $x = a$ and $x = b$. Thus,

$$G(a,\xi) = \frac{y_1(a)y_2(\xi)}{pW} = 0,$$

$$G(b,\xi) = \frac{y_1(\xi)y_2(b)}{pW} = 0.$$

Also, the Green's function is symmetric in its arguments. Interchanging the arguments gives

$$G(\xi,x) = \begin{cases} \dfrac{y_1(x)y_2(\xi)}{pW}, & a \le x \le \xi, \\ \dfrac{y_1(\xi)y_2(x)}{pW}, & \xi \le x \le b. \end{cases} \tag{6.109}$$

But a careful look at the original form shows that

$$G(x,\xi) = G(\xi,x).$$

We will make use of these properties in the next section to quickly determine the Green's functions for other boundary value problems.

Example 6.22. *Solve the boundary value problem* $y'' = x^2$, $y(0) = 0 = y(1)$ *using the boundary value Green's function.*

We first solve the homogeneous equation, $y'' = 0$. After two integrations, we have $y(x) = Ax + B$, for A and B constants to be determined.

We need one solution satisfying $y_1(0) = 0$ Thus,

$$0 = y_1(0) = B.$$

So, we can pick $y_1(x) = x$, as A is arbitrary.

The other solution has to satisfy $y_2(1) = 0$. So,

$$0 = y_2(1) = A + B.$$

This can be solved for $B = -A$. Again, A is arbitrary and we will choose $A = -1$. Thus, $y_2(x) = 1 - x$.

For this problem, $p(x) = 1$. Thus, for $y_1(x) = x$ and $y_2(x) = 1 - x$,

$$p(x)W(x) = y_1(x)y_2'(x) - y_1'(x)y_2(x) = x(-1) - 1(1 - x) = -1.$$

Note that $p(x)W(x)$ is a constant, as it should be.

Now we construct the Green's function. We have

$$G(x, \xi) = \begin{cases} -\xi(1 - x), & 0 \le \xi \le x, \\ -x(1 - \xi), & x \le \xi \le 1. \end{cases} \tag{6.110}$$

Notice the symmetry between the two branches of the Green's function. Also, the Green's function satisfies homogeneous boundary conditions: $G(0, \xi) = 0$ from the lower branch, and $G(1, \xi) = 0$ from the upper branch.

Finally, we insert the Green's function into the integral form of the solution and evaluate the integral.

$$
\begin{aligned}
y(x) &= \int_0^1 G(x, \xi) f(\xi) \, d\xi \\
&= \int_0^1 G(x, \xi) \xi^2 \, d\xi \\
&= -\int_0^x \xi(1 - x)\xi^2 \, d\xi - \int_x^1 x(1 - \xi)\xi^2 \, d\xi \\
&= -(1 - x) \int_0^x \xi^3 \, d\xi - x \int_x^1 (\xi^2 - \xi^3) \, d\xi \\
&= -(1 - x) \left[\frac{\xi^4}{4} \right]_0^x - x \left[\frac{\xi^3}{3} - \frac{\xi^4}{4} \right]_x^1 \\
&= -\frac{1}{4}(1 - x)x^4 - \frac{1}{12}x(4 - 3) + \frac{1}{12}x(4x^3 - 3x^4) \\
&= \frac{1}{12}(x^4 - x). \tag{6.111}
\end{aligned}
$$

Checking the answer, we can easily verify that $y'' = x^2$, $y(0) = 0$, and $y(1) = 0$.

6.7.2 Properties of Green's Functions

WE HAVE NOTED SOME PROPERTIES OF GREEN'S FUNCTIONS in the previous section. In this section we will elaborate on some of these properties as a tool for quickly constructing Green's functions for boundary value problems. We list five basic properties:

1. **Differential Equation:**

   The boundary value Green's function satisfies the differential equation $\frac{\partial}{\partial x}\left(p(x)\frac{\partial G(x,\xi)}{\partial x}\right) + q(x)G(x,\xi) = 0, x \neq \xi$.

   This is easily established. For $x < \xi$, we are on the second branch and $G(x,\xi)$ is proportional to $y_1(x)$. Thus, because $y_1(x)$ is a solution of the homogeneous equation, then so is $G(x,\xi)$. For $x > \xi$, we are on the first branch and $G(x,\xi)$ is proportional to $y_2(x)$. So, once again, $G(x,\xi)$ is a solution of the homogeneous problem.

2. **Boundary Conditions:**

   In the example in the previous section, we had seen that $G(a,\xi) = 0$ and $G(b,\xi) = 0$. For example, for $x = a$, we are on the second branch and $G(x,\xi)$ is proportional to $y_1(x)$. Thus, whatever condition $y_1(x)$ satisfies, $G(x,\xi)$ will satisfy. A similar statement can be made for $x = b$.

3. **Symmetry or Reciprocity:** $G(x,\xi) = G(\xi,x)$

   We had shown this reciprocity property in the previous section.

4. **Continuity of G at $x = \xi$:** $G(\xi^+,\xi) = G(\xi^-,\xi)$

   Here we define $\xi^{\pm}$ through the limits of a function as x approaches ξ from above or below. In particular,

   $$G(\xi^+,x) = \lim_{x\downarrow\xi} G(x,\xi), \quad x > \xi,$$

   $$G(\xi^-,x) = \lim_{x\uparrow\xi} G(x,\xi), \quad x < \xi.$$

   Setting $x = \xi$ in both branches, we have

   $$\frac{y_1(\xi)y_2(\xi)}{pW} = \frac{y_1(\xi)y_2(\xi)}{pW}.$$

   Therefore, we have established the continuity of $G(x,\xi)$ between the two branches at $x = \xi$.

5. **Jump Discontinuity of $\frac{\partial G}{\partial x}$ at $x = \xi$:**

   $$\frac{\partial G(\xi^+,\xi)}{\partial x} - \frac{\partial G(\xi^-,\xi)}{\partial x} = \frac{1}{p(\xi)}$$

   This case is not as obvious. We first compute the derivatives by noting which branch is involved and then evaluate the derivatives and subtract them. Thus, we have

   $$
   \begin{aligned}
   \frac{\partial G(\xi^+,\xi)}{\partial x} - \frac{\partial G(\xi^-,\xi)}{\partial x} &= -\frac{1}{pW}y_1(\xi)y_2'(\xi) + \frac{1}{pW}y_1'(\xi)y_2(\xi) \\
   &= -\frac{y_1'(\xi)y_2(\xi) - y_1(\xi)y_2'(\xi)}{p(\xi)(y_1(\xi)y_2'(\xi) - y_1'(\xi)y_2(\xi))} \\
   &= \frac{1}{p(\xi)}.
   \end{aligned}
   \tag{6.112}
   $$

Here is a summary of the properties of the boundary value Green's function based upon the previous solution:

Properties of the Green's Function

---

1. **Differential Equation:**

   $\frac{\partial}{\partial x}\left(p(x)\frac{\partial G(x,\xi)}{\partial x}\right) + q(x)G(x,\xi) = 0, x \neq \xi$

2. **Boundary Conditions:** Whatever conditions $y_1(x)$ and $y_2(x)$ satisfy, $G(x,\xi)$ will satisfy.

3. **Symmetry or Reciprocity:** $G(x,\xi) = G(\xi,x)$

4. **Continuity of G at $x = \xi$:** $G(\xi^+,\xi) = G(\xi^-,\xi)$

5. **Jump Discontinuity of $\frac{\partial G}{\partial x}$ at $x = \xi$:**

$$\frac{\partial G(\xi^+,\xi)}{\partial x} - \frac{\partial G(\xi^-,\xi)}{\partial x} = \frac{1}{p(\xi)}$$

We now show how a knowledge of these properties allows one to quickly construct a Green's function with an example.

Example 6.23. *Construct the Green's function for the problem*

$$y'' + \omega^2 y = f(x), \quad 0 < x < 1,$$

$$y(0) = 0 = y(1),$$

with $\omega \neq 0$.

I. **Find solutions to the homogeneous equation.**

A general solution to the homogeneous equation is given as

$$y_h(x) = c_1 \sin \omega x + c_2 \cos \omega x.$$

Thus, for $x \neq \xi$,

$$G(x,\xi) = c_1(\xi) \sin \omega x + c_2(\xi) \cos \omega x.$$

II. **Boundary Conditions.**

First, we have $G(0,\xi) = 0$ for $0 \leq x \leq \xi$. So,

$$G(0,\xi) = c_2(\xi) \cos \omega x = 0.$$

So,

$$G(x,\xi) = c_1(\xi) \sin \omega x, \quad 0 \leq x \leq \xi.$$

Second, we have $G(1,\xi) = 0$ for $\xi \leq x \leq 1$. So,

$$G(1,\xi) = c_1(\xi) \sin \omega + c_2(\xi) \cos \omega. = 0$$

A solution can be chosen with

$$c_2(\xi) = -c_1(\xi) \tan \omega.$$

This gives

$$G(x, \xi) = c_1(\xi) \sin \omega x - c_1(\xi) \tan \omega \cos \omega x.$$

This can be simplified by factoring out the $c_1(\xi)$ and placing the remaining terms over a common denominator. The result is

$$
\begin{aligned}
G(x, \xi) &= \frac{c_1(\xi)}{\cos \omega} \left[\sin \omega x \cos \omega - \sin \omega \cos \omega x \right] \\
&= -\frac{c_1(\xi)}{\cos \omega} \sin \omega (1 - x).
\end{aligned}
$$
(6.113)

Because the coefficient is arbitrary at this point, as can write the result as

$$G(x, \xi) = d_1(\xi) \sin \omega (1 - x), \quad \xi \leq x \leq 1.$$

We note that we could have started with $y_2(x) = \sin \omega (1 - x)$ as one of the linearly independent solutions of the homogeneous problem in anticipation that $y_2(x)$ satisfies the second boundary condition.

III. Symmetry or Reciprocity

We now impose that $G(x, \xi) = G(\xi, x)$. Up to this point we have that

$$
G(x, \xi) = \begin{cases}
c_1(\xi) \sin \omega x, & 0 \leq x \leq \xi, \\
d_1(\xi) \sin \omega (1 - x), & \xi \leq x \leq 1.
\end{cases}
$$

We can make the branches symmetric by picking the right forms for $c_1(\xi)$ and $d_1(\xi)$. We choose $c_1(\xi) = C \sin \omega (1 - \xi)$ and $d_1(\xi) = C \sin \omega \xi$. Then,

$$
G(x, \xi) = \begin{cases}
C \sin \omega (1 - \xi) \sin \omega x, & 0 \leq x \leq \xi, \\
C \sin \omega (1 - x) \sin \omega \xi, & \xi \leq x \leq 1.
\end{cases}
$$

Now the Green's function is symmetric and we still have to determine the constant C. We note that we could have gotten to this point using the Method of Variation of Parameters result where $C = \frac{1}{pW}$.

IV. Continuity of $G(x, \xi)$

We already have continuity by virtue of the symmetry imposed in the last step.

V. Jump Discontinuity in $\frac{\partial}{\partial x} G(x, \xi)$.

We still need to determine C. We can do this using the jump discontinuity in the derivative:

$$\frac{\partial G(\xi^+, \xi)}{\partial x} - \frac{\partial G(\xi^-, \xi)}{\partial x} = \frac{1}{p(\xi)}.$$

For this problem, $p(x) = 1$. Inserting the Green's function, we have

$$
\begin{aligned}
1 &= \frac{\partial G(\xi^+, \xi)}{\partial x} - \frac{\partial G(\xi^-, \xi)}{\partial x} \\
&= \frac{\partial}{\partial x} [C \sin \omega (1 - x) \sin \omega \xi]_{x=\xi} - \frac{\partial}{\partial x} [C \sin \omega (1 - \xi) \sin \omega x]_{x=\xi} \\
&= -\omega C \cos \omega (1 - \xi) \sin \omega \xi - \omega C \sin \omega (1 - \xi) \cos \omega \xi \\
&= -\omega C \sin \omega (\xi + 1 - \xi) \\
&= -\omega C \sin \omega.
\end{aligned}
$$
(6.114)

Therefore,

$$C = -\frac{1}{\omega \sin \omega}.$$

Finally, we have the Green's function:

$$G(x, \xi) = \begin{cases} -\dfrac{\sin \omega(1 - \xi) \sin \omega x}{\omega \sin \omega}, & 0 \le x \le \xi, \\[2mm] -\dfrac{\sin \omega(1 - x) \sin \omega \xi}{\omega \sin \omega}, & \xi \le x \le 1. \end{cases} \tag{6.115}$$

It is instructive to compare this result to the Variation of Parameters result.

Example 6.24. *Use the Method of Variation of Parameters to solve*

$$y'' + \omega^2 y = f(x), \quad 0 < x < 1,$$

$$y(0) = 0 = y(1), \quad \omega \neq 0.$$

We have the functions $y_1(x) = \sin \omega x$ and $y_2(x) = \sin \omega(1 - x)$ as the solutions of the homogeneous equation satisfying $y_1(0) = 0$ and $y_2(1) = 0$. We need to compute pW:

$$\begin{aligned} p(x)W(x) &= y_1(x)y_2'(x) - y_1'(x)y_2(x) \\ &= -\omega \sin \omega x \cos \omega(1 - x) - \omega \cos \omega x \sin \omega(1 - x) \\ &= -\omega \sin \omega \end{aligned} \tag{6.116}$$

Inserting this result into the Variation of Parameters result for the Green's function leads to the same Green's function as above.

6.7.3 Differential Equation for the Green's Function

As we progress in the book, we will develop a more general theory of Green's functions for ordinary and partial differential equations. Much of this theory relies on understanding that the Green's function really is the system response function to a point source. This begins with recalling that the boundary value Green's function satisfies a homogeneous differential equation for $x \neq \xi$,

$$\frac{\partial}{\partial x}\left(p(x)\frac{\partial G(x, \xi)}{\partial x} \right) + q(x)G(x, \xi) = 0, \quad x \neq \xi. \tag{6.117}$$

For $x = \xi$, we have seen that the derivative has a jump in its value. This is similar to the step, or Heaviside, function,

$$H(x) = \begin{cases} 1, & x > 0, \\ 0, & x < 0. \end{cases}$$

This function is shown in Figure 6.14 and we see that the derivative of the step function is zero everywhere except at the jump, or discontinuity. At the jump, there is an infinite slope, although technically we have learned that

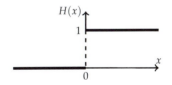

Figure 6.14: The Heaviside step function, $H(x)$.

there is no derivative at this point. We will try to remedy this situation by introducing the Dirac delta function,

$$\delta(x) = \frac{d}{dx}H(x).$$

We will show that the Green's function satisfies the differential equation

$$\frac{\partial}{\partial x}\left(p(x)\frac{\partial G(x,\xi)}{\partial x}\right) + q(x)G(x,\xi) = \delta(x-\xi). \tag{6.118}$$

However, we will first indicate why this knowledge is useful for the general theory of solving differential equations using Green's functions.

As noted, the Green's function satisfies the differential equation

The Dirac delta function is described in more detail in Section 8.4. The key property we will need here is the sifting property,

$$\int_a^b f(x)\delta(x-\xi)\,dx = f(\xi)$$

for $a < \xi < b$.

$$\frac{\partial}{\partial x}\left(p(x)\frac{\partial G(x,\xi)}{\partial x}\right) + q(x)G(x,\xi) = \delta(x-\xi) \tag{6.119}$$

and satisfies homogeneous conditions. We will use the Green's function to solve the nonhomogeneous equation

$$\frac{d}{dx}\left(p(x)\frac{dy(x)}{dx}\right) + q(x)y(x) = f(x). \tag{6.120}$$

These equations can be written in the more compact forms

$$\mathcal{L}[y] = f(x),$$
$$\mathcal{L}[G] = \delta(x-\xi). \tag{6.121}$$

Using these equations, we can determine the solution, $y(x)$, in terms of the Green's function. Multiplying the first equation by $G(x,\xi)$, the second equation by $y(x)$, and then subtracting, we have

$$G\mathcal{L}[y] - y\mathcal{L}[G] = f(x)G(x,\xi) - \delta(x-\xi)y(x).$$

Now integrate both sides from $x = a$ to $x = b$. The left-hand side becomes

$$\int_a^b [f(x)G(x,\xi) - \delta(x-\xi)y(x)]\,dx = \int_a^b f(x)G(x,\xi)\,dx - y(\xi).$$

Recall that Green's identity is given by

$$\int_a^b (u\mathcal{L}v - v\mathcal{L}u)\,dx = [p(uv' - vu')]\big|_a^b.$$

Using Green's Identity from Section 6.6.4, the right side is

$$\int_a^b (G\mathcal{L}[y] - y\mathcal{L}[G])\,dx = \left[p(x)\left(G(x,\xi)y'(x) - y(x)\frac{\partial G}{\partial x}(x,\xi)\right)\right]_{x=a}^{x=b}.$$

The general solution in terms of the boundary value Green's function with corresponding surface terms.

Combining these results and rearranging, we obtain

$$\boxed{\begin{aligned} y(\xi) &= \int_a^b f(x)G(x,\xi)\,dx \\ &\quad - \left[p(x)\left(y(x)\frac{\partial G}{\partial x}(x,\xi) - G(x,\xi)y'(x)\right)\right]_{x=a}^{x=b}. \end{aligned}} \tag{6.122}$$

We will refer to the extra terms in the solution,

$$S(b,\xi) - S(a,\xi) = \left[p(x) \left(y(x) \frac{\partial G}{\partial x}(x,\xi) - G(x,\xi)y'(x) \right) \right]_{x=a}^{x=b},$$

as the boundary, or surface, terms. Thus,

$$y(\xi) = \int_a^b f(x)G(x,\xi)\,dx - [S(b,\xi) - S(a,\xi)].$$

The result in Equation (6.122) is the key equation in determining the solution of a nonhomogeneous boundary value problem. The particular set of boundary conditions in the problem will dictate what conditions $G(x,\xi)$ must satisfy. For example, if we have the boundary conditions $y(a) = 0$ and $y(b) = 0$, then the boundary terms yield

$$
\begin{aligned}
y(\xi) &= \int_a^b f(x)G(x,\xi)\,dx - \left[p(b) \left(y(b)\frac{\partial G}{\partial x}(b,\xi) - G(b,\xi)y'(b) \right) \right] \\
&\quad + \left[p(a) \left(y(a)\frac{\partial G}{\partial x}(a,\xi) - G(a,\xi)y'(a) \right) \right] \\
&= \int_a^b f(x)G(x,\xi)\,dx + p(b)G(b,\xi)y'(b) - p(a)G(a,\xi)y'(a).
\end{aligned}
$$

$$(6.123)$$

The right-hand side will only vanish if $G(x,\xi)$ also satisfies these homogeneous boundary conditions. This then leaves us with the solution

$$y(\xi) = \int_a^b f(x)G(x,\xi)\,dx.$$

We should rewrite this as a function of x. So, we replace ξ with x and x with ξ. This gives

$$y(x) = \int_a^b f(\xi)G(\xi,x)\,d\xi.$$

However, this is not yet in the desirable form. The arguments of the Green's function are reversed. But, in this case, $G(x,\xi)$ is symmetric in its arguments. So, we can simply switch the arguments to obtain the desired result.

We can now see that the theory works for other boundary conditions. If we had $y'(a) = 0$, then the $y(a)\frac{\partial G}{\partial x}(a,\xi)$ term in the boundary terms could be made to vanish if we set $\frac{\partial G}{\partial x}(a,\xi) = 0$. So, this confirms that other boundary value problems can be posed in addition to the one elaborated upon in this chapter so far.

We can even adapt this theory to nonhomogeneous boundary conditions. We first rewrite Equation (6.122) as

$$y(x) = \int_a^b G(x,\xi)f(\xi)\,d\xi - \left[p(\xi) \left(y(\xi)\frac{\partial G}{\partial \xi}(x,\xi) - G(x,\xi)y'(\xi) \right) \right]_{\xi=a}^{\xi=b}.$$

$$(6.124)$$

Let's consider the boundary conditions $y(a) = \alpha$ and $y'(b) = \beta$. We also assume that $G(x,\xi)$ satisfies homogeneous boundary conditions

$$G(a,\xi) = 0, \quad \frac{\partial G}{\partial \xi}(b,\xi) = 0.$$

in both x and ξ because the Green's function is symmetric in its variables. Then we need only focus on the boundary terms to examine the effect on the solution. We have

$$
\begin{aligned}
S(b,x) - S(a,x) &= \left[p(b) \left(y(b) \frac{\partial G}{\partial \xi}(x,b) - G(x,b)y'(b) \right) \right] \\
&\quad - \left[p(a) \left(y(a) \frac{\partial G}{\partial \xi}(x,a) - G(x,a)y'(a) \right) \right] \\
&= -\beta p(b) G(x,b) - \alpha p(a) \frac{\partial G}{\partial \xi}(x,a). \quad (6.125)
\end{aligned}
$$

General solution satisfying the nonhomogeneous boundary conditions $y(a) = \alpha$ and $y'(b) = \beta$. Here, the Green's function satisfies homogeneous boundary conditions, $G(a,\xi) = 0$, $\frac{\partial G}{\partial \xi}(b,\xi) = 0$.

Therefore, we have the solution

$$
y(x) = \int_a^b G(x,\xi) f(\xi)\, d\xi + \beta p(b) G(x,b) + \alpha p(a) \frac{\partial G}{\partial \xi}(x,a). \quad (6.126)
$$

This solution satisfies the nonhomogeneous boundary conditions.

Example 6.25. *Solve* $y'' = x^2$, $y(0) = 1, y(1) = 2$ *using the boundary value Green's function.*

This is a modification of Example 6.22. We can use the boundary value Green's function that we found in that example:

$$
G(x,\xi) = \begin{cases} -\xi(1-x), & 0 \le \xi \le x, \\ -x(1-\xi), & x \le \xi \le 1. \end{cases} \quad (6.127)
$$

We insert the Green's function into the general solution of Equation (6.126) and use the given boundary conditions to obtain

$$
\begin{aligned}
y(x) &= \int_0^1 G(x,\xi)\xi^2\, d\xi - \left[y(\xi) \frac{\partial G}{\partial \xi}(x,\xi) - G(x,\xi)y'(\xi) \right]_{\xi=0}^{\xi=1} \\
&= \int_0^x (x-1)\xi^3\, d\xi + \int_x^1 x(\xi-1)\xi^2\, d\xi + y(0)\frac{\partial G}{\partial \xi}(x,0) - y(1)\frac{\partial G}{\partial \xi}(x,1) \\
&= \frac{(x-1)x^4}{4} + \frac{x(1-x^4)}{4} - \frac{x(1-x^3)}{3} + (x-1) - 2x \\
&= \frac{x^4}{12} + \frac{35}{12}x - 1. \quad (6.128)
\end{aligned}
$$

Of course, this problem can be solved by direct integration. The general solution is

$$
y(x) = \frac{x^4}{12} + c_1 x + c_2.
$$

Inserting this solution into each boundary condition yields the same result.

The Green's function satisfies a delta function forced differential equation.

We have seen how the introduction of the Dirac delta function in the differential equation satisfied by the Green's function, Equation (6.119), can lead to the solution of boundary value problems. The Dirac delta function also aids in the interpretation of the Green's function. We note that the Green's function is a solution of an equation in which the nonhomogeneous function is $\delta(x-\xi)$. Note that if we multiply the delta function by $f(\xi)$ and integrate, we obtain

$$
\int_{-\infty}^{\infty} \delta(x-\xi) f(\xi)\, d\xi = f(x).
$$

We can view the delta function as a unit impulse at $x = \xi$ that can be used to build $f(x)$ as a sum of impulses of different strengths, $f(\xi)$. Thus, the Green's function is the response to the impulse as governed by the differential equation and given boundary conditions.

Derivation of the jump condition for the Green's function.

In particular, the delta function forced equation can be used to derive the jump condition. We begin with the equation in the form

$$\frac{\partial}{\partial x}\left(p(x)\frac{\partial G(x,\xi)}{\partial x}\right) + q(x)G(x,\xi) = \delta(x - \xi). \tag{6.129}$$

Now integrate both sides from $\xi - \epsilon$ to $\xi + \epsilon$ and take the limit as $\epsilon \to 0$. Then,

$$\lim_{\epsilon\to 0}\int_{\xi-\epsilon}^{\xi+\epsilon}\left[\frac{\partial}{\partial x}\left(p(x)\frac{\partial G(x,\xi)}{\partial x}\right) + q(x)G(x,\xi)\right]dx = \lim_{\epsilon\to 0}\int_{\xi-\epsilon}^{\xi+\epsilon}\delta(x - \xi)\,dx$$
$$= 1. \tag{6.130}$$

Because the $q(x)$ term is continuous, the limit as $\epsilon \to 0$ of that term vanishes. Using the Fundamental Theorem of Calculus, we then have

$$\lim_{\epsilon\to 0}\left[p(x)\frac{\partial G(x,\xi)}{\partial x}\right]_{\xi-\epsilon}^{\xi+\epsilon} = 1. \tag{6.131}$$

This is the jump condition that we have been using!

6.7.4 Series Representations of Green's Functions

THERE ARE TIMES THAT IT MIGHT NOT BE SO SIMPLE to find the Green's function in the simple closed form as we have seen so far. However, there is a method for determining the Green's functions of Sturm–Liouville boundary value problems in the form of an eigenfunction expansion. We will finish our discussion of Green's functions for ordinary differential equations by showing how one obtains such series representations. (Note that we are really just repeating the steps toward developing eigenfunction expansion that we had seen in Section 6.6.7.)

We will make use of the complete set of eigenfunctions of the differential operator, $\mathcal{L}$, satisfying the homogeneous boundary conditions:

$$\mathcal{L}[\phi_n] = -\lambda_n \sigma \phi_n, \quad n = 1, 2, \ldots.$$

We want to find the particular solution y satisfying $\mathcal{L}[y] = f$ and homogeneous boundary conditions. We assume that

$$y(x) = \sum_{n=1}^{\infty} a_n \phi_n(x).$$

Inserting this into the differential equation, we obtain

$$\mathcal{L}[y] = \sum_{n=1}^{\infty} a_n \mathcal{L}[\phi_n] = -\sum_{n=1}^{\infty} \lambda_n a_n \sigma \phi_n = f.$$

This has resulted in the generalized Fourier expansion

$$f(x) = \sum_{n=1}^{\infty} c_n \sigma \phi_n(x)$$

with coefficients

$$c_n = -\lambda_n a_n.$$

We have seen how to compute these coefficients earlier in the chapter. We multiply both sides by $\phi_k(x)$ and integrate. Using the orthogonality of the eigenfunctions,

$$\int_a^b \phi_n(x)\phi_k(x)\sigma(x)\,dx = N_k \delta_{nk},$$

one obtains the expansion coefficients (if $\lambda_k \neq 0$)

$$a_k = -\frac{(f, \phi_k)}{N_k \lambda_k},$$

where $(f, \phi_k) \equiv \int_a^b f(x)\phi_k(x)\,dx$.

As before, we can rearrange the solution to obtain the Green's function. Namely, we have

$$y(x) = \sum_{n=1}^{\infty} \frac{(f, \phi_n)}{-N_n \lambda_n} \phi_n(x) = \int_a^b \underbrace{\sum_{n=1}^{\infty} \frac{\phi_n(x)\phi_n(\xi)}{-N_n \lambda_n}}_{G(x,\xi)} f(\xi)\,d\xi.$$

Therefore, we have found the Green's function as an expansion in the eigenfunctions:

Green's function as an expansion in the eigenfunctions.

$$G(x, \xi) = \sum_{n=1}^{\infty} \frac{\phi_n(x)\phi_n(\xi)}{-\lambda_n N_n}. \tag{6.132}$$

We will conclude this discussion with an example. We will solve this problem three different ways in order to summarize the methods we have used in the text.

Example 6.26. *Solve*

$$y'' + 4y = x^2, \quad x \in (0,1), \quad y(0) = y(1) = 0$$

Example using the Green's function eigenfunction expansion.

using the Green's function eigenfunction expansion.

The Green's function for this problem can be constructed fairly quickly for this problem once the eigenvalue problem is solved. The eigenvalue problem is

$$\phi''(x) + 4\phi(x) = -\lambda\phi(x),$$

where $\phi(0) = 0$ and $\phi(1) = 0$. The general solution is obtained by rewriting the equation as

$$\phi''(x) + k^2\phi(x) = 0,$$

where

$$k^2 = 4 + \lambda.$$

Solutions satisfying the boundary condition at $x = 0$ are of the form

$$\phi(x) = A \sin kx.$$

Forcing $\phi(1) = 0$ gives

$$0 = A \sin k \Rightarrow k = n\pi, \quad k = 1, 2, 3 \ldots.$$

So, the eigenvalues are

$$\lambda_n = n^2 \pi^2 - 4, \quad n = 1, 2, \ldots$$

and the eigenfunctions are

$$\phi_n = \sin n\pi x, \quad n = 1, 2, \ldots.$$

We also need the normalization constant, N_n. We have that

$$N_n = \|\phi_n\|^2 = \int_0^1 \sin^2 n\pi x = \frac{1}{2}.$$

We can now construct the Green's function for this problem using Equation (6.132):

$$G(x, \xi) = 2 \sum_{n=1}^{\infty} \frac{\sin n\pi x \sin n\pi \xi}{(4 - n^2 \pi^2)}. \tag{6.133}$$

Using this Green's function, the solution of the boundary value problem becomes

$$
\begin{aligned}
y(x) &= \int_0^1 G(x, \xi) f(\xi) \, d\xi \\
&= \int_0^1 \left(2 \sum_{n=1}^{\infty} \frac{\sin n\pi x \sin n\pi \xi}{(4 - n^2 \pi^2)} \right) \xi^2 \, d\xi \\
&= 2 \sum_{n=1}^{\infty} \frac{\sin n\pi x}{(4 - n^2 \pi^2)} \int_0^1 \xi^2 \sin n\pi \xi \, d\xi \\
&= 2 \sum_{n=1}^{\infty} \frac{\sin n\pi x}{(4 - n^2 \pi^2)} \left[\frac{(2 - n^2 \pi^2)(-1)^n - 2}{n^3 \pi^3} \right].
\end{aligned}
\tag{6.134}
$$

We can compare this solution to the one we would obtain if we did not employ Green's functions directly. The eigenfunction expansion method for solving boundary value problems, which we saw earlier, is demonstrated in the next example.

Example 6.27. *Solve*

$$y'' + 4y = x^2, \quad x \in (0, 1), \quad y(0) = y(1) = 0$$

using the eigenfunction expansion method.

We assume that the solution of this problem is in the form

$$y(x) = \sum_{n=1}^{\infty} c_n \phi_n(x).$$

Example using the eigenfunction expansion method.

Inserting this solution into the differential equation $\mathcal{L}[y] = x^2$ gives

$$
\begin{aligned}
x^2 &= \mathcal{L}\left[\sum_{n=1}^{\infty} c_n \sin n\pi x\right] \\
&= \sum_{n=1}^{\infty} c_n \left[\frac{d^2}{dx^2} \sin n\pi x + 4 \sin n\pi x\right] \\
&= \sum_{n=1}^{\infty} c_n[4 - n^2\pi^2] \sin n\pi x.
\end{aligned}
\tag{6.135}
$$

This is a Fourier sine series expansion of $f(x) = x^2$ on $[0, 1]$. Namely,

$$
f(x) = \sum_{n=1}^{\infty} b_n \sin n\pi x.
$$

In order to determine the c_n's in Equation (6.135), we will need the Fourier sine series expansion of x^2 on $[0, 1]$. Thus, we need to compute

$$
\begin{aligned}
b_n &= \frac{2}{1}\int_0^1 x^2 \sin n\pi x \\
&= 2\left[\frac{(2 - n^2\pi^2)(-1)^n - 2}{n^3\pi^3}\right], \quad n = 1, 2, \ldots.
\end{aligned}
\tag{6.136}
$$

The resulting Fourier sine series is

$$
x^2 = 2\sum_{n=1}^{\infty}\left[\frac{(2 - n^2\pi^2)(-1)^n - 2}{n^3\pi^3}\right]\sin n\pi x.
$$

Inserting this expansion in Equation (6.135), we find

$$
2\sum_{n=1}^{\infty}\left[\frac{(2 - n^2\pi^2)(-1)^n - 2}{n^3\pi^3}\right]\sin n\pi x = \sum_{n=1}^{\infty} c_n[4 - n^2\pi^2]\sin n\pi x.
$$

Due to the linear independence of the eigenfunctions, we can solve for the unknown coefficients to obtain

$$
c_n = 2\frac{(2 - n^2\pi^2)(-1)^n - 2}{(4 - n^2\pi^2)n^3\pi^3}.
$$

Therefore, the solution using the eigenfunction expansion method is

$$
\begin{aligned}
y(x) &= \sum_{n=1}^{\infty} c_n \phi_n(x) \\
&= 2\sum_{n=1}^{\infty} \frac{\sin n\pi x}{(4 - n^2\pi^2)}\left[\frac{(2 - n^2\pi^2)(-1)^n - 2}{n^3\pi^3}\right].
\end{aligned}
\tag{6.137}
$$

We note that the solution in this example is the same solution as we had obtained using the Green's function obtained in series form in the previous example.

One remaining question is the following: Is there a closed form for the Green's function and the solution to this problem? The answer is yes!

Example 6.28. *Find the closed form Green's function for the problem*

$$
y'' + 4y = x^2, \quad x \in (0, 1), \quad y(0) = y(1) = 0
$$

and use it to obtain a closed form solution to this boundary value problem.

We note that the differential operator is a special case of the example done in Section 6.7.1. Namely, we pick $\omega = 2$. The Green's function was already found in that section. For this special case, we have

Using the closed form Green's function.

$$G(x,\xi) = \begin{cases} -\dfrac{\sin 2(1-\xi)\sin 2x}{2\sin 2}, & 0 \le x \le \xi, \\ -\dfrac{\sin 2(1-x)\sin 2\xi}{2\sin 2}, & \xi \le x \le 1. \end{cases} \qquad (6.138)$$

Using this Green's function, the solution to the boundary value problem is readily computed:

$$\begin{aligned} y(x) &= \int_0^1 G(x,\xi)f(\xi)\,d\xi \\ &= -\int_0^x \frac{\sin 2(1-x)\sin 2\xi}{2\sin 2}\xi^2\,d\xi + \int_x^1 \frac{\sin 2(\xi-1)\sin 2x}{2\sin 2}\xi^2\,d\xi \\ &= -\frac{1}{4\sin 2}\left[-x^2\sin 2 + (1-\cos^2 x)\sin 2 + \sin x\cos x(1+\cos 2)\right]. \\ &= -\frac{1}{4\sin 2}\left[-x^2\sin 2 + 2\sin^2 x\sin 1\cos 1 + 2\sin x\cos x\cos^2 1\right]. \\ &= -\frac{1}{8\sin 1\cos 1}\left[-x^2\sin 2 + 2\sin x\cos 1(\sin x\sin 1 + \cos x\cos 1)\right]. \\ &= \frac{x^2}{4} - \frac{\sin x\cos(1-x)}{4\sin 1}. \qquad (6.139) \end{aligned}$$

In Figure 6.15 we show a plot of this solution along with the first five terms of the series solution. The series solution converges quickly to the closed form solution.

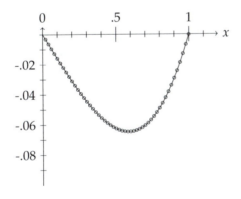

Figure 6.15: Plots of the exact solution to Example 6.26 with the first five terms of the series solution.

As one last check, we solve the boundary value problem directly, as we had done in the previous chapter.

Example 6.29. *Solve directly:*

$$y'' + 4y = x^2, \quad x \in (0,1), \quad y(0) = y(1) = 0.$$

Direct solution of the boundary value problem.

The problem has the general solution

$$y(x) = c_1\cos 2x + c_2\sin 2x + y_p(x),$$

where y_p is a particular solution of the nonhomogeneous differential equation. Using the Method of Undetermined Coefficients, we assume a solution of the form

$$y_p(x) = Ax^2 + Bx + C.$$

Inserting this guess into the nonhomogeneous equation, we have

$$2A + 4(Ax^2 + Bx + C) = x^2.$$

Thus, $B = 0, 4A = 1$, and $2A + 4C = 0$. The solution of this system is

$$A = \frac{1}{4}, \quad B = 0, \quad C = -\frac{1}{8}.$$

So, the general solution of the nonhomogeneous differential equation is

$$y(x) = c_1 \cos 2x + c_2 \sin 2x + \frac{x^2}{4} - \frac{1}{8}.$$

We next determine the arbitrary constants using the boundary conditions. We have

$$
\begin{aligned}
0 &= y(0) \\
&= c_1 - \frac{1}{8} \\
0 &= y(1) \\
&= c_1 \cos 2 + c_2 \sin 2 + \frac{1}{8}
\end{aligned}
\tag{6.140}
$$

Thus, $c_1 = \frac{1}{8}$ and

$$c_2 = -\frac{\frac{1}{8} + \frac{1}{8}\cos 2}{\sin 2}.$$

Inserting these constants into the solution, we find the same solution as before.

$$
\begin{aligned}
y(x) &= \frac{1}{8}\cos 2x - \left[\frac{\frac{1}{8} + \frac{1}{8}\cos 2}{\sin 2}\right]\sin 2x + \frac{x^2}{4} - \frac{1}{8} \\
&= \frac{(\cos 2x - 1)\sin 2 - \sin 2x(1 + \cos 2)}{8\sin 2} + \frac{x^2}{4} \\
&= \frac{(-2\sin^2 x)2\sin 1\cos 1 - \sin 2x(2\cos^2 1)}{16\sin 1\cos 1} + \frac{x^2}{4} \\
&= -\frac{(\sin^2 x)\sin 1 + \sin x\cos x(\cos 1)}{4\sin 1} + \frac{x^2}{4} \\
&= \frac{x^2}{4} - \frac{\sin x\cos(1 - x)}{4\sin 1}.
\end{aligned}
\tag{6.141}
$$

6.7.5   Nonhomogeneous Heat Equation

BOUNDARY VALUE GREEN'S FUNCTIONS DO NOT ONLY ARISE in the solution of nonhomogeneous ordinary differential equations. They are also important in arriving at the solution of nonhomogeneous partial differential equations. In this section we will show that this is the case by turning to the nonhomogeneous heat equation.

Consider the nonhomogeneous heat equation with nonhomogeneous boundary conditions:

$$u_t - k u_{xx} = h(x), \qquad 0 \le x \le L, \quad t > 0. \tag{6.142}$$

$$u(0,t) = a, \quad u(L,t) = b, \tag{6.143}$$

$$u(x,0) = f(x). \tag{6.144}$$

We are interested in finding a particular solution to this initial-boundary value problem. In fact, we can represent the solution to the general nonhomogeneous heat equation as the sum of two solutions that solve different problems.

First, we let $v(x,t)$ satisfy the homogeneous problem

$$v_t - k v_{xx} = 0, \qquad 0 \le x \le L, \quad t > 0. \tag{6.145}$$

$$v(0,t) = 0, \quad v(L,t) = 0, \tag{6.146}$$

$$v(x,0) = g(x), \tag{6.147}$$

which has homogeneous boundary conditions.

We will also need a steady-state solution to the original problem. A steady-state solution is one that satisfies $u_t = 0$. Let $w(x)$ be the steady-state solution. It satisfies the problem

$$-k w_{xx} = h(x), \qquad 0 \le x \le L, \tag{6.148}$$

$$u(0,t) = a, \quad u(L,t) = b. \tag{6.149}$$

The steady-state solution, $w(t)$, satisfies a nonhomogeneous differential equation with nonhomogeneous boundary conditions. The transient solution, $v(t)$, satisfies the homogeneous heat equation with homogeneous boundary conditions and satisfies a modified initial condition.

Now consider $u(x,t) = w(x) + v(x,t)$, the sum of the steady-state solution $w(x)$, and the transient solution $v(x,t)$. We first note that $u(x,t)$ satisfies the nonhomogeneous heat equation,

$$\begin{aligned} u_t - k u_{xx} &= (w+v)_t - (w+v)_{xx} \\ &= v_t - k v_{xx} - k w_{xx} = h(x). \end{aligned} \tag{6.150}$$

The boundary conditions are also satisfied. Evaluating, $u(x,t)$ at $x = 0$ and $x = L$, we have

$$\begin{aligned} u(0,t) &= w(0) + v(0,t) = a, \\ u(L,t) &= w(L) + v(L,t) = b. \end{aligned} \tag{6.151}$$

Finally, the initial condition gives

The transient solution satisfies

$$v(x,0) = f(x) - w(x).$$

$$u(x,0) = w(x) + v(x,0) = w(x) + g(x).$$

Thus, if we set $g(x) = f(x) - w(x)$, then $u(x,t) = w(x) + v(x,t)$ will be the solution of the nonhomogeneous boundary value problem. We already know how to solve the homogeneous problem to obtain $v(x,t)$. So, we only need to find the steady-state solution $w(x)$.

There are several methods we could use to solve Equation (6.149) for the steady-state solution. One is the Method of Variation of Parameters, which is closely related to the Green's function method for boundary value problems that we described in the previous several sections. However, we will just integrate the differential equation for the steady-state solution directly to find the solution. From this solution we will be able to "read off" the Green's function.

Integrating the steady-state equation (6.149) once yields

$$\frac{dw}{dx} = -\frac{1}{k}\int_0^x h(z)\,dz + A,$$

where we have been careful to include the integration constant, $A = w'(0)$. Integrating again, we obtain

$$w(x) = -\frac{1}{k}\int_0^x \left(\int_0^y h(z)\,dz\right) dy + Ax + B,$$

where a second integration constant has been introduced. This gives the general solution for Equation (6.149).

The boundary conditions can now be used to determine the constants. It is clear that $B = a$ for the condition at $x = 0$ to be satisfied. The second condition gives

$$b = w(L) = -\frac{1}{k}\int_0^L \left(\int_0^y h(z)\,dz\right) dy + AL + a.$$

Solving for A, we have

$$A = \frac{1}{kL}\int_0^L \left(\int_0^y h(z)\,dz\right) dy + \frac{b-a}{L}.$$

The steady-state solution.

Inserting the integration constants, the solution of the boundary value problem for the steady-state solution is then

$$w(x) = -\frac{1}{k}\int_0^x \left(\int_0^y h(z)\,dz\right) dy + \frac{x}{kL}\int_0^L \left(\int_0^y h(z)\,dz\right) dy + \frac{b-a}{L}x + a.$$

This is sufficient for an answer, but it can be written in a more compact form. In fact, we will show that the solution can be written in a way such that a Green's function can be identified.

First, we rewrite the double integrals as single integrals. We can do this using integration by parts. Consider the integral in the first term of the solution:

$$I = \int_0^x \left(\int_0^y h(z)\,dz\right) dy.$$

Setting $u = \int_0^y h(z)\,dz$ and $dv = dy$ in the standard integration by parts formula, we obtain

$$
\begin{aligned}
I &= \int_0^x \left(\int_0^y h(z)\,dz\right) dy \\
&= y\int_0^y h(z)\,dz\Big|_0^x - \int_0^x yh(y)\,dy \\
&= \int_0^x (x-y)h(y)\,dy. \tag{6.152}
\end{aligned}
$$

Thus, the double integral has now collapsed to a single integral. Replacing the integral in the solution, the steady-state solution becomes

$$w(x) = -\frac{1}{k}\int_0^x (x-y)h(y)\,dy + \frac{x}{kL}\int_0^L (L-y)h(y)\,dy + \frac{b-a}{L}x + a.$$

We can make a further simplification by combining these integrals. This can be done if the integration range, $[0, L]$, in the second integral is split into two pieces, $[0, x]$ and $[x, L]$. Writing the second integral as two integrals over these subintervals, we obtain

$$
\begin{aligned}
w(x) &= -\frac{1}{k}\int_0^x (x-y)h(y)\,dy + \frac{x}{kL}\int_0^x (L-y)h(y)\,dy \\
&\quad + \frac{x}{kL}\int_x^L (L-y)h(y)\,dy + \frac{b-a}{L}x + a.
\end{aligned}
\tag{6.153}
$$

Next, we rewrite the integrands,

$$
\begin{aligned}
w(x) &= -\frac{1}{k}\int_0^x \frac{L(x-y)}{L}h(y)\,dy + \frac{1}{k}\int_0^x \frac{x(L-y)}{L}h(y)\,dy \\
&\quad + \frac{1}{k}\int_x^L \frac{x(L-y)}{L}h(y)\,dy + \frac{b-a}{L}x + a.
\end{aligned}
\tag{6.154}
$$

It can now be seen how we can combine the first two integrals:

$$w(x) = -\frac{1}{k}\int_0^x \frac{y(L-x)}{L}h(y)\,dy + \frac{1}{k}\int_x^L \frac{x(L-y)}{L}h(y)\,dy + \frac{b-a}{L}x + a.$$

The resulting integrals now take on a similar form and this solution can be written compactly as

$$w(x) = -\int_0^L G(x,y)[-\tfrac{1}{k}h(y)]\,dy + \frac{b-a}{L}x + a,$$

where

$$
G(x,y) = \begin{cases} \dfrac{x(L-y)}{L}, & 0 \le x \le y, \\[2mm] \dfrac{y(L-x)}{L}, & y \le x \le L, \end{cases}
$$

is the Green's function for this problem.

The Green's function for the steady-state problem.

The full solution to the original problem can be found by adding to this steady-state solution a solution of the homogeneous problem,

$$u_t - ku_{xx} = 0, \qquad 0 \le x \le L, \quad t > 0. \tag{6.155}$$

$$u(0,t) = 0, \quad u(L,t) = 0, \tag{6.156}$$

$$u(x,0) = f(x) - w(x). \tag{6.157}$$

Example 6.30. *Solve the nonhomogeneous problem,*

$$u_t - u_{xx} = 10, \qquad 0 \le x \le 1, \quad t > 0. \tag{6.158}$$

$$u(0,t) = 20, \quad u(1,t) = 0, \tag{6.159}$$

$$u(x,0) = 2x(1-x). \tag{6.160}$$

In this problem, we have a rod initially at a temperature of $u(x,0) = 2x(1-x)$. The ends of the rod are maintained at fixed temperatures, and the bar is continually heated at a constant temperature, represented by the source term, 10.

First, we find the steady, state temperature, $w(x)$, satisfying

$$-w_{xx} = 10, \qquad 0 \le x \le 1. \tag{6.161}$$
$$u(0,t) = 20, \quad u(1,t) = 0. \tag{6.162}$$

Using the general solution, we have

$$w(x) = \int_0^1 10 G(x,y)\, dy - 20x + 20,$$

where

$$G(x,y) = \begin{cases} x(1-y), & 0 \le x \le y, \\ y(1-x), & y \le x \le 1. \end{cases}$$

We compute the solution

$$\begin{aligned} w(x) &= \int_0^x 10y(1-x)\, dy + \int_x^1 10x(1-y)\, dy - 20x + 20 \\ &= 5(x - x^2) - 20x + 20, \\ &= 20 - 15x - 5x^2. \end{aligned} \tag{6.163}$$

Checking this solution, it satisfies both the steady-state equation and boundary conditions.

The transient solution satisfies

$$v_t - v_{xx} = 0, \qquad 0 \le x \le 1, \quad t > 0. \tag{6.164}$$
$$v(0,t) = 0, \quad v(1,t) = 0, \tag{6.165}$$
$$v(x,0) = x(1-x) - 10. \tag{6.166}$$

Recall, that we have determined the solution of this problem as

$$v(x,t) = \sum_{n=1}^{\infty} b_n e^{-n^2 \pi^2 t} \sin n\pi x,$$

where the Fourier sine coefficients are given in terms of the initial temperature distribution,

$$b_n = 2 \int_0^1 [x(1-x) - 10] \sin n\pi x\, dx, \quad n = 1, 2, \dots.$$

Therefore, the full solution is

$$u(x,t) = \sum_{n=1}^{\infty} b_n e^{-n^2 \pi^2 t} \sin n\pi x + 20 - 15x - 5x^2.$$

Note that for large t, the transient solution tends to zero and we are left with the steady-state solution, as expected.

6.8   Appendix: Least Squares Approximation

IN THE FIRST SECTION OF THIS CHAPTER, we showed that we can expand functions over an infinite set of basis functions as

$$f(x) = \sum_{n=1}^{\infty} c_n \phi_n(x)$$

and that the generalized Fourier coefficients are given by

$$c_n = \frac{<\phi_n, f>}{<\phi_n, \phi_n>}.$$

In this section we turn to a discussion of approximating $f(x)$ by the partial sums $\sum_{n=1}^{N} c_n \phi_n(x)$ and showing that the Fourier coefficients are the best coefficients minimizing the deviation of the partial sum from $f(x)$. This will lead us to a discussion of the convergence of Fourier series.

More specifically, we set the following goal:

Goal

To find the best approximation of $f(x)$ on $[a, b]$ by $S_N(x) = \sum_{n=1}^{N} c_n \phi_n(x)$ for a set of fixed functions $\phi_n(x)$; that is, to find the expansion coefficients, c_n, such that $S_N(x)$ approximates $f(x)$ in the least squares sense.

We want to measure the deviation of the finite sum from the given function. Essentially, we want to look at the error made in the approximation. This is done by introducing the mean square deviation:

$$E_N = \int_a^b [f(x) - S_N(x)]^2 \rho(x)\, dx,$$

where we have introduced the weight function $\rho(x) > 0$. It gives us a sense as to how close the Nth partial sum is to $f(x)$.

The mean square deviation.

We want to minimize this deviation by choosing the right c_n's. We begin by inserting the partial sums and expand the square in the integrand:

$$
\begin{aligned}
E_N &= \int_a^b [f(x) - S_N(x)]^2 \rho(x)\, dx \\
&= \int_a^b \left[f(x) - \sum_{n=1}^{N} c_n \phi_n(x) \right]^2 \rho(x)\, dx \\
&= \int_a^b f^2(x) \rho(x)\, dx - 2 \int_a^b f(x) \sum_{n=1}^{N} c_n \phi_n(x) \rho(x)\, dx \\
&\quad + \int_a^b \sum_{n=1}^{N} c_n \phi_n(x) \sum_{m=1}^{N} c_m \phi_m(x) \rho(x)\, dx.
\end{aligned}
\tag{6.167}
$$

Looking at the three resulting integrals, we see that the first term is just the inner product of f with itself. The other integrations can be rewritten after interchanging the order of integration and summation. The double sum can be reduced to a single sum using the orthogonality of the ϕ_n's. Thus, we have

$$
\begin{aligned}
E_N &= <f,f> -2\sum_{n=1}^{N} c_n <f,\phi_n> + \sum_{n=1}^{N}\sum_{m=1}^{N} c_n c_m <\phi_n,\phi_m> \\
&= <f,f> -2\sum_{n=1}^{N} c_n <f,\phi_n> + \sum_{n=1}^{N} c_n^2 <\phi_n,\phi_n>.
\end{aligned} \tag{6.168}
$$

We are interested in finding the coefficients, so we will complete the square in c_n. Focusing on the last two terms, we have

$$
\begin{aligned}
&-2\sum_{n=1}^{N} c_n <f,\phi_n> + \sum_{n=1}^{N} c_n^2 <\phi_n,\phi_n> \\
&= \sum_{n=1}^{N} <\phi_n,\phi_n> c_n^2 - 2 <f,\phi_n> c_n \\
&= \sum_{n=1}^{N} <\phi_n,\phi_n> \left[c_n^2 - \frac{2 <f,\phi_n>}{<\phi_n,\phi_n>} c_n \right] \\
&= \sum_{n=1}^{N} <\phi_n,\phi_n> \left[\left(c_n - \frac{<f,\phi_n>}{<\phi_n,\phi_n>} \right)^2 - \left(\frac{<f,\phi_n>}{<\phi_n,\phi_n>} \right)^2 \right].
\end{aligned}
$$

$$\tag{6.169}$$

Up to this point, we have shown that the mean square deviation is given as

$$
E_N = <f,f> + \sum_{n=1}^{N} <\phi_n,\phi_n> \left[\left(c_n - \frac{<f,\phi_n>}{<\phi_n,\phi_n>} \right)^2 - \left(\frac{<f,\phi_n>}{<\phi_n,\phi_n>} \right)^2 \right].
$$

So, E_N is minimized by choosing

$$
c_n = \frac{<f,\phi_n>}{<\phi_n,\phi_n>}.
$$

However, these are the Fourier Coefficients. This minimization is often referred to as Minimization in Least Squares Sense.

Minimization in Least Squares Sense, Bessel's Inequality.

Inserting the Fourier Coefficients into the mean square deviation yields

$$
0 \le E_N = <f,f> - \sum_{n=1}^{N} c_n^2 <\phi_n,\phi_n>.
$$

Thus, we obtain Bessel's Inequality:

$$
<f,f> \ge \sum_{n=1}^{N} c_n^2 <\phi_n,\phi_n>.
$$

Convergence in the mean.

For convergence, we next let N get large and see if the partial sums converge to the function. In particular, we say that the infinite series converges in the mean if

$$
\int_a^b [f(x) - S_N(x)]^2 \rho(x)\,dx \to 0 \text{ as } N \to \infty.
$$

Letting N get large in Bessel's inequality shows that the sum $\sum_{n=1}^{N} c_n^2 <$ $\phi_n, \phi_n >$ converges if

$$(< f, f >= \int_a^b f^2(x)\rho(x)\,dx < \infty.$$

The space of all such f is denoted $L_\rho^2(a, b)$, the space of square integrable functions on (a, b) with weight $\rho(x)$.

From the nth term divergence test from calculus, we know that $\sum a_n$ convergence implies that $a_n \to 0$ as $n \to \infty$. Therefore, in this problem, the terms $c_n^2 < \phi_n, \phi_n >$ approach zero as n gets large. This is only possible if the c_n's go to zero as n gets large. Thus, if $\sum_{n=1}^{N} c_n \phi_n$ converges in the mean to f, then $\int_a^b [f(x) - \sum_{n=1}^{N} c_n \phi_n]^2 \rho(x)\,dx$ approaches zero as $N \to \infty$. This implies from the above derivation of Bessel's inequality that

$$< f, f > - \sum_{n=1}^{N} c_n^2 (\phi_n, \phi_n) \to 0.$$

This leads to Parseval's equality:

Parseval's equality.

$$< f, f >= \sum_{n=1}^{\infty} c_n^2 < \phi_n, \phi_n > .$$

Parseval's equality holds if and only if

$$\lim_{N \to \infty} \int_a^b \left(f(x) - \sum_{n=1}^{N} c_n \phi_n(x) \right)^2 \rho(x)\,dx = 0.$$

If this is true for every square integrable function in $L_\rho^2(a, b)$, then the set of functions $\{\phi_n(x)\}_{n=1}^{\infty}$ is said to be complete. One can view these functions as an infinite dimensional basis for the space of square integrable functions on (a, b) with weight $\rho(x) > 0$.

One can extend the above limit $c_n \to 0$ as $n \to \infty$, by assuming that $\frac{\phi_n(x)}{\|\phi_n\|}$ is uniformly bounded and that $\int_a^b |f(x)|\rho(x)\,dx < \infty$. This is the Riemann–Lebesgue Lemma, but will not be proven here.

6.9   Appendix: Fredholm Alternative Theorem

GIVEN THAT $Ly = f$, WHEN CAN ONE EXPECT to find a solution? Is it unique? These questions are answered by the Fredholm Alternative Theorem. This theorem occurs in many forms from a statement about solutions to systems of algebraic equations to solutions of boundary value problems and integral equations. The theorem comes in two parts, thus the term "alternative." Either the equation has exactly one solution for all f, or the equation has many solutions for some f's and none for the rest.

The reader is familiar with the statements of the Fredholm Alternative for the solution of systems of algebraic equations. One seeks solutions of the system $Ax = b$ for A an $n \times m$ matrix. Defining the matrix adjoint A^* through $< Ax, y >=< x, A^*y >$ for all $x, y, \in C^n$, then either

Theorem 6.2. First Alternative

   The equation $Ax = b$ has a solution if and only if $< b, v >= 0$ for all v satisfying $A^*v = 0$.

   or

Theorem 6.3. Second Alternative

   A solution of $Ax = b$, if it exists, is unique if and only if $x = 0$ is the only solution of $Ax = 0$.

The second alternative is more familiar when given in the form: The solution of a nonhomogeneous system of n equations and n unknowns is unique if the only solution to the homogeneous problem is the zero solution. Or, equivalently, A is invertible, or has nonzero determinant.

Proof. We prove the second theorem first. Assume that $Ax = 0$ for $x \neq 0$ and $Ax_0 = b$. Then $A(x_0 + \alpha x) = b$ for all α. Therefore, the solution is not unique. Conversely, if there are two different solutions, x_1 and x_2, satisfying $Ax_1 = b$ and $Ax_2 = b$, then one has a nonzero solution $x = x_1 - x_2$ such that $Ax = A(x_1 - x_2) = 0$.

   The proof of the first part of the first theorem is simple. Let $A^*v = 0$ and $Ax_0 = b$. Then we have

$$< b, v >=< Ax_0, v >=< x_0, A^*v >= 0.$$

For the second part, we assume that $< b, v >= 0$ for all v such that $A^*v = 0$. Write b as the sum of a part that is in the range of A and a part that in the space orthogonal to the range of A, $b = b_R + b_O$. Then, $0 =< b_O, Ax >=< A^*b, x >$ for all x. Thus, A^*b_O. Because $< b, v >= 0$ for all v in the nullspace of A^*, then $< b, b_O >= 0$.

   Therefore, $< b, v >= 0$ implies that

$$0 =< b, b_O >=< b_R + b_O, b_O >=< b_O, b_O > .$$

This means that $b_O = 0$, giving $b = b_R$ in the range of A. So, $Ax = b$ has a solution. $\qquad\square$

Example 6.31. *Determine the allowed forms of $\mathbf{b}$ for a solution of $A\mathbf{x} = \mathbf{b}$ to exist, where*

$$A = \begin{pmatrix} 1 & 2 \\ 3 & 6 \end{pmatrix}.$$

First note that $A^* = \overline{A}^T$. This is seen by looking at

$$
\begin{aligned}
< A\mathbf{x}, \mathbf{y} > &= < \mathbf{x}, A^*\mathbf{y} > \\
\sum_{i=1}^{n}\sum_{j=1}^{n} a_{ij} x_j \bar{y}_i &= \sum_{j=1}^{n} x_j \sum_{j=1}^{n} a_{ij} \bar{y}_i \\
&= \sum_{j=1}^{n} x_j \overline{\sum_{j=1}^{n} (\bar{a}^T)_{ji}} \, y_i.
\end{aligned}
\tag{6.170}
$$

For this example,

$$A^* = \begin{pmatrix} 1 & 3 \\ 2 & 6 \end{pmatrix}.$$

We next solve $A^*\mathbf{v} = 0$. This means, $v_1 + 3v_2 = 0$. So, the nullspace of A^* is spanned by $\mathbf{v} = (3, -1)^T$. For a solution of $A\mathbf{x} = \mathbf{b}$ to exist, $\mathbf{b}$ would have to be orthogonal to $\mathbf{v}$. Therefore, a solution exists when

$$\mathbf{b} = \alpha \begin{pmatrix} 1 \\ 3 \end{pmatrix}.$$

So, what does the Fredholm Alternative say about solutions of boundary value problems? We extend the Fredholm Alternative for linear operators. A more general statement would be:

Theorem 6.4. *If L is a bounded linear operator on a Hilbert space, then $Ly = f$ has a solution if and only if $< f, v > = 0$ for every v such that $L^\dagger v = 0$.*

The statement for boundary value problems is similar. However, we need to be careful to treat the boundary conditions in our statement. As we have seen, after several integrations by parts we have that

$$< \mathcal{L}u, v > = S(u, v) + < u, \mathcal{L}^\dagger v >,$$

where $S(u, v)$ involves the boundary conditions on u and v. Note that for nonhomogeneous boundary conditions, this term may no longer vanish.

Theorem 6.5. *The solution of the boundary value problem $\mathcal{L}u = f$ with boundary conditions $Bu = g$ exists if and only if*

$$< f, v > - S(u, v) = 0$$

for all v satisfying $\mathcal{L}^\dagger v = 0$ and $B^\dagger v = 0$.

Example 6.32. *Consider the problem*

$$u'' + u = f(x), \quad u(0) - u(2\pi) = \alpha, u'(0) - u'(2\pi) = \beta.$$

Only certain values of α and β will lead to solutions. We first note that

$$L = L^\dagger = \frac{d^2}{dx^2} + 1.$$

Solutions of

$$L^\dagger v = 0, \quad v(0) - v(2\pi) = 0, v'(0) - v'(2\pi) = 0$$

are easily found to be linear combinations of $v = \sin x$ and $v = \cos x$.
    Next, one computes

$$\begin{aligned} S(u, v) &= \left[u'v - uv' \right]_0^{2\pi} \\ &= u'(2\pi)v(2\pi) - u(2\pi)v'(2\pi) - u'(0)v(0) + u(0)v'(0). \end{aligned}$$

$$\text{(6.171)}$$

For $v(x) = \sin x$, this yields

$$S(u, \sin x) = -u(2\pi) + u(0) = \alpha.$$

Similarly,

$$S(u, \cos x) = \beta.$$

Using $< f, v > - S(u, v) = 0$, this leads to the conditions that we were seeking:

$$\int_0^{2\pi} f(x) \sin x \, dx = \alpha,$$

$$\int_0^{2\pi} f(x) \cos x \, dx = \beta.$$

Problems

1. Consider the set of vectors $(-1, 1, 1)$, $(1, -1, 1)$, $(1, 1, -1)$.

   a. Use the Gram–Schmidt process to find an orthonormal basis for R^3 using this set in the given order.

   b. What do you get if you do reverse the order of these vectors?

2. Use the Gram–Schmidt process to find the first four orthogonal polynomials satisfying the following:

   a. Interval: $(-\infty, \infty)$ Weight Function: e^{-x^2}.

   b. Interval: $(0, \infty)$ Weight Function: e^{-x}.

3. Find $P_4(x)$ using

   a. The Rodrigues Formula in Equation (6.20).

   b. The three-term recursion formula in Equation (6.22).

4. In Equations (6.35) through (6.42), we provide several identities for Legendre polynomials. Derive the results in Equations (6.36) through (6.42) as described in the text. Namely,

   a. Differentiating Equation (6.35) with respect to x, derive Equation (6.36).

   b. Derive Equation (6.37) by differentiating $g(x, t)$ with respect to x and rearranging the resulting infinite series.

   c. Combining the previous result with Equation (6.35), derive Equations (6.38) and (6.39).

   d. Adding and subtracting Equations (6.38) and (6.39), obtain Equations (6.40) and (6.41).

   e. Derive Equation (6.42) using some of the other identities.

5. Use the recursion relation (6.22) to evaluate $\int_{-1}^{1} x P_n(x) P_m(x) \, dx$, $n \leq m$.

6. Expand the following in a Fourier-Legendre series for $x \in (-1,1)$.

    a. $f(x) = x^2$.

    b. $f(x) = 5x^4 + 2x^3 - x + 3$.

    c. $f(x) = \begin{cases} -1, & -1 < x < 0, \\ 1, & 0 < x < 1. \end{cases}$

    d. $f(x) = \begin{cases} x, & -1 < x < 0, \\ 0, & 0 < x < 1. \end{cases}$

7. Use integration by parts to show $\Gamma(x+1) = x\Gamma(x)$.

8. Prove the double factorial identities:

$$(2n)!! = 2^n n!$$

and

$$(2n-1)!! = \frac{(2n)!}{2^n n!}.$$

9. Express the following as Gamma functions. Namely, noting the form $\Gamma(x+1) = \int_0^\infty t^x e^{-t} \, dt$ and using an appropriate substitution, each expression can be written in terms of a Gamma function.

    a. $\int_0^\infty x^{2/3} e^{-x} \, dx$.

    b. $\int_0^\infty x^5 e^{-x^2} \, dx$.

    c. $\int_0^1 \left[\ln \left(\frac{1}{x} \right) \right]^n \, dx$.

10. The coefficients C_k^p in the binomial expansion for $(1+x)^p$ are given by

$$C_k^p = \frac{p(p-1)\cdots(p-k+1)}{k!}.$$

    a. Write C_k^p in terms of Gamma functions.

    b. For $p = 1/2$, use the properties of Gamma functions to write $C_k^{1/2}$ in terms of factorials.

    c. Confirm your answer in part b. by deriving the Maclaurin series expansion of $(1+x)^{1/2}$.

11. The Hermite polynomials, $H_n(x)$, satisfy the following:

    i. $< H_n, H_m > = \int_{-\infty}^\infty e^{-x^2} H_n(x) H_m(x) \, dx = \sqrt{\pi} 2^n n! \delta_{n,m}$.

    ii. $H_n'(x) = 2n H_{n-1}(x)$.

    iii. $H_{n+1}(x) = 2x H_n(x) - 2n H_{n-1}(x)$.

    iv. $H_n(x) = (-1)^n e^{x^2} \frac{d^n}{dx^n} \left(e^{-x^2} \right)$.

Using these, show that

    a. $H_n'' - 2x H_n' + 2n H_n = 0$. [Use properties ii. and iii.]

    b. $\int_{-\infty}^\infty x e^{-x^2} H_n(x) H_m(x) \, dx = \sqrt{\pi} 2^{n-1} n! \left[\delta_{m,n-1} + 2(n+1)\delta_{m,n+1} \right]$.
    [Use properties i. and iii.]

c. $H_n(0) = \begin{cases} 0, & n \text{ odd,} \\ (-1)^m \frac{(2m)!}{m!}, & n = 2m. \end{cases}$ [Let $x = 0$ in iii. and iterate.

   Note from iv. that $H_0(x) = 1$ and $H_1(x) = 2x.$]

12. In Maple, one can type **simplify(LegendreP(2*n-2,0)-LegendreP(2*n,0));** to find a value for $P_{2n-2}(0) - P_{2n}(0)$. It gives the result in terms of Gamma functions. However, in Example 6.10 for Fourier-Legendre series, the value is given in terms of double factorials! So, we have

$$P_{2n-2}(0) - P_{2n}(0) = \frac{\sqrt{\pi}(4n-1)}{2\Gamma(n+1)\Gamma\left(\frac{3}{2}-n\right)} = (-1)^{n-1}\frac{(2n-3)!!}{(2n-2)!!}\frac{4n-1}{2n}.$$

You will verify that both results are the same by doing the following:

   a. Prove that $P_{2n}(0) = (-1)^n \frac{(2n-1)!!}{(2n)!!}$ using the generating function and a binomial expansion.

   b. Prove that $\Gamma\left(n + \frac{1}{2}\right) = \frac{(2n-1)!!}{2^n}\sqrt{\pi}$ using $\Gamma(x) = (x-1)\Gamma(x-1)$ and iteration.

   c. Verify the result from Maple that $P_{2n-2}(0) - P_{2n}(0) = \frac{\sqrt{\pi}(4n-1)}{2\Gamma(n+1)\Gamma\left(\frac{3}{2}-n\right)}.$

   d. Can either expression for $P_{2n-2}(0) - P_{2n}(0)$ be simplified further?

13. A solution of Bessel's equation, $x^2 y'' + xy' + (x^2 - n^2)y = 0$, can be found using the guess $y(x) = \sum_{j=0}^{\infty} a_j x^{j+n}$. One obtains the recurrence relation $a_j = \frac{-1}{j(2n+j)} a_{j-2}$. Show that for $a_0 = (n!2^n)^{-1}$, we get the Bessel function of the first kind of order n from the even values $j = 2k$:

$$J_n(x) = \sum_{k=0}^{\infty} \frac{(-1)^k}{k!(n+k)!} \left(\frac{x}{2}\right)^{n+2k}.$$

14. Use the infinite series in the Problem 13 derive the derivative identities (6.59) and (6.60):

   a. $\frac{d}{dx}[x^n J_n(x)] = x^n J_{n-1}(x).$

   b. $\frac{d}{dx}[x^{-n} J_n(x)] = -x^{-n} J_{n+1}(x).$

15. Prove the following identities based on those in the Problem 14:

   a. $J_{p-1}(x) + J_{p+1}(x) = \frac{2p}{x} J_p(x).$

   b. $J_{p-1}(x) - J_{p+1}(x) = 2J_p'(x).$

16. Use the derivative identities of Bessel functions,(6.59)-(6.60), and integration by parts to show that

$$\int x^3 J_0(x)\, dx = x^3 J_1(x) - 2x^2 J_2(x) + C.$$

17. Use the generating function to find $J_n(0)$ and $J_n'(0)$.

18. Bessel functions $J_p(\lambda x)$ are solutions of $x^2 y'' + xy' + (\lambda^2 x^2 - p^2)y = 0$. Assume that $x \in (0,1)$ and that $J_p(\lambda) = 0$ and $J_p(0)$ is finite.

a. Show that this equation can be written in the form

$$\frac{d}{dx}\left(x\frac{dy}{dx}\right) + (\lambda^2 x - \frac{p^2}{x})y = 0.$$

This is the standard Sturm–Liouville form for Bessel's equation.

b. Prove that

$$\int_0^1 x J_p(\lambda x) J_p(\mu x)\, dx = 0, \quad \lambda \neq \mu$$

by considering

$$\int_0^1 \left[J_p(\mu x)\frac{d}{dx}\left(x\frac{d}{dx}J_p(\lambda x)\right) - J_p(\lambda x)\frac{d}{dx}\left(x\frac{d}{dx}J_p(\mu x)\right)\right] dx.$$

Thus, the solutions corresponding to different eigenvalues (λ, μ) are orthogonal.

c. Prove that

$$\int_0^1 x\left[J_p(\lambda x)\right]^2 dx = \frac{1}{2}J_{p+1}^2(\lambda) = \frac{1}{2}J_p'^2(\lambda).$$

19. We can rewrite Bessel functions, $J_\nu(x)$, in a form that will allow the order to be non-integer by using the gamma function. You will need the results from Problem 12b for $\Gamma\left(k + \frac{1}{2}\right)$.

a. Extend the series definition of the Bessel function of the first kind of order ν, $J_\nu(x)$, for $\nu \geq 0$ by writing the series solution for $y(x)$ in Problem 13 using the gamma function.

b. Extend the series to $J_{-\nu}(x)$, for $\nu \geq 0$. Discuss the resulting series and what happens when ν is a positive integer.

c. Use these results to obtain the closed form expressions

$$J_{1/2}(x) = \sqrt{\frac{2}{\pi x}}\sin x,$$

$$J_{-1/2}(x) = \sqrt{\frac{2}{\pi x}}\cos x.$$

d. Use the results in part c. with the recursion formula for Bessel functions to obtain a closed form for $J_{3/2}(x)$.

20. In this problem you will derive the expansion

$$x^2 = \frac{c^2}{2} + 4\sum_{j=2}^{\infty} \frac{J_0(\alpha_j x)}{\alpha_j^2 J_0(\alpha_j c)}, \quad 0 < x < c,$$

where the $\alpha_j's$ are the positive roots of $J_1(\alpha c) = 0$, by following the below steps.

a. List the first five values of α for $J_1(\alpha c) = 0$ using Table 6.3 and Figure 6.8. [Note: Be careful in determining α_1.]

b. Show that $\|J_0(\alpha_1 x)\|^2 = \frac{c^2}{2}$. Recall that

$$\|J_0(\alpha_j x)\|^2 = \int_0^c x J_0^2(\alpha_j x)\, dx.$$

c. Show that $\|J_0(\alpha_j x)\|^2 = \frac{c^2}{2}\left[J_0(\alpha_j c)\right]^2$, $j = 2, 3, \ldots$. (This is the most involved step.) First note from Problem 18 that $y(x) = J_0(\alpha_j x)$ is a solution of

$$x^2 y'' + x y' + \alpha_j^2 x^2 y = 0.$$

i. Verify the Sturm–Liouville form of this differential equation: $(xy')' = -\alpha_j^2 xy$.

ii. Multiply the equation in part i. by $y(x)$ and integrate from $x = 0$ to $x = c$ to obtain

$$
\begin{aligned}
\int_0^c (xy')' y\, dx &= -\alpha_j^2 \int_0^c xy^2\, dx \\
&= -\alpha_j^2 \int_0^c x J_0^2(\alpha_j x)\, dx. \qquad (6.172)
\end{aligned}
$$

iii. Noting that $y(x) = J_0(\alpha_j x)$, integrate the left-hand side by parts and use the following to simplify the resulting equation:

1. $J_0'(x) = -J_1(x)$ from Equation (6.60).
2. Equation (6.63).
3. $J_2(\alpha_j c) + J_0(\alpha_j c) = 0$ from Equation (6.61).

iv. Now you should have enough information to complete part d.

d. Use the results from parts b, c, and problem 16 to derive the expansion coefficients for

$$x^2 = \sum_{j=1}^{\infty} c_j J_0(\alpha_j x)$$

in order to obtain the desired expansion.

21. Prove that if $u(x)$ and $v(x)$ satisfy the general homogeneous boundary conditions

$$
\begin{aligned}
\alpha_1 u(a) + \beta_1 u'(a) &= 0, \\
\alpha_2 u(b) + \beta_2 u'(b) &= 0 \qquad\qquad (6.173)
\end{aligned}
$$

at $x = a$ and $x = b$, then

$$p(x)[u(x)v'(x) - v(x)u'(x)]_{x=a}^{x=b} = 0.$$

22. Prove Green's identity $\int_a^b (u\mathcal{L}v - v\mathcal{L}u)\, dx = \left[p(uv' - vu')\right]\Big|_a^b$ for the general Sturm–Liouville operator $\mathcal{L}$.

23. Find the adjoint operator and its domain for $Lu = u'' + 4u' - 3u$, $u'(0) + 4u(0) = 0$, $u'(1) + 4u(1) = 0$.

24. Show that a Sturm–Liouville operator with periodic boundary conditions on $[a, b]$ is self-adjoint if and only if $p(a) = p(b)$. [Recall that periodic boundary conditions are given as $u(a) = u(b)$ and $u'(a) = u'(b)$.]

25. The Hermite differential equation is given by $y'' - 2xy' + \lambda y = 0$. Rewrite this equation in self-adjoint form. From the Sturm–Liouville form obtained, verify that the differential operator is self adjoint on $(-\infty, \infty)$. Give the integral form for the orthogonality of the eigenfunctions.

26. Find the eigenvalues and eigenfunctions of the given Sturm–Liouville problems:

    a. $y'' + \lambda y = 0$, $y'(0) = 0 = y'(\pi)$.

    b. $(xy')' + \frac{\lambda}{x}y = 0$, $y(1) = y(e^2) = 0$.

27. The eigenvalue problem $x^2 y'' - \lambda xy' + \lambda y = 0$ with $y(1) = y(2) = 0$ is not a Sturm–Liouville eigenvalue problem. Show that none of the eigenvalues are real by solving this eigenvalue problem.

28. In Example 6.20, we found a bound on the lowest eigenvalue for the given eigenvalue problem.

    a. Verify the computation in the example.

    b. Apply the method using

$$y(x) = \begin{cases} x, & 0 < x < \frac{1}{2}, \\ 1 - x, & \frac{1}{2} < x < 1. \end{cases}$$

    Is this an upper bound on λ_1?

    c. Use the Rayleigh quotient to obtain a good upper bound for the lowest eigenvalue of the eigenvalue problem: $\phi'' + (\lambda - x^2)\phi = 0$, $\phi(0) = 0$, $\phi'(1) = 0$.

29. Use the method of eigenfunction expansions to solve the problems:

    a. $y'' + 4y = x^2$, $\quad y'(0) = y'(1) = 0$.

    b. $y + 4y = x^2$ $\quad y(0) = y(1) = 0$.

30. Determine the solvability conditions for the nonhomogeneous boundary value problem: $u'(\pi/4) = \beta$.

31. Consider the problem $y'' = \sin x$, $y'(0) = 0$, $y(\pi) = 0$.

    a. Solve by direct integration.

    b. Determine the Green's function.

    c. Solve the boundary value problem using the Green's function.

    d. Change the boundary conditions to $y'(0) = 5$, $y(\pi) = -3$.

        i. Solve by direct integration.

        ii. Solve using the Green's function.

32. Consider the problem:

$$\frac{\partial^2 G}{\partial x^2} = \delta(x - x_0), \quad \frac{\partial G}{\partial x}(0, x_0) = 0, \quad G(\pi, x_0) = 0.$$

   a. Solve by direct integration.

   b. Compare this result to the Green's function in part b of Problem 31.

   c. Verify that G is symmetric in its arguments.

33. Consider the boundary value problem: $y'' - y = x$, $x \in (0, 1)$, with boundary conditions $y(0) = y(1) = 0$.

   a. Find a closed form solution without using Green's functions.

   b. Determine the closed form Green's function using the properties of Green's functions. Use this Green's function to obtain a solution of the boundary value problem.

   c. Determine a series representation of the Green's function. Use this Green's function to obtain a solution of the boundary value problem.

   d. Confirm that all of the solutions obtained give the same results.

7
Complex Representations of Functions

"He is not a true man of science who does not bring some sympathy to his studies, and expect to learn something by behavior as well as by application. It is childish to rest in the discovery of mere coincidences, or of partial and extraneous laws. The study of geometry is a petty and idle exercise of the mind, if it is applied to no larger system than the starry one. Mathematics should be mixed not only with physics but with ethics; that is mixed mathematics. The fact which interests us most is the life of the naturalist. The purest science is still biographical."—Henry David Thoreau (1817–1862)

7.1 Complex Representations of Waves

WE HAVE SEEN that we can determine the frequency content of a function $f(t)$ defined on an interval $[0, T]$ by looking for the Fourier coefficients in the Fourier series expansion

$$f(t) = \frac{a_0}{2} + \sum_{n=1}^{\infty} a_n \cos \frac{2\pi nt}{T} + b_n \sin \frac{2\pi nt}{T}.$$

The coefficients take forms like

$$a_n = \frac{2}{T} \int_0^T f(t) \cos \frac{2\pi nt}{T} \, dt.$$

However, trigonometric functions can be written in a complex exponential form. Using Euler's Formula, which was obtained using the Maclaurin expansion of e^x in Example 1.37,

$$e^{i\theta} = \cos \theta + i \sin \theta,$$

the complex conjugate is found by replacing i with $-i$ to obtain

$$e^{-i\theta} = \cos \theta - i \sin \theta.$$

Adding these expressions, we have

$$2 \cos \theta = e^{i\theta} + e^{-i\theta}.$$

Subtracting the exponentials leads to an expression for the sine function. Thus, we have the important result that sines and cosines can be written as complex exponentials:

$$\begin{aligned}
\cos\theta &= \frac{e^{i\theta} + e^{-i\theta}}{2}, \\
\sin\theta &= \frac{e^{i\theta} - e^{-i\theta}}{2i}.
\end{aligned}$$

(7.1)

So, we can write

$$\cos\frac{2\pi nt}{T} = \frac{1}{2}\left(e^{\frac{2\pi int}{T}} + e^{-\frac{2\pi int}{T}}\right).$$

Later we will see that we can use this information to rewrite the series as a sum over complex exponentials in the form

$$f(t) = \sum_{n=-\infty}^{\infty} c_n e^{\frac{2\pi int}{T}},$$

where the Fourier coefficients now take the form

$$c_n = \int_0^T f(t) e^{-\frac{2\pi int}{T}}\, dt.$$

In fact, when one considers the representation of analog signals defined over an infinite interval and containing a continuum of frequencies, one will see that Fourier series sums become integrals of complex functions and so do the Fourier coefficients. Thus, we will naturally find ourselves needing to work with functions of complex variables and perform complex integrals.

We can also develop a complex representation for waves. Recall from the discussion in Section 5.11 on finite length strings that a solution to the wave equation was given by

$$u(x,t) = \frac{1}{2}\left[\sum_{n=1}^{\infty} A_n \sin k_n(x+ct) + \sum_{n=1}^{\infty} A_n \sin k_n(x-ct)\right].$$

(7.2)

We can replace the sines with their complex forms as

$$\begin{aligned}
u(x,t) &= \frac{1}{4i}\left[\sum_{n=1}^{\infty} A_n \left(e^{ik_n(x+ct)} - e^{-ik_n(x+ct)}\right)\right. \\
&\quad \left. + \sum_{n=1}^{\infty} A_n \left(e^{ik_n(x-ct)} - e^{-ik_n(x-ct)}\right)\right].
\end{aligned}$$

(7.3)

Defining $k_{-n} = -k_n$, $n = 1, 2, \ldots$, we can rewrite this solution in the form

$$u(x,t) = \sum_{n=-\infty}^{\infty}\left[c_n e^{ik_n(x+ct)} + d_n e^{ik_n(x-ct)}\right].$$

(7.4)

Such representations are also possible for waves propagating over the entire real line. In such cases, we are not restricted to discrete frequencies and wave numbers. The sum of the harmonics will then be a sum over a continuous range, which means that the sums become integrals. So, we are led to the complex representation

$$u(x,t) = \int_{-\infty}^{\infty}\left[c(k) e^{ik(x+ct)} + d(k) e^{ik(x-ct)}\right] dk.$$

(7.5)

The forms $e^{ik(x+ct)}$ and $e^{ik(x-ct)}$ are complex representations of what are called plane waves in one dimension. The integral represents a general wave form consisting of a sum over plane waves. The Fourier coefficients in the representation can be complex valued functions, and the evaluation of the integral may be done using methods from complex analysis. We would like to be able to compute such integrals.

With the above ideas in mind, we will now take a tour of complex analysis. We will first review some facts about complex numbers and then introduce complex functions. This will lead us to the calculus of functions of a complex variable, including the differentiation and integration complex functions. This will set up the methods needed to explore Fourier transforms in the next chapter.

7.2   Complex Numbers

COMPLEX NUMBERS WERE FIRST INTRODUCED in order to solve some simple problems. The history of complex numbers only extends about 500 years. In essence, it was found that we need to find the roots of equations such as $x^2 + 1 = 0$. The solution is $x = \pm\sqrt{-1}$. Due to the usefulness of this concept, which was not realized at first, a special symbol was introduced—the imaginary unit, $i = \sqrt{-1}$. In particular, Girolamo Cardano (1501–1576) was one of the first to use square roots of negative numbers when providing solutions of cubic equations. However, complex numbers did not become an important part of mathematics or science until the late seventeenth century after people like Abraham de Moivre (1667–1754), the Bernoulli[1] family, and Euler took them seriously.

A complex number is a number of the form $z = x + iy$, where x and y are real numbers. x is called the real part of z, and y is the imaginary part of z. Examples of such numbers are $3 + 3i$, $-1i = -i$, $4i$ and 5. Note that $5 = 5 + 0i$ and $4i = 0 + 4i$.

There is a geometric representation of complex numbers in a two-dimensional plane, known as the complex plane $\mathbb{C}$. This is given by the Argand diagram as shown in Figure 7.1. Here we can think of the complex number $z = x + iy$ as a point (x, y) in the z-complex plane or as a vector. The magnitude, or length, of this vector is called the complex modulus of z, denoted by $|z| = \sqrt{x^2 + y^2}$. We can also use the geometric picture to develop a polar representation of complex numbers. From Figure 7.1 we can see that in terms of r and θ, we have that

$$
\begin{aligned}
x &= r\cos\theta, \\
y &= r\sin\theta.
\end{aligned}
\tag{7.6}
$$

Thus,

$$z = x + iy = r(\cos\theta + i\sin\theta) = re^{i\theta}. \tag{7.7}$$

So, given r and θ we have $z = re^{i\theta}$. However, given the Cartesian form,

[1] The Bernoulli's were a family of Swiss mathematicians spanning three generations. It all started with Jacob Bernoulli (1654–1705) and his brother Johann Bernoulli (1667–1748). Jacob had a son, Nicolaus Bernoulli (1687–1759) and Johann (1667–1748) had three sons, Nicolaus Bernoulli II (1695–1726), Daniel Bernoulli (1700–1872), and Johann Bernoulli II (1710–1790). The last generation consisted of Johann II's sons, Johann Bernoulli III (1747–1807) and Jacob Bernoulli II (1759–1789). Johann, Jacob and Daniel Bernoulli were the most famous of the Bernoulli's. Jacob studied with Leibniz, Johann studied under his older brother and later taught Leonhard Euler and Daniel Bernoulli, who is known for his work in hydrodynamics.

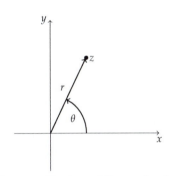

Figure 7.1: The Argand diagram for plotting complex numbers in the complex z-plane.

The complex modulus, $|z| = \sqrt{x^2 + y^2}$.

Complex numbers can be represented in rectangular (Cartesian), $z = x + iy$, or polar form, $z = re^{i\theta}$. Here we define the argument, θ, and modulus, $|z| = r$, of complex numbers.

$z = x + iy$, we can also determine the polar form, as

$$r = \sqrt{x^2 + y^2},$$

$$\tan\theta = \frac{y}{x}. \qquad (7.8)$$

Note that $r = |z|$.

Locating $1 + i$ in the complex plane, it is possible to immediately determine the polar form from the angle and length of the "complex vector." This is shown in Figure 7.2. It is obvious that $\theta = \frac{\pi}{4}$ and $r = \sqrt{2}$.

Example 7.1. *Write $z = 1 + i$ in polar form.*

If one did not see the polar form from the plot in the z-plane, then one could systematically determine the results. First, write $z = 1 + i$ in polar form, $z = re^{i\theta}$, for some r and θ.

Using the above relations between polar and Cartesian representations, we have $r = \sqrt{x^2 + y^2} = \sqrt{2}$ and $\tan\theta = \frac{y}{x} = 1$. This gives $\theta = \frac{\pi}{4}$. So, we have found that

$$1 + i = \sqrt{2}e^{i\pi/4}.$$

We can also define binary operations of addition, subtraction, multiplication, and division of complex numbers to produce a new complex number.

The addition of two complex numbers is simply done by adding the real and imaginary parts of each number. So,

$$(3 + 2i) + (1 - i) = 4 + i.$$

Subtraction is just as easy,

$$(3 + 2i) - (1 - i) = 2 + 3i.$$

We can multiply two complex numbers just like we multiply any binomials, though we now can use the fact that $i^2 = -1$. For example, we have

$$(3 + 2i)(1 - i) = 3 + 2i - 3i + 2i(-i) = 5 - i.$$

We can even divide one complex number into another one and get a complex number as the quotient. Before we do this, we need to introduce the complex conjugate, $\bar{z}$, of a complex number. The complex conjugate of $z = x + iy$, where x and y are real numbers, is given as

$$\bar{z} = x - iy.$$

Complex conjugates satisfy the following relations for complex numbers z and w and real number x.

$$\begin{aligned}
\overline{z + w} &= \bar{z} + \bar{w}, \\
\overline{zw} &= \overline{z}\overline{w}, \\
\bar{\bar{z}} &= z, \\
\bar{x} &= x. \qquad (7.9)
\end{aligned}$$

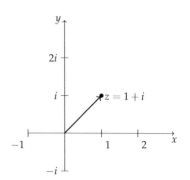

Figure 7.2: Locating $1 + i$ in the complex z-plane.

We can easily add, subtract, multiply, and divide complex numbers.

The complex conjugate of $z = x + iy$ is given as $\bar{z} = x - iy$.

One consequence is that the complex conjugate of $re^{i\theta}$ is

$$\overline{re^{i\theta}} = \overline{\cos\theta + i\sin\theta} = \cos\theta - i\sin\theta = re^{-i\theta}.$$

Another consequence is that

$$z\bar{z} = re^{i\theta}re^{-i\theta} = r^2.$$

Thus, the product of a complex number with its complex conjugate is a real number. We can also prove this result using the Cartesian form

$$z\bar{z} = (x + iy)(x - iy) = x^2 + y^2 = |z|^2.$$

Now we are in a position to write the quotient of two complex numbers in the standard form of a real plus an imaginary number.

Example 7.2. *Simplify the expression* $z = \frac{3+2i}{1-i}$.

This simplification is accomplished by multiplying the numerator and denominator of this expression by the complex conjugate of the denominator:

$$z = \frac{3+2i}{1-i} = \frac{3+2i}{1-i}\frac{1+i}{1+i} = \frac{1+5i}{2}.$$

Therefore, the quotient is a complex number and in standard form is given by $z = \frac{1}{2} + \frac{5}{2}i$.

We can also consider powers of complex numbers. For example,

$$(1+i)^2 = 2i,$$

$$(1+i)^3 = (1+i)(2i) = 2i - 2.$$

But, what is $(1+i)^{1/2} = \sqrt{1+i}$?

In general, we want to find the nth root of a complex number. Let $t = z^{1/n}$. To find t in this case is the same as asking for the solution of

$$z = t^n$$

given z. But, this is the root of an nth degree equation, for which we expect n roots. If we write z in polar form, $z = re^{i\theta}$, then we would naively compute

$$
\begin{aligned}
z^{1/n} &= \left(re^{i\theta}\right)^{1/n} \\
&= r^{1/n}e^{i\theta/n} \\
&= r^{1/n}\left[\cos\frac{\theta}{n} + i\sin\frac{\theta}{n}\right].
\end{aligned}
\tag{7.10}
$$

For example,

$$(1+i)^{1/2} = \left(\sqrt{2}e^{i\pi/4}\right)^{1/2} = 2^{1/4}e^{i\pi/8}.$$

But this is only one solution. We expected two solutions for $n = 2$.

The reason we only found one solution is that the polar representation for z is not unique. We note that

The function $f(z) = z^{1/n}$ is multivalued: $z^{1/n} = r^{1/n}e^{i(\theta+2k\pi)/n}$, $k = 0, 1, \ldots, n-1$.

$$e^{2k\pi i} = 1, \quad k = 0, \pm 1, \pm 2, \ldots.$$

So, we can rewrite z as $z = re^{i\theta}e^{2k\pi i} = re^{i(\theta + 2k\pi)}$. Now we have that

$$z^{1/n} = r^{1/n}e^{i(\theta + 2k\pi)/n}, \quad k = 0, 1, \ldots, n - 1.$$

Note that these are the only distinct values for the roots. We can see this by considering the case $k = n$. Then we find that

$$e^{i(\theta + 2\pi i n)/n} = e^{i\theta/n}e^{2\pi i} = e^{i\theta/n}.$$

So, we have recovered the $n = 0$ value. Similar results can be shown for the other k values larger than n.

Now we can finish the example we started.

Example 7.3. *Determine the square roots of $1 + i$, or $\sqrt{1 + i}$.*

As we have seen, we first write $1 + i$ in polar form: $1 + i = \sqrt{2}e^{i\pi/4}$. Then, introduce $e^{2k\pi i} = 1$ and find the roots:

$$
\begin{aligned}
(1 + i)^{1/2} &= \left(\sqrt{2}e^{i\pi/4}e^{2k\pi i} \right)^{1/2}, \quad k = 0, 1, \\
&= 2^{1/4}e^{i(\pi/8 + k\pi)}, \quad k = 0, 1, \\
&= 2^{1/4}e^{i\pi/8}, 2^{1/4}e^{9\pi i/8}. \tag{7.11}
\end{aligned}
$$

Finally, what is $\sqrt[n]{1}$? Our first guess would be $\sqrt[n]{1} = 1$. But we now know that there should be n roots. These roots are called the nth roots of unity. Using the above result with $r = 1$ and $\theta = 0$, we have that

The nth roots of unity, $\sqrt[n]{1}$.

$$\sqrt[n]{1} = \left[\cos \frac{2\pi k}{n} + i \sin \frac{2\pi k}{n} \right], \quad k = 0, \ldots, n - 1.$$

For example, we have

$$\sqrt[3]{1} = \left[\cos \frac{2\pi k}{3} + i \sin \frac{2\pi k}{3} \right], \quad k = 0, 1, 2.$$

These three roots can be written out as

$$\sqrt[3]{1} = 1, -\frac{1}{2} + \frac{\sqrt{3}}{2}i, -\frac{1}{2} - \frac{\sqrt{3}}{2}i.$$

We can locate these cube roots of unity in the complex plane. In Figure 7.3, we see that these points lie on the unit circle and are at the vertices of an equilateral triangle. In fact, all nth roots of unity lie on the unit circle and are the vertices of a regular n-gon with one vertex at $z = 1$.

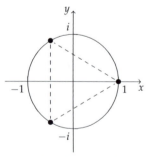

Figure 7.3: Locating the cube roots of unity in the complex z-plane.

7.3 Complex Valued Functions

WE WOULD LIKE TO NEXT EXPLORE complex functions and the calculus of complex functions. We begin by defining a function that takes complex numbers into complex numbers, $f : C \to C$. It is difficult to visualize such functions. For real functions of one variable, $f : R \to R$, we graph these functions by first drawing two intersecting copies of R and then proceed to map the domain into the range of f.

It would be more difficult to do this for complex functions. Imagine placing together two orthogonal copies of the complex plane C. One would need a four-dimensional space in order to complete the visualization. Instead, one typically uses two copies of the complex plane side by side in order to indicate how such functions behave. Over the years there have been several ways to visualize complex functions. We will describe a few of these in this chapter.

We will assume that the domain lies in the z-plane and the image lies in the w-plane. We will then write the complex function as $w = f(z)$. We show these planes in Figure 7.4 as well as the mapping between the planes.

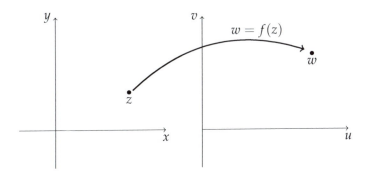

Figure 7.4: Defining a complex valued function, $w = f(z)$, on C for $z = x + iy$ and $w = u + iv$.

Letting $z = x + iy$ and $w = u + iv$, we can write the real and imaginary parts of $f(z)$:

$$w = f(z) = f(x + iy) = u(x, y) + iv(x, y).$$

We see that one can view this function as a function of z or a function of x and y. Often, we have an interest in writing out the real and imaginary parts of the function, $u(x, y)$ and $v(x, y)$, which are functions of two real variables, x and y. We will look at several functions to determine the real and imaginary parts.

Example 7.4. *Find the real and imaginary parts of $f(z) = z^2$.*

For example, we can look at the simple function $f(z) = z^2$. It is a simple matter to determine the real and imaginary parts of this function. Namely, we have

$$z^2 = (x + iy)^2 = x^2 - y^2 + 2ixy.$$

Therefore, we have that

$$u(x, y) = x^2 - y^2, \quad v(x, y) = 2xy.$$

In Figure 7.5 we show how a grid in the z-plane is mapped by $f(z) = z^2$ into the w-plane. For example, the horizontal line $x = 1$ is mapped to $u(1, y) = 1 - y^2$

and $v(1, y) = 2y$. *Eliminating the "parameter" y between these two equations, we have $u = 1 - v^2/4$. This is a parabolic curve. Similarly, the horizontal line $y = 1$ results in the curve $u = v^2/4 - 1$.*

If we look at several curves, $x =$const and $y =$const, then we get a family of intersecting parabolae, as shown in Figure 7.5.

Figure 7.5: 2D plot showing how the function $f(z) = z^2$ maps the lines $x = 1$ and $y = 1$ in the z-plane into parabolae in the w-plane.

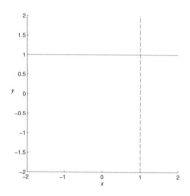
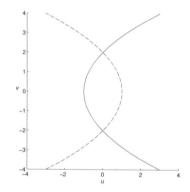

Figure 7.6: 2D plot showing how the function $f(z) = z^2$ maps a grid in the z-plane into the w-plane.

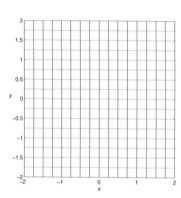
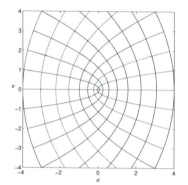

Example 7.5. *Find the real and imaginary parts of $f(z) = e^z$.*
For this case, we make use of Euler's Formula:

$$
\begin{aligned}
e^z &= e^{x+iy} \\
&= e^x e^{iy} \\
&= e^x (\cos y + i \sin y).
\end{aligned}
\tag{7.12}
$$

Thus, $u(x, y) = e^x \cos y$ and $v(x, y) = e^x \sin y$. In Figure 7.7 we show how a grid in the z-plane is mapped by $f(z) = e^z$ into the w-plane.

Example 7.6. *Find the real and imaginary parts of $f(z) = z^{1/2}$.*
We have that

$$
z^{1/2} = \sqrt{x^2 + y^2} \left(\cos(\theta + k\pi) + i \sin(\theta + k\pi) \right), \quad k = 0, 1.
\tag{7.13}
$$

Thus,

$$
u = |z| \cos(\theta + k\pi), u = |z| \cos(\theta + k\pi),
$$

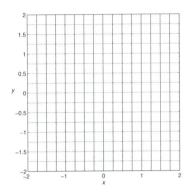 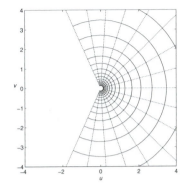

for $|z| = \sqrt{x^2 + y^2}$ and $\theta = \tan^{-1}(y/x)$. For each k-value, one has a different surface and curves of constant θ give $u/v = c_1$; and curves of constant nonzero complex modulus give concentric circles, $u^2 + v^2 = c_2$, for c_1 and c_2 constants.

Example 7.7. *Find the real and imaginary parts of $f(z) = \ln z$.*

In this case, we make use of the polar form of a complex number, $z = re^{i\theta}$. Our first thought would be to simply compute

$$\ln z = \ln r + i\theta.$$

However, the natural logarithm is multivalued, just like the square root function. Recalling that $e^{2\pi i k} = 1$ for k an integer, we have $z = re^{i(\theta + 2\pi k)}$. Therefore,

$$\boxed{\ln z = \ln r + i(\theta + 2\pi k), \qquad k = \text{integer}.}$$

The natural logarithm is a multivalued function. In fact, there are an infinite number of values for a given z. Of course, this contradicts the definition of a function that you were first taught.

Thus, one typically will only report the principal value, $\text{Log } z = \ln r + i\theta$, for θ restricted to some interval of length 2π, such as $[0, 2\pi)$. In order to account for the multivaluedness, one introduces a way to extend the complex plane so as to include all the branches. This is done by assigning a plane to each branch, using (branch) cuts along lines, and then gluing the planes together at the branch cuts to form what is called a Riemann surface. We will not elaborate upon this any further here and refer the interested reader to more advanced texts. Comparing the multivalued logarithm to the principal value logarithm, we have

$$\ln z = \text{Log } z + 2n\pi i.$$

We should note that some books use $\log z$ instead of $\ln z$. It should not be confused with the common logarithm.

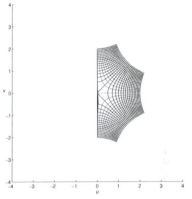

Figure 7.8: 2D plot showing how the function $f(z) = \sqrt{z}$ maps a grid in the z-plane into the w-plane.

7.3.1 *Complex Domain Coloring*

ANOTHER METHOD FOR VISUALIZING COMPLEX FUNCTIONS is domain coloring. The idea was described by Frank A. Farris. There are a few approaches to this method. The main idea is that one colors each point of the

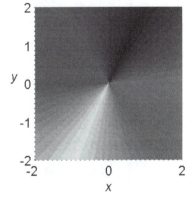

Figure 7.9: Domain coloring of the complex z-plane assigning colors to arg(z).

Figure 7.10: Domain coloring for $f(z) = z^2$. The left figure shows the phase coloring. The right figure shows the colored surface with height $|f(z)|$.

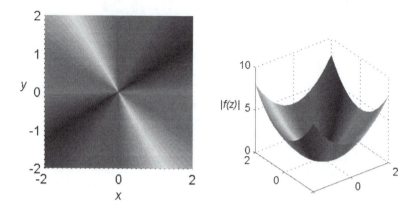

z-plane (the domain) according to $\arg(z)$ as shown in Figure 7.9. The modulus $|f(z)|$ is then plotted as a surface. Examples are shown for $f(z) = z^2$ in Figure 7.10 and $f(z) = 1/z(1-z)$ in Figure 7.11.

Figure 7.11: Domain coloring for $f(z) = 1/z(1-z)$. The left figure shows the phase coloring. The right figure shows the colored surface with height $|f(z)|$.

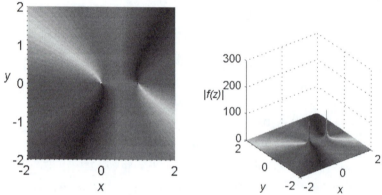

We would like to put all of this information in one plot. We can do this by adjusting the brightness of the colored domain using the modulus of the function. In the plots that follow, we use the fractional part of $\ln |z|$. In Figure 7.12 we show the effect for the z-plane using $f(z) = z$. In the figures that follow, we look at several other functions. In these plots, we have chosen to view the functions in a circular window.

One can see the rich behavior hidden in these figures. As you progress in your reading, especially after the next chapter, you should return to these figures and locate the zeros, poles, branch points, and branch cuts. A search online will lead you to other colorings and superposition of the uv grid on these figures.

As a final picture, we look at iteration in the complex plane. Consider the function $f(z) = z^2 - 0.75 - 0.2i$. Interesting figures result when studying the iteration in the complex plane. In Figure 7.15 we show $f(z)$ and $f^{20}(z)$, which is the iteration of f twenty times. It leads to an interesting coloring. What happens when one keeps iterating? Such iterations lead to the study of Julia and Mandelbrot sets. In Figure 7.16 we show six iterations of $f(z) = (1 - i/2)\sin x$.

Figure 7.12: Domain coloring for the function $f(z) = z$ showing a coloring for $\arg(z)$ and brightness based on $|f(z)|$.

Figure 7.13: Domain coloring for the function $f(z) = z^2$.

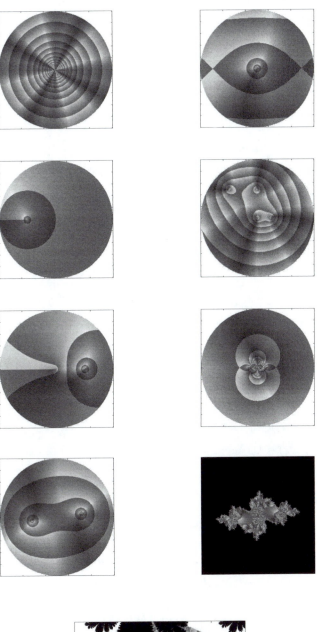

Figure 7.14: Domain coloring for several functions. On the top row, the domain coloring is shown for $f(z) = z^4$ and $f(z) = \sin z$. On the second row, plots for $f(z) = \sqrt{1+z}$ and $f(z) = \frac{1}{z(1/2-z)(z-i)(z-i+1)}$ are shown. In the last row, domain colorings for $f(z) = \ln z$ and $f(z) = \sin(1/z)$ are shown.

Figure 7.15: Domain coloring for $f(z) = z^2 - 0.75 - 0.2i$. The left figure shows the phase coloring. On the right is the plot for $f^{20}(z)$.

Figure 7.16: Domain coloring for six iterations of $f(z) = (1 - i/2) \sin x$.

The following code was used in MATLAB to produce these figures.

```
fn = @(x) (1-i/2)*sin(x);
xmin=-2; xmax=2; ymin=-2; ymax=2;
Nx=500;
Ny=500;
x=linspace(xmin,xmax,Nx);
y=linspace(ymin,ymax,Ny);
[X,Y] = meshgrid(x,y); z = complex(X,Y);
tmp=z; for n=1:6
    tmp = fn(tmp);
end Z=tmp;
XX=real(Z);
YY=imag(Z);
R2=max(max(X.^2));
R=max(max(XX.^2+YY.^2));

circle(:,:,1) = X.^2+Y.^2 < R2;
circle(:,:,2)=circle(:,:,1);
circle(:,:,3)=circle(:,:,1);

addcirc(:,:,1)=circle(:,:,1)==0;
addcirc(:,:,2)=circle(:,:,1)==0;
addcirc(:,:,3)=circle(:,:,1)==0;

warning off MATLAB:divideByZero;
hsvCircle=ones(Nx,Ny,3);
hsvCircle(:,:,1)=atan2(YY,XX)*180/pi+(atan2(YY,XX)*180/pi<0)*360;
hsvCircle(:,:,1)=hsvCircle(:,:,1)/360; lgz=log(XX.^2+YY.^2)/2;
hsvCircle(:,:,2)=0.75; hsvCircle(:,:,3)=1-(lgz-floor(lgz))/2;
hsvCircle(:,:,1) = flipud((hsvCircle(:,:,1)));

hsvCircle(:,:,2) = flipud((hsvCircle(:,:,2)));

hsvCircle(:,:,3) =flipud((hsvCircle(:,:,3)));

rgbCircle=hsv2rgb(hsvCircle);
rgbCircle=rgbCircle.*circle+addcirc;

image(rgbCircle)
axis square
set(gca,'XTickLabel',{})
set(gca,'YTickLabel',{})
```

7.4 Complex Differentiation

NEXT WE WANT TO DIFFERENTIATE COMPLEX FUNCTIONS. We generalize the definition from single variable calculus,

$$f'(z) = \lim_{\Delta z \to 0} \frac{f(z + \Delta z) - f(z)}{\Delta z}, \tag{7.14}$$

provided this limit exists.

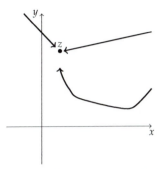

Figure 7.17: There are many paths that approach z as $\Delta z \to 0$.

The computation of this limit is similar to what one sees in multivariable calculus for limits of real functions of two variables. Letting $z = x + iy$ and $\delta z = \delta x + i \delta y$, then

$$z + \delta x = (x + \delta x) + i(y + \delta y).$$

Letting $\Delta z \to 0$ means that we get closer to z. There are many paths that one can take that will approach z. [See Figure 7.17.]

It is sufficient to look at two paths in particular. We first consider the path $y =$ constant. This horizontal path is shown in Figure 7.18. For this path, $\Delta z = \Delta x + i\Delta y = \Delta x$, as y does not change along the path. The derivative, if it exists, is then computed as

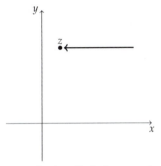

Figure 7.18: A path that approaches z with $y =$ constant.

$$
\begin{aligned}
f'(z) &= \lim_{\Delta z \to 0} \frac{f(z + \Delta z) - f(z)}{\Delta z} \\
&= \lim_{\Delta x \to 0} \frac{u(x + \Delta x, y) + iv(x + \Delta x, y) - (u(x, y) + iv(x, y))}{\Delta x} \\
&= \lim_{\Delta x \to 0} \frac{u(x + \Delta x, y) - u(x, y)}{\Delta x} + \lim_{\Delta x \to 0} i\frac{v(x + \Delta x, y) - v(x, y)}{\Delta x}.
\end{aligned}
\tag{7.15}
$$

The last two limits are easily identified as partial derivatives of real valued functions of two variables. Thus, we have shown that when $f'(z)$ exists,

$$f'(z) = \frac{\partial u}{\partial x} + i\frac{\partial v}{\partial x}. \tag{7.16}$$

A similar computation can be made if, instead, we take the vertical path, $x =$ constant, in Figure 7.17). In this case, $\Delta z = i\Delta y$ and

$$
\begin{aligned}
f'(z) &= \lim_{\Delta z \to 0} \frac{f(z + \Delta z) - f(z)}{\Delta z} \\
&= \lim_{\Delta y \to 0} \frac{u(x, y + \Delta y) + iv(x, y + \Delta y) - (u(x, y) + iv(x, y))}{i\Delta y} \\
&= \lim_{\Delta y \to 0} \frac{u(x, y + \Delta y) - u(x, y)}{i\Delta y} + \lim_{\Delta y \to 0} \frac{v(x, y + \Delta y) - v(x, y)}{\Delta y}.
\end{aligned}
\tag{7.17}
$$

Therefore,

$$f'(z) = \frac{\partial v}{\partial y} - i\frac{\partial u}{\partial y}. \tag{7.18}$$

We have found two different expressions for $f'(z)$ by following two different paths to z. If the derivative exists, then these two expressions must be the same. Equating the real and imaginary parts of these expressions, we have

$$\frac{\partial u}{\partial x} = \frac{\partial v}{\partial y}$$

$$\frac{\partial v}{\partial x} = -\frac{\partial u}{\partial y}. \tag{7.19}$$

These are known as the Cauchy–Riemann Equations[2].

[2] Augustin-Louis Cauchy (1789–1857) was a French mathematician well known for his work in analysis. Georg Friedrich Bernhard Riemann (1826–1866) was a German mathematician who made major contributions to geometry and analysis.

Theorem 7.1. $f(z)$ is holomorphic (differentiable) if and only if the Cauchy–Riemann equations are satisfied.

Example 7.8. $f(z) = z^2$.

In this case, we have already seen that $z^2 = x^2 - y^2 + 2ixy$. Therefore, $u(x,y) = x^2 - y^2$ and $v(x,y) = 2xy$. We first check the Cauchy–Riemann Equations.

$$\frac{\partial u}{\partial x} = 2x = \frac{\partial v}{\partial y}$$

$$\frac{\partial v}{\partial x} = 2y = -\frac{\partial u}{\partial y}. \tag{7.20}$$

Therefore, $f(z) = z^2$ is differentiable.

We can further compute the derivative using either Equation (7.16) or Equation (7.18). Thus,

$$f'(z) = \frac{\partial u}{\partial x} + i\frac{\partial v}{\partial x} = 2x + i(2y) = 2z.$$

This result is not surprising.

Example 7.9. $f(z) = \bar{z}$.

In this case, we have $f(z) = x - iy$. Therefore, $u(x,y) = x$ and $v(x,y) = -y$. But, $\frac{\partial u}{\partial x} = 1$ and $\frac{\partial v}{\partial y} = -1$. Thus, the Cauchy–Riemann Equations are not satisfied and we conclude that $f(z) = \bar{z}$ is not differentiable.

Another consequence of the Cauchy–Riemann Equations is that both $u(x,y)$ and $v(x,y)$ are harmonic functions. A real-valued function $u(x,y)$ is harmonic if it satisfies Laplace's Equation in 2D, $\nabla^2 u = 0$, or

$$\frac{\partial^2 u}{\partial x^2} + \frac{\partial^2 u}{\partial y^2} = 0.$$

Theorem 7.2. $f(z) = u(x,y) + iv(x,y)$ is differentiable if and only if u and v are harmonic functions.

This is easily proven using the Cauchy–Riemann Equations.

$$\frac{\partial^2 u}{\partial x^2} = \frac{\partial}{\partial x}\frac{\partial u}{\partial x}$$

$$= \frac{\partial}{\partial x}\frac{\partial v}{\partial y}$$

$$= \frac{\partial}{\partial y} \frac{\partial v}{\partial x}$$

$$= -\frac{\partial}{\partial y} \frac{\partial u}{\partial y}$$

$$= -\frac{\partial^2 u}{\partial y^2}. \qquad (7.21)$$

Example 7.10. *Is $u(x,y) = x^2 + y^2$ harmonic?*

$$\frac{\partial^2 u}{\partial x^2} + \frac{\partial^2 u}{\partial y^2} = 2 + 2 \neq 0.$$

No, it is not.

Example 7.11. *Is $u(x,y) = x^2 - y^2$ harmonic?*

$$\frac{\partial^2 u}{\partial x^2} + \frac{\partial^2 u}{\partial y^2} = 2 - 2 = 0.$$

Yes, it is.

Given a harmonic function $u(x,y)$, can one find a function $v(x,y)$ such that $f(z) = u(x,y) + iv(x,y)$ is differentiable? In this case, v is called the harmonic conjugate of u.

The harmonic conjugate function.

Example 7.12. *Find the harmonic conjugate of $u(x,y) = x^2 - y^2$ and determine $f(z) = u + iv$ such that $u + iv$ is differentiable.*

The Cauchy–Riemann Equations tell us the following about the unknown function, $v(x,y)$

$$\frac{\partial v}{\partial x} = -\frac{\partial u}{\partial y} = 2y,$$

$$\frac{\partial v}{\partial y} = \frac{\partial u}{\partial x} = 2x.$$

We can integrate the first of these equations to obtain

$$v(x,y) = \int 2y \, dx = 2xy + c(y).$$

Here, $c(y)$ is an arbitrary function of y. One can check to see that this works by simply differentiating the result with respect to x.

However, the second equation must also hold. So, we differentiate the result with respect to y to find that

$$\frac{\partial v}{\partial y} = 2x + c'(y).$$

Because we were supposed to get $2x$, we have that $c'(y) = 0$. Thus, $c(y) = k$ is a constant.

We have just shown that we get an infinite number of functions,

$$v(x,y) = 2xy + k,$$

such that

$$f(z) = x^2 - y^2 + i(2xy + k)$$

is differentiable. In fact, for $k = 0$, this is nothing other than $f(z) = z^2$.

7.5 Complex Integration

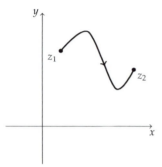

Figure 7.19: We would like to integrate a complex function $f(z)$ over the path Γ in the complex plane.

WE HAVE INTRODUCED FUNCTIONS OF A COMPLEX VARIABLE. We also established when functions are differentiable as complex functions, or holomorphic. In this chapter we will turn to integration in the complex plane. We will learn how to compute complex path integrals, or contour integrals. We will see that contour integral methods are also useful in the computation of some of the real integrals that we will face when exploring Fourier transforms in the next chapter.

7.5.1 Complex Path Integrals

IN THIS SECTION WE WILL INVESTIGATE the computation of complex path integrals. Given two points in the complex plane, connected by a path Γ as shown in Figure 7.19, we would like to define the integral of $f(z)$ along Γ,

$$\int_\Gamma f(z)\,dz.$$

A natural procedure would be to work in real variables, by writing

$$\int_\Gamma f(z)\,dz = \int_\Gamma [u(x,y) + iv(x,y)]\,(dx + idy),$$

as $z = x + iy$ and $dz = dx + idy$.

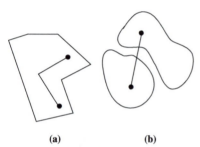

(a)　　　　**(b)**

Figure 7.20: Examples of (a) a connected set and (b) a disconnected set.

In order to carry out the integration, we then have to find a parametrization of the path and use methods from a multivariate calculus class. Namely, let u and v be continuous in domain D, and Γ a piecewise smooth curve in D. Let $(x(t), y(t))$ be a parametrization of Γ for $t_0 \leq t \leq t_1$ and $f(z) = u(x,y) + iv(x,y)$ for $z = x + iy$. Then

$$\int_\Gamma f(z)\,dz = \int_{t_0}^{t_1} [u(x(t),y(t)) + iv(x(t),y(t))]\,(\frac{dx}{dt} + i\frac{dy}{dt})dt. \qquad (7.22)$$

Here we have used

$$dz = dx + idy = \left(\frac{dx}{dt} + i\frac{dy}{dt}\right) dt.$$

Furthermore, a set D is called a domain if it is both open and connected.

Before continuing, we first define open and connected. A set D is connected if and only if for all z_1, and z_2 in D, there exists a piecewise smooth curve connecting z_1 to z_2 and lying in D. Otherwise it is called disconnected. Examples are shown in Figure 7.20

A set D is open if and only if for all z_0 in D, there exists an open disk $|z - z_0| < \rho$ in D. In Figure 7.21 we show a region with two disks.

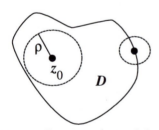

Figure 7.21: Locations of open disks inside and on the boundary of a region.

For all points on the interior of the region, one can find at least one disk contained entirely in the region. The closer one is to the boundary, the smaller the radii of such disks. However, for a point on the boundary, every such disk would contain points inside and outside the disk. Thus, an open set in the complex plane would not contain any of its boundary points.

We now have a prescription for computing path integrals. Let's see how this works with a couple of examples.

Example 7.13. *Evaluate $\int_C z^2\, dz$, where C = the arc of the unit circle in the first quadrant as shown in Figure 7.22.*

There are two ways we could carry out the parametrization. First, we note that the standard parametrization of the unit circle is

$$(x(\theta), y(\theta)) = (\cos\theta, \sin\theta), \quad 0 \le \theta \le 2\pi.$$

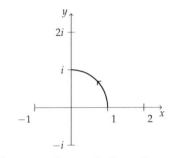

Figure 7.22: Contour for Example 7.13.

For a quarter circle in the first quadrant, $0 \le \theta \le \frac{\pi}{2}$, we let $z = \cos\theta + i\sin\theta$. Therefore, $dz = (-\sin\theta + i\cos\theta)\, d\theta$ and the path integral becomes

$$\int_C z^2\, dz = \int_0^{\frac{\pi}{2}} (\cos\theta + i\sin\theta)^2 (-\sin\theta + i\cos\theta)\, d\theta.$$

We can expand the integrand and integrate, having to perform some trigonometric integrations.

$$\int_0^{\frac{\pi}{2}} [\sin^3\theta - 3\cos^2\theta\sin\theta + i(\cos^3\theta - 3\cos\theta\sin^2\theta)]\, d\theta.$$

The reader should work out these trigonometric integrations and confirm the result. For example, you can use

$$\sin^3\theta = \sin\theta(1 - \cos^2\theta))$$

to write the real part of the integrand as

$$\sin\theta - 4\cos^2\theta\sin\theta.$$

The resulting antiderivative becomes

$$-\cos\theta + \frac{4}{3}\cos^3\theta.$$

The imaginary integrand can be integrated in a similar fashion.

While this integral is doable, there is a simpler procedure. We first note that $z = e^{i\theta}$ on C. So, $dz = ie^{i\theta}\, d\theta$. The integration then becomes

$$\begin{aligned}
\int_C z^2\, dz &= \int_0^{\frac{\pi}{2}} (e^{i\theta})^2 ie^{i\theta}\, d\theta \\
&= i\int_0^{\frac{\pi}{2}} e^{3i\theta}\, d\theta \\
&= \frac{ie^{3i\theta}}{3i}\Big|_0^{\pi/2} \\
&= -\frac{1+i}{3}.
\end{aligned} \tag{7.23}$$

Example 7.14. *Evaluate $\int_\Gamma z\, dz$, for the path $\Gamma = \gamma_1 \cup \gamma_2$ shown in Figure 7.23.*

In this problem we have a path that is a piecewise smooth curve. We can compute the path integral by computing the values along the two segments of the path and *adding the results. Let the two segments be called γ_1 and γ_2 as shown in Figure 7.23 and parametrize each path separately.*

Over γ_1 we note that $y = 0$. Thus, $z = x$ for $x \in [0,1]$. It is natural to take x as the parameter. So, we let $dz = dx$ to find

$$\int_{\gamma_1} z\, dz = \int_0^1 x\, dx = \frac{1}{2}.$$

For path γ_2, we have that $z = 1 + iy$ for $y \in [0,1]$ and $dz = i\, dy$. Inserting this parametrization into the integral, the integral becomes

$$\int_{\gamma_2} z\, dz = \int_0^1 (1 + iy)\, i\, dy = i - \frac{1}{2}.$$

Combining the results for the paths γ_1 and γ_2, we have $\int_\Gamma z\, dz = \frac{1}{2} + (i - \frac{1}{2}) = i$.

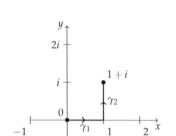

Figure 7.23: Contour for Example 7.14 with $\Gamma = \gamma_1 \cup \gamma_2$.

Example 7.15. *Evaluate $\int_{\gamma_3} z\, dz$, where γ_3 is the path shown in Figure 7.24.*

In this case, we take a path from $z = 0$ to $z = 1 + i$ along a different path than in the previous example. Let $\gamma_3 = \{(x,y) | y = x^2, x \in [0,1]\} = \{z | z = x + ix^2, x \in [0,1]\}$. Then, $dz = (1 + 2ix)\, dx$.

The integral becomes

$$
\begin{aligned}
\int_{\gamma_3} z\, dz &= \int_0^1 (x + ix^2)(1 + 2ix)\, dx \\
&= \int_0^1 (x + 3ix^2 - 2x^3)\, dx = \\
&= \left[\frac{1}{2}x^2 + ix^3 - \frac{1}{2}x^4 \right]_0^1 = i. \quad (7.24)
\end{aligned}
$$

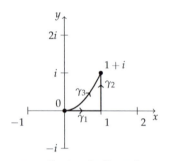

Figure 7.24: Contour for Example 7.15.

In the last case we found the same answer as obtained in Example 7.14. But we should not take this as a general rule for all complex path integrals. In fact, it is not true that integrating over different paths always yields the same results. However, when this is true, then we refer to this property as path independence. In particular, the integral $\int f(z)\, dz$ is path independent if

$$\int_{\Gamma_1} f(z)\, dz = \int_{\Gamma_2} f(z)\, dz$$

for all paths from z_1 to z_2 as shown in Figure 7.25.

We can show that if $\int f(z)\, dz$ is path independent, then the integral of $f(z)$ over all closed loops is zero:

$$\int_{\text{closed loops}} f(z)\, dz = 0.$$

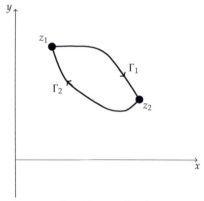

Figure 7.25: $\int_{\Gamma_1} f(z)\, dz = \int_{\Gamma_2} f(z)\, dz$ for all paths from z_1 to z_2 when the integral of $f(z)$ is path independent. A simple closed contour.

A common notation for integrating over closed loops is $\oint_C f(z)\, dz$. But first we have to define what we mean by a closed loop. A simple closed contour is a path satisfying

a. The end point is the same as the beginning point. (This makes the loop closed.)

b. The are no self-intersections. (This makes the loop simple.)

A loop in the shape of a figure eight is closed, but it is not simple.

Now consider an integral over the closed loop C shown in Figure 7.26. We pick two points on the loop, breaking it into two contours, C_1 and C_2. Then we make use of the path independence by defining C_2^- to be the path along C_2 but in the opposite direction. Then,

$\oint_C f(z)\,dz = 0$ if the integral is path independent.

$$\oint_C f(z)\,dz = \int_{C_1} f(z)\,dz + \int_{C_2} f(z)\,dz$$
$$= \int_{C_1} f(z)\,dz - \int_{C_2^-} f(z)\,dz. \qquad (7.25)$$

Assuming that the integrals from point 1 to point 2 are path independent, then the integrals over C_1 and C_2^- are equal. Therefore, we have $\oint_C f(z)\,dz = 0$.

Example 7.16. *Consider the integral $\oint_C z\,dz$ for C the closed contour shown in Figure 7.24 starting at $z = 0$ following path γ_1, then γ_2 and returning to $z = 0$. Based on the earlier examples and the fact that going backward on γ_3 introduces a negative sign, we have*

$$\oint_C z\,dz = \int_{\gamma_1} z\,dz + \int_{\gamma_2} z\,dz - \int_{\gamma_3} z\,dz = \frac{1}{2} + \left(i - \frac{1}{2}\right) - i = 0.$$

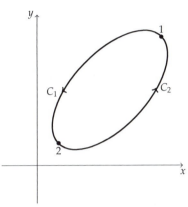

Figure 7.26: The integral $\oint_C f(z)\,dz$ around C is zero if the integral $\int_\Gamma f(z)\,dz$ is path independent.

7.5.2 Cauchy's Theorem

NEXT WE WANT TO INVESTIGATE if we can determine that integrals over simple closed contours vanish without doing all the work of parametrizing the contour. First, we need to establish the direction about which we traverse the contour. We can define the orientation of a curve by referring to the normal of the curve.

Recall that the normal is a perpendicular to the curve. There are two such perpendiculars. The above normal points outward and the other normal points toward the interior of a closed curve. We will define a positively oriented contour as one that is traversed with the outward normal pointing to the right. As one follows loops, the interior would then be on the left.

We now consider $\oint_C (u + iv)\,dz$ over a simple closed contour. This can be written in terms of two real integrals in the xy-plane.

A curve with parametrization $(x(t), y(t))$ has a normal $(n_x, n_y) = \left(-\frac{dx}{dt}, \frac{dy}{dt}\right)$.

$$\oint_C (u + iv)\,dz = \int_C (u + iv)(dx + i\,dy)$$
$$= \int_C u\,dx - v\,dy + i\int_C v\,dx + u\,dy. \qquad (7.26)$$

These integrals in the plane can be evaluated using Green's Theorem in the Plane. Recall this theorem from your last semester of calculus:

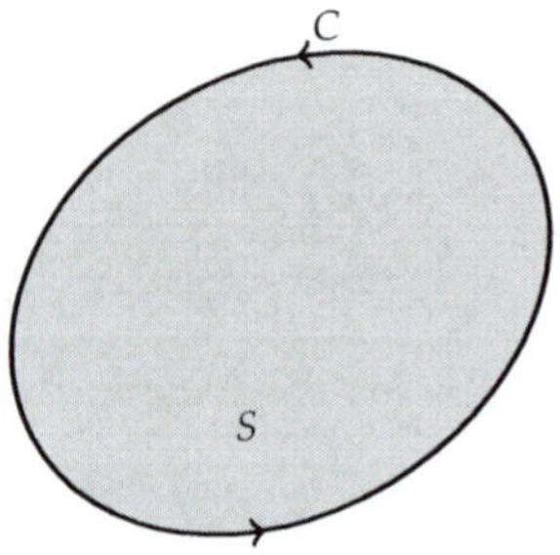

Figure 7.27: Region used in Green's Theorem.

Green's Theorem in the Plane is one of the major integral theorems of vector calculus. It was discovered by George Green (1793–1841) and published in 1828, about four years before he entered Cambridge as an undergraduate.

Green's Theorem in the Plane.

Theorem 7.3. *Let $P(x,y)$ and $Q(x,y)$ be continuously differentiable functions on and inside the simple closed curve C as shown in Figure 7.27. Denoting the enclosed region S, we have*

$$\int_C P\,dx + Q\,dy = \int\int_S \left(\frac{\partial Q}{\partial x} - \frac{\partial P}{\partial y}\right) dxdy. \tag{7.27}$$

Using Green's Theorem to rewrite the first integral in Equation (7.26), we have

$$\int_C u\,dx - v\,dy = \int\int_S \left(\frac{-\partial v}{\partial x} - \frac{\partial u}{\partial y}\right) dxdy.$$

If u and v satisfy the Cauchy–Riemann Equations (7.19), then the integrand in the double integral vanishes. Therefore,

$$\int_C u\,dx - v\,dy = 0.$$

In a similar fashion, one can show that

$$\int_C v\,dx + u\,dy = 0.$$

We have thus proven the following theorem:

Cauchy's Theorem

Theorem 7.4. *If u and v satisfy the Cauchy–Riemann Equations (7.19) inside and on the simple closed contour C, then*

$$\oint_C (u + iv)\,dz = 0. \tag{7.28}$$

Corollary $\oint_C f(z)\,dz = 0$ when f is differentiable in domain D with $C \subset D$.

Either one of these is referred to as **Cauchy's Theorem**.

Example 7.17. *Evaluate $\oint_{|z-1|=3} z^4\,dz$.*

Because $f(z) = z^4$ is differentiable inside the circle $|z-1| = 3$, this integral vanishes.

We can use Cauchy's Theorem to show that we can deform one contour into another, perhaps simpler contour.

One can deform contours into simpler ones.

Theorem 7.5. *If $f(z)$ is holomorphic between two simple closed contours, C and C', then $\oint_C f(z)\,dz = \oint_{C'} f(z)\,dz$.*

Proof. We consider the two curves C and C' as shown in Figure 7.28. Connecting the two contours with contours Γ_1 and Γ_2 (as shown in the figure), C is seen to split into contours C_1 and C_2 and C' into contours C_1' and C_2'. Note that $f(z)$ is differentiable inside the newly formed regions between the curves. Also, the boundaries of these regions are now simple closed curves.

Therefore, Cauchy's Theorem tells us that the integrals of $f(z)$ over these regions are zero.

Noting that integrations over contours opposite the positive orientation are the negative of integrals that are positively oriented, we have from Cauchy's Theorem that

$$\int_{C_1} f(z)\,dz + \int_{\Gamma_1} f(z)\,dz - \int_{C_1'} f(z)\,dz + \int_{\Gamma_2} f(z)\,dz = 0$$

and

$$\int_{C_2} f(z)\,dz - \int_{\Gamma_2} f(z)\,dz - \int_{C_2'} f(z)\,dz - \int_{\Gamma_1} f(z)\,dz = 0.$$

In the first integral, we have traversed the contours in the following order: C_1, Γ_1, C_1' backward, and Γ_2. The second integral denotes the integration over the lower region, but going backward over all contours except for C_2.

Combining these results by adding the two equations above, we have

$$\int_{C_1} f(z)\,dz + \int_{C_2} f(z)\,dz - \int_{C_1'} f(z)\,dz - \int_{C_2'} f(z)\,dz = 0.$$

Noting that $C = C_1 + C_2$ and $C' = C_1' + C_2'$, we have

$$\oint_C f(z)\,dz = \oint_{C'} f(z)\,dz,$$

as was to be proven. □

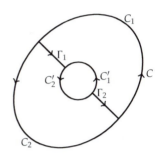

Figure 7.28: The contours needed to prove that $\oint_C f(z)\,dz = \oint_{C'} f(z)\,dz$ when $f(z)$ is holomorphic between the contours C and C'.

Example 7.18. *Compute $\oint_R \frac{dz}{z}$ for R the rectangle $[-2,2] \times [-2i, 2i]$.*

We can compute this integral by looking at four separate integrals over the sides of the rectangle in the complex plane. One simply parametrizes each line segment, performs the integration, and sums the four separate results. From Theorem 7.5, we can instead integrate over a simpler contour by deforming the rectangle into a circle as long as $f(z) = \frac{1}{z}$ is differentiable in the region bounded by the rectangle and the circle. So, using the unit circle, as shown in Figure 7.29, the integration might be easier to perform.

More specifically, Theorem 7.5 tells us that

$$\oint_R \frac{dz}{z} = \oint_{|z|=1} \frac{dz}{z}$$

The latter integral can be computed using the parametrization $z = e^{i\theta}$ for $\theta \in [0, 2\pi]$. Thus,

$$\oint_{|z|=1} \frac{dz}{z} = \int_0^{2\pi} \frac{ie^{i\theta}\,d\theta}{e^{i\theta}}$$
$$= i \int_0^{2\pi} d\theta = 2\pi i. \qquad (7.29)$$

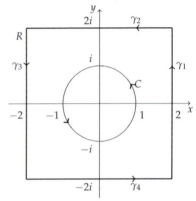

Figure 7.29: The contours used to compute $\oint_R \frac{dz}{z}$. Note that to compute the integral around R we can deform the contour to the circle C because $f(z)$ is differentiable in the region between the contours.

Therefore, we have found that $\oint_R \frac{dz}{z} = 2\pi i$ by deforming the original simple closed contour.

For fun, let's do this the long way to see how much effort was saved. We will label the contour as shown in Figure 7.30. The lower segment, γ_4, of the square can be

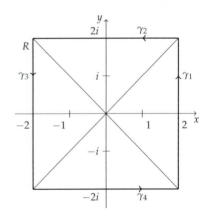

Figure 7.30: The contours used to compute $\oint_R \frac{dz}{z}$. The added diagonals are for the reader to easily see the arguments used in the evaluation of the limits when integrating over the segments of the square R.

simple parametrized by noting that along this segment, $z = x - 2i$ for $x \in [-2,2]$. Then, we have

$$
\begin{aligned}
\oint_{\gamma_4} \frac{dz}{z} &= \int_{-2}^{2} \frac{dx}{x - 2i} \\
&= \ln\left| x - 2i \right|_{-2}^{2} \\
&= \left(\ln(2\sqrt{2}) - \frac{\pi i}{4} \right) - \left(\ln(2\sqrt{2}) - \frac{3\pi i}{4} \right) \\
&= \frac{\pi i}{2}.
\end{aligned}
\tag{7.30}
$$

We note that the arguments of the logarithms are determined from the angles made by the diagonals provided in Figure 7.30.

Similarly, the integral along the top segment, $z = x + 2i$, $x \in [-2,2]$, is computed as

$$
\begin{aligned}
\oint_{\gamma_2} \frac{dz}{z} &= \int_{2}^{-2} \frac{dx}{x + 2i} \\
&= \ln\left| x + 2i \right|_{2}^{-2} \\
&= \left(\ln(2\sqrt{2}) + \frac{3\pi i}{4} \right) - \left(\ln(2\sqrt{2}) + \frac{\pi i}{4} \right) \\
&= \frac{\pi i}{2}.
\end{aligned}
\tag{7.31}
$$

The integral over the right side, $z = 2 + iy$, $y \in [-2,2]$, is

$$
\begin{aligned}
\oint_{\gamma_1} \frac{dz}{z} &= \int_{-2}^{2} \frac{i\,dy}{2 + iy} \\
&= \ln\left| 2 + iy \right|_{-2}^{2} \\
&= \left(\ln(2\sqrt{2}) + \frac{\pi i}{4} \right) - \left(\ln(2\sqrt{2}) - \frac{\pi i}{4} \right) \\
&= \frac{\pi i}{2}.
\end{aligned}
\tag{7.32}
$$

Finally, the integral over the left side, $z = -2 + iy$, $y \in [-2,2]$, is

$$
\begin{aligned}
\oint_{\gamma_3} \frac{dz}{z} &= \int_{2}^{-2} \frac{i\,dy}{-2 + iy} \\
&= \ln\left| -2 + iy \right|_{-2}^{2} \\
&= \left(\ln(2\sqrt{2}) + \frac{5\pi i}{4} \right) - \left(\ln(2\sqrt{2}) + \frac{3\pi i}{4} \right) \\
&= \frac{\pi i}{2}.
\end{aligned}
\tag{7.33}
$$

Therefore, we have that

$$
\oint_R \frac{dz}{z} = \int_{\gamma_1} \frac{dz}{z} + \int_{\gamma_2} \frac{dz}{z} + \int_{\gamma_3} \frac{dz}{z} + \int_{\gamma_4} \frac{dz}{z}
$$

$$= \frac{\pi i}{2} + \frac{\pi i}{2} + \frac{\pi i}{2} + \frac{\pi i}{2}$$
$$= 4\left(\frac{\pi i}{2}\right) = 2\pi i. \tag{7.34}$$

This gives the same answer as we found using a simple contour deformation.

The converse of Cauchy's Theorem is not true, namely $\oint_C f(z)\,dz = 0$ does not always imply that $f(z)$ is differentiable. What we do have is Morera's Theorem(Giacinto Morera, 1856–1909):

Morera's Theorem.

Theorem 7.6. *Let f be continuous in a domain D. Suppose that for every simple closed contour C in D, $\oint_C f(z)\,dz = 0$. Then f is differentiable in D.*

The proof is a bit more detailed than we need to go into here. However, this theorem is useful in the next section.

7.5.3 Analytic Functions and Cauchy's Integral Formula

IN THE PREVIOUS SECTION WE SAW that Cauchy's Theorem was useful for computing particular integrals without having to parametrize the contours or for deforming contours into simpler contours. The integrand needs to possess certain differentiability properties. In this section we will generalize the functions that we can integrate slightly so that we can integrate a larger family of complex functions. This will lead us to the Cauchy's Integral Formula, which extends Cauchy's Theorem to functions analytic in an annulus. However, first we need to explore the concept of analytic functions.

A function $f(z)$ is analytic in domain D if for every open disk $|z - z_0| < \rho$ lying in D, $f(z)$ can be represented as a power series in z_0. Namely,

$$f(z) = \sum_{n=0}^{\infty} c_n (z - z_0)^n.$$

This series converges uniformly and absolutely inside the circle of convergence, $|z - z_0| < R$, with radius of convergence R. [See the Appendix for a review of convergence.]

Because $f(z)$ can be written as a uniformly convergent power series, we can integrate it term by term over any simple closed contour in D containing z_0. In particular, we have to compute integrals like $\oint_C (z - z_0)^n\,dz$. As we will see in the homework problems, these integrals evaluate to zero for most n. Thus, we can show that for $f(z)$ analytic in D and on any closed contour C lying in D, $\oint_C f(z)\,dz = 0$. Also, f is a uniformly convergent sum of continuous functions, so $f(z)$ is also continuous. Thus, by Morera's Theorem, we have that $f(z)$ is differentiable if it is analytic. Often terms like analytic, differentiable, and holomorphic are used interchangeably, though there is a subtle distinction due to their definitions.

As examples of series expansions about a given point, we will consider series expansions and regions of convergence for $f(z) = \frac{1}{1+z}$.

There are various types of complex-valued functions.

A holomorphic function is (complex) differentiable in a neighborhood of every point in its domain.

An analytic function has a convergent Taylor series expansion in a neighborhood of each point in its domain. We see here that analytic functions are holomorphic and vice versa.

If a function is holomorphic throughout the complex plane, then it is called an entire function.

Finally, a function which is holomorphic on all of its domain except at a set of isolated poles (to be defined later), then it is called a meromorphic function.

Example 7.19. *Find the series expansion of $f(z) = \frac{1}{1+z}$ about $z_0 = 0$.*

This case is simple. From Chapter 1 we recall that $f(z)$ is the sum of a geometric series for $|z| < 1$. We have

$$f(z) = \frac{1}{1+z} = \sum_{n=0}^{\infty} (-z)^n.$$

Thus, this series expansion converges inside the unit circle ($|z| < 1$) in the complex plane.

Example 7.20. *Find the series expansion of $f(z) = \frac{1}{1+z}$ about $z_0 = \frac{1}{2}$.*

We now look into an expansion about a different point. We could compute the expansion coefficients using Taylor's formula for the coefficients. However, we can also make use of the formula for geometric series after rearranging the function. We seek an expansion in powers of $z - \frac{1}{2}$. So, we rewrite the function in a form that is a function of $z - \frac{1}{2}$. Thus,

$$f(z) = \frac{1}{1+z} = \frac{1}{1 + (z - \frac{1}{2} + \frac{1}{2})} = \frac{1}{\frac{3}{2} + (z - \frac{1}{2})}.$$

This is not quite in the form we need. It would be nice if the denominator were of the form of one plus something. [Note: This is similar to what we had seen in Example 1.36.] We can get the denominator into such a form by factoring out the $\frac{3}{2}$. Then we would have

$$f(z) = \frac{2}{3} \frac{1}{1 + \frac{2}{3}(z - \frac{1}{2})}.$$

The second factor now has the form $\frac{1}{1-r}$, which would be the sum of a geometric series with first term $a = 1$ and ratio $r = -\frac{2}{3}(z - \frac{1}{2})$ provided that $|r| < 1$. Therefore, we have found that

$$f(z) = \frac{2}{3} \sum_{n=0}^{\infty} \left[-\frac{2}{3}\left(z - \frac{1}{2}\right) \right]^n$$

for

$$\left| -\frac{2}{3}\left(z - \frac{1}{2}\right) \right| < 1.$$

This convergence interval can be rewritten as

$$\left| z - \frac{1}{2} \right| < \frac{3}{2},$$

which is a circle centered at $z = \frac{1}{2}$ with radius $\frac{3}{2}$.

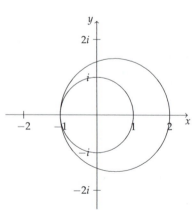

Figure 7.31: Regions of convergence for expansions of $f(z) = \frac{1}{1+z}$ about $z = 0$ and $z = \frac{1}{2}$.

In Figure 7.31 we show the regions of convergence for the power series expansions of $f(z) = \frac{1}{1+z}$ about $z = 0$ and $z = \frac{1}{2}$. We note that the first expansion gives that $f(z)$ is at least analytic inside the region $|z| < 1$. The second expansion shows that $f(z)$ is analytic in a larger region, $|z - \frac{1}{2}| < \frac{3}{2}$. We will see later that there are expansions that converge outside these regions and that some yield expansions involving negative powers of $z - z_0$.

We now present the main theorem of this section:

Cauchy Integral Formula

Theorem 7.7. *Let $f(z)$ be analytic in $|z - z_0| < \rho$ and let C be the boundary (circle) of this disk. Then,*

$$f(z_0) = \frac{1}{2\pi i} \oint_C \frac{f(z)}{z - z_0}\, dz. \qquad (7.35)$$

Proof. In order to prove this, we first make use of the analyticity of $f(z)$. We insert the power series expansion of $f(z)$ about z_0 into the integrand. Then we have

$$
\begin{aligned}
\frac{f(z)}{z - z_0} &= \frac{1}{z - z_0}\left[\sum_{n=0}^{\infty} c_n(z - z_0)^n\right] \\
&= \frac{1}{z - z_0}\left[c_0 + c_1(z - z_0) + c_2(z - z_0)^2 + \ldots\right] \\
&= \frac{c_0}{z - z_0} + \underbrace{c_1 + c_2(z - z_0) + \ldots}_{\text{analytic function}}. \qquad (7.36)
\end{aligned}
$$

As noted, the integrand can be written as

$$\frac{f(z)}{z - z_0} = \frac{c_0}{z - z_0} + h(z),$$

where $h(z)$ is an analytic function, as $h(z)$ is representable as a series expansion about z_0. We have already shown that analytic functions are differentiable, so by Cauchy's Theorem $\oint_C h(z)\, dz = 0$.

Noting also that $c_0 = f(z_0)$ is the first term of a Taylor series expansion about $z = z_0$, we have

$$\oint_C \frac{f(z)}{z - z_0}\, dz = \oint_C \left[\frac{c_0}{z - z_0} + h(z)\right] dz = f(z_0) \oint_C \frac{1}{z - z_0}\, dz.$$

We need only compute the integral $\oint_C \frac{1}{z - z_0}\, dz$ to finish the proof of Cauchy's Integral Formula. This is done by parametrizing the circle, $|z - z_0| = \rho$, as shown in Figure 7.32. This is simply done by letting

$$z - z_0 = \rho e^{i\theta}.$$

(Note that this has the right complex modulus because $|e^{i\theta}| = 1$. Then $dz = i\rho e^{i\theta}\, d\theta$. Using this parametrization, we have

$$\oint_C \frac{dz}{z - z_0} = \int_0^{2\pi} \frac{i\rho e^{i\theta}\, d\theta}{\rho e^{i\theta}} = i\int_0^{2\pi} d\theta = 2\pi i.$$

Therefore,

$$\oint_C \frac{f(z)}{z - z_0}\, dz = f(z_0) \oint_C \frac{1}{z - z_0}\, dz = 2\pi i f(z_0),$$

as was to be shown. $\qquad \square$

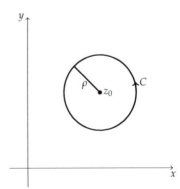

Figure 7.32: Circular contour used in proving the Cauchy Integral Formula.

Example 7.21. *Compute $\oint_{|z|=4} \frac{\cos z}{z^2 - 6z + 5} \, dz$.*

In order to apply the Cauchy Integral Formula, we need to factor the denominator, $z^2 - 6z + 5 = (z-1)(z-5)$. We next locate the zeros of the denominator. In Figure 7.33 we show the contour and the points $z = 1$ and $z = 5$. The only point inside the region bounded by the contour is $z = 1$. Therefore, we can apply the Cauchy Integral Formula for $f(z) = \frac{\cos z}{z-5}$ to the integral

$$\int_{|z|=4} \frac{\cos z}{(z-1)(z-5)} \, dz = \int_{|z|=4} \frac{f(z)}{(z-1)} \, dz = 2\pi i f(1).$$

Therefore, we have

$$\int_{|z|=4} \frac{\cos z}{(z-1)(z-5)} \, dz = -\frac{\pi i \cos(1)}{2}.$$

We have shown that $f(z_0)$ has an integral representation for $f(z)$ analytic in $|z - z_0| < \rho$. In fact, all derivatives of an analytic function have an integral representation. This is given by

$$f^{(n)}(z_0) = \frac{n!}{2\pi i} \oint_C \frac{f(z)}{(z - z_0)^{n+1}} \, dz \tag{7.37}$$

This can be proven following a derivation similar to that for the Cauchy Integral Formula. Inserting the Taylor series expansion for $f(z)$ into the integral on the right-hand side, we have

$$\oint_C \frac{f(z)}{(z - z_0)^{n+1}} \, dz = \sum_{m=0}^{\infty} c_m \oint_C \frac{(z - z_0)^m}{(z - z_0)^{n+1}} \, dz$$

$$= \sum_{m=0}^{\infty} c_m \oint_C \frac{dz}{(z - z_0)^{n-m+1}}. \tag{7.38}$$

Picking $k = n - m$, the integrals in the sum can be computed using the following result:

$$\oint_C \frac{dz}{(z - z_0)^{k+1}} = \begin{cases} 0, & k \neq 0, \\ 2\pi i, & k = 0. \end{cases} \tag{7.39}$$

The proof is left for the exercises.

The only nonvanishing integrals, $\oint_C \frac{dz}{(z-z_0)^{n-m+1}}$, occur when $k = n - m = 0$, or $m = n$. Therefore, the series of integrals collapses to one term and we have

$$\oint_C \frac{f(z)}{(z - z_0)^{n+1}} \, dz = 2\pi i c_n.$$

We finish the proof by recalling that the coefficients of the Taylor series expansion for $f(z)$ are given by

$$c_n = \frac{f^{(n)}(z_0)}{n!}.$$

Then,

$$\oint_C \frac{f(z)}{(z - z_0)^{n+1}} \, dz = \frac{2\pi i}{n!} f^{(n)}(z_0)$$

and the result follows.

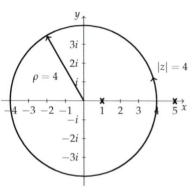

Figure 7.33: Circular contour used in computing $\oint_{|z|=4} \frac{\cos z}{z^2 - 6z + 5} \, dz$.

7.5.4   Laurent Series

UNTIL THIS POINT WE HAVE ONLY TALKED about series whose terms have nonnegative powers of $z - z_0$. It is possible to have series representations in which there are negative powers. In the previous section, we investigated expansions of $f(z) = \frac{1}{1+z}$ about $z = 0$ and $z = \frac{1}{2}$. The regions of convergence for each series was shown in Figure 7.31. Let us reconsider each of these expansions, but for values of z outside the region of convergence previously found.

Example 7.22. $f(z) = \frac{1}{1+z}$ for $|z| > 1$.

  As before, we make use of the geometric series . Because $|z| > 1$, we instead rewrite the function as

$$f(z) = \frac{1}{1+z} = \frac{1}{z} \frac{1}{1 + \frac{1}{z}}.$$

We now have the function in a form of the sum of a geometric series with first term $a = 1$ and ratio $r = -\frac{1}{z}$. We note that $|z| > 1$ implies that $|r| < 1$. Thus, we have the geometric series

$$f(z) = \frac{1}{z} \sum_{n=0}^{\infty} \left(-\frac{1}{z} \right)^n.$$

This can be re-indexed[3] as

$$f(z) = \sum_{n=0}^{\infty} (-1)^n z^{-n-1} = \sum_{j=1}^{\infty} (-1)^{j-1} z^{-j}.$$

Note that this series, which converges outside the unit circle, $|z| > 1$, has negative powers of z.

Example 7.23. $f(z) = \frac{1}{1+z}$ for $|z - \frac{1}{2}| > \frac{3}{2}$.

  As before, we express this in a form in which we can use a geometric series expansion. We seek powers of $z - \frac{1}{2}$. So, we add and subtract $\frac{1}{2}$ to the z to obtain

$$f(z) = \frac{1}{1+z} = \frac{1}{1 + (z - \frac{1}{2} + \frac{1}{2})} = \frac{1}{\frac{3}{2} + (z - \frac{1}{2})}.$$

Instead of factoring out the $\frac{3}{2}$ as we had done in Example 7.20, we factor out the $(z - \frac{1}{2})$ term. Then, we obtain

$$f(z) = \frac{1}{1+z} = \frac{1}{(z - \frac{1}{2})} \frac{1}{\left[1 + \frac{3}{2}(z - \frac{1}{2})^{-1} \right]}.$$

  Now we identify $a = 1$ and $r = -\frac{3}{2}(z - \frac{1}{2})^{-1}$. This leads to the series

$$
\begin{aligned}
f(z) &= \frac{1}{z - \frac{1}{2}} \sum_{n=0}^{\infty} \left(-\frac{3}{2}(z - \frac{1}{2})^{-1} \right)^n \\
&= \sum_{n=0}^{\infty} \left(-\frac{3}{2} \right)^n \left(z - \frac{1}{2} \right)^{-n-1}.
\end{aligned}
\tag{7.40}
$$

This converges for $|z - \frac{1}{2}| > \frac{3}{2}$ and can also be re-indexed to verify that this series involves negative powers of $z - \frac{1}{2}$.

[3] Re-indexing a series is often useful in series manipulations. In this case, we have the series

$$\sum_{n=0}^{\infty} (-1)^n z^{-n-1} = z^{-1} - z^{-2} + z^{-3} + \ldots.$$

The index is n. You can see that the index does not appear when the sum is expanded showing the terms. The summation index is sometimes referred to as a dummy index for this reason. Re-indexing allows one to rewrite the shorthand summation notation while capturing the same terms. In this example, the exponents are $-n - 1$. We can simplify the notation by letting $-n - 1 = -j$, or $j = n + 1$. Noting that $j = 1$ when $n = 0$, we get the sum $\sum_{j=1}^{\infty} (-1)^{j-1} z^{-j}$.

This leads to the following theorem:

Theorem 7.8. *Let $f(z)$ be analytic in an annulus, $R_1 < |z - z_0| < R_2$, with C a positively oriented simple closed curve around z_0 and inside the annulus as shown in Figure 7.34. Then,*

$$f(z) = \sum_{j=0}^{\infty} a_j(z - z_0)^j + \sum_{j=1}^{\infty} b_j(z - z_0)^{-j},$$

with

$$a_j = \frac{1}{2\pi i} \oint_C \frac{f(z)}{(z - z_0)^{j+1}} \, dz$$

and

$$b_j = \frac{1}{2\pi i} \oint_C \frac{f(z)}{(z - z_0)^{-j+1}} \, dz.$$

The above series can be written in the more compact form

$$f(z) = \sum_{j=-\infty}^{\infty} c_j(z - z_0)^j.$$

Such a series expansion is called a Laurent series expansion, named after its discoverer Pierre Alphonse Laurent (1813–1854).

Example 7.24. *Expand $f(z) = \frac{1}{(1-z)(2+z)}$ in the annulus $1 < |z| < 2$.*
Using partial fractions, we can write this as

$$f(z) = \frac{1}{3}\left[\frac{1}{1-z} + \frac{1}{2+z}\right].$$

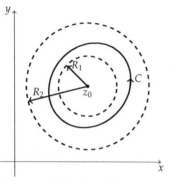

Figure 7.34: This figure shows an annulus, $R_1 < |z - z_0| < R_2$, with C a positively oriented simple closed curve around z_0 and inside the annulus.

We can expand the first fraction, $\frac{1}{1-z}$, as an analytic function in the region $|z| > 1$ and the second fraction, $\frac{1}{2+z}$, as an analytic function in $|z| < 2$. This is done as follows. First, we write

$$\frac{1}{2+z} = \frac{1}{2[1 - (-\frac{z}{2})]} = \frac{1}{2}\sum_{n=0}^{\infty}\left(-\frac{z}{2}\right)^n.$$

Then, we write

$$\frac{1}{1-z} = -\frac{1}{z[1 - \frac{1}{z}]} = -\frac{1}{z}\sum_{n=0}^{\infty}\frac{1}{z^n}.$$

Therefore, in the common region, $1 < |z| < 2$, we have that

$$\begin{aligned}
\frac{1}{(1-z)(2+z)} &= \frac{1}{3}\left[\frac{1}{2}\sum_{n=0}^{\infty}\left(-\frac{z}{2}\right)^n - \sum_{n=0}^{\infty}\frac{1}{z^{n+1}}\right] \\
&= \sum_{n=0}^{\infty}\frac{(-1)^n}{6(2^n)}z^n + \sum_{n=1}^{\infty}\frac{(-1)}{3}z^{-n}.
\end{aligned} \tag{7.41}$$

We note that this is not a Taylor series expansion due to the existence of terms with negative powers in the second sum.

Example 7.25. *Find series representations of* $f(z) = \frac{1}{(1-z)(2+z)}$ *throughout the complex plane.*

In the previous example, we found series representations of $f(z) = \frac{1}{(1-z)(2+z)}$ in the annulus $1 < |z| < 2$. However, we can also find expansions that converge for other regions. We first write

$$f(z) = \frac{1}{3}\left[\frac{1}{1-z} + \frac{1}{2+z}\right].$$

We then expand each term separately.

The first fraction, $\frac{1}{1-z}$, can be written as the sum of the geometric series

$$\frac{1}{1-z} = \sum_{n=0}^{\infty} z^n, \quad |z| < 1.$$

This series converges inside the unit circle. We indicate this by region 1 in Figure 7.35.

In the previous example, we showed that the second fraction, $\frac{1}{2+z}$, has the series expansion

$$\frac{1}{2+z} = \frac{1}{2[1-(-\frac{z}{2})]} = \frac{1}{2}\sum_{n=0}^{\infty}\left(-\frac{z}{2}\right)^n,$$

that converges in the circle $|z| < 2$. This is labeled as region 2 in Figure 7.35.

Regions 1 and 2 intersect for $|z| < 1$, so, we can combine these two series representations to obtain

$$\frac{1}{(1-z)(2+z)} = \frac{1}{3}\left[\sum_{n=0}^{\infty} z^n + \frac{1}{2}\sum_{n=0}^{\infty}\left(-\frac{z}{2}\right)^n\right], \quad |z| < 1.$$

In the annulus, $1 < |z| < 2$, we had already seen in the previous example that we needed a different expansion for the fraction $\frac{1}{1-z}$. We looked for an expansion in powers of $1/z$ that would converge for large values of z. We had found that

$$\frac{1}{1-z} = -\frac{1}{z\left(1-\frac{1}{z}\right)} = -\frac{1}{z}\sum_{n=0}^{\infty}\frac{1}{z^n}, \quad |z| > 1.$$

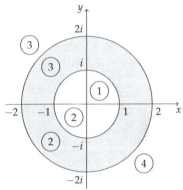

Figure 7.35: Regions of convergence for Laurent expansions of $f(z) = \frac{1}{(1-z)(2+z)}$.

This series converges in region 3 in Figure 7.35. Combining this series with the one for the second fraction, we obtain a series representation that converges in the overlap of regions 2 and 3. Thus, in the annulus $1 < |z| < 2$, we have

$$\frac{1}{(1-z)(2+z)} = \frac{1}{3}\left[\frac{1}{2}\sum_{n=0}^{\infty}\left(-\frac{z}{2}\right)^n - \sum_{n=0}^{\infty}\frac{1}{z^{n+1}}\right].$$

So far, we have series representations for $|z| < 2$. The only region not covered yet is outside this disk, $|z| > 2$. In Figure 7.35 we see that series 3, which converges in region 3, will converge in the last section of the complex plane. We just need one more series expansion for $1/(2+z)$ for large z. Factoring out a z in the denominator, we can write this as a geometric series with $r = 2/z$:

$$\frac{1}{2+z} = \frac{1}{z[\frac{2}{z}+1]} = \frac{1}{z}\sum_{n=0}^{\infty}\left(-\frac{2}{z}\right)^n.$$

This series converges for $|z| > 2$. Therefore, it converges in region 4 and the final series representation is

$$\frac{1}{(1-z)(2+z)} = \frac{1}{3}\left[\frac{1}{z}\sum_{n=0}^{\infty}\left(-\frac{2}{z}\right)^n - \sum_{n=0}^{\infty}\frac{1}{z^{n+1}}\right].$$

7.5.5 Singularities and Residue Theorem

IN THE PREVIOUS SECTION, WE FOUND that we could integrate functions satisfying some analyticity properties along contours without using detailed parametrizations around the contours. We can deform contours if the function is analytic in the region between the original and new contour. In this section we will extend our tools for performing contour integrals.

Singularities of complex functions.

The integrand in the Cauchy Integral Formula was of the form $g(z) = \frac{f(z)}{z-z_0}$, where $f(z)$ is well behaved at z_0. The point $z = z_0$ is called a singularity of $g(z)$, as $g(z)$ is not defined there. More specifically, a singularity of $f(z)$ is a point at which $f(z)$ fails to be analytic.

We can also classify these singularities. Typically, these are isolated singularities. As we saw from the proof of the Cauchy Integral Formula, $g(z) = \frac{f(z)}{z-z_0}$ has a Laurent series expansion about $z = z_0$, given by

$$g(z) = \frac{f(z_0)}{z-z_0} + f'(z_0) + \frac{1}{2}f''(z_0)(z-z_0) + \dots.$$

It is the nature of the first term that gives information about the type of singularity that $g(z)$ has. Namely, in order to classify the singularities of $f(z)$, we look at the principal part of the Laurent series of $f(z)$ about $z = z_0$,

Classification of singularities.

$\sum_{j=1}^{\infty} b_j(z-z_0)^{-j}$, which consists of the negative powers of $z - z_0$.

There are three types of singularities: removable, poles, and essential singularities. They are defined as follows:

1. If $f(z)$ is bounded near z_0, then z_0 is a removable singularity.

2. If there are a finite number of terms in the principal part of the Laurent series of $f(z)$ about $z = z_0$, then z_0 is called a pole.

3. If there are an infinite number of terms in the principal part of the Laurent series of $f(z)$ about $z = z_0$, then z_0 is called an essential singularity.

Example 7.26. *$f(z) = \frac{\sin z}{z}$ has a removable singularity at $z = 0$.*

At first it looks like there is a possible singularity at $z = 0$, as the denominator is zero at $z = 0$. However, we know from the first semester of calculus that $\lim_{z\to 0}\frac{\sin z}{z} = 1$. Furthermore, we can expand $\sin z$ about $z = 0$ and see that

$$\frac{\sin z}{z} = \frac{1}{z}\left(z - \frac{z^3}{3!} + \dots\right) = 1 - \frac{z^2}{3!} + \dots.$$

Thus, there are only nonnegative powers in the series expansion. So, $z = 0$ is a removable singularity.

Example 7.27. $f(z) = \frac{e^z}{(z-1)^n}$ *has poles at $z = 1$ for n a positive integer.*

For $n = 1$, we have $f(z) = \frac{e^z}{z-1}$. This function has a singularity at $z = 1$. The series expansion is found by expanding e^z about $z = 1$:

$$f(z) = \frac{e}{z-1}e^{z-1} = \frac{e}{z-1} + e + \frac{e}{2!}(z-1) + \cdots.$$

Note that the principal part of the Laurent series expansion about $z = 1$ only has one term, $\frac{e}{z-1}$. Therefore, $z = 1$ is a pole. Because the leading term has an exponent of -1, $z = 1$ is called a pole of order one, or a simple pole.

Simple pole.

For $n = 2$, we have $f(z) = \frac{e^z}{(z-1)^2}$. The series expansion is again found by expanding e^z about $z = 1$:

$$f(z) = \frac{e}{(z-1)^2}e^{z-1} = \frac{e}{(z-1)^2} + \frac{e}{z-1} + \frac{e}{2!} + \frac{e}{3!}(z-1) + \cdots.$$

Note that the principal part of the Laurent series has two terms involving $(z-1)^{-2}$ and $(z-1)^{-1}$. Because the leading term has an exponent of -2, $z = 1$ is called a pole of order 2, or a double pole.

Double pole.

Example 7.28. $f(z) = e^{\frac{1}{z}}$ *has an essential singularity at $z = 0$.*

In this case, we have the series expansion about $z = 0$ given by

$$f(z) = e^{\frac{1}{z}} = \sum_{n=0}^{\infty} \frac{\left(\frac{1}{z}\right)^n}{n!} = \sum_{n=0}^{\infty} \frac{1}{n!}z^{-n}.$$

We see that there are an infinite number of terms in the principal part of the Laurent series. So, this function has an essential singularity at $z = 0$.

In the above examples, we have seen poles of order 1 (a simple pole) and 2 (a double pole). In general, we can say that $f(z)$ has a pole of order k at z_0 if and only if $(z - z_0)^k f(z)$ has a removable singularity at z_0, but $(z - z_0)^{k-1} f(z)$ for $k > 0$ does not.

Poles of order k.

Example 7.29. *Determine the order of the pole of $f(z) = \cot z \csc z$ at $z = 0$.*

First we rewrite $f(z)$ in terms of sines and cosines.

$$f(z) = \cot z \csc z = \frac{\cos z}{\sin^2 z}.$$

We note that the denominator vanishes at $z = 0$.

How do we know that the pole is not a simple pole? Well, we check to see if $(z - 0)f(z)$ has a removable singularity at $z = 0$:

$$\begin{aligned}
\lim_{z\to 0}(z-0)f(z) &= \lim_{z\to 0}\frac{z\cos z}{\sin^2 z} \\
&= \left(\lim_{z\to 0}\frac{z}{\sin z}\right)\left(\lim_{z\to 0}\frac{\cos z}{\sin z}\right) \\
&= \lim_{z\to 0}\frac{\cos z}{\sin z}. \quad\quad\quad (7.42)
\end{aligned}$$

We see that this limit is undefined. So now we check to see if $(z - 0)^2 f(z)$ has a removable singularity at $z = 0$:

$$\lim_{z\to 0}(z-0)^2 f(z) = \lim_{z\to 0}\frac{z^2 \cos z}{\sin^2 z}$$

$$
= \left(\lim_{z \to 0} \frac{z}{\sin z} \right) \left(\lim_{z \to 0} \frac{z \cos z}{\sin z} \right)
$$

$$
= \lim_{z \to 0} \frac{z}{\sin z} \cos(0) = 1. \tag{7.43}
$$

In this case, we have obtained a finite, nonzero, result. So, $z = 0$ is a pole of order 2.

We could have also relied on series expansions. Expanding both the sine and cosine functions in a Taylor series expansion, we have

$$
f(z) = \frac{\cos z}{\sin^2 z} = \frac{1 - \frac{1}{2!} z^2 + \dots}{(z - \frac{1}{3!} z^3 + \dots)^2}.
$$

Factoring a z from the expansion in the denominator,

$$
f(z) = \frac{1}{z^2} \frac{1 - \frac{1}{2!} z^2 + \dots}{(1 - \frac{1}{3!} z + \dots)^2} = \frac{1}{z^2} \left(1 + O(z^2) \right),
$$

we can see that the leading term will be a $1/z^2$, indicating a pole of order 2.

We will see how knowledge of the poles of a function can aid in the computation of contour integrals. We now show that if a function, $f(z)$, has a pole of order k, then

$$
\oint_C f(z) \, dz = 2\pi i \, \mathrm{Res}[f(z); z_0],
$$

where we have defined $\mathrm{Res}[f(z); z_0]$ as the residue of $f(z)$ at $z = z_0$. In particular, for a pole of order k the residue is given by

---

Residues – Poles of Order k

$$
\mathrm{Res}[f(z); z_0] = \lim_{z \to z_0} \frac{1}{(k-1)!} \frac{d^{k-1}}{dz^{k-1}} \left[(z - z_0)^k f(z) \right]. \tag{7.44}
$$

Proof. Let $\phi(z) = (z - z_0)^k f(z)$ be an analytic function. Then $\phi(z)$ has a Taylor series expansion about z_0. As we had seen in the previous section, we can write the integral representation of any derivative of ϕ as

$$
\phi^{(k-1)}(z_0) = \frac{(k-1)!}{2\pi i} \oint_C \frac{\phi(z)}{(z - z_0)^k} \, dz.
$$

Inserting the definition of $\phi(z)$, we then have

$$
\phi^{(k-1)}(z_0) = \frac{(k-1)!}{2\pi i} \oint_C f(z) \, dz.
$$

Solving for the integral, we have the following result:

$$
\oint_C f(z) \, dz = \frac{2\pi i}{(k-1)!} \frac{d^{k-1}}{dz^{k-1}} \left[(z - z_0)^k f(z) \right]_{z=z_0}
$$

$$
\equiv 2\pi i \, \mathrm{Res}[f(z); z_0] \tag{7.45}
$$

$\square$

Note: If z_0 is a simple pole, the residue is easily computed as

$$Res[f(z); z_0] = \lim_{z \to z_0} (z - z_0) f(z).$$

The residue for a simple pole.

In fact, one can show (Problem 18) that for g and h analytic functions at z_0, with $g(z_0) \neq 0$, $h(z_0) = 0$, and $h'(z_0) \neq 0$,

$$Res\left[\frac{g(z)}{h(z)}; z_0\right] = \frac{g(z_0)}{h'(z_0)}.$$

Example 7.30. *Find the residues of $f(z) = \frac{z-1}{(z+1)^2(z^2+4)}$.*

$f(z)$ has poles at $z = -1$, $z = 2i$, and $z = -2i$. The pole at $z = -1$ is a double pole (pole of order 2). The other poles are simple poles. We compute those residues first:

$$
\begin{aligned}
Res[f(z); 2i] &= \lim_{z \to 2i} (z - 2i) \frac{z-1}{(z+1)^2(z+2i)(z-2i)} \\
&= \lim_{z \to 2i} \frac{z-1}{(z+1)^2(z+2i)} \\
&= \frac{2i-1}{(2i+1)^2(4i)} = -\frac{1}{50} - \frac{11}{100}i.
\end{aligned}
\tag{7.46}
$$

$$
\begin{aligned}
Res[f(z); -2i] &= \lim_{z \to -2i} (z + 2i) \frac{z-1}{(z+1)^2(z+2i)(z-2i)} \\
&= \lim_{z \to -2i} \frac{z-1}{(z+1)^2(z-2i)} \\
&= \frac{-2i-1}{(-2i+1)^2(-4i)} = -\frac{1}{50} + \frac{11}{100}i.
\end{aligned}
\tag{7.47}
$$

For the double pole, we have to do a little more work.

$$
\begin{aligned}
Res[f(z); -1] &= \lim_{z \to -1} \frac{d}{dz} \left[(z+1)^2 \frac{z-1}{(z+1)^2(z^2+4)} \right] \\
&= \lim_{z \to -1} \frac{d}{dz} \left[\frac{z-1}{z^2+4} \right] \\
&= \lim_{z \to -1} \frac{d}{dz} \left[\frac{z^2 + 4 - 2z(z-1)}{(z^2+4)^2} \right] \\
&= \lim_{z \to -1} \frac{d}{dz} \left[\frac{-z^2 + 2z + 4}{(z^2+4)^2} \right] \\
&= \frac{1}{25}.
\end{aligned}
\tag{7.48}
$$

Example 7.31. *Find the residue of $f(z) = \cot z$ at $z = 0$.*

We write $f(z) = \cot z = \frac{\cos z}{\sin z}$ and note that $z = 0$ is a simple pole. Thus,

$$Res[\cot z; z = 0] = \lim_{z \to 0} \frac{z \cos z}{\sin z} = \cos(0) = 1.$$

The residue of $f(z)$ at z_0 is the coefficient of the $(z - z_0)^{-1}$ term, $c_{-1} = b_1$, of the Laurent series expansion about z_0.

Another way to find the residue of a function $f(z)$ at a singularity z_0 is to look at the Laurent series expansion about the singularity. This is because the residue of $f(z)$ at z_0 is the coefficient of the $(z - z_0)^{-1}$ term, or $c_{-1} = b_1$.

Example 7.32. *Find the residue of $f(z) = \frac{1}{z(3-z)}$ at $z = 0$ using a Laurent series expansion.*

First, we need the Laurent series expansion about $z = 0$ of the form $\sum_{-\infty}^{\infty} c_n z^n$. A partial fraction expansion gives

$$f(z) = \frac{1}{z(3-z)} = \frac{1}{3}\left(\frac{1}{z} + \frac{1}{3-z}\right).$$

The first term is a power of z. The second term needs to be written as a convergent series for small z. This is given by

$$
\begin{aligned}
\frac{1}{3-z} &= \frac{1}{3(1-z/3)} \\
&= \frac{1}{3}\sum_{n=0}^{\infty}\left(\frac{z}{3}\right)^n.
\end{aligned}
\tag{7.49}
$$

Thus, we have found

$$f(z) = \frac{1}{3}\left(\frac{1}{z} + \frac{1}{3}\sum_{n=0}^{\infty}\left(\frac{z}{3}\right)^n\right).$$

The coefficient of z^{-1} can be read off to give $\text{Res}[f(z); z = 0] = \frac{1}{3}$.

Example 7.33. *Find the residue of $f(z) = z\cos\frac{1}{z}$ at $z = 0$ using a Laurent series expansion.*

In this case, $z = 0$ is an essential singularity. The only way to find residues at essential singularities is to use Laurent series. Because

$$\cos z = 1 - \frac{1}{2!}z^2 + \frac{1}{4!}z^4 - \frac{1}{6!}z^6 + \ldots,$$

then we have

$$
\begin{aligned}
f(z) &= z\left(1 - \frac{1}{2!z^2} + \frac{1}{4!z^4} - \frac{1}{6!z^6} + \ldots\right) \\
&= z - \frac{1}{2!z} + \frac{1}{4!z^3} - \frac{1}{6!z^5} + \ldots.
\end{aligned}
\tag{7.50}
$$

From the second term we have that $\text{Res}[f(z); z = 0] = -\frac{1}{2}$.

We are now ready to use residues in order to evaluate integrals.

Example 7.34. *Evaluate $\oint_{|z|=1} \frac{dz}{\sin z}$.*

We begin by looking for the singularities of the integrand. These are located at values of z for which $\sin z = 0$. Thus, $z = 0, \pm\pi, \pm 2\pi, \ldots$, are the singularities. However, only $z = 0$ lies inside the contour, as shown in Figure 7.36. We note further that $z = 0$ is a simple pole, as

$$\lim_{z\to 0}(z-0)\frac{1}{\sin z} = 1.$$

Therefore, the residue is 1 and we have

$$\oint_{|z|=1} \frac{dz}{\sin z} = 2\pi i.$$

Finding the residue at an essential singularity.

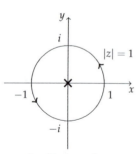

Figure 7.36: Contour for computing $\oint_{|z|=1} \frac{dz}{\sin z}$.

In general, we could have several poles of different orders. For example, we will be computing

$$\oint_{|z|=2} \frac{dz}{z^2 - 1}.$$

The integrand has singularities at $z^2 - 1 = 0$, or $z = \pm 1$. Both poles are inside the contour, as seen in Figure 7.38. One could do a partial fraction decomposition and have two integrals with one pole each integral. Then, the result could be found by adding the residues from each pole.

In general, when there are several poles, we can use the Residue Theorem:

The Residue Theorem

Theorem 7.9. *Let $f(z)$ be a function that has poles z_j, $j = 1, \ldots, N$, inside a simple closed contour C and no other singularities in this region. Then,*

$$\oint_C f(z)\, dz = 2\pi i \sum_{j=1}^{N} Res[f(z); z_j], \qquad (7.51)$$

where the residues are computed using Equation (7.44),

$$Res[f(z); z_0] = \lim_{z \to z_0} \frac{1}{(k-1)!} \frac{d^{k-1}}{dz^{k-1}} \left[(z - z_0)^k f(z) \right].$$

The proof of this theorem is based upon the contours shown in Figure 7.37. One constructs a new contour C' by encircling each pole, as shown in the figure. Then one connects a path from C to each circle. In the figure, two separated paths along the cut are shown only to indicate the direction followed on the cut. The new contour is then obtained by following C and crossing each cut as it is encountered. Then one goes around a circle in the negative sense and returns along the cut to proceed around C. The sum of the contributions to the contour integration involve two integrals for each cut, which will cancel due to the opposing directions. Thus, we are left with

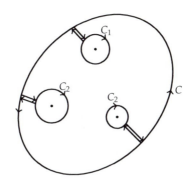

Figure 7.37: A depiction of how one cuts out poles to prove that the integral around C is the sum of the integrals around circles with the poles at the center of each.

$$\oint_{C'} f(z)\, dz = \oint_C f(z)\, dz - \oint_{C_1} f(z)\, dz - \oint_{C_2} f(z)\, dz - \oint_{C_3} f(z)\, dz = 0.$$

Of course, the sum is zero because $f(z)$ is analytic in the enclosed region, as all singularities have been cut out. Solving for $\oint_C f(z)\, dz$, one has that this integral is the sum of the integrals around the separate poles, which can be evaluated with single residue computations. Thus, the result is that $\oint_C f(z)\, dz$ is $2\pi i$ times the sum of the residues.

Example 7.35. *Evaluate $\oint_{|z|=2} \frac{dz}{z^2 - 1}$.*
  We first note that there are two poles in this integral because

$$\frac{1}{z^2 - 1} = \frac{1}{(z-1)(z+1)}.$$

In Figure 7.38 we plot the contour and the two poles, denoted by an "x." Because both poles are inside the contour, we need to compute the residues for each one. They are each simple poles, so we have

$$Res\left[\frac{1}{z^2-1};z=1\right] = \lim_{z\to 1}(z-1)\frac{1}{z^2-1}$$

$$= \lim_{z\to 1}\frac{1}{z+1} = \frac{1}{2},\qquad (7.52)$$

and

$$Res\left[\frac{1}{z^2-1};z=-1\right] = \lim_{z\to -1}(z+1)\frac{1}{z^2-1}$$

$$= \lim_{z\to -1}\frac{1}{z-1} = -\frac{1}{2}.\qquad (7.53)$$

Then,

$$\oint_{|z|=2}\frac{dz}{z^2-1} = 2\pi i\left(\frac{1}{2}-\frac{1}{2}\right) = 0.$$

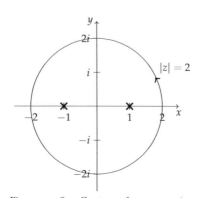

Figure 7.38:   Contour for computing $\oint_{|z|=2}\frac{dz}{z^2-1}$.

Example 7.36. *Evaluate* $\oint_{|z|=3}\frac{z^2+1}{(z-1)^2(z+2)}\,dz$.

In this example, there are two poles $z=1,-2$ inside the contour. [See Figure 7.39.] $z=1$ is a second-order pole and $z=-2$ is a simple pole. Therefore, we need to compute the residues at each pole of $f(z) = \frac{z^2+1}{(z-1)^2(z+2)}$:

$$Res[f(z);z=1] = \lim_{z\to 1}\frac{1}{1!}\frac{d}{dz}\left[(z-1)^2\frac{z^2+1}{(z-1)^2(z+2)}\right]$$

$$= \lim_{z\to 1}\left(\frac{z^2+4z-1}{(z+2)^2}\right)$$

$$= \frac{4}{9}.\qquad (7.54)$$

$$Res[f(z);z=-2] = \lim_{z\to -2}(z+2)\frac{z^2+1}{(z-1)^2(z+2)}$$

$$= \lim_{z\to -2}\frac{z^2+1}{(z-1)^2}$$

$$= \frac{5}{9}.\qquad (7.55)$$

The evaluation of the integral is found by computing $2\pi i$ times the sum of the residues:

$$\oint_{|z|=3}\frac{z^2+1}{(z-1)^2(z+2)}\,dz = 2\pi i\left(\frac{4}{9}+\frac{5}{9}\right) = 2\pi i.$$

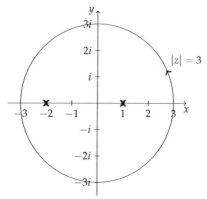

Figure 7.39:   Contour for computing $\oint_{|z|=3}\frac{z^2+1}{(z-1)^2(z+2)}\,dz$.

Example 7.37. *Compute* $\oint_{|z|=2}z^3e^{2/z}\,dz$.

In this case, $z=0$ is an essential singularity and is inside the contour. A Laurent series expansion about $z=0$ gives

$$z^3e^{2/z} = z^3\sum_{n=0}^{\infty}\frac{1}{n!}\left(\frac{2}{z}\right)^n$$

$$= \sum_{n=0}^{\infty}\frac{2^n}{n!}z^{3-n}$$

$$= z^3 + \frac{2}{2!}z^2 + \frac{4}{3!}z + \frac{8}{4!} + \frac{16}{5!z} + \dots.\qquad (7.56)$$

The residue is the coefficient of z^{-1}, *or* $Res[z^3 e^{2/z}; z = 0] = -\frac{2}{15}$. *Therefore,*

$$\oint_{|z|=2} z^3 e^{2/z}\, dz = \frac{4}{15}\pi i.$$

Example 7.38. *Evaluate* $\int_0^{2\pi} \frac{d\theta}{2+\cos\theta}$.

Here we have a real integral in which there are no signs of complex functions. In fact, we could apply simpler methods from a calculus course to do this integral, attempting to write $1 + \cos\theta = 2\cos^2\frac{\theta}{2}$. *However, we do not get very far.*

One trick, useful in computing integrals whose integrand is in the form $f(\cos\theta, \sin\theta)$, *is to transform the integration to the complex plane through the transformation* $z = e^{i\theta}$. *Then,*

$$\cos\theta = \frac{e^{i\theta} + e^{-i\theta}}{2} = \frac{1}{2}\left(z + \frac{1}{z}\right),$$

$$\sin\theta = \frac{e^{i\theta} - e^{-i\theta}}{2i} = -\frac{i}{2}\left(z - \frac{1}{z}\right).$$

Computation of integrals of functions of sines and cosines, $f(\cos\theta, \sin\theta)$.

Under this transformation, $z = e^{i\theta}$, *the integration now takes place around the unit circle in the complex plane. Noting that* $dz = ie^{i\theta}\, d\theta = iz\, d\theta$, *we have*

$$
\begin{aligned}
\int_0^{2\pi} \frac{d\theta}{2+\cos\theta} &= \oint_{|z|=1} \frac{\frac{dz}{iz}}{2 + \frac{1}{2}\left(z + \frac{1}{z}\right)} \\
&= -i \oint_{|z|=1} \frac{dz}{2z + \frac{1}{2}(z^2 + 1)} \\
&= -2i \oint_{|z|=1} \frac{dz}{z^2 + 4z + 1}.
\end{aligned}
\tag{7.57}
$$

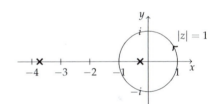

Figure 7.40: Contour for computing $\int_0^{2\pi} \frac{d\theta}{2+\cos\theta}$.

We can apply the Residue Theorem to the resulting integral. The singularities occur at the roots of $z^2 + 4z + 1 = 0$. *Using the quadratic formula, we have the roots* $z = -2 \pm \sqrt{3}$.

The location of these poles are shown in Figure 7.40. Only $z = -2 + \sqrt{3}$ *lies inside the integration contour. We will therefore need the residue of* $f(z) = \frac{-2i}{z^2+4z+1}$ *at this simple pole:*

$$
\begin{aligned}
Res[f(z); z = -2 + \sqrt{3}] &= \lim_{z\to-2+\sqrt{3}} (z - (-2+\sqrt{3}))\frac{-2i}{z^2 + 4z + 1} \\
&= -2i \lim_{z\to-2+\sqrt{3}} \frac{z - (-2+\sqrt{3})}{(z - (-2+\sqrt{3}))(z - (-2-\sqrt{3}))} \\
&= -2i \lim_{z\to-2+\sqrt{3}} \frac{1}{z - (-2-\sqrt{3})} \\
&= \frac{-2i}{-2+\sqrt{3} - (-2-\sqrt{3})} \\
&= \frac{-i}{\sqrt{3}} \\
&= \frac{-i\sqrt{3}}{3}.
\end{aligned}
\tag{7.58}
$$

Therefore, we have

$$\int_0^{2\pi} \frac{d\theta}{2+\cos\theta} = -2i \oint_{|z|=1} \frac{dz}{z^2+4z+1} = 2\pi i \left(\frac{-i\sqrt{3}}{3}\right) = \frac{2\pi\sqrt{3}}{3}. \quad (7.59)$$

Before moving on to further applications, we note that there is another way to compute the integral in the last example. Karl Theodor Wilhelm Weierstraß (1815–1897) introduced a substitution method for computing integrals involving rational functions of sine and cosine. One makes the substitution $t = \tan\frac{\theta}{2}$ and converts the integrand into a rational function of t. One can show that this substitution implies that

The Weierstraß substitution method.

$$\sin\theta = \frac{2t}{1+t^2}, \quad \cos\theta = \frac{1-t^2}{1+t^2},$$

and

$$d\theta = \frac{2dt}{1+t^2}.$$

The details are left for Problem 8 and applying the method. In order to see how it works, we will redo Example 7.3.8.

Example 7.39. *Apply the Weierstraß substitution method to compute $\int_0^{2\pi} \frac{d\theta}{2+\cos\theta}$.*

$$\begin{aligned}
\int_0^{2\pi} \frac{d\theta}{2+\cos\theta} &= \int_{-\infty}^{\infty} \frac{1}{2+\frac{1-t^2}{1+t^2}} \frac{2dt}{1+t^2} \\
&= 2\int_{-\infty}^{\infty} \frac{dt}{t^2+3} \\
&= \frac{2}{3}\sqrt{3}\left[\tan^{-1}\left(\frac{\sqrt{3}}{3}t\right)\right]_{-\infty}^{\infty} = \frac{2\pi\sqrt{3}}{3}. \quad (7.60)
\end{aligned}$$

7.5.6   Infinite Integrals

INFINITE INTEGRALS OF THE FORM $\int_{-\infty}^{\infty} f(x)\,dx$ occur often in physics. They can represent wave packets, wave diffraction, Fourier transforms, and arise in other applications. In this section, we will see that such integrals may be computed by extending the integration to a contour in the complex plane.

Recall from your calculus experience that these integrals are improper integrals, and the way that one determines if improper integrals exist, or converge, is to carefully compute these integrals using limits such as

$$\int_{-\infty}^{\infty} f(x)\,dx = \lim_{R\to\infty} \int_{-R}^{R} f(x)\,dx.$$

For example, we evaluate the integral of $f(x) = x$ as

$$\int_{-\infty}^{\infty} x\,dx = \lim_{R\to\infty} \int_{-R}^{R} x\,dx = \lim_{R\to\infty}\left(\frac{R^2}{2} - \frac{(-R)^2}{2}\right) = 0.$$

One might also be tempted to carry out this integration by splitting the integration interval, $(-\infty, 0] \cup [0, \infty)$. However, the integrals $\int_0^\infty x\, dx$ and $\int_{-\infty}^0 x\, dx$ do not exist. A simple computation confirms this.

$$\int_0^\infty x\, dx = \lim_{R\to\infty} \int_0^R x\, dx = \lim_{R\to\infty} \left(\frac{R^2}{2}\right) = \infty.$$

Therefore,

$$\int_{-\infty}^\infty f(x)\, dx = \int_{-\infty}^0 f(x)\, dx + \int_0^\infty f(x)\, dx$$

does not exist while $\lim_{R\to\infty} \int_{-R}^R f(x)\, dx$ does exist. We will be interested in computing the latter type of integral. Such an integral is called the Cauchy Principal Value Integral and is denoted with either a P, PV, or a bar through the integral:

The Cauchy Principal Value Integral.

$$P \int_{-\infty}^\infty f(x)\, dx = PV \int_{-\infty}^\infty f(x)\, dx = \!\!\!\bcancel{\int_{-\infty}^\infty} f(x)\, dx = \lim_{R\to\infty} \int_{-R}^R f(x)\, dx. \quad (7.61)$$

If there is a discontinuity in the integral, one can further modify this definition of principal value integral to bypass the singularity. For example, if $f(x)$ is continuous on $a \le x \le b$ and not defined at $x = x_0 \in [a, b]$, then

$$\int_a^b f(x)\, dx = \lim_{\epsilon\to 0} \left(\int_a^{x_0-\epsilon} f(x)\, dx + \int_{x_0+\epsilon}^b f(x)\, dx\right).$$

In our discussions we will be computing integrals over the real line in the Cauchy Principal Value sense.

Example 7.40. *Compute $\int_{-1}^1 \frac{dx}{x^3}$ in the Cauchy Principal Value sense.*
In this case, $f(x) = \frac{1}{x^3}$ is not defined at $x = 0$. So, we have

$$\begin{aligned}
\int_{-1}^1 \frac{dx}{x^3} &= \lim_{\epsilon\to 0} \left(\int_{-1}^{-\epsilon} \frac{dx}{x^3} + \int_\epsilon^1 \frac{dx}{x^3}\right) \\
&= \lim_{\epsilon\to 0} \left(-\frac{1}{2x^2}\Big|_{-1}^{-\epsilon} - \frac{1}{2x^2}\Big|_\epsilon^1\right) = 0. \quad (7.62)
\end{aligned}$$

We now proceed to the evaluation of principal value integrals using complex integration methods. We want to evaluate the integral $\int_{-\infty}^\infty f(x)\, dx$. We will extend this into an integration in the complex plane. We extend $f(x)$ to $f(z)$ and assume that $f(z)$ is analytic in the upper half plane $(\text{Im}(z) > 0)$ except at isolated poles. We then consider the integral $\int_{-R}^R f(x)\, dx$ as an integral over the interval $(-R, R)$. We view this interval as a piece of a larger contour C_R obtained by completing the contour with a semicircle Γ_R of radius R extending into the upper half plane as shown in Figure 7.41. Note, that a similar construction is sometimes needed extending the integration into the lower half plane $(\text{Im}(z) < 0)$, as we will later see.

Computation of real integrals by embedding the problem in the complex plane.

The integral around the entire contour C_R can be computed using the Residue Theorem and is related to integrations over the pieces of the contour by

$$\oint_{C_R} f(z)\, dz = \int_{\Gamma_R} f(z)\, dz + \int_{-R}^R f(z)\, dz. \quad (7.63)$$

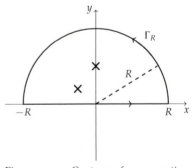

Figure 7.41: Contours for computing $P \int_{-\infty}^\infty f(x)\, dx$.

Taking the limit $R \to \infty$ and noting that the integral over $(-R, R)$ is the desired integral, we have

$$P \int_{-\infty}^{\infty} f(x)\, dx = \oint_C f(z)\, dz - \lim_{R \to \infty} \int_{\Gamma_R} f(z)\, dz, \qquad (7.64)$$

where we have identified C as the limiting contour as R gets large.

Now the key to carrying out the integration is that the second integral vanishes in the limit. This is true if $R|f(z)| \to 0$ along Γ_R as $R \to \infty$. This can be seen by the following argument. We parametrize the contour Γ_R using $z = Re^{i\theta}$. Then, when $|f(z)| < M(R)$,

$$\left| \int_{\Gamma_R} f(z)\, dz \right| = \left| \int_0^{2\pi} f(Re^{i\theta}) Re^{i\theta}\, d\theta \right|$$

$$\leq R \int_0^{2\pi} \left| f(Re^{i\theta}) \right| d\theta$$

$$< RM(R) \int_0^{2\pi} d\theta$$

$$= 2\pi R M(R). \qquad (7.65)$$

So, if $\lim_{R \to \infty} RM(R) = 0$, then $\lim_{R \to \infty} \int_{\Gamma_R} f(z)\, dz = 0$.

We now demonstrate how to use complex integration methods in evaluating integrals over real valued functions.

Example 7.41. *Evaluate $\int_{-\infty}^{\infty} \frac{dx}{1+x^2}$.*

We already know how to do this integral using calculus without complex analysis. We have that

$$\int_{-\infty}^{\infty} \frac{dx}{1+x^2} = \lim_{R \to \infty} \left(2 \tan^{-1} R \right) = 2 \left(\frac{\pi}{2} \right) = \pi.$$

We will apply the methods of this section and confirm this result. The needed contours are shown in Figure 7.42, and the poles of the integrand are at $z = \pm i$. We first write the integral over the bounded contour C_R as the sum of an integral from $-R$ to R along the real axis plus the integral over the semicircular arc in the upper half complex plane,

$$\int_{C_R} \frac{dz}{1+z^2} = \int_{-R}^{R} \frac{dx}{1+x^2} + \int_{\Gamma_R} \frac{dz}{1+z^2}.$$

Next, we let R get large.

We first note that $f(z) = \frac{1}{1+z^2}$ goes to zero fast enough on Γ_R as R gets large.

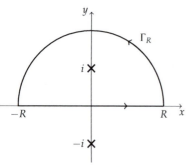

Figure 7.42: Contour for computing $\int_{-\infty}^{\infty} \frac{dx}{1+x^2}$.

$$R|f(z)| = \frac{R}{|1 + R^2 e^{2i\theta}|} = \frac{R}{\sqrt{1 + 2R^2 \cos\theta + R^4}}.$$

Thus, as $R \to \infty$, $R|f(z)| \to 0$ and $C_R \to C$. So,

$$\int_{-\infty}^{\infty} \frac{dx}{1+x^2} = \oint_C \frac{dz}{1+z^2}.$$

We need only compute the residue at the enclosed pole, $z = i$.

$$Res[f(z); z = i] = \lim_{z \to i} (z - i) \frac{1}{1+z^2} = \lim_{z \to i} \frac{1}{z+i} = \frac{1}{2i}.$$

Then, using the Residue Theorem, we have

$$\int_{-\infty}^{\infty} \frac{dx}{1+x^2} = 2\pi i \left(\frac{1}{2i}\right) = \pi.$$

Example 7.42. *Evaluate* $P \int_{-\infty}^{\infty} \frac{\sin x}{x} dx$.

For this example, the integral is unbounded at $z = 0$. Constructing the contours as before we are faced for the first time with a pole lying on the contour. We cannot ignore this fact. We can proceed with the computation by carefully going around the pole with a small semicircle of radius ϵ, as shown in Figure 7.43. Then the principal value integral computation becomes

$$P \int_{-\infty}^{\infty} \frac{\sin x}{x} dx = \lim_{\epsilon \to 0, R \to \infty} \left(\int_{-R}^{-\epsilon} \frac{\sin x}{x} dx + \int_{\epsilon}^{R} \frac{\sin x}{x} dx\right). \tag{7.66}$$

We will also need to rewrite the sine function in terms of exponentials in this integral. There are two approaches that we could take. First, we could employ the definition of the sine function in terms of complex exponentials. This gives two integrals to compute:

$$P \int_{-\infty}^{\infty} \frac{\sin x}{x} dx = \frac{1}{2i} \left(P \int_{-\infty}^{\infty} \frac{e^{ix}}{x} dx - P \int_{-\infty}^{\infty} \frac{e^{-ix}}{x} dx\right). \tag{7.67}$$

The other approach would be to realize that the sine function is the imaginary part of an exponential, $\operatorname{Im} e^{ix} = \sin x$. Then, we would have

$$P \int_{-\infty}^{\infty} \frac{\sin x}{x} dx = \operatorname{Im} \left(P \int_{-\infty}^{\infty} \frac{e^{ix}}{x} dx\right). \tag{7.68}$$

We first consider $P \int_{-\infty}^{\infty} \frac{e^{ix}}{x} dx$, which is common to both approaches. We use the contour in Figure 7.43. Then we have

$$\oint_{C_R} \frac{e^{iz}}{z} dz = \int_{\Gamma_R} \frac{e^{iz}}{z} dz + \int_{-R}^{-\epsilon} \frac{e^{iz}}{z} dz + \int_{C_\epsilon} \frac{e^{iz}}{z} dz + \int_{\epsilon}^{R} \frac{e^{iz}}{z} dz.$$

The integral $\oint_{C_R} \frac{e^{iz}}{z} dz$ vanishes because there are no poles enclosed in the contour! The sum of the second and fourth integrals gives the integral we seek as $\epsilon \to 0$ and $R \to \infty$. The integral over Γ_R will vanish as R gets large according to Jordan's Lemma.

Jordan's Lemma gives conditions such as when integrals over Γ_R will vanish as R gets large. We state a version of Jordan's Lemma here for reference and give a proof at the end of this chapter.

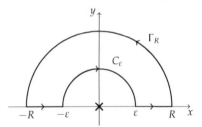

Figure 7.43: Contour for computing $P \int_{-\infty}^{\infty} \frac{\sin x}{x} dx$.

Jordan's Lemma

If $f(z)$ converges uniformly to zero as $z \to \infty$, then

$$\lim_{R \to \infty} \int_{C_R} f(z) e^{ikz} dz = 0,$$

where $k > 0$ and C_R is the upper half of the circle $|z| = R$.

A similar result applies for $k < 0$, but one closes the contour in the lower half plane. [See Section 7.5.8 for the proof of Jordan's Lemma.]

The remaining integral around the small semicircular arc must be done separately. We have

$$\int_{C_\epsilon} \frac{e^{iz}}{z}\, dz = \int_\pi^0 \frac{\exp(i\epsilon e^{i\theta})}{\epsilon e^{i\theta}} i\epsilon e^{i\theta}\, d\theta = -\int_0^\pi i \exp(i\epsilon e^{i\theta})\, d\theta.$$

Taking the limit as ϵ goes to zero, the integrand goes to i and we have

$$\int_{C_\epsilon} \frac{e^{iz}}{z}\, dz = -\pi i.$$

Note that we have not previously done integrals in which a singularity lies on the contour. One can show, as in this example, that points on the contour can be accounted for using half of a residue (times $2\pi i$). For the semicircle C_ϵ, you can verify this. The negative sign comes from going clockwise around the semicircle.

So far, we have that

$$P\int_{-\infty}^\infty \frac{e^{ix}}{x}\, dx = -\lim_{\epsilon \to 0} \int_{C_\epsilon} \frac{e^{iz}}{z}\, dz = \pi i.$$

At this point, we can get the answer using the second approach in Equation (7.68). Namely,

$$P\int_{-\infty}^\infty \frac{\sin x}{x}\, dx = \text{Im}\left(P\int_{-\infty}^\infty \frac{e^{ix}}{x}\, dx\right) = \text{Im}(\pi i) = \pi. \qquad (7.69)$$

It is instructive to carry out the first approach in Equation (7.67). We will need to compute $P\int_{-\infty}^\infty \frac{e^{-ix}}{x}\, dx$. This is done in a similar manner to the above computation, being careful with the sign changes due to the orientations of the contours as shown in Figure 7.44.

We note that the contour is closed in the lower half plane. This is because $k < 0$ in the application of Jordan's Lemma. One can understand why this is the case from the following observation. Consider the exponential in Jordan's Lemma. Let $z = z_R + iz_I$. Then,

$$e^{ikz} = e^{ik(z_R + iz_I)} = e^{-kz_I} e^{ikz_R}.$$

As $|z|$ gets large, the second factor just oscillates. The first factor would go to zero if $kz_I > 0$. So, if $k > 0$, we would close the contour in the upper half plane. If $k < 0$, then we would close the contour in the lower half plane. In the current computation, $k = -1$, so we use the lower half plane.

Working out the details, we find the same value for

$$P\int_{-\infty}^\infty \frac{e^{-ix}}{x}\, dx = \pi i.$$

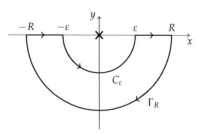

Figure 7.44: Contour in the lower half plane for computing $P\int_{-\infty}^\infty \frac{e^{-ix}}{x}\, dx$.

Finally, we can compute the original integral as

$$\begin{aligned} P\int_{-\infty}^\infty \frac{\sin x}{x}\, dx &= \frac{1}{2i}\left(P\int_{-\infty}^\infty \frac{e^{ix}}{x}\, dx - P\int_{-\infty}^\infty \frac{e^{-ix}}{x}\, dx\right)\\ &= \frac{1}{2i}(\pi i + \pi i)\\ &= \pi. \end{aligned} \qquad (7.70)$$

This is the same result as we obtained using Equation (7.68).

Example 7.43. *Evaluate $\oint_{|z|=1} \frac{dz}{z^2+1}$.*

In this example, there are two simple poles, $z = \pm i$ lying on the contour, as seen in Figure 7.45. This problem is similar to Problem 1c, except we will do it using contour integration instead of a parametrization. We bypass the two poles by drawing small semicircles around them. Because the poles are not included in the closed contour, the Residue Theorem tells us that the integral over the path vanishes. We can write the full integration as a sum over three paths: $C_\pm$ for the semicircles and C for the original contour with the poles cut out. Then we take the limit as the semicircle radii go to zero. So,

$$0 = \int_C \frac{dz}{z^2+1} + \int_{C_+} \frac{dz}{z^2+1} + \int_{C_-} \frac{dz}{z^2+1}.$$

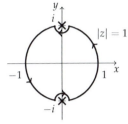

Figure 7.45: Example with poles on contour.

The integral over the semicircle around i can be done using the parametrization $z = i + \epsilon e^{i\theta}$. Then $z^2 + 1 = 2i\epsilon e^{i\theta} + \epsilon^2 e^{2i\theta}$. This gives

$$\int_{C_+} \frac{dz}{z^2+1} = \lim_{\epsilon \to 0} \int_0^{-\pi} \frac{i\epsilon e^{i\theta}}{2i\epsilon e^{i\theta} + \epsilon^2 e^{2i\theta}} \, d\theta = \frac{1}{2} \int_0^{-\pi} d\theta = -\frac{\pi}{2}.$$

As in the previous example, we note that this is just πi times the residue, $\text{Res}\left[\frac{1}{z^2+1}; z = i\right] = \frac{1}{2i}$. Because the path is traced clockwise, we find that the contribution is $-\pi i\text{Res} = -\frac{\pi}{2}$, which is what we obtained above. A similar computation will give the contribution from $z = -i$ as $\frac{\pi}{2}$. Adding these values gives the total contribution from $C_\pm$ as zero. So, the final result is that

$$\oint_{|z|=1} \frac{dz}{z^2+1} = 0.$$

Example 7.44. *Evaluate $\int_{-\infty}^{\infty} \frac{e^{ax}}{1+e^x} dx$, for $0 < a < 1$.*

In dealing with integrals involving exponentials or hyperbolic functions, it is sometimes useful to use different types of contours. This example is one such case. We will replace x with z and integrate over the contour in Figure 7.46. Letting $R \to \infty$, the integral along the real axis is the integral that we desire. The integral along the path for $y = 2\pi$ leads to a multiple of this integral because $z = x + 2\pi i$ along this path. Integration along the vertical paths vanishes as $R \to \infty$. This is captured in the following integrals:

$$\oint_{C_R} \frac{e^{az}}{1+e^z} dz = \int_{-R}^{R} \frac{e^{ax}}{1+e^x} dx + \int_0^{2\pi} \frac{e^{a(R+iy)}}{1+e^{R+iy}} dy$$
$$+ \int_R^{-R} \frac{e^{a(x+2\pi i)}}{1+e^{x+2\pi i}} dx + \int_{2\pi}^{0} \frac{e^{a(-R+iy)}}{1+e^{-R+iy}} dy \quad (7.71)$$

We can now let $R \to \infty$. For large R, the second integral decays as $e^{(a-1)R}$ and the fourth integral decays as e^{-aR}. Thus, we are left with

$$\oint_C \frac{e^{az}}{1+e^z} dz = \lim_{R \to \infty} \left(\int_{-R}^{R} \frac{e^{ax}}{1+e^x} dx - e^{2\pi i a} \int_{-R}^{R} \frac{e^{ax}}{1+e^x} dx \right)$$
$$= (1 - e^{2\pi i a}) \int_{-\infty}^{\infty} \frac{e^{ax}}{1+e^x} dx. \quad (7.72)$$

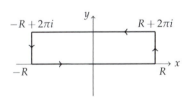

Figure 7.46: Example using a rectangular contour.

We need only evaluate the left contour integral using the Residue Theorem. The poles are found from

$$1 + e^z = 0.$$

Within the contour, this is satisfied by $z = i\pi$. So,

$$Res\left[\frac{e^{az}}{1 + e^z}; z = i\pi\right] = \lim_{z \to i\pi}(z - i\pi)\frac{e^{az}}{1 + e^z} = -e^{i\pi a}.$$

Applying the Residue Theorem, we have

$$(1 - e^{2\pi i a})\int_{-\infty}^{\infty}\frac{e^{ax}}{1 + e^x}dx = -2\pi i e^{i\pi a}.$$

Therefore, we have found that

$$\int_{-\infty}^{\infty}\frac{e^{ax}}{1 + e^x}dx = \frac{-2\pi i e^{i\pi a}}{1 - e^{2\pi i a}} = \frac{\pi}{\sin \pi a}, \quad 0 < a < 1.$$

7.5.7   Integration over Multivalued Functions

WE HAVE SEEN THAT SOME COMPLEX FUNCTIONS inherently possess multivaluedness; that is, such "functions" do not evaluate to a single value, but have many values. The key examples were $f(z) = z^{1/n}$ and $f(z) = \ln z$. The nth roots have n distinct values, and logarithms have an infinite number of values as determined by the range of the resulting arguments. We mentioned that the way to handle multivaluedness is to assign different branches to these functions, introduce a branch cut, and glue them together at the branch cuts to form Riemann surfaces. In this way, we can draw continuous paths along the Riemann surfaces as we move from one Riemann sheet to another.

Before we do examples of contour integration involving multivalued functions, let's first try to get a handle on multivaluedness in a simple case. We will consider the square root function,

$$w = z^{1/2} = r^{1/2}e^{i\left(\frac{\theta}{2} + k\pi\right)}, \quad k = 0, 1.$$

There are two branches, corresponding to each k value. If we follow a path not containing the origin, then we stay in the same branch, so the final argument (θ) will be equal to the initial argument. However, if we follow a path that encloses the origin, this will not be true. In particular, for an initial point on the unit circle, $z_0 = e^{i\theta_0}$, we have its image as $w_0 = e^{i\theta_0/2}$. However, if we go around a full revolution, $\theta = \theta_0 + 2\pi$, then

$$z_1 = e^{i\theta_0 + 2\pi i} = e^{i\theta_0},$$

but

$$w_1 = e^{(i\theta_0 + 2\pi i)/2} = e^{i\theta_0/2}e^{\pi i} \neq w_0.$$

Here we obtain a final argument (θ) that is not equal to the initial argument! Somewhere, we have crossed from one branch to another. Points, such as the origin in this example, are called branch points. Actually, there are two branch points, because we can view the closed path around the origin as a closed path around complex infinity in the compactified complex plane. However, we will not go into that at this time.

We can demonstrate this in the following figures. In Figure 7.47 we show how the points A through E are mapped from the z-plane into the w-plane under the square root function for the principal branch, $k = 0$. As we trace out the unit circle in the z-plane, we only trace out a semicircle in the w-plane. If we consider the branch $k = 1$, we then trace out a semicircle in the lower half plane, as shown in Figure 7.48 following the points from F to J.

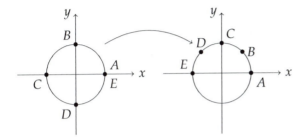

Figure 7.47: In this figure we show how points on the unit circle in the z-plane are mapped to points in the w-plane under the principal square root function.

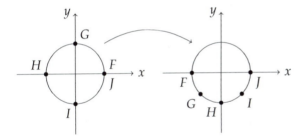

Figure 7.48: In this figure we show how points on the unit circle in the z-plane are mapped to points in the w-plane under the square root function for the second branch, $k = 1$.

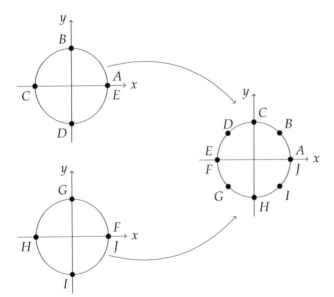

Figure 7.49: In this figure we show the combined mapping using two branches of the square root function.

We can combine these into one mapping depicting how the two complex planes corresponding to each branch provide a mapping to the w-plane. This is shown in Figure 7.49.

A common way to draw this domain, which looks like two separate complex planes, would be to glue them together. Imagine cutting each plane along the positive x-axis, extending between the two branch points, $z = 0$ and $z = \infty$. As one approaches the cut on the principal branch, one can move onto the glued second branch. Then one continues around the origin on this branch until one once again reaches the cut. This cut is glued to the principal branch in such a way that the path returns to its starting point. The resulting surface we obtain is the Riemann surface shown in Figure 7.50. Note that there is nothing that forces us to place the branch cut at a particular place. For example, the branch cut could be along the positive real axis, the negative real axis, or any path connecting the origin and complex infinity.

We now look at examples involving integrals of multivalued functions.

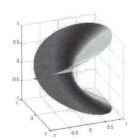

Figure 7.50: Riemann surface for $f(z) = z^{1/2}$.

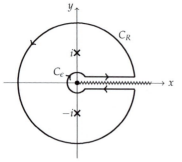

Figure 7.51: An example of a contour that accounts for a branch cut.

Example 7.45. *Evaluate* $\int_0^\infty \frac{\sqrt{x}}{1+x^2} dx$.

We consider the contour integral $\oint_C \frac{\sqrt{z}}{1+z^2} dz$.

The first thing we can see in this problem is the square root function in the integrand. Being that there is a multivalued function, we locate the branch point and determine where to draw the branch cut. In Figure 7.51 we show the contour that we will use in this problem. Note that we picked the branch cut along the positive x-axis.

We take the contour C to be positively oriented, being careful to enclose the two poles and to hug the branch cut. It consists of two circles. The outer circle C_R is a circle of radius R and the inner circle C_ϵ will have a radius of ϵ. The sought-after answer will be obtained by letting $R \to \infty$ and $\epsilon \to 0$. On the large circle we have that the integrand goes to zero fast enough as $R \to \infty$. The integral around the small circle vanishes as $\epsilon \to 0$. We can see this by parametrizing the circle as $z = \epsilon e^{i\theta}$ for $\theta \in [0, 2\pi]$:

$$\oint_{C_\epsilon} \frac{\sqrt{z}}{1+z^2} dz = \int_0^{2\pi} \frac{\sqrt{\epsilon e^{i\theta}}}{1+(\epsilon e^{i\theta})^2} i\epsilon e^{i\theta} d\theta$$
$$= i\epsilon^{3/2} \int_0^{2\pi} \frac{e^{3i\theta/2}}{1+(\epsilon^2 e^{2i\theta})} d\theta. \qquad (7.73)$$

It should now be easy to see that as $\epsilon \to 0$, this integral vanishes.

The integral above the branch cut is the one we are seeking, $\int_0^\infty \frac{\sqrt{x}}{1+x^2} dx$. The integral under the branch cut, where $z = re^{2\pi i}$, is

$$\int \frac{\sqrt{z}}{1+z^2} dz = \int_\infty^0 \frac{\sqrt{re^{2\pi i}}}{1+r^2 e^{4\pi i}} dr$$
$$= \int_0^\infty \frac{\sqrt{r}}{1+r^2} dr. \qquad (7.74)$$

We note that this is the same as that above the cut.

Up to this point, we have that the contour integral, as $R \to \infty$ and $\epsilon \to 0$, is

$$\oint_C \frac{\sqrt{z}}{1+z^2} dz = 2 \int_0^\infty \frac{\sqrt{x}}{1+x^2} dx.$$

In order to finish this problem, we need the residues at the two simple poles.

$$Res\left[\frac{\sqrt{z}}{1+z^2}; z = i\right] = \frac{\sqrt{i}}{2i} = \frac{\sqrt{2}}{4}(1+i),$$

$$Res\left[\frac{\sqrt{z}}{1+z^2}; z = -i\right] = \frac{\sqrt{-i}}{-2i} = \frac{\sqrt{2}}{4}(1-i).$$

So,

$$2 \int_0^\infty \frac{\sqrt{x}}{1+x^2} dx = 2\pi i \left(\frac{\sqrt{2}}{4}(1+i) + \frac{\sqrt{2}}{4}(1-i)\right) = \pi\sqrt{2}.$$

Finally, we have the value of the integral that we were seeking,

$$\int_0^\infty \frac{\sqrt{x}}{1+x^2} dx = \frac{\pi\sqrt{2}}{2}.$$

Example 7.46. *Compute $\int_a^\infty f(x)\, dx$ using contour integration involving logarithms.*[4]

In this example, we will apply contour integration to the integral

$$\oint_C f(z) \ln(a-z)\, dz$$

[4] This approach was originally published in Neville, E. H., 1945, Indefinite integration by means of residues. *The Mathematical Student*, **13**, 16–35, and discussed in Duffy, D. G., *Transform Methods for Solving Partial Differential Equations*, 1994.

for the contour shown in Figure 7.52.

We will assume that $f(z)$ is single valued and vanishes as $|z| \to \infty$. We will choose the branch cut to span from the origin along the positive real axis. Employing the Residue Theorem and breaking up the integrals over the pieces of the contour in Figure 7.52, we have schematically that

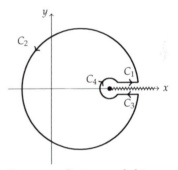

Figure 7.52: Contour needed to compute $\oint_C f(z) \ln(a-z)\, dz$.

$$2\pi i \sum Res[f(z)\ln(a-z)] = \left(\int_{C_1} + \int_{C_2} + \int_{C_3} + \int_{C_4}\right) f(z)\ln(a-z)\, dz.$$

First of all, we assume that $f(z)$ is well behaved at $z = a$ and vanishes fast enough as $|z| = R \to \infty$. Then, the integrals over C_2 and C_4 will vanish. For example, for the path C_4, we let $z = a + \epsilon e^{i\theta}$, $0 < \theta < 2\pi$. Then,

$$\int_{C_4} f(z)\ln(a-z)\, dz. = \lim_{\epsilon \to 0} \int_{2\pi}^0 f(a + \epsilon e^{i\theta}) \ln(\epsilon e^{i\theta}) i\epsilon e^{i\theta}\, d\theta.$$

If $f(a)$ is well behaved, then we only need to show that $\lim_{\epsilon \to 0} \epsilon \ln \epsilon = 0$. This is left to the reader.

Similarly, we consider the integral over C_2 as R gets large,

$$\int_{C_2} f(z)\ln(a-z)\, dz = \lim_{R \to \infty} \int_0^{2\pi} f(Re^{i\theta}) \ln(Re^{i\theta}) iRe^{i\theta}\, d\theta.$$

Thus, we need only require that

$$\lim_{R \to \infty} R \ln R |f(Re^{i\theta})| = 0.$$

Next, we consider the two straight line pieces. For C_1, the integration along the real axis occurs for $z = x$, so

$$\int_{C_1} f(z) \ln(a - z)\, dz = \int_a^\infty f(x) \ln(a - x)\, dz.$$

However, integration over C_3 requires noting that we need the branch for the logarithm such that $\ln z = \ln(a - x) + 2\pi i$. Then,

$$\int_{C_3} f(z) \ln(a - z)\, dz = \int_\infty^a f(x)[\ln(a - x) + 2\pi i]\, dz.$$

Combining these results, we have

$$
\begin{aligned}
2\pi i \sum \mathrm{Res}[f(z) \ln(a - z)] &= \int_a^\infty f(x) \ln(a - x)\, dz \\
&\quad + \int_\infty^a f(x)[\ln(a - x) + 2\pi i]\, dz. \\
&= -2\pi i \int_a^\infty f(x)\, dz. \quad (7.75)
\end{aligned}
$$

Therefore,

$$\int_a^\infty f(x)\, dx = -\sum \mathrm{Res}[f(z) \ln(a - z)].$$

Example 7.47. Compute $\int_1^\infty \frac{dx}{4x^2 - 1}$.

We can apply the previous example to this case. We see from Figure 7.53 that the two poles at $z = \pm \frac{1}{2}$ are inside contour C. So, we compute the residues of $\frac{\ln(1 - z)}{4z^2 - 1}$ at these poles and find that

$$
\begin{aligned}
\int_1^\infty \frac{dx}{4x^2 - 1} &= -\mathrm{Res}\left[\frac{\ln(1 - z)}{4z^2 - 1}; \frac{1}{2}\right] - \mathrm{Res}\left[\frac{\ln(1 - z)}{4z^2 - 1}; -\frac{1}{2}\right] \\
&= -\frac{\ln \frac{1}{2}}{4} + \frac{\ln \frac{3}{2}}{4} = \frac{\ln 3}{4}. \quad (7.76)
\end{aligned}
$$

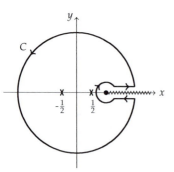

Figure 7.53: Contour needed to compute $\int_1^\infty \frac{dx}{4x^2 - 1}$.

7.5.8 Appendix: Jordan's Lemma

FOR COMPLETENESS, WE PROVE JORDAN'S LEMMA.

Theorem 7.10. *If $f(z)$ converges uniformly to zero as $z \to \infty$, then*

$$\lim_{R \to \infty} \int_{C_R} f(z) e^{ikz}\, dz = 0,$$

where $k > 0$ and C_R is the upper half of the circle $|z| = R$.

Proof. We consider the integral

$$I_R = \int_{C_R} f(z) e^{ikz}\, dz,$$

where $k > 0$ and C_R is the upper half of the circle $|z| = R$ in the complex plane. Let $z = Re^{i\theta}$ be a parametrization of C_R. Then,

$$I_R = \int_0^\pi f(Re^{i\theta}) e^{ikR\cos\theta - aR\sin\theta}\, iRe^{i\theta}\, d\theta.$$

Because

$$\lim_{|z|\to\infty} f(z) = 0, \quad 0 \leq \arg z \leq \pi,$$

then for large $|R|$, $|f(z)| < \epsilon$ for some $\epsilon > 0$. Then,

$$
\begin{aligned}
|I_R| &= \left| \int_0^\pi f(Re^{i\theta}) e^{ikR\cos\theta - aR\sin\theta} iRe^{i\theta}\, d\theta \right| \\
&\leq \int_0^\pi \left| f(Re^{i\theta}) \right| \left| e^{ikR\cos\theta} \right| \left| e^{-aR\sin\theta} \right| \left| iRe^{i\theta} \right| d\theta \\
&\leq \epsilon R \int_0^\pi e^{-aR\sin\theta}\, d\theta \\
&= 2\epsilon R \int_0^{\pi/2} e^{-aR\sin\theta}\, d\theta.
\end{aligned}
\tag{7.77}
$$

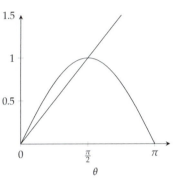

The last integral in Equation (7.77) still cannot be computed, but we can get a bound on it over the range $\theta \in [0, \pi/2]$. Note from Figure 7.54 that

$$\sin\theta \geq \frac{2}{\pi}\theta, \quad \theta \in [0, \pi/2].$$

Figure 7.54: Plots of $y = \sin\theta$ and $y = \frac{2}{\pi}\theta$ to show where $\sin\theta \geq \frac{2}{\pi}\theta$.

Therefore, we have

$$|I_R| \leq 2\epsilon R \int_0^{\pi/2} e^{-2aR\theta/\pi}\, d\theta = \frac{2\epsilon R}{2aR/\pi}(1 - e^{-aR}).$$

For large R, we have

$$\lim_{R\to\infty} |I_R| \leq \frac{\pi\epsilon}{a}.$$

So, as $\epsilon \to 0$, the integral vanishes.

$\square$

Problems

1. Write the following in standard form.

    a. $(4 + 5i)(2 - 3i)$.

    b. $(1 + i)^3$.

    c. $\frac{5+3i}{1-i}$.

2. Write the following in polar form, $z = re^{i\theta}$.

    a. $i - 1$.

    b. $-2i$.

    c. $\sqrt{3} + 3i$.

3. Write the following in rectangular form, $z = a + ib$.

    a. $4e^{i\pi/6}$.

    b. $\sqrt{2}e^{5i\pi/4}$.

    c. $(1 - i)^{100}$.

4. Find all z such that $z^4 = 16i$. Write the solutions in rectangular form, $z = a + ib$, with no decimal approximation or trig functions.

5. Show that $\sin(x + iy) = \sin x \cosh y + i \cos x \sinh y$ using trigonometric identities and the exponential forms of these functions.

6. Find all z such that $\cos z = 2$, or explain why there are none. You will need to consider $\cos(x + iy)$ and equate real and imaginary parts of the resulting expression similar to Problem 5.

7. Find the principal value of i^i. Rewrite the base, i, as an exponential first.

8. Consider the circle $|z - 1| = 1$.

    a. Rewrite the equation in rectangular coordinates by setting $z = x + iy$.

    b. Sketch the resulting circle using part a.

    c. Consider the image of the circle under the mapping $f(z) = z^2$, given by $|z^2 - 1| = 1$.

        i. By inserting $z = re^{i\theta} = r(\cos\theta + i\sin\theta)$, find the equation of the image curve in polar coordinates.

        ii. Sketch the image curve. You may need to refer to your Calculus II text for polar plots. [Maple might help.]

9. Find the real and imaginary parts of the functions:

    a. $f(z) = z^3$.

    b. $f(z) = \sinh(z)$.

    c. $f(z) = \cos\bar{z}$.

10. Find the derivative of each function in Problem 9 when the derivative exists. Otherwise, show that the derivative does not exist.

11. Let $f(z) = u + iv$ be differentiable. Consider the vector field given by $\mathbf{F} = v\mathbf{i} + u\mathbf{j}$. Show that the equations $\nabla \cdot \mathbf{F} = \mathbf{0}$ and $\nabla \times \mathbf{F} = \mathbf{0}$ are equivalent to the Cauchy–Riemann Equations. [You will need to recall from multivariable calculus the del operator, $\nabla = \mathbf{i}\frac{\partial}{\partial x} + \mathbf{j}\frac{\partial}{\partial y} + \mathbf{k}\frac{\partial}{\partial z}$.]

12. What parametric curve is described by the function

$$\gamma(t) = (t - 3) + i(2t + 1),$$

$0 \leq t \leq 2$? [Hint: What would you do if you were instead considering the parametric equations $x = t - 3$ and $y = 2t + 1$]

13. Write the equation that describes the circle of radius 3 that is centered at $z = 2 - i$ in (a) Cartesian form (in terms of x and y); (b) polar form (in terms of θ and r); (c) complex form (in terms of z, r, and $e^{i\theta}$).

14. Consider the function $u(x, y) = x^3 - 3xy^2$.

a. Show that $u(x, y)$ is harmonic; that is, $\nabla^2 u = 0$.

b. Find its harmonic conjugate, $v(x, y)$.

c. Find a differentiable function, $f(z)$, for which $u(x, y)$ is the real part.

d. Determine $f'(z)$ for the function in part c. [Use $f'(z) = \frac{\partial u}{\partial x} + i\frac{\partial v}{\partial x}$ and rewrite your answer as a function of z.]

15. Evaluate the following integrals:

a. $\int_C \bar{z}\, dz$, where C is the parabola $y = x^2$ from $z = 0$ to $z = 1 + i$.

b. $\int_C f(z)\, dz$, where $f(z) = 2z - \bar{z}$ and C is the path from $z = 0$ to $z = 2 + i$ consisting of two line segments from $z = 0$ to $z = 2$ and then $z = 2$ to $z = 2 + i$.

c. $\int_C \frac{1}{z^2 + 4}\, dz$ for C the positively oriented circle, $|z| = 2$. [Hint: Parametrize the circle as $z = 2e^{i\theta}$, multiply numerator and denominator by $e^{-i\theta}$, and put in trigonometric form.]

16. Let C be the positively oriented ellipse $3x^2 + y^2 = 9$. Define

$$F(z_0) = \int_C \frac{z^2 + 2z}{z - z_0}\, dz.$$

Find $F(2i)$ and $F(2)$. [Hint: Sketch the ellipse in the complex plane. Use the Cauchy Integral Theorem with an appropriate $f(z)$, or Cauchy's Theorem if z_0 is outside the contour.]

17. Show that

$$\int_C \frac{dz}{(z - 1 - i)^{n+1}} = \begin{cases} 0, & n \neq 0, \\ 2\pi i, & n = 0, \end{cases}$$

for C the boundary of the square $0 \leq x \leq 2, 0 \leq y \leq 2$ taken counterclockwise. [Hint: Use the fact that contours can be deformed into simpler shapes (like a circle) as long as the integrand is analytic in the region between them. After picking a simpler contour, integrate using parametrization.]

18. Show that for g and h analytic functions at z_0, with $g(z_0) \neq 0, h(z_0) = 0$, and $h'(z_0) \neq 0$,

$$\text{Res}\left[\frac{g(z)}{h(z)}; z_0\right] = \frac{g(z_0)}{h'(z_0)}.$$

19. For the following, determine if the given point is a removable singularity, an essential singularity, or a pole (indicate its order).

a. $\frac{1 - \cos z}{z^2}, \quad z = 0$.

b. $\frac{\sin z}{z^2}, \quad z = 0$.

c. $\frac{z^2 - 1}{(z - 1)^2}, \quad z = 1$.

d. $ze^{1/z}, \quad z = 0$.

e. $\cos\frac{\pi}{z - \pi}, \quad z = \pi$.

20. Find the Laurent series expansion for $f(z) = \frac{\sinh z}{z^3}$ about $z = 0$. [Hint: You need to first do a MacLaurin series expansion for the hyperbolic sine.]

21. Find series representations for all indicated regions.

  a. $f(z) = \frac{z}{z-1}$, $|z| < 1$, $|z| > 1$.

  b. $f(z) = \frac{1}{(z-i)(z+2)}$, $|z| < 1$, $1 < |z| < 2$, $|z| > 2$. [Hint: Use partial fractions to write this as a sum of two functions first.]

22. Find the residues at the given points:

  a. $\frac{2z^2+3z}{z-1}$ at $z = 1$.

  b. $\frac{\ln(1+2z)}{z}$ at $z = 0$.

  c. $\frac{\cos z}{(2z-\pi)^3}$ at $z = \frac{\pi}{2}$.

23. Consider the integral $\int_0^{2\pi} \frac{d\theta}{5-4\cos\theta}$.

  a. Evaluate this integral by making the substitution $2\cos\theta = z + \frac{1}{z}$, $z = e^{i\theta}$, and using complex integration methods.

  b. In the 1800s, Weierstrass introduced a method for computing integrals involving rational functions of sine and cosine. One makes the substitution $t = \tan\frac{\theta}{2}$ and converts the integrand into a rational function of t. Note that the integration around the unit circle corresponds to $t \in (-\infty, \infty)$.

    i. Show that
    $$\sin\theta = \frac{2t}{1+t^2}, \quad \cos\theta = \frac{1-t^2}{1+t^2}.$$

    ii. Show that
    $$d\theta = \frac{2dt}{1+t^2}.$$

    iii. Use the Weierstrass substitution to compute the above integral.

24. Do the following integrals:

  a. $\oint_{|z-i|=3} \frac{e^z}{z^2+\pi^2}\, dz.$

  b. $\oint_{|z-i|=3} \frac{z^2-3z+4}{z^2-4z+3}\, dz.$

  c. $\int_{-\infty}^{\infty} \frac{\sin x}{x^2+4}\, dx.$ [Hint: This is $\mathrm{Im}\int_{-\infty}^{\infty} \frac{e^{ix}}{x^2+4}\, dx.$]

25. Evaluate the integral $\int_0^{\infty} \frac{(\ln x)^2}{1+x^2}\, dx.$
  [Hint: Replace x with $z = e^t$ and use the rectangular contour in Figure 7.55 with $R \to \infty$.]

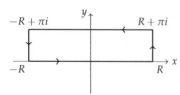

Figure 7.55: Rectangular contour for Problem 25.

26. Do the following integrals for fun!

a. For C the boundary of the square $|x| \le 2$, $|y| \le 2$,

$$\oint_C \frac{dz}{z(z-1)(z-3)^2}.$$

b.

$$\int_0^\pi \frac{\sin^2 \theta}{13 - 12 \cos \theta} \, d\theta.$$

c.

$$\int_{-\infty}^\infty \frac{dx}{x^2 + 5x + 6}.$$

d.

$$\int_0^\infty \frac{\cos \pi x}{1 - 9x^2} \, dx.$$

e.

$$\int_{-\infty}^\infty \frac{dx}{(x^2 + 9)(1-x)^2}.$$

f.

$$\int_0^\infty \frac{\sqrt{x}}{(1+x)^2} \, dx.$$

g.

$$\int_0^\infty \frac{\sqrt{x}}{(1+x)^2} \, dx.$$

8

Transform Techniques in Physics

"There is no branch of mathematics, however abstract, which may not some day be applied to phenomena of the real world."—Nikolai Lobatchevsky (1792–1856)

8.1   Introduction

SOME OF THE MOST POWERFUL TOOLS for solving problems in physics are transform methods. The idea is that one can transform the problem at hand to a new problem in a different space, hoping that the problem in the new space is easier to solve. Such transforms appear in many forms.

As we had seen in Chapter 3 and will see later in the book, the solutions of linear partial differential equations can be found by using the method of separation of variables to reduce solving partial differential equations (PDEs) to solving ordinary differential equations (ODEs). We can also use transform methods to transform the given PDE into ODEs or algebraic equations. Solving these equations, we then construct solutions of the PDE (or, the ODE) using an inverse transform. A schematic of these processes is shown below, and we will describe in this chapter how one can use Fourier and Laplace transforms to this effect.

In this chapter, we will explore the use of integral transforms. Given a function $f(x)$, we define an integral transform to a new function $F(k)$ as

$$F(k) = \int_a^b f(x)K(x,k)\,dx.$$

Here, $K(x,k)$ is called the kernel of the transform. We will concentrate specifically on Fourier transforms,

$$\hat{f}(k) = \int_{-\infty}^{\infty} f(x)e^{ikx}\,dx,$$

and Laplace transforms,

$$F(s) = \int_0^{\infty} f(t)e^{-st}\,dt.$$

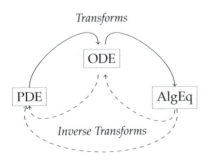

Figure 8.1: Schematic indicating that PDEs and ODEs can be transformed to simpler problems, solved in the new space, and transformed back to the original space.

8.1.1   Example 1: Linearized KdV Equation

As a relatively simple example, we consider the linearized Korteweg-de Vries (KdV) Equation:

$$u_t + cu_x + \beta u_{xxx} = 0, \quad -\infty < x < \infty. \tag{8.1}$$

This equation governs the propagation of some small-amplitude water waves. Its nonlinear counterpart has been at the center of attention for the past 40 years as a generic nonlinear wave equation.

We seek solutions that oscillate in space. So, we assume a solution of the form

$$u(x,t) = A(t)e^{ikx}. \tag{8.2}$$

The nonlinear counterpart to this equation is the Korteweg-de Vries (KdV) equation: $u_t + 6uu_x + u_{xxx} = 0$. This equation was derived by Diederik Johannes Korteweg (1848–1941) and his student Gustav de Vries (1866–1934). This equation governs the propagation of traveling waves called solitons. These were first observed by John Scott Russell (1808-1882) and were the source of a long debate on the existence of such waves. The history of this debate is interesting, and the KdV turned up as a generic equation in many other fields in the latter part of the past century, leading to many papers on nonlinear evolution equations.

Such behavior was seen in Chapters 3 and 6 for the wave equation for vibrating strings. In that case, we found plane wave solutions of the form $e^{ik(x \pm ct)}$, which we could write as $e^{i(kx \pm \omega t)}$ by defining $\omega = kc$. We further note that one often seeks complex solutions as a linear combination of such forms and then takes the real part in order to obtain physical solutions. In this case, we will find plane wave solutions for which the angular frequency $\omega = \omega(k)$ is a function of the wave number.

Inserting the guess (8.2) into the linearized KdV equation, we find that

$$\frac{dA}{dt} + i(ck - \beta k^3)A = 0. \tag{8.3}$$

Thus, we have converted the problem of seeking a solution of the partial differential equation into seeking a solution to an ordinary differential equation. This new problem is easier to solve. In fact, given an initial value, $A(0)$, we have

$$A(t) = A(0)e^{-i(ck - \beta k^3)t}. \tag{8.4}$$

Therefore, the solution of the partial differential equation is

$$u(x,t) = A(0)e^{ik(x - (c - \beta k^2)t)}. \tag{8.5}$$

We note that this solution takes the form $e^{i(kx - \omega t)}$, where

$$\omega = ck - \beta k^3.$$

A dispersion relation is an expression giving the angular frequency as a function of the wave number, $\omega = \omega(k)$.

In general, the equation $\omega = \omega(k)$ gives the angular frequency as a function of the wave number, k, and is called a dispersion relation. For $\beta = 0$, we see that c is nothing but the wave speed. For $\beta \neq 0$, the wave speed is given as

$$v = \frac{\omega}{k} = c - \beta k^2.$$

This suggests that waves with different wave numbers will travel at different speeds. Recalling that wave numbers are related to wavelengths, $k = \frac{2\pi}{\lambda}$, this means that waves with different wavelengths will travel at different speeds. For example, an initial localized wave packet will not maintain its shape. It is said to disperse, as the component waves of differing wavelengths will tend to part company.

For a general initial condition, we write the solutions to the linearized KdV as a superposition of plane waves. We can do this because the partial differential equation is linear. This should remind you of what we had done when using separation of variables. We first sought product solutions and then took a linear combination of the product solutions to obtain the general solution.

For this problem, we will sum over all wave numbers. The wave numbers are not restricted to discrete values. We instead have a continuous range of values. Thus, "summing" over k means that we have to integrate over the wave numbers. Thus, we have the general solution[1]

[1] The extra 2π has been introduced to be consistent with the definition of the Fourier transform, which is given later in the chapter.

$$u(x,t) = \frac{1}{2\pi} \int_{-\infty}^{\infty} A(k,0)e^{ik(x-(c-\beta k^2)t)}\, dk. \tag{8.6}$$

Note that we have indicated that A is a function of k. This is similar to introducing the A_n's and B_n's in the series solution for waves on a string.

How do we determine the $A(k,0)$'s? We introduce as an initial condition the initial wave profile $u(x,0) = f(x)$. Then, we have

$$f(x) = u(x,0) = \frac{1}{2\pi} \int_{-\infty}^{\infty} A(k,0)e^{ikx}\, dk. \tag{8.7}$$

Thus, given $f(x)$, we seek $A(k,0)$. In this chapter we will see that

$$A(k,0) = \int_{-\infty}^{\infty} f(x)e^{-ikx}\, dx.$$

This is what is called the Fourier transform of $f(x)$. It is just one of the so-called integral transforms that we will consider in this chapter.

In Figure 8.2 we summarize the transform scheme. One can use methods like separation of variables to solve the partial differential equation directly, evolving the initial condition $u(x,0)$ into the solution $u(x,t)$ at a later time.

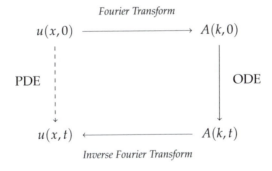

Figure 8.2: Schematic of using Fourier transforms to solve a linear evolution equation.

The transform method works as follows. Starting with the initial condition, one computes its Fourier Transform (FT) as[2]

[2] Note: The Fourier transform as used in this section and the next section are defined slightly differently than how we will define them later. The sign of the exponentials has been reversed.

$$A(k,0) = \int_{-\infty}^{\infty} f(x)e^{-ikx}\, dx.$$

Applying the transform on the partial differential equation, one obtains an ordinary differential equation satisfied by $A(k,t)$ which is simpler to solve than the original partial differential equation. Once $A(k,t)$ has been found, one then applies the Inverse Fourier Transform (IFT) to $A(k,t)$ in order to obtain the desired solution:

$$
\begin{aligned}
u(x,t) &= \frac{1}{2\pi} \int_{-\infty}^{\infty} A(k,t)e^{ikx}\, dk \\
&= \frac{1}{2\pi} \int_{-\infty}^{\infty} A(k,0)e^{ik(x-(c-\beta k^2)t)}\, dk.
\end{aligned}
\tag{8.8}
$$

8.1.2   Example 2: Free Particle Wave Function

A MORE FAMILIAR EXAMPLE IN PHYSICS comes from quantum mechanics. The Schrödinger Equation gives the wave function $\Psi(x,t)$ for a particle under the influence of forces, represented through the corresponding potential function $V(x)$. The one-dimensional, time-dependent Schrödinger Equation is given by

The one-dimensional, time-dependent Schrödinger Equation.

$$
i\hbar \Psi_t = -\frac{\hbar^2}{2m} \Psi_{xx} + V\Psi.
\tag{8.9}
$$

We consider the case of a free particle in which there are no forces, $V = 0$. Thus we have

$$
i\hbar \Psi_t = -\frac{\hbar^2}{2m} \Psi_{xx}.
\tag{8.10}
$$

Taking a hint from the study of the linearized KdV Equation, we will assume that solutions of Equation (8.10) take the form

$$
\Psi(x,t) = \frac{1}{2\pi} \int_{-\infty}^{\infty} \phi(k,t)e^{ikx}\, dk.
$$

[Here we have opted to use the more traditional notation, $\phi(k,t)$, instead of $A(k,t)$ as above.]

Inserting the expression for $\Psi(x,t)$ into Equation (8.10), we have

$$
i\hbar \int_{-\infty}^{\infty} \frac{d\phi(k,t)}{dt} e^{ikx}\, dk = -\frac{\hbar^2}{2m} \int_{-\infty}^{\infty} \phi(k,t)(ik)^2 e^{ikx}\, dk.
$$

Because this is true for all t, we can equate the integrands, giving

$$
i\hbar \frac{d\phi(k,t)}{dt} = \frac{\hbar^2 k^2}{2m} \phi(k,t).
$$

As with the previous example, we have obtained a simple ordinary differential equation. The solution of this equation is given by

$$
\phi(k,t) = \phi(k,0)e^{-i\frac{\hbar k^2}{2m}t}.
$$

Applying the inverse Fourier transform, the general solution to the time-dependent problem for a free particle is found as

$$
\Psi(x,t) = \frac{1}{2\pi} \int_{-\infty}^{\infty} \phi(k,0)e^{ik(x-\frac{\hbar k}{2m}t)}\, dk.
$$

We note that this takes the familiar form

$$\Psi(x,t) = \frac{1}{2\pi} \int_{-\infty}^{\infty} \phi(k,0)e^{i(kx-\omega t)} \, dk,$$

where the dispersion relation is found as

$$\omega = \frac{\hbar k^2}{2m}.$$

The wave speed is given as

$$v = \frac{\omega}{k} = \frac{\hbar k}{2m}.$$

As a special note, we see that this is not the particle velocity! Recall that the momentum is given as $p = \hbar k$.[3] So, this wave speed is $v = \frac{p}{2m}$, which is only half the classical particle velocity! A simple manipulation of this result will clarify the "problem."

We assume that particles can be represented by a localized wave function. This is the case if the major contributions to the integral are centered about a central wave number, k_0. Thus, we can expand $\omega(k)$ about k_0:

$$\omega(k) = \omega_0 + \omega_0'(k - k_0)t + \dots. \tag{8.11}$$

Here, $\omega_0 = \omega(k_0)$ and $\omega_0' = \omega'(k_0)$. Inserting this expression into the integral representation for $\Psi(x,t)$, we have

$$\Psi(x,t) = \frac{1}{2\pi} \int_{-\infty}^{\infty} \phi(k,0)e^{i(kx-\omega_0 t - \omega_0'(k-k_0)t - \dots)} \, dk.$$

We now make the change of variables, $s = k - k_0$, and rearrange the resulting factors to find

$$\begin{aligned}
\Psi(x,t) &\approx \frac{1}{2\pi} \int_{-\infty}^{\infty} \phi(k_0 + s,0)e^{i((k_0+s)x - (\omega_0 + \omega_0' s)t)} \, ds \\
&= \frac{1}{2\pi} e^{i(-\omega_0 t + k_0 \omega_0' t)} \int_{-\infty}^{\infty} \phi(k_0 + s,0)e^{i(k_0+s)(x - \omega_0' t)} \, ds \\
&= e^{i(-\omega_0 t + k_0 \omega_0' t)} \Psi(x - \omega_0' t, 0). \tag{8.12}
\end{aligned}$$

Summarizing: for an initially localized wave packet, $\Psi(x,0)$ with wave numbers grouped around k_0 the wave function, $\Psi(x,t)$, is a translated version of the initial wave function up to a phase factor. In quantum mechanics, we are more interested in the probability density for locating a particle, so from

$$|\Psi(x,t)|^2 = |\Psi(x - \omega_0' t, 0)|^2,$$

we see that the "velocity of the wave packet" is found to be

$$\omega_0' = \left.\frac{d\omega}{dk}\right|_{k=k_0} = \frac{\hbar k}{m}.$$

This corresponds to the classical velocity of the particle ($v_{\text{part}} = p/m$). Thus, one usually defines ω_0' to be the group velocity,

$$v_g = \frac{d\omega}{dk},$$

[3] Because $p = \hbar k$, we also see that the dispersion relation is given by

$$\omega = \frac{\hbar k^2}{2m} = \frac{p^2}{2m\hbar} = \frac{E}{\hbar}.$$

Group and phase velocities, $v_g = \frac{d\omega}{dk}$, $v_p = \frac{\omega}{k}$.

and the former velocity as the phase velocity,

$$v_p = \frac{\omega}{k}.$$

8.1.3   Transform Schemes

THESE EXAMPLES HAVE ILLUSTRATED one of the features of transform theory. Given a partial differential equation, we can transform the equation from spatial variables to wave number space, or time variables to frequency space. In the new space, the time evolution is simpler. In these cases, the evolution was governed by an ordinary differential equation. One solves the problem in the new space and then transforms back to the original space. This is depicted in Figure 8.3 for the Schrödinger Equation and was shown in Figure 8.2 for the linearized KdV Equation.

Figure 8.3:   The scheme for solving the Schrödinger Equation using Fourier transforms. The goal is to solve for $\Psi(x,t)$ given $\Psi(x,0)$. Instead of a direct solution in coordinate space (on the left side), one can first transform the initial condition, obtaining $\phi(k,0)$ in wave number space. The governing equation in the new space is found by transforming the PDE to get an ODE. This simpler equation is solved to obtain $\phi(k,t)$. Then an inverse transform yields the solution of the original equation.

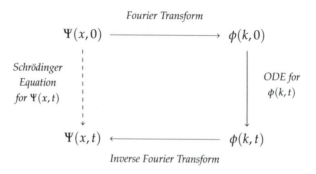

This is similar to the solution of the system of ordinary differential equations in Chapter 3, $\dot{\mathbf{x}} = A\mathbf{x}$. In that case, we diagonalized the system using the transformation $\mathbf{x} = S\mathbf{y}$. This led to a simpler system $\dot{\mathbf{y}} = \Lambda\mathbf{y}$, where $\Lambda = S^{-1}AS$. Solving for $\mathbf{y}$, we inverted the solution to obtain $\mathbf{x}$. Similarly, one can apply this diagonalization to the solution of linear algebraic systems of equations. The general scheme is shown in Figure 8.4.

Figure 8.4: This shows the scheme for solving the linear system of ODEs $\dot{\mathbf{x}} = A\mathbf{x}$. One finds a transformation between $\mathbf{x}$ and $\mathbf{y}$ of the form $\mathbf{x} = S\mathbf{y}$ that diagonalizes the system. The resulting system is easier to solve for $\mathbf{y}$. Then one uses the inverse transformation to obtain the solution to the original problem.

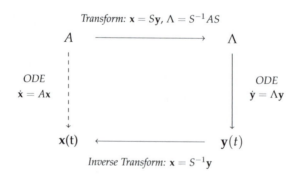

Similar transform constructions occur for many other type of problems. We will end this chapter with a study of Laplace transforms, which are useful in the study of initial value problems, particularly for linear ordinary differential equations with constant coefficients. A similar scheme for using Laplace transforms is depicted in Figure 8.30.

In this chapter, we will begin with the study of Fourier transforms. These will provide an integral representation of functions defined on the real line. Such functions can also represent analog signals. Analog signals are continuous signals that can be represented as a sum over a continuous set of frequencies, as opposed to the sum over discrete frequencies, which Fourier series were used to represent in an earlier chapter. We will then investigate a related transform, the Laplace transform, which is useful in solving initial value problems such as those encountered in ordinary differential equations.

8.2   Complex Exponential Fourier Series

BEFORE DERIVING THE FOURIER TRANSFORM, we will need to rewrite the trigonometric Fourier series representation as a complex exponential Fourier series. We first recall from Chapter 5 the trigonometric Fourier series representation of a function defined on $[-\pi, \pi]$ with period 2π. The Fourier series is given by

$$f(x) \sim \frac{a_0}{2} + \sum_{n=1}^{\infty} \left(a_n \cos nx + b_n \sin nx \right), \tag{8.13}$$

where the Fourier coefficients were found as

$$\begin{aligned}
a_n &= \frac{1}{\pi} \int_{-\pi}^{\pi} f(x) \cos nx \, dx, \quad n = 0, 1, \ldots, \\
b_n &= \frac{1}{\pi} \int_{-\pi}^{\pi} f(x) \sin nx \, dx, \quad n = 1, 2, \ldots.
\end{aligned} \tag{8.14}$$

In order to derive the exponential Fourier series, we replace the trigonometric functions with exponential functions and collect like exponential terms. This gives

$$\begin{aligned}
f(x) &\sim \frac{a_0}{2} + \sum_{n=1}^{\infty} \left[a_n \left(\frac{e^{inx} + e^{-inx}}{2} \right) + b_n \left(\frac{e^{inx} - e^{-inx}}{2i} \right) \right] \\
&= \frac{a_0}{2} + \sum_{n=1}^{\infty} \left(\frac{a_n - ib_n}{2} \right) e^{inx} + \sum_{n=1}^{\infty} \left(\frac{a_n + ib_n}{2} \right) e^{-inx}. \tag{8.15}
\end{aligned}$$

The coefficients of the complex exponentials can be rewritten by defining

$$c_n = \frac{1}{2}(a_n + ib_n), \quad n = 1, 2, \ldots. \tag{8.16}$$

This implies that

$$\bar{c}_n = \frac{1}{2}(a_n - ib_n), \quad n = 1, 2, \ldots. \tag{8.17}$$

So far, the representation is rewritten as

$$f(x) \sim \frac{a_0}{2} + \sum_{n=1}^{\infty} \bar{c}_n e^{inx} + \sum_{n=1}^{\infty} c_n e^{-inx}.$$

Re-indexing the first sum, by introducing $k = -n$, we can write

$$f(x) \sim \frac{a_0}{2} + \sum_{k=-1}^{-\infty} \bar{c}_{-k} e^{-ikx} + \sum_{n=1}^{\infty} c_n e^{-inx}.$$

Because k is a dummy index, we replace it with a new n as

$$f(x) \sim \frac{a_0}{2} + \sum_{n=-1}^{-\infty} \bar{c}_{-n} e^{-inx} + \sum_{n=1}^{\infty} c_n e^{-inx}.$$

We can now combine all the terms into a simple sum. We first define c_n for negative n's by

$$c_n = \bar{c}_{-n}, \quad n = -1, -2, \ldots.$$

Letting $c_0 = \frac{a_0}{2}$, we can write the complex exponential Fourier series representation as

$$f(x) \sim \sum_{n=-\infty}^{\infty} c_n e^{-inx}, \tag{8.18}$$

where

$$
\begin{aligned}
c_n &= \frac{1}{2}(a_n + ib_n), \quad n = 1, 2, \ldots, \\
c_n &= \frac{1}{2}(a_{-n} - ib_{-n}), \quad n = -1, -2, \ldots, \\
c_0 &= \frac{a_0}{2}.
\end{aligned}
\tag{8.19}
$$

Given such a representation, we would like to write out the integral forms of the coefficients, c_n. So, we replace the a_n's and b_n's with their integral representations and replace the trigonometric functions with complex exponential functions. Doing this, we have for $n = 1, 2, \ldots$

$$
\begin{aligned}
c_n &= \frac{1}{2}(a_n + ib_n) \\
&= \frac{1}{2}\left[\frac{1}{\pi} \int_{-\pi}^{\pi} f(x) \cos nx \, dx + \frac{i}{\pi} \int_{-\pi}^{\pi} f(x) \sin nx \, dx \right] \\
&= \frac{1}{2\pi} \int_{-\pi}^{\pi} f(x) \left(\frac{e^{inx} + e^{-inx}}{2} \right) dx + \frac{i}{2\pi} \int_{-\pi}^{\pi} f(x) \left(\frac{e^{inx} - e^{-inx}}{2i} \right) dx \\
&= \frac{1}{2\pi} \int_{-\pi}^{\pi} f(x) e^{inx} \, dx.
\end{aligned}
\tag{8.20}
$$

It is a simple matter to determine the c_n's for other values of n. For $n = 0$, we have that

$$c_0 = \frac{a_0}{2} = \frac{1}{2\pi} \int_{-\pi}^{\pi} f(x) \, dx.$$

For $n = -1, -2, \ldots$, we find that

$$c_n = \bar{c}_n = \frac{1}{2\pi} \int_{-\pi}^{\pi} f(x) \overline{e^{-inx}} \, dx = \frac{1}{2\pi} \int_{-\pi}^{\pi} f(x) e^{inx} \, dx.$$

Therefore, we have obtained the complex exponential Fourier series coefficients for all n. Now we can define the complex exponential Fourier series for the function $f(x)$ defined on $[-\pi, \pi]$ as shown below.

> ### Complex Exponential Series for $f(x)$ Defined on $[-\pi, \pi]$
>
> $$f(x) \sim \sum_{n=-\infty}^{\infty} c_n e^{-inx}, \qquad (8.21)$$
>
> $$c_n = \frac{1}{2\pi} \int_{-\pi}^{\pi} f(x) e^{inx} \, dx. \qquad (8.22)$$

We can easily extend the above analysis to other intervals. For example, for $x \in [-L, L]$ the Fourier trigonometric series is

$$f(x) \sim \frac{a_0}{2} + \sum_{n=1}^{\infty} \left(a_n \cos \frac{n\pi x}{L} + b_n \sin \frac{n\pi x}{L} \right)$$

with Fourier coefficients

$$a_n = \frac{1}{L} \int_{-L}^{L} f(x) \cos \frac{n\pi x}{L} \, dx, \quad n = 0, 1, \ldots,$$

$$b_n = \frac{1}{L} \int_{-L}^{L} f(x) \sin \frac{n\pi x}{L} \, dx, \quad n = 1, 2, \ldots.$$

This can be rewritten as an exponential Fourier series of the form

> ### Complex Exponential Series for $f(x)$ Defined on $[-L, L]$
>
> $$f(x) \sim \sum_{n=-\infty}^{\infty} c_n e^{-in\pi x/L}, \qquad (8.23)$$
>
> $$c_n = \frac{1}{2L} \int_{-L}^{L} f(x) e^{in\pi x/L} \, dx. \qquad (8.24)$$

We can now use this complex exponential Fourier series for function defined on $[-L, L]$ to derive the Fourier transform by letting L get large. This will lead to a sum over a continuous set of frequencies, as opposed to the sum over discrete frequencies, which Fourier series represent.

8.3 Exponential Fourier Transform

BOTH THE TRIGONOMETRIC AND COMPLEX EXPONENTIAL Fourier series provide us with representations of a class of functions of finite period in terms of sums over a discrete set of frequencies. In particular, for functions defined on $x \in [-L, L]$, the period of the Fourier series representation is $2L$. We can write the arguments in the exponentials, $e^{-in\pi x/L}$, in terms of the angular frequency, $\omega_n = n\pi/L$, as $e^{-i\omega_n x}$. We note that the frequencies, ν_n, are then defined through $\omega_n = 2\pi\nu_n = \frac{n\pi}{L}$. Therefore, the complex exponential series is seen to be a sum over a discrete, or countable, set of frequencies.

We would now like to extend the finite interval to an infinite interval, $x \in (-\infty, \infty)$, and to extend the discrete set of (angular) frequencies to a continuous range of frequencies, $\omega \in (-\infty, \infty)$. One can do this rigorously. It amounts to letting L and n get large and keeping $\frac{n}{L}$ fixed.

We first define $\Delta\omega = \frac{\pi}{L}$, so that $\omega_n = n\Delta\omega$. Inserting the Fourier coefficients (8.24) into Equation (8.23), we have

$$
\begin{aligned}
f(x) &\sim \sum_{n=-\infty}^{\infty} c_n e^{-in\pi x/L} \\
&= \sum_{n=-\infty}^{\infty} \left(\frac{1}{2L} \int_{-L}^{L} f(\xi) e^{in\pi\xi/L} \, d\xi \right) e^{-in\pi x/L} \\
&= \sum_{n=-\infty}^{\infty} \left(\frac{\Delta\omega}{2\pi} \int_{-L}^{L} f(\xi) e^{i\omega_n \xi} \, d\xi \right) e^{-i\omega_n x}.
\end{aligned}
\tag{8.25}
$$

Now, we let L get large, so that $\Delta\omega$ becomes small and ω_n approaches the angular frequency ω. Then,

$$
\begin{aligned}
f(x) &\sim \lim_{\Delta\omega \to 0, L \to \infty} \frac{1}{2\pi} \sum_{n=-\infty}^{\infty} \left(\int_{-L}^{L} f(\xi) e^{i\omega_n \xi} \, d\xi \right) e^{-i\omega_n x} \Delta\omega \\
&= \frac{1}{2\pi} \int_{-\infty}^{\infty} \left(\int_{-\infty}^{\infty} f(\xi) e^{i\omega\xi} \, d\xi \right) e^{-i\omega x} \, d\omega.
\end{aligned}
\tag{8.26}
$$

Definitions of the Fourier transform and the inverse Fourier transform.

Looking at this last result, we formally arrive at the definition of the Fourier transform. It is embodied in the inner integral and can be written as

$$
F[f] = \hat{f}(\omega) = \int_{-\infty}^{\infty} f(x) e^{i\omega x} \, dx.
\tag{8.27}
$$

This is a generalization of the Fourier coefficients (8.24).

Once we know the Fourier transform, $\hat{f}(\omega)$, we can reconstruct the original function, $f(x)$, using the inverse Fourier transform, which is given by the outer integration,

$$
F^{-1}[\hat{f}] = f(x) = \frac{1}{2\pi} \int_{-\infty}^{\infty} \hat{f}(\omega) e^{-i\omega x} \, d\omega.
\tag{8.28}
$$

We note that it can be proven that the Fourier transform exists when $f(x)$ is absolutely integrable, that is,

$$
\int_{-\infty}^{\infty} |f(x)| \, dx < \infty.
$$

Such functions are said to be L_1.

We combine these results below, defining the Fourier and inverse Fourier transforms and indicating that they are inverse operations of each other. We will then prove the first of the equations, Equation (8.31). [The second equation, Equation (8.32), follows in a similar way.]

The **Fourier transform** and **inverse Fourier transform** are inverse operations. Defining the Fourier transform as

$$F[f] = \hat{f}(\omega) = \int_{-\infty}^{\infty} f(x)e^{i\omega x}\, dx, \qquad (8.29)$$

and the inverse Fourier transform as

$$F^{-1}[\hat{f}] = f(x) = \frac{1}{2\pi}\int_{-\infty}^{\infty} \hat{f}(\omega)e^{-i\omega x}\, d\omega, \qquad (8.30)$$

then

$$F^{-1}[F[f]] = f(x), \qquad (8.31)$$

and

$$F[F^{-1}[\hat{f}]] = \hat{f}(\omega). \qquad (8.32)$$

Proof. The proof is carried out by inserting the definition of the Fourier transform, (Equation (8.29)), into the inverse transform definition, (Equation (8.30)), and then interchanging the orders of integration. Thus, we have

$$
\begin{aligned}
F^{-1}[F[f]] &= \frac{1}{2\pi}\int_{-\infty}^{\infty} F[f]e^{-i\omega x}\, d\omega \\
&= \frac{1}{2\pi}\int_{-\infty}^{\infty}\left[\int_{-\infty}^{\infty} f(\xi)e^{i\omega\xi}\, d\xi\right] e^{-i\omega x}\, d\omega \\
&= \frac{1}{2\pi}\int_{-\infty}^{\infty}\int_{-\infty}^{\infty} f(\xi)e^{i\omega(\xi-x)}\, d\xi d\omega \\
&= \frac{1}{2\pi}\int_{-\infty}^{\infty}\left[\int_{-\infty}^{\infty} e^{i\omega(\xi-x)}\, d\omega\right] f(\xi)\, d\xi. \qquad (8.33)
\end{aligned}
$$

In order to complete the proof, we need to evaluate the inside integral, which does not depend upon $f(x)$. This is an improper integral, so we first define

$$D_\Omega(x) = \int_{-\Omega}^{\Omega} e^{i\omega x}\, d\omega$$

and compute the inner integral as

$$\int_{-\infty}^{\infty} e^{i\omega(\xi-x)}\, d\omega = \lim_{\Omega\to\infty} D_\Omega(\xi - x).$$

We can compute $D_\Omega(x)$. A simple evaluation yields

$$
\begin{aligned}
D_\Omega(x) &= \int_{-\Omega}^{\Omega} e^{i\omega x}\, d\omega \\
&= \left.\frac{e^{i\omega x}}{ix}\right|_{-\Omega}^{\Omega} \\
&= \frac{e^{ix\Omega} - e^{-ix\Omega}}{2ix} \\
&= \frac{2\sin x\Omega}{x}. \qquad (8.34)
\end{aligned}
$$

A plot of this function is given in Figure 8.5 for $\Omega = 4$. For large Ω, the peak grows and the values of $D_\Omega(x)$ for $x \neq 0$ tend to zero as shown in

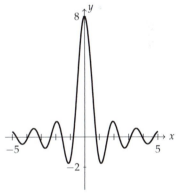

Figure 8.5: A plot of the function $D_\Omega(x)$ for $\Omega = 4$.

Figure 8.6. In fact, as x approaches 0, $D_\Omega(x)$ approaches 2Ω. For $x \neq 0$, the $D_\Omega(x)$ function tends to zero.

We further note that

$$\lim_{\Omega \to \infty} D_\Omega(x) = 0, \quad x \neq 0,$$

and $\lim_{\Omega \to \infty} D_\Omega(x)$ is infinite at $x = 0$. However, the area is constant for each Ω. In fact,

$$\int_{-\infty}^{\infty} D_\Omega(x)\, dx = 2\pi.$$

We can show this by recalling the computation in Example 7.42,

$$\int_{-\infty}^{\infty} \frac{\sin x}{x}\, dx = \pi.$$

Then,

$$
\begin{aligned}
\int_{-\infty}^{\infty} D_\Omega(x)\, dx &= \int_{-\infty}^{\infty} \frac{2\sin x\Omega}{x}\, dx \\
&= \int_{-\infty}^{\infty} 2\frac{\sin y}{y}\, dy \\
&= 2\pi.
\end{aligned}
\tag{8.35}
$$

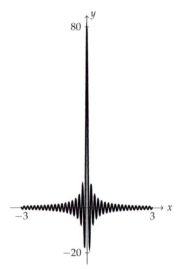

Figure 8.6: A plot of the function $D_\Omega(x)$ for $\Omega = 40$.

Another way to look at $D_\Omega(x)$ is to consider the sequence of functions $f_n(x) = \frac{\sin nx}{\pi x}$, $n = 1, 2, \ldots$. Thus we have shown that this sequence of functions satisfies the two properties,

$$\lim_{n \to \infty} f_n(x) = 0, \quad x \neq 0,$$

$$\int_{-\infty}^{\infty} f_n(x)\, dx = 1.$$

This is a key representation of such generalized functions. The limiting value vanishes at all but one point, but the area is finite.

Such behavior can be seen for the limit of other sequences of functions. For example, consider the sequence of functions

$$f_n(x) = \begin{cases} 0, & |x| > \frac{1}{n}, \\ \frac{n}{2}, & |x| \le \text{frac}1n. \end{cases}$$

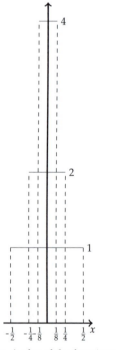

Figure 8.7: A plot of the functions $f_n(x)$ for $n = 2, 4, 8$.

This is a sequence of functions as shown in Figure 8.7. As $n \to \infty$, we find the limit is zero for $x \neq 0$ and is infinite for $x = 0$. However, the area under each member of the sequences is one. Thus, the limiting function is zero at most points but has area one.

The limit is not really a function. It is a generalized function. It is called the Dirac delta function, which is defined by

1. $\delta(x) = 0$ for $x \neq 0$.

2. $\int_{-\infty}^{\infty} \delta(x)\, dx = 1$.

Before returning to the proof that the inverse Fourier transform of the Fourier transform is the identity, we state one more property of the Dirac delta function, which we will prove in the next section. Namely, we will show that

$$\int_{-\infty}^{\infty} \delta(x - a) f(x)\, dx = f(a).$$

Returning to the proof, we now have that

$$\int_{-\infty}^{\infty} e^{i\omega(\xi - x)}\, d\omega = \lim_{\Omega \to \infty} D_\Omega(\xi - x) = 2\pi \delta(\xi - x).$$

Inserting this into Equation (8.33), we have

$$
\begin{aligned}
F^{-1}[F[f]] &= \frac{1}{2\pi} \int_{-\infty}^{\infty} \left[\int_{-\infty}^{\infty} e^{i\omega(\xi - x)}\, d\omega \right] f(\xi)\, d\xi \\
&= \frac{1}{2\pi} \int_{-\infty}^{\infty} 2\pi \delta(\xi - x) f(\xi)\, d\xi \\
&= f(x).
\end{aligned}
\tag{8.36}
$$

Thus, we have proven that the inverse transform of the Fourier transform of f is f.   □

8.4 Dirac Delta Function

IN THE PREVIOUS SECTION WE INTRODUCED the Dirac delta function, $\delta(x)$. As noted above, this is one example of what is known as a generalized function, or a distribution. Dirac had introduced this function in the 1930s in his study of quantum mechanics as a useful tool. It was later studied in a general theory of distributions and found to be more than a simple tool used by physicists. The Dirac delta function, as any distribution, only makes sense under an integral.

Two properties were used in the previous section. First, one has that the area under the delta function is one:

$$\int_{-\infty}^{\infty} \delta(x)\, dx = 1.$$

Integration over more general intervals gives

$$\int_a^b \delta(x)\, dx = \begin{cases} 1, & 0 \in [a, b], \\ 0, & 0 \notin [a, b]. \end{cases} \tag{8.37}$$

The other property that was used was the sifting property:

$$\int_{-\infty}^{\infty} \delta(x - a) f(x)\, dx = f(a).$$

This can be seen by noting that the delta function is zero everywhere, except at $x = a$. Therefore, the integrand is zero everywhere, and the only

P. A. M. Dirac (1902-1984) introduced the δ function in his book, *The Principles of Quantum Mechanics*, 4th Ed., Oxford University Press, 1958, originally published in 1930, as part of his orthogonality statement for a basis of functions in a Hilbert space, $< \xi' | \xi'' > = c\delta(\xi' - \xi'')$ in the same way we introduced discrete orthogonality using the Kronecker delta.

Properties of the Dirac delta function:

$$\int_{-\infty}^{\infty} \delta(x-a)f(x)\,dx = f(a).$$

$$\int_{-\infty}^{\infty} \delta(ax)\,dx = \frac{1}{|a|}\int_{-\infty}^{\infty}\delta(y)\,dy.$$

$$\int_{-\infty}^{\infty}\delta(f(x))\,dx = \int_{-\infty}^{\infty}\sum_{j=1}^{n}\frac{\delta(x-x_j)}{|f'(x_j)|}\,dx.$$

(For n simple roots.)

These and other properties are often written outside the integral:

$$\delta(ax) = \frac{1}{|a|}\delta(x).$$

$$\delta(-x) = \delta(x).$$

$$\delta((x-a)(x-b)) = \frac{[\delta(x-a)+\delta(x-a)]}{|a-b|}.$$

$$\delta(f(x)) = \sum_{j}\frac{\delta(x-x_j)}{|f'(x_j)|},$$

for $f(x_j)=0, f'(x_j)\neq 0$.

contribution from $f(x)$ will be from $x = a$. So, we can replace $f(x)$ with $f(a)$ under the integral. Since $f(a)$ is a constant, we have that

$$\int_{-\infty}^{\infty}\delta(x-a)f(x)\,dx = \int_{-\infty}^{\infty}\delta(x-a)f(a)\,dx$$

$$= f(a)\int_{-\infty}^{\infty}\delta(x-a)\,dx = f(a). \qquad (8.38)$$

Another property results from using a scaled argument, ax. In this case, we show that

$$\delta(ax) = |a|^{-1}\delta(x). \qquad (8.39)$$

As usual, this only has meaning under an integral sign. So, we place $\delta(ax)$ inside an integral and make a substitution $y = ax$:

$$\int_{-\infty}^{\infty}\delta(ax)\,dx = \lim_{L\to\infty}\int_{-L}^{L}\delta(ax)\,dx$$

$$= \lim_{L\to\infty}\frac{1}{a}\int_{-aL}^{aL}\delta(y)\,dy. \qquad (8.40)$$

If $a > 0$ then

$$\int_{-\infty}^{\infty}\delta(ax)\,dx = \frac{1}{a}\int_{-\infty}^{\infty}\delta(y)\,dy.$$

However, if $a < 0$ then

$$\int_{-\infty}^{\infty}\delta(ax)\,dx = \frac{1}{a}\int_{\infty}^{-\infty}\delta(y)\,dy = -\frac{1}{a}\int_{-\infty}^{\infty}\delta(y)\,dy.$$

The overall difference in a multiplicative minus sign can be absorbed into one expression by changing the factor $1/a$ to $1/|a|$. Thus,

$$\int_{-\infty}^{\infty}\delta(ax)\,dx = \frac{1}{|a|}\int_{-\infty}^{\infty}\delta(y)\,dy. \qquad (8.41)$$

Example 8.1. *Evaluate* $\int_{-\infty}^{\infty}(5x+1)\delta(4(x-2))\,dx$. *This is a straightforward integration:*

$$\int_{-\infty}^{\infty}(5x+1)\delta(4(x-2))\,dx = \frac{1}{4}\int_{-\infty}^{\infty}(5x+1)\delta(x-2)\,dx = \frac{11}{4}.$$

The first step is to write $\delta(4(x-2)) = \frac{1}{4}\delta(x-2)$. *Then, the final evaluation is given by*

$$\frac{1}{4}\int_{-\infty}^{\infty}(5x+1)\delta(x-2)\,dx = \frac{1}{4}(5(2)+1) = \frac{11}{4}.$$

Even more general than $\delta(ax)$ is the delta function $\delta(f(x))$. The integral of $\delta(f(x))$ can be evaluated, depending upon the number of zeros of $f(x)$. If there is only one zero, $f(x_1) = 0$, then one has that

$$\int_{-\infty}^{\infty}\delta(f(x))\,dx = \int_{-\infty}^{\infty}\frac{1}{|f'(x_1)|}\delta(x-x_1)\,dx.$$

This can be proven using the substitution $y = f(x)$ and is left as an exercise for the reader. This result is often written as

$$\delta(f(x)) = \frac{1}{|f'(x_1)|}\delta(x-x_1),$$

again keeping in mind that this only has meaning when placed under an integral.

Example 8.2. *Evaluate* $\int_{-\infty}^{\infty} \delta(3x - 2)x^2 \, dx$.

This is not a simple $\delta(x - a)$. So, we need to find the zeros of $f(x) = 3x - 2$. There is only one, $x = \frac{2}{3}$. Also, $|f'(x)| = 3$. Therefore, we have

$$\int_{-\infty}^{\infty} \delta(3x - 2)x^2 \, dx = \int_{-\infty}^{\infty} \frac{1}{3}\delta(x - \frac{2}{3})x^2 \, dx = \frac{1}{3}\left(\frac{2}{3}\right)^2 = \frac{4}{27}.$$

Note that this integral can be evaluated the long way using the substitution $y = 3x - 2$. Then, $dy = 3\,dx$ and $x = (y + 2)/3$. This gives

$$\int_{-\infty}^{\infty} \delta(3x - 2)x^2 \, dx = \frac{1}{3}\int_{-\infty}^{\infty} \delta(y)\left(\frac{y + 2}{3}\right)^2 \, dy = \frac{1}{3}\left(\frac{4}{9}\right) = \frac{4}{27}.$$

More generally, one can show that when $f(x_j) = 0$ and $f'(x_j) \neq 0$ for $j = 1, 2, \ldots, n$, (i.e., when one has n simple zeros), then

$$\delta(f(x)) = \sum_{j=1}^{n} \frac{1}{|f'(x_j)|}\delta(x - x_j).$$

Example 8.3. *Evaluate* $\int_{0}^{2\pi} \cos x \, \delta(x^2 - \pi^2) \, dx$.

In this case, the argument of the delta function has two simple roots. Namely, $f(x) = x^2 - \pi^2 = 0$ when $x = \pm\pi$. Furthermore, $f'(x) = 2x$. Therefore, $|f'(\pm\pi)| = 2\pi$. This gives

$$\delta(x^2 - \pi^2) = \frac{1}{2\pi}[\delta(x - \pi) + \delta(x + \pi)].$$

Inserting this expression into the integral and noting that $x = -\pi$ is not in the integration interval, we have

$$\begin{aligned}
\int_{0}^{2\pi} \cos x \, \delta(x^2 - \pi^2) \, dx &= \frac{1}{2\pi}\int_{0}^{2\pi} \cos x \, [\delta(x - \pi) + \delta(x + \pi)] \, dx \\
&= \frac{1}{2\pi}\cos\pi = -\frac{1}{2\pi}. \qquad (8.42)
\end{aligned}$$

Example 8.4. *Show* $H'(x) = \delta(x)$, *where the Heaviside function (or, step function) is defined as*

$$H(x) = \begin{cases} 0, & x < 0 \\ 1, & x > 0 \end{cases}$$

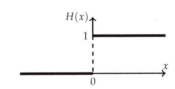

Figure 8.8: The Heaviside step function, $H(x)$.

and is shown in Figure 8.8.

Looking at the plot, it is easy to see that $H'(x) = 0$ for $x \neq 0$. In order to check that this gives the delta function, we need to compute the area integral. Therefore, we have

$$\int_{-\infty}^{\infty} H'(x) \, dx = H(x)\Big|_{-\infty}^{\infty} = 1 - 0 = 1.$$

Thus, $H'(x)$ satisfies the two properties of the Dirac delta function.

8.5 Properties of the Fourier Transform

WE NOW RETURN TO THE FOURIER TRANSFORM. Before actually computing the Fourier transform of some functions, we prove a few of the properties of the Fourier transform.

First we note that there are several forms that one may encounter for the Fourier transform. In applications, functions can either be functions of time, $f(t)$, or space, $f(x)$. The corresponding Fourier transforms are then written as

$$\hat{f}(\omega) = \int_{-\infty}^{\infty} f(t)e^{i\omega t}\, dt \tag{8.43}$$

or

$$\hat{f}(k) = \int_{-\infty}^{\infty} f(x)e^{ikx}\, dx. \tag{8.44}$$

ω is called the angular frequency and is related to the frequency ν by $\omega = 2\pi\nu$. The units of frequency are typically given in Hertz (Hz). Sometimes the frequency is denoted by f when there is no confusion. k is called the wave number. It has units of inverse length and is related to the wavelength, λ, by $k = \frac{2\pi}{\lambda}$.

We explore a few basic properties of the Fourier transform and use them in examples in the next section.

1. **Linearity:** For any functions $f(x)$ and $g(x)$ for which the Fourier transform exists and constant a, we have

$$F[f + g] = F[f] + F[g]$$

and

$$F[af] = aF[f].$$

These simply follow from the properties of integration and establish the linearity of the Fourier transform.

2. **Transform of a Derivative:** $F\left[\dfrac{df}{dx}\right] = -ik\hat{f}(k)$

Here we compute the Fourier transform (8.29) of the derivative by inserting the derivative in the Fourier integral and using integration by parts:

$$
\begin{aligned}
F\left[\frac{df}{dx}\right] &= \int_{-\infty}^{\infty} \frac{df}{dx} e^{ikx}\, dx \\
&= \lim_{L\to\infty} \left[f(x)e^{ikx}\right]_{-L}^{L} - ik\int_{-\infty}^{\infty} f(x)e^{ikx}\, dx.
\end{aligned}
\tag{8.45}
$$

The limit will vanish if we assume that $\lim_{x\to\pm\infty} f(x) = 0$. This last integral is recognized as the Fourier transform of f, proving the given property.

3. **Higher-Order Derivatives:** $F\left[\dfrac{d^n f}{dx^n}\right] = (-ik)^n \hat{f}(k)$

The proof of this property follows from the previous result, or doing several integration by parts. We will consider the case when $n = 2$. Noting that the second derivative is the derivative of $f'(x)$ and applying the previous result, we have

$$
\begin{aligned}
F\left[\frac{d^2 f}{dx^2}\right] &= F\left[\frac{d}{dx}f'\right] \\
&= -ikF\left[\frac{df}{dx}\right] = (-ik)^2 \hat{f}(k).
\end{aligned}
\tag{8.46}
$$

This result will be true if

$$
\lim_{x\to\pm\infty} f(x) = 0 \text{ and } \lim_{x\to\pm\infty} f'(x) = 0.
$$

The generalization to the transform of the nth derivative easily follows.

4. **Multiplication by x:** $F[xf(x)] = -i\dfrac{d}{dk}\hat{f}(k)$

This property can be shown by using the fact that $\frac{d}{dk}e^{ikx} = ixe^{ikx}$ and the ability to differentiate an integral with respect to a parameter.

$$
\begin{aligned}
F[xf(x)] &= \int_{-\infty}^{\infty} xf(x)e^{ikx}\, dx \\
&= \int_{-\infty}^{\infty} f(x)\frac{d}{dk}\left(\frac{1}{i}e^{ikx}\right)\, dx \\
&= -i\frac{d}{dk}\int_{-\infty}^{\infty} f(x)e^{ikx}\, dx \\
&= -i\frac{d}{dk}\hat{f}(k).
\end{aligned}
\tag{8.47}
$$

This result can be generalized to $F[x^n f(x)]$ as an exercise.

5. **Shifting Properties:** For constant a, we have the following shifting properties:

$$
f(x-a) \leftrightarrow e^{ika}\hat{f}(k),
\tag{8.48}
$$
$$
f(x)e^{-iax} \leftrightarrow \hat{f}(k-a).
\tag{8.49}
$$

Here we have denoted the Fourier transform pairs using a double arrow as $f(x) \leftrightarrow \hat{f}(k)$. These are easily proved by inserting the desired forms into the definition of the Fourier transform (8.29), or inverse Fourier transform (8.30). The first shift property (8.48) is shown by the following argument. We evaluate the Fourier transform:

$$
F[f(x-a)] = \int_{-\infty}^{\infty} f(x-a)e^{ikx}\, dx.
$$

Now perform the substitution $y = x - a$. Then,

$$
\begin{aligned}
F[f(x-a)] &= \int_{-\infty}^{\infty} f(y)e^{ik(y+a)}\, dy \\
&= e^{ika}\int_{-\infty}^{\infty} f(y)e^{iky}\, dy \\
&= e^{ika}\hat{f}(k).
\end{aligned}
\tag{8.50}
$$

The second shift property (8.49) follows in a similar way.

6. **Convolution of Functions:** We define the convolution of two functions $f(x)$ and $g(x)$ as

$$(f * g)(x) = \int_{-\infty}^{\infty} f(t)g(x-t)\, dx. \tag{8.51}$$

Then, the Fourier transform of the convolution is the product of the Fourier transforms of the individual functions:

$$F[f * g] = \hat{f}(k)\hat{g}(k). \tag{8.52}$$

We will return to the proof of this property in Section 8.6.

8.5.1 Fourier Transform Examples

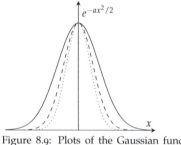

Figure 8.9: Plots of the Gaussian function $f(x) = e^{-ax^2/2}$ for $a = 1, 2, 3$.

IN THIS SECTION WE WILL COMPUTE the Fourier transforms of several functions.

Example 8.5. *Find the Fourier transform of a Gaussian, $f(x) = e^{-ax^2/2}$.*

This function, shown in Figure 8.9, is called the Gaussian function. It has many applications in areas such as quantum mechanics, molecular theory, probability, and heat diffusion. We will compute the Fourier transform of this function and show that the Fourier transform of a Gaussian is a Gaussian. In the derivation, we will introduce classic techniques for computing such integrals.

We begin by applying the definition of the Fourier transform,

$$\hat{f}(k) = \int_{-\infty}^{\infty} f(x)e^{ikx}\, dx = \int_{-\infty}^{\infty} e^{-ax^2/2 + ikx}\, dx. \tag{8.53}$$

The first step in computing this integral is to complete the square in the argument of the exponential. Our goal is to rewrite this integral so that a simple substitution will lead to a classic integral of the form $\int_{-\infty}^{\infty} e^{\beta y^2}\, dy$, which we can integrate. The completion of the square follows as usual:

$$
\begin{aligned}
-\frac{a}{2}x^2 + ikx &= -\frac{a}{2}\left[x^2 - \frac{2ik}{a}x\right] \\
&= -\frac{a}{2}\left[x^2 - \frac{2ik}{a}x + \left(-\frac{ik}{a}\right)^2 - \left(-\frac{ik}{a}\right)^2\right] \\
&= -\frac{a}{2}\left(x - \frac{ik}{a}\right)^2 - \frac{k^2}{2a}. \tag{8.54}
\end{aligned}
$$

We now put this expression into the integral and make the substitutions $y = x - \frac{ik}{a}$ and $\beta = \frac{a}{2}$.

$$
\begin{aligned}
\hat{f}(k) &= \int_{-\infty}^{\infty} e^{-ax^2/2 + ikx}\, dx \\
&= e^{-\frac{k^2}{2a}} \int_{-\infty}^{\infty} e^{-\frac{a}{2}\left(x - \frac{ik}{a}\right)^2}\, dx \\
&= e^{-\frac{k^2}{2a}} \int_{-\infty - \frac{ik}{a}}^{\infty - \frac{ik}{a}} e^{-\beta y^2}\, dy. \tag{8.55}
\end{aligned}
$$

One would be tempted to absorb the $-\frac{ik}{a}$ terms in the limits of integration. However, we know from our previous study that the integration takes place over a contour in the complex plane as shown in Figure 8.10.

In this case, we can deform this horizontal contour to a contour along the real axis because we will not cross any singularities of the integrand. So, we now safely write

$$\hat{f}(k) = e^{-\frac{k^2}{2a}} \int_{-\infty}^{\infty} e^{-\beta y^2}\, dy.$$

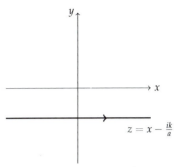

Figure 8.10: Simple horizontal contour.

The resulting integral is a classic integral and can be performed using a standard trick. Define I by[4]

$$I = \int_{-\infty}^{\infty} e^{-\beta y^2}\, dy.$$

Then,

$$I^2 = \int_{-\infty}^{\infty} e^{-\beta y^2}\, dy \int_{-\infty}^{\infty} e^{-\beta x^2}\, dx.$$

Note that we needed to change the integration variable so that we can write this product as a double integral:

$$I^2 = \int_{-\infty}^{\infty} \int_{-\infty}^{\infty} e^{-\beta(x^2+y^2)}\, dx dy.$$

[4] Here we show

$$\int_{-\infty}^{\infty} e^{-\beta y^2}\, dy = \sqrt{\frac{\pi}{\beta}}.$$

Note that we solved the $\beta = 1$ case in Example 6.11, so a simple variable transformation $z = \sqrt{\beta}y$ is all that is needed to get the answer. However, it cannot hurt to see this classic derivation again.

This is an integral over the entire xy-plane. We now transform to polar coordinates to obtain

$$
\begin{aligned}
I^2 &= \int_0^{2\pi} \int_0^{\infty} e^{-\beta r^2}\, r dr d\theta \\
&= 2\pi \int_0^{\infty} e^{-\beta r^2}\, r dr \\
&= -\frac{\pi}{\beta} \left[e^{-\beta r^2} \right]_0^{\infty} = \frac{\pi}{\beta}.
\end{aligned}
\tag{8.56}
$$

The final result is obtained by taking the square root, yielding

$$I = \sqrt{\frac{\pi}{\beta}}.$$

We can now insert this result to give the Fourier transform of the Gaussian function:

$$\hat{f}(k) = \sqrt{\frac{2\pi}{a}} e^{-k^2/2a}.
\tag{8.57}$$

Therefore, we have shown that the Fourier transform of a Gaussian is a Gaussian.

The Fourier transform of a Gaussian is a Gaussian.

Example 8.6. *Find the Fourier transform of the box, or gate, function,*

$$f(x) = \begin{cases} b, & |x| \le a, \\ 0, & |x| > a \end{cases}.$$

This function is called the box function, or gate function. It is shown in Figure 8.11. The Fourier transform of the box function is relatively easy to compute. It is given by

$$\hat{f}(k) = \int_{-\infty}^{\infty} f(x) e^{ikx}\, dx$$

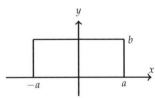

Figure 8.11: A plot of the box function in Example 8.6.

$$= \int_{-a}^{a} b e^{ikx} \, dx$$

$$= \frac{b}{ik} e^{ikx} \Big|_{-a}^{a}$$

$$= \frac{2b}{k} \sin ka. \tag{8.58}$$

We can rewrite this as

$$\hat{f}(k) = 2ab \frac{\sin ka}{ka} \equiv 2ab \operatorname{sinc} ka.$$

Here we introduced the sinc function,

$$\operatorname{sinc} x = \frac{\sin x}{x}.$$

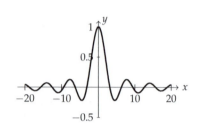

Figure 8.12: A plot of the Fourier transform of the box function in Example 8.6. This is the general shape of the sinc function.

A plot of this function is shown in Figure 8.12. This function appears often in signal analysis, and it plays a role in the study of diffraction.

We will now consider special limiting values for the box function and its transform. This will lead us to the Uncertainty Principle for signals, connecting the relationship between the localization properties of a signal and its transform.

1. *$a \to \infty$ and b fixed.*

   In this case, as a gets large, the box function approaches the constant function $f(x) = b$. At the same time, we see that the Fourier transform approaches a Dirac delta function. We had seen this function earlier when we first defined the Dirac delta function. Compare Figure 8.12 with Figure 8.5. In fact, $\hat{f}(k) = b D_a(k)$. [Recall the definition of $D_\Omega(x)$ in Equation (8.34).] So, in the limit, we obtain $\hat{f}(k) = 2\pi b \delta(k)$. This limit implies the fact that the Fourier transform of $f(x) = 1$ is $\hat{f}(k) = 2\pi\delta(k)$. As the width of the box becomes wider, the Fourier transform becomes more localized. In fact, we have arrived at the important result that

$$\int_{-\infty}^{\infty} e^{ikx} = 2\pi\delta(k). \tag{8.59}$$

2. *$b \to \infty$, $a \to 0$, and $2ab = 1$.*

   In this case, the box narrows and becomes steeper while maintaining a constant area of one. This is the way we had found a representation of the Dirac delta function previously. The Fourier transform approaches a constant in this limit. As a approaches zero, the sinc function approaches one, leaving $\hat{f}(k) \to 2ab = 1$. Thus, the Fourier transform of the Dirac delta function is one. Namely, we have

$$\int_{-\infty}^{\infty} \delta(x) e^{ikx} = 1. \tag{8.60}$$

   In this case, we have that the more localized the function $f(x)$ is, the more spread out the Fourier transform, $\hat{f}(k)$, is. We will summarize these notions in the next item by relating the widths of the function and its Fourier transform.

3. *The Uncertainty Principle:* $\Delta x \Delta k = 4\pi$.

The widths of the box function and its Fourier transform are related, as we have seen in the last two limiting cases. It is natural to define the width Δx of the box function as

$$\Delta x = 2a.$$

The width of the Fourier transform is a little trickier. This function actually extends along the entire k-axis. However, as $\hat{f}(k)$ became more localized, the central peak in Figure 8.12 became narrower. So, we define the width of this function, Δk as the distance between the first zeros on either side of the main lobe as shown in Figure 8.13. This gives

$$\Delta k = \frac{2\pi}{a}.$$

Combining these two relations, we find that

$$\Delta x \Delta k = 4\pi.$$

Thus, the more localized a signal, the less localized its transform, and vice versa. This notion is referred to as the Uncertainty Principle. For general signals, one needs to define the effective widths more carefully, but the main idea holds:

$$\Delta x \Delta k \ge c > 0.$$

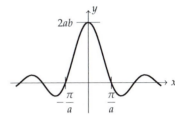

Figure 8.13: The width of the function $2ab\frac{\sin ka}{ka}$ is defined as the distance between the smallest magnitude zeros.

We now turn to other examples of Fourier transforms.

Example 8.7. *Find the Fourier transform of $f(x) = \begin{cases} e^{-ax}, & x \ge 0 \\ 0, & x < 0 \end{cases}, a > 0$.*

The Fourier transform of this function is

$$
\begin{aligned}
\hat{f}(k) &= \int_{-\infty}^{\infty} f(x) e^{ikx}\, dx \\
&= \int_{0}^{\infty} e^{ikx - ax}\, dx \\
&= \frac{1}{a - ik}.
\end{aligned}
\tag{8.61}
$$

Next, we will compute the inverse Fourier transform of this result and recover the original function.

Example 8.8. *Find the inverse Fourier transform of $\hat{f}(k) = \frac{1}{a-ik}$.*

The inverse Fourier transform of this function is

$$f(x) = \frac{1}{2\pi} \int_{-\infty}^{\infty} \hat{f}(k) e^{-ikx}\, dk = \frac{1}{2\pi} \int_{-\infty}^{\infty} \frac{e^{-ikx}}{a - ik}\, dk.$$

This integral can be evaluated using contour integral methods. We evaluate the integral

$$I = \int_{-\infty}^{\infty} \frac{e^{-ixz}}{a - iz}\, dz,$$

More formally, the Uncertainty Principle for signals is about the relation between duration and bandwidth, which are defined by $\Delta t = \frac{\|tf\|_2}{\|f\|_2}$ and $\Delta \omega = \frac{\|\omega \hat{f}\|_2}{\|\hat{f}\|_2}$, respectively, where $\|f\|_2 = \int_{-\infty}^{\infty} |f(t)|^2\, dt$ and $\|\hat{f}\|_2 = \frac{1}{2\pi} \int_{-\infty}^{\infty} |\hat{f}(\omega)|^2\, d\omega$. Under appropriate conditions, one can prove that $\Delta t \Delta \omega \ge \frac{1}{2}$. Equality holds for Gaussian signals. Werner Heisenberg (1901–1976) introduced the Uncertainty Principle into quantum physics in 1926, relating uncertainties in the position (Δx) and momentum (Δp_x) of particles. In this case, $\Delta x \Delta p_x \ge \frac{1}{2}\hbar$. Here, the uncertainties are defined as the positive square roots of the quantum mechanical variances of the position and momentum.

using Jordan's Lemma from Section 7.5.8. According to Jordan's Lemma, we need to enclose the contour with a semicircle in the upper half plane for $x < 0$ and in the lower half plane for $x > 0$, as shown in Figure 8.14.

The integrations along the semicircles will vanish and we will have

$$
\begin{aligned}
f(x) &= \frac{1}{2\pi} \int_{-\infty}^{\infty} \frac{e^{-ikx}}{a - ik} \, dk \\
&= \pm \frac{1}{2\pi} \oint_C \frac{e^{-ixz}}{a - iz} \, dz \\
&= \begin{cases} 0, & x < 0 \\ -\frac{1}{2\pi} 2\pi i \operatorname{Res}[z = -ia], & x > 0 \end{cases} \\
&= \begin{cases} 0, & x < 0 \\ e^{-ax}, & x > 0 \end{cases}.
\end{aligned}
\tag{8.62}
$$

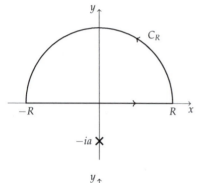
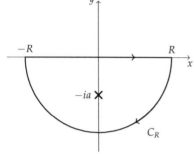

Figure 8.14: Contours for inverting $\hat{f}(k) = \frac{1}{a-ik}$.

Note that without paying careful attention to Jordan's Lemma, one might not retrieve the function from the previous example.

Example 8.9. *Find the inverse Fourier transform of $\hat{f}(\omega) = \pi \delta(\omega + \omega_0) + \pi \delta(\omega - \omega_0)$.*

We would like to find the inverse Fourier transform of this function. Instead of carrying out any integration, we will make use of the properties of Fourier transforms. Because the transforms of sums are the sums of transforms, we can look at each term individually. Consider $\delta(\omega - \omega_0)$. This is a shifted function. From the shift theorems in Equations (8.48) and (8.49) we have the Fourier transform pair

$$
e^{i\omega_0 t} f(t) \leftrightarrow \hat{f}(\omega - \omega_0).
$$

Recalling from Example 8.6 that

$$
\int_{-\infty}^{\infty} e^{i\omega t} \, dt = 2\pi \delta(\omega),
$$

we have from the shift property that

$$
F^{-1}[\delta(\omega - \omega_0)] = \frac{1}{2\pi} e^{-i\omega_0 t}.
$$

The second term can be transformed similarly. Therefore, we have

$$
F^{-1}[\pi \delta(\omega + \omega_0) + \pi \delta(\omega - \omega_0] = \frac{1}{2} e^{i\omega_0 t} + \frac{1}{2} e^{-i\omega_0 t} = \cos \omega_0 t.
$$

Example 8.10. *Find the Fourier transform of the finite wave train.*

$$
f(t) = \begin{cases} \cos \omega_0 t, & |t| \leq a, \\ 0, & |t| > a \end{cases}.
$$

For the previous example, we consider the finite wave train, which will reappear in the last chapter on signal analysis. In Figure 8.15 we show a plot of this function. A straightforward computation gives

$$
\hat{f}(\omega) = \int_{-\infty}^{\infty} f(t) e^{i\omega t} \, dt
$$

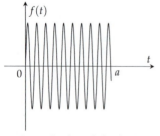

Figure 8.15: A plot of the finite wave train.

$$= \int_{-a}^{a} [\cos \omega_0 t + i \sin \omega_0 t] e^{i\omega t} \, dt$$

$$= \int_{-a}^{a} \cos \omega_0 t \cos \omega t \, dt + i \int_{-a}^{a} \sin \omega_0 t \sin \omega t \, dt$$

$$= \frac{1}{2} \int_{-a}^{a} [\cos((\omega + \omega_0)t) + \cos((\omega - \omega_0)t)] \, dt$$

$$= \frac{\sin((\omega + \omega_0)a)}{\omega + \omega_0} + \frac{\sin((\omega - \omega_0)a)}{\omega - \omega_0}. \tag{8.63}$$

8.6 Convolution Operation

IN THE LIST OF PROPERTIES OF THE FOURIER TRANSFORM, we defined the convolution of two functions, $f(x)$ and $g(x)$, to be the integral

$$(f * g)(x) = \int_{-\infty}^{\infty} f(t)g(x - t) \, dt. \tag{8.64}$$

In some sense, one is looking at a sum of the overlaps of one of the functions and all of the shifted versions of the other function. The German word for convolution is *faltung*, which means "folding" and in old texts this is referred to as the Faltung Theorem. In this section we will look into the convolution operation and its Fourier transform.

Before we get too involved with the convolution operation, it should be noted that there are really two things you need to take away from this discussion. The rest is detail. First, the convolution of two functions is a new functions as defined by Equation (8.64) when dealing with the Fourier transform. The second and most relevant is that the Fourier transform of the convolution of two functions is the product of the transforms of each function. The rest is all about the use and consequences of these two statements. In this section we will show how the convolution works and how it is useful.

The convolution is commutative.

First, we note that the convolution is commutative: $f * g = g * f$. This is easily shown by replacing $x - t$ with a new variable, $y = x - t$ and $dy = -dt$.

$$\begin{aligned}
(g * f)(x) &= \int_{-\infty}^{\infty} g(t)f(x - t) \, dt \\
&= -\int_{\infty}^{-\infty} g(x - y)f(y) \, dy \\
&= \int_{-\infty}^{\infty} f(y)g(x - y) \, dy \\
&= (f * g)(x). \tag{8.65}
\end{aligned}$$

The best way to understand the folding of the functions in the convolution is to take two functions and convolve them. The next example gives a graphical rendition followed by a direct computation of the convolution. The reader is encouraged to carry out these analyses for other functions.

Example 8.11. *Graphical convolution of the box function and a triangle function.*
In order to understand the convolution operation, we need to apply it to specific

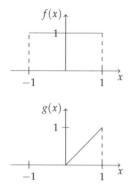

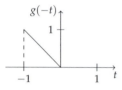

Figure 8.16: A plot of the box function $f(x)$ and the triangle function $g(x)$.

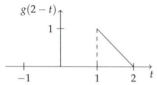

Figure 8.17: A plot of the reflected triangle function, $g(-t)$.

Figure 8.18: A plot of the reflected triangle function shifted by two units, $g(2 - t)$.

functions. We will first do this graphically for the box function

$$f(x) = \begin{cases} 1, & |x| \leq 1, \\ 0, & |x| > 1, \end{cases}$$

and the triangular function

$$g(x) = \begin{cases} x, & 0 \leq x \leq 1, \\ 0, & otherwise, \end{cases}$$

as shown in Figure 8.16.

Next, we determine the contributions to the integrand. We consider the shifted and reflected function $g(t - x)$ in Equation (8.64) for various values of t. For $t = 0$, we have $g(x - 0) = g(-x)$. This function is a reflection of the triangle function, $g(x)$, as shown in Figure 8.17.

We then translate the triangle function performing horizontal shifts by t. In Figure 8.18 we show such a shifted and reflected $g(x)$ for $t = 2$, or $g(2 - x)$.

In Figure 8.18 we show several plots of other shifts, $g(x - t)$, superimposed on $f(x)$.

The integrand is the product of $f(t)$ and $g(x - t)$ and the integral of the product $f(t)g(x - t)$ is given by the sum of the shaded areas for each value of x.

In the first plot of Figure 8.19, the area is zero, as there is no overlap of the functions. Intermediate shift values are displayed in the other plots in Figure 8.19. The value of the convolution at x is shown by the area under the product of the two functions for each value of x.

Plots of the areas of the convolution of the box and triangle functions for several values of x are given in Figure 8.18. We see that the value of the convolution integral builds up and then quickly drops to zero as a function of x. In Figure 8.20 the values of these areas is shown as a function of x.

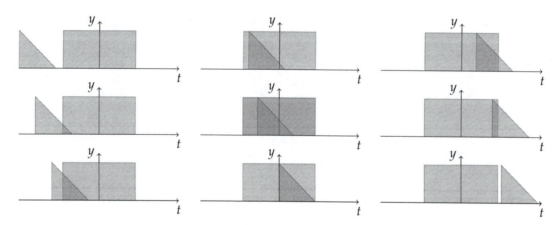

Figure 8.19: A plot of the box and triangle functions with the overlap indicated by the shaded area.

The plot of the convolution in Figure 8.20 is not easily determined using the graphical method. However, we can directly compute the convolution as shown in the next example.

Example 8.12. *Analytically find the convolution of the box function and the tri-angle function.*

The nonvanishing contributions to the convolution integral are when both $f(t)$ and $g(x-t)$ do not vanish. $f(t)$ is nonzero for $|t| \leq 1$, or $-1 \leq t \leq 1$. $g(x-t)$ is nonzero for $0 \leq x-t \leq 1$, or $x-1 \leq t \leq x$. These two regions are shown in Figure 8.21. On this region, $f(t)g(x-t) = x - t$.

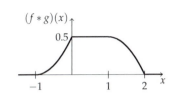

Figure 8.20: A plot of the convolution of the box and triangle functions.

Figure 8.21: Intersection of the support of $g(x)$ and $f(x)$.

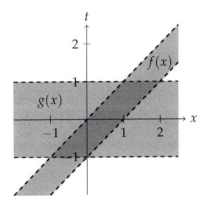

Isolating the intersection in Figure 8.22, we see in Figure 8.22 that there are three regions as shown by different shadings. These regions lead to a piecewise defined function with three different branches of nonzero values for $-1 < x < 0$, $0 < x < 1$, and $1 < x < 2$.

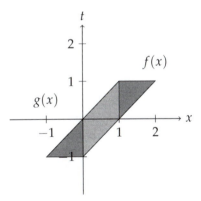

Figure 8.22: Intersection of the support of $g(x)$ and $f(x)$ showing the integration regions.

The values of the convolution can be determined through careful integration. The resulting integrals are given as

$$
\begin{aligned}
(f * g)(x) &= \int_{-\infty}^{\infty} f(t)g(x-t)\,dt \\
&= \begin{cases} \int_{-1}^{x}(x-t)\,dt, & -1 < x < 0 \\ \int_{x-1}^{x}(x-t)\,dt, & 0 < x < 1 \\ \int_{x-1}^{1}(x-t)\,dt, & 1 < x < 2 \end{cases} \\
&= \begin{cases} \frac{1}{2}(x+1)^2, & -1 < x < 0 \\ \frac{1}{2}, & 0 < x < 1 \\ \frac{1}{2}\left[1-(x-1)^2\right] & 1 < x < 2. \end{cases}
\end{aligned}
\tag{8.66}
$$

A plot of this function is shown in Figure 8.20.

8.6.1 Convolution Theorem for Fourier Transforms

IN THIS SECTION WE COMPUTE the Fourier transform of the convolution integral and show that the Fourier transform of the convolution is the product of the transforms of each function,

$$F[f * g] = \hat{f}(k)\hat{g}(k). \tag{8.67}$$

First, we use the definitions of the Fourier transform and the convolution to write the transform as

$$
\begin{aligned}
F[f * g] &= \int_{-\infty}^{\infty} (f * g)(x) e^{ikx} \, dx \\
&= \int_{-\infty}^{\infty} \left(\int_{-\infty}^{\infty} f(t) g(x - t) \, dt \right) e^{ikx} \, dx \\
&= \int_{-\infty}^{\infty} \left(\int_{-\infty}^{\infty} g(x - t) e^{ikx} \, dx \right) f(t) \, dt. \tag{8.68}
\end{aligned}
$$

We now substitute $y = x - t$ on the inside integral and separate the integrals:

$$
\begin{aligned}
F[f * g] &= \int_{-\infty}^{\infty} \left(\int_{-\infty}^{\infty} g(x - t) e^{ikx} \, dx \right) f(t) \, dt \\
&= \int_{-\infty}^{\infty} \left(\int_{-\infty}^{\infty} g(y) e^{ik(y+t)} \, dy \right) f(t) \, dt \\
&= \int_{-\infty}^{\infty} \left(\int_{-\infty}^{\infty} g(y) e^{iky} \, dy \right) f(t) e^{ikt} \, dt \\
&= \left(\int_{-\infty}^{\infty} f(t) e^{ikt} \, dt \right) \left(\int_{-\infty}^{\infty} g(y) e^{iky} \, dy \right). \tag{8.69}
\end{aligned}
$$

We see that the two integrals are just the Fourier transforms of f and g. Therefore, the Fourier transform of a convolution is the product of the Fourier transforms of the functions involved:

$$F[f * g] = \hat{f}(k)\hat{g}(k).$$

Example 8.13. *Compute the convolution of the box function of height one and width two with itself.*

Let $\hat{f}(k)$ be the Fourier transform of $f(x)$. Then, the Convolution Theorem says that $F[f * f](k) = \hat{f}^2(k)$, or

$$(f * f)(x) = F^{-1}[\hat{f}^2(k)].$$

For the box function, we have already found that

$$\hat{f}(k) = \frac{2}{k} \sin k.$$

So, we need to compute

$$
\begin{aligned}
(f * f)(x) &= F^{-1}[\frac{4}{k^2} \sin^2 k] \\
&= \frac{1}{2\pi} \int_{-\infty}^{\infty} \left(\frac{4}{k^2} \sin^2 k \right) e^{-ikx} \, dk. \tag{8.70}
\end{aligned}
$$

One way to compute this integral is to extend the computation into the complex k-plane. We first need to rewrite the integrand. Thus,

$$
\begin{aligned}
(f * f)(x) &= \frac{1}{2\pi} \int_{-\infty}^{\infty} \frac{4}{k^2} \sin^2 k\, e^{-ikx}\, dk \\
&= \frac{1}{\pi} \int_{-\infty}^{\infty} \frac{1}{k^2} [1 - \cos 2k] e^{-ikx}\, dk \\
&= \frac{1}{\pi} \int_{-\infty}^{\infty} \frac{1}{k^2} \left[1 - \frac{1}{2}(e^{ik} + e^{-ik}) \right] e^{-ikx}\, dk \\
&= \frac{1}{\pi} \int_{-\infty}^{\infty} \frac{1}{k^2} \left[e^{-ikx} - \frac{1}{2}(e^{-i(1-k)} + e^{-i(1+k)}) \right] dk. \quad (8.71)
\end{aligned}
$$

We can compute the above integrals if we know how to compute the integral

$$
I(y) = \frac{1}{\pi} \int_{-\infty}^{\infty} \frac{e^{-iky}}{k^2}\, dk.
$$

Then, the result can be found in terms of $I(y)$ as

$$
(f * f)(x) = I(x) - \frac{1}{2}[I(1-k) + I(1+k)].
$$

We consider the integral

$$
\oint_C \frac{e^{-iyz}}{\pi z^2}\, dz
$$

over the contour in Figure 8.23.

We can see that there is a double pole at $z = 0$. The pole is on the real axis. So, we will need to cut out the pole as we seek the value of the principal value integral.

Recall from Chapter 7 that

$$
\oint_{C_R} \frac{e^{-iyz}}{\pi z^2}\, dz = \int_{\Gamma_R} \frac{e^{-iyz}}{\pi z^2}\, dz + \int_{-R}^{-\epsilon} \frac{e^{-iyz}}{\pi z^2}\, dz + \int_{C_\epsilon} \frac{e^{-iyz}}{\pi z^2}\, dz + \int_{\epsilon}^{R} \frac{e^{-iyz}}{\pi z^2}\, dz.
$$

The integral $\oint_{C_R} \frac{e^{-iyz}}{\pi z^2}\, dz$ vanishes because there are no poles enclosed in the contour! The sum of the second and fourth integrals gives the integral we seek as $\epsilon \to 0$ and $R \to \infty$. The integral over Γ_R will vanish as R gets large according to Jordan's Lemma provided $y < 0$. That leaves the integral over the small semicircle.

As before, we can show that

$$
\lim_{\epsilon \to 0} \int_{C_\epsilon} f(z)\, dz = -\pi i\, \mathrm{Res}[f(z); z = 0].
$$

Therefore, we find

$$
I(y) = P \int_{-\infty}^{\infty} \frac{e^{-iyz}}{\pi z^2}\, dz = \pi i\, \mathrm{Res}\left[\frac{e^{-iyz}}{\pi z^2}; z = 0 \right].
$$

A simple computation of the residue gives $I(y) = -y$, for $y < 0$.

When $y > 0$, we need to close the contour in the lower half plane in order to apply Jordan's Lemma. Carrying out the computation, one finds $I(y) = y$, for $y > 0$. Thus,

$$
I(y) = \begin{cases} -y, & y > 0, \\ y, & y < 0, \end{cases} \quad (8.72)
$$

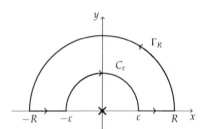

Figure 8.23:   Contour for computing $P \int_{-\infty}^{\infty} \frac{e^{-iyz}}{\pi z^2}\, dz$.

We are now ready to finish the computation of the convolution. We have to combine the integrals $I(y)$, $I(y+1)$, and $I(y-1)$, as $(f * f)(x) = I(x) - \frac{1}{2}[I(1 - k) + I(1 + k)]$. This gives different results in four intervals:

$$
\begin{aligned}
(f * f)(x) &= x - \frac{1}{2}[(x-2) + (x+2)] = 0, \quad x < -2, \\
&= x - \frac{1}{2}[(x-2) - (x+2)] = 2 + x \quad -2 < x < 0, \\
&= -x - \frac{1}{2}[(x-2) - (x+2)] = 2 - x, \quad 0 < x < 2, \\
&= -x - \frac{1}{2}[-(x-2) - (x+2)] = 0, \quad x > 2. \quad (8.73)
\end{aligned}
$$

A plot of this solution is the triangle function:

$$
(f * f)(x) = \begin{cases} 0, & x < -2 \\ 2 + x, & -2 < x < 0 \\ 2 - x, & 0 < x < 2 \\ 0, & x > 2, \end{cases} \quad (8.74)
$$

which was shown in the previous example.

Example 8.14. *Find the convolution of the box function of height one and width two with itself using a direct computation of the convolution integral.*

The nonvanishing contributions to the convolution integral are when both $f(t)$ and $f(x - t)$ do not vanish. $f(t)$ is nonzero for $|t| \leq 1$, or $-1 \leq t \leq 1$. $f(x - t)$ is nonzero for $|x - t| \leq 1$, or $x - 1 \leq t \leq x + 1$. These two regions are shown in Figure 8.25. On this region, $f(t)g(x - t) = 1$.

Figure 8.24: Plot of the regions of support for $f(t)$ and $f(x - t)$.

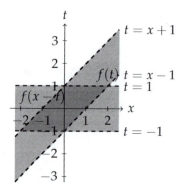

Thus, the nonzero contributions to the convolution are

$$
(f * f)(x) = \begin{cases} \int_{-1}^{x+1} dt, & 0 \leq x \leq 2, \\ \int_{x-1}^{1} dt, & -2 \leq x \leq 0, \end{cases} = \begin{cases} 2 + x, & 0 \leq x \leq 2, \\ 2 - x, & -2 \leq x \leq 0. \end{cases}
$$

Once again, we arrive at the triangle function.

In the previous section we showed the graphical convolution. For completeness, we do the same for this example. In Figure 8.25 we show the results. We see that the convolution of two box functions is a triangle function.

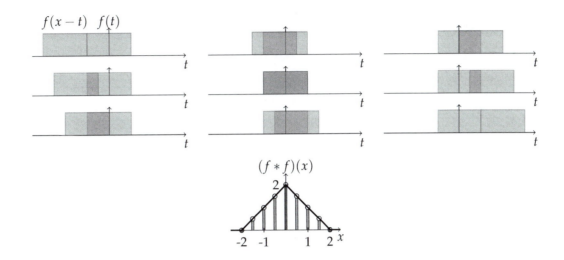

$f(x-t)$ $f(t)$

$(f * f)(x)$

Figure 8.25: A plot of the convolution of a box function with itself is a triangular function. We also show of several of the overlaps of $f(x)$ and the translations $f(x-t)$.

Example 8.15. *Show the graphical convolution of the box function of height one and width two with itself.*

Let's consider a slightly more complicated example, the convolution of two Gaussian functions.

Example 8.16. *Convolution of two Gaussian functions $f(x) = e^{-ax^2}$.*

In this example we will compute the convolution of two Gaussian functions with different widths. Let $f(x) = e^{-ax^2}$ and $g(x) = e^{-bx^2}$. A direct evaluation of the integral would be to compute

$$(f * g)(x) = \int_{-\infty}^{\infty} f(t)g(x-t)\, dt = \int_{-\infty}^{\infty} e^{-at^2 - b(x-t)^2}\, dt.$$

This integral can be rewritten as

$$(f * g)(x) = e^{-bx^2} \int_{-\infty}^{\infty} e^{-(a+b)t^2 + 2bxt}\, dt.$$

One could proceed to complete the square and finish carrying out the integration. However, we will use the Convolution Theorem to evaluate the convolution and leave the evaluation of this integral to Problem 12.

Recalling the Fourier transform of a Gaussian from Example 8.5, we have

$$\hat{f}(k) = F[e^{-ax^2}] = \sqrt{\frac{\pi}{a}} e^{-k^2/4a} \tag{8.75}$$

and

$$\hat{g}(k) = F[e^{-bx^2}] = \sqrt{\frac{\pi}{b}} e^{-k^2/4b}.$$

Denoting the convolution function by $h(x) = (f * g)(x)$, the Convolution Theorem gives

$$\hat{h}(k) = \hat{f}(k)\hat{g}(k) = \frac{\pi}{\sqrt{ab}} e^{-k^2/4a} e^{-k^2/4b}.$$

This is another Gaussian function, as seen by rewriting the Fourier transform of $h(x)$ as

$$\hat{h}(k) = \frac{\pi}{\sqrt{ab}}e^{-\frac{1}{4}\left(\frac{1}{a}+\frac{1}{b}\right)k^2} = \frac{\pi}{\sqrt{ab}}e^{-\frac{a+b}{4ab}k^2}. \qquad (8.76)$$

In order to complete the evaluation of the convolution of these two Gaussian functions, we need to find the inverse transform of the Gaussian in Equation (8.76). We can do this by looking at Equation (8.75). We have first that

$$F^{-1}\left[\sqrt{\frac{\pi}{a}}e^{-k^2/4a}\right] = e^{-ax^2}.$$

Moving the constants, we then obtain

$$F^{-1}[e^{-k^2/4a}] = \sqrt{\frac{a}{\pi}}e^{-ax^2}.$$

We now make the substitution $\alpha = \frac{1}{4a}$,

$$F^{-1}[e^{-\alpha k^2}] = \sqrt{\frac{1}{4\pi\alpha}}e^{-x^2/4\alpha}.$$

This is in the form needed to invert Equation (8.76). Thus, for $\alpha = \frac{a+b}{4ab}$, we find

$$(f*g)(x) = h(x) = \sqrt{\frac{\pi}{a+b}}e^{-\frac{ab}{a+b}x^2}.$$

8.6.2 Application to Signal Analysis

THERE ARE MANY APPLICATIONS of the convolution operation. One of these areas is the study of analog signals. An analog signal is a continuous signal and may contain either a finite or continuous set of frequencies. Fourier transforms can be used to represent such signals as a sum over the frequency content of these signals. In this section we will describe how convolutions can be used in studying signal analysis.

The first application is filtering. For a given signal, there might be some noise in the signal, or some undesirable high frequencies. For example, a device used for recording an analog signal might naturally not be able to record high frequencies. Let $f(t)$ denote the amplitude of a given analog signal and $\hat{f}(\omega)$ be the Fourier transform of this signal such as the example provided in Figure 8.26. Recall that the Fourier transform gives the frequency content of the signal.

There are many ways to filter out unwanted frequencies. The simplest would be to just drop all the high (angular) frequencies. For example, for some cutoff frequency ω_0, frequencies $|\omega| > \omega_0$ will be removed. The Fourier transform of the filtered signal would then be zero for $|\omega| > \omega_0$. This could be accomplished by multiplying the Fourier transform of the signal by a function that vanishes for $|\omega| > \omega_0$. For example, we could use the gate function

$$p_{\omega_0}(\omega) = \begin{cases} 1, & |\omega| \le \omega_0 \\ 0, & |\omega| > \omega_0 \end{cases}, \qquad (8.77)$$

as shown in Figure 8.27.

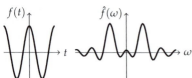

Figure 8.26: Schematic plot of a signal $f(t)$ and its Fourier transform $\hat{f}(\omega)$.

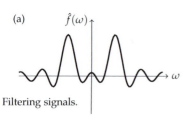

(a)

Filtering signals.

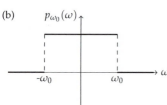

(b)

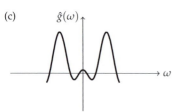

(c)

Figure 8.27: (a) Plot of the Fourier transform $\hat{f}(\omega)$ of a signal. (b) The gate function $p_{\omega_0}(\omega)$ used to filter out high frequencies. (c) The product of the functions, $\hat{g}(\omega) = \hat{f}(\omega)p_{\omega_0}(\omega)$, in (a) and (b) shows how the filters cuts out high frequencies, $|\omega| > \omega_0$.

In general, we multiply the Fourier transform of the signal by some filtering function $\hat{h}(t)$ to get the Fourier transform of the filtered signal,

$$\hat{g}(\omega) = \hat{f}(\omega)\hat{h}(\omega).$$

The new signal $g(t)$ is then the inverse Fourier transform of this product, giving the new signal as a convolution:

$$g(t) = F^{-1}[\hat{f}(\omega)\hat{h}(\omega)] = \int_{-\infty}^{\infty} h(t-\tau)f(\tau)\,d\tau. \qquad (8.78)$$

Such processes occur often in systems theory as well. One thinks of $f(t)$ as the input signal into some filtering device, which in turn produces the output, $g(t)$. The function $h(t)$ is called the impulse response. This is because it is a response to the impulse function, $\delta(t)$. In this case, one has

$$\int_{-\infty}^{\infty} h(t-\tau)\delta(\tau)\,d\tau = h(t).$$

Another application of the convolution is in windowing. This represents what happens when one measures a real signal. Real signals cannot be recorded for all values of time. Instead, data is collected over a finite time interval. If the length of time the data is collected is T, then the resulting signal is zero outside this time interval. This can be modeled in the same way as with filtering, except the new signal will be the product of the old signal with the windowing function. The resulting Fourier transform of the new signal will be a convolution of the Fourier transforms of the original signal and the windowing function.

Windowing signals.

Example 8.17. Finite Wave Train, Revisited.

We return to the finite wave train in Example 8.10 given by

$$h(t) = \begin{cases} \cos\omega_0 t, & |t| \le a, \\ 0, & |t| > a. \end{cases}$$

We can view this as a windowed version of $f(t) = \cos\omega_0 t$ obtained by multiplying $f(t)$ by the gate function

$$g_a(t) = \begin{cases} 1, & |x| \le a, \\ 0, & |x| > a. \end{cases} \qquad (8.79)$$

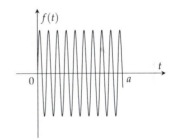

Figure 8.28: A plot of the finite wave train.

This is shown in Figure 8.28. Then, the Fourier transform is given as a convolution:

$$\begin{aligned} \hat{h}(\omega) &= (\hat{f} * \hat{g}_a)(\omega) \\ &= \frac{1}{2\pi} \int_{-\infty}^{\infty} \hat{f}(\omega - v)\hat{g}_a(v)\,dv. \end{aligned} \qquad (8.80)$$

The convolution in spectral space is defined with an extra factor of $1/2\pi$ so as to preserve the idea that the inverse Fourier transform of a convolution is the product of the corresponding signals.

Note that the convolution in frequency space requires the extra factor of $1/(2\pi)$.

We need the Fourier transforms of f and g_a in order to finish the computation. The Fourier transform of the box function was found in Example 8.6 as

$$\hat{g}_a(\omega) = \frac{2}{\omega}\sin\omega a.$$

The Fourier transform of the cosine function, $f(t) = \cos \omega_0 t$, is

$$
\begin{aligned}
\hat{f}(\omega) &= \int_{-\infty}^{\infty} \cos(\omega_0 t) e^{i\omega t}\, dt \\
&= \int_{-\infty}^{\infty} \frac{1}{2} \left(e^{i\omega_0 t} + e^{-i\omega_0 t} \right) e^{i\omega t}\, dt \\
&= \frac{1}{2} \int_{-\infty}^{\infty} \left(e^{i(\omega+\omega_0)t} + e^{i(\omega-\omega_0)t} \right) dt \\
&= \pi \left[\delta(\omega + \omega_0) + \delta(\omega - \omega_0) \right].
\end{aligned}
\tag{8.81}
$$

Note that we had earlier computed the inverse Fourier transform of this function in Example 8.9.

   Inserting these results in the convolution integral, we have

$$
\begin{aligned}
\hat{h}(\omega) &= \frac{1}{2\pi} \int_{-\infty}^{\infty} \hat{f}(\omega - v)\hat{g}_a(v)\, dv \\
&= \frac{1}{2\pi} \int_{-\infty}^{\infty} \pi \left[\delta(\omega - v + \omega_0) + \delta(\omega - v - \omega_0) \right] \frac{2}{v} \sin va\, dv \\
&= \frac{\sin(\omega + \omega_0)a}{\omega + \omega_0} + \frac{\sin(\omega - \omega_0)a}{\omega - \omega_0}.
\end{aligned}
\tag{8.82}
$$

This is the same result we had obtained in Example 8.10.

8.6.3   Parseval's Equality

The integral/sum of the (modulus) square of a function is the integral/sum of the (modulus) square of the transform.

As another example of the convolution theorem, we derive Parseval's Equality (named after Marc-Antoine Parseval (1755–1836)):

$$
\int_{-\infty}^{\infty} |f(t)|^2\, dt = \frac{1}{2\pi} \int_{-\infty}^{\infty} |\hat{f}(\omega)|^2\, d\omega.
\tag{8.83}
$$

This equality has a physical meaning for signals. The integral on the left side is a measure of the energy content of the signal in the time domain. The right side provides a measure of the energy content of the transform of the signal. Parseval's Equality, is simply a statement that the energy is invariant under the Fourier transform. Parseval's Equality is a special case of Plancherel's Formula (named after Michel Plancherel, 1885–1967).

   Let's rewrite the Convolution Theorem in its inverse form

$$
F^{-1}[\hat{f}(k)\hat{g}(k)] = (f * g)(t).
\tag{8.84}
$$

Then, by the definition of the inverse Fourier transform, we have

$$
\int_{-\infty}^{\infty} f(t - u)g(u)\, du = \frac{1}{2\pi} \int_{-\infty}^{\infty} \hat{f}(\omega)\hat{g}(\omega) e^{-i\omega t}\, d\omega.
$$

Setting $t = 0$,

$$
\int_{-\infty}^{\infty} f(-u)g(u)\, du = \frac{1}{2\pi} \int_{-\infty}^{\infty} \hat{f}(\omega)\hat{g}(\omega)\, d\omega.
\tag{8.85}
$$

Now, let $g(t) = \overline{f(-t)}$, or $f(-t) = \overline{g(t)}$. We note that the Fourier transform of $g(t)$ is related to the Fourier transform of $f(t)$:

$$
\begin{aligned}
\hat{g}(\omega) &= \int_{-\infty}^{\infty} \overline{f(-t)} e^{i\omega t}\, dt \\
&= -\int_{\infty}^{-\infty} \overline{f(\tau)} e^{-i\omega\tau}\, d\tau \\
&= \overline{\int_{-\infty}^{\infty} f(\tau) e^{i\omega\tau}\, d\tau} = \overline{\hat{f}(\omega)}.
\end{aligned}
\tag{8.86}
$$

So, inserting this result into Equation (8.85), we find that

$$
\int_{-\infty}^{\infty} f(-u)\overline{f(-u)}\, du = \frac{1}{2\pi} \int_{-\infty}^{\infty} |\hat{f}(\omega)|^2\, d\omega,
$$

which yields Parseval's Equality in the form of Equation (8.83) after substituting $t = -u$ on the left.

As noted above, the forms in Equations (8.83) and (8.85) are often referred to as the Plancherel Formula or Parseval Formula. A more commonly defined Parseval equation is that given for Fourier series. For example, for a function $f(x)$ defined on $[-\pi, \pi]$, which has a Fourier series representation, we have

$$
\frac{a_0^2}{2} + \sum_{n=1}^{\infty} (a_n^2 + b_n^2) = \frac{1}{\pi} \int_{-\pi}^{\pi} [f(x)]^2\, dx.
$$

In general, there is a Parseval identity for functions that can be expanded in a complete sets of orthonormal functions, $\{\phi_n(x)\}$, $n = 1, 2, \ldots$, which is given by

$$
\sum_{n=1}^{\infty} <f, \phi_n>^2 = \|f\|^2.
$$

Here, $\|f\|^2 = <f, f>$. The Fourier series example is just a special case of this formula.

8.7 Laplace Transform

UP TO THIS POINT WE HAVE ONLY EXPLORED Fourier exponential transforms as one type of integral transform. The Fourier transform is useful on infinite domains. However, students are often introduced to another integral transform, called the Laplace transform, in their introductory differential equations class. These transforms are defined over semi-infinite domains and are useful for solving initial value problems for ordinary differential equations.

The Fourier and Laplace transforms are examples of a broader class of transforms known as integral transforms. For a function $f(x)$ defined on an interval (a, b), we define the integral transform

$$
F(k) = \int_a^b K(x, k) f(x)\, dx,
$$

The Laplace transform is named after Pierre-Simon de Laplace (1749–1827). Laplace made major contributions, especially to celestial mechanics, tidal analysis, and probability.

Integral transform on $[a, b]$ with respect to the integral kernel, $K(x, k)$.

where $K(x,k)$ is a specified kernel of the transform. Looking at the Fourier transform, we see that the interval is stretched over the entire real axis and the kernel is of the form, $K(x,k) = e^{ikx}$. In Table 8.1 we show several types of integral transforms.

Table 8.1: A table of Common Integral Transforms.

Laplace Transform	$F(s) = \int_0^\infty e^{-sx} f(x)\,dx$
Fourier Transform	$F(k) = \int_{-\infty}^\infty e^{ikx} f(x)\,dx$
Fourier Cosine Transform	$F(k) = \int_0^\infty \cos(kx) f(x)\,dx$
Fourier Sine Transform	$F(k) = \int_0^\infty \sin(kx) f(x)\,dx$
Mellin Transform	$F(k) = \int_0^\infty x^{k-1} f(x)\,dx$
Hankel Transform	$F(k) = \int_0^\infty x J_n(kx) f(x)\,dx$

It should be noted that these integral transforms inherit the linearity of integration. Namely, let $h(x) = \alpha f(x) + \beta g(x)$, where α and β are constants. Then,

$$
\begin{aligned}
H(k) &= \int_a^b K(x,k)h(x)\,dx, \\
&= \int_a^b K(x,k)(\alpha f(x) + \beta g(x))\,dx, \\
&= \alpha \int_a^b K(x,k)f(x)\,dx + \beta \int_a^b K(x,k)g(x)\,dx, \\
&= \alpha F(x) + \beta G(x). \tag{8.87}
\end{aligned}
$$

Therefore, we have shown the linearity of the integral transforms. We have seen the linearity property used for Fourier transforms, and we will use linearity in the study of Laplace transforms.

The Laplace transform of f, $F = \mathcal{L}[f]$.

We now turn to Laplace transforms. The Laplace transform of a function $f(t)$ is defined as

$$
F(s) = \mathcal{L}[f](s) = \int_0^\infty f(t) e^{-st}\,dt, \quad s > 0. \tag{8.88}
$$

This is an improper integral and one needs

$$
\lim_{t \to \infty} f(t) e^{-st} = 0
$$

to guarantee convergence.

Laplace transforms also have proven useful in engineering for solving circuit problems and doing systems analysis. In Figure 8.29 it is shown that a signal $x(t)$ is provided as input to a linear system, indicated by $h(t)$. One is interested in the system output, $y(t)$, which is given by a convolution of the input and system functions. By considering the transforms of $x(t)$ and $h(t)$, the transform of the output is given as a product of the Laplace transforms in the s-domain. In order to obtain the output, one needs to compute a convolution product for Laplace transforms similar to the convolution operation we had seen for Fourier transforms earlier in the chapter. Of course, for us to do this in practice, we have to know how to compute Laplace transforms.

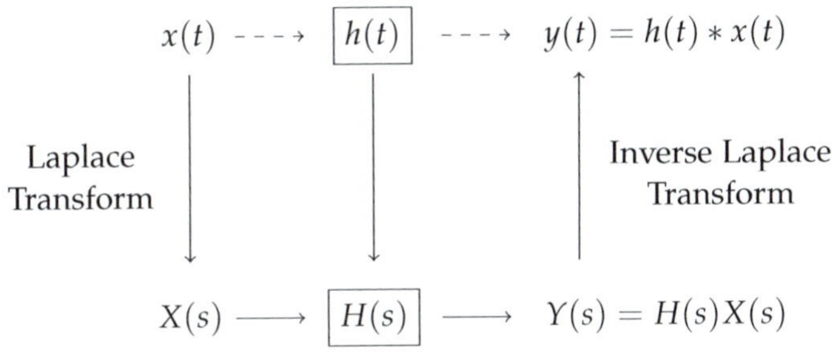

$$x(t) \dashrightarrow \boxed{h(t)} \dashrightarrow y(t) = h(t) * x(t)$$

Laplace Transform

Inverse Laplace Transform

$$X(s) \longrightarrow \boxed{H(s)} \longrightarrow Y(s) = H(s)X(s)$$

Figure 8.29: A schematic depicting the use of Laplace transforms in systems theory.

8.7.1 Properties and Examples of Laplace Transforms

IT IS TYPICAL THAT ONE MAKES USE of Laplace transforms by referring to a Table of transform pairs. A sample of such pairs is given in Table 8.2. Combining some of these simple Laplace transforms with the properties of the Laplace transform, as shown in Table 8.3, we can deal with many applications of the Laplace transform. We will first prove a few of the given Laplace transforms and show how they can be used to obtain new transform pairs. In the next section we will show how these transforms can be used to sum infinite series and to solve initial value problems for ordinary differential equations.

$f(t)$	$F(s)$	$f(t)$	$F(s)$
c	$\dfrac{c}{s}$	e^{at}	$\dfrac{1}{s-a}$, $s > a$
t^n	$\dfrac{n!}{s^{n+1}}$, $s > 0$	$t^n e^{at}$	$\dfrac{n!}{(s-a)^{n+1}}$
$\sin \omega t$	$\dfrac{\omega}{s^2 + \omega^2}$	$e^{at} \sin \omega t$	$\dfrac{\omega}{(s-a)^2 + \omega^2}$
$\cos \omega t$	$\dfrac{s}{s^2 + \omega^2}$	$e^{at} \cos \omega t$	$\dfrac{s-a}{(s-a)^2 + \omega^2}$
$t \sin \omega t$	$\dfrac{2\omega s}{(s^2 + \omega^2)^2}$	$t \cos \omega t$	$\dfrac{s^2 - \omega^2}{(s^2 + \omega^2)^2}$
$\sinh at$	$\dfrac{a}{s^2 - a^2}$	$\cosh at$	$\dfrac{s}{s^2 - a^2}$
$H(t-a)$	$\dfrac{e^{-as}}{s}$, $s > 0$	$\delta(t-a)$	e^{-as}, $a \geq 0, s > 0$

Table 8.2: Table of Selected Laplace Transform Pairs

We begin with some simple transforms. These are found by simply using the definition of the Laplace transform.

Example 8.18. *Show that $\mathcal{L}[1] = \frac{1}{s}$.*

For this example, we insert $f(t) = 1$ into the definition of the Laplace transform:

$$\mathcal{L}[1] = \int_0^\infty e^{-st}\, dt.$$

This is an improper integral, and the computation is understood by introducing an upper limit of a and then letting $a \to \infty$. We will not always write this limit, *but it will be understood that this is how one computes such improper integrals. Proceeding with the computation, we have*

$$
\begin{aligned}
\mathcal{L}[1] &= \int_0^\infty e^{-st}\,dt \\
&= \lim_{a\to\infty} \int_0^a e^{-st}\,dt \\
&= \lim_{a\to\infty} \left(-\frac{1}{s}e^{-st}\right)_0^a \\
&= \lim_{a\to\infty} \left(-\frac{1}{s}e^{-sa} + \frac{1}{s}\right) = \frac{1}{s}.
\end{aligned} \tag{8.89}
$$

Thus, we have found that the Laplace transform of 1 is $\frac{1}{s}$. This result can be extended to any constant c, using the linearity of the transform, $\mathcal{L}[c] = c\mathcal{L}[1]$. Therefore,

$$
\mathcal{L}[c] = \frac{c}{s}.
$$

Example 8.19. *Show that $\mathcal{L}[e^{at}] = \frac{1}{s-a}$, for $s > a$.*

For this example, we can easily compute the transform. Again, we only need to compute the integral of an exponential function.

$$
\begin{aligned}
\mathcal{L}[e^{at}] &= \int_0^\infty e^{at}e^{-st}\,dt \\
&= \int_0^\infty e^{(a-s)t}\,dt \\
&= \left(\frac{1}{a-s}e^{(a-s)t}\right)_0^\infty \\
&= \lim_{t\to\infty} \frac{1}{a-s}e^{(a-s)t} - \frac{1}{a-s} = \frac{1}{s-a}.
\end{aligned} \tag{8.90}
$$

Note that the previous limit was computed as $\lim_{t\to\infty} e^{(a-s)t} = 0$. This is only true if $a - s < 0$, or $s > a$. [Actually, a could be complex. In this case, we would only need s to be greater than the real part of a, $s > \mathrm{Re}(a)$.]

Example 8.20. *Show that $\mathcal{L}[\cos at] = \frac{s}{s^2+a^2}$ and $\mathcal{L}[\sin at] = \frac{a}{s^2+a^2}$.*

For these examples, we could again insert the trigonometric functions directly into the transform and integrate. For example,

$$
\mathcal{L}[\cos at] = \int_0^\infty e^{-st}\cos at\,dt.
$$

Recall how one evaluates integrals involving the product of a trigonometric function and the exponential function. One integrates by parts two times and then obtains an integral of the original unknown integral. Rearranging the resulting integral expressions, one arrives at the desired result. However, there is a much simpler way to compute these transforms.

Recall that $e^{iat} = \cos at + i\sin at$. Making use of the linearity of the Laplace transform, we have

$$
\mathcal{L}[e^{iat}] = \mathcal{L}[\cos at] + i\mathcal{L}[\sin at].
$$

Thus, transforming this complex exponential will simultaneously provide the Laplace transforms for the sine and cosine functions!

The transform is simply computed as

$$\mathcal{L}[e^{iat}] = \int_0^\infty e^{iat} e^{-st}\, dt = \int_0^\infty e^{-(s-ia)t}\, dt = \frac{1}{s-ia}.$$

Note that we could easily have used the result for the transform of an exponential, which was already proven. In this case, $s > Re(ia) = 0$.

We now extract the real and imaginary parts of the result using the complex conjugate of the denominator:

$$\frac{1}{s-ia} = \frac{1}{s-ia}\frac{s+ia}{s+ia} = \frac{s+ia}{s^2+a^2}.$$

Reading off the real and imaginary parts, we find the sought-after transforms,

$$\mathcal{L}[\cos at] = \frac{s}{s^2+a^2},$$
$$\mathcal{L}[\sin at] = \frac{a}{s^2+a^2}. \tag{8.91}$$

Example 8.21. *Show that $\mathcal{L}[t] = \frac{1}{s^2}$.*

For this example we evaluate

$$\mathcal{L}[t] = \int_0^\infty t e^{-st}\, dt.$$

This integral can be evaluated using the method of integration by parts:

$$\int_0^\infty t e^{-st}\, dt = -t\frac{1}{s}e^{-st}\Big|_0^\infty + \frac{1}{s}\int_0^\infty e^{-st}\, dt$$
$$= \frac{1}{s^2}. \tag{8.92}$$

Example 8.22. *Show that $\mathcal{L}[t^n] = \frac{n!}{s^{n+1}}$ for nonnegative integer n.*

We have seen the $n = 0$ and $n = 1$ cases: $\mathcal{L}[1] = \frac{1}{s}$ and $\mathcal{L}[t] = \frac{1}{s^2}$. We now generalize these results to nonnegative integer powers, $n > 1$, of t. We consider the integral

$$\mathcal{L}[t^n] = \int_0^\infty t^n e^{-st}\, dt.$$

Following the previous example, we again integrate by parts:[5]

$$\int_0^\infty t^n e^{-st}\, dt = -t^n\frac{1}{s}e^{-st}\Big|_0^\infty + \frac{n}{s}\int_0^\infty t^{-n}e^{-st}\, dt$$
$$= \frac{n}{s}\int_0^\infty t^{-n}e^{-st}\, dt. \tag{8.93}$$

We could continue to integrate by parts until the final integral is computed. However, look at the integral that resulted after one integration by parts. It is just the Laplace transform of t^{n-1}. So, we can write the result as

$$\mathcal{L}[t^n] = \frac{n}{s}\mathcal{L}[t^{n-1}].$$

[5] This integral can just as easily be done using differentiation. We note that

$$\left(-\frac{d}{ds}\right)^n \int_0^\infty e^{-st}\, dt = \int_0^\infty t^n e^{-st}\, dt.$$

Because

$$\int_0^\infty e^{-st}\, dt = \frac{1}{s},$$

$$\int_0^\infty t^n e^{-st}\, dt = \left(-\frac{d}{ds}\right)^n \frac{1}{s} = \frac{n!}{s^{n+1}}.$$

We compute $\int_0^\infty t^n e^{-st}\, dt$ by turning it into an initial value problem for a first-order difference equation and finding the solution using an iterative method.

This is an example of a recursive definition of a sequence. In this case, we have a sequence of integrals. Denoting

$$I_n = \mathcal{L}[t^n] = \int_0^\infty t^n e^{-st}\, dt$$

and noting that $I_0 = \mathcal{L}[1] = \frac{1}{s}$, we have the following:

$$I_n = \frac{n}{s} I_{n-1}, \quad I_0 = \frac{1}{s}. \tag{8.94}$$

This is also what is called a difference equation. It is a first-order difference equation with an "initial condition," I_0. The next step is to solve this difference equation.

Finding the solution of this first-order difference equation is easy to do using simple iteration. Note that replacing n with $n-1$, we have

$$I_{n-1} = \frac{n-1}{s} I_{n-2}.$$

Repeating the process, we find

$$
\begin{aligned}
I_n &= \frac{n}{s} I_{n-1} \\
&= \frac{n}{s}\left(\frac{n-1}{s} I_{n-2} \right) \\
&= \frac{n(n-1)}{s^2} I_{n-2} \\
&= \frac{n(n-1)(n-2)}{s^3} I_{n-3}.
\end{aligned} \tag{8.95}
$$

We can repeat this process until we get to I_0, which we know. We have to carefully count the number of iterations. We do this by iterating k times and then figure out how many steps will get us to the known initial value. A list of iterates is easily written out:

$$
\begin{aligned}
I_n &= \frac{n}{s} I_{n-1} \\
&= \frac{n(n-1)}{s^2} I_{n-2} \\
&= \frac{n(n-1)(n-2)}{s^3} I_{n-3} \\
&= \cdots \\
&= \frac{n(n-1)(n-2)\ldots(n-k+1)}{s^k} I_{n-k}.
\end{aligned} \tag{8.96}
$$

Because we know $I_0 = \frac{1}{s}$, we choose to stop at $k = n$ obtaining

$$I_n = \frac{n(n-1)(n-2)\ldots(2)(1)}{s^n} I_0 = \frac{n!}{s^{n+1}}.$$

Therefore, we have shown that $\mathcal{L}[t^n] = \frac{n!}{s^{n+1}}$.

Such iterative techniques are useful in obtaining a variety of integrals, such as $I_n = \int_{-\infty}^\infty x^{2n} e^{-x^2}\, dx$.

As a final note, one can extend this result to cases when n is not an integer. To do this, we use the Gamma function, which was discussed in Section 6.4. Recall that the Gamma function is the generalization of the factorial function and is defined as

$$\Gamma(x) = \int_0^\infty t^{x-1} e^{-t}\, dt. \tag{8.97}$$

Note the similarity to the Laplace transform of t^{x-1}:

$$\mathcal{L}[t^{x-1}] = \int_0^\infty t^{x-1} e^{-st}\, dt.$$

For $x - 1$ an integer and $s = 1$, we have that

$$\Gamma(x) = (x-1)!.$$

Thus, the Gamma function can be viewed as a generalization of the factorial, and we have shown that

$$\mathcal{L}[t^p] = \frac{\Gamma(p+1)}{s^{p+1}}$$

for $p > -1$.

Now we are ready to introduce additional properties of the Laplace transform in Table 8.3. We have already discussed the first property, which is a consequence of the linearity of integral transforms. We will prove the other properties in this and the following sections.

Laplace Transform Properties
$\mathcal{L}[af(t) + bg(t)] = aF(s) + bG(s)$
$\mathcal{L}[tf(t)] = -\dfrac{d}{ds}F(s)$
$\mathcal{L}\left[\dfrac{df}{dt}\right] = sF(s) - f(0)$
$\mathcal{L}\left[\dfrac{d^2f}{dt^2}\right] = s^2 F(s) - sf(0) - f'(0)$
$\mathcal{L}[e^{at}f(t)] = F(s-a)$
$\mathcal{L}[H(t-a)f(t-a)] = e^{-as}F(s)$
$\mathcal{L}[(f*g)(t)] = \mathcal{L}[\int_0^t f(t-u)g(u)\, du] = F(s)G(s)$

Table 8.3: Table of Selected Laplace Transform Properties

Example 8.23. *Show that* $\mathcal{L}\left[\frac{df}{dt}\right] = sF(s) - f(0)$.
We have to compute

$$\mathcal{L}\left[\frac{df}{dt}\right] = \int_0^\infty \frac{df}{dt} e^{-st}\, dt.$$

We can move the derivative off f by integrating by parts. This is similar to what we had done when finding the Fourier transform of the derivative of a function. Letting $u = e^{-st}$ and $v = f(t)$, we have

$$
\begin{aligned}
\mathcal{L}\left[\frac{df}{dt}\right] &= \int_0^\infty \frac{df}{dt} e^{-st}\, dt \\
&= \left. f(t)e^{-st} \right|_0^\infty + s \int_0^\infty f(t) e^{-st}\, dt \\
&= -f(0) + sF(s).
\end{aligned}
\tag{8.98}
$$

Here we have assumed that $f(t)e^{-st}$ vanishes for large t.

The final result is that

$$\mathcal{L}\left[\frac{df}{dt}\right] = sF(s) - f(0).$$

Example 8.24. *Show that $\mathcal{L}\left[\frac{d^2 f}{dt^2}\right] = s^2 F(s) - sf(0) - f'(0)$.*

We can compute this Laplace transform using two integrations by parts, or we could make use of the previous result. Letting $g(t) = \frac{df(t)}{dt}$, we have

$$\mathcal{L}\left[\frac{d^2 f}{dt^2}\right] = \mathcal{L}\left[\frac{dg}{dt}\right] = sG(s) - g(0) = sG(s) - f'(0).$$

But,

$$G(s) = \mathcal{L}\left[\frac{df}{dt}\right] = sF(s) - f(0).$$

So,

$$
\begin{aligned}
\mathcal{L}\left[\frac{d^2 f}{dt^2}\right] &= sG(s) - f'(0) \\
&= s\left[sF(s) - f(0)\right] - f'(0) \\
&= s^2 F(s) - sf(0) - f'(0).
\end{aligned}
\tag{8.99}
$$

We will return to the other properties in Table 8.3 after looking at a few applications.

8.8   Applications of Laplace Transforms

ALTHOUGH THE LAPLACE TRANSFORM IS A VERY USEFUL TRANSFORM, it is often encountered only as a method for solving initial value problems in introductory differential equations. In this section we will show how to solve simple differential equations. Along the way we will introduce step and impulse functions and show how the Convolution Theorem for Laplace transforms plays a role in finding solutions. However, we will first explore an unrelated application of Laplace transforms. We will see that the Laplace transform is useful in finding sums of infinite series.

8.8.1   Series Summation Using Laplace Transforms

WE SAW IN CHAPTER 5 THAT FOURIER SERIES can be used to sum series. For example, in Problem 5.13, one proves that

$$\sum_{n=1}^{\infty} \frac{1}{n^2} = \frac{\pi^2}{6}.$$

In this section we will show how Laplace transforms can be used to sum series.[6] There is an interesting history of using integral transforms to sum series. For example, Richard Feynman[7] (1918–1988) described how one can

[6] Albert D. Wheelon, *Tables of Summable Series and Integrals Involving Bessel Functions*, Holden-Day, 1968.

[7] R. P. Feynman, 1949, *Phys. Rev.* **76**, p. 769

use the Convolution Theorem for Laplace transforms to sum series with denominators that involved products. We will describe this and simpler sums in this section.

We begin by considering the Laplace transform of a known function,

$$F(s) = \int_0^\infty f(t)e^{-st}\, dt.$$

Inserting this expression into the sum $\sum_n F(n)$ and interchanging the sum and integral, we find

$$
\begin{aligned}
\sum_{n=0}^\infty F(n) &= \sum_{n=0}^\infty \int_0^\infty f(t)e^{-nt}\, dt \\
&= \int_0^\infty f(t) \sum_{n=0}^\infty \left(e^{-t}\right)^n\, dt \\
&= \int_0^\infty f(t) \frac{1}{1-e^{-t}}\, dt.
\end{aligned}
\tag{8.100}
$$

The last step was obtained using the sum of a geometric series. The key is being able to carry out the final integral as we show in the next example.

Example 8.25. *Evaluate the sum $\sum_{n=1}^\infty \frac{(-1)^{n+1}}{n}$.*

Because $\mathcal{L}[1] = 1/s$, we have

$$
\begin{aligned}
\sum_{n=1}^\infty \frac{(-1)^{n+1}}{n} &= \sum_{n=1}^\infty \int_0^\infty (-1)^{n+1} e^{-nt}\, dt \\
&= \int_0^\infty \frac{e^{-t}}{1+e^{-t}}\, dt \\
&= \int_1^2 \frac{du}{u} = \ln 2.
\end{aligned}
\tag{8.101}
$$

Example 8.26. *Evaluate the sum $\sum_{n=1}^\infty \frac{1}{n^2}$.*

This is a special case of the Riemann zeta function

$$\zeta(s) = \sum_{n=1}^\infty \frac{1}{n^s}.\tag{8.102}$$

The Riemann zeta function[8] is important in the study of prime numbers and more recently has seen applications in the study of dynamical systems. The series in this example is $\zeta(2)$. We have already seen in 5.13 that

$$\zeta(2) = \frac{\pi^2}{6}.$$

Using Laplace transforms, we can provide an integral representation of $\zeta(2)$.

The first step is to find the correct Laplace transform pair. The sum involves the function $F(n) = 1/n^2$. So, we look for a function $f(t)$ whose Laplace transform is $F(s) = 1/s^2$. We know by now that the inverse Laplace transform of $F(s) = 1/s^2$ is $f(t) = t$. As before, we replace each term in the series by a Laplace transform, exchange the summation and integration, and sum the resulting geometric series:

$$
\begin{aligned}
\sum_{n=1}^\infty \frac{1}{n^2} &= \sum_{n=1}^\infty \int_0^\infty t e^{-nt}\, dt \\
&= \int_0^\infty \frac{t}{e^t - 1}\, dt.
\end{aligned}
\tag{8.103}
$$

[8] A translation of Riemann, Bernhard (1859), "Über die Anzahl der Primzahlen unter einer gegebenen Grösse" is in H. M. Edwards (1974). *Riemann's Zeta Function.* Academic Press. Riemann had shown that the Riemann zeta function can be obtained through contour integral representation, $2\sin(\pi s)\Gamma(s) = i\oint_C \frac{(-x)^{s-1}}{e^x - 1} dx$, for a specific contour C.

So, we have that

$$\int_0^\infty \frac{t}{e^t - 1} \, dt = \sum_{n=1}^\infty \frac{1}{n^2} = \zeta(2).$$

Integrals of this type occur often in statistical mechanics in the form of Bose–Einstein integrals. These are of the form

$$G_n(z) = \int_0^\infty \frac{x^{n-1}}{z^{-1}e^x - 1} \, dx.$$

Note that $G_n(1) = \Gamma(n)\zeta(n)$.

In general, the Riemann zeta function must be tabulated through other means. In some special cases, one can use closed form expressions. For example,

$$\zeta(2n) = \frac{2^{2n-1}\pi^{2n}}{(2n)!} B_n,$$

where the B_n's are the Bernoulli numbers. Bernoulli numbers are defined through the Maclaurin series expansion

$$\frac{x}{e^x - 1} = \sum_{n=0}^\infty \frac{B_n}{n!} x^n.$$

The first few Riemann zeta functions are

$$\zeta(2) = \frac{\pi^2}{6}, \quad \zeta(4) = \frac{\pi^4}{90}, \quad \zeta(6) = \frac{\pi^6}{945}.$$

We can extend this method of using Laplace transforms to summing series whose terms take special general forms. For example, from Feynman's 1949 paper, we note that

$$\frac{1}{(a+bn)^2} = -\frac{\partial}{\partial a} \int_0^\infty e^{-s(a+bn)} \, ds.$$

This identity can be shown easily by first noting that

$$\int_0^\infty e^{-s(a+bn)} \, ds = \left[\frac{-e^{-s(a+bn)}}{a+bn} \right]_0^\infty = \frac{1}{a+bn}.$$

Now, differentiate the result with respect to a and the result follows.

The latter identity can be generalized further as

$$\frac{1}{(a+bn)^{k+1}} = \frac{(-1)^k}{k!} \frac{\partial^k}{\partial a^k} \int_0^\infty e^{-s(a+bn)} \, ds.$$

In Feynman's 1949 paper, he develops methods for handling several other general sums using the Convolution Theorem. Wheelon gives more examples of these. We will just provide one such result and an example. First, we note that

$$\frac{1}{ab} = \int_0^1 \frac{du}{[a(1-u) + bu]^2}.$$

However,

$$\frac{1}{[a(1-u) + bu]^2} = \int_0^\infty t e^{-t[a(1-u)+bu]} \, dt.$$

So, we have

$$\frac{1}{ab} = \int_0^1 du \int_0^\infty te^{-t[a(1-u)+bu]} \, dt.$$

We see in the next example how this representation can be useful.

Example 8.27. *Evaluate* $\sum_{n=0}^\infty \frac{1}{(2n+1)(2n+2)}$.

We sum this series by first letting $a = 2n + 1$ and $b = 2n + 2$ in the formula for $1/ab$. Collecting the n-dependent terms, we can sum the series leaving a double integral computation in ut-space. The details are as follows:

$$
\begin{aligned}
\sum_{n=0}^\infty \frac{1}{(2n+1)(2n+2)} &= \sum_{n=0}^\infty \int_0^1 \frac{du}{[(2n+1)(1-u) + (2n+2)u]^2} \\
&= \sum_{n=0}^\infty \int_0^1 du \int_0^\infty te^{-t(2n+1+u)} \, dt \\
&= \int_0^1 du \int_0^\infty te^{-t(1+u)} \sum_{n=0}^\infty e^{-2nt} \, dt \\
&= \int_0^\infty \frac{te^{-t}}{1 - e^{-2t}} \int_0^1 e^{-tu} \, du \, dt \\
&= \int_0^\infty \frac{te^{-t}}{1 - e^{-2t}} \frac{1 - e^{-t}}{t} \, dt \\
&= \int_0^\infty \frac{e^{-t}}{1 + e^{-t}} \, dt \\
&= \left. -\ln(1 + e^{-t}) \right|_0^\infty = \ln 2. \quad\quad (8.104)
\end{aligned}
$$

8.8.2 *Solution of ODEs Using Laplace Transforms*

ONE OF THE TYPICAL APPLICATIONS OF LAPLACE TRANSFORMS is the solution of nonhomogeneous linear constant coefficient differential equations. In the following examples we will show how this works.

The general idea is that one transforms the equation for an unknown function $y(t)$ into an algebraic equation for its transform, $Y(t)$. Typically, the algebraic equation is easy to solve for $Y(s)$ as a function of s. Then, one transforms back into t-space using Laplace transform tables and the properties of Laplace transforms. The scheme is shown in Figure 8.30.

Example 8.28. *Solve the initial value problem $y' + 3y = e^{2t}$, $y(0) = 1$.*

The first step is to perform a Laplace transform of the initial value problem. The transform of the left side of the equation is

$$\mathcal{L}[y' + 3y] = sY - y(0) + 3Y = (s + 3)Y - 1.$$

Transforming the right-hand side, we have

$$\mathcal{L}[e^{2t}] = \frac{1}{s - 2}.$$

Combining these two results, we obtain

$$(s + 3)Y - 1 = \frac{1}{s - 2}.$$

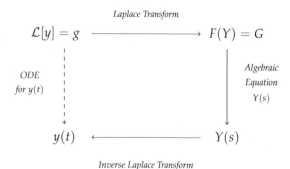

$$\mathcal{L}[y] = g \xrightarrow{\text{Laplace Transform}} F(Y) = G$$

ODE for $y(t)$

Algebraic Equation $Y(s)$

$$y(t) \longleftarrow Y(s)$$

Inverse Laplace Transform

Figure 8.30: The scheme for solving an ordinary differential equation using Laplace transforms. One transforms the initial value problem for $y(t)$ and obtains an algebraic equation for $Y(s)$. Solve for $Y(s)$ and the inverse transform gives the solution to the initial value problem.

The next step is to solve for $Y(s)$:

$$Y(s) = \frac{1}{s+3} + \frac{1}{(s-2)(s+3)}.$$

Now we need to find the inverse Laplace transform. Namely, we need to figure out what function has a Laplace transform of the above form. We will use the tables of Laplace transform pairs. Later we will show that there are other methods for carrying out the Laplace transform inversion.

The inverse transform of the first term is e^{-3t}. However, we have not seen anything that looks like the second form in the table of transforms that we have compiled, but we can rewrite the second term using a partial fraction decomposition. Let's recall how to do this.

The goal is to find constants A and B such that

$$\frac{1}{(s-2)(s+3)} = \frac{A}{s-2} + \frac{B}{s+3}. \tag{8.105}$$

This is an example of carrying out a partial fraction decomposition.

We picked this form because we know that recombining the two terms into one term will have the same denominator. We just need to make sure the numerators agree afterward. So, adding the two terms, we have

$$\frac{1}{(s-2)(s+3)} = \frac{A(s+3) + B(s-2)}{(s-2)(s+3)}.$$

Equating numerators,

$$1 = A(s+3) + B(s-2).$$

There are several ways to proceed at this point.

a. *Method 1.*

We can rewrite the equation by gathering terms with common powers of s, we have

$$(A+B)s + 3A - 2B = 1.$$

The only way that this can be true for all s is that the coefficients of the different powers of s agree on both sides. This leads to two equations for A and B:

$$A + B = 0,$$

$$3A - 2B = 1. \tag{8.106}$$

*The first equation gives $A = -B$, so the second equation becomes $-5B = 1$.
The solution is then $A = -B = \frac{1}{5}$.*

b. *Method 2.*

Because the equation $\frac{1}{(s-2)(s+3)} = \frac{A}{s-2} + \frac{B}{s+3}$ is true for all s, we can pick specific values. For $s = 2$, we find $1 = 5A$, or $A = \frac{1}{5}$. For $s = -3$, we find $1 = -5B$, or $B = -\frac{1}{5}$. Thus, we obtain the same result as Method 1, but much quicker.

c. *Method 3.*

We could just inspect the original partial fraction problem. Because the numerator has no s terms, we might guess the form

$$\frac{1}{(s-2)(s+3)} = \frac{1}{s-2} - \frac{1}{s+3}.$$

But, recombining the terms on the right-hand side, we see that

$$\frac{1}{s-2} - \frac{1}{s+3} = \frac{5}{(s-2)(s+3)}.$$

Because we were off by 5, we divide the partial fractions by 5 to obtain

$$\frac{1}{(s-2)(s+3)} = \frac{1}{5}\left[\frac{1}{s-2} - \frac{1}{s+3}\right],$$

which once again gives the desired form.

Returning to the problem, we have found that

$$Y(s) = \frac{1}{s+3} + \frac{1}{5}\left(\frac{1}{s-2} - \frac{1}{s+3}\right).$$

We can now see that the function with this Laplace transform is given by

$$y(t) = \mathcal{L}^{-1}\left[\frac{1}{s+3} + \frac{1}{5}\left(\frac{1}{s-2} - \frac{1}{s+3}\right)\right] = e^{-3t} + \frac{1}{5}\left(e^{2t} - e^{-3t}\right)$$

works. Simplifying, we have the solution of the initial value problem

$$y(t) = \frac{1}{5}e^{2t} + \frac{4}{5}e^{-3t}.$$

We can verify that we have solved the initial value problem.

$$y' + 3y = \frac{2}{5}e^{2t} - \frac{12}{5}e^{-3t} + 3\left(\frac{1}{5}e^{2t} + \frac{4}{5}e^{-3t}\right) = e^{2t}$$

and $y(0) = \frac{1}{5} + \frac{4}{5} = 1$.

Example 8.29. *Solve the initial value problem $y'' + 4y = 0$, $y(0) = 1$, $y'(0) = 3$.*

*We can probably solve this without Laplace transforms, but it is a simple exercise.
Transforming the equation, we have*

$$\begin{aligned} 0 &= s^2 Y - sy(0) - y'(0) + 4Y \\ &= (s^2 + 4)Y - s - 3. \end{aligned} \qquad (8.107)$$

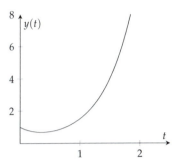

Figure 8.31: A plot of the solution to Example 8.28.

Solving for Y, we have

$$Y(s) = \frac{s+3}{s^2+4}.$$

We now ask if we recognize the transform pair needed. The denominator looks like the type needed for the transform of a sine or cosine. We just need to play with the numerator. Splitting the expression into two terms, we have

$$Y(s) = \frac{s}{s^2+4} + \frac{3}{s^2+4}.$$

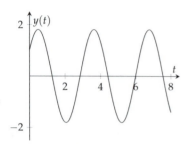

The first term is now recognizable as the transform of $\cos 2t$. *The second term is not the transform of* $\sin 2t$. *It would be if the numerator were a 2. This can be corrected by multiplying and dividing by 2:*

$$\frac{3}{s^2+4} = \frac{3}{2}\left(\frac{2}{s^2+4}\right).$$

Figure 8.32: A plot of the solution to Example 8.29.

The solution is then found as

$$y(t) = \mathcal{L}^{-1}\left[\frac{s}{s^2+4} + \frac{3}{2}\left(\frac{2}{s^2+4}\right)\right] = \cos 2t + \frac{3}{2}\sin 2t.$$

The reader can verify that this is the solution of the initial value problem.

8.8.3   Step and Impulse Functions

OFTEN, THE INITIAL VALUE PROBLEMS THAT ONE FACES in differential equations courses can be solved using either the Method of Undetermined Coefficients or the Method of Variation of Parameters. However, using the latter can be messy and involves some skill with integration. Many circuit designs can be modeled with systems of differential equations using Kirchoff's Rules. Such systems can get fairly complicated. However, Laplace transforms can be used to solve such systems, and electrical engineers have long used such methods in circuit analysis.

In this section we add a couple more transform pairs and transform properties that are useful in accounting for things like turning on a driving force, using periodic functions like a square wave, or introducing impulse forces.

We first recall the Heaviside step function, given by

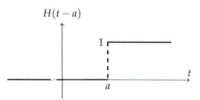

Figure 8.33: A shifted Heaviside function, $H(t-a)$.

$$H(t) = \begin{cases} 0, & t < 0, \\ 1, & t > 0. \end{cases} \qquad (8.108)$$

A more general version of the step function is the horizontally shifted step function, $H(t-a)$. This function is shown in Figure 8.33. The Laplace transform of this function is found for $a > 0$ as

$$\begin{aligned} \mathcal{L}[H(t-a)] &= \int_0^\infty H(t-a)e^{-st}\,dt \\ &= \int_a^\infty e^{-st}\,dt \\ &= \left.\frac{e^{-st}}{s}\right|_a^\infty = \frac{e^{-as}}{s}. \end{aligned} \qquad (8.109)$$

Just like the Fourier transform, the Laplace transform has two Shift Theorems involving the multiplication of the function, $f(t)$, or its transform, $F(s)$, by exponentials. The first and second Shift Properties/Theorems are given, respectively, by

The Shift Theorems.

$$\mathcal{L}[e^{at}f(t)] = F(s-a) \tag{8.110}$$
$$\mathcal{L}[f(t-a)H(t-a)] = e^{-as}F(s). \tag{8.111}$$

We prove the First Shift Theorem and leave the other proof as an exercise for the reader. Namely,

$$
\begin{aligned}
\mathcal{L}[e^{at}f(t)] &= \int_0^\infty e^{at}f(t)e^{-st}\,dt \\
&= \int_0^\infty f(t)e^{-(s-a)t}\,dt = F(s-a). \tag{8.112}
\end{aligned}
$$

Example 8.30. *Compute the Laplace transform of $e^{-at}\sin\omega t$.*

This function arises as the solution of the underdamped harmonic oscillator. We first note that the exponential multiplies a sine function. The Shift Theorem tells us that we first need the transform of the sine function. So, for $f(t) = \sin\omega t$, we have

$$F(s) = \frac{\omega}{s^2 + \omega^2}.$$

Using this transform, we can obtain the solution to this problem as

$$\mathcal{L}[e^{-at}\sin\omega t] = F(s+a) = \frac{\omega}{(s+a)^2 + \omega^2}.$$

More interesting examples can be found using piecewise defined functions. First we consider the function $H(t) - H(t-a)$. For $t < 0$, both terms are zero. In the interval $[0,a]$, the function $H(t) = 1$ and $H(t-a) = 0$. Therefore, $H(t) - H(t-a) = 1$ for $t \in [0,a]$. Finally, for $t > a$, both functions are one and therefore the difference is zero. The graph of $H(t) - H(t-a)$ is shown in Figure 8.34.

We now consider the piecewise defined function

$$g(t) = \begin{cases} f(t), & 0 \le t \le a, \\ 0, & t < 0, t > a. \end{cases}$$

This function can be rewritten in terms of step functions. We only need to multiply $f(t)$ by the above box function:

$$g(t) = f(t)[H(t) - H(t-a)].$$

We depict this in Figure 8.35.

Even more complicated functions can be written in terms of step functions. We only need to look at sums of functions of the form $f(t)[H(t-a) - H(t-b)]$ for $b > a$. This is similar to a box function. It is nonzero between a and b and has height $f(t)$.

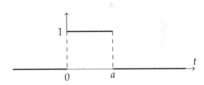

Figure 8.34: The box function, $H(t) - H(t-a)$.

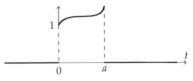

Figure 8.35: Formation of a piecewise function, $f(t)[H(t) - H(t-a)]$.

We show as an example the square wave function in Figure 8.36. It can be represented as a sum of an infinite number of boxes,

$$f(t) = \sum_{n=-\infty}^{\infty} [H(t - 2na) - H(t - (2n+1)a)],$$

for $a > 0$.

Example 8.31. *Find the Laplace Transform of a square wave "turned on" at $t = 0$.*

.

Figure 8.36: A square wave, $f(t) = \sum_{n=-\infty}^{\infty}[H(t - 2na) - H(t - (2n+1)a)]$.

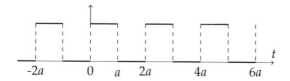

We let

$$f(t) = \sum_{n=0}^{\infty} [H(t - 2na) - H(t - (2n+1)a)], \quad a > 0.$$

Using the properties of the Heaviside function, we have

$$
\begin{aligned}
\mathcal{L}[f(t)] &= \sum_{n=0}^{\infty} \left[\mathcal{L}[H(t - 2na)] - \mathcal{L}[H(t - (2n+1)a)] \right] \\
&= \sum_{n=0}^{\infty} \left[\frac{e^{-2nas}}{s} - \frac{e^{-(2n+1)as}}{s} \right] \\
&= \frac{1 - e^{-as}}{s} \sum_{n=0}^{\infty} \left(e^{-2as} \right)^n \\
&= \frac{1 - e^{-as}}{s} \left(\frac{1}{1 - e^{-2as}} \right) \\
&= \frac{1 - e^{-as}}{s(1 - e^{-2as})}.
\end{aligned}
\tag{8.113}
$$

Note that the third line in the derivation is a geometric series. We summed this series to get the answer in a compact form as $e^{-2as} < 1$.

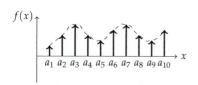

Figure 8.37: Plot representing impulse forces of height $f(a_i)$. The sum $\sum_{i=1}^{n} f(a_i)\delta(x - a_i)$ describes a general impulse function.

Other interesting examples are provided by the delta function. The Dirac delta function can be used to represent a unit impulse. Summing over a number of impulses, or point sources, we can describe a general function as shown in Figure 8.37. The sum of impulses located at points a_i, $i = 1, \ldots, n$, with strengths $f(a_i)$ would be given by

$$f(x) = \sum_{i=1}^{n} f(a_i)\delta(x - a_i).$$

A continuous sum could be written as

$$f(x) = \int_{-\infty}^{\infty} f(\xi)\delta(x - \xi)\, d\xi.$$

TRANSFORM TECHNIQUES IN PHYSICS 497

This is simply an application of the sifting property of the delta function. We will investigate a case when one would use a single impulse. While a mass on a spring is undergoing simple harmonic motion, we hit it for an instant at time $t = a$. In such a case, we could represent the force as a multiple of $\delta(t - a)$.

$$\mathcal{L}[\delta(t-a)] = e^{-as}.$$

One would then need the Laplace transform of the delta function to solve the associated initial value problem. Inserting the delta function into the Laplace transform, we find that for $a > 0$,

$$
\begin{aligned}
\mathcal{L}[\delta(t-a)] &= \int_0^\infty \delta(t-a)e^{-st}\,dt \\
&= \int_{-\infty}^\infty \delta(t-a)e^{-st}\,dt \\
&= e^{-as}.
\end{aligned}
\tag{8.114}
$$

Example 8.32. *Solve the initial value problem $y'' + 4\pi^2 y = \delta(t-2)$, $y(0) = y'(0) = 0$.*

This initial value problem models a spring oscillation with an impulse force. Without the forcing term, given by the delta function, this spring is initially at rest and not stretched. The delta function models a unit impulse at $t = 2$. Of course, we anticipate that at this time the spring will begin to oscillate. We will solve this problem using Laplace transforms.

First, we transform the differential equation:

$$s^2 Y - sy(0) - y'(0) + 4\pi^2 Y = e^{-2s}.$$

Inserting the initial conditions, we have

$$(s^2 + 4\pi^2)Y = e^{-2s}.$$

Solving for $Y(s)$, we obtain

$$Y(s) = \frac{e^{-2s}}{s^2 + 4\pi^2}.$$

We now seek the function for which this is the Laplace transform. The form of this function is an exponential times some Laplace transform, $F(s)$. Thus, we need the Second Shift Theorem because the solution is of the form $Y(s) = e^{-2s}F(s)$ for

$$F(s) = \frac{1}{s^2 + 4\pi^2}.$$

We need to find the corresponding $f(t)$ of the Laplace transform pair. The denominator in $F(s)$ suggests a sine or cosine. Because the numerator is constant, we pick sine. From the tables of transforms, we have

$$\mathcal{L}[\sin 2\pi t] = \frac{2\pi}{s^2 + 4\pi^2}.$$

So, we write

$$F(s) = \frac{1}{2\pi}\frac{2\pi}{s^2 + 4\pi^2}.$$

This gives $f(t) = (2\pi)^{-1}\sin 2\pi t$.

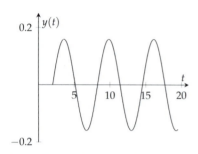

Figure 8.38: A plot of the solution to Example 8.32 in which a spring at rest experiences an impulse force at $t = 2$.

We now apply the Second Shift Theorem, $\mathcal{L}[f(t-a)H(t-a)] = e^{-as}F(s)$, or

$$
\begin{aligned}
y(t) &= \mathcal{L}^{-1}\left[e^{-2s}F(s)\right] \\
&= H(t-2)f(t-2) \\
&= \frac{1}{2\pi}H(t-2)\sin 2\pi(t-2).
\end{aligned}
\tag{8.115}
$$

This solution tells us that the mass is at rest until $t = 2$ and then begins to oscillate at its natural frequency. A plot of this solution is shown in Figure 8.38

Example 8.33. *Solve the initial value problem*

$$
y'' + y = f(t), \quad y(0) = 0, y'(0) = 0,
$$

where

$$
f(t) = \begin{cases} \cos \pi t, & 0 \le t \le 2, \\ 0, & \text{otherwise.} \end{cases}
$$

We need the Laplace transform of $f(t)$. This function can be written in terms of a Heaviside function, $f(t) = \cos \pi t\, H(t-2)$. In order to apply the Second Shift Theorem, we need a shifted version of the cosine function. We find the shifted version by noting that $\cos \pi(t-2) = \cos \pi t$. Thus, we have

$$
\begin{aligned}
f(t) &= \cos \pi t\,[H(t) - H(t-2)] \\
&= \cos \pi t - \cos \pi(t-2)H(t-2), \quad t \ge 0.
\end{aligned}
\tag{8.116}
$$

The Laplace transform of this driving term is

$$
F(s) = (1 - e^{-2s})\mathcal{L}[\cos \pi t] = (1 - e^{-2s})\frac{s}{s^2 + \pi^2}.
$$

Now we can proceed to solve the initial value problem. The Laplace transform of the initial value problem yields

$$
(s^2 + 1)Y(s) = (1 - e^{-2s})\frac{s}{s^2 + \pi^2}.
$$

Therefore,

$$
Y(s) = (1 - e^{-2s})\frac{s}{(s^2 + \pi^2)(s^2 + 1)}.
$$

We can retrieve the solution to the initial value problem using the Second Shift Theorem. The solution is of the form $Y(s) = (1 - e^{-2s})G(s)$ for

$$
G(s) = \frac{s}{(s^2 + \pi^2)(s^2 + 1)}.
$$

Then, the final solution takes the form

$$
y(t) = g(t) - g(t-2)H(t-2).
$$

We only need to find $g(t)$ in order to finish the problem. This is easily done using the partial fraction decomposition

$$
G(s) = \frac{s}{(s^2 + \pi^2)(s^2 + 1)} = \frac{1}{\pi^2 - 1}\left[\frac{s}{s^2 + 1} - \frac{s}{s^2 + \pi^2}\right].
$$

Then,

$$g(t) = \mathcal{L}^{-1}\left[\frac{s}{(s^2+\pi^2)(s^2+1)}\right] = \frac{1}{\pi^2-1}\left(\cos t - \cos \pi t\right).$$

The final solution is then given by

$$y(t) = \frac{1}{\pi^2-1}\left[\cos t - \cos \pi t - H(t-2)(\cos(t-2) - \cos \pi t)\right].$$

A plot of this solution is shown in Figure 8.39

8.9 Convolution Theorem

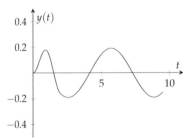

Figure 8.39: A plot of the solution to Example 8.33 in which a spring at rest experiences a piecewise defined force.

FINALLY, WE CONSIDER THE CONVOLUTION of two functions. Often, we are faced with having the product of two Laplace transforms that we know, and we seek the inverse transform of the product. For example, let's say we have obtained $Y(s) = \frac{1}{(s-1)(s-2)}$ while trying to solve an initial value problem. In this case, we could find a partial fraction decomposition. But, there are other ways to find the inverse transform, especially if we cannot perform a partial fraction decomposition. We could use the Convolution Theorem for Laplace transforms, or we could compute the inverse transform directly. We will look into these methods in the next two sections.We begin by defining the convolution.

We define the convolution of two functions defined on $[0, \infty)$ much the same way as we had done for the Fourier transform. The convolution $f * g$ is defined as

$$(f * g)(t) = \int_0^t f(u)g(t-u)\,du. \tag{8.117}$$

Note that the convolution integral has finite limits as opposed to the Fourier transform case.

The convolution operation has two important properties:

1. The convolution is commutative: $f * g = g * f$

   Proof. The key is to make a substitution $y = t - u$ in the integral. This makes f a simple function of the integration variable.

$$\begin{aligned}(g * f)(t) &= \int_0^t g(u)f(t-u)\,du \\ &= -\int_t^0 g(t-y)f(y)\,dy \\ &= \int_0^t f(y)g(t-y)\,dy \\ &= (f * g)(t). \tag{8.118}\end{aligned}$$

   $\square$

The convolution is commutative.

The Convolution Theorem for Laplace transforms.

2. The Convolution Theorem: The Laplace transform of a convolution is the product of the Laplace transforms of the individual functions:

$$\mathcal{L}[f * g] = F(s)G(s).$$

Proof. Proving this theorem takes a bit more work. We will make some assumptions that will work in many cases. First, we assume that the functions are causal, $f(t) = 0$ and $g(t) = 0$ for $t < 0$. Second, we will assume that we can interchange integrals, which needs more rigorous attention than will be provided here. The first assumption will allow us to write the finite integral as an infinite integral. Then a change of variables will allow us to split the integral into the product of two integrals that are recognized as a product of two Laplace transforms.

Carrying out the computation, we have

$$\begin{aligned} \mathcal{L}[f * g] &= \int_0^\infty \left(\int_0^t f(u)g(t-u)\,du \right) e^{-st}\,dt \\ &= \int_0^\infty \left(\int_0^\infty f(u)g(t-u)\,du \right) e^{-st}\,dt \\ &= \int_0^\infty f(u) \left(\int_0^\infty g(t-u)e^{-st}\,dt \right) du \qquad (8.119) \end{aligned}$$

Now, make the substitution $\tau = t - u$. We note that

$$int_0^\infty f(u) \left(\int_0^\infty g(t-u)e^{-st}\,dt \right) du = \int_0^\infty f(u) \left(\int_{-u}^\infty g(\tau)e^{-s(\tau+u)}\,d\tau \right) du$$

However, because $g(\tau)$ is a causal function, we have that it vanishes for $\tau < 0$ and we can change the integration interval to $[0, \infty)$. So, after a little rearranging, we can proceed to the result.

$$\begin{aligned} \mathcal{L}[f * g] &= \int_0^\infty f(u) \left(\int_0^\infty g(\tau)e^{-s(\tau+u)}\,d\tau \right) du \\ &= \int_0^\infty f(u)e^{-su} \left(\int_0^\infty g(\tau)e^{-s\tau}\,d\tau \right) du \\ &= \left(\int_0^\infty f(u)e^{-su}\,du \right) \left(\int_0^\infty g(\tau)e^{-s\tau}\,d\tau \right) \\ &= F(s)G(s). \qquad (8.120) \end{aligned}$$

$\square$

We make use of the Convolution Theorem to do the following examples.

Example 8.34. *Find* $y(t) = \mathcal{L}^{-1}\left[\frac{1}{(s-1)(s-2)} \right]$.
We note that this is a product of two functions:

$$Y(s) = \frac{1}{(s-1)(s-2)} = \frac{1}{s-1}\frac{1}{s-2} = F(s)G(s).$$

We know the inverse transforms of the factors: $f(t) = e^t$ *and* $g(t) = e^{2t}$.

Using the Convolution Theorem, we find $y(t) = (f * g)(t)$. *We compute the convolution:*

$$y(t) = \int_0^t f(u)g(t-u)\,du$$

$$= \int_0^t e^u e^{2(t-u)} \, du$$

$$= e^{2t} \int_0^t e^{-u} \, du$$

$$= e^{2t}[-e^t + 1] = e^{2t} - e^t. \tag{8.121}$$

One can also confirm this by carrying out a partial fraction decomposition.

Example 8.35. *Consider the initial value problem,* $y'' + 9y = 2\sin 3t$, $y(0) = 1$, $y'(0) = 0$.

The Laplace transform of this problem is given by

$$(s^2 + 9)Y - s = \frac{6}{s^2 + 9}.$$

Solving for $Y(s)$*, we obtain*

$$Y(s) = \frac{6}{(s^2 + 9)^2} + \frac{s}{s^2 + 9}.$$

The inverse Laplace transform of the second term is easily found as $\cos(3t)$; however, the first term is more complicated.

We can use the Convolution Theorem to find the Laplace transform of the first term. We note that

$$\frac{6}{(s^2 + 9)^2} = \frac{2}{3} \frac{3}{(s^2 + 9)} \frac{3}{(s^2 + 9)}$$

is a product of two Laplace transforms (up to the constant factor). Thus,

$$\mathcal{L}^{-1}\left[\frac{6}{(s^2 + 9)^2}\right] = \frac{2}{3}(f * g)(t),$$

where $f(t) = g(t) = \sin 3t$. Evaluating this convolution product, we have

$$
\begin{aligned}
\mathcal{L}^{-1}\left[\frac{6}{(s^2 + 9)^2}\right] &= \frac{2}{3}(f * g)(t) \\
&= \frac{2}{3} \int_0^t \sin 3u \sin 3(t - u) \, du \\
&= \frac{1}{3} \int_0^t [\cos 3(2u - t) - \cos 3t] \, du \\
&= \frac{1}{3}\left[\frac{1}{6} \sin(6u - 3t) - u \cos 3t\right]_0^t \\
&= \frac{1}{9} \sin 3t - \frac{1}{3} t \cos 3t. \tag{8.122}
\end{aligned}
$$

Combining this with the inverse transform of the second term of $Y(s)$, the solution to the initial value problem is

$$y(t) = -\frac{1}{3} t \cos 3t + \frac{1}{9} \sin 3t + \cos 3t.$$

Note that the amplitude of the solution will grow in time from the first term. You can see this in Figure 8.40. This is known as a resonance.

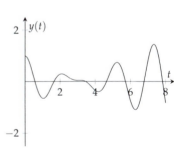

Figure 8.40: Plot of the solution to Example 8.35 showing a resonance.

Example 8.36. *Find* $\mathcal{L}^{-1}\left[\frac{6}{(s^2+9)^2}\right]$ *using partial fraction decomposition.*

If we look at Table 8.2, we see that the Laplace transform pairs with the denominator $(s^2 + \omega^2)^2$ *are*

$$\mathcal{L}[t \sin \omega t] = \frac{2\omega s}{(s^2 + \omega^2)^2}$$

and

$$\mathcal{L}[t \cos \omega t] = \frac{s^2 - \omega^2}{(s^2 + \omega^2)^2}.$$

So, we might consider rewriting a partial fraction decomposition as

$$\frac{6}{(s^2+9)^2} = \frac{A6s}{(s^2+9)^2} + \frac{B(s^2-9)}{(s^2+9)^2} + \frac{Cs+D}{s^2+9}.$$

Combining the terms on the right over a common denominator, we find

$$6 = 6As + B(s^2 - 9) + (Cs + D)(s^2 + 9).$$

Collecting like powers of s, *we have*

$$Cs^3 + (D + B)s^2 + 6As + (D - B) = 6.$$

Therefore, $C = 0$, $A = 0$, $D + B = 0$, *and* $D - B = \frac{2}{3}$. *Solving the last two equations, we find* $D = -B = \frac{1}{3}$.

Using these results, we find

$$\frac{6}{(s^2+9)^2} = -\frac{1}{3}\frac{(s^2-9)}{(s^2+9)^2} + \frac{1}{3}\frac{1}{s^2+9}.$$

This is the result we had obtained in the previous example using the Convolution Theorem.

8.10 *Inverse Laplace Transform*

UP TO THIS POINT WE HAVE SEEN that the inverse Laplace transform can be found by making use of Laplace transform tables and properties of Laplace transforms. This is typically the way Laplace transforms are taught and used in a differential equations course. One can do the same for Fourier transforms. However, in the case of Fourier transforms, we introduced an inverse transform in the form of an integral. Does such an inverse integral transform exist for the Laplace transform? Yes, it does! In this section we will derive the inverse Laplace transform integral and show how it is used.

We begin by considering a causal function $f(t)$, which vanishes for $t < 0$, and define the function $g(t) = f(t)e^{-ct}$ with $c > 0$. For $g(t)$ absolutely integrable,

A function $f(t)$ is said to be of exponential order if $\int_0^\infty |f(t)|e^{-ct}\, dt < \infty$

$$\int_{-\infty}^{\infty} |g(t)|\, dt = \int_0^{\infty} |f(t)|e^{-ct}\, dt < \infty,$$

we can write the Fourier transform,

$$\hat{g}(\omega) = \int_{-\infty}^{\infty} g(t)e^{i\omega t}dt = \int_0^{\infty} f(t)e^{i\omega t - ct}dt,$$

and the inverse Fourier transform,

$$g(t) = f(t)e^{-ct} = \frac{1}{2\pi} \int_{-\infty}^{\infty} \hat{g}(\omega)e^{-i\omega t}\, d\omega.$$

Multiplying by e^{ct} and inserting $\hat{g}(\omega)$ into the integral for $g(t)$, we find

$$f(t) = \frac{1}{2\pi} \int_{-\infty}^{\infty} \int_{0}^{\infty} f(\tau)e^{(i\omega-c)\tau}\, d\tau\, e^{-(i\omega-c)t}\, d\omega.$$

Letting $s = c - i\omega$ (so $d\omega = i\, ds$), we have

$$f(t) = \frac{i}{2\pi} \int_{c+i\infty}^{c-i\infty} \int_{0}^{\infty} f(\tau)e^{-s\tau}\, d\tau\, e^{st}\, ds.$$

Note that the inside integral is simply $F(s)$. So, we have

$$f(t) = \frac{1}{2\pi i} \int_{c-i\infty}^{c+i\infty} F(s)e^{st}\, ds. \tag{8.123}$$

The integral in the last equation is the inverse Laplace transform, called the Bromwich Integral and is named after Thomas John I'Anson Bromwich (1875–1929) . This inverse transform is not usually covered in differential equations courses because the integration takes place in the complex plane. This integral is evaluated along a path in the complex plane called the Bromwich contour. The typical way to compute this integral is to first choose c so that all poles are to the left of the contour. This guarantees that $f(t)$ is of exponential type. The contour is a closed semicircle enclosing all the poles. One then relies on a generalization of Jordan's Lemma to the second and third quadrants.[9]

Example 8.37. *Find the inverse Laplace transform of $F(s) = \frac{1}{s(s+1)}$.*
    The integral we have to compute is

$$f(t) = \frac{1}{2\pi i} \int_{c-i\infty}^{c+i\infty} \frac{e^{st}}{s(s+1)}\, ds.$$

This integral has poles at $s = 0$ and $s = -1$. The contour we will use is shown in Figure 8.41. We enclose the contour with a semicircle to the left of the path in the complex s-plane. One has to verify that the integral over the semicircle vanishes as the radius goes to infinity. Assuming that we have done this, then the result is simply obtained as $2\pi i$ times the sum of the residues. The residues in this case are

$$Res\left[\frac{e^{zt}}{z(z+1)}; z = 0\right] = \lim_{z\to 0}\frac{e^{zt}}{(z+1)} = 1$$

and

$$Res\left[\frac{e^{zt}}{z(z+1)}; z = -1\right] = \lim_{z\to -1}\frac{e^{zt}}{z} = -e^{-t}.$$

Therefore, we have

$$f(t) = 2\pi i\left[\frac{1}{2\pi i}(1) + \frac{1}{2\pi i}(-e^{-t})\right] = 1 - e^{-t}.$$

[9] Closing the contour to the left of the contour can be reasoned in a manner similar to what we saw in Jordan's Lemma. Write the exponential as $e^{st} = e^{(s_R + is_I)t} = e^{s_R t}e^{is_I t}$. The second factor is an oscillating factor and the growth in the exponential can only come from the first factor. In order for the exponential to decay as the radius of the semicircle grows, $s_R t < 0$. Since $t > 0$, we need $s < 0$ which is done by closing the contour to the left. If $t < 0$, then the contour to the right would enclose no singularities and preserve the causality of $f(t)$.

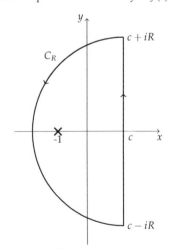

Figure 8.41: The contour used for applying the Bromwich Integral to the Laplace transform $F(s) = \frac{1}{s(s+1)}$.

We can verify this result using the Convolution Theorem or using a partial fraction decomposition. The latter method is simplest. We note that

$$\frac{1}{s(s+1)} = \frac{1}{s} - \frac{1}{s+1}.$$

The first term leads to an inverse transform of 1 and the second term gives e^{-t}. So,

$$\mathcal{L}^{-1}\left[\frac{1}{s} - \frac{1}{s+1}\right] = 1 - e^{-t}.$$

Thus, we have verified the result from doing contour integration.

Example 8.38. *Find the inverse Laplace transform of* $F(s) = \frac{1}{s(1+e^s)}$.
In this case, we need to compute

$$f(t) = \frac{1}{2\pi i}\int_{c-i\infty}^{c+i\infty} \frac{e^{st}}{s(1+e^s)}\, ds.$$

This integral has poles at complex values of s such that $1 + e^s = 0$, or $e^s = -1$. Letting $s = x + iy$, we see that

$$e^s = e^{x+iy} = e^x(\cos y + i\sin y) = -1.$$

We see $x = 0$ and y satisfies $\cos y = -1$ and $\sin y = 0$. Therefore, $y = n\pi$ for n an odd integer. Therefore, the integrand has an infinite number of simple poles at $s = n\pi i$, $n = \pm 1, \pm 3, \ldots$. It also has a simple pole at $s = 0$.

In Figure 8.42 we indicate the poles. We need to compute the residues at each pole. At $s = n\pi i$, we have

$$
\begin{aligned}
Res\left[\frac{e^{st}}{s(1+e^s)}; s = n\pi i\right] &= \lim_{s\to n\pi i} (s - n\pi i)\frac{e^{st}}{s(1+e^s)} \\
&= \lim_{s\to n\pi i} \frac{e^{st}}{se^s} \\
&= -\frac{e^{n\pi it}}{n\pi i}, \quad n \text{ odd.}
\end{aligned}
$$
(8.124)

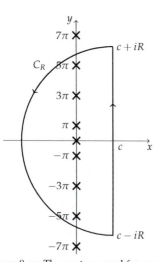

Figure 8.42: The contour used for applying the Bromwich integral to the Laplace transform $F(s) = \frac{1}{1+e^s}$.

At $s = 0$, the residue is

$$Res\left[\frac{e^{st}}{s(1+e^s)}; s = 0\right] = \lim_{s\to 0}\frac{e^{st}}{1+e^s} = \frac{1}{2}.$$

Summing the residues and noting the exponentials for $\pm n$ can be combined to form sine functions, we arrive at the inverse transform.

$$
\begin{aligned}
f(t) &= \frac{1}{2} - \sum_{n \text{ odd}} \frac{e^{n\pi it}}{n\pi i} \\
&= \frac{1}{2} - 2\sum_{k=1}^{\infty} \frac{\sin(2k-1)\pi t}{(2k-1)\pi}.
\end{aligned}
$$
(8.125)

The series in this example might look familiar. It is a Fourier sine series with odd harmonics whose amplitudes decay like $1/n$. It is a vertically shifted square wave. In fact, we had computed the Laplace transform of a general square wave in Example 8.31.

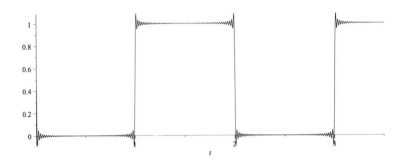

Figure 8.43: Plot of the square wave result as the inverse Laplace transform of $F(s) = \frac{1}{s(1+e^s)}$ with 50 terms.

In that example, we found

$$\mathcal{L}\left[\sum_{n=0}^{\infty}[H(t-2na) - H(t-(2n+1)a)]\right] = \frac{1-e^{-as}}{s(1-e^{-2as})}$$

$$= \frac{1}{s(1+e^{-as})}. \quad (8.126)$$

In this example, one can show that

$$f(t) = \sum_{n=0}^{\infty}[H(t-2n+1) - H(t-2n)].$$

The reader should verify that this result is indeed the square wave shown in Figure 8.43.

8.11 Transforms and Partial Differential Equations

AS ANOTHER APPLICATION OF THE TRANSFORMS, we will see that we can use transforms to solve some linear partial differential equations. We will first solve the one-dimensional heat equation and the two-dimensional Laplace equations using Fourier transforms. The transforms of the partial differential equations lead to ordinary differential equations that are easier to solve. The final solutions are then obtained using inverse transforms.

We could go further by applying a Fourier transform in space and a Laplace transform in time to convert the heat equation into an algebraic equation. We will also show that we can use a finite sine transform to solve nonhomogeneous problems on finite intervals. Along the way we will identify several Green's functions.

8.11.1 Fourier Transform and the Heat Equation

WE WILL FIRST CONSIDER THE SOLUTION OF THE HEAT EQUATION on an infinite interval using Fourier transforms. The basic scheme was discussed earlier and is outlined in Figure 8.44.

Figure 8.44: Using Fourier transforms to solve a linear partial differential equation.

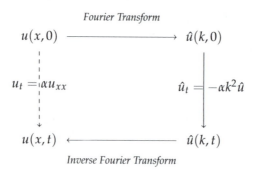

Consider the heat equation on the infinite line:

$$u_t = \alpha u_{xx}, \quad -\infty < x < \infty, t > 0,$$
$$u(x,0) = f(x), \quad -\infty < x < \infty. \tag{8.127}$$

We can Fourier transform the heat equation using the Fourier transform of $u(x,t)$,

$$\mathcal{F}[u(x,t)] = \hat{u}(k,t) = \int_{-\infty}^{\infty} u(x,t)e^{ikx}\, dx.$$

We need to transform the derivatives in the equation. First we note that

$$
\begin{aligned}
\mathcal{F}[u_t] &= \int_{-\infty}^{\infty} \frac{\partial u(x,t)}{\partial t} e^{ikx}\, dx \\
&= \frac{\partial}{\partial t} \int_{-\infty}^{\infty} u(x,t)e^{ikx}\, dx \\
&= \frac{\partial \hat{u}(k,t)}{\partial t}.
\end{aligned}
\tag{8.128}
$$

Assuming that $\lim_{|x|\to\infty} u(x,t) = 0$ and $\lim_{|x|\to\infty} u_x(x,t) = 0$ we also have that

$$
\begin{aligned}
\mathcal{F}[u_{xx}] &= \int_{-\infty}^{\infty} \frac{\partial^2 u(x,t)}{\partial x^2} e^{ikx}\, dx \\
&= -k^2 \hat{u}(k,t).
\end{aligned}
\tag{8.129}
$$

The transformed heat equation.

Therefore, the heat equation becomes

$$\frac{\partial \hat{u}(k,t)}{\partial t} = -\alpha k^2 \hat{u}(k,t).$$

This is a first-order differential equation which is readily solved as

$$\hat{u}(k,t) = A(k)e^{-\alpha k^2 t},$$

where $A(k)$ is an arbitrary function of k. The inverse Fourier transform is

$$
\begin{aligned}
u(x,t) &= \frac{1}{2\pi} \int_{-\infty}^{\infty} \hat{u}(k,t)e^{-ikx}\, dk. \\
&= \frac{1}{2\pi} \int_{-\infty}^{\infty} \hat{A}(k)e^{-\alpha k^2 t}e^{-ikx}\, dk.
\end{aligned}
\tag{8.130}
$$

We can determine $A(k)$ using the initial condition. Note that

$$\mathcal{F}[u(x,0)] = \hat{u}(k,0) = \int_{-\infty}^{\infty} f(x)e^{ikx}\,dx.$$

But we also have from the solution that,

$$u(x,0) = \frac{1}{2\pi} \int_{-\infty}^{\infty} \hat{A}(k)e^{-ikx}\,dk.$$

Comparing these two expressions for $\hat{u}(k,0)$, we see that

$$A(k) = \mathcal{F}[f(x)].$$

We note that $\hat{u}(k,t)$ is given by the product of two Fourier transforms, $\hat{u}(k,t) = A(k)e^{-\alpha k^2 t}$. So, by the Convolution Theorem, we expect that $u(x,t)$ is the convolution of the inverse transforms:

$$u(x,t) = (f * g)(x,t) = \frac{1}{2\pi} \int_{-\infty}^{\infty} f(\xi,t)g(x-\xi,t)\,d\xi,$$

where

$$g(x,t) = \mathcal{F}^{-1}[e^{-\alpha k^2 t}].$$

In order to determine $g(x,t)$, we need only recall Example 8.5. In that example, we saw that the Fourier transform of a Gaussian is a Gaussian. Namely, we found that

$$\mathcal{F}[e^{-ax^2/2}] = \sqrt{\frac{2\pi}{a}}e^{-k^2/2a},$$

or,

$$\mathcal{F}^{-1}[\sqrt{\frac{2\pi}{a}}e^{-k^2/2a}] = e^{-ax^2/2}.$$

Applying this to the current problem, we have

$$g(x) = \mathcal{F}^{-1}[e^{-\alpha k^2 t}] = \sqrt{\frac{\pi}{\alpha t}}e^{-x^2/4t}.$$

Finally, we can write the solution to the problem:

$$u(x,t) = (f * g)(x,t) = \int_{-\infty}^{\infty} f(\xi,t)\frac{e^{-(x-\xi)^2/4t}}{\sqrt{4\pi\alpha t}}\,d\xi,$$

The function in the integrand,

$$K(x,t) = \frac{e^{-x^2/4t}}{\sqrt{4\pi\alpha t}},$$

is called the heat kernel.

$K(x,t)$ is called the heat kernel.

8.11.2 Laplace's Equation on the Half Plane

WE CONSIDER A STEADY-STATE SOLUTION in two dimensions. In particular, we look for the steady-state solution, $u(x,y)$, satisfying the two-dimensional

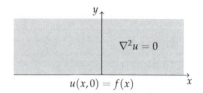

Figure 8.45: This is the domain for a semi-infinite slab with boundary value $u(x,0) = f(x)$ and governed by Laplace's equation.

Laplace equation on a semi-infinite slab with given boundary conditions as shown in Figure 8.45. The boundary value problem is given as

$$u_{xx} + u_{yy} = 0, \quad -\infty < x < \infty, \quad y > 0,$$
$$u(x,0) = f(x), \quad -\infty < x < \infty,$$
$$\lim_{y \to \infty} u(x,y) = 0, \quad \lim_{|x| \to \infty} u(x,y) = 0. \tag{8.131}$$

This problem can be solved using a Fourier transform of $u(x,y)$ with respect to x. The transform scheme for doing this is shown in Figure 8.46. We begin by defining the Fourier transform

$$\hat{u}(k,y) = \mathcal{F}[u] = \int_{-\infty}^{\infty} u(x,y)e^{ikx}\,dx.$$

We can transform Laplace's equation. We first note from the properties of Fourier transforms that

$$\mathcal{F}\left[\frac{\partial^2 u}{\partial x^2}\right] = -k^2 \hat{u}(k,y),$$

if $\lim_{|x| \to \infty} u(x,y) = 0$ and $\lim_{|x| \to \infty} u_x(x,y) = 0$. Also,

$$\mathcal{F}\left[\frac{\partial^2 u}{\partial y^2}\right] = \frac{\partial^2 \hat{u}(k,y)}{\partial y^2}.$$

Thus, the transform of Laplace's equation gives $\hat{u}_{yy} = k^2 \hat{u}$.

Figure 8.46: The transform scheme used to convert Laplace's equation to an ordinary differential equation, which is easier to solve.

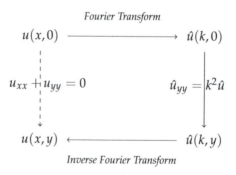

The transformed Laplace equation.

This is a simple ordinary differential equation. We can solve this equation using the boundary conditions. The general solution is

$$\hat{u}(k,y) = a(k)e^{ky} + b(k)e^{-ky}.$$

Because $\lim_{y \to \infty} u(x,y) = 0$ and k can be positive or negative, we have that $\hat{u}(k,y) = a(k)e^{-|k|y}$. The coefficient $a(k)$ can be determined using the remaining boundary condition, $u(x,0) = f(x)$. We find that $a(k) = \hat{f}(k)$ because

$$a(k) = \hat{u}(k,0) = \int_{-\infty}^{\infty} u(x,0)e^{ikx}\,dx = \int_{-\infty}^{\infty} f(x)e^{ikx}\,dx = \hat{f}(k).$$

We have found that $\hat{u}(k,y) = \hat{f}(k)e^{-|k|y}$. We can obtain the solution using the inverse Fourier transform,

$$u(x,t) = \mathcal{F}^{-1}[\hat{f}(k)e^{-|k|y}].$$

We note, that this is a product of Fourier transforms and use the Convolution Theorem for Fourier transforms. Namely, we have that $a(k) = \mathcal{F}[f]$ and $e^{-|k|y} = \mathcal{F}[g]$ for $g(x,y) = \frac{2y}{x^2+y^2}$. This last result is essentially proven in Problem 6.

Then, the Convolution Theorem gives the solution

$$
\begin{aligned}
u(x,y) &= \frac{1}{2\pi} \int_{-\infty}^{\infty} f(\xi) g(x-\xi)\, d\xi \\
&= \frac{1}{2\pi} \int_{-\infty}^{\infty} f(\xi) \frac{2y}{(x-\xi)^2 + y^2}\, d\xi.
\end{aligned}
\tag{8.132}
$$

We note for future use, that this solution is in the form

$$
u(x,y) = \int_{-\infty}^{\infty} f(\xi) G(x,\xi;y,0)\, d\xi,
$$

where

$$
G(x,\xi;y,0) = \frac{y}{\pi((x-\xi)^2 + y^2)}
$$

is the Green's function for this problem.

The Green's function for the Laplace equation.

8.11.3 Heat Equation on Infinite Interval, Revisited

WE WILL RECONSIDER THE INITIAL VALUE PROBLEM for the heat equation on an infinite interval,

$$
\begin{aligned}
u_t &= u_{xx}, \quad -\infty < x < \infty, \quad t > 0, \\
u(x,0) &= f(x), \quad -\infty < x < \infty.
\end{aligned}
\tag{8.133}
$$

We can apply both a Fourier and a Laplace transform to convert this to an algebraic problem. The general solution will then be written in terms of an initial value Green's function as

$$
u(x,t) = \int_{-\infty}^{\infty} G(x,t;\xi,0) f(\xi)\, d\xi.
$$

For the time dependence, we can use the Laplace transform; and for the spatial dependence, we use the Fourier transform. These combined transforms lead us to define

$$
\hat{u}(k,s) = \mathcal{F}[\mathcal{L}[u]] = \int_{-\infty}^{\infty} \int_{0}^{\infty} u(x,t) e^{-st} e^{ikx}\, dt dx.
$$

Applying this to the terms in the heat equation, we have

$$
\begin{aligned}
\mathcal{F}[\mathcal{L}[u_t]] &= s\hat{u}(k,s) - \mathcal{F}[u(x,0)] \\
&= s\hat{u}(k,s) - \hat{f}(k) \\
\mathcal{F}[\mathcal{L}[u_{xx}]] &= -k^2 \hat{u}(k,s).
\end{aligned}
\tag{8.134}
$$

Here we have assumed that

$$
\lim_{t\to\infty} u(x,t) e^{-st} = 0, \quad \lim_{|x|\to\infty} u(x,t) = 0, \quad \lim_{|x|\to\infty} u_x(x,t) = 0.
$$

Therefore, the heat equation can be turned into an algebraic equation for the transformed solution:

The transformed heat equation.

$$(s + k^2)\hat{u}(k, s) = \hat{f}(k),$$

or

$$\hat{u}(k, s) = \frac{\hat{f}(k)}{s + k^2}.$$

The solution to the heat equation is obtained using the inverse transforms for both the Fourier and Laplace transforms. Thus, we have

$$
\begin{aligned}
u(x, t) &= \mathcal{F}^{-1}[\mathcal{L}^{-1}[\hat{u}]] \\
&= \frac{1}{2\pi} \int_{-\infty}^{\infty} \left(\frac{1}{2\pi i} \int_{c-i\infty}^{c+\infty} \frac{\hat{f}(k)}{s + k^2} e^{st}\, ds \right) e^{-ikx}\, dk. \quad (8.135)
\end{aligned}
$$

Because the inside integral has a simple pole at $s = -k^2$, we can compute the Bromwich Integral by choosing $c > -k^2$. Thus,

$$\frac{1}{2\pi i} \int_{c-i\infty}^{c+\infty} \frac{\hat{f}(k)}{s + k^2} e^{st}\, ds = \text{Res}\left[\frac{\hat{f}(k)}{s + k^2} e^{st}; s = -k^2 \right] = e^{-k^2 t}\hat{f}(k).$$

Inserting this result into the solution, we have

$$
\begin{aligned}
u(x, t) &= \mathcal{F}^{-1}[\mathcal{L}^{-1}[\hat{u}]] \\
&= \frac{1}{2\pi} \int_{-\infty}^{\infty} \hat{f}(k) e^{-k^2 t} e^{-ikx}\, dk. \quad (8.136)
\end{aligned}
$$

This solution is of the form

$$u(x, t) = \mathcal{F}^{-1}[\hat{f}\hat{g}]$$

for $\hat{g}(k) = e^{-k^2 t}$. So, by the Convolution Theorem for Fourier transforms, the solution is a convolution:

$$u(x, t) = \int_{-\infty}^{\infty} f(\xi)g(x - \xi)\, d\xi.$$

All we need is the inverse transform of $\hat{g}(k)$.

We note that $\hat{g}(k) = e^{-k^2 t}$ is a Gaussian. Because the Fourier transform of a Gaussian is a Gaussian, we need only recall Example 8.5:

$$\mathcal{F}[e^{-ax^2/2}] = \sqrt{\frac{2\pi}{a}} e^{-k^2/2a}.$$

Setting $a = 1/2t$, this becomes

$$\mathcal{F}[e^{-x^2/4t}] = \sqrt{4\pi t}\, e^{-k^2 t}.$$

So,

$$g(x) = \mathcal{F}^{-1}[e^{-k^2 t}] = \frac{e^{-x^2/4t}}{\sqrt{4\pi t}}.$$

Inserting $g(x)$ into the solution, we have

$$
\begin{aligned}
u(x,t) &= \frac{1}{\sqrt{4\pi t}} \int_{-\infty}^{\infty} f(\xi) e^{-(x-\xi)^2/4t} \, d\xi \\
&= \int_{-\infty}^{\infty} G(x,t;\xi,0) f(\xi) \, d\xi.
\end{aligned}
\tag{8.137}
$$

Here we have identified the initial value Green's function

$$
G(x,t;\xi,0) = \frac{1}{\sqrt{4\pi t}} e^{-(x-\xi)^2/4t}.
$$

The initial value Green's function for the heat equation.

8.11.4 Nonhomogeneous Heat Equation

WE NOW CONSIDER THE NONHOMOGENEOUS HEAT EQUATION with homogeneous boundary conditions defined on a finite interval.

$$
\begin{aligned}
u_t - k u_{xx} &= h(x,t), \quad 0 \le x \le L, \quad t > 0, \\
u(0,t) &= 0, \quad u(L,t) = 0, \quad t > 0, \\
u(x,0) &= f(x), \quad 0 \le x \le .
\end{aligned}
\tag{8.138}
$$

We know that when $h(x,t) \equiv 0$, the solution takes the form

$$
u(x,t) = \sum_{n=1}^{\infty} b_n \sin \frac{n\pi x}{L}.
$$

So, when $h(x,t) \neq 0$, we might assume that the solution takes the form

$$
u(x,t) = \sum_{n=1}^{\infty} b_n(t) \sin \frac{n\pi x}{L},
$$

where the b_n's are the finite Fourier sine transform of the desired solution:

$$
b_n(t) = \mathcal{F}_s[u] = \frac{2}{L} \int_0^L u(x,t) \sin \frac{n\pi x}{L} \, dx,
$$

Note that the finite Fourier sine transform is essentially the Fourier sine transform that we encountered in Section 5.9.

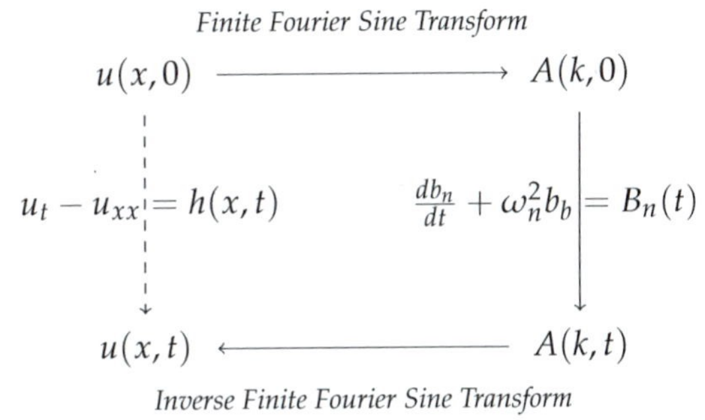

Figure 8.47: Using finite Fourier transforms to solve the heat equation by solving an ODE instead of a PDE.

The idea behind using the finite Fourier Sine Transform is to solve the given heat equation by transforming the heat equation to a simpler equation for the transform, $b_n(t)$, solve for $bn(t)$, and then do an inverse transform, that is, insert the $b_n(t)$'s back into the series representation. This is depicted in Figure 8.47. Note that we had explored a similar diagram earlier when discussing the use of transforms to solve differential equations.

First, we need to transform the partial differential equation. The finite transforms of the derivative terms are given by

$$
\begin{aligned}
\mathcal{F}_s[u_t] &= \frac{2}{L} \int_0^L \frac{\partial u}{\partial t}(x,t) \sin \frac{n\pi x}{L} \, dx \\
&= \frac{d}{dt} \left(\frac{2}{L} \int_0^L u(x,t) \sin \frac{n\pi x}{L} \, dx \right) \\
&= \frac{db_n}{dt}.
\end{aligned}
\tag{8.139}
$$

$$
\begin{aligned}
\mathcal{F}_s[u_{xx}] &= \frac{2}{L} \int_0^L \frac{\partial^2 u}{\partial x^2}(x,t) \sin \frac{n\pi x}{L} \, dx \\
&= \left[u_x \sin \frac{n\pi x}{L} \right]_0^L - \left(\frac{n\pi}{L} \right) \frac{2}{L} \int_0^L \frac{\partial u}{\partial x}(x,t) \cos \frac{n\pi x}{L} \, dx \\
&= - \left[\frac{n\pi}{L} u \cos \frac{n\pi x}{L} \right]_0^L - \left(\frac{n\pi}{L} \right)^2 \frac{2}{L} \int_0^L u(x,t) \sin \frac{n\pi x}{L} \, dx \\
&= \frac{n\pi}{L} [u(0,t) - u(L,0) \cos n\pi] - \left(\frac{n\pi}{L} \right)^2 b_n^2 \\
&= -\omega_n^2 b_n^2,
\end{aligned}
\tag{8.140}
$$

where $\omega_n = \frac{n\pi}{L}$.

Furthermore, we define

$$
H_n(t) = \mathcal{F}_s[h] = \frac{2}{L} \int_0^L h(x,t) \sin \frac{n\pi x}{L} \, dx.
$$

Then, the heat equation is transformed to

$$
\frac{db_n}{dt} + \omega_n^2 b_n = H_n(t), \quad n = 1,2,3,\ldots.
$$

This is a simple linear first-order differential equation. We can supplement this equation with the initial condition

$$
b_n(0) = \frac{2}{L} \int_0^L u(x,0) \sin \frac{n\pi x}{L} \, dx.
$$

The differential equation for b_n is easily solved using the integrating factor, $\mu(t) = e^{\omega_n^2 t}$. Thus,

$$
\frac{d}{dt} \left(e^{\omega_n^2 t} b_n(t) \right) = H_n(t) e^{\omega_n^2 t}
$$

and the solution is

$$
b_n(t) = b_n(0) e^{-\omega_n^2 t} + \int_0^t H_n(\tau) e^{-\omega_n^2 (t-\tau)} \, d\tau.
$$

The final step is to insert these coefficients (finite Fourier sine transform) into the series expansion (inverse finite Fourier sine transform) for $u(x,t)$. The result is

$$
u(x,t) = \sum_{n=1}^{\infty} b_n(0) e^{-\omega_n^2 t} \sin \frac{n\pi x}{L} + \sum_{n=1}^{\infty} \left[\int_0^t H_n(\tau) e^{-\omega_n^2 (t-\tau)} \, d\tau \right] \sin \frac{n\pi x}{L}.
$$

This solution can be written in a more compact form in order to identify the Green's function. We insert the expressions for $b_n(0)$ and $H_n(t)$ in terms of the initial profile and source term and interchange sums and integrals. This leads to

$$
\begin{aligned}
u(x,t) &= \sum_{n=1}^{\infty} \left(\frac{2}{L} \int_0^L u(\xi,0) \sin\frac{n\pi\xi}{L} \, d\xi \right) e^{-\omega_n^2 t} \sin\frac{n\pi x}{L} \\
&\quad + \sum_{n=1}^{\infty} \left[\int_0^t \left(\frac{2}{L} \int_0^L h(\xi,\tau) \sin\frac{n\pi\xi}{L} \, d\xi \right) e^{-\omega_n^2(t-\tau)} \, d\tau \right] \sin\frac{n\pi x}{L} \\
&= \int_0^L u(\xi,0) \left[\frac{2}{L} \sum_{n=1}^{\infty} \sin\frac{n\pi x}{L} \sin\frac{n\pi\xi}{L} e^{-\omega_n^2 t} \right] d\xi \\
&\quad + \int_0^t \int_0^L h(\xi,\tau) \left[\frac{2}{L} \sum_{n=1}^{\infty} \sin\frac{n\pi x}{L} \sin\frac{n\pi\xi}{L} e^{-\omega_n^2(t-\tau)} \right] \\
&= \int_0^L u(\xi,0) G(x,\xi;t,0) \, d\xi + \int_0^t \int_0^L h(\xi,\tau) G(x,\xi;t,\tau) \, d\xi d\tau.
\end{aligned}
$$
$$(8.141)$$

Here we have defined the Green's function

$$
G(x,\xi;t,\tau) = \frac{2}{L} \sum_{n=1}^{\infty} \sin\frac{n\pi x}{L} \sin\frac{n\pi\xi}{L} e^{-\omega_n^2(t-\tau)}.
$$

We note that $G(x,\xi;t,0)$ gives the initial value Green's function.

Note that at $t = \tau$,

$$
G(x,\xi;t,t) = \frac{2}{L} \sum_{n=1}^{\infty} \sin\frac{n\pi x}{L} \sin\frac{n\pi\xi}{L}.
$$

This is actually the series representation of the Dirac delta function. The Fourier sine transform of the delta function is

$$
\mathcal{F}_s[\delta(x-\xi)] = \frac{2}{L} \int_0^L \delta(x-\xi) \sin\frac{n\pi x}{L} \, dx = \frac{2}{L} \sin\frac{n\pi\xi}{L}.
$$

Then, the representation becomes

$$
\delta(x-\xi) = \frac{2}{L} \sum_{n=1}^{\infty} \sin\frac{n\pi x}{L} \sin\frac{n\pi\xi}{L}.
$$

Also, we note that

$$
\frac{\partial G}{\partial t} = -\omega_n^2 G
$$

$$
\frac{\partial^2 G}{\partial x^2} = -\left(\frac{n\pi}{L}\right)^2 G.
$$

Therefore, $G_t = G_{xx}$, at least for $\tau \neq t$ and $\xi \neq x$.

We can modify this problem by adding nonhomogeneous boundary conditions.

$$
\begin{aligned}
u_t - k u_{xx} &= h(x,t), \quad 0 \leq x \leq L, \quad t > 0, \\
u(0,t) &= A, \quad u(L,t) = B, \quad t > 0, \\
u(x,0) &= f(x), \quad 0 \leq x \leq L. \quad (8.142)
\end{aligned}
$$

One way to treat these conditions is to assume $u(x,t) = w(x) + v(x,t)$ where $v_t - kv_{xx} = h(x,t)$ and $w_{xx} = 0$. Then, $u(x,t) = w(x) + v(x,t)$ satisfies the original nonhomogeneous heat equation.

If $v(x,t)$ satisfies $v(0,t) = v(L,t) = 0$ and $w(x)$ satisfies $w(0) = A$ and $w(L) = B$, then $u(0,t) = w(0) + v(0,t) = A$ $u(L,t) = w(L) + v(L,t) = B$

Finally, we note that

$$v(x,0) = u(x,0) - w(x) = f(x) - w(x).$$

Therefore, $u(x,t) = w(x) + v(x,t)$ satisfies the original problem if

$$v_t - kv_{xx} = h(x,t), \quad 0 \leq x \leq L, \quad t > 0,$$
$$v(0,t) = 0, \quad v(L,t) = 0, \quad t > 0,$$
$$v(x,0) = f(x) - w(x), \quad 0 \leq x \leq L, \tag{8.143}$$

and

$$w_{xx} = 0, \quad 0 \leq x \leq L,$$
$$w(0) = A, \quad w(L) = B. \tag{8.144}$$

We can solve the last problem to obtain $w(x) = A + \frac{B-A}{L}x$. The solution to the problem for $v(x,t)$ is simply the problem we solved already in terms of Green's functions with the new initial condition $f(x) - A - \frac{B-A}{L}x$.

Problems

1. In this problem you will show that the sequence of functions

$$f_n(x) = \frac{n}{\pi}\left(\frac{1}{1+n^2x^2}\right)$$

approaches $\delta(x)$ as $n \to \infty$. Use the following to support your argument:

a. Show that $\lim_{n\to\infty} f_n(x) = 0$ for $x \neq 0$.

b. Show that the area under each function is one.

2. Verify that the sequence of functions $\{f_n(x)\}_{n=1}^{\infty}$, defined by $f_n(x) = \frac{n}{2}e^{-n|x|}$, approaches a delta function.

3. Evaluate the following integrals:

a. $\int_0^\pi \sin x \delta\left(x - \frac{\pi}{2}\right) dx$.

b. $\int_{-\infty}^\infty \delta\left(\frac{x-5}{3}e^{2x}\right)(3x^2 - 7x + 2) \, dx$.

c. $\int_0^\pi x^2 \delta\left(x + \frac{\pi}{2}\right) dx$.

d. $\int_0^\infty e^{-2x}\delta(x^2 - 5x + 6) \, dx$. [See Problem 4.]

e. $\int_{-\infty}^\infty (x^2 - 2x + 3)\delta(x^2 - 9) \, dx$. [See Problem 4.]

4. For the case that a function has multiple roots, $f(x_i) = 0$, $i = 1, 2, \ldots$, it can be shown that

$$\delta(f(x)) = \sum_{i=1}^n \frac{\delta(x - x_i)}{|f'(x_i)|}.$$

Use this result to evaluate $\int_{-\infty}^\infty \delta(x^2 - 5x - 6)(3x^2 - 7x + 2) \, dx$.

5. Find a Fourier series representation of the Dirac delta function, $\delta(x)$, on $[-L, L]$.

6. For $a > 0$, find the Fourier transform, $\hat{f}(k)$, of $f(x) = e^{-a|x|}$.

7. Use the result from Problem 6 plus properties of the Fourier transform to find the Fourier transform, of $f(x) = x^2 e^{-a|x|}$ for $a > 0$.

8. Find the Fourier transform, $\hat{f}(k)$, of $f(x) = e^{-2x^2 + x}$.

9. Prove the Second Shift Property in the form

$$F\left[e^{i\beta x} f(x)\right] = \hat{f}(k + \beta).$$

10. A damped harmonic oscillator is given by

$$f(t) = \begin{cases} Ae^{-\alpha t}e^{i\omega_0 t}, & t \geq 0, \\ 0, & t < 0. \end{cases}$$

    a. Find $\hat{f}(\omega)$ and

    b. the frequency distribution $|\hat{f}(\omega)|^2$.

    c. Sketch the frequency distribution.

11. Show that the convolution operation is associative: $(f * (g * h))(t) = ((f * g) * h)(t)$.

12. In this problem, you will directly compute the convolution of two Gaussian functions in two steps.

    a. Use completing the square to evaluate

$$\int_{-\infty}^{\infty} e^{-\alpha t^2 + \beta t} \, dt.$$

    b. Use the result from part a to directly compute the convolution in Example 8.16:

$$(f * g)(x) = e^{-bx^2} \int_{-\infty}^{\infty} e^{-(a+b)t^2 + 2bxt} \, dt.$$

13. You will compute the (Fourier) convolution of two box functions of the same width. Recall that the box function is given by

$$f_a(x) = \begin{cases} 1, & |x| \leq a \\ 0, & |x| > a. \end{cases}$$

Consider $(f_a * f_a)(x)$ for different intervals of x. A few preliminary sketches will help. In Figure 8.48, the factors in the convolution integrand are shown for one value of x. The integrand is the product of the first two functions. The convolution at x is the area of the overlap in the third figure. Think about how these pictures change as you vary x. Plot the resulting areas as a function of x. This is the graph of the desired convolution.

Figure 8.48: Sketch used to compute the convolution of the box function with itself. In the top figure is the box function. The middle figure shows the box shifted by x. The bottom figure indicates the overlap of the functions.

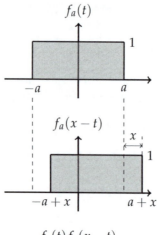

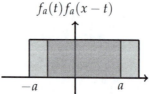

14. Define the integrals $I_n = \int_{-\infty}^{\infty} x^{2n} e^{-x^2}\, dx$. Noting that $I_0 = \sqrt{\pi}$,

    a. Find a recursive relation between I_n and I_{n-1}.

    b. Use this relation to determine I_1, I_2, and I_3.

    c. Find an expression in terms of n for I_n.

15. Find the Laplace transform of the following functions:

    a. $f(t) = 9t^2 - 7$.

    b. $f(t) = e^{5t-3}$.

    c. $f(t) = \cos 7t$.

    d. $f(t) = e^{4t} \sin 2t$.

    e. $f(t) = e^{2t}(t + \cosh t)$.

    f. $f(t) = t^2 H(t - 1)$.

    g. $f(t) = \begin{cases} \sin t, & t < 4\pi, \\ \sin t + \cos t, & t > 4\pi \end{cases}$.

    h. $f(t) = \int_0^t (t - u)^2 \sin u\, du$.

    i. $f(t) = (t + 5)^2 + t e^{2t} \cos 3t$ and write the answer in the simplest form.

16. Find the inverse Laplace transform of the following functions using the properties of Laplace transforms and the table of Laplace transform pairs.

    a. $F(s) = \dfrac{18}{s^3} + \dfrac{7}{s}$.

    b. $F(s) = \dfrac{1}{s - 5} - \dfrac{2}{s^2 + 4}$.

c. $F(s) = \dfrac{s+1}{s^2+1}$.

d. $F(s) = \dfrac{3}{s^2+2s+2}$.

e. $F(s) = \dfrac{1}{(s-1)^2}$.

f. $F(s) = \dfrac{e^{-3s}}{s^2-1}$.

g. $F(s) = \dfrac{1}{s^2+4s-5}$.

h. $F(s) = \dfrac{s+3}{s^2+8s+17}$.

17. Compute the convolution $(f*g)(t)$ (in the Laplace transform sense) and its corresponding Laplace transform $\mathcal{L}[f*g]$ for the following functions:

a. $f(t) = t^2$, $g(t) = t^3$.

b. $f(t) = t^2$, $g(t) = \cos 2t$.

c. $f(t) = 3t^2 - 2t + 1$, $g(t) = e^{-3t}$.

d. $f(t) = \delta\left(t - \frac{\pi}{4}\right)$, $g(t) = \sin 5t$.

18. For the following problems, draw the given function and find the Laplace transform in closed form.

a. $f(t) = 1 + \sum_{n=1}^{\infty}(-1)^n H(t-n)$.

b. $f(t) = \sum_{n=0}^{\infty}[H(t-2n+1) - H(t-2n)]$.

c. $f(t) = \sum_{n=0}^{\infty}(t-2n)[H(t-2n) - H(t-2n-1)] + (2n+2-t)[H(t-2n-1) - H(t-2n-2)]$.

19. Use the Convolution Theorem to compute the inverse transform of the following:

a. $F(s) = \dfrac{2}{s^2(s^2+1)}$.

b. $F(s) = \dfrac{e^{-3s}}{s^2}$.

c. $F(s) = \dfrac{1}{s(s^2+2s+5)}$.

20. Find the inverse Laplace transform in two different ways: (i) Use tables. (ii) Use the Bromwich Integral.

a. $F(s) = \dfrac{1}{s^3(s+4)^2}$.

b. $F(s) = \dfrac{1}{s^2-4s-5}$.

c. $F(s) = \dfrac{s+3}{s^2+8s+17}$.

d. $F(s) = \dfrac{s+1}{(s-2)^2(s+4)}$.

e. $F(s) = \dfrac{s^2 + 8s - 3}{(s^2 + 2s + 1)(s^2 + 1)}$.

21. Use Laplace transforms to solve the following initial value problems. Where possible, describe the solution behavior in terms of oscillation and decay.

a. $y'' - 5y' + 6y = 0$, $y(0) = 2$, $y'(0) = 0$.

b. $y'' - y = te^{2t}$, $y(0) = 0$, $y'(0) = 1$.

c. $y'' + 4y = \delta(t - 1)$, $y(0) = 3$, $y'(0) = 0$.

d. $y'' + 6y' + 18y = 2H(\pi - t)$, $y(0) = 0$, $y'(0) = 0$.

22. Use Laplace transforms to convert the following system of differential equations into an algebraic system, and find the solution of the differential equations.

$$\begin{aligned} x'' &= 3x - 6y, & x(0) &= 1, & x'(0) &= 0, \\ y'' &= x + y, & y(0) &= 0, & y'(0) &= 0. \end{aligned}$$

23. Use Laplace transforms to convert the following nonhomogeneous systems of differential equations into an algebraic system, and find the solutions of the differential equations.

a.

$$\begin{aligned} x' &= 2x + 3y + 2\sin 2t, & x(0) &= 1, \\ y' &= -3x + 2y, & y(0) &= 0. \end{aligned}$$

b.

$$\begin{aligned} x' &= -4x - y + e^{-t}, & x(0) &= 2, \\ y' &= x - 2y + 2e^{-3t}, & y(0) &= -1. \end{aligned}$$

c.

$$\begin{aligned} x' &= x - y + 2\cos t, & x(0) &= 3, \\ y' &= x + y - 3\sin t, & y(0) &= 2. \end{aligned}$$

24. Consider the series circuit in Problem 2.20 and in Figure 2.7 with $L = 1.00$ H, $R = 1.00 \times 10^2$ Ω, $C = 1.00 \times 10^{-4}$ F, and $V_0 = 1.00 \times 10^3$ V.

a. Write the second-order differential equation for this circuit.

b. Suppose that no charge is present and no current is flowing at time $t = 0$ when V_0 is applied. Use Laplace transforms to find the current and the charge on the capacitor as functions of time.

c. Replace the battery with the alternating source $V(t) = V_0 \sin 2\pi ft$ with $V_0 = 1.00 \times 10^3$ V and $f = 150$ Hz. Again, suppose that no charge is present and no current is flowing at time $t = 0$ when the AC source is applied. Use Laplace transforms to find the current and the charge on the capacitor as functions of time.

d. Plot your solutions, and describe how the system behaves over time.

25. Use Laplace transforms to sum the following series or write as a single integral.

a. $\displaystyle\sum_{n=0}^{\infty} \frac{(-1)^n}{1+2n}$.

b. $\displaystyle\sum_{n=1}^{\infty} \frac{1}{n(n+3)}$.

c. $\displaystyle\sum_{n=1}^{\infty} \frac{(-1)^n}{n(n+3)}$.

d. $\displaystyle\sum_{n=0}^{\infty} \frac{(-1)^n}{n^2-a^2}$.

e. $\displaystyle\sum_{n=0}^{\infty} \frac{1}{(2n+1)^2-a^2}$.

f. $\displaystyle\sum_{n=1}^{\infty} \frac{1}{n} e^{-an}$.

26. Use Laplace transforms to prove

$$\sum_{n=1}^{\infty} \frac{1}{(n+a)(n+b)} = \frac{1}{b-a} \int_0^1 \frac{u^a - u^b}{1-u}\, du.$$

Use this result to evaluate the following sums:

a. $\displaystyle\sum_{n=1}^{\infty} \frac{1}{n(n+1)}$.

b. $\displaystyle\sum_{n=1}^{\infty} \frac{1}{(n+2)(n+3)}$.

27. Do the following:

a. Find the first four nonvanishing terms of the Maclaurin series expansion of $f(x) = \dfrac{x}{e^x - 1}$.

b. Use the result in part a. to determine the first four nonvanishing Bernoulli numbers, B_n.

c. Use these results to compute $\zeta(2n)$ for $n = 1,2,3,4$.

28. Given the following Laplace transforms $F(s)$, find the function $f(t)$. Note that in each case there are an infinite number of poles, resulting in an infinite series representation.

a. $F(s) = \dfrac{1}{s^2(1+e^{-s})}$.

b. $F(s) = \dfrac{1}{s \sinh s}$.

c. $F(s) = \dfrac{\sinh s}{s^2 \cosh s}$.

d. $F(s) = \dfrac{\sinh(\beta\sqrt{s}x)}{s\sinh(\beta\sqrt{s}L)}$.

29. Consider the initial boundary value problem for the heat equation:

$$
\begin{aligned}
u_t &= 2u_{xx}, & 0 < t, \quad 0 \le x \le 1, \\
u(x,0) &= x(1-x), & 0 < x < 1, \\
u(0,t) &= 0, & t > 0, \\
u(1,t) &= 0, & t > 0.
\end{aligned}
$$

Use the finite transform method to solve this problem. Namely, assume that the solution takes the form $u(x,t) = \sum_{n=1}^{\infty} b_n(t)\sin n\pi x$ and obtain an ordinary differential equation for b_n and solve for the b_n's for each n.

9

Vector Analysis and EM Waves

"From a long view of the history of mankind seen from, say, ten thousand years from now, there can be little doubt that the most significant event of the 19th century will be judged as Maxwell's discovery of the laws of electrodynamics."—The Feynman Lectures on Physics (1964), Richard Feynman (1918–1988)

UP TO THIS POINT, we have mainly been confined to problems involving only one or two independent variables. In particular, the heat equation and the wave equation involved one time and one space dimension. However, we live in a world of three spatial dimensions. (However, some theoretical physicists live in worlds of many more dimensions, or at least they think so.) We will need to extend the study of the heat equation and the wave equation to three spatial dimensions.

Recall that the one-dimensional wave equation takes the form

$$\frac{\partial^2 u}{\partial t^2} = c^2 \frac{\partial^2 u}{\partial x^2}. \tag{9.1}$$

For higher-dimensional problems, we will need to generalize the $\frac{\partial^2 u}{\partial x^2}$ term. For the case of electromagnetic waves in a source-free environment, we will derive a three-dimensional wave equation for the electric and magnetic fields: It is given by

$$\frac{\partial^2 u}{\partial t^2} = c^2 \left(\frac{\partial^2 u}{\partial x^2} + \frac{\partial^2 u}{\partial y^2} + \frac{\partial^2 u}{\partial z^2} \right). \tag{9.2}$$

This is the generic form of the linear wave equation in Cartesian coordinates. It can be written in more compact form using the Laplacian,[1] ∇^2,

$$\frac{\partial^2 u}{\partial t^2} = c^2 \nabla^2 u. \tag{9.3}$$

The introduction of the Laplacian is common when generalizing to higher dimensions. In fact, we have already presented some generic ones and three-dimensional equations in Table 5.1, which we reproduce in Table 9.1. We have studied the one-dimensional wave equation, heat equation, and Schrödinger Equation. For steady-state, or equilibrium, heat flow problems, the heat equation no longer involves the time derivative. What is

[1] For the wave equation, there is the more compact notation $\Box u = 0$, where

$$\Box = \nabla^2 - \frac{1}{c^2} \frac{\partial^2}{\partial t^2}$$

is called the d'Alembertian, or d'Alembert operator. Also, the Laplacian is sometimes written as $\Delta = \nabla^2$.

left is called Laplace's equation, which we have also seen in relation to complex functions. Adding an external heat source, Laplace's equation becomes what is known as Poisson's equation.

Table 9.1: Generic Partial Differential Equations in Two and Three Spatial Dimensions

Name	2 Vars	3 D
Heat Equation	$u_t = k u_{xx}$	$u_t = k \nabla^2 u$
Wave Equation	$u_{tt} = c^2 u_{xx}$	$u_{tt} = c^2 \nabla^2 u$
Laplace's Equation	$u_{xx} + u_{yy} = 0$	$\nabla^2 u = 0$
Poisson's Equation	$u_{xx} + u_{yy} = F(x,y)$	$\nabla^2 u = F(x,y,z)$
Schrödinger's Equation	$iu_t = u_{xx} + F(x,t)u$	$iu_t = \nabla^2 u + F(x,y,z,t)u$

Using the Laplacian allows us not only to write these equations in a more compact form, but also in a coordinate-free representation. Many problems are more easily cast in other coordinate systems. For example, the propagation of electromagnetic waves in an optical fiber are naturally described in terms of cylindrical coordinates. The heat flow inside a hemispherical igloo can be described using spherical coordinates. The vibrations of a circular drumhead can be described using polar coordinates. In each of these cases, the Laplacian must be written in terms of the needed coordinate systems.

The solution of these partial differential equations can be handled using separation of variables or transform methods because these are linear partial differential equations. In the next chapter we will look at several examples of applying the separation of variables in higher dimensions. This will lead to the study of ordinary differential equations, which in turn leads to new sets of functions, other than the typical sine and cosine solutions.

In this chapter we will review some of the needed vector analysis for the derivation of the three-dimensional wave equation from Maxwell's equations. We will review the basic vector operations (the dot and cross products); define the gradient, curl, and divergence, and introduce standard vector identities that are often seen in physics courses. Equipped with these vector operations, we will derive the three-dimensional wave equations for electromagnetic waves from Maxwell's equations. We will conclude this chapter with a section on curvilinear coordinates and provide the vector differential operators for different coordinate systems.

9.1   Vector Analysis

9.1.1   A Review of Vector Products

AT THIS POINT YOU MIGHT WANT to reread the first section of Chapter 3. In that chapter, we introduced the formal definition of a vector space and some simple properties of vectors. We also discussed one of the common vector products, the dot product, which is defined as

$$\mathbf{u} \cdot \mathbf{v} = uv \cos \theta. \tag{9.4}$$

There is also a component form, which we write as

$$\mathbf{u} \cdot \mathbf{v} = u_1 v_1 + u_2 v_2 + u_3 v_3 = \sum_{k=1}^{3} u_k v_k. \qquad (9.5)$$

One of the first physical examples using a cross product is the definition of work. The work done on a body by a constant force $\mathbf{F}$ during a displacement $\mathbf{d}$ is

$$W = \mathbf{F} \cdot \mathbf{d}.$$

In the case of a nonconstant force, we have to add up the incremental contributions to the work, $dW = \mathbf{F} \cdot d\mathbf{r}$, to obtain

$$W = \int_C dW = \int_C \mathbf{F} \cdot d\mathbf{r} \qquad (9.6)$$

over the path C. Note how much this looks like a complex path integral. It is a path integral, but the path lies in a real three-dimensional space.

Another application of the dot product is the proof of the Law of Cosines. Recall that this law gives the side opposite a given angle in terms of the angle and the other two sides of the triangle:

$$c^2 = a^2 + b^2 - 2ab \cos \theta. \qquad (9.7)$$

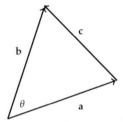

Figure 9.1: $v = r\omega$. The Law of Cosines can be derived using vectors.

Consider the triangle in Figure 9.1. We draw the sides of the triangle as vectors. Note that $\mathbf{b} = \mathbf{c} + \mathbf{a}$. Also recall that the square of the length any vector can be written as the dot product of the vector with itself. Therefore, we have

$$\begin{aligned} c^2 &= \mathbf{c} \cdot \mathbf{c} \\ &= (\mathbf{b} - \mathbf{a}) \cdot (\mathbf{b} - \mathbf{a}) \\ &= \mathbf{a} \cdot \mathbf{a} + \mathbf{b} \cdot \mathbf{b} - 2\mathbf{a} \cdot \mathbf{b} \\ &= a^2 + b^2 - 2ab \cos \theta. \end{aligned} \qquad (9.8)$$

We note that the Law of Cosines also comes up when writing out inverse square laws in many applications. Namely, the vector $\mathbf{a}$ can locate a mass, or charge, and vector $\mathbf{b}$ points to an observation point. Then, the inverse square law would involve vector $\mathbf{c}$, whose length is obtained as $\sqrt{a^2 + b^2 - 2ab \cos \theta}$. Typically, one does not have $\mathbf{a}$'s and $\mathbf{b}$'s, but something like $\mathbf{r}_1$ and $\mathbf{r}_2$, or $\mathbf{r}$ and $\mathbf{R}$. For these problems, one is interested in approximating the expression of interest in terms of ratios like $\frac{r}{R}$ for $R \gg r$. In fact, we had seen such expressions when discussing the gravitational potential for the Earth–moon system in Figure 6.5.

Another important vector product is the cross product. The cross product produces a vector, unlike the dot product, which results in a scalar. The magnitude of the cross product is given as

$$|\mathbf{a} \times \mathbf{b}| = ab \sin \theta. \qquad (9.9)$$

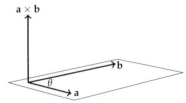

Figure 9.2: The cross product is shown. The direction is obtained using the Right Hand Rule: Curl fingers from $\mathbf{a}$ through to $\mathbf{b}$. The thumb will point in the direction of $\mathbf{a} \times \mathbf{b}$.

Being a vector, we also have to specify the direction. The cross product produces a vector that is perpendicular to both vectors $\mathbf{a}$ and $\mathbf{b}$. Thus, the vector is normal to the plane in which these vectors live. There are two possible directions. The direction taken is given by the Right Hand Rule. This is shown in Figure 9.2. The direction can also be determined using your right hand. Curl your fingers from **a** through to **b**. The thumb will point in the direction of $\mathbf{a} \times \mathbf{b}$.

One of the first occurrences of the cross product in physics is in the definition of the torque, $\boldsymbol{\tau} = \mathbf{r} \times \mathbf{F}$. Recall that the torque is the analog to the force. A net torque will cause an angular acceleration. Consider a rigid body in which a force is applied to to the body at a position **r** from the axis of rotation. (See Figure 9.3.) Then, this force produces a torque with respect to the axis. The direction of the torque is given by the Right Hand Rule. Point your fingers in the direction of **r** and rotate them toward **F**. In the figure, this would be out of the page and indicates that the bar would rotate in a counterclockwise direction if this were the only force acting on the bar.

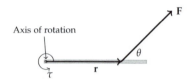

Figure 9.3: A force applied at a point located at **r** from the axis of rotation produces a torque $\boldsymbol{\tau} = \mathbf{r} \times \mathbf{F}$ with respect to the axis.

Another example is that of a body rotating about an axis as shown in Figure 9.4. We locate the body with a position vector pointing from the origin of the coordinate system to the body. The tangential velocity of the body is related to the angular velocity by a cross product $\mathbf{v} = \boldsymbol{\omega} \times \mathbf{r}$. The direction of the angular velocity is given be the Right Hand Rule. Curl the fingers of your right hand in the direction of the motion of the rotating mass. Your thumb will point in the direction of ω. Counter clockwise motion produces a positive angular velocity and clockwise will give a negative angular velocity. Note that for the origin at the center of rotation of the mass, we obtain the familiar expression $v = r\omega$.

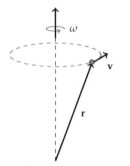

Figure 9.4: A mass rotates at an angular velocity ω about a fixed axis of rotation. The tangential velocity with respect to a given origin is given by $\mathbf{v} = \boldsymbol{\omega} \times \mathbf{r}$.

There is also a geometric interpretation of the cross product. Consider the vectors **a** and **b** in Figure 9.5. Now draw a perpendicular from the tip of **b** to vector **a**. This forms a triangle of height h. Slide the triangle over to form a rectangle of base a and height h. The area of this triangle is

$$
\begin{aligned}
A &= ah \\
&= a(b\sin\theta) \\
&= |\mathbf{a} \times \mathbf{b}|.
\end{aligned} \tag{9.10}
$$

Therefore, the magnitude of the cross product is the area of the triangle formed by the vectors **a** and **b**.

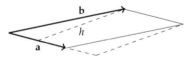

Figure 9.5: The magnitudes of the cross product give the area of the parallelogram defined by **a** and **b**.

The dot product was shown to have a simple form in terms of the components of the vectors. Similarly, we can write the cross product in component form. Recall that we can expand any vector $\mathbf{v} \in R^3$ as

$$
\mathbf{v} = \sum_{k=1}^{3} v_k \mathbf{e}_k, \tag{9.11}
$$

where the $\mathbf{e}_k$'s are the standard basis vectors.

We would like to expand the cross product of two vectors:

$$
\mathbf{u} \times \mathbf{v} = \left(\sum_{k=1}^{3} u_k \mathbf{e}_k \right) \times \left(\sum_{k=1}^{3} v_k \mathbf{e}_k \right).
$$

In order to do this, we need a few properties of the cross product.

First of all, the cross product is not commutative. In fact, it is anticommutative:

$$\mathbf{u} \times \mathbf{v} = -\mathbf{v} \times \mathbf{u}.$$

A simple consequence of this is that $\mathbf{v} \times \mathbf{v} = 0$. Just replace $\mathbf{u}$ with $\mathbf{v}$ in the Anticommutativity Rule and you have $\mathbf{v} \times \mathbf{v} = -\mathbf{v} \times \mathbf{v}$. Something that is its negative must be zero.

The cross product also satisfies distributive properties:

$$\mathbf{u} \times (\mathbf{v} + \mathbf{w}) = \mathbf{u} \times \mathbf{v} + \mathbf{u} \times \mathbf{w}$$

and

$$\mathbf{u} \times (a\mathbf{v}) = (a\mathbf{u}) \times \mathbf{v} = a(\mathbf{u} \times \mathbf{v}).$$

Thus, we can expand the cross product in terms of the components of the given vectors. A simple computation shows that $\mathbf{u} \times \mathbf{v}$ can be expressed in terms of sums over $\mathbf{e}_i \times \mathbf{e}_j$:

$$
\begin{aligned}
\mathbf{u} \times \mathbf{v} &= \left(\sum_{i=1}^{3} u_i \mathbf{e}_i \right) \times \left(\sum_{j=1}^{3} v_j \mathbf{e}_j \right) \\
&= \sum_{i=1}^{3} \sum_{j=1}^{3} u_i v_j \mathbf{e}_i \times \mathbf{e}_j.
\end{aligned}
\tag{9.12}
$$

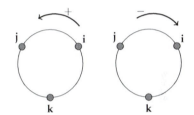

Figure 9.6: The sign for the cross product for basis vectors can be determined from a simple diagram. Arrange the vectors on a circle as above. If the needed computation goes counterclockwise, then the sign is positive. Thus, $\mathbf{j} \times \mathbf{k} = \mathbf{i}$ and $\mathbf{k} \times \mathbf{j} = -\mathbf{i}$.

The cross products of basis vectors are simple to compute. First of all, the cross products $\mathbf{e}_i \times \mathbf{e}_j$ vanish when $i = j$ by anticommutativity of the cross product. For $i \neq j$, it is not much more difficult. For the typical basis, $\{\mathbf{i}, \mathbf{j}, \mathbf{k}\}$, this is simple. Imagine computing $\mathbf{i} \times \mathbf{j}$. This is a vector of length $|\mathbf{i} \times \mathbf{j}| = |\mathbf{i}||\mathbf{j}| \sin 90° = 1$. The vector $\mathbf{i} \times \mathbf{j}$ is perpendicular to both vectors, $\mathbf{i}$ and $\mathbf{j}$. Thus, the cross product is either $\mathbf{k}$ or $-\mathbf{k}$. Using the Right Hand Rule, we have $\mathbf{i} \times \mathbf{j} = \mathbf{k}$. Similarly, we find the following:

$$
\begin{aligned}
\mathbf{i} \times \mathbf{j} = \mathbf{k}, \quad \mathbf{j} \times \mathbf{k} &= \mathbf{i}, \quad \mathbf{k} \times \mathbf{i} = \mathbf{j}, \\
\mathbf{j} \times \mathbf{i} = -\mathbf{k}, \quad \mathbf{k} \times \mathbf{j} &= -\mathbf{i}, \quad \mathbf{i} \times \mathbf{k} = -\mathbf{j}.
\end{aligned}
\tag{9.13}
$$

Inserting these results into the cross product for vectors in R^3, we have

$$\mathbf{u} \times \mathbf{v} = (u_2 v_3 - u_3 v_2)\mathbf{i} + (u_3 v_1 - u_1 v_3)\mathbf{j} + (u_1 v_2 - u_2 v_1)\mathbf{k}. \tag{9.14}$$

While this form for the cross product is correct and useful, there are other forms that help in verifying identities or making computation simpler with less memorization. However, some of these new expressions can lead to problems for the novice as dealing with indices can be daunting at first sight.

One expression that is useful for computing cross products is the familiar computation using determinants. Namely, we have that

$$\mathbf{u} \times \mathbf{v} = \begin{vmatrix} \mathbf{i} & \mathbf{j} & \mathbf{k} \\ u_1 & u_2 & u_3 \\ v_1 & v_2 & v_3 \end{vmatrix}$$

$$= \begin{vmatrix} u_2 & u_3 \\ v_2 & v_3 \end{vmatrix} \mathbf{i} - \begin{vmatrix} u_1 & u_3 \\ v_1 & v_3 \end{vmatrix} \mathbf{j} + \begin{vmatrix} u_1 & u_2 \\ v_1 & v_2 \end{vmatrix} \mathbf{k}$$

$$= (u_2 v_3 - u_3 v_2)\mathbf{i} + (u_3 v_1 - u_1 v_3)\mathbf{j} + (u_1 v_2 - u_2 v_1)\mathbf{k}.$$

$$(9.15)$$

The completely antisymmetric symbol, or permutation symbol, ϵ_{ijk}. This is also called the Levi-Civita symbol, named after the Italian mathematician Tullio Levi-Civita (1873–1941). He is known for work in tensor calculus and was the doctoral student of the inventor of tensor calculus, Gregorio Ricci-Curbastro (1853–1925).

A more compact form for the cross product is obtained by introducing the completely antisymmetric symbol, ϵ_{ijk}. This symbol is defined by the relations

$$\epsilon_{123} = \epsilon_{231} = \epsilon_{312} = 1$$

and

$$\epsilon_{321} = \epsilon_{213} = \epsilon_{132} = -1,$$

and all other combinations, like ϵ_{113}, vanish. Note that all indices must differ. Also, if the order is a cyclic permutation of $\{1, 2, 3\}$, then the value is $+1$. For this reason, ϵ_{ijk} is also called the permutation symbol or the Levi-Civita permutation symbol. We can also indicate the index permutation more generally using the following identities:

$$\epsilon_{ijk} = \epsilon_{jki} = \epsilon_{kij} = -\epsilon_{jik} = -\epsilon_{ikj} = -\epsilon_{kji}.$$

Returning to the cross product, we can introduce the standard basis $\mathbf{e}_1 = \mathbf{i}$, $\mathbf{e}_2 = \mathbf{j}$, and $\mathbf{e}_3 = \mathbf{k}$. With this notation, we have that

$$\mathbf{e}_i \times \mathbf{e}_j = \sum_{k=1}^{3} \epsilon_{ijk} \mathbf{e}_k. \qquad (9.16)$$

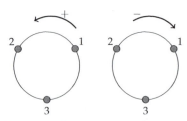

Figure 9.7: The sign for the permutation symbol can be determined from a simple cyclic diagram similar to that for the cross product. Arrange the numbers from 1 to 3 on a circle. If the needed computation goes counterclockwise, then the sign is positive, otherwise it is negative.

Example 9.1. *Compute the cross product of the basis vectors* $\mathbf{e}_2 \times \mathbf{e}_1$ *using the permutation symbol. A straightforward application of the definition of the cross product:*

$$\begin{aligned} \mathbf{e}_2 \times \mathbf{e}_1 &= \sum_{k=1}^{3} \epsilon_{21k} \mathbf{e}_k \\ &= \epsilon_{211} \mathbf{e}_1 + \epsilon_{212} \mathbf{e}_2 + \epsilon_{213} \mathbf{e}_3 \\ &= -\mathbf{e}_3. \end{aligned} \qquad (9.17)$$

It is helpful to write out enough terms in these sums until you become familiar with manipulating the indices. Note that the first two terms vanished because of repeated indices. In the last term, we used $\epsilon_{213} = -1$.

Starting with the component form of the cross product in Equation (9.12), we now write the general cross product as

$$\begin{aligned} \mathbf{u} \times \mathbf{v} &= \sum_{i=1}^{3} \sum_{j=1}^{3} u_i v_j \mathbf{e}_i \times \mathbf{e}_j \\ &= \sum_{i=1}^{3} \sum_{j=1}^{3} u_i v_j \left(\sum_{k=1}^{3} \epsilon_{ijk} \mathbf{e}_k \right) \\ &= \sum_{i,j,k=1}^{3} \epsilon_{ijk} u_i v_j \mathbf{e}_k. \end{aligned} \qquad (9.18)$$

Note that the last sum is a triple sum over the indices i, j, and k.

Example 9.2. *Let* $\mathbf{u} = 2\mathbf{i} - 3\mathbf{j}$ *and* $\mathbf{v} = \mathbf{i} + 5\mathbf{j} + 4\mathbf{k}$. *Compute* $\mathbf{u} \times \mathbf{v}$. *We can compute this easily using determinants.*

$$\mathbf{u} \times \mathbf{v} = \begin{vmatrix} \mathbf{i} & \mathbf{j} & \mathbf{k} \\ 2 & -3 & 0 \\ 1 & 5 & 4 \end{vmatrix}$$

$$= \begin{vmatrix} -3 & 0 \\ 5 & 4 \end{vmatrix}\mathbf{i} - \begin{vmatrix} 2 & 0 \\ 1 & 4 \end{vmatrix}\mathbf{j} + \begin{vmatrix} 2 & -3 \\ 1 & 5 \end{vmatrix}\mathbf{k}$$

$$= -12\mathbf{i} - 8\mathbf{j} + 13\mathbf{k}. \tag{9.19}$$

Using the permutation symbol to compute this cross product, we have

$$\begin{aligned} \mathbf{u} \times \mathbf{v} &= \epsilon_{123}u_1 v_2 \mathbf{k} + \epsilon_{231}u_2 v_3 \mathbf{i} + \epsilon_{312}u_3 v_1 \mathbf{j} \\ &\quad + \epsilon_{213}u_2 v_1 \mathbf{k} + \epsilon_{132}u_1 v_3 \mathbf{j} + \epsilon_{321}u_3 v_2 \mathbf{i} \\ &= 2(5)\mathbf{k} + (-3)4\mathbf{i} + (0)1\mathbf{j} - (-3)1\mathbf{k} - (2)4\mathbf{j} - (0)5\mathbf{i} \\ &= -12\mathbf{i} - 8\mathbf{j} + 13\mathbf{k}. \tag{9.20} \end{aligned}$$

Sometimes it is useful to note from Equation (9.18) that the kth component of the cross product is given by

$$(\mathbf{u} \times \mathbf{v})_k = \sum_{i,j=1}^{3} \epsilon_{ijk} u_i v_j.$$

In more advanced texts, or in the case of relativistic computations with tensors, the summation symbol is suppressed. For this case, one writes

$$(\mathbf{u} \times \mathbf{v})_k = \epsilon_{ijk} u_i v_j,$$

where it is understood that summation is performed over repeated indices. This is called the Einstein summation convention.

Because the cross product can be written as both a determinant,

$$\begin{aligned} \mathbf{u} \times \mathbf{v} &= \begin{vmatrix} \mathbf{i} & \mathbf{j} & \mathbf{k} \\ u_1 & u_2 & u_3 \\ v_1 & v_2 & v_3 \end{vmatrix} \\ &= \epsilon_{ij1}u_i v_j \mathbf{i} + \epsilon_{ij2}u_i v_j \mathbf{j} + \epsilon_{ij3}u_i v_j \mathbf{k}, \tag{9.21} \end{aligned}$$

and using the permutation symbol,

$$\mathbf{u} \times \mathbf{v} = \epsilon_{ijk} u_i v_j \mathbf{e}_k,$$

we can write the determinant using the Levi-Civita symbol. We start with the determinant in Equation (9.21) and replace the entries using

$$\begin{aligned} \mathbf{a}_1 &= (\mathbf{i}, \mathbf{j}, \mathbf{k}), \\ \mathbf{a}_2 &= \mathbf{u}, \\ \mathbf{a}_3 &= \mathbf{v}. \end{aligned}$$

$$\tag{9.22}$$

The Einstein summation convention is used to suppress summation notation. In general relativity, one employs raised indices, so that vector components are written in the form u^i. The convention then requires that one only sums over a combination of one lower and one upper index. Thus, we would write $\epsilon_{ijk} u^i v^j$. We will forgo the need for raised indices in most of this book.

This gives the determinant in terms of the Levi-Civita symbol.

$$\begin{vmatrix} a_{11} & a_{12} & a_{13} \\ a_{21} & a_{22} & a_{23} \\ a_{31} & a_{32} & a_{33} \end{vmatrix} = \sum_{i,j,k=1}^{3} \epsilon_{ijk} a_{1i} a_{2j} a_{3k}. \tag{9.23}$$

Here we included the triple sum in order to emphasize the hidden summations.

Example 9.3. *Compute the determinant* $\begin{vmatrix} 1 & 0 & 2 \\ 0 & -3 & 4 \\ 2 & 4 & -1 \end{vmatrix}$.

We insert the components of each row into the expression for the determinant:

$$\begin{aligned} \begin{vmatrix} 1 & 0 & 2 \\ 0 & -3 & 4 \\ 2 & 4 & -1 \end{vmatrix} &= \epsilon_{123}(1)(-3)(-1) + \epsilon_{231}(0)(4)(2) + \epsilon_{312}(2)(0)(4) \\ &\quad + \epsilon_{213}(0)(0)(-1) + \epsilon_{132}(1)(4)(4) + \epsilon_{321}(2)(-3)(2) \\ &= 3 + 0 + 0 - 0 - 14 - (-12) \\ &= 15. \end{aligned} \tag{9.24}$$

Note that if one adds copies of the first two columns, as shown in Figure 9.8, then the products of the first three diagonals, downward to the right (dashed), give the positive terms in the determinant computation and the products of the last three diagonals, downward to the left (dotted), give the negative terms.

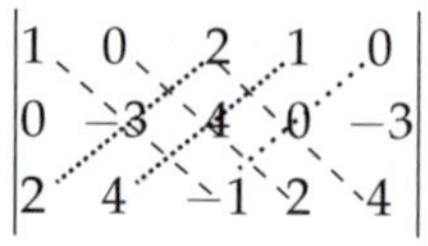

Figure 9.8: Diagram for computing determinants.

One useful identity is

$$\epsilon_{jki}\epsilon_{j\ell m} = \delta_{k\ell}\delta_{im} - \delta_{km}\delta_{i\ell},$$

Product identity satisfied by the permutation symbol, ϵ_{ijk}.

where δ_{ij} is the Kronecker delta. Note that the Einstein summation convention is used in this identity; that is, summing over j is understood. So, the left side is really a sum of three terms:

$$\epsilon_{jki}\epsilon_{j\ell m} = \epsilon_{1ki}\epsilon_{1\ell m} + \epsilon_{2ki}\epsilon_{2\ell m} + \epsilon_{3ki}\epsilon_{3\ell m}.$$

This identity is simple to understand. For nonzero values of the Levi-Civita symbol, we have to require that all indices differ for each factor on the left side of the equation: $j \neq k \neq i$ and $j \neq \ell \neq m$. Because the first two slots are the same j, and the indices only take values $1, 2$, or 3, then either $k = \ell$ or $k = m$. This will give terms with factors of $\delta_{k\ell}$ or δ_{km}. If the former is true, then there is only one possibility for the third slot, $i = m$. Thus, we have a term $\delta_{k\ell}\delta_{im}$. Similarly, the other case yields the second term on the right side of the identity. We just need to get the signs right. Obviously, changing the order of ℓ and m will introduce a minus sign. A little care will show that the identity gives the correct ordering.

Other identities involving the permutation symbol are

$$\epsilon_{mjk}\epsilon_{njk} = 2\delta_{mn},$$

$$\epsilon_{ijk}\epsilon_{ijk} = 6.$$

We will end this section by recalling triple products. There are only two ways to construct triple products. Starting with the cross product $\mathbf{b} \times \mathbf{c}$, which is a vector, we can multiply the cross product by another vector, $\mathbf{a}$, to either obtain a scalar or a vector.

In the first case, we have the triple scalar product, $\mathbf{a} \cdot (\mathbf{b} \times \mathbf{c})$. Actually, we do not need the parentheses. Writing $\mathbf{a} \cdot \mathbf{b} \times \mathbf{c}$ could only mean one thing. If we computed $\mathbf{a} \cdot \mathbf{b}$ first, we would get a scalar. Then, the result would be a multiple of $\mathbf{c}$, which is not a scalar. So, dropping the parentheses would mean that we want the triple scalar product by convention.

Let's consider the component form of this product. We will use the Einstein summation convention and the fact that the permutation symbol is cyclic in ijk. Using $\epsilon_{jki} = \epsilon_{ijk}$,

$$
\begin{aligned}
\mathbf{a} \cdot (\mathbf{b} \times \mathbf{c}) &= a_i(\mathbf{b} \times \mathbf{c})_i \\
&= \epsilon_{jki}a_ib_jc_k \\
&= \epsilon_{ijk}a_ib_jc_k \\
&= (\mathbf{a} \times \mathbf{b})_k c_k \\
&= (\mathbf{a} \times \mathbf{b}) \cdot \mathbf{c}. \tag{9.25}
\end{aligned}
$$

In order to appreciate the summation convention, here is the same computation with the explicit sums shown:

$$
\begin{aligned}
\mathbf{a} \cdot (\mathbf{b} \times \mathbf{c}) &= \sum_{i=1}^{3} a_i(\mathbf{b} \times \mathbf{c})_i \\
&= \sum_{i=1}^{3}\sum_{j=1}^{3}\sum_{k=1}^{3} \epsilon_{jki}a_ib_jc_k \\
&= \sum_{k=1}^{3}\sum_{i=1}^{3}\sum_{j=1}^{3} \epsilon_{ijk}a_ib_jc_k \\
&= \sum_{k=1}^{3} (\mathbf{a} \times \mathbf{b})_k c_k \\
&= (\mathbf{a} \times \mathbf{b}) \cdot \mathbf{c}. \tag{9.26}
\end{aligned}
$$

We have proven that

$$\mathbf{a} \cdot (\mathbf{b} \times \mathbf{c}) = (\mathbf{a} \times \mathbf{b}) \cdot \mathbf{c}.$$

Now, imagine how much writing would be involved if we had expanded everything in terms of all of the components. Well, this might be a good time to show how one proves identities using the components without the simplifying notation. So, we expand the scalar triple product $\mathbf{a} \cdot (\mathbf{b} \times \mathbf{c})$ as

$$
\begin{aligned}
\mathbf{a} \cdot (\mathbf{b} \times \mathbf{c}) &= \mathbf{a} \cdot [(b_2 c_3 - b_3 c_2)\mathbf{i} + (b_3 c_1 - b_1 c_3)\mathbf{j} + (b_1 c_2 - b_2 c_1)\mathbf{k}] \\
&= a_1(b_2 c_3 - b_3 c_2) + a_2(b_3 c_1 - b_1 c_3) + a_3(b_1 c_2 - b_2 c_1) \\
&= a_1 b_2 c_3 - a_1 b_3 c_2 + a_2 b_3 c_1 - a_2 b_1 c_3 + a_3 b_1 c_2 - a_3 b_2 c_1.
\end{aligned}
\tag{9.27}
$$

Now we stare at the expression to figure out how we can regroup the terms to obtain $(\mathbf{a} \times \mathbf{b}) \cdot \mathbf{c}$. You could just expand the terms for this expression and see if you get the same terms, or we could regroup the terms based on the components of $\mathbf{c}$. Thus,

$$
\mathbf{a} \cdot (\mathbf{b} \times \mathbf{c}) = (a_2 b_3 - a_3 b_2)c_1 + (a_3 b_1 - a_1 b_3)c_2 + (a_1 b_2 - a_2 b_1)c_3.
$$

Now, we can see that this is indeed $(\mathbf{a} \times \mathbf{b}) \cdot \mathbf{c}$.

Note that this result suggests that the triple scalar product can be computed by just computing a determinant. In particular, the third equation in Equations (9.25) gives

$$
\begin{aligned}
\mathbf{a} \cdot (\mathbf{b} \times \mathbf{c}) &= \epsilon_{ijk} a_i b_j c_k \\
&= \begin{vmatrix} a_1 & a_2 & a_3 \\ b_1 & b_2 & b_3 \\ c_1 & c_2 & c_3 \end{vmatrix}.
\end{aligned}
\tag{9.28}
$$

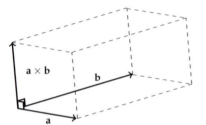

Figure 9.9: Three non-coplanar vectors define a parallelepiped. The volume is given by the triple scalar product $\mathbf{a} \cdot \mathbf{b} \times \mathbf{c}$.

There is a geometric interpretation of the scalar triple product. Consider the three vectors in Figure 9.9. If they do not all lie in a plane, then they form the sides of a parallelepiped. The cross product $\mathbf{a} \times \mathbf{b}$ gives the area of the base as we had seen earlier. The cross product is perpendicular to this base. The dot product of $\mathbf{c}$ with this cross product gives the height of the parallelepiped. So, the volume of the parallelepiped is the height times the base, or the triple scalar product. In general, one gets a signed volume, as the cross product could be pointing below the base.

The second type of triple product is the triple cross product,

$$
\mathbf{a} \times (\mathbf{b} \times \mathbf{c}) = \epsilon_{mnj} \epsilon_{ijk} a_i b_m c_n \mathbf{e}_k.
$$

In this case, we cannot drop the parentheses as this would lead to a real ambiguity. Lets think a little about this product. The vector $\mathbf{b} \times \mathbf{c}$ is a vector that is perpendicular to both $\mathbf{b}$ and $\mathbf{c}$. Computing the triple cross product would then produce a vector perpendicular to $\mathbf{a}$ and $\mathbf{b} \times \mathbf{c}$. But the latter vector is perpendicular to both $\mathbf{b}$ and $\mathbf{c}$ already. Therefore, the triple cross product must lie in the plane spanned by these vectors. In fact, there is an identity that tells us exactly the right combination of vectors $\mathbf{b}$ and $\mathbf{c}$. It is given by

The BAC-CAB rule.

$$
\boxed{\mathbf{a} \times (\mathbf{b} \times \mathbf{c}) = \mathbf{b}(\mathbf{a} \cdot \mathbf{c}) - \mathbf{c}(\mathbf{a} \cdot \mathbf{b})} \tag{9.29}
$$

This rule is called the BAC-CAB Rule because of the order of the vectors on the right side of this equation. We prove this identity later in the chapter using components. In the next example, we prove it using the Levi-Civita symbol.

Example 9.4. *Prove that $\mathbf{a} \times (\mathbf{b} \times \mathbf{c}) = \mathbf{b}(\mathbf{a} \cdot \mathbf{c}) - \mathbf{c}(\mathbf{a} \cdot \mathbf{b})$ using the Levi-Civita symbol.*

We can prove the BAC-CAB Rule using the permutation symbol and some identities. We first note that vectors can be expanded in a basis as $\mathbf{a} = a_i\mathbf{e}_i$ and $\mathbf{b} \times \mathbf{c} = (\mathbf{b} \times \mathbf{c})_j\mathbf{e}_j$. We will also need the cross products $\mathbf{e}_i \times \mathbf{e}_j = \epsilon_{ijk}\mathbf{e}_k$ and $\mathbf{b} \times \mathbf{c} = \epsilon_{mnj}b_m c_n\mathbf{e}_j$. Then, the computation proceeds as follows:

$$
\begin{aligned}
\mathbf{a} \times (\mathbf{b} \times \mathbf{c}) &= (a_i\mathbf{e}_i) \times ((\mathbf{b} \times \mathbf{c})_j\mathbf{e}_j) \\
&= a_i(\mathbf{b} \times \mathbf{c})_j(\mathbf{e}_i \times \mathbf{e}_j) \\
&= a_i(\mathbf{b} \times \mathbf{c})_j\epsilon_{ijk}\mathbf{e}_k \\
&= \epsilon_{mnj}\epsilon_{ijk}a_i b_m c_n\mathbf{e}_k
\end{aligned} \tag{9.30}
$$

In order to evaluate the product of the Levi-Civita symbols, we use the identity

$$
\epsilon_{mnj}\epsilon_{ijk} = \delta_{mk}\delta_{ni} - \delta_{mi}\delta_{nk}
$$

and the properties of the Kronecker delta functions. Thus, we obtain

$$
\begin{aligned}
\mathbf{a} \times (\mathbf{b} \times \mathbf{c}) &= \epsilon_{mnj}\epsilon_{ijk}a_i b_m c_n\mathbf{e}_k \\
&= a_i b_m c_n \left(\delta_{mk}\delta_{ni} - \delta_{mi}\delta_{nk}\right)\mathbf{e}_k \\
&= a_n b_m c_n\mathbf{e}_m - a_m b_m c_n\mathbf{e}_n \\
&= (b_m\mathbf{e}_m)(c_n a_n) - (c_n\mathbf{e}_n)(a_m b_m) \\
&= \mathbf{b}(\mathbf{a} \cdot \mathbf{c}) - \mathbf{c}(\mathbf{a} \cdot \mathbf{b}).
\end{aligned} \tag{9.31}
$$

Line 3 is obtained by noting that for the first term, $i = n$ and $k = m$. For the second term in line 3, $i = m$ and $k = n$. We then group the results by noting that

$$
b_m\mathbf{e}_m = b_1\mathbf{e}_1 + b_2\mathbf{e}_2 + b_3\mathbf{e}_3 = \mathbf{b},
$$

and

$$
c_n a_n = a_1 c_1 + a_2 c_2 + a_3 c_3 = \mathbf{a} \cdot \mathbf{c}.
$$

A similar computation simplifies the second term.

Example 9.5. *Show that*

$$
\mathbf{a} \times (\mathbf{b} \times \mathbf{c}) \neq (\mathbf{a} \times \mathbf{b}) \times \mathbf{c}.
$$

9.1.2 Differentiation and Integration of Vectors

YOU HAVE ALREADY BEEN INTRODUCED to the idea that vectors can be differentiated and integrated in your introductory physics course. These ideas are also the major theme encountered in a multivariate calculus class, or Calculus III. We review some of these topics in the next sections. We first recall the differentiation and integration of vector functions. We will consider vectors as functions of some parameter. As the parameter changes, so does the vector. The most common vectors are the position and velocity vectors for a particle in motion (Figure 9.10). The parameter in this case is naturally the time. However, we could also use the distance the particle had moved from its starting point.

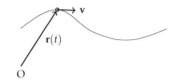

Figure 9.10: Position and velocity vectors of moving particle.

Let's begin thinking about the position vector as a function of time, $\mathbf{r}(t)$. The position vector can be written in component form as $\mathbf{r}(t) = x(t)\mathbf{i} + y(t)\mathbf{j} + x(t)\mathbf{k}$, where the components are also time dependent. The rate of change of the position is the velocity. This is determined from the derivatives of the components of the position vector:

$$
\begin{aligned}
\mathbf{v}(t) &= \frac{d\mathbf{r}}{dt} \\
&= \lim_{\Delta t \to 0} \frac{\mathbf{r}(t + \Delta t) - \mathbf{r}(t)}{\Delta t} \\
&= \frac{dx}{dt}\mathbf{i} + \frac{dy}{dt}\mathbf{j} + \frac{dz}{dt}\mathbf{k} \\
&= v_x\mathbf{i} + v_y\mathbf{k} + v_z\mathbf{k}.
\end{aligned}
\tag{9.32}
$$

The velocity vector is tangent to the path, $\mathbf{r}(t)$, as seen in Figure 9.12. The magnitude of this vector gives the speed,

$$
|\mathbf{v}| = \sqrt{\left(\frac{dx}{dt}\right)^2 + \left(\frac{dy}{dt}\right)^2 + \left(\frac{dz}{dt}\right)^2}.
$$

Moreover, the derivative of the velocity vector gives the acceleration, $\mathbf{a}(t) = \mathbf{v}'(t)$.

In general, one can differentiate an arbitrary time-dependent vector $\mathbf{v}(t) = f(t)\mathbf{i} + g(t)\mathbf{j} + h(t)\mathbf{k}$ as

$$
\frac{d\mathbf{v}}{dt} = \frac{df}{dt}\mathbf{i} + \frac{dg}{dt}\mathbf{j} + \frac{dh}{dt}\mathbf{k}.
\tag{9.33}
$$

Example 9.6. *Find the velocity and acceleration of a particle in uniform circular motion.*

A circle in the xy-plane can be parametrized as $\mathbf{r}(t) = r\cos(\omega t)\mathbf{i} + r\sin(\omega t)\mathbf{j}$. Then, the velocity is found as

$$
\mathbf{v}(t) = -r\omega\sin(\omega t)\mathbf{i} + r\omega\cos(\omega t)\mathbf{j}.
$$

Its speed is $v = r\omega$, which is easily recognized as the tangential speed. The acceleration is

$$
\mathbf{a}(t) = -\omega^2 r\cos(\omega t)\mathbf{i} - \omega^2 r\sin(\omega t)\mathbf{j}.
$$

The magnitude gives the centripetal acceleration, $a = \omega^2 r$ and the acceleration vector is pointing toward the center of the circle.

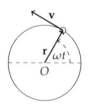

Figure 9.11: Particle on circular path.

Once one can differentiate time-dependent vectors, one can prove some standard properties. These are given in the following list but we will not prove them here.

a. $\dfrac{d}{dt}[\mathbf{u} + \mathbf{v}] = \dfrac{d\mathbf{u}}{dt} + \dfrac{d\mathbf{v}}{dt}.$

b. $\dfrac{d}{dt}[c\mathbf{u}] = c\dfrac{d\mathbf{u}}{dt}.$

c. $\dfrac{d}{dt}[f(t)\mathbf{u}] = f'(t)\mathbf{u} + f(t)\dfrac{d\mathbf{u}}{dt}.$

d. $\dfrac{d}{dt}[\mathbf{u} \cdot \mathbf{v}] = \dfrac{d\mathbf{u}}{dt} \cdot \mathbf{v} + \mathbf{u} \cdot \dfrac{d\mathbf{v}}{dt}.$

e. $\dfrac{d}{dt}[\mathbf{u} \times \mathbf{v}] = \dfrac{d\mathbf{u}}{dt} \times \mathbf{v} + \mathbf{u} \times \dfrac{d\mathbf{v}}{dt}.$

f. $\dfrac{d}{dt}[\mathbf{u}(f(t))] = \dfrac{d\mathbf{u}}{df}\dfrac{df}{dt}.$

Example 9.7. *Let* $|\mathbf{r}(t)| =$ *const. Show that* $\mathbf{r}'(t)$ *is perpendicular* $\mathbf{r}(t)$.

Because $|\mathbf{r}| =$ *const,* $|\mathbf{r}|^2 = \mathbf{r} \cdot \mathbf{r} =$ *const. Differentiating this expression, one has* $\frac{d}{dt}(\mathbf{r} \cdot \mathbf{r}) = 2\mathbf{r} \cdot \frac{d\mathbf{r}}{dt} = 0$. *Therefore,* $\mathbf{r} \cdot \frac{d\mathbf{r}}{dt} = 0$, *as was to be shown.*

In this discussion, we have referred to t as the time. However, when parametrizing spacecurves, t could represent any parameter. For example, the circle could be parametrized for t, the angle swept out along any arc of the circle, $\mathbf{r}(t) = r\cos t\,\mathbf{i} + r\sin t\,\mathbf{j}$, for $t_1 \leq t \leq t_2$. We can still differentiate with respect to this parameter. It no longer has the meaning of velocity.

Another standard parameter is that of arclength. The arclength of a path is the distance along the path from some starting point. In deriving an expression for arclength, one first considers incremental distances along paths. Moving from point (x, y, z) to point $(x + \Delta x, y + \Delta y, z + \Delta z)$, one has gone a distance of

$$\Delta s = \sqrt{(\Delta x)^2 + (\Delta y)^2 + (\Delta z)^2}.$$

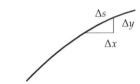

Figure 9.12: An incremental arclength, $\Delta s = \sqrt{(\Delta x)^2 + (\Delta y)^2}$.

In Figure 9.12 we see a sketch for a two-dimensional example.

Given a curve parametrized by t, such as the time, one can rewrite this as

$$\Delta s = \sqrt{\left(\dfrac{\Delta x}{\Delta t}\right)^2 + \left(\dfrac{\Delta y}{\Delta t}\right)^2 + \left(\dfrac{\Delta z}{\Delta t}\right)^2}\,\Delta t.$$

Letting Δt get small, as well as the other increments, we are led to the differential form

$$ds = \sqrt{\left(\dfrac{dx}{dt}\right)^2 + \left(\dfrac{dy}{dt}\right)^2 + \left(\dfrac{dz}{dt}\right)^2}\,dt. \qquad (9.34)$$

We note that the square root is $|\mathbf{r}'(t)|$ So,

$$ds = |\mathbf{r}'(t)|dt$$

or

$$\dfrac{ds}{dt} = |\mathbf{r}'(t)|.$$

In order to find the total arclength, we need only integrate over the parameter range,

$$s = \int_{t_1}^{t_2} |\mathbf{r}'(t)|\, dt.$$

If t is time and $\mathbf{r}(t)$ is the position vector of a particle, then $|\mathbf{r}'(t)|$ is the particle speed and we have that the distance traveled is simply an integral of the speed:

$$s = \int_{t_1}^{t_2} v\, dt.$$

If one is interested in knowing the distance traveled from point $\mathbf{r}(t_1)$ to an arbitrary point $\mathbf{r}(t)$, one can define the arclength function:

$$s(t) = \int_{t_1}^{t} |\mathbf{r}'(\tau)|\, d\tau.$$

Example 9.8. *Determine the length of the parabolic path described by* $\mathbf{r} = t\mathbf{i} + t^2\mathbf{j}$, $t \in [0,1]$.

We want to determine the length of the path, $L = \int_0^1 |\mathbf{r}'(t)|\, dt$. *First, we compute* $\mathbf{r}'(t) = \mathbf{i} + 2t\mathbf{j}$. *Then we find the magnitude,* $|\mathbf{r}'(t)| = \sqrt{1 + 4t^2}$ *and integrate:*

$$
\begin{aligned}
s &= \int_0^1 |\mathbf{r}'(t)|\, dt \\
&= \int_0^1 \sqrt{1 + 4t^2}\, dt \\
&= \left[x\sqrt{x^2 + \frac{1}{4}} + \frac{1}{4}\ln\left(x + \sqrt{x^2 + \frac{1}{4}}\right) \right]_0^1 \\
&= \frac{\sqrt{5}}{2} + \frac{1}{4}\ln(2 + \sqrt{5}).
\end{aligned}
\tag{9.35}
$$

Here we have used

$$\int \sqrt{t^2 + a^2}\, dt = \frac{1}{2}\left(t\sqrt{t^2 + a^2} + a^2 \ln(t + \sqrt{t^2 + a^2}) \right)$$

with $t = x$ *and* $a = \frac{1}{2}$.

Line integrals are defined as integrals of functions along a path, or curve, in space. Let $f(x,y,z)$ be the function, and C a parametrized path. Then, we are interested in computing $\int_C f(x,y,z)\, ds$, where s is the arclength parameter. This integral looks similar to the contour integrals that we had studied in Chapter 5. We can compute such integrals in a similar manner by introducing the parametrization:

$$\int_C f(x,y,z)\, ds = \int_C f(x(t),y(t),z(t))|\mathbf{r}'(t)|\, dt.$$

Example 9.9. *Compute* $\int_C (x^2 + y^2 + z^2)\, ds$ *for the helical path* $\mathbf{r} = (\cos t, \sin t, t)$, $t \in [0, 2\pi]$.

In order to do this integral, we have to integrate over the given range of t values. So, we replace ds with $|\mathbf{r}'(t)|dt$. In this problem, $|\mathbf{r}'(t)| = \sqrt{2}$. Also, we insert the parametric forms for $x(t) = \cos t$, $y(t) = \sin t$, and $z = t$ into $f(x,y,z)$. Thus,

$$\int_C (x^2 + y^2 + z^2)\, ds = \int_0^{2\pi} (1 + t^2)\sqrt{2}\, dt = 2\sqrt{2}\pi\left(1 + \frac{4\pi^2}{3}\right).
\tag{9.36}$$

One can also integrate vector functions. Given the vector function $\mathbf{v}(t) = f(t)\mathbf{i} + g(t)\mathbf{j} + h(t)\mathbf{k}$, we can do a straightforward term-by-term integration:

$$\int_a^b \mathbf{v}(t)\,dt = \int_a^b f(t)\,dt\mathbf{i} + \int_a^b g(t)\,dt\mathbf{j} + \int_a^b h(t)\,dt\mathbf{k}.$$

If $\mathbf{v}(t)$ is the velocity and t is the time, then

$$\int_a^b \mathbf{v}(t)\,dt = \int_a^b \frac{d\mathbf{r}}{dt}\,dt = \mathbf{r}(b) - \mathbf{r}(a).$$

We can thus interpret this integral as giving the displacement of a particle between times $t = a$ and $t = b$.

At the beginning of this chapter we recalled the concept of work: the work done on a body by a nonconstant force $\mathbf{F}$ over a path C is

$$W = \int_C \mathbf{F} \cdot d\mathbf{r}. \tag{9.37}$$

If the path is parametrized by t, then we can write $d\mathbf{r} = \frac{d\mathbf{r}}{dt}dt$. Thus, the prescription for computing line integrals such as this is

$$\int_C \mathbf{F} \cdot d\mathbf{r} = \int_C \mathbf{F} \cdot \frac{d\mathbf{r}}{dt}\,dt.$$

Note that if t is time, then

$$\int_C \mathbf{F} \cdot d\mathbf{r} = \int_C \mathbf{F} \cdot \mathbf{v}dt\,dt,$$

where $\mathbf{F} \cdot \mathbf{v}$ is the power, or energy per time.

There are other forms that line integrals can take. Let $\mathbf{F} = P(x,y,z)\mathbf{i} + Q(x,y,z)\mathbf{j} + R(x,y,z)\mathbf{k}$. Noting that $d\mathbf{r} = dx\mathbf{i} + dy\mathbf{y} + dz\mathbf{k}$, then we can write

$$\int_C \mathbf{F} \cdot d\mathbf{r} = \int_C P(x,y,z)\,dx + Q(x,y,z)\,dy + R(x,y,z)\,dz.$$

Example 9.10. *Compute the work done by the force* $\mathbf{F} = y\mathbf{i} - x\mathbf{j}$ *on a particle as it moves around the circle* $\mathbf{r} = (\cos t)\mathbf{i} + (\sin t)\mathbf{j}$, *for* $0 \leq t \leq \pi$:

$$W = \int_C \mathbf{F} \cdot d\mathbf{r} = \int_C y\,dx - x\,dy.$$

One way to complete this is to note that $dx = -\sin t\,dt$ *and* $dy = \cos t\,dt$. *Then*

$$\int_C y\,dx - x\,dy = \int_0^\pi (-\sin^2 t - \cos^2 t)\,dt = -\pi.$$

9.1.3 Div, Grad, Curl

THROUGHOUT PHYSICS WE SEE FUNCTIONS that vary in both space and time. A function $f(x,y,z,t)$ is called a scalar function when the output is a scalar, or a number. An example of such a function is the temperature. A function $\mathbf{F}(x,y,z,t)$ is called a vector (or vector valued) function if the output of the function is a vector. Let $\mathbf{v}(x,y,z,t)$ represent the velocity of a fluid at position (x, y, z) at time t. This is an example of a vector function. Typically, when we assign a number, or a vector, to every point in a domain, we refer to this as a scalar, or vector, field. In this section we discuss how fields change from one point in space to another. Namely, we look at derivatives of multivariate functions with respect to their independent variables and the meanings of these derivatives in a physical context.

In studying functions of one-variable in calculus, one is introduced to the derivative, $\frac{df}{dx}$. The derivative has several meanings. The standard mathematical meaning is that the derivative gives the slope of the graph of $f(x)$ at x. The derivative also tells us how rapidly $f(x)$ varies when x is changed by dx. Recall that dx is called a differential. We can think of the differential dx as an infinitesimal increment in x. Then, changing x by an amount dx results in a change in $f(x)$ by

$$df = \frac{df}{dx} dx.$$

We can extend this idea to functions of several variables. Consider the temperature $T(x, y, z)$ at a point in space. The change in temperature depends on the direction in which one moves in space. Extending the above relation between differentials of the dependent and independent variables, we have

$$dT = \frac{\partial T}{\partial x} dx + \frac{\partial T}{\partial y} dy + \frac{\partial T}{\partial z} dz. \tag{9.38}$$

Note that if we only changed x, keeping y and z fixed, then we recover the form $dT = \frac{dT}{dx} dx$.

Introducing the vectors

$$d\mathbf{r} = dx\mathbf{i} + dy\mathbf{j} + dz\mathbf{k}, \tag{9.39}$$

The gradient of a function,

$$\nabla T = \frac{\partial T}{\partial x}\mathbf{i} + \frac{\partial T}{\partial y}\mathbf{j} + \frac{\partial T}{\partial z}\mathbf{k},.$$

$$\nabla T \equiv \frac{\partial T}{\partial x}\mathbf{i} + \frac{\partial T}{\partial y}\mathbf{j} + \frac{\partial T}{\partial z}\mathbf{k}, \tag{9.40}$$

we can write Equation (9.38) as

$$dT = \nabla T \cdot d\mathbf{r} \tag{9.41}$$

Equation (9.40) defines the gradient of a scalar function, T. Equation (9.41) gives the change in T as one moves in the direction $d\mathbf{r}$.

Using the definition of the dot product, we also have

$$dT = |\nabla T||d\mathbf{r}| \cos\theta.$$

Note that by fixing $|d\mathbf{r}|$ and varying θ, the maximum value of dT is obtained when $\cos\theta = 1$. Thus, the maximum value of dT is in the direction of the gradient. Similarly, because $\cos\pi = -1$, the minimum value of dT is in a direction 180° from the gradient.

The greatest change in a function is in the direction of its gradient.

Example 9.11. *Plot the gradient field for the saddle* $f(x, y) = 2x^2 - y^2$.

The gradient, ∇f, *provides a vector at each point* (x, y). *Thus, the gradient generates a vector field called the gradient field. Because it points in the direction of greatest increase in a function, we could look at the level curves of the function,* $f(x, y) = c$ *and see graphically if the gradients point in a direction of increasing contours. These plots are shown in Figure 9.13.*

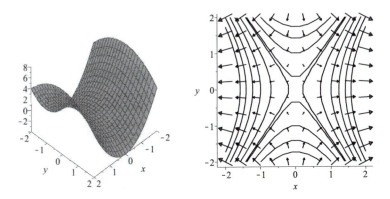

Figure 9.13: (a) Plot of the saddle, $f(x, y) = 2x^2 - y^2$. (b) Plot of the contours and the gradient field for $f(x, y) = 2x^2 - y^2$.

Notice that the gradient at each point is perpendicular to the contour passing through that point. Also, the gradient vectors point in the direction of largest increase in $f(x, y)$ *at that point. Compare the two plots for different regions and verify that this is true.*

Example 9.12. *Let* $f(x, y, z) = x^2 y + z e^{xy}$. *Compute* ∇f.

$$\nabla f = \frac{\partial f}{\partial x}\mathbf{i} + \frac{\partial f}{\partial y}\mathbf{j} + \frac{\partial f}{\partial z}\mathbf{k},$$
$$= (2xy + yze^{xy})\mathbf{i} + (x^2 + xze^{xy})\mathbf{j} + e^{xy}\mathbf{k}. \tag{9.42}$$

Example 9.13. *Find the direction of greatest change of* $f(x, y, z) = x^2 y + z e^{xy}$ *at* $(x, y, z) = (1, 1, 0)$. *What is the rate of change of* $f(x, y, z)$ *in this direction?*

The direction of greatest change is in the direction of the gradient at this point. From the last example, we have $\nabla f(1, 1, 0) = 2\mathbf{i} + \mathbf{j} + e\mathbf{k}$. *The rate of change of* $f(x, y, z)$ *in this direction is given by* $|\nabla f(1, 1, 0)| = \sqrt{2^2 + 1^2 + e^2} = \sqrt{5 + e^2}$.

From this analysis, we see that the rate of change of a function, such as $T(x, y, z,)$, depends on the direction one heads away from a given point. So, if one moves an infinitesimal distance ds in some direction $d\mathbf{r}$, then how does T change with respect to s? Another way to ask this is to ask what is the directional derivative of T in direction $\mathbf{n}$? We define this directional derivative as

$$D_{\mathbf{n}}T = \frac{dT}{ds}. \tag{9.43}$$

The directional derivative of a function, $D_{\mathbf{n}}T = \frac{dT}{ds} = \nabla T \cdot \mathbf{n}$.

We can develop an operational definition of the directional derivative. From Equation (9.41) we have

$$\frac{dT}{ds} = \nabla T \cdot \frac{d\mathbf{r}}{ds}. \tag{9.44}$$

We note that

$$\frac{d\mathbf{r}}{ds} = \left(\frac{dx}{ds}\right)\mathbf{i} + \left(\frac{dy}{ds}\right)\mathbf{j} + \left(\frac{dz}{ds}\right)\mathbf{k}$$

and

$$\left|\frac{d\mathbf{r}}{ds}\right| = \sqrt{\left(\frac{dx}{ds}\right)^2 + \left(\frac{dy}{ds}\right)^2 + \left(\frac{dz}{ds}\right)^2} = 1,$$

since $ds^2 = dx^2 + dy^2 + dz^2$. Thus, $\frac{d\mathbf{r}}{ds}$ is a unit vector pointing in the direction of interest. Defining $\mathbf{n} = \frac{d\mathbf{r}}{ds}$, the directional derivative of $T(x,y,z)$ in direction $\mathbf{n}$ can be written as

$$D_\mathbf{n} T = \nabla T \cdot \mathbf{n}. \tag{9.45}$$

Example 9.14. *Let the temperature of a rectangular plate be given by $T(x,y) = 5.0 \sin \frac{3\pi x}{2} \sin \frac{\pi y}{2}$. Determine the directional derivative at $(x,y) = (\frac{1}{2}, 1)$ in the following directions: (a) $\mathbf{i}$, (b) $3\mathbf{i} + 4\mathbf{j}$.*

For part (a), we have

$$D_\mathbf{i} T = \nabla T \cdot \mathbf{i} = \frac{\partial T}{\partial x}.$$

So,

$$D_\mathbf{i} T \bigg|_{(\frac{1}{2},1)} = \frac{15}{2} \pi \cos \frac{3\pi}{4} \sin \frac{\pi}{2} = -\frac{15}{4} \pi \sqrt{2}.$$

For part (b), the direction given is not a unit vector, $|3\mathbf{i} + 4\mathbf{j}| = 5$. Dividing by the length of the vector, we obtain a unit normal vector, $\mathbf{n} = \frac{3}{5}\mathbf{i} + \frac{4}{5}\mathbf{j}$. The directional derivative can now be computed:

$$
\begin{aligned}
D_\mathbf{n} T &= \nabla T \cdot \mathbf{n} \\
&= \frac{3}{5}\frac{\partial T}{\partial x} + \frac{4}{5}\frac{\partial T}{\partial y} \\
&= \frac{9\pi}{2} \cos \frac{3\pi x}{2} \sin \frac{\pi y}{2} + 2\pi \sin \frac{3\pi x}{2} \cos \frac{\pi y}{2}.
\end{aligned} \tag{9.46}
$$

Evaluating this result at $(x,y) = (\frac{1}{2}, 1)$, we have

$$D_\mathbf{n} T \bigg|_{(\frac{1}{2},1)} = \frac{9\pi}{2} \cos \frac{3\pi}{4} \sin \frac{\pi}{2} + 2\pi \sin \frac{3\pi}{4} \cos \frac{\pi}{2} = -\frac{9}{4}\pi\sqrt{2}.$$

From part (a) of the example, we see that

$$D_\mathbf{i} T = \frac{\partial T}{\partial x}.$$

Similarly, we have

$$D_\mathbf{j} T = \frac{\partial T}{\partial y} \quad \text{and} \quad D_\mathbf{k} T = \frac{\partial T}{\partial z}.$$

These give the rates of change along the coordinate directions and result in simple partial derivatives with respect to an independent variable keeping the other variables fixed. Thus, the directional derivative is in some sense a generalization of the partial derivative.

Gradient operator, ∇.

We can write the gradient in the form

$$\nabla T = \left(\frac{\partial}{\partial x}\mathbf{i} + \frac{\partial}{\partial y}\mathbf{j} + \frac{\partial}{\partial z}\mathbf{k}\right) T. \tag{9.47}$$

Thus, we see that the gradient can be viewed as an operator acting on T. The operator,

$$\nabla = \frac{\partial}{\partial x}\mathbf{i} + \frac{\partial}{\partial y}\mathbf{j} + \frac{\partial}{\partial z}\mathbf{k},$$

is called the *del*, or gradient, operator. It is a differential vector operator. It can act on scalar functions to produce a vector field. Recall that if the gravitational potential is given by $\Phi(\mathbf{r})$, then the gravitational force is found as $\mathbf{F} = -\nabla\Phi$.

We can also allow the del operator to act on vector fields. Recall that a vector field is simply a vector valued function. For example, a force field is a function defined at points in space indicating the force that would act on a mass placed at that location. We could denote it as $\mathbf{F}(x, y, z)$. Again, think about the gravitational force above. The force acting on a mass in the Earth's gravitational field is given by a vector field. At each point in space, one would see that the force vector takes on different magnitudes and directions, depending upon the location of the mass in space.

How can we combine the (vector) del operator and a vector field? Well, we could "multiply" them. We could either compute the dot product, $\nabla \cdot \mathbf{F}$, or we could compute the cross product $\nabla \times \mathbf{F}$. The first expression is called the divergence of the vector field, and the second is called the curl of the vector field. These are typically encountered in a third semester calculus course. In some texts they are denoted by div $\mathbf{F}$ and curl $\mathbf{F}$.

The divergence is computed the same as any other dot product. Writing the vector field in component form,

The divergence, div $\mathbf{F} = \nabla \cdot \mathbf{F}$.

$$\mathbf{F} = F_1(x, y, z)\mathbf{i} + F_2(x, y, z)\mathbf{j} + F_3(x, y, z)\mathbf{k},$$

we find the divergence is simply given as

$$
\begin{aligned}
\nabla \cdot \mathbf{F} &= \left(\frac{\partial}{\partial x}\mathbf{i} + \frac{\partial}{\partial y}\mathbf{j} + \frac{\partial}{\partial z}\mathbf{k}\right) \cdot (F_1\mathbf{i} + F_2\mathbf{j} + F_3\mathbf{k}) \\
&= \frac{\partial F_1}{\partial x} + \frac{\partial F_2}{\partial y} + \frac{\partial F_3}{\partial z}
\end{aligned}
\tag{9.48}
$$

Similarly, we can compute the curl of $\mathbf{F}$. Using the determinant form, we have

The curl $\mathbf{F} = \nabla \times \mathbf{F}$.

$$
\begin{aligned}
\nabla \times \mathbf{F} &= \left(\frac{\partial}{\partial x}\mathbf{i} + \frac{\partial}{\partial y}\mathbf{j} + \frac{\partial}{\partial z}\mathbf{k}\right) \times (F_1\mathbf{i} + F_2\mathbf{j} + F_3\mathbf{k}) \\
&= \begin{vmatrix} \mathbf{i} & \mathbf{j} & \mathbf{k} \\ \frac{\partial}{\partial x} & \frac{\partial}{\partial y} & \frac{\partial}{\partial z} \\ F_1 & F_2 & F_3 \end{vmatrix} \\
&= \left(\frac{\partial F_3}{\partial y} - \frac{\partial F_2}{\partial z}\right)\mathbf{i} + \left(\frac{\partial F_1}{\partial z} - \frac{\partial F_3}{\partial x}\right)\mathbf{j} + \left(\frac{\partial F_2}{\partial x} - \frac{\partial F_1}{\partial y}\right)\mathbf{k}.
\end{aligned}
\tag{9.49}
$$

Example 9.15. *Write the curl definition using the Levi-Civita symbol.*

We can begin with the last result and rewrite it using the standard basis, $\mathbf{e}_k$, $k = 1, 2, 3$, and coordinates $(x_1, x_2, x_3) = (x, y, z)$. Substituting these notations, *we have*

$$\nabla \times \mathbf{F} = \left(\frac{\partial F_3}{\partial x_2} - \frac{\partial F_2}{\partial x_3}\right)\mathbf{e}_1 + \left(\frac{\partial F_1}{\partial x_3} - \frac{\partial F_3}{\partial x_1}\right)\mathbf{e}_2 + \left(\frac{\partial F_2}{\partial x_1} - \frac{\partial F_1}{\partial x_2}\right)\mathbf{e}_3.$$

We can absorb the signs by inserting the appropriate Levi-Civita symbols,

$$\begin{aligned}\nabla \times \mathbf{F} &= \left(\epsilon_{231}\frac{\partial F_3}{\partial x_2} + \epsilon_{321}\frac{\partial F_2}{\partial x_3}\right)\mathbf{e}_1 + \left(\epsilon_{312}\frac{\partial F_1}{\partial x_3} + \epsilon_{132}\frac{\partial F_3}{\partial x_1}\right)\mathbf{e}_2 \\ &\quad + \left(\epsilon_{123}\frac{\partial F_2}{\partial x_1} + \epsilon_{213}\frac{\partial F_1}{\partial x_2}\right)\mathbf{e}_3.\end{aligned} \qquad (9.50)$$

Examination of this result yields

$$\nabla \times \mathbf{F} = \epsilon_{ijk}\frac{\partial F_j}{\partial x_i}\mathbf{e}_k,$$

where we have employed the Einstein summation convention.

Example 9.16. *Compute the divergence and curl of the vector field:* $\mathbf{F} = x\mathbf{i} + y\mathbf{j}$. *This vector field is shown in Figure 9.14.*

$$\nabla \cdot \mathbf{F} = \frac{\partial x}{\partial x} + \frac{\partial y}{\partial y} = 2.$$

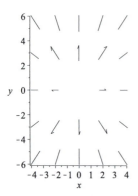

Figure 9.14: A plot of the vector field $\mathbf{F} = x\mathbf{i} + y\mathbf{j}$.

$$\begin{aligned}\nabla \times \mathbf{F} &= \begin{vmatrix} \mathbf{i} & \mathbf{j} & \mathbf{k} \\ \frac{\partial}{\partial x} & \frac{\partial}{\partial y} & \frac{\partial}{\partial z} \\ x & y & 0 \end{vmatrix} \\ &= \left(\frac{\partial y}{\partial x} - \frac{\partial x}{\partial z}\right)\mathbf{k} = 0.\end{aligned} \qquad (9.51)$$

Example 9.17. *Compute the divergence and curl of the vector field:* $\mathbf{F} = y\mathbf{i} - x\mathbf{j}$. *This vector field is shown in Figure 9.15.*

$$\nabla \cdot \mathbf{F} = \frac{\partial y}{\partial x} - \frac{\partial x}{\partial y} = 0.$$

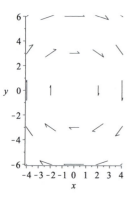

Figure 9.15: A plot of the vector field $\mathbf{F} = y\mathbf{i} - x\mathbf{j}$.

$$\begin{aligned}\nabla \times \mathbf{F} &= \begin{vmatrix} \mathbf{i} & \mathbf{j} & \mathbf{k} \\ \frac{\partial}{\partial x} & \frac{\partial}{\partial y} & \frac{\partial}{\partial z} \\ y & -x & 0 \end{vmatrix} \\ &= \left(-\frac{\partial x}{\partial x} - \frac{\partial y}{\partial y}\right)\mathbf{k} = -2.\end{aligned} \qquad (9.52)$$

These operations also have interpretations. The divergence measures how much the vector field $\mathbf{F}$ spreads from a point. When the divergence of a vector field is nonzero around a point, that is an indication that there is a source (div $\mathbf{F} > 0$) or a sink (div $\mathbf{F} < 0$). For example, $\nabla \cdot \mathbf{E} = \frac{\rho}{\epsilon_0}$ indicates that there are sources contributing to the electric i£¡led. For a single charge, the field lines are radially pointing toward (sink) or away from (source) the charge. A field in which the divergence is zero is called divergenceless, or solenoidal.

The curl is an indication of a rotational field. It is a measure of how much a field curls around a point. Consider the flow of a stream. The velocity of each element of fluid can be represented by a velocity field. If the curl of the field is nonzero, then when we drop a leaf into the stream, we will see it begin to rotate about some point. A field that has zero curl is called irrotational.

The last common differential operator is the Laplace operator. This is the common second-derivative operator, the divergence of the gradient,

$$\nabla^2 f = \nabla \cdot \nabla f.$$

It is easily computed as

$$
\begin{aligned}
\nabla^2 f &= \nabla \cdot \nabla f \\
&= \nabla \cdot \left(\frac{\partial f}{\partial x}\mathbf{i} + \frac{\partial f}{\partial y}\mathbf{j} + \frac{\partial f}{\partial z}\mathbf{k} \right) \\
&= \frac{\partial^2 f}{\partial x^2} + \frac{\partial^2 f}{\partial y^2} + \frac{\partial^2 f}{\partial z^2}.
\end{aligned}
\tag{9.53}
$$

The Laplace operator,

$$\nabla^2 f = \frac{\partial^2 f}{\partial x^2} + \frac{\partial^2 f}{\partial y^2} + \frac{\partial^2 f}{\partial z^2}.$$

Another notation for the Laplace operator is Δ. Thus, $\nabla^2 f = 0$ can be written as $\Delta f = 0$.

Example 9.18. *Compute $\nabla^2 f$ for $f(x,y,z) = x^2 + xy^2 + yz^3$.*

A simple computation gives

$$
\begin{aligned}
\nabla^2 f &= \left(\frac{\partial^2}{\partial x^2} + \frac{\partial^2}{\partial y^2} + \frac{\partial^2}{\partial z^2} \right)(x^2 + xy^2 + yz^3) \\
&= 2 + 2x + 6yz.
\end{aligned}
\tag{9.54}
$$

9.1.4 *Integral Theorems*

MAXWELL'S EQUATIONS ARE GIVEN LATER IN THIS CHAPTER in differential form and only describe electric and magnetic fields locally. At times we would like to also provide global information, or information over a finite region. In this case, one can derive various integral theorems. These are the finale in a three semester calculus sequence. These integral theorems are important and useful in deriving local conservation laws.

These theorems are all different versions of a generalized Fundamental Theorem of Calculus:

(a) $\int_a^b \frac{df}{dx}\, dx = f(b) - f(a)$, The Fundamental Theorem of Calculus in one dimension.

(b) $\int_{\mathbf{a}}^{\mathbf{b}} \nabla T \cdot d\mathbf{r} = T(\mathbf{b}) - T(\mathbf{a})$, The Fundamental Theorem of Calculus for Vector Fields.

(c) $\oint_C (P\, dx + Q\, dy) = \int_D \left(\frac{\partial Q}{\partial x} - \frac{\partial P}{\partial y} \right) dx\, dy$, Green's Theorem in the Plane.

(d) $\int_V \nabla \cdot \mathbf{F}\, dV = \oint_S \mathbf{F} \cdot d\mathbf{a}$, Gauss' Divergence Theorem.

(e) $\int_S (\nabla \times \mathbf{F}) \cdot d\mathbf{a} = \oint_C \mathbf{F} \cdot d\mathbf{r}$, Stokes' Theorem.

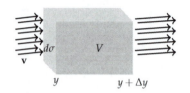

Figure 9.16: Fluid flowing through an infinitesimal rectangular box.

The connections between these integral theorems are probably more easily seen by thinking in terms of fluids. Consider a fluid with mass density $\rho(x, y, z)$ and fluid velocity $\mathbf{v}(x, y, z, t)$. We define $\mathbf{Q} = \rho \mathbf{v}$ as the mass flow rate. [Note the units are $\text{kg/m}^2/\text{s}$ indicating the flow of mass per area per time.]

Now consider the fluid flowing through an infinitesimal rectangular box as depicted in Figure 9.16. Let the fluid flow into the left face and out the right face. The rate at which the fluid mass flows in kg/s through a face can be represented by $\mathbf{Q} \cdot d\sigma$, where $d\sigma = \mathbf{n} d\sigma$ represents the differential area element normal to the face. The rate of flow across the left face is

$$\mathbf{Q} \cdot d\sigma = -Q_y \, dxdz \Big|_y ,$$

where Q_y is the y-component of $\mathbf{Q}$, and that flowing across the right face is

$$\mathbf{Q} \cdot d\sigma = Q_y \, dxdz \Big|_{y+dy}.$$

The net flow rate across the infinitesimal element is the sum of these flows:

$$Q_y dxdz \Big|_{y+dy} - Q_y dxdz \Big|_y = \frac{\partial Q_y}{\partial y} \, dxdydz.$$

Here we have used the differential relation

$$dQ_y = Q_y(y + dy) - Q_y(y) = \frac{\partial Q_y}{\partial y} \, dy.$$

A similar computation can be done for the other faces, leading to the result that the total rate of flow is $\nabla \cdot \mathbf{Q} \, d\tau$, where $d\tau = dxdydz$ is the volume element. So, the rate of flow per volume through the volume element gives

Conservation of mass equation,

$$\frac{\partial \rho}{\partial t} + \nabla \cdot \mathbf{Q} = 0.$$

$$\nabla \cdot \mathbf{Q} = -\frac{\partial \rho}{\partial t}. \tag{9.55}$$

Note that if more fluid is flowing out the right face than is flowing into the left face, then the amount of fluid inside the region will decrease. That is why the right-hand side of this equation has the negative sign.

If the fluid is incompressible, that is, $\rho = \text{const.}$, then $\nabla \cdot \mathbf{Q} = 0$, which implies $\nabla \cdot \mathbf{v} = 0$ assuming there are no sinks or sources. If there were a sink in the rectangular box, then there would be a loss of fluid not accounted for. Likewise, if a hose were inserted and fluid were supplied, then one would have a source.

If there are sinks, or sources, then the net mass due to these would contribute to an overall flow through the surrounding surface. This is captured by the equation

Gauss' Divergence Theorem

$$\underbrace{\int_V \nabla \cdot \mathbf{Q} \, d\tau}_{\text{Net mass due to sink/sources}} = \underbrace{\oint_S \mathbf{Q} \cdot \mathbf{n} \, d\sigma}_{\text{Net flow outward from } S}. \tag{9.56}$$

Dividing out the constant mass density, as $\mathbf{Q} = \rho \mathbf{v}$, this becomes

$$\int_V \nabla \cdot \mathbf{v} \, dV = \oint_S \mathbf{v} \cdot \mathbf{n} \, d\sigma. \qquad (9.57)$$

The surface integral on the right-hand side is called the flux of the vector field through surface S. This is nothing other than Gauss' Divergence Theorem.[2]

Example 9.19. *Use the Divergence Theorem to compute*

$$\int_S (x^2 dydz + y^2 dzdx + z^2 dxdy)$$

for S the surface of the unit cube, $[0,1] \times [0,1] \times [0,1]$.

We first compute the divergence of the vector $\mathbf{v} = x^2\mathbf{i} + y^2\mathbf{j} + z^2\mathbf{k}$, which we obtained from the coefficients in the given integral. Then

$$\nabla \cdot \mathbf{v} = \frac{\partial x^2}{\partial x} + \frac{\partial y^2}{\partial y} + \frac{\partial z^2}{\partial z} = 2(x + y + z).$$

Then,

$$\begin{aligned}
\int_S (x^2 dydz + y^2 dzdx + z^2 dxdy) &= \int_V 2(x + y + z) \, dV \\
&= 2\int_0^1 \int_0^1 \int_0^1 (x + y + z) \, dxdydz \\
&= 2\int_0^1 \int_0^1 (\frac{1}{2} + y + z) \, dydz \\
&= 2\int_0^1 (\frac{1}{2} + \frac{1}{2} + z) \, dz \\
&= 2(1 + \frac{1}{2}) = 3. \qquad (9.58)
\end{aligned}$$

The unit normal can be written in terms of the direction cosines,

$$\mathbf{n} = \cos\alpha\mathbf{i} + \cos\beta\mathbf{j} + \cos\gamma\mathbf{k},$$

where the angles are the directions between $\mathbf{n}$ and the coordinate axes. For example, $\mathbf{n} \cdot \mathbf{i} = \cos\alpha$. For vector $\mathbf{v} = v_1\mathbf{i} + v_2\mathbf{j} + v_3\mathbf{k}$, we have

$$\begin{aligned}
\int_S \mathbf{v} \cdot \mathbf{n} \, d\sigma &= \int_S (v_1 \cos\alpha + v_2 \cos\beta + v_3 \cos\gamma) \, d\sigma \\
&= \int_S (u_1 dydz + u_2 dzdx + u_3 dxdy). \qquad (9.59)
\end{aligned}$$

The other integral theorems are just a variation of the divergence theorem. For example, a two-dimensional version of this is obtained by considering a simply connected region, D, bounded by a simple closed curve, C. One could think of the laminar flow of a thin sheet of fluid. Then the total mass in contained in D and the net mass would be related to the next flow through the boundary, C. The integral theorem for this situation is given as

$$\int_D \nabla \cdot \mathbf{v} \, dA = \oint_C \mathbf{v} \cdot \mathbf{n} \, ds. \qquad (9.60)$$

The tangent vector to the curve at point $\mathbf{r}$ on the curve C is

$$\frac{d\mathbf{r}}{ds} = \frac{dx}{ds}\mathbf{i} + \frac{dy}{ds}\mathbf{j}.$$

[2] We should note that the Divergence Theorem holds provided $\mathbf{v}$ is a continuous vector field and has continuous partial derivatives in a domain containing V. Also, $\mathbf{n}$ is the outward normal to the surface S, as shown in Figure 9.17.

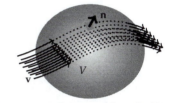

Figure 9.17: Region used in the Divergence Theorem.

Therefore, the outward normal at that point is given by

$$\mathbf{n} = \frac{dy}{ds}\mathbf{i} - \frac{dx}{ds}\mathbf{j}.$$

Letting $\mathbf{v} = Q(x,y)\mathbf{i} - P(x,y)\mathbf{j}$, the two dimensional version of the Divergence Theorem becomes

Green's Theorem in the Plane, which is a special case of Stokes' Theorem restricted to two dimensions.

$$\oint_C (P\,dx + Q\,dy) = \int_D \left(\frac{\partial Q}{\partial x} - \frac{\partial P}{\partial y}\right) dx dy. \qquad (9.61)$$

This is just Green's Theorem in the Plane.

Example 9.20. *Show that the area of the region, S, bounded by a closed curve, C, is given by*

$$A(S) = \frac{1}{2}\oint_C x\,dy - y\,dx. \qquad (9.62)$$

This is a simple application of Green's Theorem in the Plane.

$$\frac{1}{2}\oint_C x\,dy - y\,dx = \frac{1}{2}\int_S \left(\frac{\partial x}{\partial x} - \frac{\partial(-y)}{\partial y}\right) dx dy$$

$$= \int_S dx dy = A(S). \qquad (9.63)$$

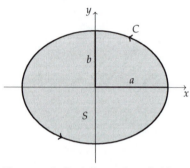

Figure 9.18: Typical area bounded by a standard ellipse C for use with application of Green's Theorem in the Plane.

Example 9.21. *Show that the area of the ellipse in Figure 9.18 with semimajor axis a and semiminor axis b is πab.*

Using the last result, we have that the area is given by

$$A(S) = \frac{1}{2}\oint_C x\,dy - y\,dx.$$

Let the equation for this ellipse be given by the standard form

$$\frac{x^2}{a^2} + \frac{y^2}{b^2} = 1.$$

In order to carry out the integration, we need the parametrization of the ellipse, $x = a\cos\theta$, $y = b\sin\theta$, for $\theta \in [0, 2\pi]$. Then, $dx = -a\sin\theta\,d\theta$, $y = b\cos\theta\,d\theta$, for $\theta \in [0, 2\pi]$. Then,

$$\begin{aligned} A(S) &= \frac{1}{2}\oint_C x\,dy - y\,dx \\ &= \frac{1}{2}\int_0^{2\pi} \left(ab\cos^2\theta + ab\sin^2\theta\right) d\theta \\ &= \frac{ab}{2}\int_0^{2\pi} d\theta = \pi ab. \end{aligned} \qquad (9.64)$$

Example 9.22. *Evaluate $\oint_C (e^x - 3y)\,dx + (e^y + 6x)\,dy$ for C the boundary of the ellipse $x^2 + 4y^2 = 4$.*

Green's Theorem in the Plane gives

$$\begin{aligned} \oint_C (e^x - 3y)\,dx + (e^y + 6x)\,dy &= \int_S \left(\frac{\partial}{\partial x}(e^y + 6x) - \frac{\partial}{\partial y}(e^x - 3y)\right) dx dy \\ &= \int_S (6 + 3)\,dx dy \\ &= 9\int_S dx dy. \end{aligned} \qquad (9.65)$$

The integral is the area of the ellipse $x^2 + 4y^2 = 4$. From the previous example, the area of an ellipse is πab. For this ellipse, $a = 2$ and $b = 1$. Therefore,

$$\oint_C (e^x - 3y)\, dx + (e^y + 6x)\, dy = 18\pi.$$

We can obtain Stokes' Theorem by applying the Divergence Theorem to the vector $\mathbf{v} \times \mathbf{n}$.

$$\int_V \nabla \cdot (\mathbf{v} \times \mathbf{n})\, d\tau = \oint_S \mathbf{n}_s \cdot (\mathbf{v} \times \mathbf{n})\, d\sigma. \qquad (9.66)$$

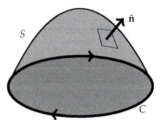

Figure 9.19: The surface S bounded by C as used in Stokes' Theorem.

Here, $\mathbf{n}_s = \mathbf{u} \times \mathbf{n}$ is a unit vector, where $\mathbf{u}$ is a unit vector tangent to the curve C, and $\mathbf{n}$ is a unit vector normal to the domain D. Noting that $(\mathbf{u} \times \mathbf{n}) \times (\mathbf{v} \times \mathbf{n}) = \mathbf{v} \cdot \mathbf{u}$ and $\nabla \cdot (\mathbf{v} \times \mathbf{n}) = \mathbf{n} \cdot \nabla \times \mathbf{v}$, then

$$\int_0^h \left(\int_D \mathbf{n} \cdot \nabla \times \mathbf{v}\, d\sigma \right) dh = \int_0^h \left(\oint_C \mathbf{v} \cdot \mathbf{u}\, ds \right) dh. \qquad (9.67)$$

Because h is arbitrary, we obtain Stokes' Theorem:

Stokes' Theorem.

$$\boxed{\int_D \mathbf{n} \cdot \nabla \times \mathbf{v}\, d\sigma = \oint_C \mathbf{v} \cdot \mathbf{u}\, ds. \qquad (9.68)}$$

Example 9.23. *Evaluate $\oint_C (z\, dx + x\, dy + y\, dz)$ for C the boundary of the triangle with vertices $(1,0,0)$, $(0,1,0)$, $(0,0,1)$ using Stokes' Theorem.*

We first identify the vector $\mathbf{v} = z\mathbf{i} + x\mathbf{j} + y\mathbf{k}$. Then we compute the curl,

$$\nabla \times \mathbf{v} = \begin{vmatrix} \mathbf{i} & \mathbf{j} & \mathbf{k} \\ \frac{\partial}{\partial x} & \frac{\partial}{\partial y} & \frac{\partial}{\partial y} \\ z & x & y \end{vmatrix}$$
$$= \mathbf{i} + \mathbf{j} + \mathbf{k}. \qquad (9.69)$$

Stokes' Theorem then gives

$$\oint_C (z\, dx + x\, dy + y\, dz) = \int_D \mathbf{n} \cdot (\mathbf{i} + \mathbf{j} + \mathbf{k})\, d\sigma,$$

where $\mathbf{n}$ is the outward normal to the surface of the triangle as shown in Figure 9.20. For a surface defined by $\phi(x, y, z) =$ const, the normal is in the direction of $\nabla \phi$. In this case, the triangle lives in the plane $x + y + z = 1$. Thus, $\phi(x, y, z) = x + y + z$ and $\nabla \phi = \mathbf{i} + \mathbf{j} + \mathbf{k}$. Thus,

For a surface defined by $\phi(x, y, z) =$ constant, the normal is in the direction of $\nabla \phi$.

$$\oint_C (z\, dx + x\, dy + y\, dz) = 3 \int_D d\sigma.$$

The remaining integral is just the area of the triangle. We can determine this area as follows. Imagine the vectors $\mathbf{a}$ and $\mathbf{b}$ pointing from $(1,0,0)$ to $(0,1,0)$ and from $(1,0,0)$ to $(0,0,1)$, respectively. So, $\mathbf{a} = \mathbf{j} - \mathbf{i}$ and $\mathbf{b} = \mathbf{k} - \mathbf{i}$.

These vectors are the sides of a parallelogram whose area is twice that of the triangle. The area of the parallelogram is given by $|\mathbf{a} \times \mathbf{b}|$. The area of the triangle is thus

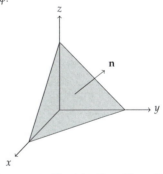

Figure 9.20: The triangle with vertices $(1,0,0), (0,1,0), (0,0,1)$.

$$\begin{aligned} \int_D d\sigma &= \frac{1}{2} |\mathbf{a} \times \mathbf{b}| \\ &= \frac{1}{2} |(\mathbf{j} - \mathbf{i}) \times (\mathbf{k} - \mathbf{i})| \\ &= \frac{1}{2} |\mathbf{i} + \mathbf{j} + \mathbf{k}| = \frac{3}{2}. \qquad (9.70) \end{aligned}$$

Finally, we have

$$\oint_C (z\,dx + x\,dy + y\,dz) = \frac{9}{2}.$$

9.1.5   *Vector Identities*

IN THIS SECTION WE WILL LIST some common vector identities and show how to prove a few of them. We will introduce two triple products and list first derivative and second derivative identities. These are useful in reducing some equations to simpler forms.

Proving these identities can be straightforward, although sometimes tedious in the more complicated cases. You should try to prove these yourself. Sometimes it is useful to write out the components on each side of the identity and see how one can fill in the needed arguments, which would provide the proofs. We will provide a couple of examples of this process.

1. Triple Products:

   (a) $\mathbf{A} \cdot (\mathbf{B} \times \mathbf{C}) = \mathbf{B} \cdot (\mathbf{C} \times \mathbf{A}) = \mathbf{C} \cdot (\mathbf{A} \times \mathbf{B})$

   (b) $\mathbf{A} \times (\mathbf{B} \times \mathbf{C}) = \mathbf{B}(\mathbf{A} \cdot \mathbf{C}) - \mathbf{C}(\mathbf{A} \cdot \mathbf{B})$

2. First Derivatives:

   (a) $\nabla(fg) = f\nabla g + g\nabla f$

   (b) $\nabla(\mathbf{A} \cdot \mathbf{B}) = \mathbf{A} \times (\nabla \times \mathbf{B}) + \mathbf{B} \times (\nabla \times \mathbf{A}) + (\mathbf{A} \cdot \nabla)\mathbf{B} + (\mathbf{B} \cdot \nabla)\mathbf{A}$

   (c) $\nabla \cdot (f\mathbf{A}) = f\nabla \cdot \mathbf{A} + \mathbf{A} \cdot \nabla f$

   (d) $\nabla \cdot (\mathbf{A} \times \mathbf{B}) = \mathbf{B} \cdot (\nabla \times \mathbf{A}) - \mathbf{A} \cdot (\nabla \times \mathbf{B})$

   (e) $\nabla \times (f\mathbf{A}) = f\nabla \times \mathbf{A} - \mathbf{A} \times \nabla f$

   (f) $\nabla \times (\mathbf{A} \times \mathbf{B}) = (\mathbf{B} \cdot \nabla)\mathbf{A} - (\mathbf{A} \cdot \nabla)\mathbf{B} + \mathbf{A}(\nabla \cdot \mathbf{B}) - \mathbf{B}(\nabla \cdot \mathbf{A})$

3. Second Derivatives:

   div curl $= 0.$

   curl grad$= \mathbf{0}.$

   (a) $\nabla \cdot (\nabla \times \mathbf{A}) = 0$

   (b) $\nabla \times (\nabla f) = 0$

   (c) $\nabla \cdot (\nabla f \times \nabla g) = 0$

   (d) $\nabla^2(fg) = f\nabla^2 g + 2\nabla f \cdot \nabla g + g\nabla^2 f$

   (e) $\nabla \cdot (f\nabla g - g\nabla f) = f\nabla^2 g - g\nabla^2 f$

   (f) $\nabla \times (\nabla \times \mathbf{A}) = \nabla(\nabla \cdot \mathbf{A}) - \nabla^2\mathbf{A}$

Example 9.24. *Prove* $\mathbf{A} \cdot (\mathbf{B} \times \mathbf{C}) = \mathbf{B} \cdot (\mathbf{C} \times \mathbf{A})$.

In such problems, one can write out the components on both sides of the identity. Using the determinant form of the triple scalar, the left-hand side becomes

$$\mathbf{A} \cdot (\mathbf{B} \times \mathbf{C}) = \begin{vmatrix} A_1 & A_2 & A_3 \\ B_1 & B_2 & B_3 \\ C_1 & C_2 & C_3 \end{vmatrix}$$

$$= A_1(B_2C_3 - B_3C_2) - A_2(B_1C_3 - B_3C_1) + A_3(B_1C_2 - B_2C_1).$$

$$(9.71)$$

Similarly, the right-hand side is given as

$$
\mathbf{B} \cdot (\mathbf{C} \times \mathbf{A}) = \begin{vmatrix} B_1 & B_2 & B_3 \\ C_1 & C_2 & C_3 \\ A_1 & A_2 & A_3 \end{vmatrix}
$$
$$
= B_1(C_2A_3 - C_3A_2) - B_2(C_1A_3 - C_3A_1) + B_3(C_1A_2 - C_2A_1).
$$
$$(9.72)$$

We can rearrange this result by separating out the components of $\mathbf{A}$.

$$
B_1(C_2A_3 - C_3A_2) - B_2(C_1A_3 - C_3A_1) + B_3(C_1A_2 - C_2A_1)
$$
$$
= A_1(B_2C_3 - B_3C_2) + A_2(B_3C_1 - B_1C_3) + A_3(B_1C_2 - B_2C_1).
$$
$$(9.73)$$

Upon inspection, we see that we obtain the same result as we had for the left-hand side.

This problem can also be solved using the completely antisymmetric symbol, ϵ_{ijk}. *Recall that the scalar triple product is given by*

$$
\mathbf{A} \cdot (\mathbf{B} \times \mathbf{C}) = \epsilon_{ijk} A_i B_j C_k.
$$

Because $\epsilon_{ijk} = \epsilon_{jki}$, *we have*

$$
\epsilon_{ijk} A_i B_j C_k = \epsilon_{jki} A_i B_j C_k = \epsilon_{jki} B_j C_k A_i.
$$

But,

$$
\epsilon_{jki} B_j C_k A_i = \mathbf{B} \cdot (\mathbf{C} \times \mathbf{A}).
$$

We have once again proven the identity. However, it took a little less work and an understanding of the antisymmetric symbol. Furthermore, you should note that this identity was proven earlier in the chapter.

Example 9.25. *Prove* $\nabla(fg) = f\nabla g + g\nabla f$. *In this problem, we compute the gradient of* fg. *Then we note that each derivative is the derivative of a product and apply the Product Rule. Carefully writing out the terms, we obtain the desired result:*

$$
\begin{aligned}
\nabla(fg) &= \frac{\partial fg}{\partial x}\mathbf{i} + \frac{\partial fg}{\partial y}\mathbf{j} + \frac{\partial fg}{\partial z}\mathbf{k} \\
&= \left(\frac{\partial f}{\partial x}\mathbf{i} + \frac{\partial f}{\partial y}\mathbf{j} + \frac{\partial f}{\partial z}\mathbf{k}\right)g + f\left(\frac{\partial g}{\partial x}\mathbf{i} + \frac{\partial g}{\partial y}\mathbf{j} + \frac{\partial g}{\partial z}\mathbf{k}\right) \\
&= f\nabla g + g\nabla f.
\end{aligned}
$$
$$(9.74)$$

9.1.6 Kepler Problem

A MAJOR TEST OF NEWTON's *Principia* was in the prediction of the return of Halley's comet in 1758. It was Edmond Halley (1656–1742) who had used Newton's Laws to take into account the effects of the larger planets on comet orbits to confirm that earlier records indicated that comets could periodically cross the night sky. It was Halley who had encouraged Newton to publish his proof that the planetary laws were the result of an inverse square law. This led to Newton's famous 1687 publication, *PhilosophiÃɇ Naturalis Principia Mathematica*, or Mathematical Principles of Natural Philosophy. Newton not only proved the law of gravitation, but introduced an axiomatic system based on his three laws of motion and derived Kepler's Laws of Planetary Motion from his theory. This was all done using geometric arguments, which were the standard of the day.

We can use the more modern rules of vector calculus to study the two-body problem, or what is referred to in classical dynamics as the Kepler problem. Recall from Section 2.10.5 that we reduced the two-body problem to a simple equation for the relative position between the bodies,

$$\ddot{\mathbf{r}} = -\mu \frac{\mathbf{r}}{r^3}, \tag{9.75}$$

where $\mu = G(m_1 + m_2)$ and $\mathbf{r}(t)$ is the position vector pointing from the central mass toward the mass rotating with respect to that mass.

Using the properties of derivatives of vector functions, we have that

$$\frac{d}{dt}(\mathbf{r} \times \dot{\mathbf{r}}) = \dot{\mathbf{r}} \times \dot{\mathbf{r}} + \mathbf{r} \times \ddot{\mathbf{r}}.$$

The first term vanishes because the cross product of a vector with itself is zero. The second term vanishes because for this problem, $\ddot{\mathbf{r}}$ is in the direction of $\mathbf{r}$. Therefore,

$$\frac{d}{dt}(\mathbf{r} \times \dot{\mathbf{r}}) = \mathbf{0}.$$

We conclude that $\mathbf{r} \times \dot{\mathbf{r}}$ is a constant vector. Because $\mathbf{r} \times \dot{\mathbf{r}} = \mathbf{r} \times \mathbf{v}$ is the angular momentum divided by the mass, we define the specific angular momentum as

$$\mathbf{h} \equiv \mathbf{r} \times \dot{\mathbf{r}}.$$

If $\mathbf{h} \neq \mathbf{0}$, then $\mathbf{h}$ is perpendicular to both $\mathbf{r}$ and $\dot{\mathbf{r}}$, because a cross product is perpendicular to each vector factor. Therefore, the orbit lies in a plane, the orbital plane.

We can choose the xy-plane to be the orbital plane and write

$$\mathbf{r} = r\cos\theta\mathbf{i} + r\sin\theta\mathbf{j}.$$

Then,

$$\dot{\mathbf{r}} = (\dot{r}\cos\theta - r\sin\theta\dot{\theta})\mathbf{i} + (\dot{r}\sin\theta + r\cos\theta\dot{\theta})\mathbf{j}.$$

Using these expressions, we obtain

$$
\begin{aligned}
\mathbf{h} &= \mathbf{r} \times \dot{\mathbf{r}} \\
&= \begin{vmatrix} \mathbf{i} & \mathbf{j} & \mathbf{k} \\ r\cos\theta & r\sin\theta & 0 \\ \dot{r}\cos\theta - r\sin\theta\dot{\theta} & \dot{r}\sin\theta + r\cos\theta\dot{\theta} & 0 \end{vmatrix} \\
&= \begin{vmatrix} r\cos\theta & r\sin\theta \\ \dot{r}\cos\theta - r\sin\theta\dot{\theta} & \dot{r}\sin\theta + r\cos\theta\dot{\theta} \end{vmatrix}\mathbf{k} \\
&= r^2\dot{\theta}\mathbf{k}.
\end{aligned} \tag{9.76}
$$

From the study of polar coordinates, we can relate the rate of change of the area swept out over time to $\dot{\theta}$. From Figure 9.21 we see that the swept area is approximately the same as the area of the shaded triangle. Because the area of a triangle is one half the base times the height,

$$\Delta A \approx \frac{1}{2} r(r \Delta \theta) = \frac{1}{2} r^2 \Delta \theta.$$

Dividing by Δt and letting $\Delta t \to 0$, we find

$$\frac{dA}{dt} = \frac{1}{2} r^2 \dot{\theta}.$$

Kepler's Second Law of Planetary Motion.

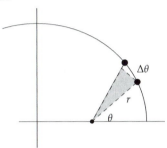

However, we have for $h = |\mathbf{h}|$ that

$$h = r^2 \dot{\theta} = 2 \frac{dA}{dt}.$$

Because h is constant, we have that $\frac{dA}{dt}$ is constant. We have just proven Kepler's Second Law of Planetary Motion. Namely, planets move in orbits such that they sweep out equal areas in equal times. In Figure 9.22 we show what this means. Area A is swept out in the same time as area B. Because the path for A is larger than for B, the speed needed to sweep out area A is less than that for area B.

Kepler's First Law of Planetary Motion says that planets move in ellipses. We can verify this by employing a few more operations with vector functions. We begin by differentiating $\dot{\mathbf{r}} \times \mathbf{h}$:

Figure 9.21: The area ΔA swept out by the mass is approximately the same as the area of the shaded triangle.

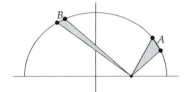

$$\begin{aligned} \frac{d}{dt}(\dot{\mathbf{r}} \times \mathbf{h}) &= \ddot{\mathbf{r}} \times \mathbf{h} \\ &= -\frac{\mu}{r^3}(\mathbf{r} \times \mathbf{h}) \\ &= -\frac{\mu}{r^3} \mathbf{r} \times (\mathbf{r} \times \dot{\mathbf{r}}). \end{aligned} \quad (9.77)$$

We can use the BAC-CAB Rule in order to simplify the triple cross product,

$$\mathbf{r} \times (\mathbf{r} \times \dot{\mathbf{r}}) = (\mathbf{r} \cdot \dot{\mathbf{r}})\mathbf{r} - (\mathbf{r} \cdot \mathbf{r})\dot{\mathbf{r}}.$$

Figure 9.22: Area A is swept out in the same time as area B. Therefore, the speed needed to sweep out area A is less than that for area B.

From the expressions for $\mathbf{r}$ and $\dot{\mathbf{r}}$, we have $\mathbf{r} \times \dot{\mathbf{r}} = r\dot{r}$. Thus, we have

$$\mathbf{r} \times (\mathbf{r} \times \dot{\mathbf{r}}) = r\dot{r}\mathbf{r} - r^2\dot{\mathbf{r}},$$

and this gives

$$\begin{aligned} \frac{d}{dt}(\dot{\mathbf{r}} \times \mathbf{h}) &= -\frac{\mu}{r^3}\left[r\dot{r}\mathbf{r} - r^2\dot{\mathbf{r}} \right] \\ &= \mu\left[\frac{\dot{\mathbf{r}}}{r} - \frac{\dot{r}}{r^2}\mathbf{r} \right] \\ &= \frac{d}{dt}\left(\mu \frac{\mathbf{r}}{r} \right). \end{aligned} \quad (9.78)$$

Integrating both sides of the equation, we have

$$\dot{\mathbf{r}} \times \mathbf{h} = \mu\left(\frac{\mathbf{r}}{r} + \mathbf{c} \right).$$

The vector on the right-hand side of this equation is the cross product of the vectors $\dot{\mathbf{r}}$ and $\mathbf{h}$. Therefore, it is perpendicular to $\mathbf{h}$ and lies in the orbital plane. Because $\mathbf{r}$ is in the orbital plane, then so is $\mathbf{c}$. Therefore, $\mathbf{c} = c_1\mathbf{i} + c_2\mathbf{j}$. We can orient the xy-plane such that $c_2 = 0$ and choose $\mathbf{c} = e\mathbf{i}$. This gives

$$\dot{\mathbf{r}} \times \mathbf{h} = \mu\left(\frac{\mathbf{r}}{r} + e\mathbf{i}\right).$$

Now we compute

$$\mathbf{r} \cdot (\dot{\mathbf{r}} \times \mathbf{h}) = \mu\mathbf{r} \cdot \left(\frac{\mathbf{r}}{r} + e\mathbf{i}\right).$$

The left side is a triple scalar product, and a little permutation of the vector gives

$$\mathbf{r} \cdot (\dot{\mathbf{r}} \times \mathbf{h}) = \mathbf{h} \cdot (\mathbf{r} \times \dot{\mathbf{h}}) = h^2.$$

The right-hand side becomes

$$\mu\mathbf{r} \cdot \left(\frac{\mathbf{r}}{r} + e\mathbf{i}\right) = \mu r(1 + e\cos\theta),$$

where θ is the angle between $\mathbf{r}$ and $\mathbf{c} = e\mathbf{i}$.

Kepler's first law in polar form.

Combining these results, we have

$$r = \frac{h^2/\mu}{1 + e\cos\theta}. \tag{9.79}$$

This is the polar equation of a conic with e the eccentricity. For $0 \le e < 1$, it describes a bound, elliptical orbit.

The periapsis, or minimum separation between the masses, is at $\theta = 0$:

$$r_{min} = \frac{h^2}{\mu(1 + e)}.$$

The periapsis is the shortest distance from a focus and the apoapsis is the largest. For orbits around the Earth, these are called perigee and apogee. For orbits around the Sun, they are called the perihelion and aphelion.

The apoapsis, or maximum separation between the masses, is at $\theta = \pi$:

$$r_{max} = \frac{h^2}{\mu(1 - e)} :$$

From these, we can find the length of the semimajor axis, a, using $2a = r_{min} + r_{max}$ to obtain

$$a = \frac{h^2}{\mu(1 - e^2)}.$$

Note that this also gives

$$\frac{h^2}{\mu} = a(1 - e^2).$$

So, the polar equation can also be written in the form

$$r = \frac{a(1 - e^2)}{1 + e\cos\theta}. \tag{9.80}$$

This allows us to write the periapsis and apoapsis as

$$r_{min} = a(1 - e) \quad r_{max} = a(1 + e).$$

We can also find the length of the semiminor axis. From the study of conics, one has that

$$e = \frac{c}{a} = \frac{\sqrt{a^2 - b^2}}{a}$$

for $a \geq b$. Solving for b, we have $b = a\sqrt{1 - e^2}$. In Figure 9.23 we indicate these parameters on the ellipse. Note that the polar coordinate system has its origin at $(c, 0)$ which is the location of the central mass. This is not the center of the ellipse as indicated by the principal axes.

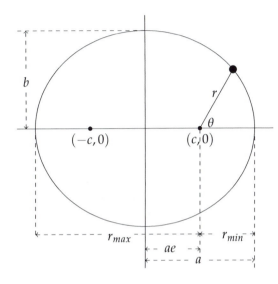

Figure 9.23: An ellipse with the ellipse parameters indicated.

In Section 2.10.5 we also used the maximum speed at the periapsis. This can be determined using the Conservation of Energy. From the equation of motion (9.75), we have

$$\begin{aligned} \ddot{x} &= -\mu \frac{x}{r^3} \\ \ddot{y} &= -\mu \frac{y}{r^3}, \end{aligned} \tag{9.81}$$

where $r = \sqrt{x^2 + y^2}$. Multiplying the first equation by $\dot{x}$, the second equation by $\dot{y}$, and adding, we have

$$\dot{x}\ddot{x} + \dot{y}\ddot{y} = \frac{\mu}{r^3}(x\dot{x} + y\dot{y}).$$

Both sides of this equation are exact derivatives, and the equation can be written as

$$\frac{d}{dt}\left(\frac{1}{2}(\dot{x}^2 + \dot{y}^2) - \frac{\mu}{r}\right) = 0.$$

Integrating yields

$$\frac{1}{2}(\dot{x}^2 + \dot{y}^2) - \frac{\mu}{r} = \text{const.}$$

The constant is usually found using the Virial Theorem, which states in this case that the time-averaged kinetic energy is related to the time-averaged potential energy,

$$< T >= -\frac{1}{2} < V > .$$

Therefore,

$$< E >= \frac{1}{2} < V >= -\frac{\mu}{2a}$$

or

$$\frac{1}{2}(\dot{x}^2 + \dot{y}^2) - \frac{\mu}{r} = -\frac{\mu}{2a}.$$

At periapsis, with $x = a(1-e)$, $y = 0$, and $\dot{x} = 0$, we have

$$\frac{1}{2}\dot{y}^2 - \frac{\mu}{a(1-e)} = -\frac{\mu}{2a}.$$

Solving for $\dot{y}$, we obtain

$$\dot{y}_{peri} = \sqrt{\frac{\mu}{a}\frac{1+e}{1-e}}.$$

Similarly, at apoapsis, we have

$$\dot{y}_{apo} = \sqrt{\frac{\mu}{a}\frac{1-e}{1+e}}.$$

It is also easy to establish conservation of (specific) angular momentum. Multiplying the first equation of motion by $\dot{y}$, the second equation by $\dot{x}$, and subtracting, we have

$$0 = \dot{y}\ddot{x} - \dot{x}\ddot{y} = \frac{d}{dt}(y\dot{x} - x\dot{y}).$$

Therefore, $y\dot{x} - x\dot{y} = $ const. We already established this in Section 2.10.5.

We can also derive Kepler's Third Law. Recall from Kepler's Second Law that

$$\frac{dA}{dt} = \frac{h}{2}.$$

Integrating, one has

$$A = \frac{h}{2}t,$$

assuming that zero area has been swept out at $t = 0$. After one complete orbit, the planet traces out the area of an ellipse in time T, the period of the motion. Recalling that the area of an ellipse is $A = \pi ab$, we have

$$\begin{aligned}
T &= \frac{2\pi ab}{h} \\
&= \frac{2\pi\sqrt{1-e^2}a^2}{\sqrt{a\mu(1-e^2)}\sqrt{\mu}\sqrt{a}} \\
&= \frac{2\pi}{\sqrt{\mu}}a^{3/2}.
\end{aligned} \tag{9.82}$$

Kepler's Third Law.

This is Kepler's Third Law, which can also be written in the form

$$T^2 = \frac{4\pi^2}{\mu}a^3.$$

9.2   Electromagnetic Waves

9.2.1   Maxwell's Equations

THERE ARE MANY APPLICATIONS leading to the equations in Table 9.1. One goal of this chapter is to derive the three-dimensional wave equation for electromagnetic waves. This derivation was first carried out by James Clerk Maxwell in 1860. At the time, much was known about the relationship between electric and magnetic fields through the work of such people as Hans Christian Ørstead (1777–1851), Michael Faraday (1791–1867), and André-Marie Ampère. Maxwell provided a mathematical formalism for these relationships, consisting of twenty partial differential equations in twenty unknowns. Later, these equations were put into more compact notations, namely in terms of quaternions, only later to be cast in vector form.

Quaternions were introduced in 1843 by William Rowan Hamilton (1805–1865) as a four-dimensional generalization of complex numbers.

In vector form, the original set of Maxwell's equations is given as

$$
\begin{aligned}
\nabla \cdot \mathbf{D} &= \rho, \\
\nabla \times \mathbf{H} &= \mu_0 \mathbf{J}_{tot}, \\
\mathbf{D} &= \epsilon \mathbf{E}, \\
\mathbf{J} &= \sigma \mathbf{E}, \\
\mathbf{J}_{tot} &= \mathbf{J} \frac{\partial \mathbf{D}}{\partial t}, \\
\nabla \cdot \mathbf{J} &= -\frac{\partial \rho}{\partial t}, \\
\mathbf{E} &= -\nabla \phi - \frac{\partial \mathbf{A}}{\partial t}, \\
\mu \mathbf{H} &= \nabla \times \mathbf{A}.
\end{aligned} \tag{9.83}
$$

Note that Maxwell expressed the electric and magnetic fields in terms of the scalar and vector potentials, ϕ and $\mathbf{A}$, respectively, as defined in the two previous equations. Here, $\mathbf{H}$ is the magnetic field, $\mathbf{D}$ is the electric displacement field, $\mathbf{E}$ is the electric field, $\mathbf{J}$ is the current density, ρ is the charge density, and σ is the conductivity.

This set of equations differs from what we typically present in physics courses. Several of these equations are defining quantities. While the potentials are part of a course in electrodynamics, they are not cast as the core set of equations now referred to as Maxwell's equations. Also, several equations are given as defining relations between the various variables, although they have some physical significance of their own, such as the continuity equation, given by $\nabla \cdot \mathbf{J} = -\frac{\partial \rho}{\partial t}$.

Furthermore, the distinction between the magnetic field strength, $\mathbf{H}$, and the magnetic flux density, $\mathbf{B}$, only becomes important in the presence of magnetic materials. Students are typically first introduced to $\mathbf{B}$ in introductory physics classes. In general, $\mathbf{B} = \mu \mathbf{H}$, where μ is the magnetic permeability of a material. In the absence of magnetic materials, $\mu = \mu_0$. In fact, in many applications of the propagation of electromagnetic waves, $\mu \approx \mu_0$.

These equations can be written in a more familiar form. The equations that we will refer to as Maxwell's equations from now on are

$$\nabla \cdot \mathbf{E} \;=\; \frac{\rho}{\epsilon_0}, \quad \text{(Gauss' Law)}$$

$$\nabla \cdot \mathbf{B} \;=\; 0$$

$$\nabla \times \mathbf{E} \;=\; -\frac{\partial \mathbf{B}}{\partial t}, \quad \text{(Faraday's Law)}$$

$$\nabla \times \mathbf{B} \;=\; \mu_0 \mathbf{J} + \mu_0 \epsilon_0 \frac{\partial \mathbf{E}}{\partial t}, \quad \text{(Maxwell-Ampère Law)} \qquad (9.84)$$

We have noted the common names attributed to each law. There are corresponding integral forms of these laws, which are often presented in an introductory physics class. The first law is Gauss' Law. It allows one to determine the electric field due to specific charge distributions. The second law typically has no name attached to it, but in some cases is called Gauss' Law for magnetic fields. It simply states that there are no free magnetic poles. The third law is Faraday's Law, indicating that changing magnetic flux induces electric potential differences. Finally, the fourth law is a modification of Ampère's Law that states that electric currents produce magnetic fields.

It should be noted that the last term in the Maxwell-Ampère Law was introduced by Maxwell. As we have seen, the divergence of the curl of any vector is zero:

The divergence of the curl of any vector is zero.

$$\nabla \cdot (\nabla \times \mathbf{V}) = 0.$$

Computing the divergence of the curl of the electric field, we find from Maxwell's equations that

$$\nabla \cdot (\nabla \times \mathbf{E}) \;=\; -\nabla \cdot \frac{\partial \mathbf{B}}{\partial t}$$

$$=\; -\frac{\partial}{\partial t} \nabla \cdot \mathbf{B} = 0, \qquad (9.85)$$

as $\nabla \cdot \mathbf{B} = 0$. Thus, the divergence of the curl of $\mathbf{E}$ vanishes as expected.

Ampère's Law in differential form.

However, before Maxwell, Ampère's Law in differential form would have been written in vector notation as

$$\nabla \times \mathbf{B} = \mu_0 \mathbf{J}.$$

The introduction of the displacement current makes Maxwell's equations mathematically consistent.

Computing the divergence of the curl of the magnetic field, we have

$$\nabla \cdot (\nabla \times \mathbf{B}) \;=\; \mu_0 \nabla \cdot \mathbf{J}$$

$$=\; -\mu_0 \frac{\partial \rho}{\partial t}. \qquad (9.86)$$

Here we made use of the Continuity Equation,

$$\frac{\partial \rho}{\partial t} + \nabla \cdot \mathbf{J} = 0. \qquad (9.87)$$

This is derived analogous to the Conservation of Mass Equation (9.55). Here, ρ is the charge density instead of the mass density and $\mathbf{Q}$, the mass flow rate, is replaced by $\mathbf{J}$, the charge flow rate or current density.

We have found that the divergence of the curl of **B** does not vanish as it should. Maxwell argued that Ampère's Law needs to account for a changing charge distribution. He introduced what he called the displacement current, $\mu_0 \epsilon_0 \frac{\partial \mathbf{E}}{\partial t}$, into the Ampère Law. Replacing **J** with

$$\mathbf{J}_{tot} = \mathbf{J} + \mu_0 \epsilon_0 \frac{\partial \mathbf{E}}{\partial t}$$

in Ampère's Law, the computation of the divergence of the curl of the magnetic field yields

$$
\begin{aligned}
\nabla \cdot (\nabla \times \mathbf{B}) &= \mu_0 \nabla \cdot \left(\mathbf{J} + \mu_0 \epsilon_0 \frac{\partial \mathbf{E}}{\partial t} \right) \\
&= -\mu_0 \frac{\partial \rho}{\partial t} + \mu_0 \epsilon_0 \frac{\partial}{\partial t} \nabla \cdot \mathbf{E} \\
&= -\mu_0 \frac{\partial \rho}{\partial t} + \mu_0 \epsilon_0 \frac{\partial}{\partial t} \left(\frac{\rho}{\epsilon_0} \right) = 0. \qquad (9.88)
\end{aligned}
$$

So, Maxwell's introduction of the displacement current was not only physically important, but it made the equations mathematically consistent.

9.2.2 Electromagnetic Wave Equation

WE ARE NOW READY TO DERIVE the wave equation for electromagnetic waves. We will consider the case of free space in which there are no free charges or currents and the waves propagate in a vacuum. We then have Maxwell's equations in the form

Maxwell's equations in a vacuum.

$$
\begin{aligned}
\nabla \cdot \mathbf{E} &= 0, \\
\nabla \cdot \mathbf{B} &= 0, \\
\nabla \times \mathbf{E} &= -\frac{\partial \mathbf{B}}{\partial t}, \\
\nabla \times \mathbf{B} &= \mu_0 \epsilon_0 \frac{\partial \mathbf{E}}{\partial t}. \qquad (9.89)
\end{aligned}
$$

We will derive the wave equation for the electric field. Consider the expression $\nabla \times (\nabla \times \mathbf{E})$. We note that the table of identities in Section 9.1.5 gives

$$\nabla \times (\nabla \times \mathbf{E}) = \nabla(\nabla \cdot \mathbf{E}) - \nabla^2 \mathbf{E}.$$

However, in the absence of charge, the divergence of **E** is zero, so we have

$$\nabla \times (\nabla \times \mathbf{E}) = -\nabla^2 \mathbf{E}. \qquad (9.90)$$

We can also use Faraday's Law on the left-hand side of this equation to obtain

$$\nabla \times (\nabla \times \mathbf{E}) = -\nabla \times \left(\frac{\partial \mathbf{B}}{\partial t} \right).$$

Interchanging the time and space derivatives, and using the Maxwell–Ampère Law, we find

$$\nabla \times (\nabla \times \mathbf{E}) = -\frac{\partial}{\partial t} (\nabla \times \mathbf{B})$$

$$= -\frac{\partial}{\partial t}\left(\epsilon_0\mu_0\frac{\partial \mathbf{E}}{\partial t}\right)$$

$$= -\epsilon_0\mu_0\frac{\partial^2 \mathbf{E}}{\partial t^2}. \tag{9.91}$$

Equations (9.92) and (9.93) are the three-dimensional wave equations for electric and magnetic fields in a vacuum.

Combining the two expressions for $\nabla \times (\nabla \times \mathbf{E})$, we have the sought-after result:

$$\epsilon_0\mu_0\frac{\partial^2 \mathbf{E}}{\partial t^2} = \nabla^2 \mathbf{E}. \tag{9.92}$$

This is the three-dimensional equation for an oscillating electric field. A similar equation can be found for the magnetic field,

$$\epsilon_0\mu_0\frac{\partial^2 \mathbf{B}}{\partial t^2} = \nabla^2 \mathbf{B}. \tag{9.93}$$

Recalling that $\epsilon_0 = 8.85 \times 10^{-12}$ C^2/Nm^2 and $\mu_0 = 4\pi \times 10^{-7}$ N/A^2, one finds that $c = 3 \times 10^8$ m/s.

One can derive more general equations. For example, we could look for waves in what are called dielectric materials. In this case, one distinguishes between free and bound charges. Free charges are charges that are free to move, such as electrons in a metal, leading to a free current density, $\mathbf{J}_f$. In materials there is also a current due to polarization effects, $\mathbf{J}_p$. Denoting the polarization by vector $\mathbf{P}$, these are related by

$$\mathbf{J}_p = \frac{\partial \mathbf{P}}{\partial t}.$$

The polarization can also lead to a local concentration of charge, ρ_p, obeying the Continuity Equation [see Equation (9.87)]

$$\frac{\partial \rho_p}{\partial t} + \nabla \cdot \mathbf{J_p} = 0.$$

Inserting the definition of J_p, we find that

$$\rho_p = -\nabla \cdot \mathbf{P}.$$

The total charge density is then $\rho = \rho_f + \rho_p$.

Inserting these expressions into Maxwell's equations, we have

$$\begin{aligned} \nabla \cdot \mathbf{E} &= \frac{\rho_f}{\epsilon_0} - \frac{\nabla \cdot \mathbf{P}}{\epsilon_0}, \\ \nabla \cdot \mathbf{B} &= 0 \\ \nabla \times \mathbf{E} &= -\frac{\partial \mathbf{B}}{\partial t}, \\ \nabla \times \frac{\mathbf{B}}{\mu_0} &= \epsilon_0\frac{\partial \mathbf{E}}{\partial t} + \frac{\partial \mathbf{P}}{\partial t} + \mathbf{J}_f. \end{aligned} \tag{9.94}$$

Gauss's Law can be written as

$$\nabla \cdot (\epsilon_0\mathbf{E} + \mathbf{P}) = \rho_f.$$

Therefore, the electric field displacement is often used: $\mathbf{D} \equiv \epsilon_0\mathbf{E} + \mathbf{P}$. In the case that the polarization is proportional to the electric field, the medium is called linear. Writing $\mathbf{P} = \epsilon_0\chi\mathbf{E}$, then

$$\mathbf{D} = (1+\chi)\epsilon_0\mathbf{E} \equiv \epsilon\mathbf{E}.$$

Here, χ is called the susceptibility and ϵ is the electric permittivity.

Similarly, one can introduce the magnetization in materials and define $\mathbf{H} = \frac{\mathbf{B}}{\mu}$ where μ is the magnetic permeability of the material. Then, the wave speed in a vacuum, c, is replaced by the wave speed in the medium, v. It is given by

$$v = \frac{1}{\sqrt{\epsilon\mu}} = \frac{c}{n}.$$

Here, n is the index of refraction, $n = \sqrt{\frac{\epsilon\mu}{\epsilon_0\mu_0}} = \sqrt{1+\bar{\zeta}}$. In many materials, $\mu \approx \mu_0$ and we will not go further into the magnetization properties. Introducing the dielectric constant, $\kappa = \frac{\epsilon}{\epsilon_0}$, one finds that the index of refraction is $n \approx \sqrt{\kappa}$.

The wave equations derived in this section lead to many of the properties of the electric and magnetic fields. We can also study systems in which these waves are confined, such as waveguides. In such cases, we can impose boundary conditions and determine what modes are allowed to propagate within certain structures, such as optical fibers. We will take up some of these examples in the next chapter.

9.2.3 Potential Functions and Helmholtz's Theorem

ANOTHER APPLICATION OF THE USE OF VECTOR ANALYSIS for studying electromagnetism is that of potential theory. In this section we describe the use of a scalar potential, $\phi(\mathbf{r},t)$ and a vector potential, $\mathbf{A}(\mathbf{r},t)$ to solve problems in electromagnetic theory. Helmholtz's Theorem says that a vector field is uniquely determined by knowing its divergence and its curl. Combining this result with the definitions of the electric and magnetic potentials, we will show that solutions of Maxwell's equations for the electric and magnetic fields can be found by simply solving a set of Poisson equations, $\nabla^2 u = f$, for the certain potential functions.

In the case of static fields, we have from Maxwell's equations that

Hermann Ludwig Ferdinand von Helmholtz (1821–1894) made many contributions to physics. There are several theorems named after him.

Electric and magnetic potentials.

$$\nabla \cdot \mathbf{B} = 0, \quad \nabla \times \mathbf{E} = 0.$$

We saw earlier in this chapter that the curl of a gradient is zero and the divergence of a curl is zero. This suggests that $\mathbf{E}$ is the gradient of a scalar function and $\mathbf{B}$ is the curl of a vector function:

$$\mathbf{E} = -\nabla\phi,$$

$$\mathbf{B} = \nabla \times \mathbf{A}.$$

ϕ is called the electric potential, and $\mathbf{A}$ is called the magnetic potential.

The remaining Maxwell equations are

$$\nabla \cdot \mathbf{E} = \frac{\rho}{\epsilon_0}, \quad \nabla \times \mathbf{B} = \mu_0\mathbf{J}.$$

Inserting the potential functions, we have

$$\nabla^2\phi = -\frac{\rho}{\epsilon_0}, \quad \nabla \times (\nabla \times \mathbf{A}) = \mu_0\mathbf{J}.$$

Thus, ϕ satisfies a Poisson equation, which is a simple partial differential equation that can be solved using separation of variables, or other techniques.

The equation satisfied by the magnetic potential looks a little more complicated. However, we can use the identity

$$\nabla \times (\nabla \times \mathbf{A}) = \nabla(\nabla \cdot \mathbf{A}) - \nabla^2 \mathbf{A}.$$

If $\nabla \cdot \mathbf{A} = 0$, then we find that

$$\nabla^2 \mathbf{A} = -\mu_0 \mathbf{J}.$$

Thus, the components of the magnetic potential also satisfy Poisson equations!

It turns out that requiring $\nabla \cdot \mathbf{A} = 0$ is not as restrictive as one might first think. Potential functions are not unique. For example, adding a constant to a potential function will still give the same fields. For example,

$$\nabla(\phi + c) = \nabla\phi = -\mathbf{E}.$$

This is not too alarming because it is the field that is physical and not the potential. In the case of the magnetic potential, adding the gradient of some field gives the same magnetic field, $\nabla \times (\mathbf{A} + \nabla\psi) = \nabla \times \mathbf{A} = \mathbf{B}$. So, we can choose ψ such that the new magnetic potential is divergenceless, $\nabla \cdot \mathbf{A} = 0$. A particular choice of the scalar and vector potentials is a called a gauge, and the process is called fixing, or choosing, a gauge. The choice of $\nabla \cdot \mathbf{A} = 0$ is called the Coulomb gauge.

Coulomb gauge: $\nabla \cdot \mathbf{A} = 0$.

If the fields are dynamic, that is, functions of time, then the magnetic potential also contributes to the electric field. In this case, we have

$$\mathbf{E} = -\nabla\phi - \frac{\partial \mathbf{A}}{\partial t},$$

$$\mathbf{B} = \nabla \times \mathbf{A}.$$

Thus, two of Maxwell's equations are automatically satisfied:

$$\nabla \cdot \mathbf{B} = 0, \quad \nabla \times \mathbf{E} = -\frac{\partial \mathbf{B}}{\partial t}.$$

The other two equations become

$$\nabla \cdot \mathbf{E} = \frac{\rho}{\epsilon_0} \Rightarrow \nabla^2\phi + \frac{\partial}{\partial t}\nabla \cdot \mathbf{A} = -\frac{\rho}{\epsilon_0},$$

and

$$\nabla \times \mathbf{B} = \mu_0 \mathbf{J} + \mu_0\epsilon_0 \frac{\partial \mathbf{E}}{\partial t} \Rightarrow$$

$$\nabla(\nabla \cdot \mathbf{A}) - \nabla^2 \mathbf{A} = \mu_0 \mathbf{J} - \frac{1}{c^2}\frac{\partial}{\partial t}\left(\nabla\phi + \frac{\partial \mathbf{A}}{\partial t}\right).$$

Rearranging, we have

$$\left(\nabla^2 - \frac{1}{c^2}\frac{\partial^2}{\partial t^2}\right)\mathbf{A} - \nabla\left(\nabla \cdot \mathbf{A} + \frac{1}{c^2}\frac{\partial\phi}{\partial t}\right) = -\mu_0 \mathbf{J}.$$

If we choose the Lorentz gauge, by requiring

$$\nabla \cdot \mathbf{A} + \frac{1}{c^2} \frac{\partial \phi}{\partial t} = 0,$$

then

$$\left(\nabla^2 - \frac{1}{c^2} \frac{\partial^2}{\partial t^2} \right) \phi = -\frac{\rho}{\epsilon_0},$$

$$\left(\nabla^2 - \frac{1}{c^2} \frac{\partial^2}{\partial t^2} \right) \mathbf{A} = -\mu_0 \mathbf{J}.$$

Thus, the potential functions satisfy nonhomogeneous wave equations, which can be solved with standard methods as one will see in a course in electrodynamics.

The above introduction of potentials to describe the electric and magnetic fields is a special case of Helmholtz's Theorem for vectors. This theorem states that any "nice" vector field in three dimensions can be resolved into the sum of a curl-free (irrotational) and divergence-free (solenoidal) vector field. This is called the Helmholtz decomposition. Namely, given any nice vector field $\mathbf{v}$, we can write it as

$$\mathbf{v} = \underbrace{-\nabla \phi}_{\text{irrotational}} + \underbrace{\nabla \times \mathbf{A}}_{\text{solenoidal}}.$$

Given

$$\nabla \cdot \mathbf{v} = \rho, \quad \nabla \times \mathbf{v} = \mathbf{F},$$

then one has

$$\nabla^2 \phi = \rho$$

and

$$\nabla (\nabla \cdot \mathbf{A}) - \nabla^2 \mathbf{A} = \mathbf{F}.$$

Forcing $\nabla \cdot \mathbf{A} = 0$,

$$\nabla^2 \mathbf{A} = -\mathbf{F}.$$

Thus, one obtains Poisson equations for ϕ and $\mathbf{A}$. This is just a generalization of the above procedure, which was a special case for static electromagnetic fields.

9.3 Curvilinear Coordinates

IN ORDER TO STUDY SOLUTIONS OF THE WAVE EQUATION, the heat equation, or even Schrödinger's Equation in different geometries, we need to see how differential operators, such as the Laplacian, appear in these geometries. The most common coordinate systems arising in physics are polar coordinates, cylindrical coordinates, and spherical coordinates. These reflect the common geometrical symmetries often encountered in physics.

In such systems, it is easier to describe boundary conditions and to make use of these symmetries. For example, specifying that the electric potential

Lorentz gauge: $\nabla \cdot \mathbf{A} + \frac{1}{c^2} \frac{\partial \phi}{\partial t} = 0$.

In relativity, one uses the d'Alembertian, $\Box \equiv \nabla^2 - \frac{1}{c^2} \frac{\partial^2}{\partial t^2}$. The equations for the potentials can be written compactly as

$$\Box \phi = -\frac{\rho}{\epsilon_0},$$

and

$$\Box \mathbf{A} = -\mu_0 \mathbf{J}.$$

is 10.0 V on a spherical surface of radius one, we would say $\phi(x, y, z) = 10$ for $x^2 + y^2 + z^2 = 1$. However, if we use spherical coordinates, (r, θ, ϕ), then we would say $\phi(r, \theta, \phi) = 10$ for $r = 1$, or $\phi(1, \theta, \phi) = 10$. This is a much simpler representation of the boundary condition.

However, this simplicity in boundary conditions leads to a more complicated-looking partial differential equation in spherical coordinates. In this section we will consider general coordinate systems and how the differential operators are written in the new coordinate systems. In the next chapter we will solve some of these new problems.

We begin by introducing the general coordinate transformations between Cartesian coordinates and the more general curvilinear coordinates. Let the Cartesian coordinates be designated by (x_1, x_2, x_3) and the new coordinates by (u_1, u_2, u_3). We will assume that these are related through the transformations

$$
\begin{aligned}
x_1 &= x_1(u_1, u_2, u_3), \\
x_2 &= x_2(u_1, u_2, u_3), \\
x_3 &= x_3(u_1, u_2, u_3).
\end{aligned}
\tag{9.95}
$$

Thus, given the curvilinear coordinates (u_1, u_2, u_3) for a specific point in space, we can determine the Cartesian coordinates (x_1, x_2, x_3) of that point. We will assume that we can invert this transformation: Given the Cartesian coordinates, one can determine the corresponding curvilinear coordinates.

In the Cartesian system, we can assign an orthogonal basis, $\{\mathbf{i}, \mathbf{j}, \mathbf{k}\}$. As a particle traces out a path in space, one locates its position by the coordinates (x_1, x_2, x_3). Picking x_2 and x_3 constant, the particle lies on the curve $x_1 =$ value of the x_1 coordinate. This line lies in the direction of the basis vector $\mathbf{i}$. We can do the same with the other coordinates and essentially map out a grid in three-dimensional space, as sown in Figure 9.24. All of the x_i-curves intersect at each point orthogonally, and the basis vectors $\{\mathbf{i}, \mathbf{j}, \mathbf{k}\}$ lie along the grid lines and are mutually orthogonal. We would like to mimic this construction for general curvilinear coordinates. Requiring the orthogonality of the resulting basis vectors leads to orthogonal curvilinear coordinates.

As for the Cartesian case, we consider u_2 and u_3 constant. This leads to a curve parametrized by $u_1 : \mathbf{r} = x_1(u_1)\mathbf{i} + x_2(u_1)\mathbf{j} + x_3(u_1)\mathbf{k}$. We call this the u_1-curve. Similarly, when u_1 and u_3 are constant, we obtain a u_2-curve and for u_1 and u_2 constant, we obtain a u_3-curve. We will assume that these curves intersect such that each pair of curves intersects orthogonally, as seen in Figure 9.25. Furthermore, we will assume that the unit tangent vectors to these curves form a right-handed system similar to the $\{\mathbf{i}, \mathbf{j}, \mathbf{k}\}$ systems for Cartesian coordinates. We will denote these as $\{\hat{\mathbf{u}}_1, \hat{\mathbf{u}}_2, \hat{\mathbf{u}}_3\}$.

We can determine these tangent vectors from the coordinate transformations. Consider the position vector as a function of the new coordinates,

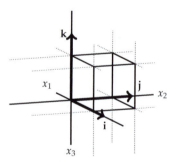

Figure 9.24: Plots of x_i-curves forming an orthogonal Cartesian grid.

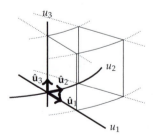

Figure 9.25: Plots of general u_i-curves forming an orthogonal grid.

$$
\mathbf{r}(u_1, u_2, u_3) = x_1(u_1, u_2, u_3)\mathbf{i} + x_2(u_1, u_2, u_3)\mathbf{j} + x_3(u_1, u_2, u_3)\mathbf{k}.
$$

Then, the infinitesimal change in position is given by

$$dr = \frac{\partial \mathbf{r}}{\partial u_1}du_1 + \frac{\partial \mathbf{r}}{\partial u_2}du_2 + \frac{\partial \mathbf{r}}{\partial u_3}du_3 = \sum_{i=1}^{3} \frac{\partial \mathbf{r}}{\partial u_i}du_i.$$

We note that the vectors $\frac{\partial \mathbf{r}}{\partial u_i}$ are tangent to the u_i-curves. Thus, we define the unit tangent vectors

$$\hat{\mathbf{u}}_i = \frac{\frac{\partial \mathbf{r}}{\partial u_i}}{\left|\frac{\partial \mathbf{r}}{\partial u_i}\right|}.$$

Solving for the original tangent vector, we have

$$\frac{\partial \mathbf{r}}{\partial u_i} = h_i\hat{\mathbf{u}}_i,$$

where

$$h_i \equiv \left|\frac{\partial \mathbf{r}}{\partial u_i}\right|.$$

The scale factors, $h_i \equiv \left|\frac{\partial \mathbf{r}}{\partial u_i}\right|$.

The h_i's are called the scale factors for the transformation. The infinitesimal change in position in the new basis is then given by

$$d\mathbf{r} = \sum_{i=1}^{3} h_i u_i \hat{\mathbf{u}}_i.$$

Example 9.26. *Determine the scale factors for the polar coordinate transformation. The transformation for polar coordinates is*

$$x = r\cos\theta, \quad y = r\sin\theta.$$

Here we note that $x_1 = x$, $x_2 = y$, $u_1 = r$, and $u_2 = \theta$. The u_1-curves are curves with $\theta = $ const. Thus, these curves are radial lines. Similarly, the u_2-curves have $r = $ const. These curves are concentric circles about the origin as shown in Figure 9.26.

The unit vectors are easily found. We will denote them by $\hat{\mathbf{u}}_r$ and $\hat{\mathbf{u}}_\theta$. We can determine these unit vectors by first computing $\frac{\partial \mathbf{r}}{\partial u_i}$. Let

$$\mathbf{r} = x(r,\theta)\mathbf{i} + y(r,\theta)\mathbf{j} = r\cos\theta\mathbf{i} + r\sin\theta\mathbf{j}.$$

Then,

$$\begin{aligned} \frac{\partial \mathbf{r}}{\partial r} &= \cos\theta\mathbf{i} + \sin\theta\mathbf{j}, \\ \frac{\partial \mathbf{r}}{\partial \theta} &= -r\sin\theta\mathbf{i} + r\cos\theta\mathbf{j}. \end{aligned} \tag{9.96}$$

The first vector already is a unit vector. So,

$$\hat{\mathbf{u}}_r = \cos\theta\mathbf{i} + \sin\theta\mathbf{j}.$$

The second vector has length r because $|-r\sin\theta\mathbf{i} + r\cos\theta\mathbf{j}| = r$. Dividing $\frac{\partial \mathbf{r}}{\partial \theta}$ by r, we have

$$\hat{\mathbf{u}}_\theta = -\sin\theta\mathbf{i} + \cos\theta\mathbf{j}.$$

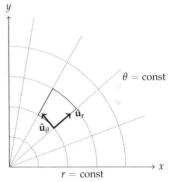

Figure 9.26: Plots an orthogonal polar grid.

We can see that these vectors are orthogonal ($\hat{\mathbf{u}}_r \cdot \hat{\mathbf{u}}_\theta = 0$) and form a right-hand system. That they form a right-hand system can be seen by either drawing the vectors or computing the cross product:

$$(\cos\theta\mathbf{i} + \sin\theta\mathbf{j}) \times (-\sin\theta\mathbf{i} + \cos\theta\mathbf{j}) = \cos^2\theta\,\mathbf{i}\times\mathbf{j} - \sin^2\theta\,\mathbf{j}\times\mathbf{i}$$
$$= \mathbf{k}. \qquad (9.97)$$

Because

$$\frac{\partial\mathbf{r}}{\partial r} = \hat{\mathbf{u}}_r,$$
$$\frac{\partial\mathbf{r}}{\partial\theta} = r\hat{\mathbf{u}}_\theta,$$

The scale factors are $h_r = 1$ and $h_\theta = r$.

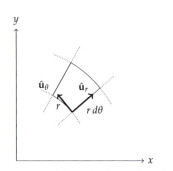

Figure 9.27: Infinitesimal area in polar coordinates.

Once we know the scale factors, we have that

$$d\mathbf{r} = \sum_{i=1}^{3} h_i\,du_i\,\hat{\mathbf{u}}_i.$$

The infinitesimal arclength is then given by the Euclidean line element

$$ds^2 = d\mathbf{r} \cdot d\mathbf{r} = \sum_{i=1}^{3} h_i^2\,du_i^2$$

when the system is orthogonal. The h_i^2 are referred to as the metric coefficients.

Example 9.27. *Verify that $d\mathbf{r} = dr\,\hat{\mathbf{u}}_r + r\,d\theta\,\hat{\mathbf{u}}_\theta$ directly from $\mathbf{r} = r\cos\theta\mathbf{i} + r\sin\theta\mathbf{j}$ and obtain the Euclidean line element for polar coordinates.*

We begin by computing

$$d\mathbf{r} = d(r\cos\theta\mathbf{i} + r\sin\theta\mathbf{j})$$
$$= (\cos\theta\mathbf{i} + \sin\theta\mathbf{j})\,dr + r(-\sin\theta\mathbf{i} + \cos\theta\mathbf{j})\,d\theta$$
$$= dr\,\hat{\mathbf{u}}_r + r\,d\theta\,\hat{\mathbf{u}}_\theta. \qquad (9.98)$$

This agrees with the form $d\mathbf{r} = \sum_{i=1}^{3} h_i\,du_i\,\hat{\mathbf{u}}_i$ when the scale factors for polar coordinates are inserted.

The line element is found as

$$ds^2 = d\mathbf{r} \cdot d\mathbf{r}$$
$$= (dr\,\hat{\mathbf{u}}_r + r\,d\theta\,\hat{\mathbf{u}}_\theta) \cdot (dr\,\hat{\mathbf{u}}_r + r\,d\theta\,\hat{\mathbf{u}}_\theta)$$
$$= dr^2 + r^2\,d\theta^2. \qquad (9.99)$$

This is the Euclidean line element in polar coordinates.

Also, along the u_i-curves,

$$d\mathbf{r} = h_i\,du_i\,\hat{\mathbf{u}}_i, \quad \text{(no summation)}.$$

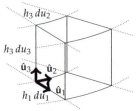

Figure 9.28: Infinitesimal volume element with sides of length $h_i\,du_i$.

This can be seen in Figure 9.28 by focusing on the u_1 curve. Along this curve, u_2 and u_3 are constant. So, $du_2 = 0$ and $du_3 = 0$. This leaves $d\mathbf{r} = h_1\,du_1\,\hat{\mathbf{u}}_1$ along the u_1-curve. Similar expressions hold along the other two curves.

We can use this result to investigate infinitesimal volume elements for general coordinate systems as shown in Figure 9.28. At a given point (u_1, u_2, u_3), we can construct an infinitesimal parallelepiped of sides $h_i du_i$, $i = 1, 2, 3$. This infinitesimal parallelepiped has a volume of size

$$dV = \left| \frac{\partial \mathbf{r}}{\partial u_1} \cdot \frac{\partial \mathbf{r}}{\partial u_2} \times \frac{\partial \mathbf{r}}{\partial u_3} \right| du_1 du_2 du_3.$$

The triple scalar product can be computed using determinants, and the resulting determinant is called the Jacobian and is given by

$$
\begin{aligned}
J &= \left| \frac{\partial(x_1, x_2, x_3)}{\partial(u_1, u_2, u_3)} \right| \\
&= \left| \frac{\partial \mathbf{r}}{\partial u_1} \cdot \frac{\partial \mathbf{r}}{\partial u_2} \times \frac{\partial \mathbf{r}}{\partial u_3} \right| \\
&= \left| \begin{matrix} \frac{\partial x_1}{\partial u_1} & \frac{\partial x_2}{\partial u_1} & \frac{\partial x_3}{\partial u_1} \\ \frac{\partial x_1}{\partial u_2} & \frac{\partial x_2}{\partial u_2} & \frac{\partial x_3}{\partial u_2} \\ \frac{\partial x_1}{\partial u_3} & \frac{\partial x_2}{\partial u_3} & \frac{\partial x_3}{\partial u_3} \end{matrix} \right|.
\end{aligned}
\tag{9.100}
$$

Therefore, the volume element can be written as

$$dV = J\, du_1 du_2 du_3 = \left| \frac{\partial(x_1, x_2, x_3)}{\partial(u_1, u_2, u_3)} \right| du_1 du_2 du_3.$$

Example 9.28. *Determine the volume element for cylindrical coordinates (r, θ, z), given by*

$$x = r \cos \theta, \tag{9.101}$$

$$y = r \sin \theta, \tag{9.102}$$

$$z = z. \tag{9.103}$$

Here, we have $(u_1, u_2, u_3) = (r, \theta, z)$ as displayed in Figure 9.29. Then, the Jacobian is given by

$$
\begin{aligned}
J &= \left| \frac{\partial(x, y, z)}{\partial(r, \theta, z)} \right| \\
&= \left| \begin{matrix} \frac{\partial x}{\partial r} & \frac{\partial y}{\partial r} & \frac{\partial z}{\partial r} \\ \frac{\partial x}{\partial \theta} & \frac{\partial y}{\partial \theta} & \frac{\partial z}{\partial \theta} \\ \frac{\partial x}{\partial z} & \frac{\partial y}{\partial z} & \frac{\partial z}{\partial z} \end{matrix} \right| \\
&= \left| \begin{matrix} \cos \theta & \sin \theta & 0 \\ -r \sin \theta & r \cos \theta & 0 \\ 0 & 0 & 1 \end{matrix} \right| \\
&= r
\end{aligned}
\tag{9.104}
$$

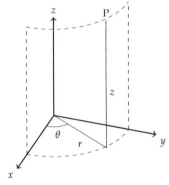

Figure 9.29: Cylindrical coordinate system.

Thus, the volume element is given as

$$dV = r\, dr\, d\theta\, dz.$$

This result should be familiar from multivariate calculus.

Another approach is to consider the geometry of the infinitesimal volume element. The directed edge lengths are given by $d\mathbf{s}_i = h_i du_i \hat{\mathbf{u}}_i$, as seen in Figure 9.25. The infinitesimal area element of for the face in direction $\hat{\mathbf{u}}_k$ is found from a simple cross product,

$$d\mathbf{A}_k = d\mathbf{s}_i \times d\mathbf{s}_j = h_i h_j du_i du_j \hat{\mathbf{u}}_i \times \hat{\mathbf{u}}_j.$$

Because these are unit vectors, the areas of the faces of the infinitesimal volumes are $dA_k = h_i h_j du_i du_j$.

The infinitesimal volume is then obtained as

$$dV = |d\mathbf{s}_k \cdot d\mathbf{A}_k| = h_i h_j h_k du_i du_j du_k |\hat{\mathbf{u}}_i \cdot (\hat{\mathbf{u}}_k \times \hat{\mathbf{u}}_j)|.$$

Thus, $dV = h_1 h_2 h_3 du_1 du_1 du_3$. Of course, this should not be a surprise because

$$J = \left| \frac{\partial \mathbf{r}}{\partial u_1} \cdot \frac{\partial \mathbf{r}}{\partial u_2} \times \frac{\partial \mathbf{r}}{\partial u_3} \right| = |h_1 \hat{\mathbf{u}}_1 \cdot h_2 \hat{\mathbf{u}}_2 \times h_3 \hat{\mathbf{u}}_3| = h_1 h_2 h_3.$$

Example 9.29. *For polar coordinates, determine the infinitesimal area element.*

In an earlier example, we found the scale factors for polar coordinates as $h_r = 1$ and $h_\theta = r$. Thus, $dA = h_r h_\theta\, dr d\theta = r\, dr d\theta$. Also, the previous example for cylindrical coordinates will yield similar results if we already know the scales factors without having to compute the Jacobian directly. Furthermore, the area element perpendicular to the z-coordinate gives the polar coordinate system result.

Next we will derive the forms of the gradient, divergence, and curl in curvilinear coordinates using several of the identities in Section 9.1.5. The results are given here for quick reference.

Gradient, divergence, and curl in orthogonal curvilinear coordinates.

$$
\begin{aligned}
\nabla \phi &= \sum_{i=1}^{3} \frac{\hat{\mathbf{u}}_i}{h_i} \frac{\partial \phi}{\partial u_i} \\
&= \frac{\hat{\mathbf{u}}_1}{h_1} \frac{\partial \phi}{\partial u_1} + \frac{\hat{\mathbf{u}}_2}{h_2} \frac{\partial \phi}{\partial u_2} + \frac{\hat{\mathbf{u}}_3}{h_3} \frac{\partial \phi}{\partial u_3}. & (9.105)\\
\nabla \cdot \mathbf{F} &= \frac{1}{h_1 h_2 h_3} \left(\frac{\partial}{\partial u_1}(h_2 h_3 F_1) + \frac{\partial}{\partial u_2}(h_1 h_3 F_2) + \frac{\partial}{\partial u_3}(h_1 h_2 F_3) \right). \\
& & (9.106)\\
\nabla \times \mathbf{F} &= \frac{1}{h_1 h_2 h_3} \begin{vmatrix} h_1 \hat{\mathbf{u}}_1 & h_2 \hat{\mathbf{u}}_2 & h_3 \hat{\mathbf{u}}_3 \\ \frac{\partial}{\partial u_1} & \frac{\partial}{\partial u_2} & \frac{\partial}{\partial u_3} \\ F_1 h_1 & F_2 h_2 & F_3 h_3 \end{vmatrix}. & (9.107)\\
\nabla^2 \phi &= \frac{1}{h_1 h_2 h_3} \left(\frac{\partial}{\partial u_1} \left(\frac{h_2 h_3}{h_1} \frac{\partial \phi}{\partial u_1} \right) + \frac{\partial}{\partial u_2} \left(\frac{h_1 h_3}{h_2} \frac{\partial \phi}{\partial u_2} \right) \right. \\
& \left. \quad + \frac{\partial}{\partial u_3} \left(\frac{h_1 h_2}{h_3} \frac{\partial \phi}{\partial u_3} \right) \right). & (9.108)
\end{aligned}
$$

Derivation of the gradient form.

We begin the derivations of these formulae by looking at the gradient, $\nabla \phi$, of the scalar function $\phi(u_1, u_2, u_3)$. We recall that the gradient operator appears in the differential change of a scalar function,

$$d\phi = \nabla \phi \cdot d\mathbf{r} = \sum_{i=1}^{3} \frac{\partial \phi}{\partial u_i} du_i.$$

Because

$$dr = \sum_{i=1}^{3} h_i du_i \hat{u}_i, \qquad (9.109)$$

we also have that

$$d\phi = \nabla\phi \cdot dr = \sum_{i=1}^{3} (\nabla\phi)_i h_i du_i.$$

Comparing these two expressions for $d\phi$, we determine that the components of the del operator can be written as

$$(\nabla\phi)_i = \frac{1}{h_i} \frac{\partial\phi}{\partial u_i}$$

and thus the gradient is given by

$$\nabla\phi = \frac{\hat{u}_1}{h_1} \frac{\partial\phi}{\partial u_1} + \frac{\hat{u}_2}{h_2} \frac{\partial\phi}{\partial u_2} + \frac{\hat{u}_3}{h_3} \frac{\partial\phi}{\partial u_3}. \qquad (9.110)$$

Next we compute the divergence:

Derivation of the divergence form.

$$\nabla \cdot \mathbf{F} = \sum_{i=1}^{3} \nabla \cdot (F_i \hat{u}_i).$$

We can do this by computing the individual terms in the sum. We will compute $\nabla \cdot (F_1 \hat{u}_1)$.

Using Equation (9.110), we have that

$$\nabla u_i = \frac{\hat{u}_i}{h_i}.$$

Then

$$\nabla u_2 \times \nabla u_3 = \frac{\hat{u}_2 \times \hat{u}_3}{h_2 h_3} = \frac{\hat{u}_1}{h_2 h_3}.$$

Solving for $\hat{u}_1$ gives

$$\hat{u}_1 = h_2 h_3 \nabla u_2 \times \nabla u_3.$$

Inserting this result into $\nabla \cdot (F_1 \hat{u}_1)$ and using the vector identity 2c from Section 9.1.5,

$$\nabla \cdot (f\mathbf{A}) = f\nabla \cdot \mathbf{A} + \mathbf{A} \cdot \nabla f,$$

we have

$$\begin{aligned} \nabla \cdot (F_1 \hat{u}_1) &= \nabla \cdot (F_1 h_2 h_3 \nabla u_2 \times \nabla u_3) \\ &= \nabla (F_1 h_2 h_3) \cdot \nabla u_2 \times \nabla u_3 + F_1 h_2 h_2 \nabla \cdot (\nabla u_2 \times \nabla u_3). \end{aligned}$$
$$(9.111)$$

The second term of this result vanishes by vector identity 3c,

$$\nabla \cdot (\nabla f \times \nabla g) = 0.$$

Because $\nabla u_2 \times \nabla u_3 = \frac{\hat{u}_1}{h_2 h_3}$, the first term can be evaluated as

$$\nabla \cdot (F_1 \hat{u}_1) = \nabla (F_1 h_2 h_3) \cdot \frac{\hat{u}_1}{h_2 h_3} = \frac{1}{h_1 h_2 h_3} \frac{\partial}{\partial u_1} (F_1 h_2 h_3).$$

Similar computations can be carried out for the remaining components, leading to the sought expression for the divergence in curvilinear coordinates:

$$\nabla \cdot \mathbf{F} = \frac{1}{h_1 h_2 h_3} \left(\frac{\partial}{\partial u_1}(h_2 h_3 F_1) + \frac{\partial}{\partial u_2}(h_1 h_3 F_2) + \frac{\partial}{\partial u_3}(h_1 h_2 F_3) \right). \quad (9.112)$$

Example 9.30. *Write the divergence operator in cylindrical coordinates.*

In this case, we have

$$
\begin{aligned}
\nabla \cdot \mathbf{F} &= \frac{1}{h_r h_\theta h_z} \left(\frac{\partial}{\partial r}(h_\theta h_z F_r) + \frac{\partial}{\partial \theta}(h_r h_z F_\theta) + \frac{\partial}{\partial \theta}(h_r h_\theta F_z) \right) \\
&= \frac{1}{r} \left(\frac{\partial}{\partial r}(r F_r) + \frac{\partial}{\partial \theta}(F_\theta) + \frac{\partial}{\partial \theta}(r F_z) \right) \\
&= \frac{1}{r}\frac{\partial}{\partial r}(r F_r) + \frac{1}{r}\frac{\partial}{\partial \theta}(F_\theta) + \frac{\partial}{\partial \theta}(F_z). \quad (9.113)
\end{aligned}
$$

Derivation of the curl form.

We now turn to the curl operator. In this case, we need to evaluate

$$\nabla \times \mathbf{F} = \sum_{i=1}^{3} \nabla \times (F_i \hat{\mathbf{u}}_i).$$

Again we focus on one term, $\nabla \times (F_1 \hat{\mathbf{u}}_1)$. Using the vector identity 2e,

$$\nabla \times (f\mathbf{A}) = f \nabla \times \mathbf{A} - \mathbf{A} \times \nabla f,$$

we have

$$
\begin{aligned}
\nabla \times (F_1 \hat{\mathbf{u}}_1) &= \nabla \times (F_1 h_1 \nabla u_1) \\
&= F_1 h_1 \nabla \times \nabla u_1 - \nabla (F_1 h_1) \times \nabla u_1. \quad (9.114)
\end{aligned}
$$

The curl of the gradient vanishes, leaving

$$\nabla \times (F_1 \hat{\mathbf{u}}_1) = \nabla (F_1 h_1) \times \nabla u_1.$$

Since $\nabla u_1 = \frac{\hat{\mathbf{u}}_1}{h_1}$, we have

$$
\begin{aligned}
\nabla \times (F_1 \hat{\mathbf{u}}_1) &= \nabla (F_1 h_1) \times \frac{\hat{\mathbf{u}}_1}{h_1} \\
&= \left(\sum_{i=1}^{3} \frac{\hat{\mathbf{u}}_i}{h_i} \frac{\partial (F_1 h_1)}{\partial u_i} \right) \times \frac{\hat{\mathbf{u}}_1}{h_1} \\
&= \frac{\hat{\mathbf{u}}_2}{h_3 h_1} \frac{\partial (F_1 h_1)}{\partial u_3} - \frac{\hat{\mathbf{u}}_3}{h_1 h_2} \frac{\partial (F_1 h_1)}{\partial u_2}. \quad (9.115)
\end{aligned}
$$

The other terms can be handled in a similar manner. The overall result is that

$$
\begin{aligned}
\nabla \times \mathbf{F} &= \frac{\hat{\mathbf{u}}_1}{h_2 h_3} \left(\frac{\partial (h_3 F_3)}{\partial u_2} - \frac{\partial (h_2 F_2)}{\partial u_3} \right) + \frac{\hat{\mathbf{u}}_2}{h_1 h_3} \left(\frac{\partial (h_1 F_1)}{\partial u_3} - \frac{\partial (h_3 F_3)}{\partial u_1} \right) \\
&\quad + \frac{\hat{\mathbf{u}}_3}{h_1 h_2} \left(\frac{\partial (h_2 F_2)}{\partial u_1} - \frac{\partial (h_1 F_1)}{\partial u_2} \right) \quad (9.116)
\end{aligned}
$$

This can be written more compactly as

$$\nabla \times \mathbf{F} = \frac{1}{h_1 h_2 h_3} \begin{vmatrix} h_1 \hat{\mathbf{u}}_1 & h_2 \hat{\mathbf{u}}_2 & h_3 \hat{\mathbf{u}}_3 \\ \frac{\partial}{\partial u_1} & \frac{\partial}{\partial u_2} & \frac{\partial}{\partial u_3} \\ F_1 h_1 & F_2 h_2 & F_3 h_3 \end{vmatrix} \quad (9.117)$$

Example 9.31. *Write the curl operator in cylindrical coordinates.*

$$\nabla \times \mathbf{F} = \frac{1}{r} \begin{vmatrix} \hat{\mathbf{e}}_r & r\hat{\mathbf{e}}_\theta & \hat{\mathbf{e}}_z \\ \frac{\partial}{\partial r} & \frac{\partial}{\partial \theta} & \frac{\partial}{\partial z} \\ F_r & rF_\theta & F_z \end{vmatrix}$$

$$= \left(\frac{1}{r} \frac{\partial F_z}{\partial \theta} - \frac{\partial F_\theta}{\partial z} \right) \hat{\mathbf{e}}_r + \left(\frac{\partial F_r}{\partial z} - \frac{\partial F_z}{\partial r} \right) \hat{\mathbf{e}}_\theta$$

$$+ \frac{1}{r} \left(\frac{\partial (rF_\theta)}{\partial r} - \frac{\partial F_r}{\partial \theta} \right) \hat{\mathbf{e}}_z. \tag{9.118}$$

Finally, we turn to the Laplacian. In the next chapter we will solve higher-dimensional problems in various geometric settings such as the wave equation, the heat equation, and Laplace's equation. These all involve knowing how to write the Laplacian in different coordinate systems. Because $\nabla^2 \phi = \nabla \cdot \nabla \phi$, we need only combine the results from Equations (9.110) and (9.112), respectively, for the gradient and the divergence in curvilinear coordinates. This is straightforward and gives

$$\nabla^2 \phi = \frac{1}{h_1 h_2 h_3} \left(\frac{\partial}{\partial u_1} \left(\frac{h_2 h_3}{h_1} \frac{\partial \phi}{\partial u_1} \right) + \frac{\partial}{\partial u_2} \left(\frac{h_1 h_3}{h_2} \frac{\partial \phi}{\partial u_2} \right) \right.$$

$$\left. + \frac{\partial}{\partial u_3} \left(\frac{h_1 h_2}{h_3} \frac{\partial \phi}{\partial u_3} \right) \right). \tag{9.119}$$

The results of rewriting the standard differential operators in cylindrical and spherical coordinates are shown in Problems 31 and 32. In particular, the Laplacians are given as

Cylindrical Coordinates:

$$\nabla^2 f = \frac{1}{r} \frac{\partial}{\partial r} \left(r \frac{\partial f}{\partial r} \right) + \frac{1}{r^2} \frac{\partial^2 f}{\partial \theta^2} + \frac{\partial^2 f}{\partial z^2} \tag{9.120}$$

Spherical Coordinates:

$$\nabla^2 f = \frac{1}{\rho^2} \frac{\partial}{\partial \rho} \left(\rho^2 \frac{\partial f}{\partial \rho} \right) + \frac{1}{\rho^2 \sin \theta} \frac{\partial}{\partial \theta} \left(\sin \theta \frac{\partial f}{\partial \theta} \right) + \frac{1}{\rho^2 \sin^2 \theta} \frac{\partial^2 f}{\partial \phi^2} \tag{9.121}$$

These forms will be used in the next chapter for the solution of Laplace's equation, the heat equation, and the wave equation in these coordinate systems.

9.4 Tensors

MAXWELL'S EQUATIONS HAVE BEEN WRITTEN IN VECTOR FORM as four equations. Vectors are not only useful because they are a compact way to write these equations, but vectors also have the property that their length is invariant, or does not change, under rotation of the underlying coordinate system. This notion of invariance under transformations can be extended to other objects. When a system involves non-isotropic systems, one usually encounters what are called tensors.

As an example, an isotropic distribution of mass would be a spherical distribution. No matter what direction one looks, the distribution looks the same. We had seen this notion earlier when discussing Friedmann's cosmological model of the universe. A simpler model is a spherical ball. However, an ellipsoidal distribution is slightly non-isotropic. In general, the distribution of matter can be described using the inertia tensor, which to most students is simply a 3×3 matrix:

$$I = \begin{pmatrix} I_{xx} & I_{xy} & I_{xz} \\ I_{yx} & I_{yy} & I_{yz} \\ I_{zx} & I_{zy} & I_{zz} \end{pmatrix}.$$

The entries are the moments of inertia about the origin for a continuous distribution of mass:

$$
\begin{aligned}
I_{xx} &= \int y^2 + z^2 \, dm, \\
I_{yy} &= \int x^2 + z^2 \, dm, \\
I_{zz} &= \int x^2 + y^2 \, dm, \\
I_{xy} &= I_{yx} = -\int xy \, dm, \\
I_{yz} &= I_{zy} = -\int yz \, dm, \\
I_{zx} &= I_{xz} = -\int xz \, dm.
\end{aligned}
\tag{9.122}
$$

These are used to describe the total angular momentum, $\mathbf{L} = I\boldsymbol{\omega}$, where $\boldsymbol{\omega}$ is a column vector of the components of angular velocity. The components of the moment of inertia can be written more compactly using index notation as

$$I_{ij} = \int (r^2 \delta_{ij} - x_i x_j) \rho \, dV.$$

Under a rotation of the coordinate system, the angular momentum equation becomes $\mathbf{L}' = I'\boldsymbol{\omega}'$. The relation to the original quantities is found using simple rules from Chapter 3:

$$
\begin{aligned}
\mathbf{L}' &= I'\boldsymbol{\omega}' \\
\hat{R}_\theta \mathbf{L} &= I' \hat{R}_\theta \boldsymbol{\omega}.
\end{aligned}
\tag{9.123}
$$

Therefore,

$$\mathbf{L} = \hat{R}_\theta^{-1} I' \hat{R}_\theta \boldsymbol{\omega}.$$

Therefore, the moment of inertia changes under a similarity transformation,

$$I = \hat{R}_\theta^{-1} I' \hat{R}_\theta.$$

If the resulting matrix is diagonal, then we have diagonalized the moment of inertia tensor,

$$I = \begin{pmatrix} I_x & 0 & 0 \\ 0 & I_y & 0 \\ 0 & 0 & I_z \end{pmatrix}.$$

The diagonal entries are called the principal moments of inertia.

There are other tensors of importance, such as the polarization tensor, describing the anisotropy of the response of a dielectric to an applied electric field. However, the occurrence of tensors is most noted in the classical theory of general relativity with the appearance of the metric tensor which describes the curvature of spacetime and the energy-momentum tensor and curvature tensor, which appear in Einstein's equation of general relativity.

Tensors are more than matrices. They are a generalization of scalars and vectors and are defined in the way they transform under coordinate transformations.

Let's consider a a set of coordinates x^j, for $j = 1, \ldots, n$ and a transformation to new coordinates, $x^{i'} = x^{i'}(x^j)$, $i, j = 1, \ldots, n$, as a function of the old coordinates. We assume that the transformation is invertible, $x^j = x^j(x_{i'})$. Then we note that

$$dx^{i'} = \sum_{j=1}^{n} \frac{\partial x^{i'}}{\partial x^j} dx^j = \frac{\partial x^{i'}}{\partial x^j} dx^j, \tag{9.124}$$

$$\frac{\partial}{\partial x^{i'}} = \sum_{j=1}^{n} \frac{\partial x^j}{\partial x^{i'}} \frac{\partial}{\partial x^j} = \frac{\partial x^j}{\partial x^{i'}} \frac{\partial}{\partial x^j}. \tag{9.125}$$

Here we have introduced the Einstein summation convention.

We note that dx^j and $\frac{\partial}{\partial x^j}$ transform differently under a coordinate transformation. Quantities that transform like dx^j are called contravariant quantities. Quantities that transform like $\frac{\partial}{\partial x^j}$ are called covariant quantities.

There are different types of tensors. The simplest type of tensor is one in which there is no change under the transformation. Thus, $T' = T$. This is true for scalars. These are called rank zero tensors.

Let's denote

$$\Lambda^i_j = \frac{\partial x'^i}{\partial x^j}.$$

Then, a quantity that transforms like

$$V^i \to V'^i = \Lambda^i_j V^j$$

is called a contravariant vector, or a contravariant tensor of rank one. Similarly, a covariant vector, or a covariant tensor of rank one, transforms like

$$V^i \to V'_i = V_j (\Lambda^{-1})^j_i.$$

We can extend this to higher rank tensors. A contravariant second-rank tensor transforms like

$$T^{\alpha\beta} \to T'^{\alpha\beta} = \Lambda^\alpha_\gamma \Lambda^\beta_\delta T^{\gamma\delta}.$$

A covariant second-rank tensor transforms like

$$T_{\alpha\beta} \to T'_{\alpha\beta} = T_{\gamma\delta} (\Lambda^{-1})^\gamma_\alpha (\Lambda^{-1})^\delta_\beta.$$

Finally, one can mix contravariant and covariant indices. For example,

$$T^\alpha_\beta \to T'^\alpha_\beta = \Lambda^\alpha_\gamma T^\gamma_\delta (\Lambda^{-1})^\delta_\beta.$$

A special tensor is the Kronecker delta.

$$
\begin{aligned}
\delta_{\beta}^{\prime \alpha} &= \Lambda_{\gamma}^{\alpha} \delta_{\delta}^{\gamma} (\Lambda^{-1})_{\beta}^{\delta} \\
&= \Lambda_{\gamma}^{\alpha} (\Lambda^{-1})_{\beta}^{\gamma} \\
&= \delta_{\beta}^{\alpha}.
\end{aligned}
\tag{9.126}
$$

Linear combinations of tensors.

There are some simple operations one can perform with tensors. First, a linear combination of tensors is a tensor. Consider a linear combination of R_{β}^{α} and S_{β}^{α}, $T_{\beta}^{\alpha} = a R_{\beta}^{\alpha} + b S_{\beta}^{\alpha}$. Then,

$$
\begin{aligned}
T_{\beta}^{\prime \alpha} &= a R_{\beta}^{\prime \alpha} + b S_{\beta}^{\prime \alpha} \\
&= a \Lambda_{\gamma}^{\alpha} \Lambda_{\beta}^{\delta} R_{\delta}^{\gamma} + b \Lambda_{\gamma}^{\alpha} \Lambda_{\beta}^{\delta} S_{\delta}^{\gamma} \\
&= \Lambda_{\gamma}^{\alpha} \Lambda_{\beta}^{\delta} (a R_{\delta}^{\gamma} + b S_{\delta}^{\gamma}) \\
&= \Lambda_{\gamma}^{\alpha} \Lambda_{\beta}^{\delta} T_{\delta}^{\gamma}.
\end{aligned}
\tag{9.127}
$$

Direct product of tensors.

The direct product of two tensors A_{β}^{α} and B^{γ} is given by $T_{\beta}^{\alpha\gamma} = A_{\beta}^{\alpha} B^{\gamma}$. It transforms as

$$
T_{\beta}^{\prime \alpha \gamma} = \Lambda_{\delta}^{\alpha} \Lambda_{\beta}^{\epsilon} \Lambda_{\rho}^{\gamma} T_{\epsilon}^{\delta\rho}.
$$

Contraction of a tensor.

The contraction operation on a tensor is a sum over an upper and lower index. Thus, $T^{\alpha\gamma} = T_{\beta}^{\alpha\gamma\beta}$. Here we note that the repeated index implies summation over the index. Thus,

$$
T_{\beta}^{\alpha\gamma\beta} = \sum_{\beta=1}^{n} T_{\beta}^{\alpha\gamma\beta}.
$$

The transformation becomes

$$
T^{\prime \alpha \gamma} = \Lambda_{\delta}^{\alpha} \Lambda_{\delta}^{\gamma} \delta_{\mu}^{\rho} T_{\rho}^{\delta\sigma\mu}
$$

Differentiation of a tensor.

Differentiation of tensors is also possible. The derivative of a second-rank tensor gives a third-rank tensor, denoted $T\alpha\beta_{,\alpha}$. Thus, we define

$$
T_{\beta}^{\prime \alpha \gamma \beta} = \Lambda_{\delta}^{\alpha} \Lambda_{\sigma}^{\gamma} \delta_{\mu}^{\beta} T_{\rho}^{\delta\sigma\mu}.
$$

It transforms as

$$
T_{,\alpha}^{\prime \beta \gamma} = \frac{\partial}{\partial x^{\prime \alpha}} T^{\prime \beta \gamma} = \Lambda_{\alpha}^{\delta} \frac{\partial}{\partial x^{\delta}} \Lambda_{\sigma}^{\beta} \Lambda_{\rho}^{\gamma} T^{\sigma\rho}.
$$

Problems

1. Compute $\mathbf{u} \times \mathbf{v}$ using the permutation symbol. Verify your answer by computing these products using traditional methods.

    a. $\mathbf{u} = 2\mathbf{i} - 3\mathbf{k}$, $\mathbf{v} = 3\mathbf{i} - 2\mathbf{j}$.

    b. $\mathbf{u} = \mathbf{i} + \mathbf{j} + \mathbf{k}$, $\mathbf{v} = \mathbf{i} - \mathbf{k}$.

    c. $\mathbf{u} = 5\mathbf{i} + 2\mathbf{j} - 3\mathbf{k}$, $\mathbf{v} = \mathbf{i} - 4\mathbf{j} + 2\mathbf{k}$.

2. Compute the following determinants using the permutation symbol. Verify your answer.

a. $\begin{vmatrix} 3 & 2 & 0 \\ 1 & 4 & -2 \\ -1 & 4 & 3 \end{vmatrix}$

b. $\begin{vmatrix} 1 & 2 & 2 \\ 4 & -6 & 3 \\ 2 & 3 & 1 \end{vmatrix}$

3. For the given expressions, write out all values for $i, j = 1, 2, 3$.

   a. ϵ_{i2j}.

   b. ϵ_{i13}.

   c. $\epsilon_{ij1}\epsilon_{i32}$.

4. Show that

   a. $\delta_{ii} = 3$.

   b. $\delta_{ij}\epsilon_{ijk} = 0$

   c. $\epsilon_{imn}\epsilon_{jmn} = 2\delta_{ij}$.

   d. $\epsilon_{ijk}\epsilon_{ijk} = 6$.

5. Show that the vector $(\mathbf{a} \times \mathbf{b}) \times (\mathbf{c} \times \mathbf{d})$ lies on the line of intersection of the two planes: (1) the plane containing $\mathbf{a}$ and $\mathbf{b}$ and (2) the plane containing $\mathbf{c}$ and $\mathbf{d}$.

6. Prove the following vector identities:

   a. $(\mathbf{a} \times \mathbf{b}) \cdot (\mathbf{c} \times \mathbf{d}) = (\mathbf{a} \cdot \mathbf{c})(\mathbf{b} \cdot \mathbf{d}) - (\mathbf{a} \cdot \mathbf{d})(\mathbf{b} \cdot \mathbf{c})$.

   b. $(\mathbf{a} \times \mathbf{b}) \times (\mathbf{c} \times \mathbf{d}) = (\mathbf{a} \cdot \mathbf{b} \times \mathbf{d})\mathbf{c} - (\mathbf{a} \cdot \mathbf{b} \times \mathbf{c})\mathbf{d}$.

7. Use Problem 6a to prove that $|\mathbf{a} \times \mathbf{b}| = ab \sin \theta$.

8. A particle moves on a straight line, $\mathbf{r} = t\mathbf{u}$, from the center of a disk. If the disk is rotating with angular velocity ω, then $\mathbf{u}$ rotates. Let $\mathbf{u} = (\cos \omega t)\mathbf{i} + (\sin \omega t)\mathbf{j}$.

   a. Determine the velocity, $\mathbf{v}$.

   b. Determine the acceleration, $\mathbf{a}$.

   c. Describe the resulting acceleration terms identifying the centripetal acceleration and Coriolis acceleration.

9. Compute the gradient of the following:

   a. $f(x,y) = x^2 - y^2$.

   b. $f(x,y,z) = yz + xy + xz$.

   c. $f(x,y) = \tan^{-1}\left(\frac{y}{x}\right)$.

   d. $f(x,y,z) = 2y^x \cos z - 5 \sin z \cos y$.

10. Find the directional derivative of the given function at the indicated point in the given direction.

  a. $f(x,y) = x^2 - y^2$, $(3,2)$, $\mathbf{u} = \mathbf{i} + \mathbf{j}$.

  b. $f(x,y) = \frac{y}{x}$, $(2,1)$, $\mathbf{u} = 3\mathbf{i} + 4\mathbf{j}$.

  c. $f(x,y,z) = x^2 + y^2 + z^2$, $(1,0,2)$, $\mathbf{u} = 2\mathbf{i} - \mathbf{j}$.

11. Zaphod Beeblebrox was in trouble after the infinite improbability drive caused the Heart of Gold, the spaceship Zaphod had stolen when he was President of the Galaxy, to appear between a small insignificant planet and its hot sun. The temperature of the ship's hull is given by $T(x,y,z) = e^{-k(x^2+y^2+z^2)}$ Nivleks. He is currently at $(1,1,1)$, in units of globs, and $k = 2$ globs^{-2}. (Check the *Hitchhikers Guide* for the current conversion of globs to kilometers and Nivleks to Kelvin.)

  a. In what direction should he proceed so as to decrease the temperature the quickest?

  b. If the Heart of Gold travels at e^6 globs per second, then how fast will the temperature decrease in the direction of fastest decline?

12. A particle moves under the force field $\mathbf{F} = -\nabla V$, where the potential function is given by $V(x,y,z) = x^3 + y^3 - 3xy + 5$. Find the equilibrium points of $\mathbf{F}$ and determine if the equilibria are stable or unstable.

13. For the given vector field, find the divergence and curl of the field.

  a. $\mathbf{F} = x\mathbf{i} + y\mathbf{j}$.

  b. $\mathbf{F} = \frac{y}{r}\mathbf{i} - \frac{x}{r}\mathbf{j}$, for $r = \sqrt{x^2 + y^2}$.

  c. $\mathbf{F} = x^2 y\mathbf{i} + z\mathbf{j} + xyz\mathbf{k}$.

14. Write the following using ϵ_{ijk} notation and simplify if possible.

  a. $\mathbf{C} \times (\mathbf{A} \times (\mathbf{A} \times \mathbf{C}))$.

  b. $\nabla \times (\nabla \times \mathbf{A})$.

  c. $\nabla \times \nabla \phi$.

15. Prove the identities:

  a. $\nabla \cdot (\nabla \times \mathbf{A}) = 0$.

  b. $\nabla \cdot (f\nabla g - g\nabla f) = f\nabla^2 g - g\nabla^2 f$.

  c. $\nabla r^n = nr^{n-2}\mathbf{r}, \quad n \geq 2$.

16. For $\mathbf{r} = x\mathbf{i} + y\mathbf{j} + z\mathbf{k}$ and $r = |\mathbf{r}|$, simplify the following.

  a. $\nabla \times (\mathbf{k} \times \mathbf{r})$.

  b. $\nabla \cdot \left(\frac{\mathbf{r}}{r}\right)$.

  c. $\nabla \times \left(\frac{\mathbf{r}}{r}\right)$.

  d. $\nabla \cdot \left(\frac{\mathbf{r}}{r^3}\right)$.

e. $\nabla \times \left(\frac{\mathbf{r}}{r^3} \right)$.

17. Newton's Law of Gravitation gives the gravitational force between two masses as

$$\mathbf{F} = -\frac{GmM}{r^3} \mathbf{r}.$$

a. Prove that $\mathbf{F}$ is irrotational.

b. Find a scalar potential for $\mathbf{F}$.

18. Consider a constant electric dipole moment $\mathbf{p}$ at the origin. It produces an electric potential of $\phi = \frac{\mathbf{p} \cdot \mathbf{r}}{4\pi\epsilon_0 r^3}$ outside the dipole. Noting that $\mathbf{E} = -\nabla\phi$, find the electric field at $\mathbf{r}$.

19. In fluid dynamics, the Euler equations govern inviscid fluid flow and provide quantitative statements on the conservation of mass, momentum, and energy. The continuity equation is given by

$$\frac{\partial \rho}{\partial t} + \nabla \cdot (\rho \mathbf{v}) = 0,$$

where $\rho(x,y,z,t)$ is the mass density and $\mathbf{v}(x,y,z,t)$ is the fluid velocity. The momentum equations are given by

$$\frac{\partial \rho \mathbf{v}}{\partial t} + \mathbf{v} \cdot \nabla(\rho \mathbf{v}) = \mathbf{f} - \nabla p.$$

Here, $p(x,y,z,t)$ is the pressure and $\mathbf{f}$ is the external force per volume.

a. Show that the continuity equation can be rewritten as

$$\frac{\partial \rho}{\partial t} + \rho \nabla \cdot (\mathbf{v}) + \mathbf{v} \cdot \nabla \rho = 0.$$

b. Prove the identity $\frac{1}{2}\nabla v^2 = \mathbf{v} \cdot \nabla \mathbf{v}$ for irrotational $\mathbf{v}$.

c. Assume that

- the external forces are conservative ($\mathbf{f} = -\rho\nabla\phi$),
- the velocity field is irrotational ($\nabla \times \mathbf{v} = \mathbf{0}$),
- the fluid is incompressible ($\rho =$ const), and
- the flow is steady, $\frac{\partial \mathbf{v}}{\partial t} = 0$.

Under these assumptions, prove Bernoulli's Principle:

$$\frac{1}{2}v^2 + \phi + \frac{p}{\rho} = \text{const.}$$

20. Find the lengths of the following curves:

a. $y(x) = x$ for $x \in [0,2]$.

b. $(x,y,z) = (t, \ln t, 2\sqrt{2}t)$ for $1 \le t \le 2$.

c. $y(x) = \cosh x$, $x \in [-2,2]$. (Recall the hanging chain example from classical dynamics.)

21. Consider the integral $\int_C y^2\,dx - 2x^2\,dy$. Evaluate this integral for the following curves:

    a. C is a straight line from $(0,2)$ to $(1,1)$.

    b. C is the parabolic curve $y = x^2$ from $(0,0)$ to $(2,4)$.

    c. C is the circular path from $(1,0)$ to $(0,1)$ in a clockwise direction.

22. Evaluate $\int_C (x^2 - 2xy + y^2)\,ds$ for the curve $x(t) = 2\cos t$, $y(t) = 2\sin t$, $0 \le t \le \pi$.

23. Prove that the magnetic flux density, **B**, satisfies the wave equation.

24. Let C be a closed curve and D the enclosed region. Prove the identity

$$\int_C \phi \nabla \phi \cdot \mathbf{n}\,ds = \int_D (\phi \nabla^2 \phi + \nabla \phi \cdot \nabla \phi)\,dA.$$

25. Let S be a closed surface and V the enclosed volume. Prove Green's first and second identities, respectively.

    a. $\int_S \phi \nabla \psi \cdot \mathbf{n}\,dS = \int_V (\phi \nabla^2 \psi + \nabla \phi \cdot \nabla \psi)\,dV$.

    b. $\int_S [\phi \nabla \psi - \psi \nabla \phi] \cdot \mathbf{n}\,dS = \int_V (\phi \nabla^2 \psi - \psi \nabla^2 \phi)\,dV$.

26. Let C be a closed curve and D the enclosed region. Prove Green's identities in two dimensions.

    a. First prove

$$\int_D (v \nabla \cdot \mathbf{F} + \mathbf{F} \cdot \nabla v)\,dA = \int_C (v\mathbf{F}) \cdot d\mathbf{s}.$$

    b. Let $\mathbf{F} = \nabla u$ and obtain Green's first identity,

$$\int_D (v \nabla^2 u + \nabla u \cdot \nabla v)\,dA = \int_C (v \nabla u) \cdot d\mathbf{s}.$$

    c. Use Green's first identity to prove Green's second identity,

$$\int_D (u \nabla^2 v - v \nabla^2 u)\,dA = \int_C (u \nabla v - v \nabla u) \cdot d\mathbf{s}.$$

27. Compute the work done by the force $\mathbf{F} = (x^2 - y^2)\mathbf{i} + 2xy\mathbf{j}$ in moving a particle counterclockwise around the boundary of the rectangle $R = [0,3] \times [0,5]$.

28. Compute the following integrals:

    a. $\int_C (x^2 + y)\,dx + (3x + y^3)\,dy$ for C the ellipse $x^2 + 4y^2 = 4$.

    b. $\int_S (x - y)\,dydz + (y^2 + z^2)\,dzdx + (y - x^2)\,dxdy$ for S the positively oriented unit sphere.

    c. $\int_C (y - z)\,dx + (3x + z)\,dy + (x + 2y)\,dz$, where C is the curve of intersection between $z = 4 - x^2 - y^2$ and the plane $x + y + z = 0$.

    d. $\int_C x^2 y\,dx - xy^2\,dy$ for C a circle of radius 2 centered about the origin.

e. $\int_S x^2 y \, dydz + 3y^2 \, dzdx - 2xz^2 \, dxdy$, where S is the surface of the cube $[-1,1] \times [-1,1] \times [-1,1]$.

29. Use Stokes' Theorem to evaluate the integral

$$\int_C -y^3 \, dx + x^3 \, dy - z^3 \, dz$$

for C the (positively oriented) curve of intersection between the cylinder $x^2 + y^2 = 1$ and the plane $x + y + z = 1$.

30. Use Stokes' Theorem to derive the integral form of Faraday's law,

$$\int_C \mathbf{E} \cdot d\mathbf{s} = -\frac{\partial}{\partial t} \int\int_D \mathbf{B} \cdot d\mathbf{S}$$

from the differential form of Maxwell's equations.

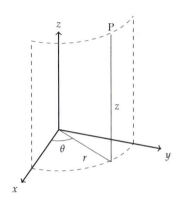

Figure 9.30: Cylindrical coordinate system.

31. For cylindrical coordinates,

$$\begin{aligned} x &= r \cos\theta, \\ y &= r \sin\theta, \\ z &= z, \end{aligned}$$ (9.128)

find the scale factors and derive the following expressions:

$$\nabla f = \frac{\partial f}{\partial r} \hat{\mathbf{e}}_r + \frac{1}{r} \frac{\partial f}{\partial \theta} \hat{\mathbf{e}}_\theta + \frac{\partial f}{\partial z} \hat{\mathbf{e}}_z.$$ (9.129)

Note that it is customary to write the basis as $\{\hat{\mathbf{e}}_r, \hat{\mathbf{e}}_\theta, \hat{\mathbf{e}}_z\}$ instead of $\{\hat{\mathbf{u}}_1, \hat{\mathbf{u}}_2, \hat{\mathbf{u}}_3\}$.

$$\nabla \cdot \mathbf{F} = \frac{1}{r} \frac{\partial(rF_r)}{\partial r} + \frac{1}{r} \frac{\partial F_\theta}{\partial \theta} + \frac{\partial F_z}{\partial z}$$ (9.130)

$$\nabla \times \mathbf{F} = \left(\frac{1}{r}\frac{\partial F_z}{\partial \theta} - \frac{\partial F_\theta}{\partial z}\right) \hat{\mathbf{e}}_r + \left(\frac{\partial F_r}{\partial z} - \frac{\partial F_z}{\partial r}\right) \hat{\mathbf{e}}_\theta + \frac{1}{r}\left(\frac{\partial(rF_\theta)}{\partial r} - \frac{\partial F_r}{\partial \theta}\right) \hat{\mathbf{e}}_z.$$ (9.131)

$$\nabla^2 f = \frac{1}{r}\frac{\partial}{\partial r}\left(r\frac{\partial f}{\partial r}\right) + \frac{1}{r^2}\frac{\partial^2 f}{\partial \theta^2} + \frac{\partial^2 f}{\partial z^2}.$$ (9.132)

32. For spherical coordinates,

$$\begin{aligned} x &= \rho \sin\theta \cos\phi, \\ y &= \rho \sin\theta \sin\phi, \\ z &= \rho \cos\theta, \end{aligned}$$ (9.133)

find the scale factors and derive the following expressions:

$$\nabla f = \frac{\partial f}{\partial \rho} \hat{\mathbf{e}}_\rho + \frac{1}{\rho}\frac{\partial f}{\partial \theta} \hat{\mathbf{e}}_\theta + \frac{1}{\rho \sin\theta}\frac{\partial f}{\partial \phi} \hat{\mathbf{e}}_\phi.$$ (9.134)

$$\nabla \cdot \mathbf{F} = \frac{1}{\rho^2}\frac{\partial(\rho^2 F_\rho)}{\partial \rho} + \frac{1}{\rho \sin\theta}\frac{\partial(\sin\theta F_\theta)}{\partial \theta} + \frac{1}{\rho \sin\theta}\frac{\partial F_\phi}{\partial \phi}.$$ (9.135)

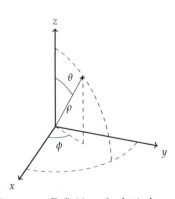

Figure 9.31: Definition of spherical coordinates for Problem 32.

$$\nabla \times \mathbf{F} = \frac{1}{\rho \sin \theta} \left(\frac{\partial (\sin \theta F_\phi)}{\partial \theta} - \frac{\partial F_\theta}{\partial \phi} \right) \hat{\mathbf{e}}_\rho + \frac{1}{\rho} \left(\frac{1}{\sin \theta} \frac{\partial F_\rho}{\partial \phi} - \frac{\partial (\rho F_\phi)}{\partial \rho} \right) \hat{\mathbf{e}}_\theta,$$
$$+ \frac{1}{\rho} \left(\frac{\partial (\rho F_\theta)}{\partial \rho} - \frac{\partial F_\rho}{\partial \theta} \right) \hat{\mathbf{e}}_\phi. \tag{9.136}$$

$$\nabla^2 f = \frac{1}{\rho^2} \frac{\partial}{\partial \rho} \left(\rho^2 \frac{\partial f}{\partial \rho} \right) + \frac{1}{\rho^2 \sin \theta} \frac{\partial}{\partial \theta} \left(\sin \theta \frac{\partial f}{\partial \theta} \right) + \frac{1}{\rho^2 \sin^2 \theta} \frac{\partial^2 f}{\partial \phi^2}. \tag{9.137}$$

33. The moments of inertia for a system of point masses are given by sums instead of integrals. For example, $I_{xx} = \sum_i m_i(y_i^2 + z_i^2)$ and $I_{xy} = -\sum_i m_i x_i y_i$. Find the inertia tensor about the origin for $m_1 = 2.0$ kg at $(1.0, 0, 1.0)$, $m_2 = 5.0$ kg at $(1.0, -1.0, 0)$, and $m_3 = 1.0$ kg at $(1.0, 1.0, 1.0)$ where the coordinate units are in meters.

34. Consider the octant of a uniform sphere of density 5 grams per cubic centimeter and radius a lying in the first octant.

   a. Find the inertia tensor about the origin.

   b. What are the principal moments of inertia?

   c Find the principal axes of inertia, that is, the eigenvectors of the inertia tensor.

35. Let T^α be a contravariant vector and S_α be a covariant vector.

   a. Show that $R_\beta = g_{\alpha\beta} T^\alpha$ is a covariant vector.

   b. Show that $R^\beta = g^{\alpha\beta} S_\alpha$ is a contravariant vector.

36. Show that $T^{\alpha\beta\gamma\delta\rho} S_{\beta\rho}$ is a tensor. What is its rank?

37. The line element in terms of the metric tensor, $g_{\alpha\beta}$ is given by

$$ds^2 = g_{\alpha\beta} \, dx^\alpha dx^\beta.$$

Show that the transformed metric for the transformation $x'^\alpha = x'^\alpha(x^\beta)$ is given by

$$g'_{\gamma\delta} = g_{\alpha\beta} \frac{\partial x^\alpha}{\partial x'^\gamma} \frac{\partial x^\beta}{\partial x'^\delta}$$

10

Extrema and Variational Calculus

"The reader will find no figures in this work. The methods which I set forth do not require either constructions or geometrical or mechanical reasonings: but only algebraic operations, subject to a regular and uniform rule of procedure."—Joseph-Louis Lagrange (1736–1813) Preface to *Méchanique Analytique*.

10.1   Stationary and Extreme Values of Functions

YOU MAY RECALL LOOKING FOR MAXIMA AND MINIMA of functions in calculus. The extreme values seem to say little more than where the local maxima and minima of a function are, or where the highest and lowest points on a mountain range might be. However, the field of optimization goes much further than simple functions. In this chapter we will be asking what functions $y(x)$ will minimize, or maximize, integrals of the form

$$I = \int_a^b f(x, y(x), y'(x)) \, dx, \qquad (10.1)$$

where a, b, $f(a)$, $f(b)$, and $f(x)$ are given. $y(x)$ could be a path and I the length of the path. Such a problem lies in the realm of the calculus of variations.

Some of the most fundamental principles in physics are based on seeking extrema of this type. For example, Fermat's principle states that light takes the path of least time. Newton discussed the Principle of Least Resistance, or Least Effort, along which one seeks a path of minimal resistance. Such integrals also lead to determining the shapes of soap bubbles, hanging chains, or the path a particle might take near a black hole.

There are many problems that are simply stated and that have led to interesting mathematics and physics. For example, we will consider problems such as

1. What curve connecting two given points has the shortest length?

2. Given two points, what path will a particle follow giving the shortest time?

3. Find the minimum surface of revolution passing through two given points.

In this chapter we will explore some of these problems. However, before we discuss the principles leading to applications of the calculus of variations in physics, we will first recall how one finds the extrema of functions of one or more variables.

10.1.1 Functions of One Variable

ONE LEARNS IN CALCULUS HOW TO FIND the minima and maxima of a function of one variable. We will review the process of finding critical points and characterizing them for functions of several variables. We will also consider the case when there are additional imposed constraints.

A function $f(x)$ has a local minimum at $x = a$ if $f(x) \geq f(a)$ for points near $x = a$; that is, $f(x)$ has a local minimum at $x = a$ if for $\epsilon > 0$, $f(x) \geq f(a)$ whenever $|x - a| < \epsilon$. Similarly, $f(x)$ has a local maximum at $x = a$ if for $\epsilon > 0$, $f(x) \leq f(a)$ whenever $|x - a| < \epsilon$. The minima, or maxima, collectively referred to as extrema, are global extrema if the inequality holds over the entire interval of interest.

For functions of one variable, it is easy to locate minima and maxima if one graphs the function such as in Figure 10.1. These are the local peaks and valleys in the graph of the function. In calculus one learns that at extrema, the tangent line is horizontal, or the slope of tangent line is zero. Thus, if a function $f(x)$ has a local extremum at $x = a$, then $f'(a) = 0$.

However, it is not always the case that when $f'(a) = 0$, one has an extremum. Such points where the derivative vanishes are called critical points and one needs to classify these critical points in order to determine if these points are local minima or maxima. This leads to the second derivative test.

The second derivative test naturally results from considering the Taylor series expansion of $f(x)$ near $x = a$. The first terms of the expansion are given by

$$f(x) = f(a) + f'(a)(x - a) + \frac{1}{2!}f''(a)(x - a)^2 + \frac{1}{3!}f^{(3)}(a)(x - a)^3 + \dots.$$

$$(10.2)$$

Keeping the first two terms, one has the equation for the tangent line of $f(x)$ at $x = a$,

$$f(x) \approx f(a) + f'(a)(x - a).$$

This is known as a linearization of $f(x)$ at $x = a$. Defining the increments in x and f by $\Delta x = x - a$ and $\Delta f = f(x) - f(a)$, we have the relation

$$\Delta f \approx f'(a)\Delta x.$$

As Δx approaches zero, we arrive at the relation between differentials,

$$df = f'(a)\,dx.$$

If $x = a$ is a critical point of $f(x)$, then the linear approximation becomes $f(x) = f(a)$. While the graph of this line is horizontal as shown in Figure

---

Local maxima and local minima.

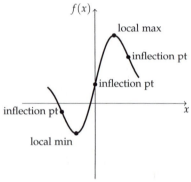

Figure 10.1: Locating local maxima, local minima, and inflection points.

Linear approximation of $f(x)$.

10.2, it does not say much about points near $x = a$. So, we should include the next term of the Taylor series.

The next approximation, the second-order Taylor polynomial, is given by

$$f(x) \approx f(a) + f'(a)(x - a) + \frac{1}{2!}f''(a)(x - a)^2.$$

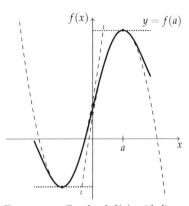

This is easily recognizable as a quadratic function in x whose graph is a parabola. Thus, we have a quadratic approximation of $f(x)$. If $x = a$ is a critical point of $f(x)$, then this approximation leads to

$$f(x) = f(a) + \frac{1}{2!}f''(a)(x - a)^2.$$

Figure 10.2: Graph of $f(x)$ with linear and parabolic approximations shown at the extrema.

Expanding the quadratic term, we have

$$f(x) = \frac{1}{2}f''(a)x^2 - f''(a)ax + \frac{1}{2}f''(a)a^2 + f(a).$$

If $f''(a) \neq 0$, then the coefficient of x^2 determines if the parabola opens upward, or downward. More formally, the function has a positive concavity (concave up, $f''(a) > 0$) or a negative concavity (concave down, $f''(a) < 0$) at $x = a$. Examples of quadratic approximations at local extrema are shown in Figure 10.2.

Quadratic approximation of $f(x)$.

Because the quadratic function is an approximation of $f(x)$ near $x = a$, a positive concavity indicates that there is a local minimum at $x = a$. Similarly, a negative concavity indicates that there is a local maximum at $x = a$.

If $f''(a) = 0$, then the approximation reduces to $f(x) \approx f(a)$, and we need more terms from the series expansion. Thus,

$$f(x) = f(a) + \frac{1}{3!}f^{(3)}(a)(x - a)^3 + \frac{1}{4!}f^{(4)}(a)(x - a)^4 + \dots.$$

In these cases, we would need information about higher-order derivatives of $f(x)$ at $x = a$. If the third derivative does not vanish, then we do not have a local minimum or maximum. However, if it does vanish, maybe the next term can lead to a determination of a local extremum.

Typically in calculus, the test for extrema stops at the second derivative. Thus, the classification of a critical point leads to the following second derivative test, based on the above analysis:

The second derivative test.

Let $x = a$ be a critical point of $f(x)$, i.e., $f'(a) = 0$. Then, if $f(x)$ is twice differentiable,

1. If $f''(a) > 0$, then $x = a$ is a local minimum.

2. If $f''(a) > 0$, then $x = a$ is a local maximum.

3. If $f''(x) = 0$, then there is no concavity at $x = a$.

Inflection points.

Note that in the third case, nothing can be said and further investigation is needed. In fact, when the concavity changes sign near a point, that point is called a point of inflection. An example is shown in Figure 10.1. One need not have a vanishing first derivative to have an inflection point. One only needs that the lowest-order nonvanishing derivative is of odd order greater than two. If the lowest-order nonvanishing derivative is even, then the point is sometimes called an undulation point. The simplest examples of these cases are $f(x) = x^3$ and $f(x) = x^4$. In both cases, $f''(0) = 0$. However, the concavity changes sign for the cubic function and not for the quartic function. If $x = a$ is an inflection point such that $f'(a) = 0$, then it is called a saddle point.

Example 10.1. *For $f(x) = 2\sin x + \sin 2x + 1$ locate and classify any extrema for $x \in [-2, 2]$.*

The critical points are determined by setting $f'(x) = 0$. Namely,

$$
\begin{aligned}
0 &= 2\cos x + 2\cos 2x \\
&= 2\cos x + 2(2\cos^2 x - 1) \\
&= 2(2\cos^2 x + \cos x - 1) \\
&= 2(2\cos x - 1)(\cos x + 1)
\end{aligned}
\tag{10.3}
$$

Therefore,

$$
\cos x = \frac{1}{2}, \quad \text{or } \cos x = -1.
$$

On the interval $[-2, 2]$, the second equation does not hold. However, the first equation is satisfied for $x = \pm\frac{\pi}{3}$.

The second derivative is given by

$$
f''(x) = -2\sin x - 4\sin 2x.
$$

At the critical points we have $f''\left(\frac{\pi}{3}\right) = -3\sqrt{3}$ and $f''\left(-\frac{\pi}{3}\right) = 3\sqrt{3}$. Therefore, there is a local maximum at $x = \frac{\pi}{3}$ and a local maximum at $x = -\frac{\pi}{3}$. The values of these extrema are $f\left(\frac{\pi}{3}\right) = 1 + \frac{1}{2}\sqrt{3} \approx 3.598$ and $f\left(-\frac{\pi}{3}\right) = 1 - \frac{1}{2}\sqrt{3} \approx -1.598$.

If one is seeking global extrema on the interval, then one also needs to valuate the function at the endpoints. Because $f(2) = 2.0618$ and $f(-2) = -0.0618$, the extrema are also global extrema on the given interval.

As a last note, there is also an inflection point on this curve. In Figure 10.1 we actually show a sketch of this function. In order to locate the inflection point, we set the second derivative equal to zero. Then,

$$
\begin{aligned}
0 &= f''(x) \\
&= -2\sin x - 4\sin 2x \\
&= -2\sin x(1 + 4\cos x).
\end{aligned}
\tag{10.4}
$$

Solutions on $[-2, 2]$ are $x = 0, \pm\cos^{-1}(-\frac{1}{4})$. These are indicated in Figure 10.1. Furthermore, we note that the first derivative does not vanish at these points.

10.1.2   *Functions of Several Variables*

A SIMILAR ANALYSIS OF EXTREMA CAN BE CARRIED OUT for functions of several variables. We begin with functions of two variables $f(x, y)$. As with the case of functions of one variable, we consider the Taylor series approximation in order to define and classify critical points. Instead of looking for tangent lines of slope zero, we now imagine horizontal tangent planes.

The tangent plane approximation to a function of two variables $f(x, y)$ at a point (x_0, y_0) is given by the equation of the plane

$$f(x, y) \approx f(x_0, y_0) + f_x(x_0, y_0)(x - x_0) + f_y(x_0, y_0)(y - y_0) \qquad (10.5)$$

where f_x and f_y denote partial derivatives of f with respect to the independent variables. This can be written a little more compactly using the gradient operation:

$$f(x, y) \approx f(x_0, y_0) + \nabla f(x_0, y_0) \cdot \Delta \mathbf{x}, \qquad (10.6)$$

where $\Delta \mathbf{x} = (x - x_0, y - y_0)^T$ and $\nabla f = \langle f_x, f_y \rangle$.

Example 10.2. *Find the tangent plane to the surface $z = \frac{x^2}{a^2} + \frac{y^2}{b^2}$ at the point (x_0, y_0).*

We first compute the first-order derivatives and evaluate them at (x_0, y_0) :

$$f_x(x_0, y_0) = \frac{2x_0}{a^2}, \quad f_y(x_0, y_0) = \frac{2y_0}{b^2}.$$

Then,

$$\begin{aligned}
f(x, y) &\approx f(x_0, y_0) + f_x(x_0, y_0)(x - x_0) + f_y(x_0, y_0)(y - y_0) \\
&= \frac{x_0^2}{a^2} + \frac{y_0^2}{b^2} + \frac{2x_0}{a^2}(x - x_0) + \frac{2y_0}{b^2}(y - y_0) \\
&= \frac{x_0}{a^2}(2x - x_0) + \frac{y_0}{b^2}(2y - y_0). \qquad (10.7)
\end{aligned}$$

Example 10.3. *Find the tangent plane to the surface $z = x^3 - 3xy^2$, a monkey saddle, at the point $(2, 1, 2)$.*

We first find that $z(2, 1) = 2$ and

$$\nabla z = \langle 3x^2 - 3y^2, -6xy \rangle \big|_{(2,1)} = \langle 9, -12 \rangle.$$

Therefore, the equation of the tangent plane is

$$\begin{aligned}
f(x, y) &\approx f(x_0, y_0) + f_x(x_0, y_0)(x - x_0) + f_y(x_0, y_0)(y - y_0) \\
&= 2 + 9(x - 2) - 12(y - 1) \\
&= 9x - 12y - 4. \qquad (10.8)
\end{aligned}$$

This is shown in Figure 10.4

The second-order approximation involves second-order derivatives and quadratic powers in x and y.

The tangent plane approximation or linear approximation of a function of two variables.

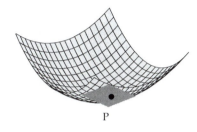

Figure 10.3: Tangent plane to a surface, $z = f(x, y)$.

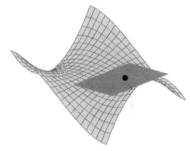

Figure 10.4: The equation of the tangent plane to the monkey saddle $z = x^3 - 3xy^2$ at $(2, 1, 2)$.

$$f(x,y) \approx f(x_0, y_0) + \nabla f(x_0, y_0) \cdot \Delta \mathbf{x} + \frac{1}{2} \left[f_{xx}(x_0, y_0)(x - x_0)^2 \right.$$
$$\left. + \; 2 f_{xy}(x_0, y_0)(x - x_0)(y - y_0) + f_{yy}(x_0, y_0)(y - y_0)^2 \right]. \tag{10.9}$$

Defining the Hessian matrix

$$H(f) = \begin{pmatrix} \frac{\partial^2 f}{\partial x^2} & \frac{\partial^2 f}{\partial x \partial y} \\ \frac{\partial^2 f}{\partial x \partial y} & \frac{\partial^2 f}{\partial y^2} \end{pmatrix},$$

we can rewrite the second-order approximation as

$$f(x,y) \approx f(x_0, y_0) + \nabla f(x_0, y_0) \cdot \Delta \mathbf{x} + \frac{1}{2} \Delta \mathbf{x}^T H(f(x_0, y_0)) \Delta \mathbf{x}. \tag{10.10}$$

The second-order approximation of a function of two variables.

Example 10.4. *Find the second-order approximation to the monkey saddle* $z(x,y) = x^3 - 3xy^2$ *at* $(2, 1, 2)$.

In a previous example, we found the linear approximation

$$z \approx 2 + 9(x - 2) - 12(y - 1) = 9x - 12y - 4.$$

This is the equation of the tangent plane at the point $(2, 1, 2)$ *on the surface. We next seek the quadratic approximation at this point. In order to compute that term, we need to find the Hessian matrix at* $(2, 1)$.

$$H(z(2,1)) = \begin{pmatrix} \frac{\partial^2 f}{\partial x^2} & \frac{\partial^2 f}{\partial x \partial y} \\ \frac{\partial^2 f}{\partial x \partial y} & \frac{\partial^2 f}{\partial y^2} \end{pmatrix}_{(2,1)} = \begin{pmatrix} 6x & -6y \\ -6y & -6x \end{pmatrix}_{(2,1)} = \begin{pmatrix} 12 & -6 \\ -6 & -12 \end{pmatrix}.$$

Letting $\Delta x = x - 2$ *and* $\Delta y = y - 1$, *the quadratic term then becomes*

$$\frac{1}{2} \Delta \mathbf{x}^T H \Delta \mathbf{x} = \begin{pmatrix} \Delta x & \Delta y \end{pmatrix} \begin{pmatrix} 6 & -3 \\ -3 & -6 \end{pmatrix} \begin{pmatrix} \Delta x \\ \Delta y \end{pmatrix}$$

$$= \begin{pmatrix} \Delta x & \Delta y \end{pmatrix} \begin{pmatrix} 6\Delta x - 3\Delta y \\ -3\Delta x - 6\Delta y \end{pmatrix}$$

$$= \Delta x(6\Delta x - 3\Delta y) + \Delta y(-3\Delta x - 6\Delta y)$$

$$= 6\Delta x^2 - 6\Delta x \Delta y - 6\Delta y^2. \tag{10.11}$$

Therefore, the quadratic approximation is

$$z(x,y) \approx -9x - 12y - 4 + 6[(x-2)^2 - (x-2)(y-1) - (y-1)^2]$$
$$= 6x^2 - 6xy - 6y^2 - 9x + 12y + 2. \tag{10.12}$$

Figure 10.5: Fit of the quadratic approximation to the monkey saddle.

Now let's return to the idea of extrema. A point $\mathbf{x}_0$ is a local minimum of $f(\mathbf{x})$ if there is a neighborhood of $\mathbf{x}_0$ such that for all points $\mathbf{x}$ in the neighborhood, $f(\mathbf{x}) \geq f(\mathbf{x}_0)$. Similarly, local maxima can be defined and such points are collectively called local (or relative) extrema.

As in the case of functions of a single variable, we need to find the critical points. A point $\mathbf{x}_0$ is a critical point of $f(\mathbf{x})$ if $\nabla f(\mathbf{x}_0) = 0$. A point that is not a local extremum is a saddle point. Once one has located the critical points, they need to be classified. For functions of one variable, we used the second derivative test. This was geometrically connected to the concavity of the quadratic approximation. For multivariate functions, we have a similar picture. In particular, for functions of two variables, the quadratic approximation can be graphed as quadric surface. Because we are only interested in functions of the form $z = f(x, y)$, we would obtain quadric surfaces of the form

$$z = Ax^2 + +2Bxy + Cy^2 + Dx + Ey + F.$$

As with conics in Section 3.3.3, we can apply translations to obtain

$$z = A(x-a)^2 + 2B(x-a)(y-b) + C(y-b)^2 + G.$$

The nature of the quadric surface depends on the constants A, B, and C. For example, the surfaces $z = x^2 + 3y^2$ and $z = 9 - x^2 - y^2$ are elliptical paraboloids. The surface $z = x^2 - 3y^2$ is a hyperbolic paraboloid. When seeking local extrema, we would expect that locally the quadratic approximation would be the equation of an elliptical paraboloid. In fact, this depends on the determinant of the matrix of quadric coefficients,

$$M = \begin{pmatrix} A & B \\ B & C \end{pmatrix},$$

which is $\det(M) = AC - B^2$. We will show this using these quadrics because they are already in diagonalized form as quadratics. [See Section 3.3.3.]

Example 10.5. *Evaluate* $\det(M)$ *and find the eigenvalues for the surfaces* $z = x^2 + 3y^2$, $z = 9 - x^2 - y^2$, *and* $z = x^2 - 3y^2$.

a. $z = x^2 + 3y^2$ *is the equation of an elliptical paraboloid that opens upward.*

$$M = \begin{pmatrix} 1 & 0 \\ 0 & 3 \end{pmatrix}.$$

$\det(M) = 3$, *and* $\lambda = 1, 3$.

b. $z = 9 - x^2 - y^2$ *is the equation of an elliptical paraboloid that opens downward.*

$$M = \begin{pmatrix} -1 & 0 \\ 0 & -1 \end{pmatrix}.$$

$\det(M) = 1$, *and* $\lambda = -1$.

c. $z = x^2 - 3y^2$ *is the equation of an hyperbolic paraboloid, or saddle.*

$$M = \begin{pmatrix} 1 & 0 \\ 0 & -3 \end{pmatrix}.$$

$\det(M) = -3$, *and* $\lambda = 1, -3$.

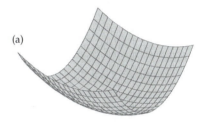
(a)

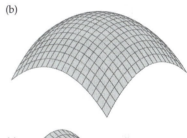
(b)

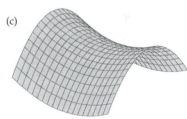
(c)

Figure 10.6: Plots of the elliptical paraboloids (a) $z = x^2 + 3y^2$, (b) $z = 9 - x^2 - y^2$, and the saddle (c) $z = x^2 - 3y^2$.

From these simple examples, we see that $\det(M) > 0$ for the elliptical paraboloids and that $\det(M) < 0$ for the saddle. In the more general case, M is a real symmetric matrix and can be diagonalized. The similarity transformation results from a rotation of the general quadratic so as to eliminate B, making the matrix diagonal. In fact, the determinant is invariant under this similarity transformation, and thus we would expect that the determinant $AC - B^2$ can be used to classify the surfaces. Namely,

1. $AC - B^2 > 0$ implies the surface is an elliptical paraboloid; and,

2. $AC - B^2 < 0$ implies the surface is a saddle.

In terms of the eigenvalues of M, we have that

1. $\lambda_1 \lambda_2 > 0$ implies the surface is an elliptical paraboloid; and,

2. $\lambda_1 \lambda_2 < 0$ implies the surface is a saddle.

Furthermore, the eigenvalues need to have the same sign for elliptical paraboloids. Positive eigenvalues will result in upward-pointing paraboloids.

The connection to the quadratic approximation of a given surface near a critical point is that the matrix M is proportional to the Hessian matrix. We can show this by considering the critical point (x_0, y_0) of $f(x, y)$. Then, the quadratic approximation at the critical point is given by

$$
\begin{aligned}
f(x, y) \approx \ & f(x_0, y_0) + \frac{1}{2} \Big[f_{xx}(x_0, y_0)(x - x_0)^2 \\
& + 2 f_{xy}(x_0, y_0)(x - x_0)(y - y_0) + f_{yy}(x_0, y_0)(y - y_0)^2 \Big]
\end{aligned}
$$
(10.13)

and the matrix of quadratic coefficients is given by

$$
M = \frac{1}{2} \begin{pmatrix} \frac{\partial^2 f}{\partial x^2} & \frac{\partial^2 f}{\partial x \partial y} \\ \frac{\partial^2 f}{\partial x \partial y} & \frac{\partial^2 f}{\partial y^2} \end{pmatrix}_{(x_0, y_0)}.
$$

The determinant $\det(M) = \frac{1}{2}\det(H)$. Therefore, the conditions on the determinant translate into the standard second derivative test for functions of two variables.

The second derivative test for function of two variables.

Classify the critical points (x_0, y_0) of $f(x, y) : \nabla f(\mathbf{x}_0) = 0$.
Compute

$$
\det(H) = \left[\frac{\partial^2 f}{\partial x^2} \frac{\partial^2 f}{\partial y^2} - \left(\frac{\partial^2 f}{\partial x \partial y} \right)^2 \right]_{(x_0, y_0)}.
$$

Then, if

a. $\det(H) > 0$, and

  If $f_{xx} > 0$, the critical point is a local minimum, or

  If $f_{xx} < 0$, the critical point is a local maximum.

b. $\det(H) < 0$, then the critical point is a saddle point.

c. $\det(H) = 0$, then the test says nothing.

Example 10.6. *Find and classify the critical points for the surface*

$$f(x,y) = 3x^3 + y^2 - 9x - 6y + 1.$$

We first set $\nabla f = 0$, or

$$\frac{\partial f}{\partial x} = 9x^2 - 9 = 0,$$

$$\frac{\partial f}{\partial y} = 2y - 6 = 0. \qquad (10.14)$$

Solving these equations gives $y = 3$ and $x = \pm 1$. So, there are critical points at $(1, 3, -14)$ and $(-1, 3, -2)$. These are shown in Figure 10.7.

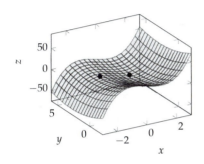

Figure 10.7: Graph of $f(x,y) = 3x^3 + y^2 - 9x - 6y + 1$, which has a saddle point and a local minimum at the indicated points.

At each critical point we need to apply the second derivative test. First we need the Hessian matrix,

$$H = \begin{pmatrix} \frac{\partial^2 f}{\partial x^2} & \frac{\partial^2 f}{\partial x \partial y} \\ \frac{\partial^2 f}{\partial x \partial y} & \frac{\partial^2 f}{\partial y^2} \end{pmatrix} = \begin{pmatrix} 18x & 0 \\ 0 & 2 \end{pmatrix}.$$

Inserting (x_0, y_0) for each case into $\det(H) = 36x$, we have

   a. At $(1, 3, -14)$ $\det(H) > 0$. Because $f_{xx}(1, 3) > 0$, then the critical point is a local minimum.

   b. At $(-1, 3, -2)$ $\det(H) < 0$. So, this critical point is a saddle point.

For a function of more than two variables, the classification depends on the eigenvalues of the Hessian matrix evaluated at the critical point. We assume that

$$\nabla f(\mathbf{x}_0) = 0 \quad \text{and} \quad \det(H(f(\mathbf{x}_0))) \neq 0.$$

Then,

Classification using eigenvalues of $H(f(\mathbf{x}_0))$.

   a. If the eigenvalues of $H(f(\mathbf{x}_0))$ are all positive, then $f(\mathbf{x})$ has a local minimum at $\mathbf{x} = \mathbf{x}_0$.

   b. If the eigenvalues of $H(f(\mathbf{x}_0))$ are all negative, then $f(\mathbf{x})$ has a local maximum at $\mathbf{x} = \mathbf{x}_0$.

   c. If $H(f(\mathbf{x}_0))$ has at least one positive eigenvalues and at least one negative eigenvalue, then $f(\mathbf{x})$ has a saddle point at $\mathbf{x} = \mathbf{x}_0$.

Example 10.7. *Find and classify the critical points of the function*

$$f(x,y,z) = x^3 - xy + y^2 - 2y + z^3 - 3z + 1.$$

First, we note that

$$\nabla f = \langle 3x^2 - y, -x + 2y - 2, 3z^2 - 3 \rangle.$$

Setting $\nabla f = 0$, we find the four critical points $(-\frac{1}{2}, \frac{3}{4}, \pm 1), (\frac{2}{3}, \frac{4}{3}, \pm 1)$.
The Hessian is given by

$$H = \begin{pmatrix} 6x & -1 & 0 \\ -1 & 2 & 0 \\ 0 & 0 & 6z \end{pmatrix}.$$

Evaluations at each critical point leads to the results in Table 10.1

Table 10.1: Critical Points and Eigenvalues of the Hessian Matrix for Each Point

Critical Point	Eigenvalues	Classification
$(-\frac{1}{2}, \frac{3}{4}, 1)$	$6, -\frac{1}{2} \pm \frac{\sqrt{29}}{2}$	Saddle
$(-\frac{1}{2}, \frac{3}{4}, -1)$	$-6, -\frac{1}{2} \pm \frac{\sqrt{29}}{2}$	Saddle
$(\frac{2}{3}, \frac{4}{3}, 1)$	$6, 3 \pm \sqrt{2}$	Local minimum
$(\frac{2}{3}, \frac{4}{3}, -1)$	$-6, 3 \pm \sqrt{2}$	Saddle

10.1.3   Linear Regression

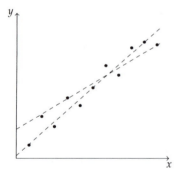

Figure 10.8: Possible linear relations for a given set of data.

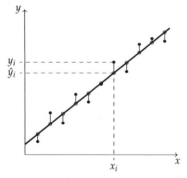

Figure 10.9: Depiction of residual errors, $e_i = y_i - \hat{y}_i$ between data and linear fit.

A COMMON TOOL IN DATA ANALYSIS is linear regression and in this section we use minimization techniques in order to find a best fit curve to a set of data. In particular, we collect a set of data (x_i, y_i) for $i = 1, \ldots, N$. We plot the data and it seems that there is a linear relationship between the variables, $y = ax + b$. A sample scatterplot is shown in Figure 10.8. The goal is to determine the slope and intercept of the best fit line.

One way you can find the best fit line is to take a straight edge and align it with the data to get an approximate line. Your lab partner repeats this process with his own copy of the data set and obtains a different line that looks like it fits the data. These two lines most likely lie between two extreme lines such as those shown in Figure 10.8 You and your lab partner determine the slopes and intercepts of your lines and find that they might be close, but not exact. Whose line is the best one? You can repeat this several times and come up with an average slope and intercept and even get an estimate of the error in these quantities.

Well, this is one way to proceed. However, you probably have not actually found the best fit line using this method. Instead, you rely on using a statistical package, mathematics software, built-in functions in a calculator, or a spreadsheet program like MS Excel. The built-in functions can perform a least squares best fit. In this section we will describe the minimization process behind finding the best fit line.

The basic idea is to minimize the sum of the squares of the difference between the y-values on the linear approximation and the y-values in the data set. In Figure 10.9 we show a data set and the linear fit. Let the equation of the line be $y = ax + b$. Using the x-values from the data set, the model predicts the values

$$\hat{y}_i = ax_i + b, \quad , i = 1, \ldots, N.$$

The data points (x_i, y_i) and the predicted points $(x_i, \hat{y}_i)$ are shown in Figure 10.9. The vertical deviation of the data point to the line can easily be seen. These differences are called the residuals, $e_i = y_i - \hat{y}_i$. We assume that a plot of the residuals shows no trends or patterns. Their distribution should be somewhat random. Otherwise, a different model is needed.

We need a measure of the total deviation of the fit from the data. A common measure is to add the squares of the residuals. If we only added the residuals, which are assumed to be randomly distributed about zero, then we would get zero as the total distance the data is above the regression line is the same as that below the line. We avoid this cancellation by first squaring the residuals and adding these results. This sum can be written as

$$E(a, b) = \sum_{i=1}^{N} (y_i - \hat{y}_i)^2.$$

We are seeking the best fit line to the data with $\hat{y}_i = ax_i + b$. So, the sum of squares of the deviations between the data and the line is

$$E(a, b) = \sum_{i=1}^{N} [y_i - (ax_i + b)]^2.$$

The goal is to find the point (a, b) that minimizes this total deviation, $E(a, b)$.

We have already seen that the condition for determining the extrema of a function of two variables requires the solution of the equations

$$\frac{\partial E}{\partial a} = 0, \quad \frac{\partial E}{\partial b} = 0.$$

For the least squares error, this means that

$$0 = \frac{\partial E}{\partial a} = 2 \sum_{i=1}^{N} (y_i - (ax_i + b))(-x_i),$$

$$0 = \frac{\partial E}{\partial b} = 2 \sum_{i=1}^{N} (y_i - (ax_i + b))(-1). \qquad (10.15)$$

Expanding the sums, these equations become

$$a \sum_{i=1}^{N} x_i^2 + b \sum_{i=1}^{N} x_i = \sum_{i=1}^{N} x_i y_i,$$

$$a \sum_{i=1}^{N} x_i + bN = \sum_{i=1}^{N} y_i, \qquad (10.16)$$

In statistics

$$\sigma_y^2 = \frac{E(a, b)}{N - 2}$$

is the standard deviation of the y_i-values about the best fit line.

Normal equations.

where we have used

$$\sum_{i=1}^{N} 1 = N.$$

In numerical analysis, this set of equations is referred to as the normal equations. It is a pair of linear equations in a and b and can easily be solved to give

$$a = \frac{N \sum_{i=1}^{N} x_i y_i - \left(\sum_{i=1}^{N} x_i\right)\left(\sum_{i=1}^{N} y_i\right)}{N \sum_{i=1}^{N} x_i^2 - \left(\sum_{i=1}^{N} x_i\right)^2},$$

$$b = \frac{\left(\sum_{i=1}^{N} x_i^2\right)\left(\sum_{i=1}^{N} y_i\right) - \left(\sum_{i=1}^{N} x_i y_i\right)\left(\sum_{i=1}^{N} x_i\right)}{N \sum_{i=1}^{N} x_i^2 - \left(\sum_{i=1}^{N} x_i\right)^2}. \tag{10.17}$$

These equations look messy, but we can simplify them by defining the averages

$$\overline{x} = \frac{1}{N} \sum_{i=1}^{N} x_i,$$

$$\overline{y} = \frac{1}{N} \sum_{i=1}^{N} y_i,$$

$$\overline{xy} = \frac{1}{N} \sum_{i=1}^{N} x_i y_i,$$

$$\overline{x^2} = \frac{1}{N} \sum_{i=1}^{N} x_i^2. \tag{10.18}$$

Then, the best fit line, $y = ax + b$, has the slope, a, and intercept, b, as

$$a = \frac{\overline{xy} - \overline{x}\,\overline{y}}{\overline{x^2} - (\overline{x})^2}$$

$$b = \frac{\overline{x^2}\,\overline{y} - \overline{xy}\,\overline{x}}{\overline{x^2} - (\overline{x})^2}. \tag{10.19}$$

Further exploration of this example of determining extrema of functions of several variables is an aside useful to those interested in linear regression. With this said, we will apply linear regression to a simple set of data.

Example 10.8. *Find the slope and intercept of the best fit line to the data in Table 10.2.*

A standard introductory physics experiment is to verify Hooke's Law for an ideal spring. One places different masses on the spring and measures the elongation of the spring. In Table 10.2 is a sample set of data. From the masses one obtains the weight, and the plot in Figure 10.10 shows that the set of points are linearly related.

x (m)	m (kg)	F (N)
0.100	0.1110	1.089
0.200	0.2081	2.042
0.300	0.2723	2.671
0.400	0.3583	3.515
0.500	0.4491	4.406

Table 10.2: Data for the Elongation (x) of a Spring as a Function of Mass (m)

As we know, Hooke's Law is given by the restoring force $F_s = -kx$, where x is the displacement from the spring's equilibrium. The amount of weight, F, that keeps a given mass on a spring in equilibrium is

$$F = kx.$$

This is what is seen in the data plot in Figure 10.10. We show this by adding a best fit line. We now use the above expressions to compute the slope and intercept for this line.

As shown in Table 10.3, we add the columns of products x^2 and xy and then compute the needed averages. From these we find

$$a = \frac{\overline{xy} - \overline{x}\,\overline{y}}{\overline{x^2} - (\overline{x})^2} = \frac{.9860 - 0.300(2.745)}{0.110 - 0.300^2} = 8.107,$$

$$b = \frac{\overline{x^2}\overline{y} - \overline{xy}\,\overline{x}}{\overline{x^2} - (\overline{x})^2} = \frac{0.110(2.745) - 0.9860}{0.110 - 0.300^2} = 0.3125. \qquad (10.20)$$

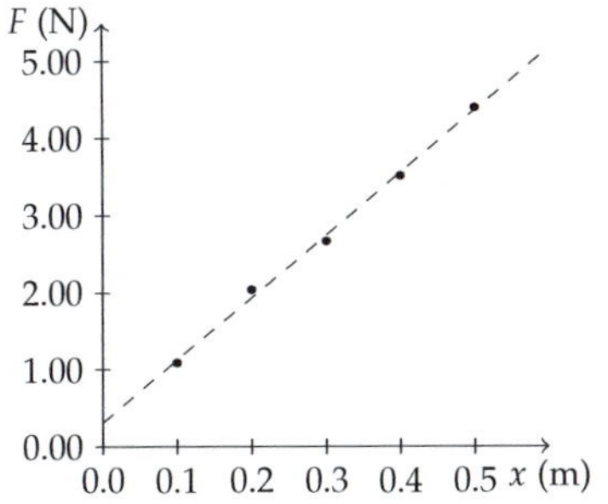

Figure 10.10: Plot of data and best fit line.

x (m)	F (N)	x^2 (m²)	xy (Nm)
0.1	1.089	0.01	0.1089
0.2	2.042	0.04	0.4084
0.3	2.671	0.09	0.8013
0.4	3.515	0.16	1.4060
0.5	4.406	0.25	2.2030
$\overline{x}$ (m)	$\overline{F}$ (N)	$\overline{x^2}$ (m²)	$\overline{xy}$ (Nm)
0.300	2.745	0.110	0.9860

Table 10.3: Averages for the Elongation Data

Thus, the final result is $F = 8.107x + 0.3125$.

The equations in (10.19) for the slope and intercept of the best fit line can be written more compactly by borrowing terms from statistics. First, we define the mean, or average, of N numbers as

Mean, μ_x.

$$\mu_x = \frac{1}{N} \sum_{i=1}^{N} x_i.$$

Thus, $\overline{x} = \mu_x$ and $\overline{y} = \mu_y$.

Variance, Var(x).

The variance is defined as the average of the squared differences from the mean. Thus,

$$\text{Var}(x) = \frac{1}{N} \sum_{i=1}^{N} (x_i - \mu_x)^2.$$

Standard deviation, σ_x.

While this is a measure in the spread of the data, it does not have the same units as the original variable. So, taking the square root, we obtain the standard deviation,

$$\sigma_x = \sqrt{\text{Var}(x)}.$$

Due to this relationship, the variance of x is often written as σ_x^2. It is handy to note that the variance can be written in a slightly different form.

$$
\begin{aligned}
\text{Var}(x) &= \frac{1}{N}\sum_{i=1}^{N}(x_i - \mu_x)^2 \\
&= \frac{1}{N}\sum_{i=1}^{N}(x_i^2 - 2x_i\mu_x + \mu_x^2) \\
&= \frac{1}{N}\sum_{i=1}^{N}x_i^2 - 2\mu_x\frac{1}{N}\sum_{i=1}^{N}x_i + \mu_x^2\frac{1}{N}\sum_{i=1}^{N}1 \\
&= \overline{x^2} - 2\mu_x^2 + \mu_x^2 \\
&= \overline{x^2} - \bar{x}^2,
\end{aligned}
\tag{10.21}
$$

The variance is the mean of the squares minus the square of the mean.

The covariance, $\text{Cov}(x, y)$.

or the variance is the mean of the squares minus the square of the mean. Note that this is precisely the form of the denominators in the expressions for the slope and intercept of the best fit line.

Another useful statistical quantity is the covariance. The covariance is a measure of the relationship between two variables. It is defined as

$$\text{Cov}(x, y) = \frac{1}{N}\sum_{i=1}^{N}(x_i - \mu_x)(y_i - \mu_y).$$

Following a similar rearrangement of terms as was carried out for the variance, one can show that

$$\text{Cov}(x, y) = \overline{xy} - \bar{x}\bar{y} = \mu_{xy} - \mu_x\mu_y,$$

which is the numerator in the equation for the slope.

So, we can write the slope compactly as

$$a = \frac{\text{Cov}(x, y)}{\text{Var}(x)} \tag{10.22}$$

and the intercept is given by

$$
\begin{aligned}
b &= \frac{\overline{x^2}\bar{y} - \overline{xy}\,\bar{x}}{\overline{x^2} - (\bar{x})^2} \\
&= \frac{(\overline{x^2} - \bar{x}^2)\bar{y} - (\overline{xy} - \bar{x}\bar{y})\bar{x}}{\text{Var}(x)} \\
&= \bar{y} - \frac{\text{Cov}(x, y)}{\text{Var}(x)}\bar{x} \\
&= \bar{y} - a\bar{x}.
\end{aligned}
\tag{10.23}
$$

In summary, we compute the slope using the variance and covariance and then compute the intercept by rewriting the equation of the best fit line in terms of the mean of the x and y values.

$$a = \frac{\text{Cov}(x,y)}{\text{Var}(x)}, \quad b = \bar{y} - a\bar{x}.$$

Example 10.9. *Recalculate the slope and intercept for the spring data.*

In Table 10.4 we compute the variance and covariance for the data in the previous example. From this data we find the same results as before:

$$a = \frac{\text{Cov}(x,y)}{\text{Var}(x)} = 8.107,$$

$$b = \bar{y} - a\bar{x} = 0.3125.$$

x (m)	F (N)	$(x-\bar{x})^2$ (m^2)	$(x-\bar{x})(y-\bar{y})$ (Nm)
0.1	1.089	0.04	0.3311
0.2	2.042	0.01	0.0703
0.3	2.671	0.00	0.0000
0.4	3.515	0.01	0.0770
0.5	4.406	0.04	0.3323

$\bar{x}$ (m)	$\bar{F}$ (N)	$\text{Var}(x)$ (m^2)	$\text{Cov}(x,y)$ (Nm)
0.300	2.745	0.02	0.1621

Table 10.4: Computation of Variance and Covariance for the Elongation Data

One can use built-in functions in various software packages to compute best fit curves. For example, MATLAB has a built-in function for doing a polynomial fit to data. For the current data set, the code for typing in the arrays and getting the slope and intercept is given by

```
>> x=[.1 .2 .3 .4 .5];
>> y=[1.089 2.042 2.6710 3.5150 4.4060];
>> polyfit(x,y,1)
ans =

    8.1070    0.3125
```

A more common tool is found in Microsoft's Excel program. One can enter the data in columns and do a scatterplot. Selecting the data and adding a trendline with a displayed equation give a visual representation of the fit and the best fit line with the same slope and intercept as above. In fact, one can do a little better. The LINEST function can be used to produce not only the slope and intercept, but also the standard errors in the slope and intercept.

The steps to follow are

1. Enter **=LINEST(y-values, x-values,1,TRUE)** where you want to place the results.

2. Select a 2-by-5 set of cells

3. Hit **F2** to display the equations

4. Enter **CTRL-SHIFT-ENTER**, to fill the array.

The resulting numbers are shown in Table 10.5. The annotation in the first row and column is provided to describe what the data is for the sample data from the previous example.

Table 10.5: Output from Excel's LINEST Function

	Slope	Intercept	
Value	8.1070	0.3125	Value
se_{slope}	0.2677	0.0888	se_{int}
R^2	0.9967	0.0846	se_y
F	917.2292	3.0000	df
SSR	6.5723	0.0215	SSE

It is useful in practice to know how these numbers are computed and how they can be used. In Table 10.6 we have the usual data columns plus the next to the last row has selected sums of the columns and the last row has selected averages of columns. We have added a column for the estimates $\hat{y}$ based on the best fit, $\hat{y} = ax + b$, the squares of the residuals, $(y - \hat{y})^2$, and the squares of the estimates minus the mean,$(\hat{y} - \mu_y)^2$.

Table 10.6: Example of Spreadsheet Columns for Computing SSE, SSR, and SST.

x	F	x^2	$(x - \mu_x)^2$	$(y - \mu_y)^2$	$\hat{y}$	$(y - \hat{y})^2$	$(\hat{y} - \mu_y)^2$
0.1	1.089	0.010	0.04	2.741	1.123	0.001	2.629
0.2	2.042	0.040	0.01	0.494	1.934	0.012	0.657
0.3	2.671	0.090	0.00	0.005	2.745	0.005	0.000
0.4	3.515	0.160	0.01	0.594	3.555	0.002	0.657
0.5	4.406	0.250	0.04	2.760	4.366	0.002	2.629
1.5		0.550	0.10	6.594		0.022	6.572
0.3	2.745		0.02	1.319			

The sums of the squares of error, SSE; residual, SSR; and total, SST.

The sums of the squares of the errors,

$$SSE = \sum_i (y_i - \hat{y}_i)^2 = 0.0215$$

and the sums of the squares of the residuals,

$$SSR = \sum_i (\hat{y}_i - \mu_y)^2 = 6.572$$

are computed. The total sums of squares is given by

$$SST = SSE + SSR = N\text{Var}(y) = 6.594.$$

df stands for the number of degrees of freedom and is given by the number of samples, N, less the number of parameters (slope and intercept). In this case, $df = N - 2$. Knowing these values, one can compute most of the entries given by LINEST:

$$R^2 = \frac{SSR}{SST} = 0.997,$$

$$se_y = \sqrt{\frac{SSE}{N-2}} = 0.085,$$

$$se_{slope} = \frac{se_y}{\sqrt{N\text{Var}(y)}} = 0.267,$$

$$se_{int} = \frac{se_y}{\sqrt{N(1 - \bar{x}^2/\overline{x^2})}} = 0.089.$$

The coefficient of determination.

The coefficient of determination, or R^2-value, is often used as an indicator of how good the fit is. The closer to one it is, the better. se_{slope} and se_{int} are the standard errors in the parameters and are useful in determining the error in the slope and intercept. More training in statistics allows one to determine confidence intervals, accept or reject hypotheses, or apply errors bars on the results.

In MATLAB one can generate the same data. The code for doing this using the **polyfit** function is below:

```
clear

% Enter Data
x=[.1 .2 .3 .4 .5];
y=[1.089 2.042 2.6710 3.5150 4.4060];
figure(1)

% Polynomial Fit
n = 1;                       % n = deg polynomial fit
[m,b]=polyfit(x,y,n);        % m = [slope intercept]

% Plot (xfit, yfit) for model
xfit = linspace(0,.6,20);
yfit = polyval(m,xfit);
plot(x,y,'ok',xfit,yfit);

% Residuals
ymodel = polyval(m,x);       % predicted values
resid  = y - ymodel;         % residuals

% R^2 Value (Rsq)
dev = y - mean(y);           % deviations - measure of spread
SST = sum(dev.^2);           % total variation accounted for
SSE = sum(resid.^2);         % variation NOT accounted for
Rsq = 1 - SSE/SST;           % percent of error explained

% F-Statistic
SSR = SST - SSE;             % variation of residual
dfe = length(x) - 1 - n;     % degrees of freedom
MSE = SSE/dfe;               % mean-square error of residuals
MSR = SSR/n;                 % mean-square error for regression
```

```
f = MSR/MSE;                    % f-statistic

% t-Statistics for coeffs from b in polyfit
R = b.R;
covm = (R'*R)\eye(n+1);         % The covariance matrix
d = diag(covm)';                % ROW vector of the diagonal elements
MSE = (b.normr^2)/b.df;         % variance of the residuals
se = sqrt(MSE*d);               % the standard errors
t = m./se;                      % observed T-value
```

Force the intercept to be zero.

There are some other issues that can be explored based upon the last example. First, one often is interested in forcing the intercept to be zero, known as regression through the origin. So, we would need to fit the data to the line $\hat{y} = ax$. There is only one parameter and the results, which the reader can confirm, are that

$$a = \frac{\sum_{i=1}^{N} x_i y_i}{\sum_{i=1}^{N} x_i^2} = \frac{\overline{xy}}{\overline{x^2}}.$$

This can be done in MATLAB along with an R^2 computation.

```
% Enter Data
x=[.1 .2 .3 .4 .5];
y=[1.089 2.042 2.6710 3.5150 4.4060];

% Plot (xfit, yfit) for model
xfit = linspace(0,.6,20);

% Force Intercept to zero - or [ones(length(x'),1) x']\y' w/intercept
 m=x'\y';

% Plot (xfit, yfit) for model
yfit = m*xfit;
plot(xfit,yfit,x,y,'o');

% R^2 Value
ymodel = m*x;                   % predicted values
resid  = y - ymodel;            % residuals
dev = y - mean(y);              % deviations - measure of spread
SST = sum(dev.^2);              % total variation accounted for
SSE = sum(resid.^2);            % variation NOT accounted for
Rsq = 1 - SSE/SST;              % percent of error explained
```

However, there is some debate as to how to evaluate the statistics in this case. A more appropriate value for R^2 is given by

```
Rsq2=sum(ymodel)/sum(y);
```

However, it should be noted that Excel's output agrees with *Rsq* above.

Finally, other common fits in physics include fitting data to exponential, $y = ae^{bx}$, or power, $y = ax^b$, functions. A standard approach would be to linearize the data using semilog or log-log plots. Taking the logarithm of each function, one has, respectively,

$$\ln y = \ln a + bx,$$

$$\ln y = \ln a + b \ln x.$$

These are now linear in the parameters, and linear regression can be performed on the adjusted data. Though this is not a least squares regression, one can get estimates of the parameters. We demonstrate this with an example.

Example 10.10. *Fit the atmospheric density to an exponential function.*

The atmospheric density as a function of altitude is shown in Figure 2.31. It appears to be exponential. So, we can seek an expression of the form $\rho = ae^{bh}$. We select eleven points from the plot as shown in the side note and in Figure 10.11.

A semilog plot would give a plot of the function

$$\ln \rho = \ln a + bh.$$

For exponential data, this would yield a straight line with slope b and intercept $\ln a$ *as shown in Figure 10.11. Using Excel, we find the best fit line is given by*

$$\ln \rho = -1.486h + 0.453.$$

Carrying out the exponential fit directly in Excel without linearization yields the same result.

A similar analysis can be done in MATLAB. However, one can also do a nonlinear regression using the function **fminsearch**. *The code implementing both methods is given below:*

```
clear

% Exponential fit using fminsearch
  h = 0:4:40
  D = density2(h*1000);

% Define exponential function
f = @(x,p) p(1)*exp(x.*p(2));

% define the error function.
  errf = @(p,x,y) sum((y(:)-f(x(:),p)).^2);

% an initial guess of the parameters
  p0 = [mean(D) (max(D)-min(D)) (max(h) - min(h))/2];

% search for solution
```

Exponential and power laws.

The data used for the exponential fit is given by

h (km)	D (kg/m3)
0	1.2216
4	0.8174
8	0.5244
12	0.3110
16	0.1660
20	0.0886
24	0.0473
28	0.0241
32	0.0127
36	0.0069
40	0.0039

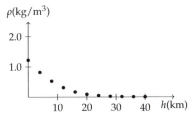

Figure 10.11: Plot of atmospheric density vs altitude.

```
P = fminsearch(errf,p0,[],h,D);

% plot the result
h2 = linspace(0,40,100);
figure(1)
plot(h,D,'ko',h2,f(h2,P),'b-')

% Compare to linearization approach
LD=log(D);
X=h'; Y=LD';
B = [ones(length(X),1) X] \ Y;
Y2=B(2)*h2+B(1);
figure(2)
semilogy(h2,exp(Y2),'b')
hold
semilogy(h,D,'ko')
hold

figure(3)
plot(h,density2(h*1000),'ko')
hold
plot(h2,exp(Y2),'k')
plot(h2,f(h2,P),'k--')
hold
xlabel('h (km)')
ylabel('\rho (kg/m^3)')

% Nonlinear function
display([num2str(P(1)) ' exp(' num2str(P(2)) '*x)'])
% Linearization function
display([num2str(exp(B(1))) ' exp(' num2str(B(2)) '*x)
```

The output from these routines gives

```
1.2484 exp(-0.117*x)
1.5748 exp(-0.14866*x)
```

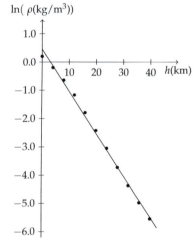

Figure 10.12: Semilog plot of atmospheric density versus altitude with best fit line.

*It is clear that the two methods give different results. In Figure 10.13 we plot the exponential fits of atmospheric density as a function of altitude. The solid curve is obtained from using linearization plus linear regression. The dashed line is obtained using MATLAB's **fminsearch** function and the latter looks like a better fit. Also, the linearization result agrees with the results from using Excel.*

Example 10.11. *Find a power law fit to the drag coefficient as a function of Reynolds number for $10 < Re < 1,000$.*

In Section 2.10.4 we investigated the problem of the falling raindrop. There we saw that Edwards, Wilder, and Scime (2001) used an approximation to the drag coefficient of $C_D = 12Re^{-1/2}$ for Reynolds number Re. In Section 2.10.7 we *discussed some of the models that have been used to study air drag as a function of Reynolds number. One of the models that is claimed to work for $Re < 100$ was given by Schiller-Naumann (1933),*

$$C_D = \frac{24}{Re}\left(1 + 0.15Re^{0.687}\right).$$

A plot of this function is shown in Figure 10.14. The data seems to look like a simple function over this range with a horizontal asymptote near $C_D = 0.4$.

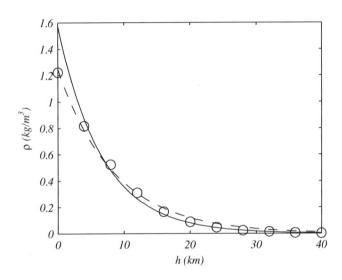

Figure 10.13: Exponential fits of atmospheric density as a function of altitude. The solid curve is from using linearization plus linear regression. The dashed line is obtained using MATLAB's fminsearch function.

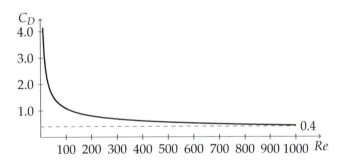

Figure 10.14: The drag coefficient as a function of Reynolds number as given by Schiller–Naumann (1933).

We can try to fit this formula to a function of the form

$$C_D = ax^b$$

*with parameters a and b. Using MATLAB's **fminsearch** again, we use the code*

```
x=linspace(10,1000,100);
y=24./x.*(1+0.15.*x.^.687);  %Schiller-Naumann
f = @(x,p) p(1).*x.^p(2);
errf = @(p,x,y) sum((y(:)-f(x(:),p)).^2);
p0 = [mean(y) (max(y)-min(y)) (max(x) - min(x))/2];
P = fminsearch(errf,p0,[],x,y);
```

```
x2 = linspace(10,1000,100);
figure(3)
plot(x,y,'ok','MarkerSize',3)
hold on
plot(x2,f(x2,P),'k-');
xlabel('Re')
ylabel('C_D')
hold off
display([num2str(P(1)) ' x^' num2str(P(2))])
```

Here we selected 100 points for the data set. The result is a fit to this data set with parameters $a = 11.9285$ and $b = -0.49975$, or $C_D = 11.9 RE^{-0.500}$. This is consistent with Edwards, Wilder, and Scime (2001). In Figure 10.15 we show how well this fits with the data.

Figure 10.15: A power law fit for the drag coefficient as a function of Reynolds number using a nonlinear regression.

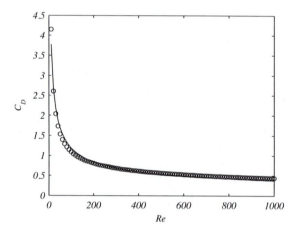

Another approach would be to do a linear fit to the data and extract the parameters. In this example, the points are sampled with logarithmic spacing. The MATLAB code for this is given below.

Figure 10.16: A power law fit for the drag coefficient as a function of Reynolds number using linearization and linear regression.

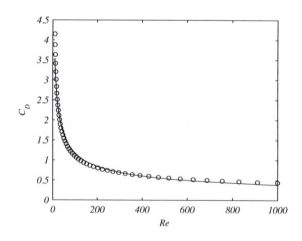

```
x=logspace(1,3);
y=24./x.*(1+0.15.*x.^.687);  %Schiller-Naumann
[m,b]=polyfit(log(x),log(y),1);
```

The result is a fit to this data set with parameters $a = 10.7016$ and $b = -0.47963$, or $C_D = 10.7Re^{-0.480}$. The fit is shown in Figure 10.16.

10.1.4   Lagrange Multipliers and Constraints

SOMETIMES WE ARE INTERESTED IN LOCATING EXTREMA under some given constraints. For example, we might want to find the extrema of a function $f(x,y)$ subject to conditions on the variables in the form $g(x,y) =$ constant.

Example 10.12. *Find the minimum and maximum values of $f(x,y) = x+y+2$ when $x^2 + y^2 = 1$. Geometrically, a plot of the surface $z = f(x,y)$ is the plane $z = x+y+2$. It should be obvious that $f(x,y)$ can be as large, or as small, as one desires. Mathematically, there are no critical points, since $\nabla f = \langle 1,1 \rangle \neq 0$ for any point on the plane. But, the story is different when a constraint is introduced.*

The points in the plane whose x and y components are related by $x^2 + y^2 = 1$ form an ellipse in the plane. The shadow of this ellipse on the xy-plane is the unit circle. In Figure 10.17 the solid ellipse indicates the points in the plane satisfying the constraint. The dashed curve is the constraint curve.

We can determine the location of the maxima and minima by reducing this problem to an optimization problem for a function of one variable by using the constraint. Solving for y in the constraint, $y = \pm\sqrt{1-x^2}$. Inserting this result into $z = x+y+2$, we have

$$z = x+y+2 = x+2 \pm \sqrt{1-x^2}.$$

We can determine the critical points by setting $z'(x) = 0$, or

$$1 \mp \frac{x}{\sqrt{1-x^2}} = 0.$$

Solving this equation, we have $x = \pm\frac{1}{\sqrt{2}}$. This gives $y = \pm\frac{1}{\sqrt{2}}$ and

$$z = x+y+2 = 2 \pm \frac{2}{\sqrt{2}} = \pm\sqrt{2}.$$

[Careful analysis shows that x and y have the same sign.]

In the previous example we solved the optimization of a function of two variables with a constraint by converting it to an optimization problem for a function of one variable. However, it is not always possible to do this. So, we need to develop another method by looking at the relationship between the constraint and the level curves of $f(x,y)$.

In Figure 10.18 we show the level curves of $f(x,y) = x+y+2$ and the constraint $x^2 + y^2 = 1$. Two level curves appear to be tangent to the unit circle.

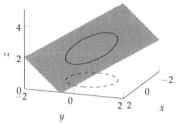

Figure 10.17: Figure showing the plane $z = x+y+2$ and the domain constraint $x^2 + y^2 = 1$. The solid ellipse indicates the points in the plane satisfying the constraint. The projection is the equation of the constraint.

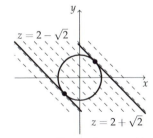

Figure 10.18: The level curves (dashed) of $f(x,y) = x+y+2$ and the constraint $x^2 + y^2 = 1$. The solid level curves correspond to $z = 2\sqrt{2}$.

Example 10.13. *Show that the level curves $z = 2 \pm \sqrt{2}$ are tangent to the unit circle.*

We consider the general level curves

$$x + y + 2 = c.$$

If these curves intersect the unit circle, then there are at most two points of intersection. Solving for $y = c - 2 - x$, and substituting into the equation of the circle,

$$x^2 + (c - 2 - x)^2 = 1.$$

Expanding,

$$2x^2 - 2(c - 2)x + [(c - 2)^2 - 1] = 0.$$

A level curve will intersect at one point provided the discriminant vanishes. Thus,

$$0 = 4(c - 2)^2 - 8[(c - 2)^2 - 1] = 8 - 4(c - 2)^2.$$

Therefore, the level curve is given by $z = c = 2 \pm \sqrt{2}$. Furthermore, $x = (2 - c)/2 = \mp\sqrt{2}/2$.

We can generalize this picture to other problems. We seek to find the extrema of $f(x, y)$ subject to $g(x, y) = c$. In Figure 10.19 we show the level curves of f, given by $f(x, y) = k$ and the constraint curve. Superimposed on two level curves and the constraint curve are the gradients of the functions. We see that the gradient ∇g is normal to the constraint curve at each point, as indicated by the arrows in Figure 10.19 emanating from the constraint curve.

As we saw in the previous chapter, the gradient ∇f points in the direction of largest increase in f. Let's consider the level curve passing through point b. All points on this curve have the same value for $f(x, y)$. The gradients on this curve all point upward. Thus, the next level curves have values that are greater than the value at point b. All the intersections of the constraint curve with those level curves will give $f(x, y)$ greater values than that at b. A similar argument can be provided at point a. Values on level curves below the curve that a is on will give lower values for $f(x, y)$. Any larger values lie on the curves just outside the constraint region.

The two small black circles at points a and b indicate the locations (x, y) of the extreme values of $f(x, y)$ on the constraint curve. We note that at both points, the gradients ∇f and ∇g are parallel. This means that the gradients are multiples of each other for (x, y) the sought location of the extrema. We can write this result as $\nabla f = \lambda \nabla g$, where the constant λ is called a Lagrange multiplier.[1]

In this example we have a function of two variables. There are two unknowns, x and y, for locating the extrema. Now, we also have the unknown Lagrange multiplier. However, the gradient relation only provides us with two equations from the two components of the gradient:

$$\frac{\partial f}{\partial x} = \lambda \frac{\partial g}{\partial x}, \quad \frac{\partial f}{\partial y} = \lambda \frac{\partial g}{\partial y}. \tag{10.24}$$

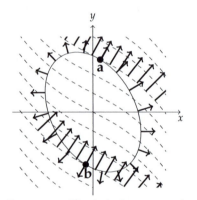

Figure 10.19: The dashed curves are the level curves for $z = f(x, y)$. The solid closed curve is a plot of the constraint, $g(x, y) = c$. Several gradient vectors for these functions are shown. The two small circles at points **a** and **b** indicate the locations of the extreme values of $f(x, y)$ on the constraint curve.

[1] Joseph Louis Lagrange introduced what are now called Lagrange multipliers in his treatise on analytical mechanics in 1788.

We can use the constraint equation as a third equation so that we will have three equations and three unknowns. Thus, the problem of finding the extrema for a function $f(x,y)$ subject to the constraint $g(x,y) = c$, now amounts to solving the system of equations

$$
\begin{aligned}
\nabla f &= \lambda \nabla g, \\
g(x,y) &= c \qquad\qquad (10.25)
\end{aligned}
$$

This can easily be extended to functions of several variables and even several constraints. For example, we consider a function of three variables, $f(x,y,z)$. If in addition there are two constraints, we add another Lagrange multiplier and we will have to solve a system of five equations for five unknowns, $\{x,y,z,\lambda_1,\lambda_2\}$, in the form

$$
\begin{aligned}
\nabla f &= \lambda_1 \nabla g_1 + \lambda_2 \nabla g_2, \\
g_1(x,y) &= c_1, \\
g_2(x,y) &= c_2. \qquad\qquad (10.26)
\end{aligned}
$$

Example 10.14. *Show that the Lagrange multiplier method gives the same locations of the extrema of $f(x,y) = x + y + 2$ subject to the constraint $x^2 + y^2 = 1$ as in the previous example.*

We begin by writing out $\nabla f = \lambda \nabla g$ as

$$
\begin{aligned}
\frac{\partial f}{\partial x} &= \lambda \frac{\partial g}{\partial x}, \\
\frac{\partial f}{\partial y} &= \lambda \frac{\partial g}{\partial y}. \qquad\qquad (10.27)
\end{aligned}
$$

Therefore,

$$
\begin{aligned}
1 &= 2\lambda x, \\
1 &= 2\lambda y. \qquad\qquad (10.28)
\end{aligned}
$$

Solving for x in the first equation and y in the second equation, we have

$$
x = y = \frac{1}{2\lambda}.
$$

Substituting these values into the constraint, $g(x,y) = x^2 + y^2 = 1$, we have

$$
\left(\frac{1}{2\lambda}\right)^2 + \left(\frac{1}{2\lambda}\right)^2 = 1.
$$

Solving for $\lambda = \pm\frac{1}{\sqrt{2}}$. Therefore, $x = y = \pm\frac{\sqrt{2}}{2}$ as before.

Example 10.15. *Find the extrema of the function $f(x,y) = x^2y - y^3$ restricted to the unit disk, $x^2 + y^2 \leq 1$.*

The graph of the function $f(x,y) = x^2y - y^3$ is a monkey saddle. The only critical point is $(0,0)$ and it is a saddle point. Thus, the extrema would have to come from the constraint.

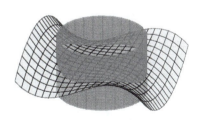

Figure 10.20: A cylindrical cookie cutter cutting away the outside region from the monkey saddle.

The restriction to the unit disk can be pictured if one thinks of $f(x, y)$ as a sheet of dough and it is cut with a cylindrical cookie cutter $x^2 + y^2 = 1$. This is depicted in Figure 10.20.

We can use Lagrange multipliers to do the problem. We identify the functions $f(x, y) = x^2 y - y^3$ and $g(x, y) = x^2 + y^2$. Then, the Lagrange multiplier condition is

$$\nabla f = \lambda \nabla g,$$
$$\langle 2xy, x^2 - 3y^2 \rangle = \lambda \langle 2x, 2y \rangle. \tag{10.29}$$

Thus, the system of equations that we need to solve is

$$2xy = 2\lambda x,$$
$$x^2 - 3y^2 = 2\lambda y,$$
$$x^2 + y^2 = 1. \tag{10.30}$$

Typically, one simplifies the system and looks for an easy implication. For example, the first equation gives

$$x(y - \lambda) = 0.$$

We see that either $x = 0$ or $y = \lambda$. Inserting these into the remaining equations leads to two cases.

For $x = 0$,

$$-3y^2 = 2\lambda y,$$
$$y^2 = 1. \tag{10.31}$$

This easily yields that $y = \pm 1$. So, we obtain the candidates $(x, y) = (0, 1), (0, -1)$. [Note that we need not solve for λ because we do not need λ.] The corresponding points on the surface are

$$(x, y, f(x, y)) = (0, 1, -1), (0, -1, 1).$$

For $y = \lambda$, we have

$$x^2 - 3\lambda^2 = 2\lambda^2,$$
$$x^2 + \lambda^2 = 1. \tag{10.32}$$

The first equation gives $x^2 = 5\lambda^2$. Inserting this result into the last equation, we have $\lambda^2 = \frac{1}{6}$. So, $y = \lambda = \pm \frac{1}{\sqrt{6}}$, and $x = \pm \frac{5}{\sqrt{6}}$. Therefore, we have four points:

$$(x, y) = \left(\sqrt{\frac{5}{6}}, \pm \frac{1}{\sqrt{6}} \right), \left(-\sqrt{\frac{5}{6}}, \pm \frac{1}{\sqrt{6}} \right).$$

Evaluating $f(x, y) = x^2 y - y^3$ at these points, we find $f(x, y) = \frac{5}{6}y - \frac{1}{6}y = \frac{2}{3}y$ for all four points. Therefore, the corresponding points on the surface are

$$(x, y, f(x, y)) = \left(\pm \sqrt{\frac{5}{6}}, \frac{1}{\sqrt{6}}, \frac{2}{3\sqrt{6}} \right), \left(\pm \sqrt{\frac{5}{6}}, \frac{1}{\sqrt{6}}, -\frac{2}{3\sqrt{6}} \right).$$

From these six solutions for (x, y) of the Lagrange multiplier conditions, the maximum value of f is $f = 1$ at $(0, -1)$ and the minimum value of f is $f = -1$ at $(0, 1)$.

Example 10.16. *Find the largest rectangle inscribed in the ellipse $ax^2 + by^2 = c$.*

Consider a rectangle of width $2x$ and height $2y$ is inscribed in the ellipse shown in Figure 10.21. The area of the rectangle is

$$A(x,y) = \ell w = 4xy.$$

The corners of the rectangle lie on the ellipse and satisfy the constraint

$$g(x,y) = ax^2 + by^2 = c.$$

Thus, we see the maximum of $A(x,y)$ subject to the constraint $g(x,y) = c$. The Lagrange multiplier condition then becomes $\nabla A = \lambda \nabla g$ and the system of equations that we need to solve are given as

$$
\begin{aligned}
4y &= 2a\lambda x, \\
4x &= 2b\lambda y, \\
c &= ax^2 + by^2.
\end{aligned}
\tag{10.33}
$$

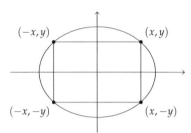

Figure 10.21: A rectangle of width $2x$ and height $2y$ is inscribed in the ellipse $ax^2 + by^2 = c$.

The first two equations are coupled in x and y, and we can eliminate one of these variables. Solving the second equation for

$$x = \frac{1}{2}b\lambda y,$$

and inserting this into the first equation, we have

$$(4 - ab\lambda^2)y = 0.$$

This equation is true if $y = 0$ or $4 - ab\lambda^2 = 0$. Since $y = 0$ gives at most a line segment, we have

$$\lambda = \pm\sqrt{\frac{4}{ab}}.$$

Without loss of generality, we pick the positive solution.

Inserting λ into the second equation, we can write y in terms of x,

$$y = \sqrt{\frac{a}{b}}x.$$

The third equation becomes

$$c = ax^2 + by^2 = 2ax^2.$$

The positive solution is $x = \sqrt{\frac{c}{2a}}$. Then, $y = \sqrt{\frac{c}{2b}}$. Therefore, we have found that maximum area is given by

$$A = 4xy = 4\sqrt{\frac{c}{2a}}\sqrt{\frac{c}{2b}} = \frac{2c}{\sqrt{ab}}$$

where

$$x = \sqrt{\frac{c}{2a}}, \quad y = \sqrt{\frac{c}{2b}}.$$

Example 10.17. *Find the distribution with maximum entropy.*

Ludwig Boltzmann (1844–1906), who contributed to our understanding of kinetic theory, the Second Law of Thermodynamics, and more generally statistical physics, has written on his tombstone

$$S = k \log W.$$

This equation relates the entropy to the number of microstates corresponding to a thermodynamic state of a system.

In statistical mechanics, one thinks of a macrostate of a system as an ensemble of systems in different microstates. This is then represented by a distribution function that gives the probability that a system chosen from the ensemble is in a particular microstate. Boltzmann then tells us that the entropy is related to the distribution functions, f_j, by the equation

$$S = -k \sum_j f_j \ln f_j,$$

where k is Boltzmann's constant and the probability distribution functions sum to one,

$$\sum_j f_j = 1.$$

The sum is over all the states in the ensemble. It is this form of entropy that is related to the disorder of a system.

For now we will not be too rigorous but will assume that the number of microstates is finite. We now restate the problem: What probability distribution will give the maximum entropy?

We want to maximize

$$S(f_1, \ldots, f_N) = -k \sum_{j=1}^{N} f_j \ln f_j$$

subject to the constraint

$$g(f_1, \ldots, f_N) = \sum_{j=1}^{N} f_j = 1.$$

Using the Lagrange Multiplier Method, we then have

$$\frac{\partial}{\partial f_i} \left[S(f_1, \ldots, f_N) - \lambda g(f_1, \ldots, f_N) \right] = 0, \quad i = 1, \ldots, N,$$

or

$$\frac{\partial}{\partial f_i} \left(-k \sum_{j=1}^{N} f_j \ln f_j - \lambda \sum_{j=1}^{N} f_j \right) = 0, \quad i = 1, \ldots, N.$$

This implies,

$$-k(\ln f_i + 1) + \lambda = 0, \quad i = 1, \ldots, N.$$

Solving for f_i, we have

$$f_i = e^{\lambda/k - 1}, \quad i = 1, \ldots, N.$$

Thus, the distribution for each state in the ensemble is the same function.

We still need to find the Lagrange multiplier. This can be done using the constraint equation. Thus,

$$\sum_{j=1}^{N} f_j = \sum_{j=1}^{N} e^{\lambda/k-1} = N e^{\lambda/k-1} = 1.$$

So,

$$f_i = e^{\lambda/k-1} = \frac{1}{N}, \quad j = 1, \ldots, N$$

and $\lambda = k(\ln f_j + 1) = k(1 - \ln N)$. Therefore, the distribution with the greatest entropy is the uniform distribution.

10.2   Calculus of Variations

10.2.1   Introduction

NEWTON FORMULATED THE LAWS OF MOTION in his 1687 volumes, collectively called the *Philosophiae Naturalis Principia Mathematica,* or simply the *Principia.* However, Newton's development was geometrical and is not how we see classical dynamics presented when we first learn mechanics. The laws of mechanics are what are now considered analytical mechanics, in which classical dynamics is presented in a more elegant way. It is based upon variational principles, whose foundations began with the work of Euler and Lagrange and have been refined by other now-famous figures in the eighteenth and nineteenth centuries.

Euler coined the term the *calculus of variations* in 1756, though it is also called variational calculus. The goal is to find minima or maxima of functions of the form $f : M \rightarrow R$, where M can be a set of numbers, functions, paths, curves, surfaces, etc. Interest in extrema problems in classical mechanics began near the end of the seventeenth century with Newton and Leibniz. In the *Principia,* Newton was interested in the least resistance of a surface of revolution as it moves through a fluid.

Seeking extrema at the time was not new, as the Egyptians knew that the shortest path between two points is a straight line and that a circle encloses the largest area for a given perimeter. Heron, an Alexandrian scholar, determined that light travels along the shortest path. This problem was later taken up by Willibrord Snellius (1580–1626) after whom Snell's law of refraction is named. Pierre de Fermat (1601/7/8-1665), the same for whom Fermat's Last Theorem is named, derived the law of reflection from his principle of least time. Namely, light always travels along a path for which the time taken is a minimum. This led to his principle of geometric optics, which was used to prove Snell's Law of Refraction.

By the seventeenth century, mathematicians were interested in paths of quickest descent. Galileo noted ,"From the preceding it is possible to infer that the path of quickest descent from one point to another is not the shortest path, namely, a straight line, but the arc of a circle." The curves that interested people of that time were the cycloid, the tautochrone (or, isochrone), and the brachistochrone. These are simply defined in terms of simple motions along curves.

- The cycloid is the curve traced by a point on the circumference of a wheel that rolls without slipping.

- The brachistochrone is the curve along which a free-sliding particle will descend more quickly between two given points than on any other AB-curve.

- The tautochrone is the curve along which a particle has a descent time independent of the initial position.

Christian Huygens (1629–1695) had shown in 1659 that a particle sliding on a cycloid undergoes simple harmonic motion with a period independent of the starting point. He also showed in 1673 that the solution of the tautochrone problem is the cycloid. Having studied Huygens solutions, Johann Bernoulli (1667–1748), investigated the brachistochrone problem and offered a challenge, specifically aimed at his older brother Jacob (1654–1705). In July 1696, Johann wrote to his mathematical contemporaries that he had solved the problem and challenged them to do so.

If two points A and B are given in a vertical plane, to assign to a mobile particle M the path AMB along which, descending under its own weight, it passes from the point A to the point B in the briefest time.—Smith, D.E. 1929, p. 644.

Johann Bernoulli's solution was published the following year. At that time, five others solved the problem (Jacob Bernoulli, Gottfried Leibniz, Isaac Newton, Ehrenfried Walther von Tschirnhaus, and Guillaume de L'Hopital.) The shape of the curve is that of a cycloid, as we will see later.

It was the study of these problems that led Johann Bernoulli to study variational problems. It turns out that Leonhard Euler was living with his brother Jacob but studied under Johann Bernoulli. He began a systematic study of extreme value problems and was aware of developments by Joseph Louis Lagrange. Euler introduced a condition on the path in the form of differential equations, which we later introduce as Euler's Equation. He took the Principle of Least Action and put it on firm ground. However, it was Lagrange who was to apply the calculus of variations to mechanics as the foundation of analytical mechanics.

Another important player in the development of minimum principles was Pierre Louis Maupertuis (1698–1759). This Principle of Least Action, which he believed was the basis for mechanics, was mostly philosophical but could be summarized as saying that all bodies move with minimum effort. The measure of effort is the action, which has units of energy times time. Maupertuis was a student of Johann Bernoulli, corresponded with Euler, and formulated his Principle of Least Action in 1744. It was based on his thoughts about the reflection and refraction of light. In that same year,

Euler added to the ideas that natural phenomena obey laws of maxima or minima. In 1746, Maupertuis went on further to use the Principle of Least Action to support his religious views. This sparked some heated discussions with scientists and philosophers of the day. Also, a dispute as to the originator of the Principle of Least Action emerged as Leibniz was thought by some to be the first to introduce these ideas.

Euler was the first to describe the Principle of Least Action on a firm mathematical basis. Lagrange further developed the principle and published examples of its use in dynamics. He introduced the variation of functions and derived the Euler-Lagrange equations. In 1867 Lagrange generalized the principle of least action basing his work on the conservation of energy and d'Alembert's Principle of Virtual Work. Further modifications by Adrien-Marie Legendre (1752–1833), Carl Gustav Jacob Jacobi (1804–1851), William Rowan Hamilton (1805–1865), and others, led to other formulations that applied to nonstationary constraints and nonconservative forces.

10.2.2   Variational Problems

THE TERM *calculus of variations* WAS FIRST COINED by Euler in 1756 as a description of the method that Joseph Louis Lagrange had introduced the previous year. The method was since expanded and studied by Euler, Hamilton, and others. As noted at the beginning of the chapter, the main idea is to determine which functions $y(x)$ will minimize, or maximize, integrals of the form

$$J[y] = \int_a^b f(x, y(x), y'(x)) \, dx, \tag{10.34}$$

where a, b, $f(a)$, $f(b)$, and $f(x)$ are given. Integrals like $J[y]$ are called functionals. This is a mapping from a function space to a scalar. J takes a function and spits out a number.

An example of a functional is the length of a curve in the plane. As we will recall, the length of a curve $y = y(x)$ from $x = x_1$ to $x = x_2$ is given by

$$L[y] = \int_{x_1}^{x_2} \sqrt{1 + \left(\frac{dy}{dx}\right)^2} \, dx.$$

This maps a given curve $y(x)$ to a number, L.

We are interested in finding the extrema of such functionals. We will not formally determine if the extrema are minima or maxima. However, in most cases, it is clear which type of extrema comes out of the analysis. Further analysis of the second variation can be found elsewhere, such as Lanczos' book on variational calculus. For interesting problems, historical principles have led to the formulation of problems in the calculus of variations, such as the Principles of Least Time, Least Action, Least Effort, or the shortest distance between geometric points on a surface. We will explore some standard examples leading to finding the extrema of functionals.

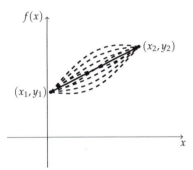

Figure 10.22: Possible paths between two fixed points.

Example 10.18. *Find the functional needed to determine the shortest distance between two given points in the plane.*

We have all heard that the shortest distance between points is a straight line. Of course, this statement needs some clarification. We will consider that the possible paths lie in a two-dimensional plane as shown in Figure 10.22. We then consider the length of a curve connecting two points in the plane. Let the points be given by (x_1, y_1) and (x_2, y_2). We assume that the functional form of the curve is given by $y = f(x)$.

Recall from the previous chapter that for a parametrized path $\mathbf{r}(t)$, the length of a curve is given by

$$L = \int_C ds = \int_{t_1}^{t_2} |\mathbf{r}'(t)|\, dt.$$

If we let the parameter be $t = x$, then

$$\mathbf{r}(x) = x\mathbf{i} + y(x)\mathbf{j}$$

and

$$\mathbf{r}'(x) = \mathbf{i} + y'(x)\mathbf{j}.$$

So, the length of the curve $y = y(x)$ from $x = x_1$ to $x = x_2$ is

$$
\begin{aligned}
L[y] &= \int_{x_1}^{x_2} |\mathbf{r}'(x)|\, dx \\
&= \int_{x_1}^{x_2} \sqrt{1 + \left(\frac{dy}{dx}\right)^2}\, dx. \quad (10.35)
\end{aligned}
$$

Example 10.19. *Find the functional needed to determine the path of fastest descent for a bead sliding frictionlessly down a wire under the influence of gravity from rest.*

We will let the shape of the wire path be given by $y = y(x)$ from $(x_1, y_1) = (0, 0)$ to (x_2, y_2). Also, we recall that the speed is given by

$$v = \frac{ds}{dt},$$

where $s = s(t)$ is the arclength as measured from the origin. Thus, the time of descent is given by the integral

$$T[y] = \int_C dt = \int_C \frac{ds}{v}.$$

We need to parametrize the path and determine the limits of integration before using this functional. We will assume, contrary to the usual choice, that downward is positive; i.e., $y(x) > 0$. Because the bead is undergoing free fall under gravity, the speed is found in terms of the height, $y(x)$:

$$v = \sqrt{2gy}.$$

This is a simple consequence of the conservation of mechanical energy,

$$T + U = \frac{1}{2}mv^2 - mgy = 0.$$

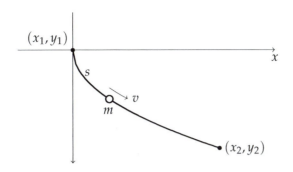

Figure 10.23: A bead of mass m sliding frictionlessly down a wire of shape $y = y(x)$ under the influence of gravity.

Because $v = v(y)$, we should write the differential arclength as

$$ds = \sqrt{dx^2 + dy^2} = \sqrt{1 + \left(\frac{dx}{dy}\right)^2}\, dy$$

and think of the path described by the function $x = x(y)$. Therefore, the time functional becomes

$$T[x] = \int_0^{y_2} \sqrt{\frac{1 + [x'(y)]^2}{2gy}}\, dy.$$

   The solution to the problem is to find the path $x(y)$ that makes this functional stationary.

10.2.3   Euler Equation

IN THE PREVIOUS EXAMPLES WE HAVE REDUCED THE PROBLEMS to finding functions $y = y(x)$ that extremize functionals of the form

$$J[y] = \int_a^b F(x, y, y')\, dx$$

for F twice continuous (C^2) in all variables. Formally, we say that J is stationary,

Stationary functionals.

$$\delta J[y] = 0,$$

at the function $y = y(x)$, or that the function $y = y(x)$ that extremizes $J[y]$ satisfies

Euler's Equation.

$$\frac{\partial F}{\partial y} - \frac{d}{dx}\left(\frac{\partial F}{\partial y'}\right) = 0. \qquad (10.36)$$

This is called the Euler's Equation. We will derive Euler's Equation and then show how it is used for some common examples.

   The idea is to consider all paths connected to the two fixed points and finding the path that is an extremum of $J[y]$. In fact, we need only consider parametizing paths near the optimal path and writing the problem in a form that we can use with the methods for local extrema of real functions.

Let's consider the paths $y(x; \epsilon) = u(x) + \epsilon\eta(x)$ near the optimal path $u = u(x)$ with $\eta(a) = \eta(b) = 0$ and η is C^2. Then, we consider the functional

$$J[y] = J[u + \epsilon\eta] \equiv \phi(\epsilon).$$

We note that if $J[u + \epsilon\eta]$ has a local extremum at u, then u is a stationary function for J. This will occur when

$$\frac{d\phi}{d\epsilon}\bigg|_{\epsilon=0} = 0.$$

We compute this derivative and find

$$
\begin{aligned}
\frac{d\phi}{d\epsilon} &= \frac{d}{d\epsilon} \int_a^b F(x, u + \epsilon\eta, u' + \epsilon\eta')\, dx \\
&= \int_a^b \frac{\partial}{\partial \epsilon} F(x, u + \epsilon\eta, u' + \epsilon\eta')\, dx \\
&= \int_a^b \left[\frac{\partial F}{\partial y} \frac{\partial y}{\partial \epsilon} + \frac{\partial F}{\partial y'} \frac{\partial y'}{\partial \epsilon} \right] dx \\
&= \int_a^b \left[\frac{\partial F}{\partial y}\eta(x) + \frac{\partial F}{\partial y'}\eta'(x) \right] dx.
\end{aligned}
\tag{10.37}
$$

We can perform an integration by parts on the second integral in order to move the derivative off of $\eta'(x)$. This is accomplished by setting $u(x) = \frac{\partial F}{\partial y'}$ and $dv = \eta'(x)\, dx$ in the integration by parts formula. Then,

$$\int_a^b \frac{\partial F}{\partial y'}\eta'(x)\, dx = \eta(x)\frac{\partial F}{\partial y'}\bigg|_a^b - \int_a^b \frac{d}{dx}\left(\frac{\partial F}{\partial y'} \right)\eta(x)\, dx.$$

The first terms vanish because $\eta(a) = \eta(b) = 0$.

This leaves

$$
\begin{aligned}
\frac{d\phi}{d\epsilon} &= \int_a^b \left[\frac{\partial F}{\partial y}\eta(x) + \frac{\partial F}{\partial y'}\eta'(x) \right] dx \\
&= \int_a^b \left[\frac{\partial F}{\partial y} - \frac{d}{dx}\left(\frac{\partial F}{\partial y'} \right) \right]\eta(x)\, dx.
\end{aligned}
\tag{10.38}
$$

Evaluation at $\epsilon = 0$ gives

$$\int_a^b \left[\frac{\partial F}{\partial y} - \frac{d}{dx}\left(\frac{\partial F}{\partial y'} \right) \right]\eta(x)\, dx = 0.$$

Noting that $\eta(x)$ is an arbitrary function and that this integral vanishes for all $\eta(x)$, we can say that the integrand vanishes,

$$\frac{\partial F}{\partial y} - \frac{d}{dx}\left(\frac{\partial F}{\partial y'} \right) = 0,$$

for all $x \in [a, b]$. This is Euler's Equation, (10.36).

Because $F = F(x, y(x), y'(x))$, one can prove a second form of Euler's Equation. We first note for the Chain Rule that

$$\frac{dF}{dx} = \frac{\partial F}{\partial x} + \frac{\partial F}{\partial y} \frac{\partial y}{\partial x} + \frac{\partial F}{\partial y'} \frac{\partial y'}{\partial x}.$$

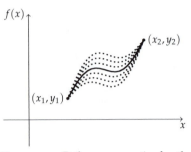

$f(x)$

(x_2, y_2)

(x_1, y_1)

x

Figure 10.24: Paths near an optimal path between two fixed points.

Now we insert

$$\frac{\partial F}{\partial y} = \frac{d}{dx}\left(\frac{\partial F}{\partial y'}\right)$$

from Euler's Equation to find

$$
\begin{aligned}
\frac{dF}{dx} &= \frac{\partial F}{\partial x} + \frac{\partial F}{\partial y}\frac{\partial y}{\partial x} + \frac{\partial F}{\partial y'}\frac{\partial y'}{\partial x} \\
&= \frac{\partial F}{\partial x} + \frac{d}{dx}\left(\frac{\partial F}{\partial y'}\right)y' + \frac{\partial F}{\partial y'}\frac{\partial y'}{\partial x} \\
&= \frac{\partial F}{\partial x} + \frac{d}{dx}\left(\frac{\partial F}{\partial y'}y'\right).
\end{aligned}
\tag{10.39}
$$

Rearranging this result, we obtain

The second form of Euler's Equation.

$$\boxed{\frac{\partial F}{\partial x} - \frac{d}{dx}\left(F - \frac{\partial F}{\partial y'}y'\right) = 0. \tag{10.40}}$$

This is the second form of Euler's Equation.

The second form of Euler's Equation is handy when $\frac{\partial F}{\partial x} = 0$, or when $F = F(y, y')$ is independent of x. In this case, we have

Special Cases:

$$F(x,y) \quad \Rightarrow \quad \frac{\partial F}{\partial y} = 0$$

$$F(x,y') \quad \Rightarrow \quad \frac{\partial F}{\partial y'} = c$$

$$F(y,y') \quad \Rightarrow \quad F - \frac{\partial F}{\partial y'}y' = c$$

$$F(y') \quad \Rightarrow \quad y'' = 0$$

$$F - \frac{\partial F}{\partial y'}y' = c,$$

where c is an arbitrary constant.

There are other special cases. In Euler's Equation if $F = F(x, y')$,

$$\frac{\partial F}{\partial y'} = c,$$

and when $F = F(x, y)$, we have

$$\frac{\partial F}{\partial y} = 0.$$

Finally, when $F = F(y')$, Euler's Second Equation implies

$$y'' = 0,$$

or $y(x) = c_1 + c_2 x$.

Example 10.20. *Find the path with the shortest distance between two points in a plane.*

Shortest distance between points.

In Example 10.18, we derived the functional

$$L[y] = \int_{x_1}^{x_2} \sqrt{1 + \left(\frac{dy}{dx}\right)^2}\, dx,$$

which gives the length of the path $y = y(x)$ from $x = a$ to $x = b$. The path that makes this functional stationary satisfies the Euler equation with

$$F(x, y, y') = \sqrt{1 + \left(\frac{dy}{dx}\right)^2}.$$

Because $F(x, y, y') = F(y')$ is independent of x and y, Euler's second form yields

$$y(x) = c_1 + c_2 x.$$

This is obviously a straight line path, and the constants can be found using the prescribed values of $y(x)$ at $x = a$ and $x = b$.

Example 10.21. *The brachistochrone problem. Find the path of fastest descent for a bead sliding frictionlessly down a wire under the influence of gravity.*

The brachistochrone problem.

In Example 10.23 we found that the time functional for a bead sliding down a wire of shape $x = x(y)$ is given by

$$T[x] = \int_0^{y_2} \sqrt{\frac{1 + [x'(y)]^2}{2gy}} \, dy.$$

In this case, we can factor out $(2g)^{-1/2}$ and define

$$F(y, x, x') = \sqrt{\frac{1 + [x'(y)]^2}{y}}.$$

In this problem, the roles of x and y have been switched; that is, $x = x(y)$. So, we need to first rewrite Euler's Equation in the form

$$\frac{\partial F}{\partial x} - \frac{d}{dy} \left(\frac{\partial F}{\partial x'} \right) = 0.$$

Because $F(y, x, x') = F(y, x')$ is independent of x,

$$\frac{d}{dy} \left(\frac{\partial F}{\partial x'} \right) = 0.$$

Therefore,

$$\frac{\partial F}{\partial x'} = c,$$

where c is a constant.

Inserting $F(y, x, x')$, we have

$$\frac{x'}{\sqrt{(1 + [x'(y)]^2)y}} = c.$$

Squaring both sides of the equation,

$$\frac{x'^2}{(1 + [x'(y)]^2)y} = c^2,$$

and rearranging, we obtain

$$x'^2 = \frac{c^2 y}{1 - c^2 y}.$$

Taking the square root, we have the first-order, separable, differential equation

$$\frac{dx}{dy} = \sqrt{\frac{c^2 y}{1 - c^2 y}}.$$

Separating variables and integrating, yields

$$x(y) = \int \sqrt{\frac{c^2 y}{1 - c^2 y}}\, dy.$$

We can use a trigonometric substitution to evaluate this integral. We let

$$c^2 y = \sin^2 \theta = \frac{1}{2}(1 - \cos 2\theta),$$

and

$$c^2 dy = 2 \sin \theta \cos \theta\, d\theta.$$

Then,

$$
\begin{aligned}
x(y) &= \int \sqrt{\frac{c^2 y}{1 - c^2 y}}\, dy \\
&= \int \sqrt{\frac{\sin^2 \theta}{1 - \sin^2 \theta}} \frac{2}{c^2} \sin \theta \cos \theta\, d\theta \\
&= \frac{2}{c^2} \int \sin^2 \theta\, d\theta \\
&= \frac{1}{c^2}\left(\theta - \frac{1}{2}\sin 2\theta\right) + k,
\end{aligned}
\qquad (10.41)
$$

where k is an integration constant.

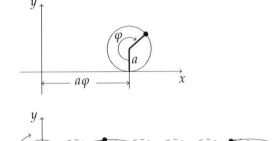

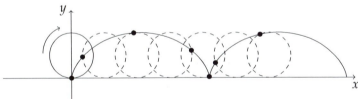

Figure 10.25: A point on a rolling wheel that does not slip traces out a cycloid. The motion consists of a translation of the center of mass, $v = at$, plus a rotation about the center of mass, $x = -a \sin \varphi$, $y = -a \cos \varphi$. The no-slip condition requires $t = \varphi$.

Summarizing, we assumed that

$$y = \frac{1}{2c^2}(1 - \cos 2\theta),$$

and found

$$x = \frac{1}{c^2}\left(\theta - \frac{1}{2}\sin 2\theta\right) + k.$$

This is a set of parametric equations for the curve along which the bead travels in minimum time. We can place these equations in standard form by defining $a = \frac{1}{2c^2}$ and $\varphi = 2\theta$, obtaining

$$
\begin{aligned}
x &= a(\varphi - \sin \varphi) + k, \\
y &= a(1 - \cos \varphi).
\end{aligned}
\qquad (10.42)
$$

We have identified a family of curves, but have not used the initial and final points on the curve. The initial point is $(x, y) = (0, 0)$. From $y = a(1 - \cos \varphi) = 0$, we choose $\varphi = 0$ at the starting point. Inserting this result into the expression for x, we get $k = 0$. This gives

Equations for a cycloid.

$$\begin{aligned} x &= a(\varphi - \sin \varphi), \\ y &= a(1 - \cos \varphi). \end{aligned} \qquad (10.43)$$

The set of parametric equations in the brachistochrone example describe the cycloid curve in Figure 10.25. A cycloid curve is traced out by a point on a rolling wheel that does not slip. One can derive this motion by combining a translation of the center of mass, $x = vt$, plus a rotation about the center of mass, $x = -a \sin \omega t$, $y = -a \cos \omega t$. The no slip condition requires that $v = a\omega$. If we set $v = a$, then $\omega = 1$ and the equations of the cycloid take the form $x = a(t - \sin t)$ and $y = a(1 - \cos t)$. Setting $t = \varphi$, we obtain the cycloid solution in the bead problem.

Figure 10.26: Plot of a brachistochrone path on top of a cycloid.

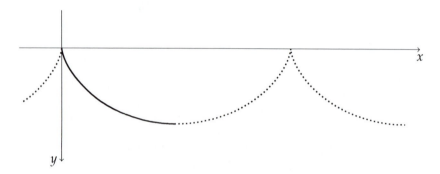

Example 10.22. *Plot solutions to the brachistochrone problem.*

While we have found the desired shape of the wire in the brachistochrone problem, we have not actually specified the final point (x_2, y_2). Namely, for what values of a and φ will the curve go through (x_2, y_2)? Inserting these values into the parametric equations, we have

$$\begin{aligned} x_2 &= a(\varphi - \sin \varphi), \\ y_2 &= a(1 - \cos \varphi). \end{aligned} \qquad (10.44)$$

Looking at the ratio of these equations, we can eliminate a :

$$\frac{y_2}{x_2} = \frac{1 - \cos \varphi}{\varphi - \sin \varphi}$$

This is a transcendental relation for determining φ. In Figure 10.27 the right-hand side of the equation is graphed with two cases shown, $\frac{y_2}{x_2} = 2$ and $\frac{y_2}{x_2} = \frac{1}{2}$.

x_2	y_2	φ	a
1.000	2.000	1.401	2.406
1.000	1.000	2.412	0.573
1.000	1.500	1.786	1.236
1.500	1.500	2.412	0.859
2.000	1.000	3.508	0.517

Table 10.7: Data for the brachistoschrone problem

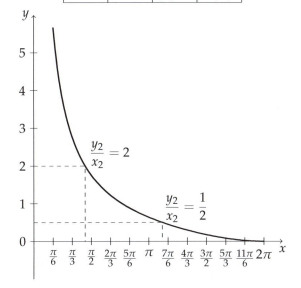

Figure 10.27: Determination of φ for a given endpoint.

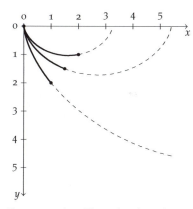

Using a root finding method, one can find φ. Inserting this value into one of the parametric equations, a is determined. In Table 10.7 we show these results. Three of these curves are shown in Figure 10.28. We note that the location of the points on each curve is different and is generally not at the bottom of the curve.

Figure 10.28: Three brachistochrone curves from the origin to the points (1,2), (2,1), (1.5,1.5).
Minimal surface area.

Example 10.23. *Find the function $y(x)$ that gives the minimum surface area of the surface obtained from rotating $y(x)$ about the x-axis.*

Another type of problem is that of determining the function that gives the minimum surface area for the surface of revolution. Consider a function $y(x)$ defined for $x \in [a, b]$. Recall that the surface of revolution is found by rotating $y(x)$ about the x-axis as shown in Figure 10.29.

One can find an expression for the surface area by looking at the slice of thickness dx in Figure 10.29. The infinitesimal area of the rim of the slice is found to be the product of the perimeter length and the width of the band, ds. Thus,

$$dA = 2\pi y(x)\, ds.$$

We also need to write ds in terms of dx.

$$ds = \sqrt{dx^2 + dy^2} = \sqrt{1 + \left(\frac{dy}{dx}\right)^2}\, dx.$$

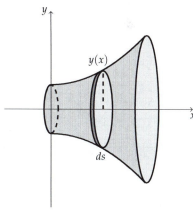

Figure 10.29: The surface of revolution generated by rotating $y(x)$ about the x-axis.

Thus, the total area becomes the functional

$$A[y] = 2\pi \int_a^b y(x) \sqrt{1 + \left(\frac{dy}{dx}\right)^2} \, dx.$$

In order to determine the minimum surface area generated by $y = y(x)$, we can apply Euler's condition to the function

$$F(x, y, y') = y\sqrt{1 + y'^2}.$$

Because $F(x, y, y') = F(y, y')$ is independent of x, we can employ Euler's second form, leading to

$$F - \frac{\partial F}{\partial y'} y' = c.$$

Inserting F, we obtain

$$y\sqrt{1 + y'^2} - \frac{yy'^2}{\sqrt{1 + y'^2}} = c.$$

Simplifying, we have

$$\frac{y}{\sqrt{1 + y'^2}} = c.$$

Finally, solving for y', we again obtain a first-order, separable differential equation,

$$c\frac{dy}{dx} = \sqrt{y^2 - c^2}.$$

We can carry out the integration using a hyperbolic function substitution, $y = c \cosh u$, $dy = c \sinh u \, du$. Thus,

$$
\begin{aligned}
x &= \int \frac{c}{\sqrt{y^2 - c^2}} \, dy \\
&= c \int du \\
&= c(u + k) = c\left(\cosh^{-1}\frac{y}{c} + k\right).
\end{aligned}
\tag{10.45}
$$

Thus, the curve shape is given by

$$y(x) = c \cosh\left(\frac{x}{c} + k\right),$$

where c and k can, in principle, be determined from the endpoints of the curve at $x = a$ and $x = b$.

This curve is called a catenary and the surface of revolution generated by the catenary is called a catenoid. For $k = 0$ and $c = 1$, the catenoid is shown in Figure 10.30.

10.2.4   Isoperimetic Problems

ACCORDING VIRGIL'S *Aeneid*, Princess Dido fled the ancient Phoenician city of Tyre after her brother, Pygmalion, murdered her husband. She went

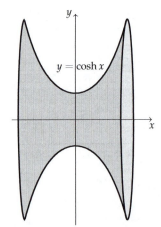

Figure 10.30: The catenoid is a minimal surface of revolution created by the rotation of the catenary curve, $y(x) = c \cosh x$.

to the Mediterranean coast of Africa and asked for a piece of land. She agreed to pay for as much land as she could enclose with an ox-hide. She cut up the ox-hide into long, thin strips, tied the strips together, and used that to enclose the largest area of land. This land supposedly became Carthage, and Dido became its Queen.

This is a classic example of what is called an isoperimetric problem. Namely, one seeks to find the shape of the closed curve of fixed length that encloses the maximum area. This is just one example of a class of variational problems with constraints. Namely, one seeks the path that makes the integral

$$J[y] = \int_a^b F(x, y, y')\, dx$$

stationary, leaving

$$I[y] = \int_a^b G(x, y, y')\, dx = c$$

invariant.

We can derive the needed Euler Equation for this problem. Let $u(x)$ be the path that leaves $J[y]$ stationary. Then we consider the variations

$$y(x) = u(x) + \epsilon(\alpha \eta(x) + \beta \xi(x)),$$

where the values at the endpoints are the same. Namely, we assume that $\eta(a) = \eta(b) = 0$ and $\xi(a) = \xi(b) = 0$. The second perturbation of $u(x)$ is introduced so as to manage the constraint.

We now consider the function

$$J[y] = J[u + \epsilon(\alpha \eta + \beta \xi)] \equiv \phi(\epsilon).$$

As before, the condition for stationarity is

$$\frac{d\phi}{d\epsilon}\bigg|_{\epsilon=0} = 0.$$

The derivation proceeds by differentiating under the integral and performing a integration by parts,

$$
\begin{aligned}
\frac{d\phi}{d\epsilon} &= \frac{d}{d\epsilon} \int_a^b F(x, u + \epsilon(\alpha\eta + \beta\xi), u' + \epsilon(\alpha\eta' + \beta\xi'))\, dx \\
&= \int_a^b \frac{\partial}{\partial \epsilon} F(x, u + \epsilon(\alpha\eta + \beta\xi), u' + \epsilon(\alpha\eta' + \beta\xi'))\, dx \\
&= \int_a^b \left[\frac{\partial F}{\partial y} \frac{\partial y}{\partial \epsilon} + \frac{\partial F}{\partial y'} \frac{\partial y'}{\partial \epsilon} \right] dx \\
&= \int_a^b \left[\frac{\partial F}{\partial y}(\alpha\eta + \beta\xi) + \frac{\partial F}{\partial y'}(\alpha\eta' + \beta\xi') \right] dx \\
&= \int_a^b \left[\frac{\partial F}{\partial y} - \frac{d}{dx}\left(\frac{\partial F}{\partial y'} \right) \right] (\alpha\eta + \beta\xi)\, dx. \qquad (10.46)
\end{aligned}
$$

This time we cannot simply equate a factor of the integrand to zero, as the perturbation $\alpha\eta(x) + \beta\xi(x)$ is not an arbitrary function. Using the constraint, we can separate η from ξ. We first note that because

$$I[y] = I[u + \epsilon(\alpha\eta + \beta\xi)] \equiv \psi(\epsilon)$$

is constant, then

$$\left.\frac{d\psi}{d\epsilon}\right|_{\epsilon=0} = 0.$$

Following the derivation for $\phi(\epsilon)$, we have

$$\frac{d\psi}{d\epsilon} = \int_a^b \left[\frac{\partial G}{\partial y} - \frac{d}{dx}\left(\frac{\partial G}{\partial y'}\right)\right](\alpha\eta + \beta\xi)\,dx.$$

Setting $\epsilon = 0$, we have

$$\int_a^b \left[\frac{\partial F}{\partial y} - \frac{d}{dx}\left(\frac{\partial F}{\partial y'}\right)\right](\alpha\eta + \beta\xi)\,dx = 0,$$

$$\int_a^b \left[\frac{\partial G}{\partial y} - \frac{d}{dx}\left(\frac{\partial G}{\partial y'}\right)\right](\alpha\eta + \beta\xi)\,dx = 0. \qquad (10.47)$$

In each equation, we can solve for $\frac{\alpha}{\beta}$, obtaining

$$\frac{\alpha}{\beta} = \frac{\int_a^b \left[\frac{\partial F}{\partial y} - \frac{d}{dx}\left(\frac{\partial F}{\partial y'}\right)\right]\xi\,dx}{\int_a^b \left[\frac{\partial F}{\partial y} - \frac{d}{dx}\left(\frac{\partial F}{\partial y'}\right)\right]\eta\,dx}$$

$$= \frac{\int_a^b \left[\frac{\partial G}{\partial y} - \frac{d}{dx}\left(\frac{\partial G}{\partial y'}\right)\right]\xi\,dx}{\int_a^b \left[\frac{\partial G}{\partial y} - \frac{d}{dx}\left(\frac{\partial G}{\partial y'}\right)\right]\eta\,dx}. \qquad (10.48)$$

Eliminating $\frac{\alpha}{\beta}$ and rearranging, we have

$$\frac{\int_a^b \left[\frac{\partial F}{\partial y} - \frac{d}{dx}\left(\frac{\partial F}{\partial y'}\right)\right]\eta\,dx}{\int_a^b \left[\frac{\partial G}{\partial y} - \frac{d}{dx}\left(\frac{\partial G}{\partial y'}\right)\right]\eta\,dx} = \frac{\int_a^b \left[\frac{\partial F}{\partial y} - \frac{d}{dx}\left(\frac{\partial F}{\partial y'}\right)\right]\xi\,dx}{\int_a^b \left[\frac{\partial G}{\partial y} - \frac{d}{dx}\left(\frac{\partial G}{\partial y'}\right)\right]\xi\,dx}$$

Because the left expression a function of η and the right expression is a function of ξ, we conclude that these two functions must be a constant, λ. Therefore, upon setting the left expression equal to λ, we have

$$\int_a^b \left[\frac{\partial(F - \lambda G)}{\partial y} - \frac{d}{dx}\left(\frac{\partial(F - \lambda G)}{\partial y'}\right)\right]\eta\,dx = 0.$$

Now we can use the fact that $\eta(x)$ is an arbitrary function and conclude that the path that makes the integral

$$J[y] = \int_a^b F(x, y, y')\,dx$$

stationary, leaving

$$I[y] = \int_a^b G(x, y, y')\,dx = c$$

invariant, is the same path that leaves $F(x, y, y') - \lambda G(x, y, y')$ stationary for some constant λ without any constraint. The path can be determined from the solution of the Euler Equation

$$\frac{\partial(F - \lambda G)}{\partial y} - \frac{d}{dx}\left(\frac{\partial(F - \lambda G)}{\partial y'}\right) = 0.$$

Example 10.24. *Maximize the area under a curve between the points $(0,0)$ and $(1,0)$ constrained to a fixed length.*

Thus, we seek to maximize the area

$$A[y] = \int_0^1 y(x)\,dx$$

subject to

$$L[y] = \int_0^1 \sqrt{1 + [y'(x)]^2}\,dx = \ell.$$

Introducing a Lagrange multiplier, we form the function

$$F(x, y, y'; \lambda) = y(x) + \lambda\sqrt{1 + [y'(x)]^2}.$$

The Euler Equation,

$$\frac{\partial F}{\partial y} - \frac{d}{dx}\left(\frac{\partial F}{\partial y'}\right) = 0,$$

becomes

$$-1 + \frac{d}{dx}\left(\lambda\frac{y'}{\sqrt{1 + y'^2}}\right) = 0,$$

or

$$\lambda\frac{y'}{\sqrt{1 + y'^2}} = x - c_1.$$

Solving for y', we have

$$y' = \pm\frac{x - c_1}{\sqrt{\lambda^2 - (x - c_1)^2}}$$

This is easily integrated to obtain

$$y = \mp\sqrt{\lambda^2 - (x - c_1)^2} + c_2.$$

Rearranging, the equation for the curve is

$$(x - c_1)^2 + (y - c_2)^2 = \lambda^2.$$

This is easily recognizable as the equation of a circular arc.

There are three unknowns, c_1, c_2, and λ. Using the end points and the length of the curve, one can determine these constants. The curve passes through the points $(0,0)$ and $(1,0)$. Therefore, the equation of the arc demands that

$$\begin{aligned} c_1^2 + c_2^2 &= \lambda^2, \\ (1 - c_1)^2 + c_2^2 &= \lambda^2. \end{aligned} \tag{10.49}$$

Subtracting one equation from the other, we have

$$c_1^2 = (1 - c_1)^2.$$

Thus, $c_1 = \frac{1}{2}$.

Next, we will determine λ. We know that for a given length ℓ, we have

$$
\begin{aligned}
\ell &= \int_0^1 \sqrt{1 + [y'(x)]^2}\, dx \\
&= \int_0^1 \frac{\lambda}{\sqrt{\lambda^2 - (x - c_1)^2}}\, dx \\
&= \lambda \sin^{-1}\left(\frac{1 - c_1}{\lambda}\right) + \lambda \sin^{-1}\left(\frac{c_1}{\lambda}\right),
\end{aligned} \qquad (10.50)
$$

where we have used $y' = \pm \frac{x - c_1}{\sqrt{\lambda^2 - (x - c_1)^2}}$. Because we know $c_1 = 1 - c_1 = \frac{1}{2}$,

$$
\ell = 2\lambda \sin^{-1}\left(\frac{1}{2\lambda}\right).
$$

We can now determine c_2, because $c_1^2 + c_2^2 = \lambda^2$. In Table 10.8 we show a few values of the parameters with the lengths and areas of the arcs.

Table 10.8: Parameters for Arcs Passing through $(0,0)$ and $(1,0)$ with $\sin \ell = 0.5$, Where ℓ is the Length of the Arc

c_1	λ	ℓ	c_2	Area
0.6180	2.0247	0.5	1.9280	3.9283
1.6180	2.3272	0.5	1.6727	3.6866
−0.6180	2.3272	0.5	2.2436	4.2575
−1.6180	2.9771	0.5	2.4990	4.5483

Figure 10.31: Arcs through the points $(0,0)$ and $(1,0)$ that maximize the area under the arc for a fixed length of the arc, $\ell = 2\lambda \sin^{-1}(1/2\lambda)$.

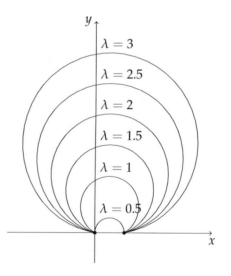

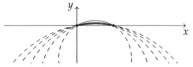

Figure 10.32: Arcs through the points $(0,0)$ and $(1,0)$ that maximize the area under the arc for a fixed length of the arc, $\ell = 2\lambda \sin^{-1}(1/2\lambda)$ with $c_2 \leq 0$. These arcs are pieces of circles that are reflections of those in Figure 10.31.

In Figure 10.31 we show examples of arcs for various values of λ as found in Table 10.8. These correspond to $c_2 > 0$. For $c_2 < 0$, the arcs correspond to the smaller cut arcs in Figure 10.32 and are shown in Figure 10.32.

Note that the arc corresponding to $c_2 = 0$ occurs for $\lambda = c_1 = 0.5$. In this case one obtains a circle centered at $(0.5, 0)$ of radius $\lambda = 0.5$. Thus, the length of the arc is half the circumference, $\ell = \frac{\pi}{2}$, and the area under the arc is half the area of the circle, $A = \frac{\pi}{8}$, as seen in Table 10.8.

Another approach to such problems is to work with parametrized paths, such as $\{(x(t), y(t)) | t \in [a, b]\}$. In these cases, one has to find Euler equations based on integrals of the form

$$J[x, y] = \int_a^b F(t, x(t), y(y), \dot{x}(t), \dot{y}(t)) \, dt.$$

As before, we look at variations of the path about a path that makes $J[x, y]$ stationary. Let that path be represented by $u(t), v(t)$. Then we consider the variations as functions of t in the form

$$x(t) = u(t) + \epsilon \eta(t),$$
$$y(t) = v(t) + \epsilon \xi(t), \tag{10.51}$$

where $\eta(a) = \eta(b) = 0$ and $\xi(a) = \xi(b) = 0$. We can now proceed as before.

We consider the function

$$\phi(\epsilon) = \int_a^b F(t, u + \epsilon \eta, v + \epsilon \xi, \dot{u} + \epsilon \dot{\eta}, \dot{v} + \epsilon \dot{\xi}) \, dt$$

and seek conditions such that $\frac{d\phi}{d\epsilon}\big|_{\epsilon=0}$. Computing the derivative, we have

$$
\begin{aligned}
\frac{d\phi}{d\epsilon} &= \frac{d}{d\epsilon} \int_a^b F(t, u + \epsilon \eta, v + \epsilon \xi, \dot{u} + \epsilon \dot{\eta}, \dot{v} + \epsilon \dot{\xi}) \, dt \\
&= \int_a^b \frac{\partial}{\partial \epsilon} F(t, u + \epsilon \eta, v + \epsilon \xi, \dot{u} + \epsilon \dot{\eta}, \dot{v} + \epsilon \dot{\xi}) \, dt \\
&= \int_a^b \left[\frac{\partial f}{\partial x} \frac{dx}{d\epsilon} + \frac{\partial f}{\partial \dot{x}} \frac{d\dot{x}}{d\epsilon} + \frac{\partial f}{\partial y} \frac{dy}{d\epsilon} + \frac{\partial f}{\partial \dot{y}} \frac{d\dot{y}}{d\epsilon} \right] dt \\
&= \int_a^b \left[\frac{\partial f}{\partial x} \eta + \frac{\partial f}{\partial \dot{x}} \dot{\eta} + \frac{\partial f}{\partial y} \xi + \frac{\partial f}{\partial \dot{y}} \dot{\xi} \right] dt. \tag{10.52}
\end{aligned}
$$

As before, we perform integration by parts to remove the derivatives from η and ξ using the boundary conditions on η and ξ to obtain

$$\int_a^b \left[\left(\frac{\partial f}{\partial x} - \frac{d}{dt} \left(\frac{\partial f}{\partial \dot{x}} \right) \right) \eta + \left(\frac{\partial f}{\partial y} - \frac{d}{dt} \left(\frac{\partial f}{\partial \dot{y}} \right) \right) \xi \right] dt = 0.$$

Because $\eta(t)$ and $\xi(t)$ are arbitrary functions, their coefficients in the integral have to vanish. For example, we consider the cases where $\xi(t) = 0$. This leaves

$$\int_a^b \left[\frac{\partial f}{\partial x} - \frac{d}{dt} \left(\frac{\partial f}{\partial \dot{x}} \right) \right] \eta \, dt = 0.$$

Then, for $\eta(t)$ arbitrary, we conclude that

$$\frac{\partial f}{\partial x} - \frac{d}{dt} \left(\frac{\partial f}{\partial \dot{x}} \right) = 0.$$

Similarly, we can look at cases when $\eta(t) = 0$ and obtain a second Euler Equation,

$$\frac{\partial f}{\partial y} - \frac{d}{dt} \left(\frac{\partial f}{\partial \dot{y}} \right) = 0.$$

Thus, we have obtained a pair of Euler equations for extremizing $J[x, y]$ along a parametrized path. This is summarized below:

The path $\{(x(t), y(t)) | t \in [a, b]\}$ for which

$$J[x, y] = \int_a^b F(t, x(t), y(y), \dot{x}(t), \dot{y}(t)) \, dt$$

is an extremum that satisfies the two Euler equations

$$\frac{\partial f}{\partial x} - \frac{d}{dt}\left(\frac{\partial f}{\partial \dot{x}}\right) = 0,$$

$$\frac{\partial f}{\partial y} - \frac{d}{dt}\left(\frac{\partial f}{\partial \dot{y}}\right) = 0. \tag{10.53}$$

This is easily generalized to higher-dimensional paths, $\mathbf{r}(t) = (x(t), y(t), z(t))$.

Example 10.25. *Determine the closed curve with a given fixed length that encloses the largest possible area.*

As seen in the previous chapter, the area enclosed by curve C is given by

$$A = \frac{1}{2}\oint_C x \, dy - y \, dx.$$

This was verified as an example of Green's Theorem in the Plane, as

$$\oint_C x \, dy - y \, dx = \int_S \left(\frac{\partial x}{\partial x} - \frac{\partial(-y)}{\partial y}\right) dx dy = 2\int_S dx dy.$$

If the curve is parametrized as $x = x(t)$ and $y = y(t)$, then

$$A = \frac{1}{2}\oint_C (x\dot{y} - y\dot{x}) \, dt.$$

Also, the length of the curve is given as

$$L = \oint_C \sqrt{\dot{x}^2 + \dot{y}^2} \, dt.$$

The problem can be mathematically stated as finding the maximum of $A[x, y]$ subject to the constraint $L[x, y]$. This is equivalent to finding the path for which $I[x, y] = A[x, y] + \lambda L[x, y]$ is stationary for some constant λ. Thus, we consider the integral

$$I[x, y] = \oint_C \left(\frac{1}{2}(x\dot{y} - y\dot{x}) + \lambda\sqrt{\dot{x}^2 + \dot{y}^2}\right) dt.$$

We define the integrand

$$F(t, x, \dot{x}, y, \dot{y}) = \frac{1}{2}(x\dot{y} - y\dot{x}) + \lambda\sqrt{\dot{x}^2 + \dot{y}^2}.$$

The variational problem states that F satisfies a pair of Euler-Lagrange equations:

$$\frac{\partial F}{\partial x} - \frac{d}{dt}\left(\frac{\partial F}{\partial \dot{x}}\right) = 0,$$

$$\frac{\partial F}{\partial y} - \frac{d}{dt}\left(\frac{\partial F}{\partial \dot{y}}\right) = 0. \tag{10.54}$$

Inserting F, we have

$$\frac{1}{2}\dot{y} - \frac{d}{dt}\left(-\frac{1}{2}y + \frac{\lambda\dot{x}}{\sqrt{\dot{x}^2 + \dot{y}^2}}\right) = 0,$$

$$-\frac{1}{2}\dot{x} - \frac{d}{dt}\left(\frac{1}{2}x + \frac{\lambda\dot{y}}{\sqrt{\dot{x}^2 + \dot{y}^2}}\right) = 0. \qquad (10.55)$$

Each of these equations is a perfect derivative and can be integrated. The results are

$$y - \frac{\lambda\dot{x}}{\sqrt{\dot{x}^2 + \dot{y}^2}} = c_1,$$

$$x + \frac{\lambda\dot{y}}{\sqrt{\dot{x}^2 + \dot{y}^2}} = c_2, \qquad (10.56)$$

where c_1 and c_2 are two arbitrary constants. Rearranging, squaring both equations, and adding, we have

$$\begin{aligned}(x - c_2)^2 + (y - c_1)^2 &= \frac{\lambda^2\dot{x}^2}{\dot{x}^2 + \dot{y}^2} + \frac{\lambda^2\dot{y}^2}{\dot{x}^2 + \dot{y}^2}\\ &= \lambda^2. \qquad (10.57)\end{aligned}$$

Therefore, the maximum area is enclosed by a circle.

10.3 Hamilton's Principle

IN CLASSICAL DYNAMICS, AN IMPORTANT APPLICATION of the calculus of variations is the derivation of the equations of motion for masses subject to forces. This is based upon Hamilton's Principle, which states that the path of a particle, $\mathbf{r}(t)$, between initial position $\mathbf{r}(t_0)$ and final position $\mathbf{r}(t_1)$ is such to make the action integral stationary, $\delta J(\mathbf{r}) = 0$, where

The action integral.

$$J(\mathbf{r}) = \int_{t_0}^{t_1} \mathcal{L}(t, \mathbf{r}, \dot{\mathbf{r}})\, dt$$

and $\mathcal{L}(t, \mathbf{r}, \dot{\mathbf{r}}) = T(\dot{\mathbf{r}}) - V(t, \mathbf{r})$ is called the Lagrangian function. Here, T is the kinetic energy and V is the potential energy.

Consider a particle moving along a path $\mathbf{r}(t) = x(t)\mathbf{i} + y(t)\mathbf{j} + z(t)\mathbf{k}$ under a force $\mathbf{F} = -\nabla V(x, y, z)$. It has a kinetic energy given by

$$T = \frac{1}{2}m\dot{x}^2 + \frac{1}{2}m\dot{y}^2 + \frac{1}{2}m\dot{z}^2.$$

The Lagrangian function is then

$$\mathcal{L} = \frac{1}{2}m\dot{x}^2 + \frac{1}{2}m\dot{y}^2 + \frac{1}{2}m\dot{z}^2 - V(x, y, z).$$

The Lagrangian.

We note that the Lagrangian is defined on a path parametrized by time. So, we can apply the variational methods from the previous section to the action integral. We then obtain three Euler-Lagrange equations,

$$
\begin{aligned}
0 &= \frac{d}{dt}\left(\frac{\partial \mathcal{L}}{\partial \dot{x}}\right) - \frac{\partial \mathcal{L}}{\partial x}, \\
0 &= \frac{d}{dt}\left(\frac{\partial \mathcal{L}}{\partial \dot{y}}\right) - \frac{\partial \mathcal{L}}{\partial y}, \\
0 &= \frac{d}{dt}\left(\frac{\partial \mathcal{L}}{\partial \dot{z}}\right) - \frac{\partial \mathcal{L}}{\partial z}.
\end{aligned}
\tag{10.58}
$$

These equations simplify when we insert the kinetic energy,

$$
\begin{aligned}
0 &= \frac{d}{dt}\left(m\dot{x}\right) + \frac{\partial V}{\partial x}, \\
0 &= \frac{d}{dt}\left(m\dot{y}\right) + \frac{\partial V}{\partial y}, \\
0 &= \frac{d}{dt}\left(m\dot{z}\right) + \frac{\partial V}{\partial z},
\end{aligned}
\tag{10.59}
$$

leading to

$$
\begin{aligned}
m\ddot{x} &= -\frac{\partial V}{\partial x} = F_x, \\
m\ddot{y} &= -\frac{\partial V}{\partial y} = F_y, \\
m\ddot{z} &= -\frac{\partial V}{\partial z} = F_z.
\end{aligned}
\tag{10.60}
$$

We recognize these equations as the three components of Newton's Second Law of Motion, $\mathbf{F} = m\mathbf{a}$ when $\mathbf{F} = -\nabla V$. Therefore, the Euler-Lagrange equations lead to the equations of motion.

Example 10.26. *Find the equation of motion of a mass moving in one dimension under the potential $V(x) = \frac{1}{2}kx^2$.*

The Lagrangian for this problem is given by

$$
\mathcal{L} = \frac{1}{2}\dot{x}^2 - \frac{1}{2}kx^2.
$$

The Euler-Lagrange equation becomes

$$
\begin{aligned}
0 &= \frac{d}{dt}\left(\frac{\partial \mathcal{L}}{\partial \dot{x}}\right) - \frac{\partial \mathcal{L}}{\partial x} \\
0 &= \frac{d}{dt}(m\dot{x}) + kx \\
m\ddot{x} &= -kx.
\end{aligned}
\tag{10.61}
$$

Of course, this is just the equation of motion for one-dimensional simple harmonic motion.

Example 10.27. *Find the equation of motion for a pendulum.*

We consider a point mass m hanging on a string of length L from a support as shown in Figure 10.33. The Cartesian coordinates for the position of this mass at time t are $(x(t), y(t))$. However, these components are related by

$$
x^2(t) + y^2(t) = L^2.
$$

Due to this constraint, there really is only one degree of freedom of motion. As usual, we choose the motion to be described by the angle that the pendulum string makes with the vertical, θ. The relation to the Cartesian coordinate system as shown in Figure 10.33 is

$$x(t) = L \sin \theta,$$
$$y(t) = L \cos \theta. \tag{10.62}$$

This gives the kinetic energy as

$$T = \frac{1}{2}m\dot{x}^2 + \frac{1}{2}m\dot{y}^2 = \frac{1}{2}mL^2\dot{\theta}^2.$$

The potential energy is

$$V = mgh = mgL(1 - \cos\theta).$$

So, the Lagrangian is

$$\mathcal{L}(t, \theta, \dot{\theta}) = \frac{1}{2}mL^2\dot{\theta}^2 - mgL(1 - \cos\theta).$$

The equations of motion are obtained using the Euler-Lagrange equation

$$0 = \frac{d}{dt}\left(\frac{\partial \mathcal{L}}{\partial \dot{\theta}}\right) - \frac{\partial \mathcal{L}}{\partial \theta}$$
$$0 = mL^2\ddot{\theta} + mgL\sin\theta. \tag{10.63}$$

Dividing by mL^2, we obtain the familiar equation for the nonlinear pendulum [see Chapters 2 and 4],

$$\ddot{\theta} + \frac{g}{L}\sin\theta = 0.$$

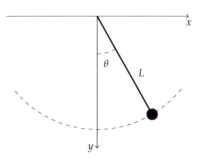

Figure 10.33: A simple pendulum of mass m attached to a string of length L.

The power of using Hamilton's Principle in classical dynamics is when one can apply it to more complex systems involving complicated geometries or many particles. In the previous example, we saw that a coordinate constraint can reduce the number of degrees of freedom, and therefore reduce the number of dynamic variables needed in the problem. In general, a system of N particles can be described by $3N$ degrees of freedom corresponding to the total number of coordinates involved. However, there might be some constraints, or relations, between these coordinates, such as restricting some of the particles to a fixed separation or to stay on a given plane (like the floor).

If these constraints are related through k equations

$$g_j(\mathbf{r}_1, \ldots, \mathbf{r}_N, t) = c_j, \quad j = 1, \ldots, k,$$

then the constraints are called holonomic constraints. There are then $3N - k$ independent degrees of freedom.

For example, the mass of a simple pendulum is constrained to a vertical plane and is further constrained to a fixed distance from the pivot. Assuming, as we had, that the pendulum swings in the xy-plane, then the equations of constraint are

$$z = 0,$$
$$x^2 + y^2 = L^2. \tag{10.64}$$

So, the three degrees of freedom of a free particle less these two constraints yields one independent degree of freedom. In our previous example, this degree of freedom was defined as the angle θ.

Constraints that are not holonomic are called nonholonomic constraints. Either there are no simple equations relating the coordinates, or there are equations that not only relate the coordinates, but also involve the velocities. We will not address nonholonomic constraints here, but refer the reader to the numerous texts on classical dynamics.

Having established the number of independent degrees of freedom, we would like to describe the system in $3N - k$ variables. These new coordinates, which can describe the configuration of the system, are called generalized coordinates. These are typically denoted $q_1, \ldots, q_n$, where $n = 3N - k$. For holonomic constraints, then

$$\mathbf{r}_j = \mathbf{r}_j(q_1, \ldots, q_n, t), \quad j = 1, \ldots, 3N.$$

Along with the generalized coordinates, one defines the generalized velocities as $\dot{q}_j$, $j = 1, \ldots, n$. Then one can show that the Lagrangian can be written in terms of the generalized coordinates and velocities,

$$\mathcal{L} = T(q_j, \dot{q}_j, t) - V(q_j, t).$$

This result depends on basic assumptions of having only holonomic constraints and conservative forces. Departure from these assumptions is beyond the scope of this discussion.

Hamilton's Principle now takes the form

$$\delta \int_{t_0}^{t_1} \mathcal{L}(q_1, \dot{q}_2, \ldots, q_n, \dot{q}_n, t) \, dt.$$

As noted earlier, there are different nomenclatures used in the literature. Many make no distinctions between the Euler equations and the Euler-Lagrange equations. Some even just refer to this set of equations as Lagrange's equations. This set of equations was first derived by Lagrange for mechanical systems in his treatise in 1788.

Applying the methods of variational calculus results in the system of equations

$$\frac{d}{dt}\left(\frac{\partial \mathcal{L}}{\partial \dot{q}_j}\right) - \frac{\partial \mathcal{L}}{\partial q_j} = 0, \quad j = 1, \ldots, n. \tag{10.65}$$

These are the Euler-Lagrange equations in generalized coordinates. They reduce to the expressions used in the previous examples with an appropriate selection of generalized coordinates and generalized velocities.

Example 10.28. *Find the equations for motion for two masses, one on a frictionless inclined plane supported by a second mass hanging from a rope on a pulley.*

In Figure 10.34 are shown two masses hanging from a pulley and connected by a rope of length ℓ. Mass m_1 is sitting on the incline a distance x from the pulley. Mass m_2 is hanging a distance $\ell - x$ from the pulley. In this problem, the motion of the masses depends on one degree of freedom that is captured by the variable x.

Taking a reference of zero potential energy from the pulley, the potential energies for each mass are given by

$$\begin{aligned} V_1 &= -m_1 g h = -m_1 g x \sin\theta, \\ V_2 &= m_2 g(x - \ell). \end{aligned} \tag{10.66}$$

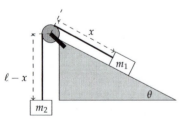

Figure 10.34: A mass on an inclined plane is attached to a second mass via a pulley at the top of the inclined plane.

The kinetic energy is

$$T = \frac{1}{2}(m_1 + m_2)\dot{x}^2.$$

Therefore, the Lagrangian is given as

$$\mathcal{L} = \frac{1}{2}(m_1 + m_2)\dot{x}^2 + m_1 g x \sin\theta + m_2 g(x - \ell).$$

The equations of motion are obtained using the Euler-Lagrange Equation

$$\begin{aligned}
0 &= \frac{d}{dt}\left(\frac{\partial\mathcal{L}}{\partial\dot{x}}\right) - \frac{\partial\mathcal{L}}{\partial x} \\
&= (m_1 + m_2)\ddot{x} + (m_1 \sin\theta - m_2)g.
\end{aligned} \tag{10.67}$$

Solving for the acceleration, we have

$$\ddot{x} = \frac{m_2 - m_1 \sin\theta}{m_1 + m_2}g.$$

We note that this is a constant acceleration. In the case that $m_2 = 0$, we recover the acceleration of a block sliding frictonlessly down an incline.

Example 10.29. *Find the Lagrangian for the double pendulum.*

Consider the double pendulum depicted in Figure 10.35. Mass m_1 is set up as a simple pendulum with length ℓ_1 One then attaches a second mass, m_2, to the first mass on a string of length ℓ_2. In this problem, we will set up the Lagrangian for this system.

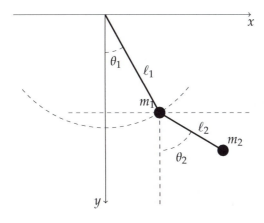

Figure 10.35: The double pendulum consisting of two masses attached to a support.

The idea is simple. We have two masses, each with a kinetic energy $T = \frac{1}{2}mv^2$ and gravitational potential energy $V = mgh$. We simply need to write these energies in terms of the two independent variables in the problem, θ_1 and θ_2. Note that we are assuming that the masses move in the same plane, giving one constraint each. Also, mass m_1 is constrained to be a distance ℓ_1 from the pivot and mass m_2 is constrained to be a distance ℓ_2 from mass m_1. So, the number of degrees of freedom is $3N - 4 = 2$.

From Figure 10.35 it is easy to see that the position of the first mass is given by

$$(x_1, y_1) = (\ell_1 \sin\theta_1, -\ell_1 \cos\theta_1)$$

and the position of the second mass is

$$(x_2, y_2) = (\ell_1 \sin\theta_1 + \ell_2 \sin\theta_2, -\ell_1 \cos\theta_1 - \ell_2 \cos\theta_2).$$

The potential energies depend on the vertical positions. We will take the potential energy to be zero at the pivot, $V = mgy$. So,

$$
\begin{aligned}
V &= m_1 g y_1 + m_2 g y_2 \\
&= m_1 g(-\ell_1 \cos\theta_1) + m_2 g(-\ell_1 \cos\theta_1 - \ell_2 \cos\theta_2) \\
&= -(m_1 + m_2) g \ell_1 \cos\theta_1 - m_2 g \ell_2 \cos\theta_2. \quad (10.68)
\end{aligned}
$$

In order to write the kinetic energy contributions, we need the velocities

$$
\begin{aligned}
(\dot{x}_1, \dot{y}_1) &= (\ell_1 \cos\theta_1, \ell_1 \sin\theta_1)\dot{\theta}_1, \\
(\dot{x}_2, \dot{y}_2) &= (\ell_1 \cos\theta_1 \dot{\theta}_1 + \ell_2 \cos\theta_2 \dot{\theta}_2, \ell_1 \sin\theta_1 \dot{\theta}_1 + \ell_2 \sin\theta_2 \dot{\theta}_2).
\end{aligned}
$$

$$(10.69)$$

Thus,

$$
\begin{aligned}
v_1^2 &= \dot{x}_1^2 + \dot{y}_1^2 = \ell_1^2 \dot{\theta}_1^2, \\
v_2^2 &= \dot{x}_2^2 + \dot{y}_2^2 \\
&= (\ell_1 \cos\theta_1 \dot{\theta}_1 + \ell_2 \cos\theta_2 \dot{\theta}_2)^2 + (\ell_1 \sin\theta_1 \dot{\theta}_1 + \ell_2 \sin\theta_2 \dot{\theta}_2)^2 \\
&= \ell_1^2 \dot{\theta}_1^2 + \ell_2^2 \dot{\theta}_2^2 + 2\ell_1 \ell_2 (\cos\theta_1 \cos\theta_2 + \sin\theta_1 \sin\theta_2)\dot{\theta}_1 \dot{\theta}_2 \\
&= \ell_1^2 \dot{\theta}_1^2 + \ell_2^2 \dot{\theta}_2^2 + 2\ell_1 \ell_2 \cos(\theta_1 - \theta_2)\dot{\theta}_1 \dot{\theta}_2. \quad (10.70)
\end{aligned}
$$

Thus, the total kinetic energy is

$$
\begin{aligned}
T &= \frac{1}{2} m_1 v_1^2 + \frac{1}{2} m_2 v_2^2 \\
&= \frac{1}{2} m_1 \ell_1^2 \dot{\theta}_1^2 + \frac{1}{2} m_2 \left[\ell_1^2 \dot{\theta}_1^2 + \ell_2^2 \dot{\theta}_2^2 + 2\ell_1 \ell_2 \cos(\theta_1 - \theta_2)\dot{\theta}_1 \dot{\theta}_2 \right]. (10.71)
\end{aligned}
$$

Therefore, the Lagrangian is given as

$$
\begin{aligned}
\mathcal{L} &= T - V \\
&= \frac{1}{2} m_1 \ell_1^2 \dot{\theta}_1^2 + \frac{1}{2} m_2 \left[\ell_1^2 \dot{\theta}_1^2 + \ell_2^2 \dot{\theta}_2^2 + 2\ell_1 \ell_2 \cos(\theta_1 - \theta_2)\dot{\theta}_1 \dot{\theta}_2 \right] \\
&\quad - \left[(m_1 + m_2) \ell_1 \cos\theta_1 + m_2 g \ell_2 \cos\theta_2 \right]. \quad (10.72)
\end{aligned}
$$

Example 10.30. *Find the equations of motion for a bead free to move without friction on a spinning wire.*

The bead on the wire is shown in Figure 10.36. The height of the bead from the bottom of the wire is $h = R(1 - \cos\theta)$. Thus, we can take the potential energy as

$$V = mgh = mgR(1 - \cos\theta).$$

The kinetic energy has two contributions due to the bead moving on the wire and the wire spinning with angular frequency Ω. In each case, there is a contribution of the form $\frac{1}{2} mv^2$. In the first case, the tangential velocity on the wire is $v_T = R\dot{\theta}$. The second contribution is the result of the bead rotating into the plane of the wire *about a circle of radius $\rho = R\sin\theta$ as depicted by the dashed circle in Figure 10.36. Thus, this velocity contribution is given by $v_\Omega = \rho\Omega = \Omega R\sin\theta$. Therefore, the total kinetic energy is*

$$T = \frac{1}{2}mR^2(\dot\theta^2 + \Omega^2\sin^2\theta).$$

The Lagrangian, $\mathcal{L} = T - V$, is then found as

$$\mathcal{L} = \frac{1}{2}mR^2(\dot\theta^2 + \Omega^2\sin^2\theta) - mgR(1 - \cos\theta).$$

Inserting this expression into the Euler-Lagrange equations, we have

$$
\begin{aligned}
0 &= \frac{d}{dt}\left(\frac{\partial\mathcal{L}}{\partial\dot\theta}\right) - \frac{\partial\mathcal{L}}{\partial\theta} \\
&= mR^2\ddot\theta - mR^2\Omega^2\sin\theta\cos\theta - mgR\sin\theta.
\end{aligned}
\tag{10.73}
$$

Solving for the angular acceleration, we have

$$\ddot\theta = \left(\Omega^2\cos\theta - \frac{g}{R}\right)\sin\theta.$$

In the following discussion we will take $\frac{g}{R} = \alpha\Omega^2$. Then, the equation of motion becomes

$$\ddot\theta = \Omega^2(\cos\theta - \alpha)\sin\theta.$$

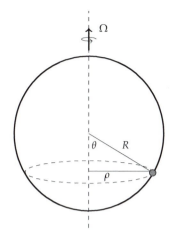

Figure 10.36: A bead on a rotating wire.

In the previous example, we found that the motion of the bead on a rotating wire lead to a nonlinear second-order differential equation for $\theta(t)$. There are various ways one can get information about the solutions of this equation. For example, if the bead moves for only small angles, we have the approximation

$$\ddot\theta + \Omega^2(\alpha - 1)\theta \approx 0.$$

This is simply the equation for simple harmonic motion with period of oscillation

$$T = \frac{2\pi}{\sqrt{\alpha - 1}\,\Omega}$$

provided $\alpha > 1$.

However, for larger angles, the situation is more complicated. We can investigate this behavior using some tools from earlier chapters.

Example 10.31. *Find and classify the equilibria of $\ddot\theta + \Omega^2(\alpha - 1)\theta \approx 0$.*

We can write this nonlinear equation as two first-order equations and look for equilibria by introducing $\omega = \dot\theta$. Thus, we have

$$
\begin{aligned}
\dot\theta &= \omega, \\
\dot\omega &= \Omega^2(\cos\theta - \alpha)\sin\theta.
\end{aligned}
\tag{10.74}
$$

Using the methods for studying nonlinear systems, we seek the equilibrium solutions. The equilibrium solutions occur for

$$
\begin{aligned}
0 &= \omega, \\
0 &= \Omega^2(\cos\theta - \alpha)\sin\theta.
\end{aligned}
\tag{10.75}
$$

$\omega = 0$ *just means that the bead is not sliding up or down the wire. The more interesting part of the solution comes from the second equation in the system. Either $\sin\theta = 0$ or $\cos\theta = \alpha$.*

We next consider the stability of the solutions corresponding to the theta-values satisfying these equations. For this analysis, we will need to evaluate the Jacobian of the system. The Jacobian is given by

$$J(\theta) = \begin{pmatrix} 0 & 1 \\ \mu & 0 \end{pmatrix},$$

where $\mu = \Omega^2(\cos^2\theta - \sin^2\theta - \alpha\cos\theta)$. The eigenvalues of the Jacobian matrix are given by $\lambda = \pm\sqrt{\mu}$. So, when μ is positive for a given equilibrium point, the equilibrium point is an unstable saddle point. When μ is negative, then the equilibrium point is a center. There are three cases to consider.

For the first case, $\sin\theta = 0$. When the bead is at the bottom of the wire ($\theta = 0, \pm 2\pi$, etc.), $\mu = \Omega^2(1 - \alpha)$. When $\alpha > 1$, then the bead is stable and small oscillations are possible about this point. However, when $\alpha < 1$, μ is positive and then these equilibrium points are unstable saddle points.

When the bead is at the top of the wire, $\theta = \pm\pi, \pm 3\pi$, etc., then $\mu = \Omega^2(1 + \alpha)$ is always positive for positive α.. Thus, the point at the top of the loop is always a saddle point.

The third case is when $\cos\theta_0 = \alpha$ for the equilibrium point $\theta = \theta_0$. In this case, $\mu = -\Omega^2\sin^2\theta_0 = \Omega^2(\alpha^2 - 1)$ is always negative. So, when this equilibrium point exists, that is, when $\alpha = \frac{g}{R\Omega^2} < 1$, the equilibrium point is a center and it is possible to have stable oscillations about a point on the rotating wire away from the bottom! For example, if $R = 20$ cm, then $\Omega > 7$ cps.

Figure 10.37: The phase portrait with several orbits for the bead problem with $\alpha = 1.5$ and $\Omega = 10$.

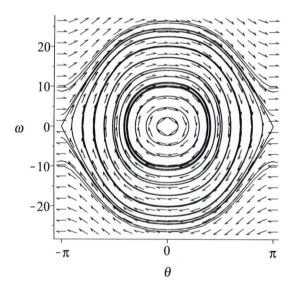

Example 10.32. *Plot solutions of the equation $\ddot\theta + \Omega^2(\alpha - 1)\theta \approx 0$.*

Another tool we can use to study the qualitative behavior of motion of the bead is to use a phase portrait, which we used in Chapters 2 and 4. In Figure 10.37 we show plots for several initial conditions for α = 1.5 and Ω = 10. Because α > 1, the origin is a stable center. In fact, we see the familiar pattern from the nonlinear pendulum. Extending the plot outside the interval [−π, π], we see in Figure 10.38 the emergence of centers and saddle points at the right locations. Also, there are unbounded orbits for large enough energies.

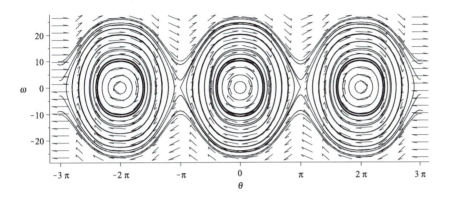

Figure 10.38: The phaseportrait for several periods for the bead problem with α = 1.5 and Ω = 10.

These figures were generated using Maple. One first loads the needed library and sets up the system.

```
> restart; with(DEtools);
> sys := {diff(x(t), t) = y(t),
      diff(y(t), t) = Omega^2*(cos(x(t))-alpha)*sin(x(t))};
> alpha := 1.5; Omega := 10
```

The built-in package for drawing phase portraits can now be called. An example with a minimal number of options is given by

```
phaseportrait(sys, [x(t), y(t)], t = -1 .. 2.5,
   [[x(0) = .1, y(0) = 0], [x(0) = .4, y(0) = 0],
   [x(0) = .7, y(0) = 0]])
```

Figure 10.37 was generated with the slightly more complicated form:

```
phaseportrait(sys, [x(t), y(t)], t = -3 .. 3,
   [seq([x(0) = 2*k*Pi*(1/20), y(0) = 0], k = -20 .. 20),
   seq([x(0) = 2*k*Pi*(1/20), y(0) = -10], k = -20 .. 20),
   seq([x(0) = 2*k*Pi*(1/20), y(0) = 10], k = -20 .. 20)],
   arrows = thin, color = black,
   dirfield = [20, 20], axesfont = [TIMES, ROMAN, 16],
   labelfont = [TIMES, ROMAN, 16, Italic], axes = framed,
   linecolor = black, thickness = 1, numpoints = 300,
   labels = ['theta', 'omega'],
   tickmarks = [spacing(Pi), default], x = -Pi .. Pi)
```

In Figure 10.39 we show plots for several initial conditions for α = 0.5 and Ω = 10. Because α > 1, the origin is an unstable saddle point. This phase portrait also shows the existence of centers, which occur for $\cos\theta_0 = 0.5$, or $\theta_0 = \frac{\pi}{3}$.

Figure 10.39: The phase portrait for several initial conditions for the bead problem with $\alpha = 0.5$ and $\Omega = 10$.

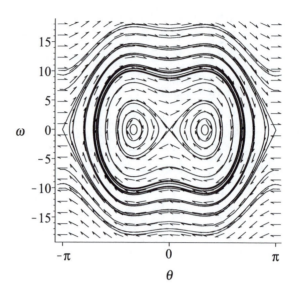

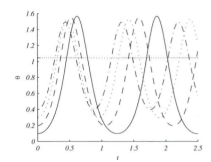

Figure 10.40: Solutions to several initial conditions in the bead problem with $\alpha = 0.5$ and $\Omega = 10$. The initial conditions are given by $\omega = 0$, and $\theta = 0.1, 0.2, 0.3, 0.4$ with the first solution given the solid curve and the others follow (dashed, dotted, dash-dot).

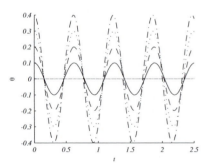

Figure 10.41: Solutions to several initial conditions in the bead problem with $\alpha = 2.0$ and $\Omega = 10$. The initial conditions are given by $\omega = 0$, and $\theta = 0.1, 0.2, 0.3, 0.4$ with the first solution given the solid curve and the others follow (dashed, dotted, dash-dot).

One can also use other packages to generate these figures. In MATLAB, we define the function

```
function dy=beadf(t,y);
global Omega alpha

dy=[y(2);Omega^2*(cos(y(1))-alpha)*sin(y(1))];
```

We can then call up **ode45** or any other numerical solver and plot solutions as a function of time. The typical MATLAB routine is given by

```
clear
global Omega alpha

alpha=.5;
Omega=10;

tspan=[0 5];
y0=[.1;0];
[t,y]=ode45(@beadf,tspan,y0);
plot(t,y(:,1),'k')
```

The solutions as a function of time for several initial conditions for $\alpha = 0.5, 2.0$ are shown in Figures 10.40 and 10.41, respectively.

One can also plot the solutions on top of the direction field in order to display a phase portrait. The code is given below and a sample is shown in Figure 10.42.

```
clear
global Omega alpha

alpha=.5;
```

```
Omega=10;
tspan=[0 5];

figure(2)
[theta,omega]=meshgrid(-pi:.2:pi,-20:1:20);
v1=omega;
v2=Omega^2*(cos(theta)-alpha).*sin(theta);
quiver(theta,omega,v1,v2,'k')
hold on

for k=.1:.1:3
    y0=[k;0];
    [t,y]=ode45(@beadf,tspan,y0);
    plot(y(:,1),y(:,2),'k')
end

axis([-pi pi -20 20])
xlabel('\theta')
ylabel('\omega')
hold off
```

10.4 Geodesics

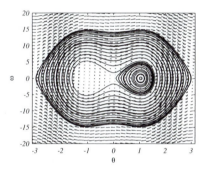

Figure 10.42: The phaseportrait generated by MATLAB for several initial conditions for the bead problem with $\alpha = 0.5$ and $\Omega = 10$.

THE SHORTEST DISTANCE BETWEEN TWO POINTS is a straight line. Well, this is true in a Euclidean plane. But, what is the shortest distance between two points on a sphere? What path should light follow as it passes the sun? Paths that are the shortest distances between points on a curved surface or events in curved spacetime are called geodesics. In general, we can set up an appropriate integral to compute these distances along paths and seek the paths that render the integral stationary. We begin by looking at the geodesics on a sphere.

Example 10.33. *Determine the shortest distance between two points on the surface of a sphere of radius R.*

The length of a curve between points A and B is generally given by

$$L = \int_{S_A}^{S_B} ds.$$

We need to represents ds on the sphere. We can do this by recalling from the previous chapter that for orthogonal curvilinear coordinates.

$$ds^2 = d\mathbf{r} \cdot d\mathbf{r} = \sum_{i=1}^{3} h_i^2 du_i^2,$$

where h_i are the scale factors and u_i are the generalized coordinates. For spherical coordinates, we have

$$d\mathbf{r} = d\rho d\mathbf{e}_\rho + \rho d\theta d\mathbf{e}_\theta + \rho \sin \theta d\phi d\mathbf{e}_\phi.$$

On the surface of a sphere of radius $\rho = R$, $d\rho = 0$ and $h_\phi = R\sin\theta$, $h_\theta = R$. So,

$$ds^2 = R^2 d\theta^2 + R^2 \sin^2\theta d\phi^2.$$

Thinking of the path in the form $\phi(\theta)$, the length of the curve then becomes

$$
\begin{aligned}
L &= \int_{S_A}^{S_B} ds \\
&= R\int_{S_A}^{S_B} \sqrt{d\theta^2 + \sin^2\theta d\phi^2} \\
&= R\int_{S_A}^{S_B} \sqrt{1 + \sin^2\theta \left(\frac{d\phi}{d\theta}\right)^2}\, d\theta.
\end{aligned}
\tag{10.76}
$$

We can apply Euler's Equation to the integrand $F(\theta, \phi, \phi') = \sqrt{1 + \sin^2\theta\phi'^2}$,

$$\frac{\partial F}{\partial \phi} - \frac{d}{d\theta}\left(\frac{\partial F}{\partial \phi'}\right) = 0.$$

We first note that F is independent of ϕ. So,

$$\frac{d}{d\theta}\left(\frac{\partial F}{\partial \phi'}\right) = 0,$$

or

$$\frac{\sin^2\theta\phi'}{\sqrt{1 + \sin^2\theta\phi'^2}} = c.$$

Solving for ϕ',

$$\frac{d\phi}{d\theta} = \frac{c}{\sin\theta\sqrt{\sin^2\theta - c^2}}.$$

One can solve for $\phi(\theta)$ by integrating:

$$
\begin{aligned}
\phi &= \int \frac{c\,d\theta}{\sin\theta\sqrt{\sin^2\theta - c^2}} \\
&= \int \frac{c\,d\theta}{\sin^2\theta\sqrt{1 - c^2\csc^2\theta}}.
\end{aligned}
\tag{10.77}
$$

Letting $u = \cot\theta$ and $du = -\csc^2\theta\,d\theta$, the integral becomes

$$
\begin{aligned}
\phi &= \int \frac{c\,d\theta}{\sin^2\theta\sqrt{1 - c^2\csc^2\theta}} \\
&= \int \frac{-c\,du/\csc^2\theta}{\sin^2\theta\sqrt{1 - c^2(1 + u^2)}} \\
&= \int \frac{-c\,du}{\sqrt{(1 - c^2) - u^2}}.
\end{aligned}
\tag{10.78}
$$

Letting $u = (1 - c^2)\sin t$, $du = (1 - c^2)\cos t\,dt$,

$$
\begin{aligned}
\phi &= \int \frac{-c\,du}{\sqrt{(1 - c^2) - u^2}} \\
&= \frac{-c}{\sqrt{1 - c^2}}\int dt
\end{aligned}
$$

$$= \frac{-c}{\sqrt{1-c^2}}t + \phi_0$$

$$= \frac{-c}{\sqrt{1-c^2}}\sin^{-1}u + \phi_0$$

$$= \frac{-c}{\sqrt{1-c^2}}\sin^{-1}(\cot\theta) + \phi_0. \qquad (10.79)$$

Thus,

$$\cot\theta = a\sin(\phi - \phi_0),$$

where $a = \frac{\sqrt{1-c^2}}{c}$.

While this gives the implicit relation between ϕ and θ, it may not be clear what the geodesics look like. However, writing this relation in Cartesian coordinates is useful. Multiplying the equation by $\sin\theta$ and expanding, we have

$$\cos\theta = a\sin\phi_0\sin\theta\cos\phi - a\cos\phi_0\sin\theta\sin\phi.$$

Recalling the transformation for spherical coordinates from the previous chapter,

$$x = \rho\sin\theta\cos\phi,$$
$$y = \rho\sin\theta\sin\phi,$$
$$z = \rho\cos\theta, \qquad (10.80)$$

we need only multiply the equation by ρ and obtain

$$z = (a\sin\phi_0)x - (a\cos\phi_0)y.$$

This is the equation of a plane in Cartesian coordinates that passes through the origin. Such planes intersect the surface of the sphere in great circles. Thus, geodesics on a sphere lie on great circles. In Figure 10.43 we plot the geodesic between the points $(\theta, \phi) = \left(\frac{\pi}{6}, \frac{5\pi}{18}\right)$ and $(\theta, \phi) = \left(\frac{5\pi}{9}, -\frac{2\pi}{9}\right)$.

Figure 10.43: The geodesic between the two points $(\theta, \phi) = \left(\frac{\pi}{6}, \frac{5\pi}{18}\right)$ and $(\theta, \phi) = \left(\frac{5\pi}{9}, -\frac{2\pi}{9}\right)$.

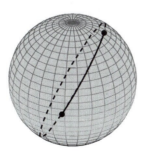

We can generalize the search for geodesics in curved space using the general line element. For a three-dimensional space described by the coordinates (u^1, u^2, u^3), the line element is given by

$$ds^2 = g_{ij}du^i du^j, \quad i, j = 1, 2, 3.$$

Here, repeated indices obey the Einstein summation convention as seen in the previous chapter. Because

$$d\mathbf{r} = \frac{\partial \mathbf{r}}{\partial u^i} du^i,$$

then the metric elements are given by

$$g_{ij} = \frac{\partial \mathbf{r}}{\partial u^i} \cdot \frac{\partial \mathbf{r}}{\partial u^j}.$$

Recall that for an orthogonal system,

$$ds^2 = \sum_{i=1}^{3} h_i^2 (du^i)^2.$$

Therefore, the metric g is given by a diagonal matrix,

$$(g_{ij}) = \begin{pmatrix} h1^2 & 0 & 0 \\ 0 & h_2^2 & 0 \\ 0 & 0 & h_3^2 \end{pmatrix}.$$

For example, the line element in spherical coordinates is given by

$$ds^2 = d\rho^2 + \rho^2 d\theta^2 + \rho^2 \sin^2 \theta d\phi^2,$$

and the corresponding metric g is given by the matrix

$$(g_{ij}) = \begin{pmatrix} 1 & 0 & 0 \\ 0 & \rho^2 & 0 \\ 0 & 0 & \rho^2 \sin^2 \theta \end{pmatrix}.$$

The length of a curve in generalized coordinates is given by the forms

$$
\begin{aligned}
L &= \int_{S_a}^{S_b} ds \\
&= \int_{S_a}^{S_b} \sqrt{g_{ij} du^i du^j} \\
&= \int_{\sigma_a}^{\sigma_b} \sqrt{g_{ij} \frac{du^i}{d\sigma} \frac{du^j}{d\sigma}} \, d\sigma,
\end{aligned}
\tag{10.81}
$$

where σ is the parameter used in the parametrization of the path $(u^i(\sigma))$. For an orthogonal system, the expanded form of this would be

$$L = \int_{\sigma_a}^{\sigma_b} \sqrt{g_{11} \left(\frac{du^1}{d\sigma}\right)^2 + g_{22} \left(\frac{du^2}{d\sigma}\right)^2 + g_{33} \left(\frac{du^3}{d\sigma}\right)^2} \, d\sigma.$$

The integrand takes the form

$$F(u^1, \dot{u}^1, u^2, \dot{u}^2, u^3, \dot{u}^3) = \sqrt{g_{ij} \frac{du^i}{d\sigma} \frac{du^j}{d\sigma}}.$$

Thus, the Euler equations are given by

$$\frac{\partial F}{\partial u^k} - \frac{d}{d\sigma}\left(\frac{\partial F}{\partial \dot{u}^k}\right) = 0, \quad k = 1, 2, 3.$$

Before embarking on the general theory of geodesics, we consider applying the Euler-Lagrange equations in order to determine geodesics equations.

Example 10.34. *Geodesics for polar coordinates in the plane.*

For polar coordinates, the line element is given by

$$ds^2 = dr^2 + r^2 d\phi^2.$$

The length of the curve is given as

$$L = \int_a^b \sqrt{\left(\frac{dr}{d\sigma}\right)^2 + r^2 \left(\frac{d\phi}{d\sigma}\right)^2}\, d\sigma.$$

We apply the Euler equations for

$$F\left(r, \frac{dr}{d\sigma}, \phi, \frac{d\phi}{d\sigma}\right) = \sqrt{\left(\frac{dr}{d\sigma}\right)^2 + r^2 \left(\frac{d\phi}{d\sigma}\right)^2}$$

Thus,

$$\begin{aligned}
0 &= \frac{\partial F}{\partial r} - \frac{d}{d\sigma}\left(\frac{\partial F}{\partial(\partial r/\partial\sigma)}\right) \\
&= \frac{r}{F}\left(\frac{d\phi}{d\sigma}\right)^2 - \frac{d}{d\sigma}\left(\frac{1}{F}\frac{dr}{d\sigma}\right)
\end{aligned} \tag{10.82}$$

$$\begin{aligned}
0 &= \frac{\partial F}{\partial \phi} - \frac{d}{d\sigma}\left(\frac{\partial F}{\partial(\partial\phi/\partial\sigma)}\right) \\
&= -\frac{d}{d\sigma}\left(\frac{1}{F} r^2 \frac{d\phi}{d\sigma}\right).
\end{aligned} \tag{10.83}$$

Noting that $\frac{ds}{d\sigma} = F$, and therefore, $\frac{d}{d\sigma} = F\frac{d}{ds}$, these equations become

$$\begin{aligned}
r\left(\frac{d\phi}{ds}\right)^2 - \frac{d^2 r}{ds^2} &= 0, \\
\frac{d}{ds}\left(r^2 \frac{d\phi}{ds}\right) &= 0.
\end{aligned} \tag{10.84}$$

These are the geodesic equations for geodesics in the plane written in polar coordinates. Solving this coupled set of equations leads to the geodesic paths. See the problem set.

Example 10.35. *Determine the geodesic equations for the two-dimensional sphere of radius R.*

This problem was actually considered earlier in Example 10.43. In that problem, the line element was given by

$$ds^2 = R^2 d\theta^2 + R^2 \sin^2\theta d\phi^2.$$

Then, we consider

$$F(\theta, \phi, \phi') = \sqrt{\left(\frac{d\theta}{d\sigma}\right)^2 + \sin^2\theta\left(\frac{d\phi}{d\sigma}\right)^2},$$

We apply Euler's equation and find

$$\frac{\sin\theta\cos\theta}{F}\left(\frac{d\phi}{d\sigma}\right)^2 - \frac{d}{d\sigma}\left(\frac{1}{F}\frac{d\theta}{d\sigma}\right) = 0,$$

$$-\frac{d}{d\sigma}\left(\frac{\sin^2\theta}{F}\frac{d\phi}{d\sigma}\right) = 0. \qquad (10.85)$$

Again, using $\frac{ds}{d\sigma} = F$, these give the geodesic equations

$$\sin\theta\cos\theta\left(\frac{d\phi}{ds}\right)^2 - \frac{d^2\theta}{ds^2} = 0,$$

$$\frac{d}{ds}\left(\sin^2\theta\frac{d\phi}{ds}\right) = 0. \qquad (10.86)$$

We can extend this process to seeking geodesics in a four-dimensional spacetime. For Minkowski spacetimes, the line element takes the form

$$ds^2 = -(cdt)^2 + dx^2 + dy^2 + dz^2,$$

where c is the speed of light. This line element is related to the proper time, τ, the time measured by a clock carried by the observer. Then,

$$d\tau^2 = -\frac{ds^2}{c^2}.$$

In relativity, one seeks the world line of free test particles moving between two timelike separated points. This particular path extremizes the proper time.

We begin with the line element

$$ds^2 = g_{\alpha\beta}\,dx^\alpha dx^\beta, \qquad (10.87)$$

where $g_{\alpha\beta}$ is the metric with $\alpha, \beta = 0, 1, 2, 3$. Also, we are using the Einstein summation convention in that we sum over repeated indices that occur as a subscript and superscript pair. In order to find the geodesic equation, we use the Variational Principle which states that freely falling test particles follow a path between two fixed points in spacetime that extremizes the proper time τ.

The proper time is defined by $d\tau^2 = -ds^2$. (We are assuming that $c = 1$.) So, formally, we have

$$\tau_{AB} = \int_A^B \sqrt{-ds^2} = \int_A^B \sqrt{-g_{\alpha\beta}\,dx^\alpha dx^\beta}.$$

In order to write this as an integral that we can compute, we consider a parametrized worldline, $x^\alpha = x^\alpha(\sigma)$, where the parameter $\sigma = 0$ at point A and $\sigma = 1$ at point B. Then we write

$$\tau_{AB} = \int_0^1 \left[-g_{\alpha\beta}\frac{dx^\alpha}{d\sigma}\frac{dx^\beta}{d\sigma}\right] = \int_0^1 L\left[x^\alpha, \frac{dx^\alpha}{d\sigma}\right]. \qquad (10.88)$$

Here we have introduced the Lagrangian $L\left[x^\alpha, \frac{dx^\alpha}{d\sigma}\right]$.

We note also that

$$L = \frac{d\tau}{d\sigma}.$$

Therefore, for functions $f = f(\tau(\sigma))$, we have

$$\frac{df}{d\sigma} = \frac{df}{d\tau}\frac{d\tau}{d\sigma} = L\frac{df}{d\tau}.$$

We will use this later to change derivatives with respect to our arbitrary parameter σ to derivatives with respect to the proper time τ.

Using variational methods as seen in this chapter, we obtain the Euler-Lagrange equations in the form

$$\frac{\partial L}{\partial x^\gamma} - \frac{d}{d\sigma}\left(\frac{\partial L}{\partial(dx^\gamma/d\sigma)}\right) = 0. \tag{10.89}$$

We carefully compute these derivatives for the general metric. First we find

$$\begin{aligned}
\frac{\partial L}{\partial x^\gamma} &= -\frac{1}{2L}\frac{\partial g_{\alpha\beta}}{\partial x^\gamma}\frac{dx^\alpha}{d\sigma}\frac{dx^\beta}{d\sigma} \\
&= -\frac{L}{2}\frac{\partial g_{\alpha\beta}}{\partial x^\gamma}\frac{dx^\alpha}{d\tau}\frac{dx^\beta}{d\tau},
\end{aligned} \tag{10.90}$$

The σ derivatives have been converted to τ derivatives.

Now we compute

$$\begin{aligned}
\frac{\partial L}{\partial(dx^\gamma/d\sigma)} &= -\frac{1}{2L}g_{\alpha\beta}\left(\delta^\alpha_\gamma\frac{dx^\beta}{d\sigma} + \frac{dx^\alpha}{d\sigma}\delta^\beta_\gamma\right) \\
&= -\frac{1}{2L}\left(g_{\gamma\beta}\frac{dx^\beta}{d\sigma} + g_{\alpha\gamma}\frac{dx^\alpha}{d\sigma}\right) \\
&= -\frac{1}{L}g_{\alpha\gamma}\frac{dx^\alpha}{d\sigma}.
\end{aligned} \tag{10.91}$$

In the previous step we used the symmetry of the metric and the fact that α and β are dummy indices.

We differentiate the last result to obtain

$$\begin{aligned}
-\frac{d}{d\sigma}\left(\frac{\partial L}{\partial(dx^\gamma/d\sigma)}\right) &= \frac{d}{d\sigma}\left(\frac{1}{L}g_{\alpha\gamma}\frac{dx^\alpha}{d\sigma}\right) \\
&= L\frac{d}{d\tau}\left(g_{\alpha\gamma}\frac{dx^\alpha}{d\tau}\right) \\
&= L\left[g_{\alpha\gamma}\frac{d^2x^\alpha}{d\tau^2} + \frac{dg_{\alpha\gamma}}{d\tau}\frac{dx^\alpha}{d\tau}\right] \\
&= L\left[g_{\alpha\gamma}\frac{d^2x^\alpha}{d\tau^2} + \frac{dg_{\alpha\gamma}}{dx^\beta}\frac{dx^\beta}{d\tau}\frac{dx^\alpha}{d\tau}\right] \\
&= L\left[g_{\alpha\gamma}\frac{d^2x^\alpha}{d\tau^2} + \frac{1}{2}\left(\frac{dg_{\alpha\gamma}}{dx^\beta} + \frac{dg_{\gamma\beta}}{dx^\alpha}\right)\frac{dx^\beta}{d\tau}\frac{dx^\alpha}{d\tau}\right].
\end{aligned} \tag{10.92}$$

Again, we have used the symmetry of the metric and the re-indexing of repeated indices. Also, we have eliminated appearances of L by changing to derivatives with respect to the proper time.

So far, we have found that

$$
\begin{aligned}
0 &= \frac{\partial L}{\partial x^\gamma} - \frac{d}{d\sigma}\left(\frac{\partial L}{\partial(dx^\gamma/d\sigma)}\right) \\
&= L\left[g_{\alpha\gamma}\frac{d^2x^\alpha}{d\tau^2} + \frac{1}{2}\left(\frac{dg_{\alpha\gamma}}{dx^\beta} + \frac{dg_{\gamma\beta}}{dx^\alpha}\right)\frac{dx^\beta}{d\tau}\frac{dx^\alpha}{d\tau}\right] - \frac{L}{2}\frac{\partial g_{\alpha\beta}}{\partial x^\gamma}\frac{dx^\alpha}{d\tau}\frac{dx^\beta}{d\tau}.
\end{aligned}
$$

Rearranging the terms on the right-hand side and changing the dummy index α to δ, we have

$$
\begin{aligned}
g_{\alpha\gamma}\frac{d^2x^\alpha}{d\tau^2} &= \frac{1}{2}\frac{\partial g_{\alpha\beta}}{\partial x^\gamma}\frac{dx^\alpha}{d\tau}\frac{dx^\beta}{d\tau} - \frac{1}{2}\left(\frac{dg_{\alpha\gamma}}{dx^\beta} + \frac{dg_{\gamma\beta}}{dx^\alpha}\right)\frac{dx^\alpha}{d\tau}\frac{dx^\beta}{d\tau} \\
&= -\frac{1}{2}\left[\frac{dg_{\alpha\gamma}}{dx^\beta} + \frac{dg_{\gamma\beta}}{dx^\alpha} - \frac{dg_{\alpha\beta}}{dx^\gamma}\right]\frac{dx^\alpha}{d\tau}\frac{dx^\beta}{d\tau} \\
&= -\frac{1}{2}\left[\frac{dg_{\delta\gamma}}{dx^\beta} + \frac{dg_{\gamma\beta}}{dx^\delta} - \frac{dg_{\delta\beta}}{dx^\gamma}\right]\frac{dx^\delta}{d\tau}\frac{dx^\beta}{d\tau} \\
&\equiv -g_{\alpha\gamma}\Gamma^\alpha_{\delta\beta}\frac{dx^\delta}{d\tau}\frac{dx^\beta}{d\tau}. \tag{10.93}
\end{aligned}
$$

We have found the geodesic equation,

$$
\frac{d^2x^\alpha}{d\tau^2} + \Gamma^\alpha_{\delta\beta}\frac{dx^\delta}{d\tau}\frac{dx^\beta}{d\tau} = 0, \tag{10.94}
$$

where the Christoffel symbols satisfy

$$
g_{\alpha\gamma}\Gamma^\alpha_{\delta\beta} = \frac{1}{2}\left[\frac{dg_{\delta\gamma}}{dx^\beta} + \frac{dg_{\gamma\beta}}{dx^\delta} - \frac{dg_{\delta\beta}}{dx^\gamma}\right]. \tag{10.95}
$$

This is a linear system of equations for the Christoffel symbols. If the metric is diagonal in the coordinate system, then the computation is relatively simple as there is only one term on the left side of Equation (10.95). In general, one needs to use the matric inverse of $g_{\alpha\beta}$. Also, we should note that the Christoffel symbol is symmetric in the lower indices,

$$
\Gamma^\alpha_{\delta\beta} = \Gamma^\alpha_{\beta\delta}.
$$

We can solve for the Christoffel symbols by introducing the inverse of the metric, $g^{\alpha\beta}$, satisfying

$$
g_{\mu\gamma}g_{\alpha\gamma} = \delta^\mu_\alpha.
$$

Here, δ^μ_α is the usual Kronecker delta, which vanishes for $\mu \neq \alpha$ and is one otherwise. Then,

$$
g^{\mu\gamma}g_{\alpha\gamma}\Gamma^\alpha_{\delta\beta} = \Gamma^\mu_{\delta\beta}.
$$

Therefore,

$$
\Gamma^\mu_{\delta\beta} = \frac{1}{2}g^{\mu\gamma}\left[\frac{dg_{\delta\gamma}}{dx^\beta} + \frac{dg_{\gamma\beta}}{dx^\delta} - \frac{dg_{\delta\beta}}{dx^\gamma}\right].
$$

Example 10.36. *Find the Christoffel symbols for the surface of a sphere.*

This is an example of how the general geodesic computation can be used for Riemannian metrics. First, we look at the geodesics found in Example 10.35:

$$
\begin{aligned}
\sin\theta\cos\theta\left(\frac{d\phi}{ds}\right)^2 - \frac{d^2\theta}{ds^2} &= 0 \\
\frac{d}{ds}\left(\sin^2\theta\frac{d\phi}{ds}\right) &= 0. \tag{10.96}
\end{aligned}
$$

Solving for the second-order derivatives, we have

$$\frac{d^2\theta}{ds^2} - \sin\theta\cos\theta\left(\frac{d\phi}{ds}\right)^2 = 0,$$

$$\frac{d^2\phi}{ds^2} + 2\cot\theta\frac{d\theta}{ds}\frac{d\phi}{ds} = 0. \qquad (10.97)$$

The expanded forms of the geodesic equations for θ and ϕ are

$$\frac{d^2\theta}{ds^2} + \Gamma^\theta_{\theta\theta}\frac{d\theta}{ds}\frac{d\theta}{ds} + \Gamma^\theta_{\theta\phi}\frac{d\theta}{ds}\frac{d\phi}{ds} + \Gamma^\theta_{\phi\theta}\frac{d\phi}{ds}\frac{d\theta}{ds} + \Gamma^\theta_{\phi\phi}\frac{d\phi}{ds}\frac{d\phi}{ds} = 0, \qquad (10.98)$$

$$\frac{d^2\phi}{ds^2} + \Gamma^\phi_{\theta\theta}\frac{d\theta}{ds}\frac{d\theta}{ds} + \Gamma^\phi_{\theta\phi}\frac{d\theta}{ds}\frac{d\phi}{ds} + \Gamma^\phi_{\phi\theta}\frac{d\phi}{ds}\frac{d\theta}{ds} + \Gamma^\phi_{\phi\phi}\frac{d\phi}{ds}\frac{d\phi}{ds} = 0. \qquad (10.99)$$

Comparing these with the geodesics, we have that

$$\Gamma^\theta_{\phi\phi} = -\sin\theta\cos\theta, \quad \Gamma^\phi_{\theta\phi} = \Gamma^\phi_{\phi\theta} = \cot\theta,$$

and the rest of the Christoffel symbols vanish.

Now we use Equation (10.95) to compute the Christoffel symbols. We first note that the metric is given by

$$g = \begin{pmatrix} 1 & 0 \\ 0 & \sin^2\theta \end{pmatrix}.$$

Because g is diagonal, the coefficients are relatively easy to find. In $g_{\alpha\gamma}\Gamma^\alpha_{\delta\beta}$, we note that $\alpha = \gamma$ for nonvanishing contributions from the metric. So, this gives for $\gamma = \theta$

$$g_{\theta\theta}\Gamma^\theta_{\delta\beta} = \frac{1}{2}\left[\frac{dg_{\delta\theta}}{dx^\beta} + \frac{dg_{\theta\beta}}{dx^\delta} - \frac{dg_{\delta\beta}}{dx^\theta}\right]. \qquad (10.100)$$

Because g is independent of ϕ, all ϕ derivatives will vanish. Also, $g_{\phi\phi}$ is the only coefficient that depends on θ. So, the only time the right side of the equation does not vanish is for $\delta = \beta = \phi$. This leaves

$$\Gamma^\theta_{\phi\phi} = -\frac{dg_{\phi\phi}}{dx^\theta} = -\sin\theta\cos\theta$$

Similarly, for $\gamma = \phi$ we have

$$g_{\phi\phi}\Gamma^\phi_{\delta\beta} = \frac{1}{2}\left[\frac{dg_{\delta\phi}}{dx^\beta} + \frac{dg_{\phi\beta}}{dx^\delta} - \frac{dg_{\delta\beta}}{dx^\phi}\right]. \qquad (10.101)$$

Using the same arguments about the derivative of the metric elements, we find one of β or δ is ϕ and the other is θ. For example,

$$g_{\phi\phi}\Gamma^\phi_{\phi\theta} = \frac{1}{2}\left[\frac{dg_{\phi\phi}}{dx^\theta} + \frac{dg_{\phi\theta}}{dx^\phi} - \frac{dg_{\phi\theta}}{dx^\phi}\right],$$

$$\sin^2\theta\,\Gamma^\phi_{\phi\theta} = \frac{1}{2}(2\sin\theta\cos\theta),$$

$$\Gamma^\phi_{\phi\theta} = \cot\theta. \qquad (10.102)$$

Because $\Gamma^\phi_{\phi\theta} = \Gamma^\phi_{\theta\phi}$, we have obtained the same results based on reading the geodesic equation.

Problems

1. For each problem, locate the critical points and classify each one using the second derivative test.

    a. $f(x,y) = (x+y)^2$.

    b. $f(x,y) = x^2y + xy^2$.

    c. $f(x,y) = x^4y + xy^4 - xy$.

    d. $f(x,y) = x^2 - 3xy + 2x + 10y + 6y^2 + 12$.

    e. $f(x,y) = (x^2 - y^2)e^{-y}$.

2. For each problem, locate the critical points and evaluate the Hessian matrix at each critical point.

    a. $f(x,y) = (x+y)^2$.

    b. $f(x,y) = x^2y + xy^2$.

    c. $f(x,y) = x^4y + xy^4 - xy$.

    d. $f(x,y,z) = xy + xz + yz$.

    e. $f(x,y,z) = x^2 + y^2 + xz + 2z^2$.

3. Find the absolute maxima and minima of the function $f(x,y) = x^2 + xy + y^2$ on the unit circle.

4. A thin plate has a temperature distribution of $T(x,y) = x^2 - y^3 - x^2y + y + 20$ for $0 \leq x, y, \leq 2$. Find the coldest and hottest points on the plate.

5. Find the extrema of the given function subject to the given constraint.

    a. $f(x,y) = (x+y)^2$, $x^2 + y = 1$.

    b. $f(x,y) = x^2y + xy^2$, $x^2 + y^2 = 2$.

    c. $f(x,y) = 2x + 3y$, $3x^2 + 2y^2 = 3$.

    d. $f(x,y,z) = x^2 + y^2 + z^2$, $xyz = 1$.

    e. $f(x,y,z) = xy + yx$, $x^2 + y^2 = 1$, $xz = 1$.

6. A particle moves under the force field $\mathbf{F} = -\nabla V$, where the potential function is given by $V(x,y) = x^3 + y^3 - 3xy + 5$. Find the equilibrium points of $\mathbf{F}$ and determine if the equilibria are stable or unstable.

7. For each of the following, find a path that extremizes the given integral.

a. $\int_1^2 (y'^2 + 2yy' + y^2)dy$, $y(1) = 0, y(2) = 1$.

b. $\int_0^2 y^2(1 - y'^2)dy$, $y(0) = 1, y(2) = 2$.

c. $\int_{-1}^1 5y'^2 + 2yy'dy$, $y(-1) = 1, y(1) = 0$.

8. In 1929 E. Hubble confirmed the linear dependence of the velocity on the distance by using the observed values of these quantities. He proposed that the observed radial velocity v, and the estimated distance d, satisfied the equation $v = H_0 d$. This equation is known as Hubble's Law and the constant H_0 is known as Hubble's constant.

Consider the set of data in Table 1, which was extracted from Hubble's 1929 paper to find the proposed linear relationship between the radial velocities and distances of several nebulae.

"A Relation Between Distance and Radial Velocity Among Extra-Galactic Nebulae", E. Hubble, *Proc. Nat. Acad.* **15**, 168 (1929).

   a. Plot the velocity vs. distance data and find the best fir line through the data using your favorite graphing application.

   b. Carry out the computation of the means of the distance and velocities, variance, and covariance for the data set.

   c. Determine the slope, the intercept, and the correlation coefficient for fit. Plot this line on the graph of the data.

   d. From the previous results, compute the sums of squares of the errors, the residuals and the total sums of squares.

   e. Compute the coefficient of determinatin and the standard errors in the slope and the intercept.

   f. What linear relationship did you find between the velocities and the distances?

   g. What is your estimate of Hubble's constant? [Show that kilometers per second per megaparsec has units of $1/\text{time}$]. How does this compare to the currently accepted value of Hubble's constant?

   h. Estimate the age of the universe (in years) using your data. How does this compare to the current estimates?

9. For each of the following, find a path that extremizes the given integral.

   a. $\int_1^2 (y'^2 + 2yy' + y^2)\, dy$, $y(1) = 0$, $y(2) = 1$.

   b. $\int_0^3 y^2 (1 - y'^2)^2\, dy$, $y(0) = 1$, $y(2) = 2$.

   c. $\int_{-1}^1 5y'^2 + 2yy'\, dy$, $y(-1) = 1$, $y(1) = 0$.

10. A bead slides frictionlessly down a wire from the point $(1,0)$ to $(0,0)$.

   a. Determine the equation of the path (shape of the wire) such that the bead takes the leat time between the given points.

   b. How much time did it take the bead to follow this path?

   c. If the path between these points is given by the line $y = x$, then how much time does the bead take?

   d. If the path between these points is given by $y = x^2$, then how much time does the bead take? Explain your answer.

11. A light ray travels from point A in a medium with index of refraction n_1 toward point B in a medium with index of refraction n_2. Assume that the

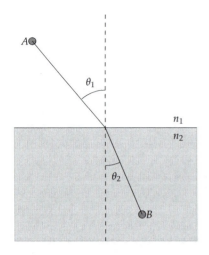

Figure 10.44: A light ray travels between two media from point A toward point B.

rays travel in straight lines and bend at a point at the interface as shown in Figure 10.44. The time functional is

$$T[y] = \int \frac{ds}{v}.$$

a. Write the time functional in terms of the travel path $y(x)$.

b. Apply Fermat's Principle of least time to write the Euler Equation for this functional.

c. Solve the equation in part b. and show that $n \sin \theta$ is a constant.

d. Let the point at which the light is incident to the interface be at $(x, 0)$. Write an expression for the total time to travel from point A at (x_1, y_1) to point B at (x_2, y_2) in terms of the indices of refraction and the coordinates $x, x_1, x_2,$ and y_2.

e. Treating the time as a function of x, minimize this function as a function of one variable and derive Snell's law of refraction.

12. Given the cylinder defined by $x^2 + y^2 = 4$, find the path of shortest length connecting the given points.

a. $(2, 0, 0)$ and $(0, 2, 5)$.

b. $(2, 0, 0)$ and $(2, 0, 5)$.

13. The shape of a hanging chain between the points $(-a, b)$ and (a, b) is such that the gravitational potential energy

$$V[y] = \rho g \int_{-a}^{a} y \sqrt{1 + y'^2}\, dx$$

is minimized subject to the length of the chain remaining constant,

$$L[y] = \int_{-a}^{a} \sqrt{1 + y'^2}\, dx.$$

Find the shape, $y(x)$ of the hanging chain.

14. In Example 10.34, the geodesic equations for geodesics in the plane in polar coordinates were found as

$$r \left(\frac{d\phi}{ds} \right)^2 - \frac{d^2 r}{ds^2} = 0,$$

$$\frac{d}{ds} \left(r^2 \frac{d\phi}{ds} \right) = 0. \qquad (10.103)$$

a. Solve this set of equations for $r = r(s)$ and $\phi = \phi(s)$.

b. Prove that the solutions obtained in part a. are the familiar straight lines in the plane.

15. Use a Lagrange multiplier to find the curve $x(y)$ of length $L = \pi$ on the interval $[0, 1]$ which maximizes the integral $I = \int_0^1 y(x)\, dx$ and pass through the points $(0, 0)$ and $(1, 0)$.

16. A mass m lies on a table and is connected to a string of length ℓ as shown in Figure 10.45. The string passes through a hole in the table and is connected to another mass M that is hanging in the air. We assume that the string remains taught and that mass M can only move vertically.

   a. The Lagrangian for this setup is

$$\mathcal{L} = \frac{1}{2}M\dot{r}^2 + \frac{1}{2}m(\dot{r}^2 + r^2\dot{\theta}^2) + Mg(\ell - r),$$

   where r and θ are polar coordinates describing where mass m is on the table with respect to the hole. Explain why the terms in the Lagrangian are appropriate.

   b. Derive the equations of motion for $r(t)$ and $\theta(t)$.

   c. What angular velocity is needed for mass m to maintain uniform circular motion?

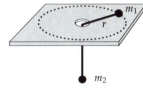

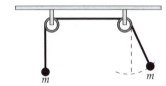

Figure 10.45: Two masses connected by a string.

17. An inclined plane of mass M and inclination θ lies on a frictionless horizontal surface. A small block of mass m is placed carefully on the plane. Find the acceleration of the inclined plane using the Euler–Lagrange equations for the indcated horizontal coordinates in Figure 10.46

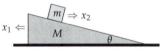

Figure 10.46: The weight of a small block on an inclined plane.

18. Two similar masses are connected by a string of fixed length and hung over two pulleys that are the same height as depicted in Figure 10.47. One mass is set into pendular motion and the other only moves vertically. Find the equations of motion of the length and angle from vertical of the swinging pendulum.

Figure 10.47: Two masses are connected to two pulleys and one is set into motion.

11

Problems in Higher Dimensions

"Equations of such complexity as are the equations of the gravitational field can be found only through the discovery of a logically simple mathematical condition that determines the equations completely or at least almost completely."

"What I have to say about this book can be found inside this book."
—*Albert Einstein (1879–1955)*

IN THIS CHAPTER WE WILL EXPLORE several examples of the solution of initial-boundary value problems involving higher spatial dimensions. These are described by higher-dimensional partial differential equations, such as the ones presented in Table 9.1 in the previous chapter. The spatial domains of the problems span many different geometries, which will necessitate the use of rectangular, polar, cylindrical, or spherical coordinates.

We will solve many of these problems using the method of separation of variables, which we first saw in Chapter 5. Using separation of variables will result in a system of ordinary differential equations for each problem. Adding the boundary conditions, we will need to solve a variety of eigenvalue problems. The product solutions that result will involve trigonometric or some of the special functions that we had encountered in Chapter 6. These methods are used in solving the hydrogen atom and other problems in quantum mechanics and in electrostatic problems in electrodynamics. We will bring to this discussion many of the tools from earlier in this book showing how much of what we have seen can be used to solve some generic partial differential equations that describe oscillation and diffusion type problems.

As we proceed through the examples in this chapter, we will see some common features. For example, the two key equations that we have studied are the heat equation and the wave equation. For higher-dimensional problems, these take the form

$$u_t = k\nabla^2 u, \tag{11.1}$$

$$u_{tt} = c^2\nabla^2 u. \tag{11.2}$$

We can separate out the time dependence in each equation. Inserting a guess of $u(\mathbf{r}, t) = \phi(\mathbf{r})T(t)$ into the heat and wave equations, we obtain

$$T'\phi = kT\nabla^2\phi, \tag{11.3}$$

$$T''\phi = c^2 T \nabla^2 \phi. \tag{11.4}$$

Dividing each equation by $\phi(\mathbf{r})T(t)$, we can separate the time and space dependence just as we had in Chapter 5. In each case, we find that a function of time equals a function of the spatial variables. Thus, these functions must be constant functions. We set these equal to the constant $-\lambda$ and find the respective equations

$$\frac{1}{k}\frac{T'}{T} = \frac{\nabla^2\phi}{\phi} = -\lambda, \tag{11.5}$$

$$\frac{1}{c^2}\frac{T''}{T} = \frac{\nabla^2\phi}{\phi} = -\lambda. \tag{11.6}$$

The sign of λ is chosen because we expect decaying solutions in time for the heat equation and oscillations in time for the wave equation and will pick $\lambda > 0$.

The respective equations for the temporal functions $T(t)$ are given by

$$T' = -\lambda k T, \tag{11.7}$$

$$T'' + c^2\lambda T = 0. \tag{11.8}$$

These are easily solved, as we had seen in Chapter 5. We have

$$T(t) = T(0)e^{-\lambda k t}, \tag{11.9}$$

$$T(t) = a\cos\omega t + b\sin\omega t, \quad \omega = c\sqrt{\lambda}, \tag{11.10}$$

where $T(0)$, a, and b are integration constants, and ω is the angular frequency of vibration.

In both cases, the spatial equation is of the same form:

The Helmholtz Equation.

$$\nabla^2\phi + \lambda\phi = 0. \tag{11.11}$$

The Helmholtz Equation is named after Hermann Ludwig Ferdinand von Helmholtz (1821–1894). He was both a physician and a physicist, and made significant contributions in physiology, optics, acoustics, and electromagnetism.

This equation is called the Helmholtz equation. For one dimensional problems, which we have already solved, the Helmholtz equation takes the form $\phi'' + \lambda\phi = 0$. We had to impose the boundary conditions and found that there were a discrete set of eigenvalues, λ_n, and associated eigenfunctions, ϕ_n.

In higher-dimensional problems, we need to further separate out the spatial dependence. We will again use the boundary conditions to find the eigenvalues, λ, and eigenfunctions, $\phi(\mathbf{r})$, for the Helmholtz equation, though the eigenfunctions will be labeled with more than one index. The resulting boundary value problems are often second-order ordinary differential equations, that can be set up as Sturm–Liouville problems. We know from Chapter 6 that such problems possess an orthogonal set of eigenfunctions. These can then be used to construct a general solution from the product solutions that may involve elementary, or special, functions, such as Legendre polynomials and Bessel functions.

We will begin our study of higher-dimensional problems by considering the vibrations of two-dimensional membranes. First we will solve the problem of a vibrating rectangular membrane and then we will turn our attention to a vibrating circular membrane. The rest of the chapter will be devoted to the study of other two- and three-dimensional problems possessing cylindrical or spherical symmetry.

11.1 Vibrations of Rectangular Membranes

OUR FIRST EXAMPLE WILL BE THE STUDY of the vibrations of a rectangular membrane. You can think of this as a drumhead with a rectangular cross section as shown in Figure 11.1. We stretch the membrane over the drumhead and fasten the material to the boundary of the rectangle. The height of the vibrating membrane is described by its height from equilibrium, $u(x, y, t)$. This problem is a much simpler example of higher-dimensional vibrations than that possessed by the oscillating electric and magnetic fields in the previous chapter.

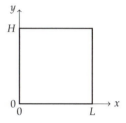

Figure 11.1: The rectangular membrane of length L and width H. There are fixed boundary conditions along the edges.

Example 11.1. *The vibrating rectangular membrane.*

The problem is given by the two-dimensional wave equation in Cartesian coordinates,

$$u_{tt} = c^2(u_{xx} + u_{yy}), \quad t > 0, 0 < x < L, 0 < y < H, \quad (11.12)$$

a set of boundary conditions,

$$u(0, y, t) = 0, \quad u(L, y, t) = 0, \quad t > 0, \quad 0 < y < H,$$
$$u(x, 0, t) = 0, \quad u(x, H, t) = 0, \quad t > 0, \quad 0 < x < L, \quad (11.13)$$

and a pair of initial conditions (as the equation is second order in time),

$$u(x, y, 0) = f(x, y), \quad u_t(x, y, 0) = g(x, y). \quad (11.14)$$

The first step is to separate the variables: $u(x, y, t) = X(x)Y(y)T(t)$. Inserting the guess $u(x, y, t)$ into the wave equation, we have

$$X(x)Y(y)T''(t) = c^2 \left(X''(x)Y(y)T(t) + X(x)Y''(y)T(t) \right).$$

Dividing by both $u(x, y, t)$ and c^2, we obtain

$$\underbrace{\frac{1}{c^2}\frac{T''}{T}}_{\text{Function of } t} = \underbrace{\frac{X''}{X} + \frac{Y''}{Y}}_{\text{Function of } x \text{ and } y} = -\lambda. \quad (11.15)$$

We see that we have a function of t that equals a function of x and y. Thus, both expressions are constant. We expect oscillations in time, so we choose the constant λ to be positive, $\lambda > 0$. (Note: As usual, the primes mean differentiation with respect to the specific dependent variable. So, there should be no ambiguity.)

These lead to two equations:

$$T'' + c^2\lambda T = 0, \quad (11.16)$$

and

$$\frac{X''}{X} + \frac{Y''}{Y} = -\lambda. \tag{11.17}$$

We note that the spatial equation is just the separated form of Helmholtz's Equation with $\phi(x,y) = X(x)Y(y)$.

The first equation is easily solved. We have

$$T(t) = a\cos\omega t + b\sin\omega t, \tag{11.18}$$

where

$$\omega = c\sqrt{\lambda}. \tag{11.19}$$

This is the angular frequency in terms of the separation constant, or eigenvalue. It leads to the frequency of oscillations for the various harmonics of the vibrating membrane as

$$\nu = \frac{\omega}{2\pi} = \frac{c}{2\pi}\sqrt{\lambda}. \tag{11.20}$$

Once we know λ, we can compute these frequencies.

Next we solve the spatial equation. We need to carry out another separation of variables. Rearranging the spatial equation, we have

$$\underbrace{\frac{X''}{X}}_{\text{Function of } x} = \underbrace{-\frac{Y''}{Y} - \lambda}_{\text{Function of } y} = -\mu. \tag{11.21}$$

Here we have a function of x equal to a function of y. So, the two expressions are constant, which we indicate with a second separation constant, $-\mu < 0$. We pick the sign in this way because we expect oscillatory solutions for $X(x)$. This leads to two equations:

$$X'' + \mu X = 0,$$
$$Y'' + (\lambda - \mu)Y = 0. \tag{11.22}$$

We now impose the boundary conditions. We have $u(0,y,t) = 0$ for all $t > 0$ and $0 < y < H$. This implies that $X(0)Y(y)T(t) = 0$ for all t and y in the domain. This is only true if $X(0) = 0$. Similarly, from the other boundary conditions, we find that $X(L) = 0$, $Y(0) = 0$, and $Y(H) = 0$. We note that homogeneous boundary conditions are important in carrying out this process. Nonhomogeneous boundary conditions could be imposed just like we had in Section 6.7.5, but we still need the solutions for homogeneous boundary conditions before tackling the more general problems.

In summary, the boundary value problems we need to solve are

$$X'' + \mu X = 0, \quad X(0) = 0, X(L) = 0,$$
$$Y'' + (\lambda - \mu)Y = 0, \quad Y(0) = 0, Y(H) = 0. \tag{11.23}$$

We have seen boundary value problems of these forms in Chapter 5. The solutions of the first eigenvalue problem are

$$X_n(x) = \sin\frac{n\pi x}{L}, \quad \mu_n = \left(\frac{n\pi}{L}\right)^2, \quad n = 1, 2, 3, \ldots.$$

The second eigenvalue problem is solved in the same manner. The differences from the first problem are that the "eigenvalue" is $\lambda - \mu$, the independent variable is y, and the interval is $[0, H]$. Thus, we can quickly write the solutions as

$$Y_m(y) = \sin \frac{m\pi x}{H}, \quad \lambda - \mu_m = \left(\frac{m\pi}{H}\right)^2, \quad m = 1, 2, 3, \ldots.$$

At this point we need to be careful about the indexing of the separation constants. So far, we have seen that μ depends on n and that the quantity $\kappa = \lambda - \mu$ depends on m. Solving for λ, we should write $\lambda_{nm} = \mu_n + \kappa_m$, or

$$\lambda_{nm} = \left(\frac{n\pi}{L}\right)^2 + \left(\frac{m\pi}{H}\right)^2, \quad n, m = 1, 2, \ldots. \tag{11.24}$$

Because $\omega = c\sqrt{\lambda}$, we have that the discrete frequencies of the harmonics are given by

$$\omega_{nm} = c\sqrt{\left(\frac{n\pi}{L}\right)^2 + \left(\frac{m\pi}{H}\right)^2}, \quad n, m = 1, 2, \ldots. \tag{11.25}$$

The harmonics for the vibrating rectangular membrane are given by

$$\nu_{nm} = \frac{c}{2}\sqrt{\left(\frac{n}{L}\right)^2 + \left(\frac{m}{H}\right)^2},$$

for $n, m = 1, 2, \ldots.$

We have successfully carried out the separation of variables for the wave equation for the vibrating rectangular membrane. The product solutions can be written as

$$u_{nm} = (a \cos \omega_{nm} t + b \sin \omega_{nm} t) \sin \frac{n\pi x}{L} \sin \frac{m\pi y}{H}, \tag{11.26}$$

and the most general solution is written as a linear combination of the product solutions,

$$u(x, y, t) = \sum_{n,m} (a_{nm} \cos \omega_{nm} t + b_{nm} \sin \omega_{nm} t) \sin \frac{n\pi x}{L} \sin \frac{m\pi y}{H}.$$

However, before we carry the general solution any further, we will first concentrate on the two-dimensional harmonics of this membrane.

For the vibrating string the nth harmonic corresponds to the function $\sin \frac{n\pi x}{L}$ and several are shown in Figure 11.2. The various harmonics correspond to the pure tones supported by the string. These then lead to the corresponding frequencies that one would hear. The actual shapes of the harmonics are sketched by locating the nodes, or places on the string that do not move.

In the same way, we can explore the shapes of the harmonics of the vibrating membrane. These are given by the spatial functions

$$\phi_{nm}(x, y) = \sin \frac{n\pi x}{L} \sin \frac{m\pi y}{H}. \tag{11.27}$$

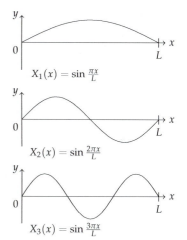

Figure 11.2: The first harmonics of the vibrating string.

Instead of nodes, we will look for the nodal curves, or nodal lines. These are the points (x, y) at which $\phi_{nm}(x, y) = 0$. Of course, these depend on the indices n and m.

A discussion of the nodal lines.

For example, when $n = 1$ and $m = 1$, we have

$$\sin \frac{\pi x}{L} \sin \frac{\pi y}{H} = 0.$$

Figure 11.3: The first few modes of the vibrating rectangular membrane. The dashed lines show the nodal lines indicating the points that do not move for the particular mode. Compare these with the nodal lines to the 3D view in Figure 11.1

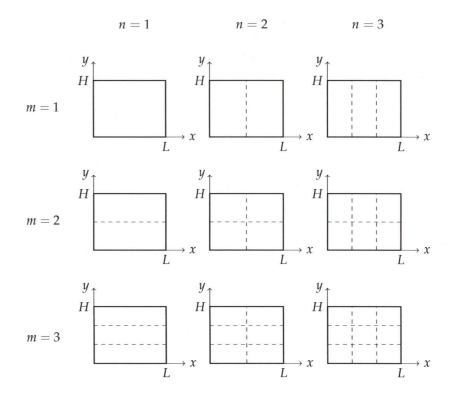

These are zero when either

$$\sin \frac{\pi x}{L} = 0, \quad \text{or} \quad \sin \frac{\pi y}{H} = 0.$$

Of course, this can only happen for $x = 0, L$ and $y = 0, H$. Thus, there are no interior nodal lines.

When $n = 2$ and $m = 1$, we have $y = 0, H$ and

$$\sin \frac{2\pi x}{L} = 0,$$

or, $x = 0, \frac{L}{2}, L$. Thus, there is one interior nodal line at $x = \frac{L}{2}$. These points stay fixed during the oscillation, and all other points oscillate on either side of this line. A similar solution shape results for the (1,2)-mode; that is, $n = 1$ and $m = 2$.

In Figure 11.3 we show the nodal lines for several modes for $n, m = 1, 2, 3$ with different columns corresponding to different n-values while the rows are labeled with different m-values. The blocked regions appear to vibrate independently. A better view is the three-dimensional view depicted in Figure 11.1. The frequencies of vibration are easily computed using the formula for ω_{nm}.

For completeness, we now return to the general solution and apply the initial conditions. The general solution is given by a linear superposition of the product solutions. There are two indices to sum over. Thus, the general

The general solution for the vibrating rectangular membrane.

solution is

$$u(x, y, t) = \sum_{n=1}^{\infty} \sum_{m=1}^{\infty} (a_{nm} \cos \omega_{nm} t + b_{nm} \sin \omega_{nm} t) \sin \frac{n\pi x}{L} \sin \frac{m\pi y}{H}, \quad (11.28)$$

$m = 1$ $m = 2$ $m = 3$

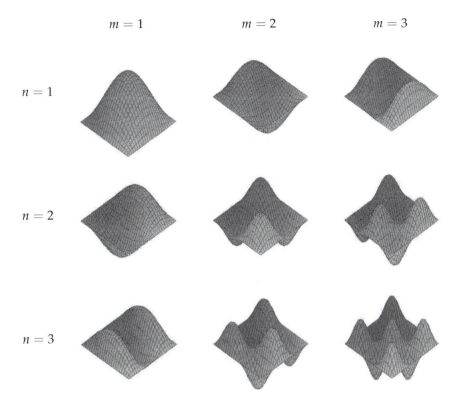

$n = 1$

$n = 2$

$n = 3$

Table 11.1: A Three-Dimensional View of the Vibrating Rectangular Membrane for the Lowest Modes. Compare these images with the nodal lines in Figure 11.3

where

$$\omega_{nm} = c\sqrt{\left(\frac{n\pi}{L}\right)^2 + \left(\frac{m\pi}{H}\right)^2}. \tag{11.29}$$

The first initial condition is $u(x,y,0) = f(x,y)$. Setting $t = 0$ in the general solution, we obtain

$$f(x,y) = \sum_{n=1}^{\infty} \sum_{m=1}^{\infty} a_{nm} \sin\frac{n\pi x}{L} \sin\frac{m\pi y}{H}. \tag{11.30}$$

This is a double Fourier sine series. The goal is to find the unknown coefficients a_{nm}.

The coefficients a_{nm} can be found knowing what we already know about Fourier sine series. We can write the initial condition as the single sum

$$f(x,y) = \sum_{n=1}^{\infty} A_n(y) \sin\frac{n\pi x}{L}, \tag{11.31}$$

where

$$A_n(y) = \sum_{m=1}^{\infty} a_{nm} \sin\frac{m\pi y}{H}. \tag{11.32}$$

These are two Fourier sine series. Recalling from Chapter 5 that the coefficients of Fourier sine series can be computed as integrals, we have

$$\begin{aligned} A_n(y) &= \frac{2}{L} \int_0^L f(x,y) \sin\frac{n\pi x}{L}\, dx, \\ a_{nm} &= \frac{2}{H} \int_0^H A_n(y) \sin\frac{m\pi y}{H}\, dy. \end{aligned} \tag{11.33}$$

Inserting the integral for $A_n(y)$ into that for a_{nm}, we have an integral representation for the Fourier coefficients in the double Fourier sine series,

$$a_{nm} = \frac{4}{LH} \int_0^H \int_0^L f(x,y) \sin \frac{n\pi x}{L} \sin \frac{m\pi y}{H} dx dy. \qquad (11.34)$$

The Fourier coefficients for the double Fourier sine series.

We can carry out the same process for satisfying the second initial condition, $u_t(x,y,0) = g(x,y)$, for the initial velocity of each point. Inserting the general solution into this initial condition, we obtain

$$g(x,y) = \sum_{n=1}^\infty \sum_{m=1}^\infty b_{nm} \omega_{nm} \sin \frac{n\pi x}{L} \sin \frac{m\pi y}{H}. \qquad (11.35)$$

Again, we have a double Fourier sine series. But now we can quickly determine the Fourier coefficients using the above expression for a_{nm} to find that

$$b_{nm} = \frac{4}{\omega_{nm} LH} \int_0^H \int_0^L g(x,y) \sin \frac{n\pi x}{L} \sin \frac{m\pi y}{H} dx dy. \qquad (11.36)$$

The full solution of the vibrating rectangular membrane.

This completes the full solution of the vibrating rectangular membrane problem. Namely, we have obtained the following solution:

$$u(x,y,t) = \sum_{n=1}^\infty \sum_{m=1}^\infty (a_{nm} \cos \omega_{nm} t + b_{nm} \sin \omega_{nm} t) \sin \frac{n\pi x}{L} \sin \frac{m\pi y}{H},$$

$$(11.37)$$

where

$$a_{nm} = \frac{4}{LH} \int_0^H \int_0^L f(x,y) \sin \frac{n\pi x}{L} \sin \frac{m\pi y}{H} dx dy, \qquad (11.38)$$

$$b_{nm} = \frac{4}{\omega_{nm} LH} \int_0^H \int_0^L g(x,y) \sin \frac{n\pi x}{L} \sin \frac{m\pi y}{H} dx dy, \qquad (11.39)$$

and the angular frequencies are given by

$$\omega_{nm} = c \sqrt{\left(\frac{n\pi}{L}\right)^2 + \left(\frac{m\pi}{H}\right)^2} \qquad (11.40)$$

11.2 Vibrations of a Kettle Drum

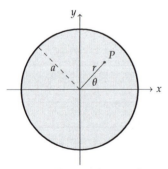

Figure 11.4: The circular membrane of radius a. A general point on the membrane is given by the distance from the center, r, and the angle, θ. There are fixed boundary conditions along the edge at $r = a$.

IN THIS SECTION WE CONSIDER the vibrations of a circular membrane of radius a as shown in Figure 11.4. Again we are looking for the harmonics of the vibrating membrane, but with the membrane fixed around the circular boundary given by $x^2 + y^2 = a^2$. However, expressing the boundary condition in Cartesian coordinates is awkward. Namely, we can only write $u(x,y,t) = 0$ for $x^2 + y^2 = a^2$. It is more natural to use polar coordinates as indicated in Figure 11.4. Let the height of the membrane be given by $u = u(r,\theta,t)$ at time t and position (r,θ). Now the boundary condition is given as $u(a,\theta,t) = 0$ for all $t > 0$ and $\theta \in [0,2\pi]$.

Before solving the initial-boundary value problem, we have to cast the full problem in polar coordinates. This means that we need to rewrite the Laplacian in r and θ. To do so would require that we know how to transform derivatives in x and y into derivatives with respect to r and θ. Using the results from Section 9.3 on curvilinear coordinates, we know that the Laplacian can be written in polar coordinates. In fact, we could use the results from Problem 31 in Chapter 9 for cylindrical coordinates for functions that are z-independent, $f = f(r, \theta)$. Then we would have

$$\nabla^2 f = \frac{1}{r} \frac{\partial}{\partial r} \left(r \frac{\partial f}{\partial r} \right) + \frac{1}{r^2} \frac{\partial^2 f}{\partial \theta^2}.$$

Derivation of Laplacian in polar coordinates.

We can obtain this result using a more direct approach, namely applying the Chain Rule in higher dimensions. First recall the transformations between polar and Cartesian coordinates:

$$x = r \cos \theta, \quad y = r \sin \theta$$

and

$$r = \sqrt{x^2 + y^2}, \quad \tan \theta = \frac{y}{x}.$$

Now, consider a function $f = f(x(r, \theta), y(r, \theta)) = g(r, \theta)$. (Technically, once we transform a given function of Cartesian coordinates, we obtain a new function g of the polar coordinates. Many texts do not rigorously distinguish between the two functions.) Thinking of $x = x(r, \theta)$ and $y = y(r, \theta)$, we have from the Chain Rule for functions of two variables:

$$\begin{aligned}
\frac{\partial f}{\partial x} &= \frac{\partial g}{\partial r} \frac{\partial r}{\partial x} + \frac{\partial g}{\partial \theta} \frac{\partial \theta}{\partial x} \\
&= \frac{\partial g}{\partial r} \frac{x}{r} - \frac{\partial g}{\partial \theta} \frac{y}{r^2} \\
&= \cos \theta \frac{\partial g}{\partial r} - \frac{\sin \theta}{r} \frac{\partial g}{\partial \theta}.
\end{aligned} \qquad (11.41)$$

Here we have used

$$\frac{\partial r}{\partial x} = \frac{x}{\sqrt{x^2 + y^2}} = \frac{x}{r};$$

and

$$\frac{\partial \theta}{\partial x} = \frac{d}{dx} \left(\tan^{-1} \frac{y}{x} \right) = \frac{-y/x^2}{1 + \left(\frac{y}{x}\right)^2} = -\frac{y}{r^2}.$$

Similarly,

$$\begin{aligned}
\frac{\partial f}{\partial y} &= \frac{\partial g}{\partial r} \frac{\partial r}{\partial y} + \frac{\partial g}{\partial \theta} \frac{\partial \theta}{\partial y} \\
&= \frac{\partial g}{\partial r} \frac{y}{r} + \frac{\partial g}{\partial \theta} \frac{x}{r^2} \\
&= \sin \theta \frac{\partial g}{\partial r} + \frac{\cos \theta}{r} \frac{\partial g}{\partial \theta}.
\end{aligned} \qquad (11.42)$$

The 2D Laplacian can now be computed as

$$\frac{\partial^2 f}{\partial x^2} + \frac{\partial^2 f}{\partial y^2} = \cos \theta \frac{\partial}{\partial r} \left(\frac{\partial f}{\partial x} \right) - \frac{\sin \theta}{r} \frac{\partial}{\partial \theta} \left(\frac{\partial f}{\partial x} \right)$$

$$
+ \sin\theta \frac{\partial}{\partial r}\left(\frac{\partial f}{\partial y}\right) + \frac{\cos\theta}{r}\frac{\partial}{\partial\theta}\left(\frac{\partial f}{\partial y}\right)
$$

$$
= \cos\theta\frac{\partial}{\partial r}\left(\cos\theta\frac{\partial g}{\partial r} - \frac{\sin\theta}{r}\frac{\partial g}{\partial\theta}\right)
$$

$$
- \frac{\sin\theta}{r}\frac{\partial}{\partial\theta}\left(\cos\theta\frac{\partial g}{\partial r} - \frac{\sin\theta}{r}\frac{\partial g}{\partial\theta}\right)
$$

$$
+ \sin\theta\frac{\partial}{\partial r}\left(\sin\theta\frac{\partial g}{\partial r} + \frac{\cos\theta}{r}\frac{\partial g}{\partial\theta}\right)
$$

$$
+ \frac{\cos\theta}{r}\frac{\partial}{\partial\theta}\left(\sin\theta\frac{\partial g}{\partial r} + \frac{\cos\theta}{r}\frac{\partial g}{\partial\theta}\right)
$$

$$
= \cos\theta\left(\cos\theta\frac{\partial^2 g}{\partial r^2} + \frac{\sin\theta}{r^2}\frac{\partial g}{\partial\theta} - \frac{\sin\theta}{r}\frac{\partial^2 g}{\partial r\partial\theta}\right)
$$

$$
- \frac{\sin\theta}{r}\left(\cos\theta\frac{\partial^2 g}{\partial\theta\partial r} - \frac{\sin\theta}{r}\frac{\partial^2 g}{\partial\theta^2} - \sin\theta\frac{\partial g}{\partial r} - \frac{\cos\theta}{r}\frac{\partial g}{\partial\theta}\right)
$$

$$
+ \sin\theta\left(\sin\theta\frac{\partial^2 g}{\partial r^2} + \frac{\cos\theta}{r}\frac{\partial^2 g}{\partial r\partial\theta} - \frac{\cos\theta}{r^2}\frac{\partial g}{\partial\theta}\right)
$$

$$
+ \frac{\cos\theta}{r}\left(\sin\theta\frac{\partial^2 g}{\partial\theta\partial r} + \frac{\cos\theta}{r}\frac{\partial^2 g}{\partial\theta^2} + \cos\theta\frac{\partial g}{\partial r} - \frac{\sin\theta}{r}\frac{\partial g}{\partial\theta}\right)
$$

$$
= \frac{\partial^2 g}{\partial r^2} + \frac{1}{r}\frac{\partial g}{\partial r} + \frac{1}{r^2}\frac{\partial^2 g}{\partial\theta^2}
$$

$$
= \frac{1}{r}\frac{\partial}{\partial r}\left(r\frac{\partial g}{\partial r}\right) + \frac{1}{r^2}\frac{\partial^2 g}{\partial\theta^2}.
$$

(11.43)

The last form often occurs in texts because it is in the form of a Sturm-Liouville operator. Also, it agrees with the result from using the Laplacian written in cylindrical coordinates as given in Problem 31 of Chapter 9.

Now that we have written the Laplacian in polar coordinates, we can pose the problem of a vibrating circular membrane.

Example 11.2. *The vibrating circular membrane.*

This problem is given by a partial differential equation,[1]

$$
u_{tt} = c^2\left[\frac{1}{r}\frac{\partial}{\partial r}\left(r\frac{\partial u}{\partial r}\right) + \frac{1}{r^2}\frac{\partial^2 u}{\partial\theta^2}\right], \tag{11.44}
$$

$$
t > 0, \quad 0 < r < a, \quad -\pi < \theta < \pi;
$$

the boundary condition,

$$
u(a,\theta,t) = 0, \quad t > 0, \quad -\pi < \theta < \pi; \tag{11.45}
$$

and the initial conditions,

$$
\begin{aligned}
u(r,\theta,0) &= f(r,\theta), \quad 0 < r < a, -\pi < \theta < \pi, \\
u_t(r,\theta,0) &= g(r,\theta),, \quad 0 < r < a, -\pi < \theta < \pi. \tag{11.46}
\end{aligned}
$$

Now we are ready to solve this problem using separation of variables. As before, we can separate out the time dependence. Let $u(r,\theta,t) = T(t)\phi(r,\theta)$.

[1] Here we state the problem of a vibrating circular membrane. We have chosen $-\pi < \theta < \pi$, but could have just as easily used $0 < \theta < 2\pi$. The symmetric interval about $\theta = 0$ will make the use of boundary conditions simpler.

As usual, $T(t)$ can be written in terms of sines and cosines. This leads to the Helmholtz Equation,

$$\nabla^2 \phi + \lambda \phi = 0.$$

We now separate the Helmholtz Equation by letting $\phi(r,\theta) = R(r)\Theta(\theta)$. This gives

$$\frac{1}{r}\frac{\partial}{\partial r}\left(r\frac{\partial R\Theta}{\partial r}\right) + \frac{1}{r^2}\frac{\partial^2 R\Theta}{\partial \theta^2} + \lambda R\Theta = 0. \qquad (11.47)$$

Dividing by $u = R\Theta$, as usual, leads to

$$\frac{1}{rR}\frac{d}{dr}\left(r\frac{dR}{dr}\right) + \frac{1}{r^2\Theta}\frac{d^2\Theta}{d\theta^2} + \lambda = 0. \qquad (11.48)$$

The third term is a constant. The first term is a function of r. However, the middle term involves both r and θ. This can be remedied by multiplying the equation by r^2. Rearranging the resulting equation, we can separate out the θ-dependence from the radial dependence. Letting μ be another separation constant, we have

$$\frac{r}{R}\frac{d}{dr}\left(r\frac{dR}{dr}\right) + \lambda r^2 = -\frac{1}{\Theta}\frac{d^2\Theta}{d\theta^2} = \mu. \qquad (11.49)$$

This gives us two ordinary differential equations:

$$\frac{d^2\Theta}{d\theta^2} + \mu\Theta = 0,$$

$$r\frac{d}{dr}\left(r\frac{dR}{dr}\right) + (\lambda r^2 - \mu)R = 0. \qquad (11.50)$$

Let's consider the first of these equations. It should look familiar by now. For $\mu > 0$, the general solution is

$$\Theta(\theta) = a\cos\sqrt{\mu}\theta + b\sin\sqrt{\mu}\theta.$$

The next step typically is to apply the boundary conditions in θ. However, when we look at the given boundary conditions in the problem, we do not see anything involving θ. This is a case for which the boundary conditions that are needed are implied and not stated outright.

We can determine the hidden boundary conditions by making some observations. Let's consider the solution corresponding to the endpoints $\theta = \pm\pi$. We note that at these θ-values, we are at the same physical point for any $r < a$. So, we would expect the solution to have the same value at $\theta = -\pi$ as it has at $\theta = \pi$. Namely, the solution is continuous at these physical points. Similarly, we expect the slope of the solution to be the same at these points. This can be summarized using the boundary conditions

$$\Theta(\pi) = \Theta(-\pi), \quad \Theta'(\pi) = \Theta'(-\pi).$$

The boundary conditions in θ are periodic boundary conditions.

Such boundary conditions are called periodic boundary conditions.

Let's apply these conditions to the general solution for $\Theta(\theta)$. First, we set $\Theta(\pi) = \Theta(-\pi)$ and use the symmetries of the sine and cosine functions to obtain

$$a\cos\sqrt{\mu}\pi + b\sin\sqrt{\mu}\pi = a\cos\sqrt{\mu}\pi - b\sin\sqrt{\mu}\pi.$$

This implies that

$$\sin \sqrt{\mu} \pi = 0.$$

This can only be true for $\sqrt{\mu} = m$, for $m = 0, 1, 2, 3, \ldots$. Therefore, the eigenfunctions are given by

$$\Theta_m(\theta) = a \cos m\theta + b \sin m\theta, \quad m = 0, 1, 2, 3, \ldots.$$

For the other half of the periodic boundary conditions, $\Theta'(\pi) = \Theta'(-\pi)$, we have that

$$-am \sin m\pi + bm \cos m\pi = am \sin m\pi + bm \cos m\pi.$$

But, this gives no new information because this equation boils down to $bm = bm$.

To summarize what we know at this point, we have found the general solutions to the temporal and angular equations. The product solutions will have various products of $\{\cos \omega t, \sin \omega t\}$ and $\{\cos m\theta, \sin m\theta\}_{m=0}^{\infty}$. We also know that $\mu = m^2$ and $\omega = c\sqrt{\lambda}$.

We still need to solve the radial equation. Inserting $\mu = m^2$, the radial equation has the form

$$r \frac{d}{dr} \left(r \frac{dR}{dr} \right) + (\lambda r^2 - m^2) R = 0. \tag{11.51}$$

Expanding the derivative term, we have

$$r^2 R''(r) + r R'(r) + (\lambda r^2 - m^2) R(r) = 0. \tag{11.52}$$

The reader should recognize this differential equation from Equation (6.66). It is a Bessel Equation with bounded solutions $R(r) = J_m(\sqrt{\lambda} r)$.

Recall that there are two linearly independent solutions of this second-order equation: $J_m(\sqrt{\lambda} r)$, the Bessel function of the first kind of order m, and $N_m(\sqrt{\lambda} r)$, the Bessel function of the second kind of order m, or Neumann functions. Plots of these functions are shown in Figures 6.8 and 6.9. So, we have the general solution of the radial equation is

$$R(r) = c_1 J_m(\sqrt{\lambda} r) + c_2 N_m(\sqrt{\lambda} r).$$

Now we are ready to apply the boundary conditions to the radial factor in the product solutions. Looking at the original problem, we find only one condition: $u(a, \theta, t) = 0$ for $t > 0$ and $-\pi < < \pi$. This implies that $R(a) = 0$. But where is the second condition?

This is another unstated boundary condition. Look again at the plots of the Bessel functions. Notice that the Neumann functions are not well behaved at the origin. Do you expect that the solution will become infinite at the center of the drum? No, the solutions should be finite at the center. So, this observation leads to the second boundary condition. Namely, $|R(0)| < \infty$. This implies that $c_2 = 0$.

Now we are left with

$$R(r) = J_m(\sqrt{\lambda} r).$$

We have set $c_1 = 1$ for simplicity. We can apply the vanishing condition at $r = a$. This gives

$$J_m(\sqrt{\lambda}a) = 0.$$

Looking again at the plots of $J_m(x)$, we see that there are an infinite number of zeros, but they are not as easy as π! In Table 11.2 we list the nth zeros of J_m, which were first seen in Table 6.3.

Table 11.2: The Zeros of Bessel Functions, $J_m(j_{mn}) = 0$

n	$m = 0$	$m = 1$	$m = 2$	$m = 3$	$m = 4$	$m = 5$
1	2.405	3.832	5.136	6.380	7.588	8.771
2	5.520	7.016	8.417	9.761	11.065	12.339
3	8.654	10.173	11.620	13.015	14.373	15.700
4	11.792	13.324	14.796	16.223	17.616	18.980
5	14.931	16.471	17.960	19.409	20.827	22.218
6	18.071	19.616	21.117	22.583	24.019	25.430
7	21.212	22.760	24.270	25.748	27.199	28.627
8	24.352	25.904	27.421	28.908	30.371	31.812
9	27.493	29.047	30.569	32.065	33.537	34.989

Let's denote the nth zero of $J_m(x)$ by j_{mn}. Then, the boundary condition tells us that

$$\sqrt{\lambda}a = j_{mn}, \quad m = 0, 1, \ldots, \quad n = 1, 2, \ldots.$$

This gives us the eigenvalues as

$$\lambda_{mn} = \left(\frac{j_{mn}}{a}\right)^2, \quad m = 0, 1, \ldots, \quad n = 1, 2, \ldots.$$

Thus, the radial function satisfying the boundary conditions is

$$R_{mn}(r) = J_m\left(\frac{j_{mn}}{a}r\right).$$

We are finally ready to write out the product solutions for the vibrating circular membrane. They are given by

Product solutions for the vibrating circular membrane.

$$u(r, \theta, t) = \left\{\begin{array}{c} \cos\omega_{mn}t \\ \sin\omega_{mn}t \end{array}\right\} \left\{\begin{array}{c} \cos m\theta \\ \sin m\theta \end{array}\right\} J_m\left(\frac{j_{mn}}{a}r\right). \tag{11.53}$$

Here we have indicated choices with the braces, leading to four different types of product solutions. Also, the angular frequency depends on the zeros of the Bessel functions,

$$\omega_{mn} = \frac{j_{mn}}{a}c, \quad m = 0, 1, \ldots, \quad n = 1, 2, \ldots.$$

As with the rectangular membrane, we are interested in the shapes of the harmonics. So, we consider the spatial solution ($t = 0$)

$$\phi(r, \theta) = (\cos m\theta)J_m\left(\frac{j_{mn}}{a}r\right).$$

Figure 11.5: The first few modes of the vibrating circular membrane. The dashed lines show the nodal lines indicating the points that do not move for the particular mode. Compare these nodal lines with the three-dimensional images in Figure 11.3.

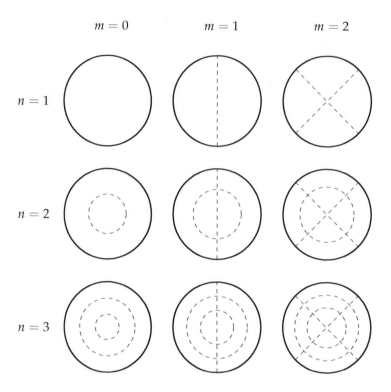

Including the solutions involving $\sin m\theta$ will only rotate these modes. The nodal curves are given by $\phi(r, \theta) = 0$. This can be satisfied if $\cos m\theta = 0$, or $J_m\left(\frac{j_{mn}}{a} r\right) = 0$. The various nodal curves that result are shown in Figure 11.5.

For the angular part, we easily see that the nodal curves are radial lines, $\theta = $const. For $m = 0$, there are no solutions, as $\cos m\theta = 1$ for $m = 0$. In Figure 11.5, this is seen by the absence of radial lines in the first column.

For $m = 1$, we have $\cos \theta = 0$. This implies that $\theta = \pm\frac{\pi}{2}$. These values give the vertical line as shown in the second column in Figure 11.5. For $m = 2$, $\cos 2\theta = 0$ implies that $\theta = \frac{\pi}{4}, \frac{3\pi}{4}$. This results in the two lines shown in the last column of Figure 11.5.

We can also consider the nodal curves defined by the Bessel functions. We seek values of r for which $\frac{j_{mn}}{a} r$ is a zero of the Bessel function and lies in the interval $[0, a]$. Thus, we have

$$\frac{j_{mn}}{a} r = j_{mj}, \quad 1 \le j \le n,$$

or

$$r = \frac{j_{mj}}{j_{mn}} a, \quad 1 \le j \le n.$$

These will give circles of these radii with $j_{mj} \le j_{mn}$, or $j \le n$. For $m = 0$ and $n = 1$, there is only one zero and $r = a$. In fact, for all $n = 1$ modes, there is only one zero giving $r = a$. Thus, the first row in Figure 11.5 shows no interior nodal circles.

For a three-dimensional view, one can look at Figure 11.3. Imagine that the various regions are oscillating independently and that the points on the nodal curves are not moving.

$$n = 1 \qquad n = 2 \qquad n = 3$$

$m = 0$

$m = 1$

$m = 2$

Table 11.3: A three dimensional view of the vibrating circular membrane for the lowest modes. Compare these images with the nodal line plots in Figure 11.5.

We should note that the nodal circles are not evenly spaced and that the radii can be computed relatively easily. For the $n = 2$ modes, we have two circles, $r = a$ and $r = \frac{j_{m1}}{j_{m2}}a$, as shown in the second row of Figure 11.5. For $m = 0$,

$$r = \frac{2.405}{5.520}a \approx 0.4357a$$

for the inner circle. For $m = 1$,

$$r = \frac{3.832}{7.016}a \approx 0.5462a,$$

and for $m = 2$,

$$r = \frac{5.136}{8.417}a \approx 0.6102a.$$

For $n = 3$, we obtain circles of radii

$$r = a, \quad r = \frac{j_{m1}}{j_{m3}}a, \text{ and } r = \frac{j_{m2}}{j_{m3}}a.$$

For $m = 0$,

$$r = a, \quad \frac{5.520}{8.654}a \approx 0.6379a, \quad \frac{2.405}{8.654}a \approx 0.2779a.$$

Similarly, for $m = 1$,

$$r = a, \quad \frac{3.832}{10.173}a \approx 0.3767a, \quad \frac{7.016}{10.173}a \approx 0.6897a,$$

and for $m = 2$,

$$r = a, \quad \frac{5.136}{11.620}a \approx 0.4420a, \quad \frac{8.417}{11.620}a \approx 0.7224a.$$

Example 11.3. *Vibrating Annulus*

More complicated vibrations can be dreamt up for this geometry. Consider an annulus in which the drum is formed from two concentric circular cylinders and the membrane is stretched between the two with an annular cross section as shown in Figure 11.6. The separation would follow as before except now the boundary conditions are that the membrane is fixed around the two circular boundaries. In this case, we cannot toss out the Neumann functions because the origin is not part of the drum head.

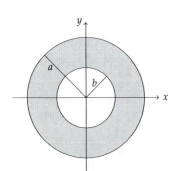

Figure 11.6: An annular membrane with radii a and $b > a$. There are fixed boundary conditions along the edges at $r = a$ and $r = b$.

The domain for this problem is shown in Figure 11.6, and the problem is given by the partial differential equation

$$u_{tt} = c^2 \left[\frac{1}{r} \frac{\partial}{\partial r} \left(r \frac{\partial u}{\partial r} \right) + \frac{1}{r^2} \frac{\partial^2 u}{\partial \theta^2} \right], \tag{11.54}$$

$$t > 0, \quad b < r < a, \quad -\pi < \theta < \pi,$$

the boundary conditions,

$$u(b,\theta,t) = 0, \quad u(a,\theta,t) = 0, \quad t > 0, \quad -\pi < \theta < \pi, \tag{11.55}$$

and the initial conditions,

$$\begin{aligned} u(r,\theta,0) &= f(r,\theta), \quad b < r < a, -\pi < \theta < \pi, \\ u_t(r,\theta,0) &= g(r,\theta), \quad b < r < a, -\pi < \theta < \pi. \end{aligned} \tag{11.56}$$

Because we cannot dispose of the Neumann functions, the product solutions take the form

$$u(r,\theta,t) = \left\{ \begin{array}{c} \cos \omega t \\ \sin \omega t \end{array} \right\} \left\{ \begin{array}{c} \cos m\theta \\ \sin m\theta \end{array} \right\} R_m(r), \tag{11.57}$$

where

$$R_m(r) = c_1 J_m(\sqrt{\lambda} r) + c_2 N_m(\sqrt{\lambda} r)$$

and $\omega = c\sqrt{\lambda}$, $m = 0,1,\ldots$.

For this problem, the radial boundary conditions are that the membrane is fixed at $r = a$ and $r = b$. Taking $b < a$, we then have to satisfy the conditions

$$\begin{aligned} R(a) &= c_1 J_m(\sqrt{\lambda} a) + c_2 N_m(\sqrt{\lambda} a) = 0, \\ R(b) &= c_1 J_m(\sqrt{\lambda} b) + c_2 N_m(\sqrt{\lambda} b) = 0. \end{aligned} \tag{11.58}$$

This leads to two homogeneous equations for c_1 and c_2. The coefficient determinant of this system has to vanish if there are to be nontrivial solutions. This gives the eigenvalue equation for λ:

$$J_m(\sqrt{\lambda} a) N_m(\sqrt{\lambda} b) - J_m(\sqrt{\lambda} b) N_m(\sqrt{\lambda} a) = 0.$$

There are an infinite number of zeros of the function

$$F(\lambda) = \lambda : J_m(\sqrt{\lambda} a) N_m(\sqrt{\lambda} b) - J_m(\sqrt{\lambda} b) N_m(\sqrt{\lambda} a).$$

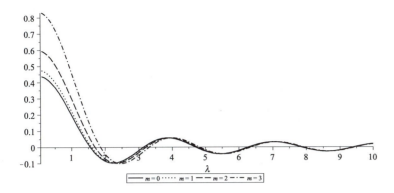

Figure 11.7: Plot of the function

$$F(\lambda) = J_m(\sqrt{\lambda}a)N_m(\sqrt{\lambda}b) - J_m(\sqrt{\lambda}b)N_m(\sqrt{\lambda}a$$

for $a = 4$ and $b = 2$ and $m = 0, 1, 2, 3$.

In Figure 11.7 we show a plot of $F(\lambda)$ for $a = 4$, $b = 2$ and $m = 0, 1, 2, 3$.

This eigenvalue equation needs to be solved numerically. Choosing $a = 2$ and $b = 4$, we have for the first few modes

$$
\begin{aligned}
\sqrt{\lambda_{mn}} &\approx 1.562, 3.137, 4.709, \quad m = 0 \\
&\approx 1.598, 3.156, 4.722, \quad m = 1 \\
&\approx 1.703, 3.214, 4.761, \quad m = 2.
\end{aligned}
\tag{11.59}
$$

Note, because $\omega_{mn} = c\sqrt{\lambda_{mn}}$, these numbers essentially give us the frequencies of oscillation.

For these particular roots, we can solve for c_1 and c_2 up to a multiplicative constant. A simple solution is to set

$$c_1 = N_m(\sqrt{\lambda_{mn}}b), \quad c_2 = J_m(\sqrt{\lambda_{mn}}b).$$

This leads to the basic modes of vibration,

$$R_{mn}(r)\Theta_m(\theta) = \cos m\theta \left(N_m(\sqrt{\lambda_{mn}}b)J_m(\sqrt{\lambda_{mn}}r) - J_m(\sqrt{\lambda_{mn}}b)N_m(\sqrt{\lambda_{mn}}r) \right),$$

for $m = 0, 1, \ldots,$ and $n = 1, 2, \ldots.$ In Figure 11.4 we show various modes for the particular choice of annular membrane dimensions, $a = 2$ and $b = 4$.

11.3   Laplace's Equation in 2D

ANOTHER OF THE GENERIC PARTIAL DIFFERENTIAL EQUATIONS is Laplace's Equation, $\nabla^2 u = 0$. This equation first appeared in the chapter on complex variables when we discussed harmonic functions. Another example is the electric potential for electrostatics. As we described Chapter 9, for static electromagnetic fields,

$$\nabla \cdot \mathbf{E} = \rho/\epsilon_0, \quad \mathbf{E} = \nabla\phi.$$

In regions devoid of charge, these equations yield the Laplace Equation $\nabla^2\phi = 0$.

Table 11.4: A Three-Dimensional View of the Vibrating Annular Membrane for the Lowest Modes

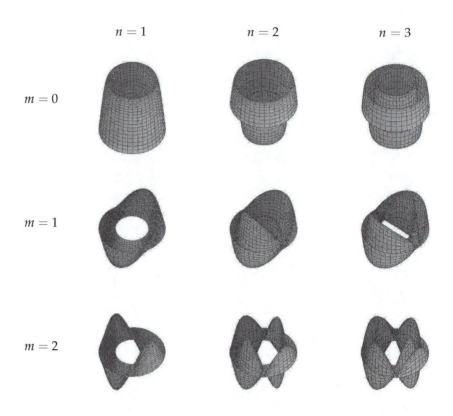

Another example comes from studying temperature distributions. Consider a thin rectangular plate with the boundaries set at fixed temperatures. Temperature changes of the plate are governed by the heat equation. The solution of the heat equation subject to these boundary conditions is time dependent. In fact, after a long period of time, the plate will reach thermal equilibrium. If the boundary temperature is zero, then the plate temperature decays to zero across the plate. However, if the boundaries are maintained at a fixed nonzero temperature, which means energy is being put into the system to maintain the boundary conditions, the internal temperature may reach a nonzero equilibrium temperature. Reaching thermal equilibrium means that asymptotically in time, the solution becomes time independent. Thus, the equilibrium state is a solution of the time-independent heat equation, which is another Laplace Equation, $\nabla^2 u = 0$.

Thermodynamic equilibrium, $\nabla^2 u = 0$.

As another example, we could look at fluid flow. For an incompressible flow, $\nabla \cdot \mathbf{v} = 0$. If the flow is irrotational, then $\nabla \times \mathbf{v} = 0$. We can introduce a velocity potential, $\mathbf{v} = \nabla \phi$. Thus, $\nabla \times \mathbf{v}$ vanishes by a vector identity and $\nabla \cdot \mathbf{v} = 0$ implies $\nabla^2 \phi = 0$. So, once again, we obtain Laplace's Equation.

Incompressible, irrotational fluid flow, $\nabla^2 \phi = 0$, for velocity $\mathbf{v} = \nabla \phi$.

In this section we will look at examples of Laplace's Equation in two dimensions. The solutions in these examples could be examples from any of the applications in the above physical situations and the solutions can be applied appropriately.

Example 11.4. *Equilibrium Temperature Distribution for a Rectangular Plate*

Let's consider Laplace's Equation in Cartesian coordinates,

$$u_{xx} + u_{yy} = 0, \quad 0 < x < L, \quad 0 < y < H$$

with the boundary conditions

$$u(0,y) = 0, \quad u(L,y) = 0, \quad u(x,0) = f(x), \quad u(x,H) = 0.$$

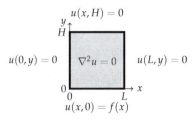

Figure 11.8: In this figure we show the domain and boundary conditions for the example of determining the equilibrium temperature distribution for a rectangular plate.

The boundary conditions are shown in Figure 11.8

As with the heat and wave equations, we can solve this problem using the method of separation of variables. Let $u(x,y) = X(x)Y(y)$. Then, Laplace's Equation becomes

$$X''Y + XY'' = 0$$

and we can separate the x- and y-dependent functions and introduce a separation constant, λ,

$$\frac{X''}{X} = -\frac{Y''}{Y} = -\lambda.$$

Thus, we are led to two differential equations:

$$\begin{aligned} X'' + \lambda X &= 0, \\ Y'' - \lambda Y &= 0. \end{aligned} \tag{11.60}$$

From the boundary condition $u(0,y) = 0, u(L,y) = 0$, we have $X(0) = 0, X(L) = 0$. So, we have the usual eigenvalue problem for $X(x)$,

$$X'' + \lambda X = 0, \quad X(0) = 0, X(L) = 0.$$

The solutions to this problem are given by

$$X_n(x) = \sin \frac{n\pi x}{L}, \quad \lambda_n = \left(\frac{n\pi}{L}\right)^2, \quad n = 1, 2, \dots.$$

The general solution of the equation for $Y(y)$ is given by

$$Y(y) = c_1 e^{\sqrt{\lambda} y} + c_2 e^{-\sqrt{\lambda} y}.$$

The boundary condition $u(x,H) = 0$ implies $Y(H) = 0$. So, we have

$$c_1 e^{\sqrt{\lambda} H} + c_2 e^{-\sqrt{\lambda} H} = 0.$$

Thus,

$$c_2 = -c_1 e^{2\sqrt{\lambda} H}.$$

Inserting this result into the expression for $Y(y)$, we have

$$\begin{aligned} Y(y) &= c_1 e^{\sqrt{\lambda} y} - c_1 e^{2\sqrt{\lambda} H} e^{-\sqrt{\lambda} y} \\ &= c_1 e^{\sqrt{\lambda} H} \left(e^{-\sqrt{\lambda} H} e^{\sqrt{\lambda} y} - e^{\sqrt{\lambda} H} e^{-\sqrt{\lambda} y} \right) \\ &= c_1 e^{\sqrt{\lambda} H} \left(e^{-\sqrt{\lambda}(H-y)} - e^{\sqrt{\lambda}(H-y)} \right) \\ &= -2c_1 e^{\sqrt{\lambda} H} \sinh \sqrt{\lambda}(H-y). \end{aligned} \tag{11.61}$$

Note: Having carried out this computation, we can now see that it would be better to guess this form in the future. So, for $Y(H) = 0$, one would guess a solution $Y(y) = \sinh \sqrt{\lambda}(H-y)$. For $Y(0) = 0$, one would guess a solution $Y(y) = \sinh \sqrt{\lambda} y$. Similarly, if $Y'(H) = 0$, one would guess a solution $Y(y) = \cosh \sqrt{\lambda}(H-y)$.

Because we already know the values of the eigenvalues λ_n from the eigenvalue problem for $X(x)$, we have that the y-dependence is given by

$$Y_n(y) = \sinh \frac{n\pi(H-y)}{L}.$$

So, the product solutions are given by

$$u_n(x,y) = \sin \frac{n\pi x}{L} \sinh \frac{n\pi(H-y)}{L}, \quad n = 1,2,\ldots.$$

These solutions satisfy Laplace's Equation and the three homogeneous boundary conditions in the problem.

The remaining boundary condition, $u(x,0) = f(x)$, still must be satisfied. Inserting $y = 0$ in the product solutions does not satisfy the boundary condition unless $f(x)$ is proportional to one of the eigenfunctions $X_n(x)$. So, we first write the general solution as a linear combination of the product solutions:

$$u(x,y) = \sum_{n=1}^{\infty} a_n \sin \frac{n\pi x}{L} \sinh \frac{n\pi(H-y)}{L}. \tag{11.62}$$

Now we apply the boundary condition, $u(x,0) = f(x)$, to find that

$$f(x) = \sum_{n=1}^{\infty} a_n \sinh \frac{n\pi H}{L} \sin \frac{n\pi x}{L}. \tag{11.63}$$

Defining $b_n = a_n \sinh \frac{n\pi H}{L}$, this becomes

$$f(x) = \sum_{n=1}^{\infty} b_n \sin \frac{n\pi x}{L}. \tag{11.64}$$

We see that the determination of the unknown coefficients, b_n, is simply done by recognizing that this is a Fourier sine series. The Fourier coefficients are easily found as

$$b_n = \frac{2}{L} \int_0^L f(x) \sin \frac{n\pi x}{L} \, dx. \tag{11.65}$$

Because $a_n = b_n / \sinh \frac{n\pi H}{L}$, we can finish solving the problem. The solution is

$$u(x,y) = \sum_{n=1}^{\infty} a_n \sin \frac{n\pi x}{L} \sinh \frac{n\pi(H-y)}{L}, \tag{11.66}$$

where

$$a_n = \frac{2}{L \sinh \frac{n\pi H}{L}} \int_0^L f(x) \sin \frac{n\pi x}{L} \, dx. \tag{11.67}$$

Example 11.5. *Equilibrium Temperature Distribution for a Rectangular Plate for General Boundary Conditions*

A more general problem is to seek solutions to Laplace's Equation in Cartesian coordinates,

$$u_{xx} + u_{yy} = 0, \quad 0 < x < L, 0 < y < H,$$

with non-zero boundary conditions on more than one side of the domain,

$$u(0,y) = g_1(y), \quad u(L,y) = g_2(y), \quad 0 < y < H,$$

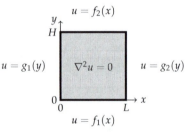

Figure 11.9: In this figure we show the domain and general boundary conditions for the example of determining the equilibrium temperature distribution for a rectangular plate.

$$u(x,0) = f_1(x), \quad u(x,H) = f_2(x), \quad 0 < x < L.$$

These boundary conditions are shown in Figure 11.9

The problem with this example is that none of the boundary conditions are homogeneous. This means that the corresponding eigenvalue problems will not have the homogeneous boundary conditions that Sturm–Liouville Theory in Section 6.6 needs. However, we can express this problem in terms of four different problems with nonhomogeneous boundary conditions on only one side of the rectangle.

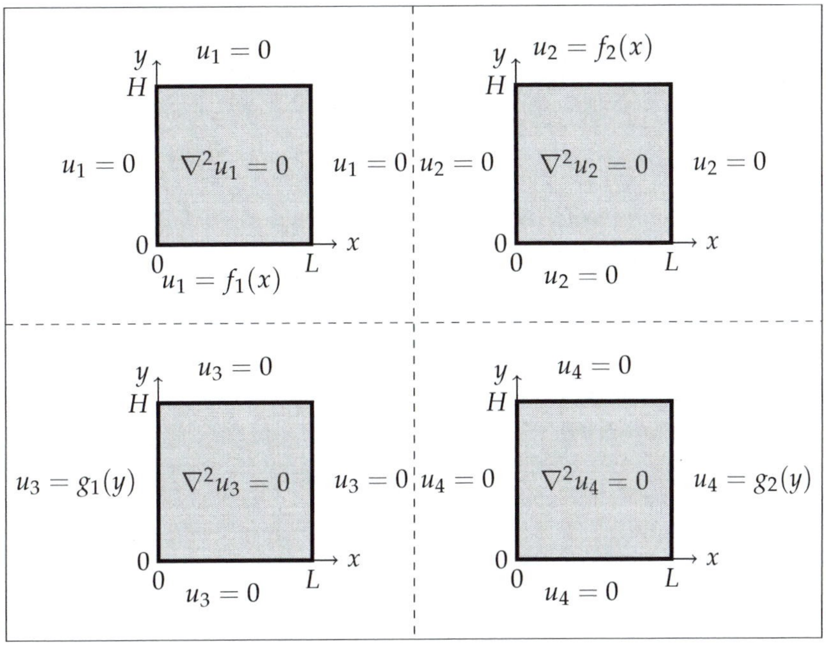

Figure 11.10: The general boundary value problem for a rectangular plate can be written as the sum of these four separate problems.

In Figure 11.10 we show how the problem can be broken up into four separate problems for functions $u_i(x,y)$, $i = 1, \ldots, 4$. Because the boundary conditions and Laplace's Equation are linear, the solution to the general problem is simply the sum of the solutions to these four problems:

$$u(x,y) = u_1(x,y) + u_2(x,y) + u_3(x,y) + u_4(x,y).$$

Then, this solution satisfies Laplace's Equation,

$$\nabla^2 u(x,y) = \nabla^2 u_1(x,y) + \nabla^2 u_2(x,y) + \nabla^2 u_3(x,y) + \nabla^2 u_4(x,y) = 0,$$

and the boundary conditions. For example, using the boundary conditions defined in Figure 11.10, we have for $y = 0$,

$$u(x,0) = u_1(x,0) + u_2(x,0) + u_3(x,0) + u_4(x,0) = f_1(x).$$

The other boundary conditions can also be shown to hold.

We can solve each of the problems in Figure 11.10 quickly, based on the solution we obtained in the previous example. The solution for $u_1(x,y)$, that satisfies the boundary conditions

$$u_1(0,y) = 0, \quad u_1(L,y) = 0, \quad 0 < y < H,$$

$$u_1(x,0) = f_1(x), \quad u_1(x,H) = 0, \quad 0 < x < L,$$

is the easiest to write down. It is given by

$$u_1(x,y) = \sum_{n=1}^{\infty} a_n \sin \frac{n\pi x}{L} \sinh \frac{n\pi(H-y)}{L}. \tag{11.68}$$

where

$$a_n = \frac{2}{L \sinh \frac{n\pi H}{L}} \int_0^L f_1(x) \sin \frac{n\pi x}{L} \, dx. \tag{11.69}$$

For the boundary conditions

$$u_2(0,y) = 0, \quad u_2(L,y) = 0, \quad 0 < y < H,$$

$$u_2(x,0) = 0, \quad u_2(x,H) = f_2(x), \quad 0 < x < L.$$

The boundary conditions for $X(x)$ are $X(0) = 0$ and $X(L) = 0$. So, we get the same form for the eigenvalues and eigenfunctions as before:

$$X_n(x) = \sin \frac{n\pi x}{L}, \quad \lambda_n = \left(\frac{n\pi}{L}\right)^2, n = 1,2,\ldots.$$

The remaining homogeneous boundary condition is now $Y(0) = 0$. Recalling that the equation satisfied by $Y(y)$ is

$$Y'' - \lambda Y = 0,$$

we can write the general solution as

$$Y(y) = c_1 \cosh \sqrt{\lambda} y + c_2 \sinh \sqrt{\lambda} y.$$

Requiring $Y(0) = 0$, we have $c_1 = 0$, or

$$Y(y) = c_2 \sinh \sqrt{\lambda} y.$$

Then, the general solution is

$$u_2(x,y) = \sum_{n=1}^{\infty} b_n \sin \frac{n\pi x}{L} \sinh \frac{n\pi y}{L}. \tag{11.70}$$

We now force the nonhomogeneous boundary condition, $u_2(x,H) = f_2(x)$,

$$f_2(x) = \sum_{n=1}^{\infty} b_n \sin \frac{n\pi x}{L} \sinh \frac{n\pi H}{L}. \tag{11.71}$$

Once again, we have a Fourier sine series. The Fourier coefficients are given by

$$b_n = \frac{2}{L \sinh \frac{n\pi H}{L}} \int_0^L f_2(x) \sin \frac{n\pi x}{L} \, dx. \tag{11.72}$$

Next we turn to the problem with the boundary conditions

$$u_3(0,y) = g_1(y), \quad u_3(L,y) = 0, \quad 0 < y < H,$$

$$u_3(x,0) = 0, \quad u_3(x,H) = 0, \quad 0 < x < L.$$

In this case, the pair of homogeneous boundary conditions $u_3(x,0) = 0$, $u_3(x,H) = 0$ lead to solutions

$$Y_n(y) = \sin \frac{n\pi y}{H}, \quad \lambda_n = -\left(\frac{n\pi}{H}\right)^2, \quad n = 1, 2 \ldots.$$

The condition $u_3(L,0) = 0$ gives $X(x) = \sinh \frac{n\pi(L-x)}{H}$.

The general solution satisfying the homogeneous conditions is

$$u_3(x,y) = \sum_{n=1}^{\infty} c_n \sin \frac{n\pi y}{H} \sinh \frac{n\pi(L-x)}{H}. \tag{11.73}$$

Applying the nonhomogeneous boundary condition, $u_3(0,y) = g_1(y)$, we obtain the Fourier sine series

$$g_1(y) = \sum_{n=1}^{\infty} c_n \sin \frac{n\pi y}{H} \sinh \frac{n\pi L}{H}. \tag{11.74}$$

The Fourier coefficients are found as

$$c_n = \frac{2}{H \sinh \frac{n\pi L}{H}} \int_0^H g_1(y) \sin \frac{n\pi y}{H} \, dy. \tag{11.75}$$

Finally, we can find the solution

$$u_4(0,y) = 0, \quad u_4(L,y) = g_2(y), \quad 0 < y < H,$$

$$u_4(x,0) = 0, \quad u_4(x,H) = 0, \quad 0 < x < L.$$

Following the above analysis, we find the general solution

$$u_4(x,y) = \sum_{n=1}^{\infty} d_n \sin \frac{n\pi y}{H} \sinh \frac{n\pi x}{H}. \tag{11.76}$$

The nonhomogeneous boundary condition, $u(L,y) = g_2(y)$, is satisfied if

$$g_2(y) = \sum_{n=1}^{\infty} d_n \sin \frac{n\pi y}{H} \sinh \frac{n\pi L}{H}. \tag{11.77}$$

The Fourier coefficients, d_n, are given by

$$d_n = \frac{2}{H \sinh \frac{n\pi L}{H}} \int_0^H g_1(y) \sin \frac{n\pi y}{H} \, dy. \tag{11.78}$$

The solution to the general problem is given by the sum of these four solutions.

$$u(x,y) = \sum_{n=1}^{\infty} \left[\left(a_n \sinh \frac{n\pi(H-y)}{L} + b_n \sinh \frac{n\pi y}{L} \right) \sin \frac{n\pi x}{L} \right.$$
$$\left. + \left(c_n \sinh \frac{n\pi(L-x)}{H} + d_n \sinh \frac{n\pi x}{H} \right) \sin \frac{n\pi y}{H} \right], \tag{11.79}$$

where the coefficients are given by the above Fourier integrals.

Example 11.6. *Laplace's Equation on a Disk*

We now turn to solving Laplace's Equation on a disk of radius a as shown in Figure 11.11. Laplace's Equation in polar coordinates is given by

$$\frac{1}{r}\frac{\partial}{\partial r}\left(r\frac{\partial u}{\partial r}\right) + \frac{1}{r^2}\frac{\partial^2 u}{\partial \theta^2} = 0, \quad 0 < r < a, \quad -\pi < \theta < \pi. \tag{11.80}$$

The boundary conditions are given as

$$u(a,\theta) = f(\theta), \quad -\pi < \theta < \pi, \tag{11.81}$$

plus periodic boundary conditions in θ.

Separation of variable proceeds as usual. Let $u(r,\theta) = R(r)\Theta(\theta)$. Then

$$\frac{1}{r}\frac{\partial}{\partial r}\left(r\frac{\partial(R\Theta)}{\partial r}\right) + \frac{1}{r^2}\frac{\partial^2(R\Theta)}{\partial\theta^2} = 0, \tag{11.82}$$

or

$$\Theta\frac{1}{r}(rR')' + \frac{1}{r^2}R\Theta'' = 0. \tag{11.83}$$

Diving by $u(r,\theta) = R(r)\Theta(\theta)$, multiplying by r^2, and rearranging, we have

$$\frac{r}{R}(rR')' = -\frac{\Theta''}{\Theta} = \lambda. \tag{11.84}$$

Because this equation gives a function of r equal to a function of θ, we set the equation equal to a constant. Thus, we have obtained two differential equations, which can be written as

$$\begin{aligned} r(rR')' - \lambda R &= 0, & (11.85) \\ \Theta'' + \lambda\Theta &= 0. & (11.86) \end{aligned}$$

We can solve the second equation subject to the periodic boundary conditions in the θ variable. The reader should be able to confirm that

$$\Theta(\theta) = a_n \cos n\theta + b_n \sin n\theta, \quad \lambda = n^2, n = 0, 1, 2, \dots$$

is the solution. Note that the n = 0 case just leads to a constant solution.

Inserting $\lambda = n^2$ into the radial equation, we find

$$r^2 R'' + rR' - n^2 R = 0.$$

This is a Cauchy-Euler type of ordinary differential equation. Recall that we solve such equations by guessing a solution of the form $R(r) = r^m$. This leads to the characteristic equation $m^2 - n^2 = 0$. Therefore, $m = \pm n$. So,

$$R(r) = c_1 r^n + c_2 r^{-n}.$$

Because we expect finite solutions at the origin, $r = 0$, we can set $c_2 = 0$. Thus, the general solution is

$$u(r,\theta) = \frac{a_0}{2} + \sum_{n=1}^{\infty}(a_n \cos n\theta + b_n \sin n\theta)\, r^n. \tag{11.87}$$

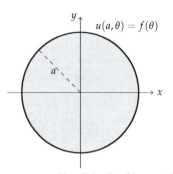

Figure 11.11: The disk of radius *a* with boundary condition along the edge at *r = a*.

PROBLEMS IN HIGHER DIMENSIONS 671

Note that we have taken the constant term out of the sum and put it into a familiar form.

Now we can impose the remaining boundary condition, $u(a, \theta) = f(\theta)$, or

$$f(\theta) = \frac{a_0}{2} + \sum_{n=1}^{\infty} (a_n \cos n\theta + b_n \sin n\theta) \, a^n. \tag{11.88}$$

This is a Fourier trigonometric series. The Fourier coefficients can be determined using the results from Chapter 4:

$$a_n = \frac{1}{\pi a^n} \int_{-\pi}^{\pi} f(\theta) \cos n\theta \, d\theta, \quad n = 0, 1, \dots, \tag{11.89}$$

$$b_n = \frac{1}{\pi a^n} \int_{-\pi}^{\pi} f(\theta) \sin n\theta \, d\theta \quad n = 1, 2 \dots. \tag{11.90}$$

11.3.1 Poisson Integral Formula

WE CAN PUT THE SOLUTION FROM THE PREVIOUS EXAMPLE in a more compact form by inserting the Fourier coefficients into the general solution. Doing this, we have

$$
\begin{aligned}
u(r, \theta) &= \frac{a_0}{2} + \sum_{n=1}^{\infty} (a_n \cos n\theta + b_n \sin n\theta) \, r^n \\
&= \frac{1}{2\pi} \int_{-\pi}^{\pi} f(\phi) \, d\phi \\
&\quad + \frac{1}{\pi} \int_{-\pi}^{\pi} \sum_{n=1}^{\infty} [\cos n\phi \cos n\theta + \sin n\phi \sin n\theta] \left(\frac{r}{a}\right)^n f(\phi) \, d\phi \\
&= \frac{1}{\pi} \int_{-\pi}^{\pi} \left[\frac{1}{2} + \sum_{n=1}^{\infty} \cos n(\theta - \phi) \left(\frac{r}{a}\right)^n \right] f(\phi) \, d\phi. \tag{11.91}
\end{aligned}
$$

The term in the brackets can be summed. We note that

$$
\begin{aligned}
\cos n(\theta - \phi) \left(\frac{r}{a}\right)^n &= \text{Re}\left(e^{in(\theta-\phi)} \left(\frac{r}{a}\right)^n \right) \\
&= \text{Re}\left(\frac{r}{a} e^{i(\theta-\phi)} \right)^n. \tag{11.92}
\end{aligned}
$$

Therefore,

$$\sum_{n=1}^{\infty} \cos n(\theta - \phi) \left(\frac{r}{a}\right)^n = \text{Re}\left(\sum_{n=1}^{\infty} \left(\frac{r}{a} e^{i(\theta-\phi)} \right)^n \right).$$

The right-hand side of this equation is a geometric series with common ratio of $\frac{r}{a} e^{i(\theta-\phi)}$, which is also the first term of the series. Because $\left| \frac{r}{a} e^{i(\theta-\phi)} \right| = \frac{r}{a} < 1$, the series converges. Summing the series, we obtain

$$
\begin{aligned}
\sum_{n=1}^{\infty} \left(\frac{r}{a} e^{i(\theta-\phi)} \right)^n &= \frac{\frac{r}{a} e^{i(\theta-\phi)}}{1 - \frac{r}{a} e^{i(\theta-\phi)}} \\
&= \frac{r e^{i(\theta-\phi)}}{a - r e^{i(\theta-\phi)}}. \tag{11.93}
\end{aligned}
$$

We need to rewrite this result so that we can easily take the real part. Thus, we multiply and divide by the complex conjugate of the denominator to obtain

$$
\sum_{n=1}^{\infty} \left(\frac{r}{a} e^{i(\theta - \phi)} \right)^n = \frac{re^{i(\theta - \phi)}}{a - re^{i(\theta - \phi)}} \frac{a - re^{-i(\theta - \phi)}}{a - re^{-i(\theta - \phi)}}
$$

$$
= \frac{are^{-i(\theta - \phi)} - r^2}{a^2 + r^2 - 2ar\cos(\theta - \phi)}. \tag{11.94}
$$

The real part of the sum is given as

$$
\mathrm{Re}\left(\sum_{n=1}^{\infty} \left(\frac{r}{a} e^{i(\theta - \phi)} \right)^n \right) = \frac{ar\cos(\theta - \phi) - r^2}{a^2 + r^2 - 2ar\cos(\theta - \phi)}.
$$

Therefore, the factor in the brackets under the integral in Equation (11.91) is

$$
\frac{1}{2} + \sum_{n=1}^{\infty} \cos n(\theta - \phi) \left(\frac{r}{a} \right)^n = \frac{1}{2} + \frac{ar\cos(\theta - \phi) - r^2}{a^2 + r^2 - 2ar\cos(\theta - \phi)}
$$

$$
= \frac{a^2 - r^2}{2(a^2 + r^2 - 2ar\cos(\theta - \phi))}. \tag{11.95}
$$

Thus, we have shown that the solution of Laplace's Equation on a disk of radius a with boundary condition $u(a, \theta) = f(\theta)$ can be written in the closed form

$$
u(r, \theta) = \frac{1}{2\pi} \int_{-\pi}^{\pi} \frac{a^2 - r^2}{a^2 + r^2 - 2ar\cos(\theta - \phi)} f(\phi)\, d\phi. \tag{11.96}
$$

This result is called the Poisson Integral Formula and

$$
K(\theta, \phi) = \frac{a^2 - r^2}{a^2 + r^2 - 2ar\cos(\theta - \phi)}
$$

is called the Poisson kernel.

Example 11.7. *Evaluate the solution in Equation (11.96) at the center of the disk. We insert $r = 0$ into the solution (11.96) to obtain*

$$
u(0, \theta) = \frac{1}{2\pi} \int_{-\pi}^{\pi} f(\phi)\, d\phi.
$$

Recalling that the average of a function $g(x)$ on $[a, b]$ is given by

$$
g_{ave} = \frac{1}{b - a} \int_{a}^{b} g(x)\, dx,
$$

we see that the value of the solution u at the center of the disk is the average of the boundary values. This is sometimes referred to as the Mean Value Theorem.

11.4   Three-Dimensional Cake Baking

IN THE REST OF THE CHAPTER WE WILL EXTEND our studies to three-dimensional problems. In this section we will solve the heat equation as we look at examples of baking cakes.

We consider cake batter, which is at room temperature of $T_i = 80°$F. It is placed into an oven, also at a fixed temperature, $T_b = 350°$F. For simplicity, we will assume that the thermal conductivity and cake density are constant. Of course, this is not quite true. However, it is an approximation that simplifies the model. We will consider two cases, one in which the cake is a rectangular solid, such as baking it in a $13'' \times 9'' \times 2''$ baking pan. The other case will lead to a cylindrical cake, such as you would obtain from a round cake pan.

Assuming that the heat constant k is indeed constant and the temperature is given by $T(\mathbf{r}, t)$, we begin with the heat equation in three dimensions:

$$\frac{\partial T}{\partial t} = k \nabla^2 T. \tag{11.97}$$

We will need to specify initial and boundary conditions. Let T_i be the initial batter temperature, $T(x, y, z, 0) = T_i$.

We choose the boundary conditions to be fixed at the oven temperature T_b. However, these boundary conditions are not homogeneous and would lead to problems when carrying out separation of variables. This is easily remedied by subtracting the oven temperature from all temperatures involved and defining $u(\mathbf{r}, t) = T(\mathbf{r}, t) - T_b$. The heat equation then becomes

$$\frac{\partial u}{\partial t} = k \nabla^2 u \tag{11.98}$$

with initial condition

$$u(\mathbf{r}, 0) = T_i - T_b.$$

The boundary conditions are now homogeneous. We cannot be any more specific than this until we specify the geometry.

Example 11.8. *Temperature of a Rectangular Cake*

We will consider a rectangular cake with dimensions $0 \le x \le W$, $0 \le y \le L$, and $0 \le z \le H$ as shown in Figure 11.12. For this problem, we seek solutions of the heat equation plus the conditions

$$
\begin{aligned}
u(x, y, z, 0) &= T_i - T_b, \\
u(0, y, z, t) = u(W, y, z, t) &= 0, \\
u(x, 0, z, t) = u(x, L, z, t) &= 0, \\
u(x, y, 0, t) = u(x, y, H, t) &= 0.
\end{aligned}
$$

Using the method of separation of variables, we seek solutions of the form

$$u(x, y, z, t) = X(x)Y(y)Z(z)G(t). \tag{11.99}$$

Substituting this form into the heat equation, we get

$$\frac{1}{k}\frac{G'}{G} = \frac{X''}{X} + \frac{Y''}{Y} + \frac{Z''}{Z}. \tag{11.100}$$

Setting these expressions equal to $-\lambda$, we get

$$\frac{1}{k}\frac{G'}{G} = -\lambda \quad and \quad \frac{X''}{X} + \frac{Y''}{Y} + \frac{Z''}{Z} = -\lambda. \tag{11.101}$$

This discussion of cake baking is adapted from R. Wilkinson's thesis work. That in turn was inspired by work done by Dr. Olszewski,(2006) From baking a cake to solving the diffusion equation. *American Journal of Physics* 74(6).

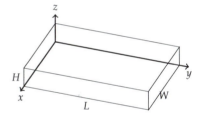

Figure 11.12: The dimensions of a rectangular cake.

Therefore, the equation for $G(t)$ is given by

$$G' + k\lambda G = 0.$$

We further have to separate out the functions of x, y, and z. We anticipate that the homogeneous boundary conditions will lead to oscillatory solutions in these variables. Therefore, we expect that separation of variables will lead to the eigenvalue problems

$$
\begin{aligned}
X'' + \mu^2 X &= 0, \quad X(0) = X(W) = 0, \\
Y'' + \nu^2 Y &= 0, \quad Y(0) = Y(L) = 0, \\
Z'' + \kappa^2 Z &= 0, \quad Z(0) = Z(H) = 0.
\end{aligned}
\tag{11.102}
$$

Noting that

$$\frac{X''}{X} = -\mu^2, \quad \frac{Y''}{Y} = -\nu^2, \quad \frac{Z''}{Z} = -\kappa^2,$$

we find from the heat equation that the separation constants are related:

$$\lambda^2 = \mu^2 + \nu^2 + \kappa^2.$$

We could have gotten to this point quicker by writing the first separated equation labeled with the separation constants as

$$\underbrace{\frac{1}{k}\frac{G'}{G}}_{-\lambda} = \underbrace{\frac{X''}{X}}_{-\mu} + \underbrace{\frac{Y''}{Y}}_{-\nu} + \underbrace{\frac{Z''}{Z}}_{-\kappa}.$$

Then, we can read off the eigenvalues problems and determine that $\lambda^2 = \mu^2 + \nu^2 + \kappa^2$.

From the boundary conditions, we get product solutions for $u(x, y, z, t)$ in the form

$$u_{mn\ell}(x, y, z, t) = \sin \mu_m x \sin \nu_n y \sin \kappa_\ell z \, e^{-\lambda_{mn\ell} k t},$$

for

$$\lambda_{mnl} = \mu_m^2 + \nu_n^2 + \kappa_\ell^2 = \left(\frac{m\pi}{W}\right)^2 + \left(\frac{n\pi}{L}\right)^2 + \left(\frac{\ell\pi}{H}\right)^2, \quad m, n, \ell = 1, 2, \ldots.$$

The general solution is a linear combination of all the product solutions, summed over three different indices,

$$u(x, y, z, t) = \sum_{m=1}^{\infty} \sum_{n=1}^{\infty} \sum_{\ell=1}^{\infty} A_{mnl} \sin \mu_m x \sin \nu_n y \sin \kappa_\ell z \, e^{-\lambda_{mn\ell} k t}, \tag{11.103}$$

where the $A_{mn\ell}$'s are arbitrary constants.

We can use the initial condition $u(x, y, z, 0) = T_i - T_b$ to determine the $A_{mn\ell}$'s. We find

$$T_i - T_b = \sum_{m=1}^{\infty} \sum_{n=1}^{\infty} \sum_{\ell=1}^{\infty} A_{mnl} \sin \mu_m x \sin \nu_n y \sin \kappa_\ell z. \tag{11.104}$$

This is a triple Fourier sine series.

We can determine these coefficients in a manner similar to how we handled double Fourier sine series earlier in the chapter. Defining

$$b_m(y,z) = \sum_{n=1}^{\infty} \sum_{\ell=1}^{\infty} A_{mnl} \sin \nu_n y \sin \kappa_\ell z,$$

we obtain a simple Fourier sine series:

$$T_i - T_b = \sum_{m=1}^{\infty} b_m(y,z) \sin \mu_m x. \tag{11.105}$$

The Fourier coefficients can then be found as

$$b_m(y,z) = \frac{2}{W} \int_0^W (T_i - T_b) \sin \mu_m x \, dx.$$

Using the same technique for the remaining sine series and noting that $T_i - T_b$ is constant, we can determine the general coefficients A_{mnl} by carrying out the needed integrations:

$$\begin{aligned}
A_{mnl} &= \frac{8}{WLH} \int_0^H \int_0^L \int_0^W (T_i - T_b) \sin \mu_m x \sin \nu_n y \sin \kappa_\ell z \, dx dy dz \\
&= (T_i - T_b) \frac{8}{\pi^3} \left[\frac{\cos\left(\frac{m\pi x}{W}\right)}{m} \right]_0^W \left[\frac{\cos\left(\frac{n\pi y}{L}\right)}{n} \right]_0^L \left[\frac{\cos\left(\frac{\ell\pi z}{H}\right)}{\ell} \right]_0^H \\
&= (T_i - T_b) \frac{8}{\pi^3} \left[\frac{\cos m\pi - 1}{m} \right] \left[\frac{\cos n\pi - 1}{n} \right] \left[\frac{\cos \ell\pi - 1}{\ell} \right] \\
&= (T_i - T_b) \frac{8}{\pi^3} \begin{cases} 0, & \text{for at least one } m, n, \ell \text{ even}, \\ \left[\frac{-2}{m}\right]\left[\frac{-2}{n}\right]\left[\frac{-2}{\ell}\right], & \text{for } m, n, \ell \text{ all odd}. \end{cases}
\end{aligned}$$

Because only the odd multiples yield non-zero $A_{mn\ell}$, we let $m = 2m' - 1$, $n = 2n' - 1$, and $\ell = 2\ell' - 1$ for $m', n', \ell' = 1, 2, \ldots$. The expansion coefficients can now be written in the simpler form

$$A_{mnl} = \frac{64(T_b - T_i)}{(2m'-1)(2n'-1)(2\ell'-1)\pi^3}.$$

Substituting this result into the general solution and dropping the primes, we find

$$u(x,y,z,t) = \frac{64(T_b - T_i)}{\pi^3} \sum_{m=1}^{\infty} \sum_{n=1}^{\infty} \sum_{\ell=1}^{\infty} \frac{\sin \mu_m x \sin \nu_n y \sin \kappa_\ell z \, e^{-\lambda_{mn\ell}kt}}{(2m-1)(2n-1)(2\ell-1)},$$

where

$$\lambda_{mn\ell} = \left(\frac{(2m-1)\pi}{W} \right)^2 + \left(\frac{(2n-1)\pi}{L} \right)^2 + \left(\frac{(2\ell-1)\pi}{H} \right)^2$$

for $m, n, \ell = 1, 2, \ldots$.

Recalling that the solution to the physical problem is

$$T(x,y,z,t) = u(x,y,z,t) + T_b,$$

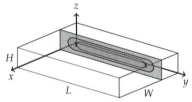

Figure 11.13: Rectangular cake showing a vertical slice.

Figure 11.14: Temperature evolution for a $13'' \times 9'' \times 2''$ cake shown as vertical slices at the indicated length in feet.

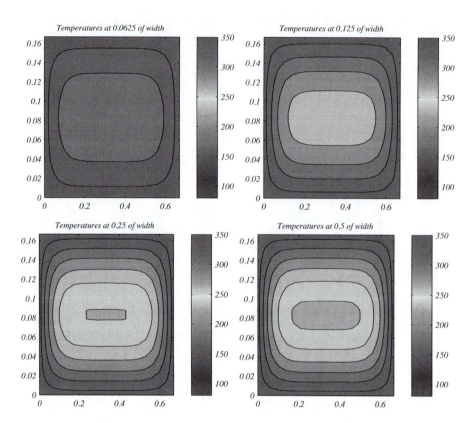

the final solution is given by

$$T(x,y,z,t) = T_b + \frac{64(T_b - T_i)}{\pi^3} \sum_{m=1}^{\infty} \sum_{n=1}^{\infty} \sum_{\ell=1}^{\infty} \frac{\sin \hat{\mu}_m x \sin \hat{\nu}_n y \sin \hat{\kappa}_\ell z \, e^{-\hat{\lambda}_{mn\ell} kt}}{(2m-1)(2n-1)(2\ell-1)}.$$

We show some temperature distributions in Figure 11.14. Because we cannot capture the entire cake, we show vertical slices such as depicted in Figure 11.13. Vertical slices are taken at the positions and times indicated for a $13'' \times 9'' \times 2''$ cake. Obviously, this is not accurate because the cake consistency is changing, and this will affect the parameter k. A more realistic model would be to allow $k = k(T(x,y,z,t))$. However, such problems are beyond the simple methods described in this book.

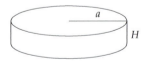

Figure 11.15: Geometry for a cylindrical cake.

Example 11.9. *Circular Cakes*

In this case, the geometry of the cake is cylindrical as shown in Figure 11.15. Therefore, we need to express the boundary conditions and heat equation in cylindrical coordinates. Also, we will assume that the solution, $u(r,z,t) = T(r,z,t) - T_b$, is independent of θ due to axial symmetry. This gives the heat equation in θ-independent cylindrical coordinates as

$$\frac{\partial u}{\partial t} = k \left(\frac{1}{r} \frac{\partial}{\partial r} \left(r \frac{\partial u}{\partial r} \right) + \frac{\partial^2 u}{\partial z^2} \right), \tag{11.106}$$

where $0 \leq r \leq a$ and $0 \leq z \leq Z$. The initial condition is

$$u(r,z,0) = T_i - T_b,$$

and the homogeneous boundary conditions on the side, top, and bottom of the cake are

$$u(a, z, t) = 0,$$
$$u(r, 0, t) = u(r, Z, t) = 0.$$

Again, we seek solutions of the form $u(r, z, t) = R(r)H(z)G(t)$. Separation of variables leads to

$$\frac{1}{k}\frac{G'}{G} = \underbrace{\frac{1}{rR}\frac{d}{dr}\left(rR'\right)}_{-\mu^2} + \underbrace{\frac{H''}{H}}_{-\nu^2}. \qquad (11.107)$$

Here we have indicated the separation constants, which lead to three ordinary differential equations. These equations and the boundary conditions are

$$G' + k\lambda G = 0,$$
$$\frac{d}{dr}\left(rR'\right) + \mu^2 rR = 0, \quad R(a) = 0, \quad R(0) \text{ is finite,}$$
$$H'' + \nu^2 H = 0, \quad H(0) = H(Z) = 0. \qquad (11.108)$$

We further note that the separation constants are related by $\lambda = \mu^2 + \nu^2$.

We can easily write the solutions for $G(t)$ and $H(z)$:

$$G(t) = Ae^{-\lambda kt}$$

and

$$H_n(z) = \sin\frac{n\pi z}{Z}, \quad n = 1, 2, 3, \ldots,$$

where $\nu = \frac{n\pi}{Z}$. Recalling from the rectangular case that only odd terms arise in the Fourier sine series coefficients for the constant initial condition, we proceed by rewriting $H(z)$ as

$$H_n(z) = \sin\frac{(2n-1)\pi z}{Z}, \quad n = 1, 2, 3, \ldots, \qquad (11.109)$$

with $\nu = \frac{(2n-1)\pi}{Z}$.

The radial equation can be written in the form

$$r^2 R'' + rR' + \mu^2 r^2 R = 0.$$

*This is a Bessel equation of the first kind of order zero that we had seen in Section **??**. Therefore, the general solution is a linear combination of Bessel functions of the first and second kind,*

$$R(r) = c_1 J_0(\mu r) + c_2 N_0(\mu r). \qquad (11.110)$$

Because $R(r)$ is bounded at $r = 0$ and $N_0(\mu r)$ is not well behaved at $r = 0$, we set $c_2 = 0$. Up to a constant factor, the solution becomes

$$R(r) = J_0(\mu r). \qquad (11.111)$$

The boundary condition $R(a) = 0$ gives the eigenvalues as

$$\mu_m = \frac{j_{0m}}{a}, \quad m = 1, 2, 3, \ldots,$$

where j_{0m} is the mth roots of the zeroeth-order Bessel function, $J_0(j_{0m}) = 0$.
Therefore, we have found the product solutions

$$H_n(z)R_m(r)G(t) = \sin \frac{(2n-1)\pi z}{Z} J_0\left(\frac{r}{a}j_{0m}\right) e^{-\lambda_{nm}kt}, \tag{11.112}$$

where $m = 1, 2, 3, \ldots, n = 1, 2, \ldots$. Combining the product solutions, the general solution is found as

$$u(r, z, t) = \sum_{n=1}^{\infty} \sum_{m=1}^{\infty} A_{nm} \sin \frac{(2n-1)\pi z}{Z} J_0\left(\frac{r}{a}j_{0m}\right) e^{-\lambda_{nm}kt} \tag{11.113}$$

with

$$\lambda_{nm} = \left(\frac{(2n-1)\pi}{Z}\right)^2 + \left(\frac{j_{0m}}{a}\right)^2,$$

for $n, m = 1, 2, 3, \ldots$.
Inserting the solution into the constant initial condition, we have

$$T_i - T_b = \sum_{n=1}^{\infty} \sum_{m=1}^{\infty} A_{nm} \sin \frac{(2n-1)\pi z}{Z} J_0\left(\frac{r}{a}j_{0m}\right).$$

This is a double Fourier series but it involves a Fourier-Bessel expansion. Writing

$$b_n(r) = \sum_{m=1}^{\infty} A_{nm} J_0\left(\frac{r}{a}j_{0m}\right),$$

the condition becomes

$$T_i - T_b = \sum_{n=1}^{\infty} b_n(r) \sin \frac{(2n-1)\pi z}{Z}.$$

As seen previously, this is a Fourier sine series and the Fourier coefficients are given by

$$\begin{aligned} b_n(r) &= \frac{2}{Z} \int_0^Z (T_i - T_b) \sin \frac{(2n-1)\pi z}{Z} dz \\ &= \frac{2(T_i - T_b)}{Z} \left[-\frac{Z}{(2n-1)\pi} \cos \frac{(2n-1)\pi z}{Z}\right]_0^Z \\ &= \frac{4(T_i - T_b)}{(2n-1)\pi}. \end{aligned}$$

We insert this result into the Fourier-Bessel series,

$$\frac{4(T_i - T_b)}{(2n-1)\pi} = \sum_{m=1}^{\infty} A_{nm} J_0\left(\frac{r}{a}j_{0m}\right),$$

and recall from Section ?? that we can determine the Fourier coefficients A_{nm} using the Fourier-Bessel series,

$$f(x) = \sum_{n=1}^{\infty} c_n J_p(j_{pn}\frac{x}{a}), \tag{11.114}$$

where the Fourier-Bessel coefficients are found as

$$c_n = \frac{2}{a^2 \left[J_{p+1}(j_{pn})\right]^2} \int_0^a x f(x) J_p(j_{pn}\frac{x}{a}) dx. \tag{11.115}$$

Comparing these series expansions, we have

$$A_{nm} = \frac{2}{a^2 J_1^2(j_{0m})} \frac{4(T_i - T_b)}{(2n-1)\pi} \int_0^a J_0(\mu_m r) r\, dr. \qquad (11.116)$$

In order to evaluate $\int_0^a J_0(\mu_m r) r\, dr$, we let $y = \mu_m r$ and get

$$
\begin{aligned}
\int_0^a J_0(\mu_m r) r\, dr &= \int_0^{\mu_m a} J_0(y) \frac{y}{\mu_m} \frac{dy}{\mu_m} \\
&= \frac{1}{\mu_m^2} \int_0^{\mu_m a} J_0(y) y\, dy \\
&= \frac{1}{\mu_m^2} \int_0^{\mu_m a} \frac{d}{dy}\left(y J_1(y)\right) dy \\
&= \frac{1}{\mu_m^2} (\mu_m a) J_1(\mu_m a) = \frac{a^2}{j_{0m}} J_1(j_{0m}). \qquad (11.117)
\end{aligned}
$$

Here we have made use of the identity $\frac{d}{dx}\left(x J_1(x)\right) = J_0(x)$ from Section 6.5.

Substituting the result of this integral computation into the expression for A_{nm}, we find

$$A_{nm} = \frac{8(T_i - T_b)}{(2n-1)\pi} \frac{1}{j_{0m} J_1(j_{0m})}.$$

Substituting this result into the original expression for $u(r,z,t)$, gives

$$u(r,z,t) = \frac{8(T_i - T_b)}{\pi} \sum_{n=1}^{\infty} \sum_{m=1}^{\infty} \sin\frac{(2n-1)\pi z}{Z} \frac{J_0(\frac{r}{a} j_{0m}) e^{-\lambda_{nm}kt}}{(2n-1)\, j_{0m} J_1(j_{0m})}.$$

Therefore, $T(r,z,t)$ is found as

$$T(r,z,t) = T_b + \frac{8(T_i - T_b)}{\pi} \sum_{n=1}^{\infty} \sum_{m=1}^{\infty} \frac{\sin\frac{(2n-1)\pi z}{Z}}{(2n-1)} \frac{J_0(\frac{r}{a} j_{0m}) e^{-\lambda_{nm}kt}}{j_{0m} J_1(j_{0m})},$$

where

$$\lambda_{nm} = \left(\frac{(2n-1)\pi}{Z}\right)^2 + \left(\frac{j_{0m}}{a}\right)^2, \quad n,m = 1,2,3,\ldots.$$

We have therefore found the general solution for the three-dimensional heat equation in cylindrical coordinates with constant diffusivity. Similar to the solutions shown in Figure 11.14 of the previous section, we show in Figure 11.17 the temperature evolution throughout a standard 9″ round cake pan. These are vertical slices similar to what is depicted in Figure 11.16.

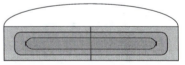

Figure 11.16: Depiction of a sideview of a vertical slice of a circular cake.

Again, one could generalize this example to considerations of other types of cakes with cylindrical symmetry. For example, there are muffins, Boston steamed bread that is steamed in tall cylindrical cans. One could also consider an annular pan, such as a bundt cake pan. In fact, such problems extend beyond baking cakes to possible heating molds in manufacturing.

11.5 *Laplace's Equation and Spherical Symmetry*

WE HAVE SEEN THAT LAPLACE'S EQUATION, $\nabla^2 u = 0$, arises in electrostatics as an equation for electric potential outside a charge distribution and it

Figure 11.17: Temperature evolution for a standard 9″ cake shown as vertical slices through the center.

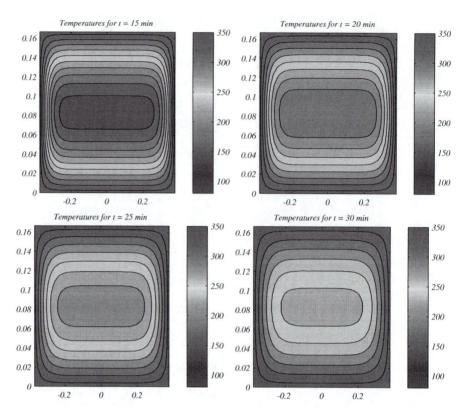

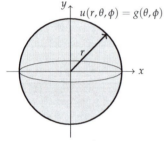

Figure 11.18: A sphere of radius r with the boundary condition $u(r,\theta,\phi) = g(\theta,\phi)$.

occurs as the equation governing equilibrium temperature distributions. As we had seen in the previous chapter, Laplace's Equation generally occurs in the study of potential theory, which also includes the study of gravitational and fluid potentials. The equation is named after Pierre-Simon Laplace (1749–1827), who had studied the properties of this equation. Solutions of Laplace's Equation are called harmonic functions.

Example 11.10. *Solve Laplace's equation in spherical coordinates.*

We seek solutions of this equation inside a sphere of radius r subject to the boundary condition as shown in Figure 11.18. The problem is given by Laplace's Equation in spherical coordinates[2]

² The Laplacian in spherical coordinates is given in Problem 32 in Chapter 8.

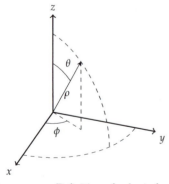

Figure 11.19: Definition of spherical coordinates (ρ,θ,ϕ). Note that there are different conventions for labeling spherical coordinates. This labeling is used often in physics.

$$\frac{1}{\rho^2}\frac{\partial}{\partial\rho}\left(\rho^2\frac{\partial u}{\partial\rho}\right) + \frac{1}{\rho^2\sin\theta}\frac{\partial}{\partial\theta}\left(\sin\theta\frac{\partial u}{\partial\theta}\right) + \frac{1}{\rho^2\sin^2\theta}\frac{\partial^2 u}{\partial\phi^2} = 0, \qquad (11.118)$$

where $u = u(\rho,\theta,\phi)$.

The boundary conditions are given by

$$u(r,\theta,\phi) = g(\theta,\phi), \quad 0 < \phi < 2\pi, \quad 0 < \theta < \pi,$$

and the periodic boundary conditions

$$u(\rho,\theta,0) = u(\rho,\theta,2\pi), \quad u_\phi(\rho,\theta,0) = u_\phi(\rho,\theta,2\pi),$$

where $0 < \rho < \infty$, and $0 < \theta < \pi$.

As before, we perform a separation of variables by seeking product solutions of the form $u(\rho,\theta,\phi) = R(\rho)\Theta(\theta)\Phi(\phi)$. Inserting this form into the Laplace Equation, we obtain

$$\frac{\Theta\Phi}{\rho^2}\frac{d}{d\rho}\left(\rho^2\frac{dR}{d\rho}\right) + \frac{R\Phi}{\rho^2\sin\theta}\frac{d}{d\theta}\left(\sin\theta\frac{d\Theta}{d\theta}\right) + \frac{R\Theta}{\rho^2\sin^2\theta}\frac{d^2\Phi}{d\phi^2} = 0. \quad (11.119)$$

Multiplying this equation by ρ^2 and dividing by $R\Theta\Phi$ yields

$$\frac{1}{R}\frac{d}{d\rho}\left(\rho^2\frac{dR}{d\rho}\right) + \frac{1}{\sin\theta\,\Theta}\frac{d}{d\theta}\left(\sin\theta\frac{d\Theta}{d\theta}\right) + \frac{1}{\sin^2\theta\,\Phi}\frac{d^2\Phi}{d\phi^2} = 0. \quad (11.120)$$

Note that the first term is the only term depending upon ρ. Thus, we can separate out the radial part. However, there is still more work to do on the other two terms, which give the angular dependence. Thus, we have

$$-\frac{1}{R}\frac{d}{d\rho}\left(\rho^2\frac{dR}{d\rho}\right) = \frac{1}{\sin\theta\,\Theta}\frac{d}{d\theta}\left(\sin\theta\frac{d\Theta}{d\theta}\right) + \frac{1}{\sin^2\theta\,\Phi}\frac{d^2\Phi}{d\phi^2} = -\lambda, \quad (11.121)$$

where we have introduced the first separation constant. This leads to two equations:

$$\frac{d}{d\rho}\left(\rho^2\frac{dR}{d\rho}\right) - \lambda R = 0 \qquad (11.122)$$

and

$$\frac{1}{\sin\theta\,\Theta}\frac{d}{d\theta}\left(\sin\theta\frac{d\Theta}{d\theta}\right) + \frac{1}{\sin^2\theta\,\Phi}\frac{d^2\Phi}{d\phi^2} = -\lambda. \qquad (11.123)$$

The final separation can be performed by multiplying the last equation by $\sin^2\theta$, rearranging the terms, and introducing a second separation constant:

$$\frac{\sin\theta}{\Theta}\frac{d}{d\theta}\left(\sin\theta\frac{d\Theta}{d\theta}\right) + \lambda\sin^2\theta = -\frac{1}{\Phi}\frac{d^2\Phi}{d\phi^2} = \mu. \qquad (11.124)$$

From this expression we can determine the differential equations satisfied by $\Theta(\theta)$ and $\Phi(\phi)$:

$$\sin\theta\frac{d}{d\theta}\left(\sin\theta\frac{d\Theta}{d\theta}\right) + (\lambda\sin^2\theta - \mu)\Theta = 0 \qquad (11.125)$$

and

$$\frac{d^2\Phi}{d\phi^2} + \mu\Phi = 0. \qquad (11.126)$$

Equation (11.123) is a key equation which occurs when studying problems possessing spherical symmetry. It is an eigenvalue problem for $Y(\theta,\phi) = \Theta(\theta)\Phi(\phi)$, $LY = -\lambda Y$, where

$$L = \frac{1}{\sin\theta}\frac{\partial}{\partial\theta}\left(\sin\theta\frac{\partial}{\partial\theta}\right) + \frac{1}{\sin^2\theta}\frac{\partial^2}{\partial\phi^2}.$$

The eigenfunctions of this operator are referred to as spherical harmonics.

We now have three ordinary differential equations to solve. These are the radial equation (11.122) and the two angular equations (11.125) and (11.126). We note that all three are in Sturm-Liouville form. We will solve each eigenvalue problem subject to appropriate boundary conditions.

The simplest of these differential equations is Equation (11.126) for $\Phi(\phi)$. We have seen equations of this form many times, and the general solution is a linear combination of sines and cosines. Furthermore, in this problem, $u(\rho,\theta,\phi)$ is periodic in ϕ:

$$u(\rho,\theta,0) = u(\rho,\theta,2\pi), \quad u_\phi(\rho,\theta,0) = u_\phi(\rho,\theta,2\pi).$$

Because these conditions hold for all ρ and θ, we must require that $\Phi(\phi)$ satisfy the periodic boundary conditions

$$\Phi(0) = \Phi(2\pi), \quad \Phi'(0) = \Phi'(2\pi).$$

The eigenfunctions and eigenvalues for Equation (11.126) are then found as

$$\Phi(\phi) = \{\cos m\phi, \sin m\phi\}, \quad \mu = m^2, \quad m = 0, 1, \ldots. \tag{11.127}$$

Next we turn to solving Equation, (11.125). We first transform this equation in order to identify the solutions. Let $x = \cos\theta$. Then the derivatives with respect to θ transform as

$$\frac{d}{d\theta} = \frac{dx}{d\theta}\frac{d}{dx} = -\sin\theta\frac{d}{dx}.$$

Letting $y(x) = \Theta(\theta)$ and noting that $\sin^2\theta = 1 - x^2$, Equation (11.125) becomes

$$\frac{d}{dx}\left((1-x^2)\frac{dy}{dx}\right) + \left(\lambda - \frac{m^2}{1-x^2}\right)y = 0. \tag{11.128}$$

We further note that $x \in [-1, 1]$, as can be easily confirmed by the reader.

This is a Sturm-Liouville eigenvalue problem. The solutions consist of a set of orthogonal eigenfunctions. For the special case that $m = 0$, Equation (11.128) becomes

$$\frac{d}{dx}\left((1-x^2)\frac{dy}{dx}\right) + \lambda y = 0. \tag{11.129}$$

In a course in differential equations, one learns to seek solutions of this equation in the form

$$y(x) = \sum_{n=0}^{\infty} a_n x^n.$$

This leads to the recursion relation

$$a_{n+2} = \frac{n(n+1) - \lambda}{(n+2)(n+1)}a_n.$$

Setting $n = 0$ and seeking a series solution, one finds that the resulting series does not converge for $x = \pm 1$. This is remedied by choosing $\lambda = \ell(\ell+1)$ for $\ell = 0, 1, \ldots$, leading to the differential equation

$$\frac{d}{dx}\left((1-x^2)\frac{dy}{dx}\right) + \ell(\ell+1)y = 0. \tag{11.130}$$

We saw this equation in Chapter 6 in the form

$$(1-x^2)y'' - 2xy' + \ell(\ell+1)y = 0.$$

The solutions of this differential equation are Legendre polynomials, denoted by $P_\ell(x)$.

For the more general case, $m \neq 0$, the differential equation (11.128) with $\lambda = \ell(\ell+1)$ becomes

Associated Legendre functions.

$$\frac{d}{dx}\left((1-x^2)\frac{dy}{dx}\right) + \left(\ell(\ell+1) - \frac{m^2}{1-x^2}\right)y = 0. \tag{11.131}$$

The solutions of this equation are called the associated Legendre functions. The two linearly independent solutions are denoted by $P_\ell^m(x)$ and $Q_\ell^m(x)$ The latter functions are not well behaved at $x = \pm 1$, corresponding to the north and south poles of the original problem. So, we can throw out these solutions in many physical cases, leaving

$$\Theta(\theta) = P_\ell^m(\cos\theta)$$

as the needed solutions. In Table 11.5 we list a few of these.

	$P_n^m(x)$	$P_n^m(\cos\theta)$
$P_0^0(x)$	1	1
$P_1^0(x)$	x	$\cos\theta$
$P_1^1(x)$	$-(1-x^2)^{\frac{1}{2}}$	$-\sin\theta$
$P_2^0(x)$	$\frac{1}{2}(3x^2-1)$	$\frac{1}{2}(3\cos^2\theta-1)$
$P_2^1(x)$	$-3x(1-x^2)^{\frac{1}{2}}$	$-3\cos\theta\sin\theta$
$P_2^2(x)$	$3(1-x^2)$	$3\sin^2\theta$
$P_3^0(x)$	$\frac{1}{2}(5x^3-3x)$	$\frac{1}{2}(5\cos^3\theta-3\cos\theta)$
$P_3^1(x)$	$-\frac{3}{2}(5x^2-1)(1-x^2)^{\frac{1}{2}}$	$-\frac{3}{2}(5\cos^2\theta-1)\sin\theta$
$P_3^2(x)$	$15x(1-x^2)$	$15\cos\theta\sin^2\theta$
$P_3^3(x)$	$-15(1-x^2)^{\frac{3}{2}}$	$-15\sin^3\theta$

Table 11.5: Associated Legendre Functions, $P_n^m(x)$

The associated Legendre functions are related to the Legendre polynomials by[3]

$$P_\ell^m(x) = (-1)^m(1-x^2)^{m/2}\frac{d^m}{dx^m}P_\ell(x), \tag{11.132}$$

for $\ell = 0, 1, 2,,\ldots$ and $m = 0, 1, \ldots, \ell$. We further note that $P_\ell^0(x) = P_\ell(x)$, as one can see in the table. Because $P_\ell(x)$ is a polynomial of degree ℓ, then for $m > \ell$, $\frac{d^m}{dx^m}P_\ell(x) = 0$ and $P_\ell^m(x) = 0$.

Furthermore, because the differential equation only depends on m^2, $P_\ell^{-m}(x)$ is proportional to $P_\ell^m(x)$. One normalization is given by

$$P_\ell^{-m}(x) = (-1)^m\frac{(\ell-m)!}{(\ell+m)!}P_\ell^m(x).$$

The associated Legendre functions also satisfy the orthogonality condition

$$\int_{-1}^1 P_\ell^m(x)P_{\ell'}^m(x)\,dx = \frac{2}{2\ell+1}\frac{(\ell+m)!}{(\ell-m)!}\delta_{\ell\ell'}. \tag{11.133}$$

Orthogonality relation.

The last differential equation we need to solve is the radial equation. With $\lambda = \ell(\ell+1)$, $\ell = 0, 1, 2, \ldots$, the radial equation (11.122) can be written as

$$\rho^2 R'' + 2\rho R' - \ell(\ell+1)R = 0. \tag{11.134}$$

The radial equation is a Cauchy-Euler type of equation. So, we can guess the form of the solution to be $R(\rho) = \rho^s$, where s is a yet to be determined constant. Inserting this guess into the radial equation, we obtain the characteristic equation

$$s(s+1) = \ell(\ell+1).$$

Solving for s, we have

$$s = \ell, -(\ell+1).$$

[3] The factor of $(-1)^m$ is known as the Condon–Shortley phase and is useful in quantum mechanics in the treatment of agular momentum. It is sometimes omitted by some.

Thus, the general solution of the radial equation is

$$R(\rho) = a\rho^\ell + b\rho^{-(\ell+1)}.$$ (11.135)

We would normally apply boundary conditions at this point. The boundary condition $u(r, \theta, \phi) = g(\theta, \phi)$ is not a homogeneous boundary condition, so we will need to hold off using it until we have the general solution to the three-dimensional problem. However, we do have a hidden condition. Because we are interested in solutions inside the sphere, we need to consider what happens at $\rho = 0$. Note that $\rho^{-(\ell+1)}$ is not defined at the origin. Because the solution is expected to be bounded at the origin, we can set $b = 0$. So, in the current problem, we have established that

$$R(\rho) = a\rho^\ell.$$

When seeking solutions outside the sphere, one considers the boundary condition $R(\rho) \to 0$ as $\rho \to \infty$. In this case, $R(\rho) = \rho^{-(\ell+1)}$.

We have carried out the full separation of Laplace's Equation in spherical coordinates. The product solutions consist of the forms

$$u(\rho, \theta, \phi) = \rho^\ell P_\ell^m(\cos\theta)\cos m\phi$$

and

$$u(\rho, \theta, \phi) = \rho^\ell P_\ell^m(\cos\theta)\sin m\phi$$

for $\ell = 0, 1, 2, \ldots$ and $m = 0, \pm 1, \ldots, \pm\ell$. These solutions can be combined to give a complex representation of the product solutions as

$$u(\rho, \theta, \phi) = \rho^\ell P_\ell^m(\cos\theta)e^{im\phi}.$$

[4] While this appears to be a complex-valued solution, it can be rewritten as a sum over real functions. The inner sum contains terms for both $m = k$ and $m = -k$. Adding these contributions, we have that

$$a_{\ell k}\rho^\ell P_\ell^k(\cos\theta)e^{ik\phi} + a_{\ell(-k)}\rho^\ell P_\ell^{-k}(\cos\theta)e^{-ik\phi}$$

can be rewritten as

$$(A_{\ell k}\cos k\phi + B_{\ell k}\sin k\phi)\rho^\ell P_\ell^k(\cos\theta).$$

The general solution is then given as a linear combination of these product solutions. As there are two indices, we have a double sum:[4]

$$u(\rho, \theta, \phi) = \sum_{\ell=0}^{\infty} \sum_{m=-\ell}^{\ell} a_{\ell m}\rho^\ell P_\ell^m(\cos\theta)e^{im\phi}.$$ (11.136)

Example 11.11. *Laplace's Equation with Azimuthal Symmetry*

As a simple example, we consider the solution of Laplace's Equation in which there is azimuthal symmetry. Let

$$u(r, \theta, \phi) = g(\theta) = 1 - \cos 2\theta.$$

This function is zero at the poles and has a maximum at the equator. So, this could be a crude model of the temperature distribution of the Earth with zero temperature at the poles and a maximum near the equator.

In problems in which there is no ϕ-dependence, only the $m = 0$ terms of the general solution survives. Thus, we have that

$$u(\rho, \theta, \phi) = \sum_{\ell=0}^{\infty} a_\ell \rho^\ell P_\ell(\cos\theta).$$ (11.137)

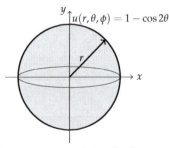

Figure 11.20: A sphere of radius r with the boundary condition

$$u(r, \theta, \phi) = 1 - \cos 2\theta.$$

Here we have used the fact that $P_\ell^0(x) = P_\ell(x)$. We just need to determine the unknown expansion coefficients, a_ℓ. Imposing the boundary condition at $\rho = r$, we are led to

$$g(\theta) = \sum_{\ell=0}^{\infty} a_\ell r^\ell P_\ell(\cos\theta).$$ (11.138)

This is a Fourier-Legendre series representation of $g(\theta)$. Because the Legendre polynomials are an orthogonal set of eigenfunctions, we can extract the coefficients.

In Chapter 6 we had proven that

$$\int_0^{\pi} P_n(\cos\theta)P_m(\cos\theta)\sin\theta\,d\theta = \int_{-1}^{1} P_n(x)P_m(x)\,dx = \frac{2}{2n+1}\delta_{nm}.$$

So, multiplying the expression for $g(\theta)$ by $P_m(\cos\theta)\sin\theta$ and integrating, we obtain the expansion coefficients:

$$a_\ell = \frac{2\ell+1}{2r^\ell}\int_0^{\pi} g(\theta)P_\ell(\cos\theta)\sin\theta\,d\theta. \tag{11.139}$$

Sometimes it is easier to rewrite $g(\theta)$ as a polynomial in $\cos\theta$ and avoid the integration. For this example, we see that

$$
\begin{aligned}
g(\theta) &= 1 - \cos 2\theta \\
&= 2\sin^2\theta \\
&= 2 - 2\cos^2\theta. \tag{11.140}
\end{aligned}
$$

Thus, setting $x = \cos\theta$ and $G(x) = g(\theta(x))$, we have $G(x) = 2 - 2x^2$.

We seek the form

$$G(x) = c_0 P_0(x) + c_1 P_1(x) + c_2 P_2(x),$$

where $P_0(x) = 1$, $P_1(x) = x$, and $P_2(x) = \frac{1}{2}(3x^2 - 1)$. Because $G(x) = 2 - 2x^2$ does not have any x terms, we know that $c_1 = 0$. So,

$$2 - 2x^2 = c_0(1) + c_2\frac{1}{2}(3x^2 - 1) = c_0 - \frac{1}{2}c_2 + \frac{3}{2}c_2 x^2.$$

By observation we have $c_2 = -\frac{4}{3}$ and thus $c_0 = 2 + \frac{1}{2}c_2 = \frac{4}{3}$. Therefore, $G(x) = \frac{4}{3}P_0(x) - \frac{4}{3}P_2(x)$.

We have found the expansion of $g(\theta)$ in terms of Legendre polynomials,

$$g(\theta) = \frac{4}{3}P_0(\cos\theta) - \frac{4}{3}P_2(\cos\theta). \tag{11.141}$$

Therefore, the nonzero coefficients in the general solution become

$$a_0 = \frac{4}{3}, \quad a_2 = \frac{4}{3}\frac{1}{r^2},$$

and the rest of the coefficients are zero. Inserting these into the general solution, we have the final solution

$$
\begin{aligned}
u(\rho,\theta,\phi) &= \frac{4}{3}P_0(\cos\theta) - \frac{4}{3}\left(\frac{\rho}{r}\right)^2 P_2(\cos\theta) \\
&= \frac{4}{3} - \frac{2}{3}\left(\frac{\rho}{r}\right)^2(3\cos^2\theta - 1). \tag{11.142}
\end{aligned}
$$

11.5.1 Spherical Harmonics

THE SOLUTIONS OF THE ANGULAR PARTS OF THE PROBLEM are often combined into one function of two variables, as problems with spherical symmetry arise often, leaving the main differences between such problems confined to the radial equation. These functions are referred to as spherical harmonics, $Y_{\ell m}(\theta,\phi)$, which are defined with a special normalization as

$Y_{\ell m}(\theta,\phi)$, are the spherical harmonics. Spherical harmonics are important in applications from atomic electron configurations to gravitational fields, planetary magnetic fields, and the cosmic microwave background radiation.

$$Y_{\ell m}(\theta, \phi) = (-1)^m \sqrt{\frac{2\ell+1}{4\pi} \frac{(\ell-m)!}{(\ell+m)!}} P_\ell^m(\cos\theta) e^{im\phi}. \qquad (11.143)$$

These satisfy the simple orthogonality relation

$$\int_0^\pi \int_0^{2\pi} Y_{\ell m}(\theta,\phi) Y_{\ell' m'}^*(\theta,\phi) \sin\theta \, d\phi \, d\theta = \delta_{\ell\ell'} \delta_{mm'}.$$

As seen earlier in the chapter, the spherical harmonics are eigenfunctions of the eigenvalue problem $LY = -\lambda Y$, where

$$L = \frac{1}{\sin\theta} \frac{\partial}{\partial\theta} \left(\sin\theta \frac{\partial}{\partial\theta} \right) + \frac{1}{\sin^2\theta} \frac{\partial^2}{\partial\phi^2}.$$

This operator appears in many problems in which there is spherical symmetry, such as obtaining the solution of Schrödinger's Equation for the hydrogen atom, as we will see later. Therefore, it is customary to plot spherical harmonics. Because the $Y_{\ell m}$'s are complex functions, one typically plots either the real part or the modulus squared. One rendition of $|Y_{\ell m}(\theta,\phi)|^2$ is shown in Table 11.6 for $\ell, m = 0, 1, 2, 3$.

Table 11.6: The First Few Spherical Harmonics, $|Y_{\ell m}(\theta,\phi)|^2$

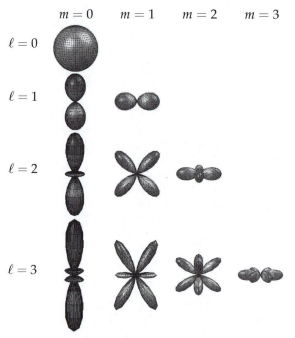

We could also look for the nodal curves of the spherical harmonics like we had for vibrating membranes. Such surface plots on a sphere are shown in Table 11.7. The colors provide for the amplitude of the $|Y_{\ell m}(\theta,\phi)|^2$. We can match these with the shapes in Table 11.6 by coloring the plots with some of the same colors as shown in Table 11.7. However, by plotting just the sign of the spherical harmonics, as in Table 11.8, we can pick out the nodal curves much more easily.

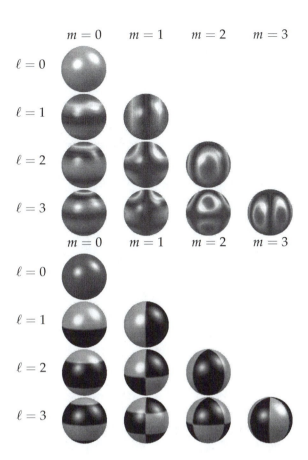

Table 11.7: Spherical Harmonic Contours for $|Y_{\ell m}(\theta,\phi)|^2$

Table 11.8: In these figures we show the nodal curves of $|Y_{\ell m}(\theta,\phi)|^2$. Along the first column ($m = 0$) are the zonal harmonics seen as ℓ horizontal circles. Along the top diagonal ($m = \ell$) are the sectional harmonics. These look like orange sections formed from m vertical circles. The remaining harmonics are tesseral harmonics. They look like a checkerboard pattern formed from intersections of $\ell - m$ horizontal circles and m vertical circles.

Spherical, or surface, harmonics can be further grouped into zonal, sectoral, and tesseral harmonics. Zonal harmonics correspond to the $m = 0$ modes. In this case, one seeks nodal curves for which $P_\ell(\cos\theta) = 0$. Solutions of this equation lead to constant θ values such that $\cos\theta$ is a zero of the Legendre polynomial, $P_\ell(x)$. The zonal harmonics correspond to the first column in Table 11.8. Because $P_\ell(x)$ is a polynomial of degree ℓ, the zonal harmonics consist of ℓ latitudinal circles.

Sectoral, or meridional, harmonics result for the case that $m = \pm\ell$. For this case, we note that $P_\ell^{\pm\ell}(x) \propto (1 - x^2)^{m/2}$. This function vanishes for $x = \pm 1$, or $\theta = 0, \pi$. Therefore, the spherical harmonics can only produce nodal curves for $e^{im\phi} = 0$. Thus, one obtains the meridians satisfying the condition $A\cos m\phi + B\sin m\phi = 0$. Solutions of this equation are of the form $\phi = $ constant. These modes can be seen in Table 11.8 in the top diagonal and can be described as m circles passing through the poles, or longitudinal circles.

Tesseral harmonics consist of the rest of the modes, which typically look like a checkerboard glued to the surface of a sphere. Examples can be seen in the pictures of nodal curves, such as Table 11.8. Looking in Table 11.8 along the diagonals going downward from left to right, one can see the same number of latitudinal circles. In fact, there are $\ell - m$ latitudinal nodal curves in these figures

Figure 11.21: Zonal harmonics, $\ell = 1$, $m = 0$.

Figure 11.22: Zonal harmonics, $\ell = 2$, $m = 0$.

Figure 11.23: Sectoral harmonics, $\ell = 2$, $m = 2$.

Figure 11.24: Tesseral harmonics, $\ell = 3$, $m = 1$.

Figure 11.25: Sectoral harmonics, $\ell = 3$, $m = 3$.

Figure 11.26: Tesseral harmonics, $\ell = 4$, $m = 3$.

In summary, the spherical harmonics have several representations, as show in Tables 11.7 and 11.8. Note that there are ℓ nodal lines, m meridional curves, and $\ell - m$ horizontal curves in these figures. The plots in Table 11.6 are the typical plots shown in physics for discussion of the wavefunctions of the hydrogen atom. Those in Table 11.7 are useful for describing gravitational or electric potential functions, temperature distributions, or wave modes on a spherical surface. The relationships between these pictures and the nodal curves can be better understood by comparing respective plots. Several modes were separated out in Figures 11.21 through 11.26 to make this comparison easier.

11.6   Schrödinger Equation in Spherical Coordinates

ANOTHER IMPORTANT EIGENVALUE PROBLEM IN PHYSICS is the Schrödinger Equation. The time-dependent Schrödinger Equation is given by

$$i\hbar \frac{\partial \Psi}{\partial t} = -\frac{\hbar^2}{2m} \nabla^2 \Psi + V\Psi. \tag{11.144}$$

Here, $\Psi(\mathbf{r}, t)$ is the wave function, which determines the quantum state of a particle of mass m subject to a (time independent) potential, $V(\mathbf{r})$. From Planck's constant, h, one defines $\hbar = \frac{h}{2\pi}$. The probability of finding the particle in an infinitesimal volume, dV, is given by $|\Psi(\mathbf{r}, t)|^2\, dV$, assuming the wave function is normalized,

$$\int_{\text{all space}} |\Psi(\mathbf{r}, t)|^2\, dV = 1.$$

One can separate out the time dependence by assuming a special form, $\Psi(\mathbf{r}, t) = \psi(\mathbf{r})e^{-iEt/\hbar}$, where E is the energy of the particular stationary state solution, or product solution. Inserting this form into the time-dependent equation, one finds that $\psi(\mathbf{r})$ satisfies the time-independent Schrödinger Equation,

$$-\frac{\hbar^2}{2m} \nabla^2 \psi + V\psi = E\psi. \tag{11.145}$$

Assuming that the potential depends only on the distance from the origin, $V = V(\rho)$, we can further separate out the radial part of this solution using spherical coordinates. Recall that the Laplacian in spherical coordinates is given by

$$\nabla^2 = \frac{1}{\rho^2} \frac{\partial}{\partial \rho} \left(\rho^2 \frac{\partial}{\partial \rho} \right) + \frac{1}{\rho^2 \sin\theta} \frac{\partial}{\partial \theta} \left(\sin\theta \frac{\partial}{\partial \theta} \right) + \frac{1}{\rho^2 \sin^2\theta} \frac{\partial^2}{\partial \phi^2}. \tag{11.146}$$

Then, the time-independent Schrödinger Equation can be written as

$$\begin{aligned}
-\frac{\hbar^2}{2m} &\left[\frac{1}{\rho^2} \frac{\partial}{\partial \rho} \left(\rho^2 \frac{\partial \psi}{\partial \rho} \right) + \frac{1}{\rho^2 \sin\theta} \frac{\partial}{\partial \theta} \left(\sin\theta \frac{\partial \psi}{\partial \theta} \right) + \frac{1}{\rho^2 \sin^2\theta} \frac{\partial^2 \psi}{\partial \phi^2} \right] \\
&= [E - V(\rho)]\psi. \tag{11.147}
\end{aligned}$$

Let's continue with the separation of variables. Assuming that the wave function takes the form $\psi(\rho, \theta, \phi) = R(\rho)Y(\theta, \phi)$, we obtain

$$- \frac{\hbar^2}{2m} \left[\frac{Y}{\rho^2} \frac{d}{d\rho} \left(\rho^2 \frac{dR}{d\rho} \right) + \frac{R}{\rho^2 \sin\theta} \frac{\partial}{\partial\theta} \left(\sin\theta \frac{\partial Y}{\partial\theta} \right) + \frac{R}{\rho^2 \sin^2\theta} \frac{\partial^2 Y}{\partial\phi^2} \right]$$
$$= RY[E - V(\rho)]\psi. \tag{11.148}$$

Dividing by $\psi = RY$, multiplying by $-\frac{2m\rho^2}{\hbar^2}$, and rearranging, we have

$$\frac{1}{R} \frac{d}{d\rho} \left(\rho^2 \frac{dR}{d\rho} \right) - \frac{2m\rho^2}{\hbar^2} [V(\rho) - E] = -\frac{1}{Y} LY,$$

where

$$L = \frac{1}{\sin\theta} \frac{\partial}{\partial\theta} \left(\sin\theta \frac{\partial}{\partial\theta} \right) + \frac{1}{\sin^2\theta} \frac{\partial^2}{\partial\phi^2}.$$

We have a function of ρ equal to a function of the angular variables. So, we set each side equal to a constant. We will judiciously write the separation constant as $\ell(\ell + 1)$. The resulting equations are then

$$\frac{d}{d\rho} \left(\rho^2 \frac{dR}{d\rho} \right) - \frac{2m\rho^2}{\hbar^2} [V(\rho) - E] R = \ell(\ell + 1)R, \tag{11.149}$$

$$\frac{1}{\sin\theta} \frac{\partial}{\partial\theta} \left(\sin\theta \frac{\partial Y}{\partial\theta} \right) + \frac{1}{\sin^2\theta} \frac{\partial^2 Y}{\partial\phi^2} = -\ell(\ell + 1)Y. \tag{11.150}$$

The second of these equations should look familiar from the previous section. This is the equation for spherical harmonics,

$$Y_{\ell m}(\theta, \phi) = \sqrt{\frac{2\ell + 1}{2} \frac{(\ell - m)!}{(\ell + m)!}} P_\ell^m e^{im\phi}. \tag{11.151}$$

So, any further analysis of the problem depends upon the choice of potential, $V(\rho)$, and the solution of the radial equation. For this, we turn to the determination of the wave function for an electron in orbit about a proton.

Solution of the hydrogen problem.

Example 11.12. *The Hydrogen Atom: $\ell = 0$ States*

Historically, the first test of the Schrödinger Equation was the determination of the energy levels in a hydrogen atom. This is modeled by an electron orbiting a proton. The potential energy is provided by the Coulomb potential,

$$V(\rho) = -\frac{e^2}{4\pi\epsilon_0\rho}.$$

Thus, the radial equation becomes

$$\frac{d}{d\rho} \left(\rho^2 \frac{dR}{d\rho} \right) + \frac{2m\rho^2}{\hbar^2} \left[\frac{e^2}{4\pi\epsilon_0\rho} + E \right] R = \ell(\ell + 1)R. \tag{11.152}$$

Before looking for solutions, we need to simplify the equation by absorbing some of the constants. One way to do this is to make an appropriate change of variables. Let $\rho = ar$. Then, by the Chain Rule we have

$$\frac{d}{d\rho} = \frac{dr}{d\rho} \frac{d}{dr} = \frac{1}{a} \frac{d}{dr}.$$

Under this transformation, the radial equation becomes

$$\frac{d}{dr}\left(r^2\frac{du}{dr}\right) + \frac{2ma^2r^2}{\hbar^2}\left[\frac{e^2}{4\pi\epsilon_0 ar} + E\right]u = \ell(\ell+1)u, \qquad (11.153)$$

where $u(r) = R(\rho)$. Expanding the second term,

$$\frac{2ma^2r^2}{\hbar^2}\left[\frac{e^2}{4\pi\epsilon_0 ar} + E\right]u = \left[\frac{mae^2}{2\pi\epsilon_0\hbar^2}r + \frac{2mEa^2}{\hbar^2}r^2\right]u,$$

we see that we can define

$$a = \frac{2\pi\epsilon_0\hbar^2}{me^2}, \qquad (11.154)$$

$$\epsilon = -\frac{2mEa^2}{\hbar^2}$$

$$= -\frac{2(2\pi\epsilon_0)^2\hbar^2}{me^4}E. \qquad (11.155)$$

Using these constants, the radial equation becomes

$$\frac{d}{dr}\left(r^2\frac{du}{dr}\right) + ru - \ell(\ell+1)u = \epsilon r^2 u. \qquad (11.156)$$

Expanding the derivative and dividing by r^2,

$$u'' + \frac{2}{r}u' + \frac{1}{r}u - \frac{\ell(\ell+1)}{r^2}u = \epsilon u. \qquad (11.157)$$

The first two terms in this differential equation came from the Laplacian. The third term came from the Coulomb potential. The fourth term can be thought to contribute to the potential and is attributed to angular momentum. Thus, ℓ is called the angular momentum quantum number. This is an eigenvalue problem for the radial eigenfunctions $u(r)$ and energy eigenvalues ϵ.

The solutions of this equation are determined in a quantum mechanics course. In order to get a feeling for the solutions, we will consider the zero angular momentum case, $\ell = 0$:

$$u'' + \frac{2}{r}u' + \frac{1}{r}u = \epsilon u. \qquad (11.158)$$

Even this equation is one we have not encountered in this book. Let's see if we can find some of the solutions.

First, we consider the behavior of the solutions for large r. For large r, the second and third terms on the left-hand side of the equation are negligible. So, we have the approximate equation

$$u'' - \epsilon u = 0. \qquad (11.159)$$

Therefore, the solutions behave like $u(r) = e^{\pm\sqrt{\epsilon}r}$ for large r. For bounded solutions, we choose the decaying solution.

This suggests that solutions take the form $u(r) = v(r)e^{-\sqrt{\epsilon}r}$ for some unknown function, $v(r)$. Inserting this guess into Equation (11.158) gives an equation for $v(r)$:

$$rv'' + 2\left(1 - \sqrt{\epsilon}r\right)v' + (1 - 2\sqrt{\epsilon})v = 0. \qquad (11.160)$$

Next we seek a series solution to this equation. Let

$$v(r) = \sum_{k=0}^{\infty} c_k r^k.$$

Inserting this series into Equation (11.160), we have

$$\sum_{k=1}^{\infty} [k(k-1) + 2k] c_k r^{k-1} + \sum_{k=1}^{\infty} [1 - 2\sqrt{\epsilon}(k+1)] c_k r^k = 0.$$

We can re-index the dummy variable in each sum. Let $k = m$ in the first sum and $k = m - 1$ in the second sum. We then find that

$$\sum_{k=1}^{\infty} \left[m(m+1)c_m + (1 - 2m\sqrt{\epsilon})c_{m-1} \right] r^{m-1} = 0.$$

Because this has to hold for all $m \geq 1$,

$$c_m = \frac{2m\sqrt{\epsilon} - 1}{m(m+1)} c_{m-1}.$$

Further analysis indicates that the resulting series leads to unbounded solutions unless the series terminates. This is only possible if the numerator, $2m\sqrt{\epsilon} - 1$, vanishes for $m = n$, $n = 1, 2 \ldots$. Thus,

$$\epsilon = \frac{1}{4n^2}.$$

Because ϵ is related to the energy eigenvalue, E, we have

$$E_n = -\frac{me^4}{2(4\pi\epsilon_0)^2 \hbar^2 n^2}.$$

Inserting the values for the constants, this gives

$$E_n = -\frac{13.6 \, eV}{n^2}.$$

This is the well-known set of energy levels for the hydrogen atom.

Energy levels for the hydrogen atom.

The corresponding eigenfunctions are polynomials, because the infinite series was forced to terminate. We could obtain these polynomials by iterating the recursion equation for the c_m's. However, we will instead rewrite the radial equation (11.160).

Let $x = 2\sqrt{\epsilon}r$ and define $y(x) = v(r)$. Then

$$\frac{d}{dr} = 2\sqrt{\epsilon}\frac{d}{dx}.$$

This gives

$$2\sqrt{\epsilon}xy'' + (2-x)2\sqrt{\epsilon}y' + (1 - 2\sqrt{\epsilon})y = 0.$$

Rearranging, we have

$$xy'' + (2-x)y' + \frac{1}{2\sqrt{\epsilon}}(1 - 2\sqrt{\epsilon})y = 0.$$

Noting that $2\sqrt{\epsilon} = \frac{1}{n}$, this equation becomes

$$xy'' + (2-x)y' + (n-1)y = 0. \qquad (11.161)$$

The resulting equation is well known. It takes the form

$$xy'' + (\alpha + 1 - x)y' + ny = 0. \tag{11.162}$$

Solutions of this equation are the associated Laguerre polynomials. The solutions are denoted by $L_n^{\alpha}(x)$. They can be defined in terms of the Laguerre polynomials,

$$L_n(x) = e^x \left(\frac{d}{dx} \right)^n (e^{-x} x^n).$$

The associated Laguerre polynomials are defined as

$$L_{n-m}^m(x) = (-1)^m \left(\frac{d}{dx} \right)^m L_n(x).$$

Note: The Laguerre polynomials were first encountered in Problem 2 in Chapter 5 as an example of a classical orthogonal polynomial defined on $[0, \infty)$ with weight $w(x) = e^{-x}$. Some of these polynomials are listed in Table 11.9 and several Laguerre polynomials are shown in Figure 11.27.

The associated Laguerre polynomials are named after the French mathematician Edmond Laguerre (1834-1886).

Comparing Equation (11.161) with Equation (11.162), we find that $y(x) = L_{n-1}^1(x)$.

Table 11.9: Associated Laguerre Functions, $L_n^m(x)$ (Note that $L_n^0(x) = L_n(x)$.)

	$L_n^m(x)$
$L_0^0(x)$	1
$L_1^0(x)$	$1 - x$
$L_2^0(x)$	$\frac{1}{2}(x^2 - 4x + 2)$
$L_3^0(x)$	$\frac{1}{6}(-x^3 + 9x^2 - 18x + 6)$
$L_0^1(x)$	1
$L_1^1(x)$	$2 - x$
$L_2^1(x)$	$\frac{1}{2}(x^2 - 6x + 6)$
$L_3^1(x)$	$\frac{1}{6}(-x^3 + 3x^2 - 36x + 24)$
$L_0^2(x)$	1
$L_1^2(x)$	$3 - x$
$L_2^2(x)$	$\frac{1}{2}(x^2 - 8x + 12)$
$L_3^2(x)$	$\frac{1}{12}(-2x^3 + 30x^2 - 120x + 120)$

In most derivations in quantum mechanics $a = \frac{a_0}{2}$, where $a_0 = \frac{4\pi\epsilon_0\hbar^2}{me^2}$ is the Bohr radius and $a_0 = 5.2917 \times 10^{-11}$m.

In summary, we have made the following transformations:

1. $R(\rho) = u(r)$, $\rho = ar$.

2. $u(r) = v(r)e^{-\sqrt{\epsilon}r}$.

3. $v(r) = y(x) = L_{n-1}^1(x)$, $x = 2\sqrt{\epsilon}r$.

Therefore,

$$R(\rho) = e^{-\sqrt{\epsilon}\rho/a} L_{n-1}^1(2\sqrt{\epsilon}\rho/a).$$

However, we also found that $2\sqrt{\epsilon} = 1/n$. So,

$$R(\rho) = e^{-\rho/2na} L_{n-1}^1(\rho/na).$$

In Figure 11.28 we show a few of these solutions.

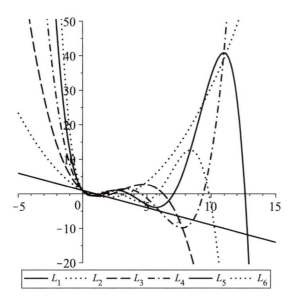

Figure 11.27: Plots of the first few Laguerre polynomials.

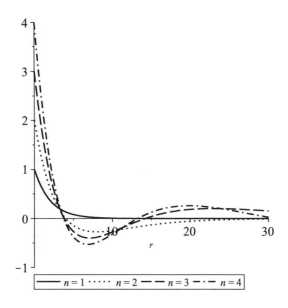

Figure 11.28: Plots of $R(\rho)$ for $a = 1$ and $n = 1, 2, 3, 4$ for the $\ell = 0$ states.

Example 11.13. *Find the $\ell \geq 0$ solutions of the radial equation.*

For the general case, for all $\ell \geq 0$, we need to solve the differential equation

$$u'' + \frac{2}{r}u' + \frac{1}{r}u - \frac{\ell(\ell+1)}{r^2}u = \epsilon u. \tag{11.163}$$

Instead of letting $u(r) = v(r)e^{-\sqrt{\epsilon}r}$, we let

$$u(r) = v(r)r^{\ell}e^{-\sqrt{\epsilon}r}.$$

This leads to the differential equation

$$rv'' + 2(\ell+1 - \sqrt{\epsilon}r)v' + (1 - 2(\ell+1)\sqrt{\epsilon})v = 0. \tag{11.164}$$

As before, we let $x = 2\sqrt{\epsilon}r$ to obtain

$$xy'' + 2\left[\ell+1 - \frac{x}{2}\right]v' + \left[\frac{1}{2\sqrt{\epsilon}} - \ell(\ell+1)\right]v = 0.$$

Noting that $2\sqrt{\epsilon} = 1/n$, we have

$$xy'' + 2[2(\ell+1) - x]v' + (n - \ell(\ell+1))v = 0.$$

We see that this is once again in the form of the associate Laguerre equation and the solutions are

$$y(x) = L_{n-\ell-1}^{2\ell+1}(x).$$

So, the solution to the radial equation for the hydrogen atom is given by

$$\begin{aligned}
R(\rho) &= r^{\ell}e^{-\sqrt{\epsilon}r}L_{n-\ell-1}^{2\ell+1}(2\sqrt{\epsilon}r) \\
&= \left(\frac{\rho}{2na}\right)^{\ell}e^{-\rho/2na}L_{n-\ell-1}^{2\ell+1}\left(\frac{\rho}{na}\right). \tag{11.165}
\end{aligned}$$

Interpretations of these solutions will be left for your quantum mechanics course.

11.7   Solution of the 3D Poisson Equation

WE RECALL FROM ELECTROSTATICS THAT THE GRADIENT OF THE ELECTRIC POTENTIAL gives the electric field, $\mathbf{E} = -\nabla\phi$. However, we also have from Gauss' Law for electric fields that $\nabla \cdot \mathbf{E} = \frac{\rho}{\epsilon_0}$, where $\rho(\mathbf{r})$ is the charge distribution at position $\mathbf{r}$. Combining these equations, we arrive at Poisson's Equation for the electric potential:

Poisson's Equation for the electric potential.

$$\nabla^2\phi = -\frac{\rho}{\epsilon_0}.$$

We note that Poisson's Equation also arises in Newton's Theory of Gravitation for the gravitational potential in the form $\nabla^2\phi = -4\pi G\rho$, where ρ is the matter density.

Poisson's Equation for the gravitational potential.

We consider Poisson's Equation in the form

$$\nabla^2\phi(\mathbf{r}) = -4\pi f(\mathbf{r})$$

for **r** defined throughout all space. We will seek a solution for the potential function using a three-dimensional Fourier transform. In the electrostatic problem, $f = \rho(\mathbf{r})/4\pi\epsilon_0$; and the gravitational problem has $f = G\rho(\mathbf{r})$

The Fourier transform can be generalized to three dimensions as

$$\hat{\phi}(\mathbf{k}) = \int_V \phi(\mathbf{r})e^{i\mathbf{k}\cdot\mathbf{r}}\,d^3r,$$

where the integration is over all space, V, $d^3r = dxdydz$, and **k** is a three-dimensional wave number, $\mathbf{k} = k_x\mathbf{i} + k_y\mathbf{j} + k_z\mathbf{k}$. The inverse Fourier transform can then be written as

$$\phi(\mathbf{r}) = \frac{1}{(2\pi)^3}\int_{V_k} \hat{\phi}(\mathbf{k})e^{-i\mathbf{k}\cdot\mathbf{r}}\,d^3k,$$

Three-dimensional Fourier transform.

where $d^3k = dk_x dk_y dk_z$ and V_k is all of k-space.

The Fourier transform of the Laplacian follows from computing Fourier transforms of any derivatives that are present. Assuming that ϕ and its gradient vanish for large distances, then

$$\mathcal{F}[\nabla^2\phi] = -(k_x^2 + k_y^2 + k_z^2)\hat{\phi}(\mathbf{k}).$$

Defining $k^2 = k_x^2 + k_y^2 + k_z^2$, then Poisson's Equation becomes the algebraic equation

$$k^2\hat{\phi}(\mathbf{k}) = 4\pi\hat{f}(\mathbf{k}).$$

Solving for $\hat{\phi}(\mathbf{k})$, we have

$$\phi(\mathbf{k}) = \frac{4\pi}{k^2}\hat{f}(\mathbf{k}).$$

The solution to Poisson's Equation is then determined from the inverse Fourier transform,

$$\phi(\mathbf{r}) = \frac{4\pi}{(2\pi)^3}\int_{V_k} \hat{f}(\mathbf{k})\frac{e^{-i\mathbf{k}\cdot\mathbf{r}}}{k^2}\,d^3k. \qquad (11.166)$$

First we will consider an example of a point charge (or mass in the gravitational case) at the origin. We will set $f(\mathbf{r}) = f_0\delta^3(\mathbf{r})$ in order to represent a point source. For a unit point charge, $f_0 = 1/4\pi\epsilon_0$.

Here we have introduced the three-dimensional Dirac delta function which, like the one-dimensional case, vanishes outside the origin and satisfies a unit volume condition,

The three-dimensional Dirac delta function, $\delta^3(\mathbf{r} - \mathbf{r}_0)$.

$$\int_V \delta^3(\mathbf{r})\,d^3r = 1.$$

Also, there is a sifting property, which takes the form

$$\int_V \delta^3(\mathbf{r} - \mathbf{r}_0)f(\mathbf{r})\,d^3r = f(\mathbf{r}_0).$$

In Cartesian coordinates,

$$\delta^3(\mathbf{r}) = \delta(x)\delta(y)\delta(z),$$

$$\int_V \delta^3(\mathbf{r})\, d^3 r = \int_{-\infty}^{\infty}\int_{-\infty}^{\infty}\int_{-\infty}^{\infty} \delta(x)\delta(y)\delta(z)\, dxdydz = 1,$$

and

$$\int_{-\infty}^{\infty}\int_{-\infty}^{\infty}\int_{-\infty}^{\infty} \delta(x-x_0)\delta(y-y_0)\delta(z-z_0)f(x,y,z)\, dxdydz = f(x_0, y_0, z_0).$$

One can define similar delta functions operating in two dimensions and n dimensions.

We can also transform the Cartesian form into curvilinear coordinates. Recall from the Section 9.3 that the volume element in curvilinear coordinates is

$$d^3 r = dxdydz = h_1 h_2 h_3\, du_1 du_2 du_3.$$

Here we have used the notation from Section 9.3. This gives

$$\int_V \delta^3(\mathbf{r})\, d^3 r = \int_V \delta^3(\mathbf{r})\, h_1 h_2 h_3\, du_1 du_2 du_3 = 1.$$

Therefore,

$$
\begin{aligned}
\delta^3(\mathbf{r}) &= \frac{\delta(u_1)}{\left|\frac{\partial \mathbf{r}}{\partial u_1}\right|}\frac{\delta(u_2)}{\left|\frac{\partial \mathbf{r}}{\partial u_2}\right|}\frac{\delta(u_3)}{\left|\frac{\partial \mathbf{r}}{\partial u_2}\right|} \\
&= \frac{1}{h_1 h_2 h_3}\delta(u_1)\delta(u_2)\delta(u_3).
\end{aligned}
\tag{11.167}
$$

So, for cylindrical coordinates,

$$\delta^3(\mathbf{r}) = \frac{1}{r}\delta(r)\delta(\theta)\delta(z).$$

Example 11.14. *Find the solution of Poisson's Equation for a point source of the form $f(\mathbf{r}) = f_0 \delta^3(\mathbf{r})$.*

The solution is found by inserting the Fourier transform of this source into Equation (11.166) and carrying out the integration. The transform of $f(\mathbf{r})$ is found as

$$\hat{f}(\mathbf{k}) = \int_V f_0 \delta^3(\mathbf{r})e^{i\mathbf{k}\cdot\mathbf{r}}\, d^3 r = f_0.$$

Inserting $\hat{f}(\mathbf{k})$ into the inverse transform in Equation (11.166) and carrying out the integration using spherical coordinates in k-space, we find

$$
\begin{aligned}
\phi(\mathbf{r}) &= \frac{4\pi}{(2\pi)^3}\int_{V_k} f_0 \frac{e^{-i\mathbf{k}\cdot\mathbf{r}}}{k^2}\, d^3 k \\
&= \frac{f_0}{2\pi^2}\int_0^{2\pi}\int_0^{\pi}\int_0^{\infty} \frac{e^{-ikx\cos\theta}}{k^2} k^2 \sin\theta\, dkd\theta d\phi \\
&= \frac{f_0}{\pi}\int_0^{\pi}\int_0^{\infty} e^{-ikx\cos\theta}\sin\theta\, dkd\theta \\
&= \frac{f_0}{\pi}\int_0^{\infty}\int_{-1}^{1} e^{-ikxy}\, dkdy, \quad y = \cos\theta, \\
&= \frac{2f_0}{\pi r}\int_0^{\infty} \frac{\sin z}{z}\, dz = \frac{f_0}{r}.
\end{aligned}
\tag{11.168}
$$

If the last example is applied to a unit point charge, then $f_0 = 1/4\pi\epsilon_0$. So, the electric potential outside a unit point charge located at the origin becomes

$$\phi(\mathbf{r}) = \frac{1}{4\pi\epsilon_0 r}.$$

This is the form familiar from introductory physics.

Also, by setting $f_0 = 1$, we have also shown in the previous example that

$$\nabla^2 \left(\frac{1}{r}\right) = -4\pi\delta^3(\mathbf{r}).$$

Because $\nabla \left(\frac{1}{r}\right) = -\frac{\mathbf{r}}{r^3}$, then we have also shown that

$$\nabla \cdot \left(\frac{\mathbf{r}}{r^3}\right) = 4\pi\delta^3(\mathbf{r}).$$

Example 11.15. *Find the Green's Function for Poisson's Equation*

For general $f(\mathbf{r})$, we note that the solution (11.166) of Poisson's Equation can be viewed as the inverse Fourier transform

$$\phi(\mathbf{r}) = \mathcal{F}^{-1}\left[\frac{4\pi}{k^2}\hat{f}(\mathbf{k})\right].$$

This is the inverse Fourier transform of a product of Fourier transforms $\hat{f}(\mathbf{k})$ and $\hat{g}(\mathbf{k}) = \frac{4\pi}{k^2}$. Therefore, we can apply the Convolution Theorem to rewrite the solution.

From the previous example, we have that $g(\mathbf{r}) = 1/r = 1/|\mathbf{r}|$. So, $\phi(\mathbf{r})$ is given by the convolution

$$\begin{aligned}
\phi(\mathbf{r}) &= (f * g)(\mathbf{r}) \\
&= \int_V f(\mathbf{r}')g(\mathbf{r} - \mathbf{r}')\, d^3r' \\
&= \int_V \frac{f(\mathbf{r}')}{|\mathbf{r} - \mathbf{r}'|}\, d^3r'.
\end{aligned} \tag{11.169}$$

In this solution, we can identify the Green's Function as

Green's Function for Poisson's Equation in three dimensions.

$$G(\mathbf{r}, \mathbf{r}') = \frac{1}{|\mathbf{r} - \mathbf{r}'|}$$

and write the solution in the form

$$\phi(\mathbf{r}) = \int_V G(\mathbf{r}, \mathbf{r}')f(\mathbf{r}')\, d^3r'.$$

11.7.1 Green's Functions for the 2D Poisson Equation

IN THIS SECTION WE CONSIDER the two-dimensional Poisson Equation with Dirichlet boundary conditions. We consider the problem

$$\begin{aligned}
\nabla^2 u &= f, \quad \text{in } D, \\
u &= g, \quad \text{on } C,
\end{aligned} \tag{11.170}$$

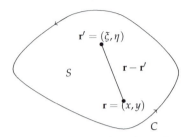

Figure 11.29: Domain for solving Poisson's equation.

for the domain in Figure 11.29

We seek to solve this problem using a Green's Function. As in earlier discussions, the Green's Function satisfies the differential equation and homogeneous boundary conditions. The associated problem is given by

$$\nabla^2 G = \delta(\xi - x, \eta - y), \quad \text{in } D,$$
$$G \equiv 0, \quad \text{on } C. \tag{11.171}$$

However, we need to be careful as to which variables appear in the differentiation. Many times, we just make the adjustment after the derivation of the solution, assuming that the Green's Function is symmetric in its arguments. However, this is not always the case and depends on things such as the self-adjointedness of the problem. Thus, we will assume that the Green's Function satisfies

$$\nabla_{r'}^2 G = \delta(\xi - x, \eta - y),$$

where the notation $\nabla_{r'}$ means differentiation with respect to the variables ξ and η. Thus,

$$\nabla_{r'}^2 G = \frac{\partial^2 G}{\partial \xi^2} + \frac{\partial^2 G}{\partial \eta^2}.$$

With this notation in mind, we now apply Green's second identity for two dimensions from Problem 26 in Chapter 9. We have

$$\int_D (u\nabla_{r'}^2 G \qquad -G\nabla_{r'}^2 u) \, dA'$$
$$= \int_C (u\nabla_{r'}G - G\nabla_{r'}u) \cdot ds'. \tag{11.172}$$

Inserting the differential equations, the left-hand side of the equation becomes

$$\int_D (u\nabla_{r'}^2 G - G\nabla_{r'}^2 u) \, dA' = \int_D (u(\xi, \eta)\delta(\xi - x, \eta - y) - G(x, y; \xi, \eta)f(\xi, \eta)) \, d\xi d\eta$$
$$= u(x, y) - \int_D G(x, y; \xi, \eta)f(\xi, \eta) \, d\xi d\eta. \tag{11.173}$$

Using the boundary conditions, $u(\xi, \eta) = g(\xi, \eta)$ on C and $G(x, y; \xi, \eta) = 0$ on C, the right-hand side of the equation becomes

$$\int_C (u\nabla_{r'}G - G\nabla_{r'}u) \cdot ds' = \int_C g(\xi, \eta)\nabla_{r'}G \cdot ds'. \tag{11.174}$$

Solving for $u(x, y)$, we have the solution written in terms of the Green's function,

$$u(x, y) = \int_D G(x, y; \xi, \eta)f(\xi, \eta) \, d\xi d\eta + \int_C g(\xi, \eta)\nabla_{r'}G \cdot ds'.$$

Now we need to find the Green's function. We find the Green's functions for several examples.

Example 11.16. *Find the two-dimensional Green's Function for the antisymmetric Poisson Equation; that is, we seek solutions that are θ-independent.*

The problem we need to solve in order to find the Green's function involves writing the Laplacian in polar coordinates,

$$v_{rr} + \frac{1}{r}v_r = \delta(r).$$

For $r \neq 0$, this is a Cauchy-Euler type of differential equation. The general solution is $v(r) = A \ln r + B$.

Due to the singularity at $r = 0$, we integrate over a domain in which a small circle of radius ϵ is cut form the plane and apply the two-dimensional Divergence Theorem. In particular, we have

$$
\begin{aligned}
1 &= \int_{D_\epsilon} \delta(r)\, dA \\
&= \int_{D_\epsilon} \nabla^2 v\, dA \\
&= \int_{C_\epsilon} \nabla v \cdot d\mathbf{s} \\
&= \int_{C_\epsilon} \frac{\partial v}{\partial r}\, dS = 2\pi A. \qquad (11.175)
\end{aligned}
$$

Therefore, $A = 1/2\pi$. We note that B is arbitrary, so we will take $B = 0$ in the remaining discussion.

Using this solution for a source of the form $\delta(\mathbf{r} - \mathbf{r}')$, we obtain the Green's function for Poisson's Equation as

$$G(\mathbf{r}, \mathbf{r}') = \frac{1}{2\pi} \ln|\mathbf{r} - \mathbf{r}'|.$$

Example 11.17. *Find the Green's function for the infinite plane.*

From Figure 11.29 we have $|\mathbf{r} - \mathbf{r}'| = \sqrt{(x-\xi)^2 + (y-\eta)^2}$. Therefore, the Green's function from the previous example gives

$$G(x, y, \xi, \eta) = \frac{1}{4\pi} \ln((\xi - x)^2 + (\eta - y)^2).$$

Green's function for the infinite plane.

Example 11.18. *Find the Green's function for the half plane, $\{(x, y) | y > 0\}$.*

This problem can be solved using the result for the Green's function for the infinite plane. We use the method of images to construct a function such that $G = 0$ on the boundary, $y = 0$. Namely, we use the image of the point (x, y) with respect to the x-axis, $(x, -y)$.

Imagine that the Green's function $G(x, y, \xi, \eta)$ represents a point charge at (x, y) and $G(x, y, \xi, \eta)$ provides the electric potential, or response, at (ξ, η). This single charge cannot yield a zero potential along the x-axis ($y=0$). One needs an additional charge to yield a zero equipotential line. This is shown in Figure 11.30.

The positive charge has a source of $\delta(\mathbf{r} - \mathbf{r}')$ at $\mathbf{r} = (x, y)$ and the negative charge is represented by the source $-\delta(\mathbf{r}^* - \mathbf{r}')$ at $\mathbf{r}^* = (x, -y)$. We construct the Green's functions at these two points and introduce a negative sign for the negative image source. Thus, we have

$$G(x, y, \xi, \eta) = \frac{1}{4\pi}\ln((\xi - x)^2 + (\eta - y)^2) - \frac{1}{4\pi}\ln((\xi - x)^2 + (\eta + y)^2).$$

Green's function for the half plane.

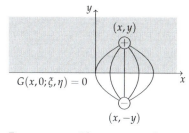

Figure 11.30: The source and image source for the Green's function for the half plane. Imagine two opposite charges forming a dipole. The electric field lines are depicted indicating that the electric potential, or Green's function, is constant along $y = 0$.

These functions satisfy the differential equation and the boundary condition

$$G(x,0,\xi,\eta) = \frac{1}{4\pi} \ln((\xi-x)^2 + (\eta)^2) - \frac{1}{4\pi} \ln((\xi-x)^2 + (\eta)^2) = 0.$$

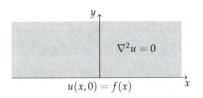

Figure 11.31:   This is the domain for a semi-infinite slab with boundary value $u(x,0) = f(x)$ and governed by Laplace's Equation.

Example 11.19. *Solve the homogeneous version of the problem; that is, solve Laplace's Equation on the half plane with a specified value on the boundary.*

We want to solve the problem

$$\nabla^2 u = 0, \quad in\ D,$$
$$u = g, \quad on\ C. \tag{11.176}$$

From the previous analysis, the solution takes the form

$$u(x,y) = \int_C g\nabla G \cdot \mathbf{n}\, ds = \int_C g \frac{\partial G}{\partial n}\, ds.$$

Because

$$G(x,y,\xi,\eta) = \frac{1}{4\pi} \ln((\xi-x)^2 + (\eta-y)^2) - \frac{1}{4\pi} \ln((\xi-x)^2 + (\eta+y)^2),$$

$$\frac{\partial G}{\partial n} = \frac{\partial G(x,y,\xi,\eta)}{\partial \eta}\bigg|_{\eta=0} = \frac{1}{\pi}\frac{y}{(\xi-x)^2 + y^2}.$$

We have arrived at the same surface Green's function as we had found in Example 8.11.2 and the solution is

$$u(x,y) = \frac{1}{\pi}\int_{-\infty}^{\infty} \frac{y}{(x-\xi)^2 + y^2} f(\xi)\, d\xi.$$

11.8   Green's Functions for Partial Differential Equations

11.8.1   Introduction

WE HAVE SEEN THROUGHOUT THE COURSE that Green's functions are the solutions of a differential equation representing the effect of a point impulse on either source terms, or initial and boundary conditions. The Green's function is obtained from transform methods or as an eigenfunction expansion. In the text we have occasionally rewritten solutions of differential equations in term's of Green's functions. We will first provide a few of these examples and then present a compilation of Green's functions for generic partial differential equations.

For example, in Section 5.13, we wrote the solution of the one-dimensional heat equation as

$$u(x,t) = \int_0^L G(x,\xi;t,0)f(\xi)\, d\xi,$$

where

$$G(x,\xi;t,0) = \frac{2}{L}\sum_{n=1}^{\infty} \sin\frac{n\pi x}{L} \sin\frac{n\pi\xi}{L} e^{\lambda_n kt},$$

and the solution of the wave equation as

$$u(x,t) = \int_0^L G_c(x,\xi,t,0)f(\xi)\,d\xi + \int_0^L G_s(x,\xi,t,0)g(\xi)\,d\xi,$$

where

$$G_c(x,\xi,t,0) = \frac{2}{L}\sum_{n=1}^{\infty} \sin\frac{n\pi x}{L}\sin\frac{n\pi\xi}{L}\cos\frac{n\pi ct}{L},$$

$$G_s(x,\xi,t,0) = \frac{2}{L}\sum_{n=1}^{\infty} \sin\frac{n\pi x}{L}\sin\frac{n\pi\xi}{L}\frac{\sin\frac{n\pi ct}{L}}{n\pi c/L}.$$

We note that setting $t = 0$ in $G_c(x,\xi;t,0)$, we obtain

$$G_c(x,\xi,0,0) = \frac{2}{L}\sum_{n=1}^{\infty} \sin\frac{n\pi x}{L}\sin\frac{n\pi\xi}{L}.$$

This is the Fourier sine series representation of the Dirac delta function, $\delta(x-\xi)$. Similarly, if we differentiate $G_s(x,\xi,t,0)$ with repsect to t and set $t = 0$, we once again obtain the Fourier sine series representation of the Dirac delta function.

It is also possible to find a closed form expression for Green's functions, which we had done for the heat equation on the infinite interval:

$$u(x,t) = \int_{-\infty}^{\infty} G(x,t;\xi,0)f(\xi)\,d\xi,$$

where

$$G(x,t;\xi,0) = \frac{e^{-(x-\xi)^2/4t}}{\sqrt{4\pi t}},$$

and for Poisson's Equation,

$$\phi(\mathbf{r}) = \int_V G(\mathbf{r},\mathbf{r}')f(\mathbf{r}')\,d^3r',$$

where the three-dimensional Green's function is given by

$$G(\mathbf{r},\mathbf{r}') = \frac{1}{|\mathbf{r}-\mathbf{r}'|}.$$

We can construct Green's functions for other problems that we have seen in this book. For example, the solution of the two-dimensional wave equation on a rectangular membrane was found in Equation (11.37) as

$$u(x,y,t) = \sum_{n=1}^{\infty}\sum_{m=1}^{\infty}(a_{nm}\cos\omega_{nm}t + b_{nm}\sin\omega_{nm}t)\sin\frac{n\pi x}{L}\sin\frac{m\pi y}{H},$$

$$\tag{11.177}$$

where

$$a_{nm} = \frac{4}{LH}\int_0^H\int_0^L f(x,y)\sin\frac{n\pi x}{L}\sin\frac{m\pi y}{H}\,dxdy, \tag{11.178}$$

$$b_{nm} = \frac{4}{\omega_{nm}LH}\int_0^H\int_0^L g(x,y)\sin\frac{n\pi x}{L}\sin\frac{m\pi y}{H}\,dxdy, \tag{11.179}$$

where the angular frequencies are given by

$$\omega_{nm} = c\sqrt{\left(\frac{n\pi}{L}\right)^2 + \left(\frac{m\pi}{H}\right)^2}. \tag{11.180}$$

Rearranging the solution, we have

$$u(x,y,t) = \int_0^H \int_0^L \left[G_c(x,y;\xi,\eta;t,0)f(\xi,\eta) + G_s(x,y;\xi,\eta;t,0)g(\xi,\eta)\right] d\xi d\eta,$$

where

$$G_c(x,y;\xi,\eta;t,0) = \frac{4}{LH} \sum_{n=1}^{\infty} \sum_{m=1}^{\infty} \sin\frac{n\pi x}{L} \sin\frac{n\pi\xi}{L} \sin\frac{m\pi y}{H} \sin\frac{m\pi\eta}{H} \cos\omega_{nm}t$$

and

$$G_s(x,y;\xi,\eta;t,0) = \frac{4}{LH} \sum_{n=1}^{\infty} \sum_{m=1}^{\infty} \sin\frac{n\pi x}{L} \sin\frac{n\pi\xi}{L} \sin\frac{m\pi y}{H} \sin\frac{m\pi\eta}{H} \frac{\sin\omega_{nm}t}{\omega_{nm}}.$$

Once again, we note that setting $t = 0$ in $G_c(x,\xi;t,0)$ and setting $t = 0$ in $\frac{\partial G_c(x,\xi;t,0)}{\partial t}$, we obtain a Fourier series representation of the Dirac delta function in two dimensions:

$$\delta(x-\xi)\delta(y-\eta) = \frac{4}{LH} \sum_{n=1}^{\infty} \sum_{m=1}^{\infty} \sin\frac{n\pi x}{L} \sin\frac{n\pi\xi}{L} \sin\frac{m\pi y}{H} \sin\frac{m\pi\eta}{H}.$$

Another example was the solution of the two-dimensional Laplace Equation on a disk given by Equation (11.87). We found that

$$u(r,\theta) = \frac{a_0}{2} + \sum_{n=1}^{\infty} \left(a_n \cos n\theta + b_n \sin n\theta\right) r^n. \tag{11.181}$$

$$a_n = \frac{1}{\pi a^n} \int_{-\pi}^{\pi} f(\theta) \cos n\theta \, d\theta, \quad n = 0, 1, \ldots, \tag{11.182}$$

$$b_n = \frac{1}{\pi a^n} \int_{-\pi}^{\pi} f(\theta) \sin n\theta \, d\theta \quad n = 1, 2 \ldots. \tag{11.183}$$

We saw that this solution can be written as

$$u(r,\theta) = \int_{-\pi}^{\pi} G(\theta,\phi;r,a)f(\phi) \, d\phi,$$

where the Green's function could be summed giving the Poisson kernel

$$G(\theta,\phi;r,a) = \frac{1}{2\pi} \frac{a^2 - r^2}{a^2 + r^2 - 2ar\cos(\theta - \phi)}.$$

We had also investigated the nonhomogeneous heat equation in Section 8.11.4,

$$u_t - ku_{xx} = h(x,t), \quad 0 \le x \le L, \quad t > 0.$$
$$u(0,t) = 0, \quad u(L,t) = 0, \quad t > 0,$$
$$u(x,0) = f(x), \quad 0 \le x \le . \tag{11.184}$$

We found that the solution of the heat equation is given by

$$u(x,t) = \int_0^L f(\xi)G(x,\xi;t,0)d\xi + \int_0^t \int_0^L h(\xi,\tau)G(x,\xi;t,\tau)\, d\xi d\tau,$$

where

$$G(x,\xi;t,\tau) = \frac{2}{L}\sum_{n=1}^{\infty} \sin\frac{n\pi x}{L} \sin\frac{n\pi\xi}{L}e^{-\omega_n^2(t-\tau)}.$$

Note that setting $t = \tau$, we again get a Fourier sine series representation of the Dirac delta function.

In general, Green's functions based on eigenfunction expansions over eigenfunctions of Sturm-Liouville eigenvalue problems are a common way to construct Green's functions. For example, surface and initial value Green's functions are constructed in terms of a modification of delta function representations modified by factors that make the Green's function a solution of the given differential equations and a factor taking into account the boundary or initial condition plus a restoration of the delta function when applied to the condition. Examples with an indication of these factors are shown below.

1. Surface Green's Function: Cube $[0,a] \times [0,b] \times [0,c]$:

$$g(x,y,z;x',y',c) = \sum_{\ell,n} \underbrace{\frac{2}{a}\sin\frac{\ell\pi x}{a} \sin\frac{\ell\pi x'}{a} \frac{2}{b}\sin\frac{n\pi y}{b} \sin\frac{n\pi y'}{b}}_{\delta-\text{function}} \left[\underbrace{\sinh\gamma_{\ell n}z}_{\text{D.E.}} / \underbrace{\sinh\gamma_{\ell n}c}_{\text{restore }\delta}\right].$$

2. Surface Green's Function: Sphere $[0,a] \times [0,\pi] \times [0,2\pi]$:

$$g(r,\phi,\theta;a,\phi',\theta') = \sum_{\ell,m} \underbrace{Y_\ell^{m*}(\psi'\,\theta')Y_\ell^{m*}(\psi\,\theta)}_{\delta-\text{function}} \left[\underbrace{r^\ell}_{\text{D.E.}} / \underbrace{a^\ell}_{\text{restore }\delta}\right].$$

3. Initial Value Green's Function: 1D Heat Equation on $[0,L]$, $k_n = \frac{n\pi}{L}$:

$$g(x,t;x',t_0) = \sum_n \underbrace{\frac{2}{L}\sin\frac{n\pi x}{L} \sin\frac{n\pi x'}{L}}_{\delta-\text{function}} \left[\underbrace{e^{-a^2 k_n^2 t}}_{\text{D.E.}} / \underbrace{e^{-a^2 k_n^2 t_0}}_{\text{restore }\delta}\right].$$

4. Initial Value Green's Function: 1D Heat Equation on infinite domain:

$$g(x,t;x',0) = \underbrace{\frac{1}{2\pi}\int_{-\infty}^{\infty} dk e^{ik(x-x')}}_{\delta-\text{function}} \underbrace{e^{-a^2 k^2 t}}_{\text{D.E.}} = \frac{e^{-(x-x')^2/4a^2 t}}{\sqrt{4\pi a^2 t}}.$$

We can extend this analysis to a more general theory of Green's functions. This theory is based upon Green's Theorems, or identities.

1. Green's First Theorem:

$$\oint_S \varphi\nabla\chi \cdot \hat{\mathbf{n}}\, dS = \int_V (\nabla\varphi \cdot \nabla\chi + \varphi\nabla^2\chi)\, dV.$$

This is easily proven starting with the identity

$$\nabla \cdot (\varphi \nabla \chi) = \nabla \varphi \cdot \nabla \chi + \varphi \nabla^2 \chi,$$

integrating over a volume of space and using Gauss' Integral Theorem.

2. Green's Second Theorem:

$$\int_V (\varphi \nabla^2 \chi - \chi \nabla^2 \varphi) \, dV = \oint_S (\varphi \nabla \chi - \chi \nabla \varphi) \cdot \hat{\mathbf{n}} \, dS.$$

This is proven by interchanging φ and χ in the first theorem and subtracting the two versions of the theorem.

The next step is to let $\varphi = u$ and $\chi = G$. Then,

$$\int_V (u \nabla^2 G - G \nabla^2 u) \, dV = \oint_S (u \nabla G - G \nabla u) \cdot \hat{\mathbf{n}} \, dS.$$

As we had seen earlier for Poisson's Equation, inserting the differential equation yields

$$u(x, y) = \int_V Gf \, dV + \oint_S (u \nabla G - G \nabla u) \cdot \hat{\mathbf{n}} \, dS.$$

If we have the Green's function, we only need to know the source term and boundary conditions in order to obtain the solution to a given problem.

In the next sections we provide a summary of these ideas as applied to some generic partial differential equations.[5]

11.8.2   Laplace's Equation: $\nabla^2 \psi = 0$.

1. **Boundary Conditions:**

   (a) *Dirichlet* – ψ is given on the surface.

   (b) *Neumann* – $\hat{\mathbf{n}} \cdot \nabla \psi = \frac{\partial \psi}{\partial n}$ is given on the surface.

   Note: Boundary conditions can be Dirichlet on part of the surface and Neumann on part. If they are Neumann on the whole surface, then the Divergence Theorem requires the constraint

   $$\int \frac{\partial \psi}{\partial n} \, dS = 0.$$

2. **Solution by Surface Green's Function, $g(\vec{\mathbf{r}}, \vec{\mathbf{r}}')$:**

   (a) Dirichlet conditions:

   $$\nabla^2 g_D(\vec{\mathbf{r}}, \vec{\mathbf{r}}') = 0,$$
   $$g_D(\vec{\mathbf{r}}_s, \vec{\mathbf{r}}_s') = \delta^{(2)}(\vec{\mathbf{r}}_s - \vec{\mathbf{r}}_s'),$$
   $$\psi(\vec{\mathbf{r}}) = \int g_D(\vec{\mathbf{r}}, \vec{\mathbf{r}}_s') \psi(\vec{\mathbf{r}}_s') \, dS'.$$

   (b) Neumann conditions:

   $$\nabla^2 g_N(\vec{\mathbf{r}}, \vec{\mathbf{r}}') = 0,$$
   $$\frac{\partial g_N}{\partial n}(\vec{\mathbf{r}}_s, \vec{\mathbf{r}}_s') = \delta^{(2)}(\vec{\mathbf{r}}_s - \vec{\mathbf{r}}_s'),$$
   $$\psi(\vec{\mathbf{r}}) = \int g_N(\vec{\mathbf{r}}, \vec{\mathbf{r}}_s') \frac{\partial \psi}{\partial n}(\vec{\mathbf{r}}_s') \, dS'.$$

   Note: Use of g is readily generalized to any number of dimensions.

11.8.3 Homogeneous Time-Dependent Equations

1. **Typical Equations:**

   (a) Diffusion/Heat Equation $\nabla^2\Psi = \frac{1}{a^2}\frac{\partial}{\partial t}\Psi$.

   (b) Schrödinger Equation $-\nabla^2\Psi + U\Psi = i\frac{\partial}{\partial t}\Psi$.

   (c) Wave Equation $\nabla^2\Psi = \frac{1}{c^2}\frac{\partial^2}{\partial t^2}\Psi$.

   (d) General form: $\mathcal{D}\Psi = \mathcal{T}\Psi$.

2. **Initial Value Green's Function,** $g(\vec{r},\vec{r}';t,t')$:

   (a) **Homogeneous Boundary Conditions**

      i. Diffusion, or Schrödinger Equation (First-order in time), $\mathcal{D}g = \mathcal{T}g$.

      $$\Psi(\vec{r},t) = \int g(\vec{r},\vec{r}';t,t_0)\Psi(\mathbf{r}',t_0)\,d^3\mathbf{r}',$$

      where

      $$g(\mathbf{r},\mathbf{r}';t_0,t_0) = \delta(\mathbf{r}-\mathbf{r}'),$$

      $g(\mathbf{r}_s)$ satisfies homogeneous boundary conditions.

      ii. Wave Equation

      $$\Psi(\mathbf{r},t) = \int[g_c(\mathbf{r},\mathbf{r}';t,t_0)\Psi(\mathbf{r}',t_0) + g_s(\mathbf{r},\mathbf{r}';t,t_0)\dot{\Psi}(\mathbf{r}',t_0)]d^3\mathbf{r}'.$$

      The first two properties in (a) above hold, but

      $$g_c(\mathbf{r},\mathbf{r}';t_0,t_0) = \delta(\mathbf{r}-\mathbf{r}')$$

      $$\dot{g}_s(\mathbf{r},\mathbf{r}';t_0,t_0) = \delta(\mathbf{r}-\mathbf{r}')$$

      Note: For the diffusion and Schrödinger equations the initial condition is Dirichlet in time. For the wave equation the initial condition is Cauchy, where Ψ and $\dot{\Psi}$ are given.

   (b) **Inhomogeneous, Time Independent (steady) Boundary Conditions (B.C.'s)**

      i. Solve Laplace's Equation, $\nabla^2\psi_s = 0$, for inhomogeneous B.C.'s

      ii. Solve homogeneous, time-dependent equation for

      $$\Psi_t(\mathbf{r},t) \text{ satisfying } \Psi_t(\mathbf{r},t_0) = \Psi(\mathbf{r},t_0) - \psi_s(\mathbf{r}).$$

      iii. Then $\Psi(\mathbf{r},t) = \Psi_t(\mathbf{r},t) + \psi_s(\mathbf{r})$.

      Note: Ψ_t is the *transient part* and ψ_s is the *steady-state part*.

3. **Time Dependent Boundary Conditions with Homogeneous Initial Conditions (I.C.'s)**

   (a) Use the Boundary Value Green's function, $h(\mathbf{r},\mathbf{r}'_s;t,t')$, which is similar to the surface Green's function in a previous section:

      $$\Psi(\mathbf{r},t) = \int_{t_0}^{\infty} h_D(\mathbf{r},\mathbf{r}'_s;t,t')\Psi(\mathbf{r}'_s,t')\,dt',$$

      or

      $$\Psi(\mathbf{r},t) = \int_{t_0}^{\infty} \frac{\partial h_N}{\partial n}(\mathbf{r},\mathbf{r}'_s;t,t')\Psi(\mathbf{r}'_s,t')\,dt'.$$

(b) Properties of $h(\mathbf{r}, \mathbf{r}'_s; t, t')$:

$$\mathcal{D}h = \mathcal{T}h$$

$$h_D(\mathbf{r}_s, \mathbf{r}'_s; t, t') = \delta(t - t'), \quad \text{or} \frac{\partial h_N}{\partial n}(\mathbf{r}_s, \mathbf{r}'_s; t, t') = \delta(t - t'),$$

$$h(\mathbf{r}, \mathbf{r}'_s; t, t') = 0, \quad t' > t, \text{ (causality).}$$

(c) **Note:** For inhomogeneous I.C.,

$$\Psi = \int g\Psi(\mathbf{r}', t_0) + \int dt' h_D \Psi(\mathbf{r}'_s, t') \, d^3\mathbf{r}'.$$

11.8.4 Inhomogeneous Steady-State Equation

1. **Poisson's Equation:**

$$\nabla^2 \psi(\mathbf{r}, t) = f(\mathbf{r}), \qquad \psi(\mathbf{r}_s) \quad \text{or} \quad \frac{\partial \psi}{\partial n}(\mathbf{r}_s) \quad \text{given.}$$

(a) Green's Theorem:

$$\int [\psi(\mathbf{r}')\nabla'^2 G(\mathbf{r}, \mathbf{r}') - G(\mathbf{r}, \mathbf{r}')\nabla'^2 \psi(\mathbf{r}')] \, d^3\mathbf{r}'$$

$$= \int [\psi(\mathbf{r}')\nabla' G(\mathbf{r}, \mathbf{r}') - G(\mathbf{r}, \mathbf{r}')\nabla' \psi(\mathbf{r}')] \cdot d\vec{S}',$$

where ∇' denotes differentiation with respect to r'.

(b) Properties of $G(\mathbf{r}, \mathbf{r}')$:

   i. $\nabla'^2 G(\mathbf{r}, \mathbf{r}') = \delta(\mathbf{r} - \mathbf{r}')$.

   ii. $G|_s = 0$ or $\frac{\partial G}{\partial n'}|_s = 0$.

   iii. Solution:

$$\psi(\mathbf{r}) = \int G(\mathbf{r}, \mathbf{r}') f(\mathbf{r}') \, d^3\mathbf{r}'$$

$$+ \int [\psi(\mathbf{r}')\nabla' G(\mathbf{r}, \mathbf{r}') - G(\mathbf{r}, \mathbf{r}')\nabla' \psi(\mathbf{r}')] \cdot d\vec{S}'.$$

$$(11.185)$$

(c) For the case of pure Neumann B.C.'s, the Divergence Theorem leads to the constraint

$$\int \nabla \psi \cdot d\vec{S} = \int f \, d^3\mathbf{r}.$$

If there are pure Neumann conditions and S is finite and $\int f \, d^3\mathbf{r} \neq 0$ by symmetry, then $\vec{\hat{n}}' \cdot \nabla' G|_s \neq 0$ and the Green's function method is much more complicated to solve.

(d) From the above result:

$$\vec{\hat{n}}' \cdot \nabla' G(\mathbf{r}, \mathbf{r}'_s) = g_D(\mathbf{r}, \mathbf{r}'_s)$$

or

$$G_N(\mathbf{r}, \mathbf{r}'_s) = -g_N(\mathbf{r}, \mathbf{r}'_s).$$

It is often simpler to use G for $\int d^3\mathbf{r}'$ and g for $\int d\vec{S}'$, separately.

(e) G satisfies a reciprocity property, $G(\mathbf{r}, \mathbf{r}') = G(\mathbf{r}', \mathbf{r})$ for either Dirichlet or Neumann boundary conditions.

(f) $G(\mathbf{r}, \mathbf{r}')$ can be considered as a potential at $\mathbf{r}$ due to a point charge $q = -1/4\pi$ at $\mathbf{r}'$, with all surfaces being grounded conductors.

11.8.5 Inhomogeneous, Time-Dependent Equations

1. **Diffusion/Heat Flow** $\nabla^2 \Psi - \frac{1}{a^2}\dot{\Psi} = f(\mathbf{r}, t)$:

    (a)

$$[\nabla^2 - \frac{1}{a^2}\frac{\partial}{\partial t}]G(\mathbf{r}, \mathbf{r}'; t, t') = [\nabla'^2 + \frac{1}{a^2}\frac{\partial}{\partial t'}]G(\mathbf{r}, \mathbf{r}'; t, t')$$
$$= \delta(\mathbf{r} - \mathbf{r}')\delta(t - t'). \quad (11.186)$$

    (b) Green's Theorem in four dimensions $(\mathbf{r}, t)$ yields

$$\Psi(\mathbf{r}, t) = \int\int_{t_0}^{\infty} G(\mathbf{r}, \mathbf{r}'; t, t') f(\mathbf{r}', t')\, dt'd^3\mathbf{r}' - \frac{1}{a^2}\int G(\mathbf{r}, \mathbf{r}'; t, t_0)\Psi(\mathbf{r}', t_0)\, d^3\mathbf{r}'$$
$$+ \int_{t_0}^{\infty}\int [\Psi(\mathbf{r}'_s, t)\nabla' G_D(\mathbf{r}, \mathbf{r}'_s; t, t') - G_N(\mathbf{r}, \mathbf{r}'_s; t, t')\nabla'\Psi(\mathbf{r}'_s, t')] \cdot \vec{dS}' dt'.$$

    (c) Either $G_D(\mathbf{r}'_s) = 0$ or $G_N(\mathbf{r}'_s) = 0$ on S at any point $\mathbf{r}'_s$.

    (d) $\hat{\mathbf{n}}' \cdot \nabla' G_D(\mathbf{r}'_s) = h_D(\mathbf{r}'_s)$, $G_N(\mathbf{r}'_s) = -h_N(\mathbf{r}'_s)$, and $-\frac{1}{a^2}G(\mathbf{r}, \mathbf{r}'; t, t_0) = g(\mathbf{r}, \mathbf{r}'; t, t_0)$.

2. **Wave Equation** $\nabla^2 \Psi - \frac{1}{c^2}\frac{\partial^2 \Psi}{\partial^2 t} = f(\mathbf{r}, t)$:

    (a)

$$[\nabla^2 - \frac{1}{c^2}\frac{\partial^2}{\partial t^2}]G(\mathbf{r}, \mathbf{r}'; t, t') = [\nabla'^2 - \frac{1}{c^2}\frac{\partial^2}{\partial t^2}]G(\mathbf{r}, \mathbf{r}'; t, t')$$
$$= \delta(\mathbf{r} - \mathbf{r}')\delta(t - t'). \quad (11.187)$$

    (b) Green's Theorem in four dimensions $(\mathbf{r}, t)$ yields

$$\Psi(\mathbf{r}, t) = \int\int_{t_0}^{\infty} G(\mathbf{r}, \mathbf{r}'; t, t') f(\mathbf{r}', t')\, dt'd^3\mathbf{r}'$$
$$- \frac{1}{c^2}\int [G(\mathbf{r}, \mathbf{r}'; t, t_0)\frac{\partial}{\partial t'}\Psi(\mathbf{r}', t_0) - \Psi(\mathbf{r}', t_0)\frac{\partial}{\partial t'}G(\mathbf{r}, \mathbf{r}'; t, t_0)]\, d^3\mathbf{r}'$$
$$+ \int_{t_0}^{\infty}\int [\Psi(\mathbf{r}'_s, t)\nabla' G_D(\mathbf{r}, \mathbf{r}'_s; t, t') - G_N(\mathbf{r}, \mathbf{r}'_s; t, t')\nabla'\psi(\mathbf{r}'_s, t')] \cdot \vec{dS}' dt'.$$

    (c) Cauchy initial conditions are given: $\Psi(t_0)$ and $\dot{\Psi}(t_0)$.

    (d) The wave and diffusion equations satisfy a causality condition $G(t, t') = 0$, $\quad t' > t$.

Problems

1. A rectangular plate $0 \le x \le L$ $0 \le y \le H$ with heat diffusivity constant k is insulated on the edges $y = 0, H$ and is kept at constant zero temperature on the other two edges. Assuming an initial temperature of $u(x, y, 0) = f(x, y)$, use separation of variables to find the general solution.

2. Solve the following problem:

$$u_{xx} + u_{yy} + u_{zz} = 0, \quad 0 < x < 2\pi, \quad 0 < y < \pi, \quad 0 < z < 1,$$
$$u(x,y,0) = \sin x \sin y, \quad u(x,y,z) = 0 \text{ on other faces.}$$

3. Consider Laplace's Equation on the unit square, $u_{xx} + u_{yy} = 0$, $0 \le x, y \le 1$. Let $u(0,y) = 0, u(1,y) = 0$ for $0 < y < 1$ and $u_y(x,0) = 0$ for $0 < y < 1$. Carry out the needed separation of variables and write the product solutions satisfying these boundary conditions.

4. Consider a cylinder of height H and radius a.

    a. Write Laplace's Equation for this cylinder in cylindrical coordinates.

    b. Carry out the separation of variables and obtain the three ordinary differential equations that result from this problem.

    c. What kind of boundary conditions could be satisfied in this problem in the independent variables?

5. Consider a square drum of side s and a circular drum of radius a.

    a. Rank the modes corresponding to the first six frequencies for each.

    b. Write each frequency (in Hz) in terms of the fundamental (i.e., the lowest frequency.)

    c. What must the lengths of the sides of the square drum be to have the same fundamental frequency? (Assume that $c = 1.0$ for each one.)

6. We presented the full solution of the vibrating rectangular membrane in Equation (11.37). Finish the solution to the vibrating circular membrane by writing out a similar full solution.

7. A copper cube 10.0 cm on a side is heated to $100°C$. The block is placed on a surface that is kept at $0°C$. The sides of the block are insulated, so the normal derivatives on the sides are zero. Heat flows from the top of the block to the air governed by the gradient $u_z = -10°C/m$. Determine the temperature of the block at its center after 1.0 minutes. Note that the thermal diffusivity is given by $k = \frac{K}{\rho c_p}$, where K is the thermal conductivity, ρ is the density, and c_p is the specific heat capacity.

8. Consider a spherical balloon of radius a. Small deformations on the surface can produce waves on the balloon's surface.

    a. Write the wave equation in spherical polar coordinates. (Note: ρ is constant!)

    b. Carry out a separation of variables and find the product solutions for this problem.

    c. Describe the nodal curves for the first six modes.

    d. For each mode determine the frequency of oscillation (in Hz) assuming $c = 1.0$ m/s and $a = 0.02$ m.

9. Consider a circular cylinder of radius $R = 4.00$ cm and height $H = 20.0$ cm that obeys the steady-state heat equation

$$u_{rr} + \frac{1}{r}u_r + u_{zz}.$$

Find the temperature distribution, $u(r,z)$, given that $u(r,0°) = 0°C$, $u(r,20°) = 20°C$, and heat is lost through the sides due to Newton's Law of Cooling:

$$[u_r + hu]_{r=4} = 0,$$

for $h = 1.0$ cm^{-1}.

10. The spherical surface of a homogeneous ball of radius 1.0 is maintained at zero temperature. It has an initial temperature distribution $u(\rho,0) = 100°C$. Assuming a heat diffusivity constant k, find the temperature throughout the sphere, $u(\rho,\theta,\phi,t)$.

11. Determine the steady-state temperature of a spherical ball maintained at the temperature

$$u(x,y,z) = x^2 + 2y^2 + 3z^2, \quad \rho = 1.$$

[Hint: Rewrite the problem in spherical coordinates and use the properties of spherical harmonics.]

12. Find the Green's function for the homogeneous fixed values on the boundary of the quarter plane $x > 0$, $y > 0$, for Poisson's equation using the infinite plane Green's function for Poisson's Equation. Use the method of images.

13. Find the Green's function for the one-dimensional heat equation with boundary conditions $u(0,t) = 0$, $u_x(L,t)$, $t > 0$.

14. Consider Laplace's Equation on the rectangular plate in Figure 11.8. Construct the Green's function for this problem.

15. Construct the Green's function for Laplace's Equation in the spherical domain in Figure 11.18.

Appendix
Review of Sequences and Infinite Series

"Once you eliminate the impossible, whatever remains, no matter how improbable, must be the truth."—Sherlock Holmes (by Sir Arthur Conan Doyle, 1859–1930)

IN THIS APPENDIX WE WILL REVIEW and extend some of the concepts and definitions related to infinite series that you might have seen previously in your calculus class [1] [2] [3]. Working with infinite series can be a little tricky and we need to understand some of the basics before moving on to the study of series of trigonometric functions.

For example, one can show that the infinite series

$$S = 1 - \frac{1}{2} + \frac{1}{3} - \frac{1}{4} + \frac{1}{5} - \cdots$$

converges to $\ln 2$. However, the terms can be rearranged to give

$$1 + \left(\frac{1}{3} - \frac{1}{2} + \frac{1}{5}\right) + \left(\frac{1}{7} - \frac{1}{4} + \frac{1}{9}\right) + \left(\frac{1}{11} - \frac{1}{6} + \frac{1}{13}\right) + \cdots = \frac{3}{2}\ln 2.$$

In fact, other rearrangements can be made to give any desired sum!

Other problems with infinite series can occur. Try to sum the following infinite series to find that

$$\sum_{k=2}^{\infty} \frac{\ln k}{k^2} \sim 0.937548\ldots.$$

A sum of even as many as a million terms only gives convergence to four or five decimal places.

The series

$$\frac{1}{x} - \frac{1}{x^2} + \frac{2!}{x^3} - \frac{3!}{x^4} + \frac{4!}{x^5} - \cdots, \quad x > 0$$

diverges for all x. So, you might think this divergent series is useless. However, truncation of this divergent series leads to an approximation of the integral

$$\int_0^{\infty} \frac{e^{-t}}{x+t}\, dt, \quad x > 0.$$

So, can we make sense out of any of these, or other manipulations, of infinite series? We will not answer all these questions, but we will go back and review what you have seen in your calculus classes.

The material in this appendix is a review of material covered in a standard course in calculus with some additional notions from advanced calculus. It is provided as a review before encountering the notion of Fourier series and their convergence as seen in Chapter 4.

[1] G. B. Thomas and R. L. Finney. *Calculus and Analytic Geometry*. Addison-Wesley Press, Cambridge, MA, ninth edition, 1995; and

[2] J. Stewart. *Calculus: Early Transcendentals*. Brooks Cole, sixth edition, 2007; and

[3] K. Kaplan. *Advanced Calculus*. Addison Wesley Publishing Company, Reading, MA fourth edition, 1991

As we will see,

$$\ln(1+x) = x - \frac{x}{2} + \frac{x}{3} - \cdots.$$

So, inserting $x = 1$ yields the first result—at least formally! It was shown in Cowen, Davidson, and Kaufman (in *The American Mathematical Monthly*, Vol. 87, No. 10. (Dec., 1980), pp. 817–819) that expressions like

$$\begin{aligned} f(x) &= \frac{1}{2}\left[\ln\frac{1+x}{1-x} + \ln(1-x^4)\right] \\ &= \frac{1}{2}\ln\left[(1+x)^2(1+x^2)\right] \end{aligned}$$

lead to alternate sums of the rearrangement of the alternating harmonic series. See Problem 6.

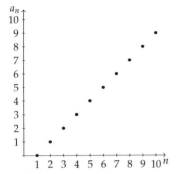

Figure A.1: Plot of the terms of the sequence $a_n = n - 1$, $n = 1, 2, \ldots, 10$.

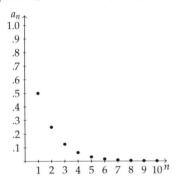

Figure A.2: Plot of the terms of the sequence $a_n = \dfrac{1}{2^n}$, for $n = 1, 2, \ldots, 10$.

[4] Leonardo Pisano Fibonacci (c.1170–c.1250) is best known for this sequence of numbers. This sequence is the solution of a problem in one of his books: *A certain man put a pair of rabbits in a place surrounded on all sides by a wall. How many pairs of rabbits can be produced from that pair in a year if it is supposed that every month each pair begets a new pair which from the second month on becomes productive* http://www-history.mcs.st-and.ac.uk

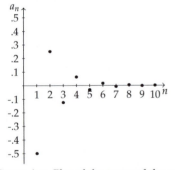

Figure A.3: Plot of the terms of the sequence $a_n = \dfrac{(-1)^n}{2^n}$, for $n = 1, 2, \ldots, 10$.

A.1   Sequences of Real Numbers

WE BEGIN WITH THE DEFINITIONS for sequences and series of numbers. A sequence is a function whose domain is the set of positive integers, $a(n)$, $n \in N$ [$N = \{1, 2, \ldots.\}$].

Examples are

1. $a(n) = n$ yields the sequence $\{1, 2, 3, 4, 5, \ldots\}$,

2. $a(n) = 3n$ yields the sequence $\{3, 6, 9, 12, \ldots\}$.

However, one typically uses subscript notation and not functional notation: $a_n = a(n)$. We then call a_n the nth term of the sequence. Furthermore, we will denote sequences by $\{a_n\}_{n=1}^{\infty}$. Sometimes we will only give the nth term of the sequence and will assume that $n \in N$ unless otherwise noted.

Another way to define a particular sequence is recursively. A recursive sequence is defined in two steps:

1. The value of first term (or first few terms) is given.

2. A rule, or recursion formula, to determine later terms from earlier ones is given.

Example A.1. *A typical example is given by the Fibonacci*[4] *sequence. It can be defined by the recursion formula $a_{n+1} = a_n + a_{n-1}$, $n \geq 2$ and the starting values of $a_1 = 0$ and $a_1 = 1$. The resulting sequence is $\{a_n\}_{n=1}^{\infty} = \{0, 1, 1, 2, 3, 5, 8, \ldots\}$. Writing the general expression for the nth term is possible, but it is not as simply stated. Recursive definitions are often useful in doing computations for large values of n.*

A.2   Convergence of Sequences

NEXT WE ARE INTERESTED IN THE BEHAVIOR OF SEQUENCES as n gets large. For the sequence defined by $a_n = n - 1$, we find the behavior as shown in Figure A.1. Notice that as n gets large, a_n also gets large. This sequence is said to be divergent.

On the other hand, the sequence defined by $a_n = \frac{1}{2^n}$ approaches a limit as n gets large. This is depicted in Figure A.2. Another related series, $a_n = \frac{(-1)^n}{2^n}$, is shown in Figure A.3 and it is also seen to approach 0. The latter sequence is called an alternating sequence because the signs alternate from term to term. The terms in the sequence are $\{-\frac{1}{2}, \frac{1}{4}, -\frac{1}{8}, \ldots\}$

The last two sequences are said to converge. In general, a sequence a_n converges to the number L if to every positive number ϵ there corresponds an integer N such that for all n,

$$n > N \Rightarrow |a - L| < \epsilon.$$

If no such number exists, then the sequence is said to diverge.

In Figures A.4 and A.5 we see what this means. For the sequence given by $a_n = \frac{(-1)^n}{2^n}$, we see that the terms approach $L = 0$. Given an $\epsilon > 0$, we ask for what value of N the nth terms ($n > N$) lie in the interval $[L - \epsilon, L + \epsilon]$. In these figures, this interval is depicted by a horizontal band. We see that for convergence, sooner, or later, the tail of the sequence ends up entirely within this band.

If a sequence $\{a_n\}_{n=1}^{\infty}$ converges to a limit L, then we write either $a_n \to L$ as $n \to \infty$ or $\lim_{n \to \infty} a_n = L$. For example, we have already seen in Figure A.3 that $\lim_{n \to \infty} \frac{(-1)^n}{2^n} = 0$.

A.3   Limit Theorems

ONCE WE HAVE DEFINED THE NOTION of convergence of a sequence to some limit, then we can investigate the properties of the limits of sequences. Here we list a few general limit theorems and some special limits, which arise often.

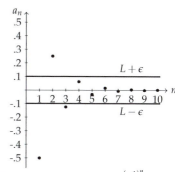

Figure A.4: Plot of $a_n = \frac{(-1)^n}{2^n}$ for $n = 1 \ldots 10$. Picking $\epsilon = 0.1$, one sees that the tail of the sequence lies between $L + \epsilon$ and $L - \epsilon$ for $n > 3$.

Limit Theorem

Theorem A.1. *Consider two convergent sequences $\{a_n\}$ and $\{b_n\}$ and a number k. Assume that $\lim_{n \to \infty} a_n = A$ and $\lim_{n \to \infty} b_n = B$. Then we have*

1. $\lim_{n \to \infty}(a_n \pm b_n) = A \pm B$.

2. $\lim_{n \to \infty}(k b_n) = kB$.

3. $\lim_{n \to \infty}(a_n b_n) = AB$.

4. $\lim_{n \to \infty} \frac{a_n}{b_n} = \frac{A}{B}, \quad B \neq 0$.

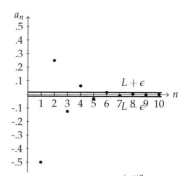

Figure A.5: Plot of $a_n = \frac{(-1)^n}{2^n}$ for $n = 1 \ldots 10$. Picking $\epsilon = 0.015$, one sees that the tail of the sequence lies between $L + \epsilon$ and $L - \epsilon$ for $n > 4$.

Some special limits are given next. These are generally first encountered in a second course in calculus.

Special Limits

Theorem A.2. *The following are special cases:*

1. $\lim_{n \to \infty} \frac{\ln n}{n} = 0$.

2. $\lim_{n \to \infty} n^{\frac{1}{n}} = 1$.

3. $\lim_{n \to \infty} x^{\frac{1}{n}} = 1, \quad x > 0$.

4. $\lim_{n \to \infty} x^n = 0, \quad |x| < 1$.

5. $\lim_{n \to \infty} \left(1 + \frac{x}{n}\right)^n = e^x$.

6. $\lim_{n \to \infty} \frac{x^n}{n!} = 0$.

The proofs generally are straightforward. For example, one can prove the first limit by first realizing that $\lim_{n \to \infty} \frac{\ln n}{n} = \lim_{x \to \infty} \frac{\ln x}{x}$. This limit in its current form is indeterminate as x gets large ($x \to \infty$) because the

L'Hopital's Rule is used often in computing limits. We recall this powerful rule here as a reference for the reader.

Theorem A.3. *Let c be a finite number or $c = \infty$. If $\lim_{x \to c} f(x) = 0$ and $\lim_{x \to c} g(x) = 0$, then*

$$\lim_{x \to c} \frac{f(x)}{g(x)} = \lim_{x \to c} \frac{f'(x)}{g'(x)}.$$

If $\lim_{x \to c} f(x) = \infty$ and $\lim_{x \to c} g(x) = \infty$, then

$$\lim_{x \to c} \frac{f(x)}{g(x)} = \lim_{x \to c} \frac{f'(x)}{g'(x)}.$$

[5] We should note that we are assuming something about limits of composite functions. Let a and b be real numbers. Suppose f and g are continuous functions, $\lim_{x \to a} f(x) = f(a)$ and $\lim_{x \to b} g(x) = b$, and $g(b) = a$. Then,

$$\begin{aligned} \lim_{x \to b} f(g(x)) &= f\left(\lim_{x \to b} g(x) \right) \\ &= f(g(b)) = f(a). \end{aligned}$$

numerator and the denominator get large for large x. In such cases, one employs L'Hopital's Rule. We find that

$$\lim_{x \to \infty} \frac{\ln x}{x} = \lim_{x \to \infty} \frac{1/x}{1} = 0.$$

The second limit in Theorem A.2 can be proven by first looking at

$$\lim_{n \to \infty} \ln n^{1/n} = \lim_{n \to \infty} \frac{1}{n} \ln n = 0$$

from the previous limit case. Now, if $\lim_{n \to \infty} \ln f(n) = 0$, then $\lim_{n \to \infty} f(n) = e^0 = 1$. Thus proving the second limit.[5]

The third limit can be done similarly. The reader is left to confirm the other limits. We finish this section with a few selected examples.

Example A.2. *Evaluate $\lim_{n \to \infty} \frac{n^2 + 2n + 3}{n^3 + n}$.*
Divide the numerator and denominator by n^2. Then,

$$\lim_{n \to \infty} \frac{n^2 + 2n + 3}{n^3 + n} = \lim_{n \to \infty} \frac{1 + \frac{2}{n} + \frac{3}{n^2}}{n + \frac{1}{n}} = \lim_{n \to \infty} \frac{1}{n} = 0.$$

Another approach to this type of problem is to consider the behavior of the numerator and denominator as $n \to \infty$. As n gets large, the numerator behaves like n^2, as $2n + 3$ becomes negligible for large enough n. Similarly, the denominator behaves like n^3 for large n. Thus,

$$\lim_{n \to \infty} \frac{n^2 + 2n + 3}{n^3 + n} = \lim_{n \to \infty} \frac{n^2}{n^3} = 0.$$

Example A.3. *Evaluate $\lim_{n \to \infty} \frac{\ln n^2}{n}$.*
Rewriting $\frac{\ln n^2}{n} = \frac{2 \ln n}{n}$, we find from identity 1 of Theorem A.2 that

$$\lim_{n \to \infty} \frac{\ln n^2}{n} = 2 \lim_{n \to \infty} \frac{\ln n}{n} = 0.$$

Example A.4. *Evaluate $\lim_{n \to \infty} (n^2)^{\frac{1}{n}}$.*
To compute this limit, we rewrite

$$\lim_{n \to \infty} (n^2)^{\frac{1}{n}} = \lim_{n \to \infty} (n)^{\frac{1}{n}} (n)^{\frac{1}{n}} = 1,$$

using identity 2 of Theorem A.2.

Example A.5. *Evaluate $\lim_{n \to \infty} (\frac{n-2}{n})^n$.*
This limit can be written as

$$\lim_{n \to \infty} \left(\frac{n-2}{n} \right)^n = \lim_{n \to \infty} \left(1 + \frac{(-2)}{n} \right)^n = e^{-2}.$$

Here we used identity 5 of Theorem A.2.

A.4 Infinite Series

IN THIS SECTION WE INVESTIGATE the meaning of infinite series, which are infinite sums of the form

$$a_1 + a_2 + a_2 + \ldots . \tag{A.1}$$

A typical example is the infinite series

$$1 + \frac{1}{2} + \frac{1}{4} + \frac{1}{8} + \ldots . \tag{A.2}$$

How would one evaluate this sum? We begin by just adding the terms. For example,

$$
\begin{aligned}
1 + \frac{1}{2} &= \frac{3}{2}, \\
1 + \frac{1}{2} + \frac{1}{4} &= \frac{7}{4}, \\
1 + \frac{1}{2} + \frac{1}{4} + \frac{1}{8} &= \frac{15}{8}, \\
1 + \frac{1}{2} + \frac{1}{4} + \frac{1}{8} + \frac{1}{16} &= \frac{31}{16}, \ldots .
\end{aligned}
\tag{A.3}
$$

The values tend to a limit. We can see this graphically in Figure A.6.

In general, we want to make sense out of Equation (A.1). As with the example, we look at a sequence of partial sums . Thus, we consider the sums

$$
\begin{aligned}
s_1 &= a_1, \\
s_2 &= a_1 + a_2, \\
s_3 &= a_1 + a_2 + a_3, \\
s_4 &= a_1 + a_2 + a_3 + a_4, \ldots .
\end{aligned}
\tag{A.4}
$$

In general, we define the nth partial sum as

$$s_n = a_1 + a_2 + \ldots + a_n.$$

If the infinite series (A.1) is to make any sense, then the sequence of partial sums should converge to some limit. We define this limit to be the sum of the infinite series, $S = \lim_{n \to \infty} s_n$. If the sequence of partial sums converges to the limit L as n gets large, then the infinite series is said to have the sum L.

We will use the compact summation notation

$$\sum_{n=1}^{\infty} a_n = a_1 + a_2 + \ldots + a_n + \ldots .$$

Here, n will be referred to as the index and it may start at values other than $n = 1$.

There is story described in E.T. Bell's "Men of Mathematics" about Carl Friedrich Gauß (1777–1855). Gauß' third grade teacher needed to occupy the students, so she asked the class to sum the first 100 integers thinking that this would occupy the students for a while. However, Gauß was able to do so in practically no time. He recognized the sum could be written as $(1 + 100) + (2 + 99) + \ldots (50 + 51) = 50(101)$. This sum is a special case of

$$\sum_{k=1}^{n} k = \frac{n(n+1)}{2}.$$

This is an example of an arithmetic progression that is a finite sum of terms.

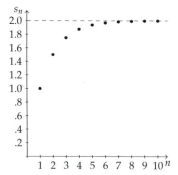

Figure A.6: Plot of $s_n = \sum_{k=1}^{n} \frac{1}{2^{k-1}}$ for $n = 1 \ldots 10$.

A.5   Convergence Tests

GIVEN A GENERAL INFINITE SERIES, it would be nice to know if it converges, or not. Often, we are only interested in the convergence and not the actual sum as it is often difficult to determine the sum even if the series converges. In this section we will review some of the standard tests for convergence, which you should have seen in Calculus II.

First, we have the nth Term Divergence Test. This is motivated by two examples:

1. $\sum_{n=0}^{\infty} 2^n = 1 + 2 + 4 + 8 + \dots$.
2. $\sum_{n=1}^{\infty} \frac{n+1}{n} = \frac{2}{1} + \frac{3}{2} + \frac{4}{3} + \dots$.

In the first example, it is easy to see that each term is getting larger and larger, and thus the partial sums will grow without bound. In the second case, each term is bigger than one. Thus, the series will be bigger than adding the same number of ones as there are terms in the sum. Obviously, this series will also diverge.

The nth Term Divergence Test.

This leads to the nth Term Divergence Test:

Theorem A.4. *If* $\lim a_n \neq 0$ *or if this limit does not exist, then* $\sum_n a_n$ *diverges.*

This theorem does not imply that just because the terms are getting smaller, the series will converge. Otherwise, we would not need any other convergence theorems.

For the next theorems, we will assume that the series has nonnegative terms.

The Comparison Test.

Comparison Test

The series $\sum a_n$ converges if there is a convergent series $\sum c_n$ such that $a_n \leq c_n$ for all $n > N$ for some N. The series $\sum a_n$ diverges if there is a divergent series $\sum d_n$ such that $d_n \leq a_n$ for all $n > N$ for some N.

This is easily seen. In the first case, we have

$$a_n \leq c_n, \forall n > N.$$

Summing both sides of the inequality, we have

$$\sum_n a_n \leq \sum_n c_n.$$

If $\sum c_n$ converges, $\sum c_n < \infty$, the $\sum a_n$ converges as well. A similar argument applies for the divergent series case.

For this test, one has to dream up a second series for comparison. Typically, this requires some experience with convergent series. Often it is better to use other tests first, if possible.

Example A.6. *Determine if* $\sum_{n=0}^{\infty} \frac{1}{3^n}$ *converges using the Comparison Test.*

We already know that $\sum_{n=0}^{\infty} \frac{1}{2^n}$ *converges. So, we compare these two series. In the above notation, we have* $a_n = \frac{1}{3^n}$ *and* $c_n = \frac{1}{2^n}$. *Because* $\frac{1}{2^n} \leq \frac{1}{3^n}$ *for* $n \geq 0$ *and* $\sum_{n=0}^{\infty} \frac{1}{2^n}$ *converges, then* $\sum_{n=0}^{\infty} \frac{1}{3^n}$ *converges by the Comparison Test.*

Limit Comparison Test

If $\lim_{n\to\infty} \frac{a_n}{b_n}$ is finite, then $\sum a_n$ and $\sum b_n$ converge together or diverge together.

Example A.7. *Determine if $\sum_{n=1}^{\infty} \frac{2n+1}{(n+1)^2}$ converges using the Limit Comparison Test.*

In order to establish the convergence, or divergence, of this series, we look to see how the terms $a_n = \frac{2n+1}{(n+1)^2}$ behave for large n. As n gets large, the numerator behaves like $2n$ and the denominator behaves like n^2. Therefore, a_n behaves like $\frac{2n}{n^2} = \frac{2}{n}$. The factor of 2 does not really matter.

This leads us to compare the infinite series $\sum_{n=1}^{\infty} \frac{2n+1}{(n+1)^2}$ with the series $\sum_{n=1}^{\infty} \frac{1}{n}$. Then,

$$\lim_{n\to\infty} \frac{a_n}{b_n} = \lim_{n\to\infty} \frac{2n^2+n}{(n+1)^2} = 2.$$

We can conclude that these two series both converge or both diverge.

If we knew the behavior of the second series, then we could finish the problem. Using the next test, we will prove that $\sum_{n=1}^{\infty} \frac{1}{n}$ diverges. Therefore, $\sum_{n=1}^{\infty} \frac{2n+1}{(n+1)^2}$ also diverges by the Limit Comparison Test. Another example of this test is given in Example A.9.

Integral Test

The Integral Test.

Consider the infinite series $\sum_{n=1}^{\infty} a_n$, where $a_n = f(n)$. Then, $\sum_{n=1}^{\infty} a_n$ and $\int_1^{\infty} f(x)\,dx$ both converge or both diverge. Here we mean that the integral converges or diverges as an improper integral.

Example A.8. *Does the harmonic series, $\sum_{n=1}^{\infty} \frac{1}{n}$, converge?*

We are interested in the convergence or divergence of the infinite series $\sum_{n=1}^{\infty} \frac{1}{n}$ that we saw in the Limit Comparison Test example. This infinite series is famous and is called the harmonic series. The plot of the partial sums is given in Figure A.7. It appears that the series could possibly converge or diverge. It is hard to tell graphically.

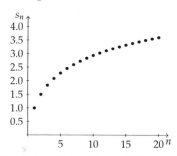

Figure A.7: Plot of the partial sums of the harmonic series $\sum_{n=1}^{\infty} \frac{1}{n}$.

In this case, we can use the Integral Test. In Figure A.8 we plot $f(x) = \frac{1}{x}$ and at each integer n we plot a box from n to $n+1$ of height $\frac{1}{n}$. We can see from the figure that the total area of the boxes is greater than the area under the curve. Because the area of each box is $\frac{1}{n}$, we have that

$$\int_1^{\infty} \frac{dx}{x} < \sum_{n=1}^{\infty} \frac{1}{n}.$$

But, we can compute the integral

$$\int_1^{\infty} \frac{dx}{x} = \lim_{x\to\infty}(\ln x) = \infty.$$

Thus, the integral diverges. However, the infinite series is larger than this! So, the harmonic series diverges by the Integral Test.

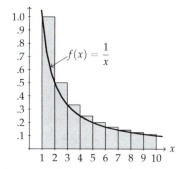

Figure A.8: Plot of $f(x) = \frac{1}{x}$ and boxes of height $\frac{1}{n}$ and width 1.

The Integral Test provides us with the convergence behavior for a class of infinite series called a *p*-series. These series are of the form $\sum_{n=1}^{\infty} \frac{1}{n^p}$.

Recalling that the improper integrals $\int_1^\infty \frac{dx}{x^p}$ converge for $p > 1$ and diverge otherwise, we have the p-test :

$$\sum_{n=1}^\infty \frac{1}{n^p}[-2cm] converges \, for \, p > 1$$

p-series and p-test.

and diverges otherwise.

Example A.9. *Does the series $\sum_{n=1}^\infty \frac{n+1}{n^3-2}$ converge?*

We first note that as n gets large, the general term behaves like $\frac{1}{n^2}$ because the numerator behaves like n and the denominator behaves like n^3. So, we expect that this series behaves like the series $\sum_{n=1}^\infty \frac{1}{n^2}$. Thus, by the Limit Comparison Test,

$$\lim_{n\to\infty} \frac{n+1}{n^3-2}(n^2) = 1.$$

These series both converge, or both diverge. However, we know that $\sum_{n=1}^\infty \frac{1}{n^2}$ converges by the p-test because $p = 2$. Therefore, the original series converges.

The Ratio Test.

Ratio Test

Consider the series $\sum_{n=1}^\infty a_n$ for $a_n > 0$. Let

$$\rho = \lim_{n\to\infty} \frac{a_{n+1}}{a_n}.$$

Then, the behavior of the infinite series can be determined from the conditions

$$\rho < 1, \quad converges$$
$$\rho > 1, \quad diverges.$$

Example A.10. *Use the Ratio Test to determine the convergence of $\sum_{n=1}^\infty \frac{n^{10}}{10^n}$.*

We compute

$$\begin{aligned} \rho &= \lim_{n\to\infty} \frac{a_{n+1}}{a_n} \\ &= \lim_{n\to\infty} \frac{(n+1)^{10}}{n^{10}} \frac{10^n}{10^{n+1}} \\ &= \lim_{n\to\infty} \left(1 + \frac{1}{n}\right)^{10} \frac{1}{10} \\ &= \frac{1}{10} < 1. \end{aligned}$$

(A.5)

Therefore, the series is said to converge by the Ratio Test.

Example A.11. *Use the Ratio Test to determine the convergence of $\sum_{n=1}^\infty \frac{3^n}{n!}$.*

[6] *In this case we make use of the fact that[6] $(n+1)! = (n+1)n!$. We compute*

$$\begin{aligned} \rho &= \lim_{n\to\infty} \frac{a_{n+1}}{a_n} \\ &= \lim_{n\to\infty} \frac{3^{n+1}}{3^n} \frac{n!}{(n+1)!} \\ &= \lim_{n\to\infty} \frac{3}{n+1} = 0 < 1. \end{aligned}$$

(A.6)

[6] Note that the Ratio Test works when factorials are involved because using $(n+1)! = (n+1)n!$ helps to reduce the needed ratios into something manageable.

This series also converges by the Ratio Test.

nth Root Test

Consider the series $\sum_{n=1}^{\infty} a_n$ for $a_n > 0$. Let

$$\rho = \lim_{n \to \infty} a_n^{1/n}.$$

Then the behavior of the infinite series can be determined using

$$\rho < 1, \quad \text{converges}$$
$$\rho > 1, \quad \text{diverges}.$$

Example A.12. *Use the nth Root Test to determine the convergence of $\sum_{n=0}^{\infty} e^{-n}$.*

We use the nth Root Test: $\lim_{n \to \infty} \sqrt[n]{a_n} = \lim_{n \to \infty} e^{-1} = e^{-1} < 1$. Thus, this series converges by the nth Root Test.[7]

[7] Note that the Root Test works when there are no factorials and simple powers are involved. In such cases special limit rules help in the evaluation.

Example A.13. *Use the nth Root Test to determine the convergence of $\sum_{n=1}^{\infty} \frac{n^n}{2^{n^2}}$.*
This series also converges by the nth Root Test.

$$\lim_{n \to \infty} \sqrt[n]{a_n} = \lim_{n \to \infty} \left(\frac{n^n}{2^{n^2}} \right)^{1/n} = \lim_{n \to \infty} \frac{n}{2^n} = 0 < 1.$$

Absolute and Conditional Convergence

We next turn to series that have both positive and negative terms. We can toss out the signs by taking absolute values of each of the terms. We note that since $a_n \leq |a_n|$, we have

$$- \sum_{n=1}^{\infty} |a_n| \leq \sum_{n=1}^{\infty} a_n \leq \sum_{n=1}^{\infty} |a_n|.$$

If the sum $\sum_{n=1}^{\infty} |a_n|$ converges, then the original series converges. For example, if $\sum_{n=1}^{\infty} |a_n| = S$, then by this inequality, $-S \leq \sum_{n=1}^{\infty} a_n \leq S$.

Convergence of the series $\sum |a_n|$ is useful because we can use the previous tests to establish convergence of such series. Thus, we say that a series converges absolutely if $\sum_{n=1}^{\infty} |a_n|$ converges. If a series converges, but does not converge absolutely, then it is said to converge conditionally.

Conditional and absolute convergence.

Example A.14. *Show that the series $\sum_{n=1}^{\infty} \frac{\cos \pi n}{n^2}$ converges absolutely.*
This series converges absolutely because $\sum_{n=1}^{\infty} |a_n| = \sum_{n=1}^{\infty} \frac{1}{n^2}$ is a p-series with $p = 2$.

Finally, there is one last test that we recall from your introductory calculus class. We consider the alternating series, given by $\sum_{n=1}^{\infty} (-1)^{n+1} a_n$. The convergence of an alternating series is determined from Leibniz's Theorem.[8]

Convergence of alternating series.

[8] Gottfried Wilhelm Leibniz (1646–1716) developed calculus independently of Sir Isaac Newton (1643–1727).

Theorem A.5. *The series $\sum_{n=1}^{\infty} (-1)^{n+1} a_n$ converges if*

1. *a_n's are positive.*

2. *$a_n \geq a_{n+1}$ for all n.*

3. *$a_n \to 0$.*

The first condition guarantees that we have alternating signs in the series. The next condition says that the magnitude of the terms gets smaller, and the last condition imposes further that the terms approach zero.

Example A.15. *Establish the type of convergence of the alternating harmonic series,* $\sum_{n=1}^{\infty} \frac{(-1)^{n+1}}{n}$.

First of all, this series is an alternating series. The a_n's in Leibniz's Theorem are given by $a_n = \frac{1}{n}$. Condition 2 for this case is

$$\frac{1}{n} \geq \frac{1}{n+1}.$$

This is certainly true, as condition 2 just means that the terms are not getting bigger as n increases. Finally, condition 3 says that the terms are in fact going to zero as n increases. This is true in this example. Therefore, the alternating harmonic series converges by Leibniz's Theorem.

Note: The alternating harmonic series converges conditionally, because the series of absolute values $\sum_{n=1}^{\infty} \left| \frac{(-1)^{n+1}}{n} \right| = \sum_{n=1}^{\infty} \frac{1}{n}$ gives the (divergent) harmonic series. So, the alternating harmonic series does not converge absolutely.

Example A.16. *Determine the type of convergence of the series $\sum_{n=0}^{\infty} \frac{(-1)^n}{2^n}$.*

$\sum_{n=0}^{\infty} \frac{(-1)^n}{2^n}$ *also passes the conditions of Leibniz's Theorem. It should be clear that the terms of this alternating series are getting smaller and approach zero. Furthermore, this series converges absolutely!*

A.6   Sequences of Functions

OUR IMMEDIATE GOAL IS TO PREPARE for studying Fourier series, which are series whose terms are functions. So, in this section we begin to discuss series of functions and the convergence of such series. Once more, we will need to resort to the convergence of the sequence of partial sums. This means we really need to start with sequences of functions. A sequence of functions is simply a set of functions $f_n(x)$, $n = 1, 2, \ldots$ defined on a common domain D. A frequently used example will be the sequence of functions $\{1, x, x^2, \ldots\}, x \in [-1, 1]$.

Evaluating each sequence of functions at a given value of x, we obtain a sequence of real numbers. As before, we can ask if this sequence converges. Doing this for each point in the domain D, we then ask if the resulting collection of limits defines a function on D. More formally, this leads us to the idea of pointwise convergence.

Pointwise convergence of sequences of functions.

A sequence of functions f_n converges pointwise on D to a limit g if

$$\lim_{n \to \infty} f_n(x) = g(x)$$

for each $x \in D$. More formally, we write that

$$\lim_{n \to \infty} f_n = g \text{ (pointwise on } D)$$

if given $x \in D$ and $\epsilon > 0$, there exists an integer N such that

$$|f_n(x) - g(x)| < \epsilon, \quad \forall n \geq N.$$

The symbol ∀ means "for all."

Example A.17. *Consider the sequence of functions*

$$f_n(x) = \frac{1}{1 + nx}, \quad |x| < \infty, \quad n = 1, 2, 3, \ldots.$$

The limits depends on the value of x. We consider two cases, $x = 0$ and $x \neq 0$.

1. $x = 0$. *Here* $\lim_{n \to \infty} f_n(0) = \lim_{n \to \infty} 1 = 1$.

2. $x \neq 0$. *Here* $\lim_{n \to \infty} f_n(x) = \lim_{n \to \infty} \frac{1}{1 + nx} = 0$.

Therefore, we can say that $f_n \to g$ pointwise for $|x| < \infty$, where

$$g(x) = \begin{cases} 0, & x \neq 0, \\ 1, & x = 0. \end{cases} \tag{A.7}$$

We also note that for a sequence that converges pointwise, N generally depends on both x and ϵ, $N(x, \epsilon)$. We will show this by example.

Example A.18. *Consider the functions $f_n(x) = x^n, x \in [0, 1], n = 1, 2, \ldots$.*
We recall that the definition for pointwise convergence suggests that for each x, we seek an N such that $|f_n(x) - g(x)| < \epsilon, \forall n \geq N$. This is not at first easy to see. So, we will provide some simple examples showing how N can depend on both x and ϵ.

1. $x = 0$. *Here we have $f_n(0) = 0$ for all n. So, given $\epsilon > 0$, we seek an N such that $|f_n(0) - 0| < \epsilon, \forall n \geq N$. Inserting $f_n(0) = 0$, we have $0 < \epsilon$. Because this is true for all n, we can pick $N = 1$.*

2. $x = \frac{1}{2}$. *In this case, we have $f_n(\frac{1}{2}) = \frac{1}{2^n}$, for $n = 1, 2, \ldots$.*
As n gets large, $f_n \to 0$. So, given $\epsilon > 0$, we seek N such that

$$\left| \frac{1}{2^n} - 0 \right| < \epsilon, \quad \forall n \geq N.$$

This result means that $\frac{1}{2^n} < \epsilon$.
Solving the inequality for n, we have

$$n > N \geq -\frac{\ln \epsilon}{\ln 2}.$$

We choose $N \geq -\frac{\ln \epsilon}{\ln 2}$. Thus, our choice of N depends on ϵ.
For, $\epsilon = 0.1$, this gives

$$N \geq -\frac{\ln 0.1}{\ln 2} = \frac{\ln 10}{\ln 2} \approx 3.32.$$

So, we pick $N = 4$ and we have $n > N = 4$.

3. $x = \frac{1}{10}$. *This case can be examined like the previous example.*
We have $f_n(\frac{1}{10}) = \frac{1}{10^n}$, for $n = 1, 2, \ldots$. This leads to $N \geq -\frac{\ln \epsilon}{\ln 10}$. For $\epsilon = 0.1$, this gives $N \geq 1$, or $n > 1$.

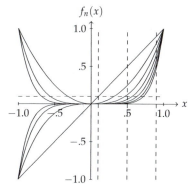

Figure A.9: Plot of $f_n(x) = x^n$ showing how N depends on $x = 0, 0.1, 0.5, 0.9$ (the vertical lines) and $\epsilon = 0.1$ (the horizontal line). Look at the intersection of a given vertical line with the horizontal line and determine N from the number of curves not under the intersection point.

4. $x = \frac{9}{10}$. *This case can be examined like the previous two examples. We have* $f_n(\frac{9}{10}) = (\frac{9}{10})^n$, *for* $n = 1, 2, \ldots$. *So, given an* $\epsilon > 0$, *we seek an N such that* $(\frac{9}{10})^n < \epsilon$ *for all* $n > N$. *Therefore,*

$$n > N \geq \frac{\ln \epsilon}{\ln \left(\frac{9}{10}\right)}.$$

For $\epsilon = 0.1$, *we have* $N \geq 21.85$, *or* $n > N = 22$.

So, for these cases, we have shown that N can depend on both x and ϵ. These cases are shown in Figure A.9.

There are other questions that can be asked about sequences of functions. Let the sequence of functions f_n be continuous on D. If the sequence of functions converges pointwise to g on D, then we can ask the following.

1. Is g continuous on D?

2. If each f_n is integrable on $[a, b]$, then does

$$\lim_{n \to \infty} \int_a^b f_n(x)\, dx = \int_a^b g(x)\, dx?$$

3. If each f_n is differentiable at c, then does

$$\lim_{n \to \infty} f_n'(c) = g'(c)?$$

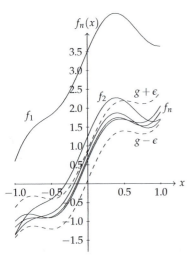

Figure A.10: For uniform convergence, as n gets large, $f_n(x)$ lies in the band $g(x) - \epsilon, g(x) - \epsilon$.

Uniform convergence of sequences of functions.

It turns out that pointwise convergence is not enough to provide an affirmative answer to any of these questions. Though we will not prove it here, what we will need is uniform convergence.

Consider a sequence of functions $\{f_n(x)\}_{n=1}^\infty$ on D. Let $g(x)$ be defined for $x \in D$. Then the sequence converges uniformly on D, or

$$\lim_{n \to \infty} f_n = g \text{ uniformly on } D,$$

if given $\epsilon > 0$, there exists an N such that

$$|f_n(x) - g(x)| < \epsilon, \quad \forall n \geq N \text{ and } \forall x \in D.$$

This definition almost looks like the definition for pointwise convergence. However, the seemingly subtle difference lies in the fact that N does not depend upon x. The sought N works for all x in the domain. As seen in Figure A.10 as n gets large, $f_n(x)$ lies in the band $(g(x) - \epsilon, g(x) + \epsilon)$.

Example A.19. *Does the sequence of functions $f_n(x) = x^n$ converge uniformly on $[0, 1]$?*

Note that in this case as n gets large, $f_n(x)$ does not lie in the band $(g(x) - \epsilon, g(x) + \epsilon)$ as seen in Figure A.11. Therefore, this sequence of functions does not converge uniformly on $[-11]$.

Example A.20. *Does the sequence of functions $f_n(x) = \cos(nx)/n^2$ converge uniformly on $[-1, 1]$?*

For this example, we plot the first several members of the sequence in Figure A.12. We can see that eventually $(n \geq N)$ members of this sequence do lie inside a band of width ϵ about the limit $g(x) \equiv 0$ for all values of x. Thus, this sequence of functions will converge uniformly to the limit.

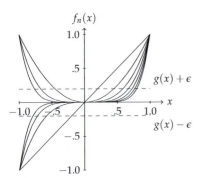

Figure A.11: Plot of $f_n(x) = x^n$ on [-1,1] for $n = 1, \ldots, 10$ and $g(x) \pm \epsilon$ for $\epsilon = 0.2$.

Finally, we should note that if a sequence of functions is uniformly convergent, then it converges pointwise. However, the examples should bear out that the converse is not true.

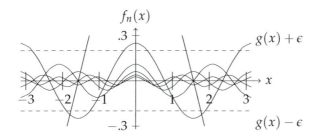

Figure A.12: Plot of $f_n(x) = \cos(nx)/n^2$ on $[-\pi, \pi]$ for $n = 1 \dots 10$ and $g(x) \pm \epsilon$ for $\epsilon = 0.2$.

A.7 Infinite Series of Functions

WE NOW TURN OUR ATTENTION TO INFINITE SERIES of functions, which will form the basis of our study of Fourier series. An infinite series of functions is given by $\sum_{n=1}^{\infty} f_n(x)$, $x \in D$. Using powers of x, an example of an infinite series of functions might be $\sum_{n=1}^{\infty} x^n$, $x \in [-1, 1]$. In order to investigate the convergence of this series, we could substitute values for x and determine if the resulting series of numbers converges. In general, to investigate the convergence of infinite series of functions, we would consider the Nth partial sums

$$s_N(x) = \sum_{n=1}^{N} f_n(x)$$

and ask if this sequence converges. We will begin to answer this question by defining pointwise and uniform convergence of infinite series of functions.

The infinite series $\sum f_j(x)$ converges pointwise to $f(x)$ on D if given $x \in D$, and $\epsilon > 0$, there exists and N such that

$$|f(x) - s_n(x)| < \epsilon$$

for all $n > N$.

Pointwise convergence of an infinite series.

The infinite series $\sum f_j(x)$ converges uniformly to $f(x)$ on D given $\epsilon > 0$, there exists an N such that

$$|f(x) - s_n(x)| < \epsilon$$

for all $n > N$ and all $x \in D$.

Uniform convergence f an infinite series.

Again, we state without proof the following important properties of uniform convergence for infinite series of functions:

1. Uniform convergence implies pointwise convergence.

2. If f_n is continuous on D, and $\sum_n^{\infty} f_n$ converges uniformly to f on D, then f is continuous on D.

Uniform convergence gives nice properties under some additional conditions, such as being able to integrate, or differentiate, term by term.

3. If f_n is continuous on $[a, b] \subset D$, $\sum_n^\infty f_n$ converges uniformly on D to g, and $\int_a^b f_n(x)\, dx$ exists, then

$$\sum_n^\infty \int_a^b f_n(x)\, dx = \int_a^b \sum_n^\infty f_n(x)\, dx = \int_a^b g(x)\, dx.$$

4. If f_n' is continuous on $[a, b] \subset D$, $\sum_n^\infty f_n$ converges pointwise to g on D, and $\sum_n^\infty f_n'$ converges uniformly on D, then

$$\sum_n^\infty f_n'(x) = \frac{d}{dx}\left(\sum_n^\infty f_n(x)\right) = g'(x)$$

for $x \in (a, b)$.

[9] Karl Theodor Wilhelm Weierstraß (1815–1897) was a German mathematician who may be thought of as the "father of analysis."

Because uniform convergence of series gives so much, like term by term integration and differentiation, we would like to be able to recognize when we have a uniformly convergent series. One test for such convergence is the Weierstraß M-Test[9].

Theorem A.6. Weierstraß M-Test *Let* $\{f_n\}_{n=1}^\infty$ *be a sequence of functions on* D. *If* $|f_n(x)| \le M_n$, *for* $x \in D$ *and* $\sum_{n=1}^\infty M_n$ *converges, then* $\sum_{n=1}^\infty f_n$ *converges uniformly on* D.

Proof. First, we note that for $x \in D$,

$$\sum_{n=1}^\infty |f_n(x)| \le \sum_{n=1}^\infty M_n.$$

Because by the assumption that $\sum_{n=1}^\infty M_n$ converges, we have that $\sum_{n=1}^\infty f_n$ converges absolutely on D. Therefore, $\sum_{n=1}^\infty f_n$ converges pointwise on D. So, we let $\sum_{n=1}^\infty f_n = g$.

We now want to prove that this convergence is, in fact, uniform. So, given an $\epsilon > 0$, we need to find an N such that

$$\left| g(x) - \sum_{j=1}^n f_j(x) \right| < \epsilon$$

if $n \ge N$ for all $x \in D$. Therefore, for any $x \in D$, we find a bound on $\left| g(x) - \sum_{j=1}^n f_j(x) \right|$:

$$
\begin{aligned}
\left| g(x) - \sum_{j=1}^n f_j(x) \right| &= \left| \sum_{j=1}^\infty f_j(x) - \sum_{j=1}^n f_j(x) \right| \\
&= \left| \sum_{j=n+1}^\infty f_j(x) \right| \\
&\le \sum_{j=n+1}^\infty |f_j(x)|, \quad \text{by the triangle inequality} \\
&\le \sum_{j=n+1}^\infty M_j. \quad\quad\quad\quad\quad\quad\quad\quad\text{(A.8)}
\end{aligned}
$$

Now, the sum over the M_j's is convergent, so we can choose N such that

$$\sum_{j=n+1}^{\infty} M_j < \epsilon, \quad n \geq N.$$

Combining these results, we have

$$\left| g(x) - \sum_{j=1}^{n} f_j(x) \right| \leq \sum_{j=n+1}^{\infty} M_j < \epsilon$$

for all $n \geq N$ and $x \in D$. Thus, we conclude that the series $\sum f_j$ converges uniformly to g, which we write $\sum f_j \rightarrow g$ uniformly on D. $\qquad\square$

We now give an example of how to use the Weierstraß M-Test.

Example A.21. *Show that the series $\sum_{n=1}^{\infty} \frac{\cos nx}{n^2}$ converges uniformly on $[-\pi, \pi]$.*

Each term of the infinite series is bounded by $\left| \frac{\cos nx}{n^2} \right| = \frac{1}{n^2} \equiv M_n$. We also know that $\sum_{n=1}^{\infty} M_n = \sum_{n=1}^{\infty} \frac{1}{n^2} < \infty$. Thus, we can conclude that the original series converges uniformly, as it satisfies the conditions of the Weierstraß M-Test.

A.8 Special Series Expansions

EXAMPLES OF INFINITE SERIES ARE geometric series, power series, and binomial series. These are discussed more fully in Sections 1.1.6, 1.1.7, and 1.1.8, respectively. These series are defined as follows:

1. The sum of the geometric series exists for $|r| < 1$ and is given by

   The geometric series.

   $$\sum_{n=0}^{\infty} ar^n = \frac{a}{1-r}, \qquad |r| < 1. \tag{A.9}$$

   A power series expansion about $x = a$ with coefficient sequence c_n is given by $\sum_{n=0}^{\infty} c_n(x-a)^n$.

2. A Taylor series expansion of $f(x)$ about $x = a$ is the series

   Taylor series expansion.

   $$f(x) \sim \sum_{n=0}^{\infty} c_n(x-a)^n, \tag{A.10}$$

   where

   $$c_n = \frac{f^{(n)}(a)}{n!}. \tag{A.11}$$

3. A Maclaurin series expansion of $f(x)$ is a Taylor series expansion of $f(x)$ about $x = 0$, or

   Maclaurin series expansion.

   $$f(x) \sim \sum_{n=0}^{\infty} c_n x^n, \tag{A.12}$$

   where

   $$c_n = \frac{f^{(n)}(0)}{n!}. \tag{A.13}$$

Some common expansions are provided in Table A.1.

Table A.1: Common Mclaurin Series Expansions

	Series Expansions You Should Know	
e^x	$= \quad 1 + x + \dfrac{x^2}{2} + \dfrac{x^3}{3!} + \dfrac{x^4}{4!} + \dots$	$= \quad \displaystyle\sum_{n=0}^{\infty} \dfrac{x^n}{n!}$
$\cos x$	$= \quad 1 - \dfrac{x^2}{2} + \dfrac{x^4}{4!} - \dots$	$= \quad \displaystyle\sum_{n=0}^{\infty} (-1)^n \dfrac{x^{2n}}{(2n)!}$
$\sin x$	$= \quad x - \dfrac{x^3}{3!} + \dfrac{x^5}{5!} - \dots$	$= \quad \displaystyle\sum_{n=0}^{\infty} (-1)^n \dfrac{x^{2n+1}}{(2n+1)!}$
$\cosh x$	$= \quad 1 + \dfrac{x^2}{2} + \dfrac{x^4}{4!} + \dots$	$= \quad \displaystyle\sum_{n=0}^{\infty} \dfrac{x^{2n}}{(2n)!}$
$\sinh x$	$= \quad x + \dfrac{x^3}{3!} + \dfrac{x^5}{5!} + \dots$	$= \quad \displaystyle\sum_{n=0}^{\infty} \dfrac{x^{2n+1}}{(2n+1)!}$
$\dfrac{1}{1-x}$	$= \quad 1 + x + x^2 + x^3 + \dots$	$= \quad \displaystyle\sum_{n=0}^{\infty} x^n$
$\dfrac{1}{1+x}$	$= \quad 1 - x + x^2 - x^3 + \dots$	$= \quad \displaystyle\sum_{n=0}^{\infty} (-x)^n$
$\tan^{-1} x$	$= \quad x - \dfrac{x^3}{3} + \dfrac{x^5}{5} - \dfrac{x^7}{7} + \dots$	$= \quad \displaystyle\sum_{n=0}^{\infty} (-1)^n \dfrac{x^{2n+1}}{2n+1}$
$\ln(1+x)$	$= \quad x - \dfrac{x^2}{2} + \dfrac{x^3}{3} - \dots$	$= \quad \displaystyle\sum_{n=1}^{\infty} (-1)^{n+1} \dfrac{x^n}{n}$

4. The binomial series indexbinomial series is a special Maclaurin series. Namely, it is the series expansion of $(1+x)^p$ for p a nonnegative integer.

We also considered the convergence of power series $\sum_{n=0}^{\infty} c_n (x-a)^n$. For $x = a$, the series obviously converges. In general, if $\sum_{n=0}^{\infty} c_n (b-a)^n$ converges for $b \neq a$, then $\sum_{n=0}^{\infty} c_n (x-a)^n$ converges absolutely for all x satisfying $|x - a| < |b - a|$.

This leads to the three possibilities:

1. $\sum_{n=0}^{\infty} c_n (x-a)^n$ may only converge at $x = a$.

2. $\sum_{n=0}^{\infty} c_n (x-a)^n$ may converge for all real numbers.

3. $\sum_{n=0}^{\infty} c_n (x-a)^n$ converges for $|x - a| < R$ and diverges for $|x - a| > R$.

Interval and radius of convergence.

The number R is called the radius of convergence of the power series and $(a - R, a + R)$ is called the interval of convergence. Convergence at the endpoints of this interval must be tested for each power series.

Finally, we have the special case of the general binomial expansion for $(1+x)^p$ for p real. The series is given by the

The binomial expansion.

$$(1+x)^p = \sum_{r=0}^{\infty} \frac{p(p-1)\cdots(p-r+1)}{r!} x^r, \quad |x| < 1. \qquad \text{(A.14)}$$

The binomial approximation.

In practice, one only needs the first few terms for $|x| \ll 1$. Then

$$(1+x)^p \approx 1 + px + \frac{p(p-1)}{2} x^2 |x| \ll 1. \qquad \text{(A.15)}$$

A.9   Order of Sequences and Functions

OFTEN WE ARE INTERESTED IN COMPARING the rates of convergence of sequences or asymptotic behavior of functions. This is also useful in approximation theory. We begin with the comparison of sequences and introduce *big-Oh* notation. We will then extend this to functions of continuous variables.

Big-Oh notation.

Let $\{a_n\}$ and $\{b_n\}$ be two sequences. Then, if there are numbers N and K (independent of N) such that

$$\left|\frac{a_n}{b_n}\right| < K \quad \text{whenever} \quad n > N,$$

we say that a_n is *of the order of b_n*. We write this as

$$a_n = O(b_n) \quad \text{as} \quad n \to \infty$$

and say a_n is "big O" of b_n.

Example A.22. *Consider the sequences given by $a_n = \frac{2n+1}{3n^2+2}$ and $b_n = \frac{1}{n}$.*
In this case, we consider the ratio

$$\left|\frac{a_n}{b_n}\right| = \left|\frac{\frac{2n+1}{3n^2+2}}{\frac{1}{n}}\right| = \left|\frac{2n^2+n}{3n^2+2}\right|.$$

We want to find a bound on the last expression as n gets large. We divide the numerator and denominator by n^2 and find that

$$\left|\frac{a_n}{b_n}\right| = \left|\frac{2+1/n}{3+2/n^2}\right|.$$

Further dividing out a 2/3, we find

$$\left|\frac{a_n}{b_n}\right| = \frac{2}{3}\left|\frac{1+1/2n}{1+2/3n^2}\right|.$$

The last expression is largest for $n = 3$. This gives

$$\left|\frac{a_n}{b_n}\right| = \frac{2}{3}\left|\frac{1+1/2n}{1+2/3n^2}\right| \le \frac{2}{3}\left|\frac{1+1/6}{1+2/27}\right| = \frac{21}{29}.$$

Considering the function $f(x) = \frac{2x^2+x}{3x^2+2}$, setting $f'(x) = 0$, we find the maximum is actually at $x = \frac{1}{3}(4 + \sqrt{22}) \approx 2.897$. Also, inserting the first few integers yields the sequence $\{0.6000, 0.7143, 0.7241, 0.7200, 0.7143, \dots\}$. In both cases, this supports choosing $n = 3$ in the example.

Thus, for $n > 3$, we have that

$$\left|\frac{a_n}{b_n}\right| \le \frac{21}{29} < 1 \equiv K.$$

We then conclude from the definition of big-Oh that

$$a_n = O(b_n) = O\left(\frac{1}{n}\right).$$

In practice, one is often given a sequence like a_n, but the second simpler sequence needs to be found by looking at the large n behavior of a_n.

Referring to the previous example, we are given $a_n = \frac{2n+1}{3n^2+2}$. We look at the large n behavior. The numerator behaves like $2n$ and the denominator behaves like $3n^2$. Thus, $a_n = \frac{2n+1}{3n^2+2} \sim \frac{2n}{3n^2} = \frac{2}{3n}$ for large n. Therefore, we say that $a_n = O(\frac{1}{n})$ for large n. Note that we are only interested in the n-dependence and not the multiplicative constant because $\frac{1}{n}$ and $\frac{2}{3n}$ have the same growth rate.

In a similar way, we can compare functions. We modify our definition of big-Oh for functions of a continuous variable: $f(x)$ is *of the order of* $g(x)$, or $f(x) = O(g(x))$, as $x \to x_0$ if

$$\lim_{x \to x_0} \left| \frac{f(x)}{g(x)} \right| < K$$

for some finite nonzero constant K independent of x_0.

Example A.23. *Show that*

$$\cos x - 1 + \frac{x^2}{2} = O(x^4) \quad \text{as } x \to 0.$$

This should be apparent from the Taylor series expansion for $\cos x$,

$$\cos x = 1 - \frac{x^2}{2} + O(x^4) \text{ as } x \to 0.$$

However, we will show that $\cos x - 1 + \frac{x^2}{2}$ is of the order of $O(x^4)$ using the above definition.

We need to compute

$$\lim_{x \to 0} \left| \frac{\cos x - 1 + \frac{x^2}{2}}{x^4} \right|.$$

The numerator and denominator separately go to zero, so we have an indeterminate form. This suggests that we need to apply L'Hopital's Rule. (See Theorem A.3.) In fact, we apply it several times to find that

$$\lim_{x \to 0} \left| \frac{\cos x - 1 + \frac{x^2}{2}}{x^4} \right| = \lim_{x \to 0} \left| \frac{-\sin x + x}{4x^3} \right|$$

$$= \lim_{x \to 0} \left| \frac{-\cos x + 1}{12x^2} \right|$$

$$= \lim_{x \to 0} \left| \frac{\sin x}{24x} \right| = \frac{1}{24}.$$

Thus, for any number $K > \frac{1}{24}$, we have that

$$\lim_{x \to 0} \left| \frac{\cos x - 1 + \frac{x^2}{2}}{x^4} \right| < K.$$

We conclude that

$$\cos x - 1 + \frac{x^2}{2} = O(x^4) \quad \text{as } x \to 0.$$

Example A.24. *Determine the order of $f(x) = (x^3 - x)^{1/3} - x$ as $x \to \infty$. We can use a binomial expansion to write the first term in powers of x. However, because $x \to \infty$, we want to write $f(x)$ in powers of $\frac{1}{x}$, so that we can neglect higher-order powers. We can do this by first factoring out the x^3 :*

$$
\begin{aligned}
(x^3 - x)^{1/3} - x &= x\left(1 - \frac{1}{x^2}\right)^{1/3} - x \\
&= x\left(1 - \frac{1}{3x^2} + O\left(\frac{1}{x^4}\right)\right) - x \\
&= -\frac{1}{3x} + O\left(\frac{1}{x^3}\right).
\end{aligned}
\tag{A.16}
$$

Now we can see from the first term on the right that $(x^3 - x)^{1/3} - x = O\left(\frac{1}{x}\right)$ as $x \to \infty$.

Problems

1. For those sequences that converge, find the limit $\lim_{n\to\infty} a_n$.

    a. $a_n = \frac{n^2+1}{n^3+1}$.

    b. $a_n = \frac{3n+1}{n+2}$.

    c. $a_n = \left(\frac{3}{n}\right)^{1/n}$.

    d. $a_n = \frac{2n^2+4n^3}{n^3+5\sqrt{2+n^6}}$.

    e. $a_n = n \ln\left(1 + \frac{1}{n}\right)$.

    f. $a_n = n \sin\left(\frac{1}{n}\right)$.

    g. $a_n = \frac{(2n+3)!}{(n+1)!}$.

2. Find the sum for each of the series:

    a. $\sum_{n=0}^{\infty} \frac{(-1)^n 3}{4^n}$.

    b. $\sum_{n=2}^{\infty} \frac{2}{5^n}$.

    c. $\sum_{n=0}^{\infty} \left(\frac{5}{2^n} + \frac{1}{3^n}\right)$.

    d. $\sum_{n=1}^{\infty} \frac{3}{n(n+3)}$.

3. Determine if the following converge, or diverge, using one of the convergence tests. If the series converges, is it absolute or conditional?

    a. $\sum_{n=1}^{\infty} \frac{n+4}{2n^3+1}$.

    b. $\sum_{n=1}^{\infty} \frac{\sin n}{n^2}$.

    c. $\sum_{n=1}^{\infty} \left(\frac{n}{n+1}\right)^{n^2}$.

    d. $\sum_{n=1}^{\infty} (-1)^n \frac{n-1}{2n^2-3}$.

    e. $\sum_{n=1}^{\infty} \frac{\ln n}{n}$.

    f. $\sum_{n=1}^{\infty} \frac{100^n}{n^{200}}$.

g. $\sum_{n=1}^{\infty}(-1)^n \frac{n}{n+3}$.

h. $\sum_{n=1}^{\infty}(-1)^n \frac{\sqrt{5n}}{n+1}$.

4. Do the following:

    a. Compute: $\lim_{n\to\infty} n\ln\left(1 - \frac{3}{n}\right)$.

    b. Use L'Hopital's Rule to evaluate $L = \lim_{x\to\infty}\left(1 - \frac{4}{x}\right)^x$. [Hint: Consider $\ln L$.]

    c. Determine the convergence of $\sum_{n=1}^{\infty}\left(\frac{n}{3n+2}\right)^{n^2}$.

    d. Sum the series $\sum_{n=1}^{\infty}\left[\tan^{-1}n - \tan^{-1}(n+1)\right]$ by first writing the Nth partial sum and then computing $\lim_{N\to\infty} s_N$.

5. Consider the sum $\sum_{n=1}^{\infty}\frac{1}{(n+2)(n+1)}$.

    a. Use an appropriate convergence test to show that this series converges.

    b. Verify that

$$\sum_{n=1}^{\infty}\frac{1}{(n+2)(n+1)} = \sum_{n=1}^{\infty}\left(\frac{n+1}{n+2} - \frac{n}{n+1}\right).$$

    c. Find the nth partial sum of the series $\sum_{n=1}^{\infty}\left(\frac{n+1}{n+2} - \frac{n}{n+1}\right)$ and use it to determine the sum of the resulting *telescoping* series.

6. Recall that the alternating harmonic series converges conditionally.

    a. From the Taylor series expansion for $f(x) = \ln(1+x)$, inserting $x = 1$ gives the alternating harmonic series. What is the sum of the alternating harmonic series?

    b Because the alternating harmonic series does not converge absolutely, a rearrangement of the terms in the series will result in series whose sums vary. One such rearrangement in alternating p positive terms and n negative terms leads to the following sum[10]:

$$\frac{1}{2}\ln\frac{4p}{n} = \underbrace{\left(1 + \frac{1}{3} + \cdots + \frac{1}{2p-1}\right)}_{p\ \text{terms}} - \underbrace{\left(\frac{1}{2} + \frac{1}{4} + \cdots + \frac{1}{2n}\right)}_{n\ \text{terms}}$$

$$+ \underbrace{\left(\frac{1}{2p+1} + \cdots + \frac{1}{4p-1}\right)}_{p\ \text{terms}} - \underbrace{\left(\frac{1}{2n+2} + \cdots + \frac{1}{4n}\right)}_{n\ \text{terms}} + \cdots.$$

Find rearrangements of the alternating harmonic series to give the following sums; that is, determine p and n for the given expression and write down the above series explicitly; that is, determine p and n leading to the following sums.

    i. $\frac{5}{2}\ln 2$.

    ii. $\ln 8$.

[10] This is discussed by Lawrence H. Riddle in the *Kenyon Math. Quarterly*, 1(2), 6–21.

iii. 0.

iv. A sum that is close to π.

7. Determine the radius and interval of convergence of the following infinite series:

a. $\sum_{n=1}^{\infty} (-1)^n \frac{(x-1)^n}{n}$.

b. $\sum_{n=1}^{\infty} \frac{x^n}{2^n n!}$.

c. $\sum_{n=1}^{\infty} \frac{1}{n} \left(\frac{x}{5}\right)^n$

d. $\sum_{n=1}^{\infty} (-1)^n \frac{x^n}{\sqrt{n}}$.

8. Find the Taylor series centered at $x = a$ and its corresponding radius of convergence for the given function. In most cases, you need not employ the direct method of computation of the Taylor coefficients.

a. $f(x) = \sinh x$, $a = 0$.

b. $f(x) = \sqrt{1+x}$, $a = 0$.

c. $f(x) = xe^x$, $a = 1$.

d. $f(x) = \frac{x-1}{2+x}$, $a = 1$.

9. Test for pointwise and uniform convergence on the given set. [The Weierstraß M-Test might be helpful.]

a. $f(x) = \sum_{n=1}^{\infty} \frac{\ln nx}{n^2}$, $x \in [1,2]$.

b. $f(x) = \sum_{n=1}^{\infty} \frac{1}{3^n} \cos \frac{x}{2^n}$ on R.

10. Consider Gregory's expansion

$$\tan^{-1} x = x - \frac{x^3}{3} + \frac{x^5}{5} - \cdots = \sum_{k=0}^{\infty} \frac{(-1)^k}{2k+1} x^{2k+1}.$$

a. Derive Gregory's expansion using the definition

$$\tan^{-1} x = \int_0^x \frac{dt}{1+t^2},$$

expanding the integrand in a Maclaurin series, and integrating the resulting series term by term.

b. From this result, derive Gregory's series for π by inserting an appropriate value for x in the series expansion for $\tan^{-1} x$.

11. Use deMoivre's Theorem to write $\sin^3 \theta$ in terms of $\sin \theta$ and $\sin 3\theta$. [Hint: Focus on the imaginary part of $e^{3i\theta}$.]

12. Evaluate the following expressions at the given point. Use your calculator and your computer (such as Maple). Then use series expansions to find an approximation to the value of the expression to as many places as you trust.

a. $\frac{1}{\sqrt{1+x^3}} - \cos x^2$ at $x = 0.015$.

b. $\ln \sqrt{\frac{1+x}{1-x}} - \tan x$ at $x = 0.0015$.

c. $f(x) = \frac{1}{\sqrt{1+2x^2}} - 1 + x^2$ at $x = 5.00 \times 10^{-3}$.

d. $f(R, h) = R - \sqrt{R^2 + h^2}$ for $R = 1.374 \times 10^3$ km and $h = 1.00$ m.

e. $f(x) = 1 - \frac{1}{\sqrt{1-x}}$ for $x = 2.5 \times 10^{-13}$.

13. Determine the order, $O(x^p)$, of the following functions. You may need to use series expansions in powers of x when $x \to 0$, or series expansions in powers of $1/x$ when $x \to \infty$.

a. $\sqrt{x(1-x)}$ as $x \to 0$.

b. $\frac{x^{5/4}}{1-\cos x}$ as $x \to 0$.

c. $\frac{x}{x^2-1}$ as $x \to \infty$.

d. $\sqrt{x^2 + x} - x$ as $x \to \infty$.

Bibliography

M. J. Ablowitz and A. S. Fokas. *Complex Variables: Introduction and Applications*. Cambridge University Press, Cambridge, U.K.; New York, 2003.

M. Abromowitz and I. A. Stregum. *Handbook of Mathematical Functions, with Formulas, Graphs, and Mathematical Tables*. Dover Publications, New York, 1965.

G. E. Andrews, R. Askey, and R. Roy. *Special Functions*. Cambridge University Press, Cambridge, U.K. ; New York, 1999.

G. Arfken. *Mathematical Methods for Physicists*. Academic Press, New York second edition, 1970.

E. T. Bell. *Men of Mathematics*. Fireside Books, New York, 1965.

C. M. Bender and S. A. Orszag. *Advanced Mathematical Methods for Scientists and Engineers*. McGraw-Hill, New York, 1978.

M. L. Boas. *Mathematical Methods in the Physical Sciences*. John Wiley & Sons, Inc. New York, third edition, 2006.

M. Braun. *Differential Equations and Their Applications: An Introduction to Applied Mathematics*. Springer-Verlag, New York, 1975.

J. W. Brown and R. V. Churchill. *Complex Variables and Applications*. McGraw-Hill, New York, seventh edition, 2004.

R. L. Burden and J. D. Faires. *Numerical Analysis*. PWS Publishing Company, Boston, MA, 5th edition, 2004.

E. Butkov. *Mathematical Physics*. Addison Wesley Publishing Company, Reading, MA, 1968.

T. L. Chow. *Mathematical Methods for Physicists, A Concise Introduction*. Cambridge University Press, Cambridge, U.K. ; New York, 2002.

R. V. Churchill. *Introduction to Complex Variables and Applications*. McGraw-Hill, New York, 1948.

C. C. Cowen, K. R. Davidson, and R. P. Kaufman. Rearranging the alternating harmonic series. *The American Mathematical Monthly*, 87(10):817–819, 1980.

J. W. Dettman. *Applied Complex Variables*. Dover Publications, New York, 2010.

J. Franklin. *Classical Electromagetism*. Addison-Wesley Press, Cambridge, MA, 2005.

H. Goldstein. *Classical mechanics*. Addison-Wesley Press, Cambridge, MA, 1950.

I. M. Gradshtein, I. M. Ryzhik, and A. Jeffrey. *Table of Integrals, Series, and Products*. Academic Press, New York, 1980.

D. J. Griffiths. *Introduction to Quantum Mechanics*. Prentice Hall, Upper Saddle River, NJ, 1994.

D. J. Griffiths. *Introduction to Electrodynamics*. Prentice Hall, Upper Saddle River, NJ, 1999.

R. Haberman. *Applied Partial Differential Equations with Fourier Series and Boundary Value Problems*. Pearson Education, Inc., Upper Saddle River, NJ, fourth edition, 2004.

J. B. Hartle. *Gravity, An Introduction to Einstein's General Relativity*. Addison Wesley Publishing Company, San Fransisco, CA, 2003.

S. Hassani. *Foundations of Mathematical Physics*. Allyn and Bacon, Needham Heights, MA, 1991.

H. Jeffreys and B. S. Jeffreys. *Methods of Mathematical Physics*. Cambridge University Press, Cambridge, U.K. ; New York, 1972.

A. J. Jerri. *Integral and Discrete Transforms with Applications and Error Analysis*. Marcel Dekker, Inc. New York, 1992.

A. J. Jerri. *Introduction to Integral Equatons*. Sampling Publishing, Potsdam, New York, second edition, 1999.

K. Kaplan. *Advanced Calculus*. Addison Wesley Publishing Company, Reading, MA fourth edition, 1991.

J. P. Keener. *Principles of Applied Mathematics, Transformation ad Approximation*. Perseus Books, Cambridge, MA, second edition, 2000.

C. Lanczos. *The Variational Principles of Mechanics*. Dover Publications, New York, fourth edition, 1986.

S. M. Lea. *Mathematics for Physicists*. Brooks/Cole, 2004.

J. B. Marion and S. T. Thornton. *Classical Dynamics of Particles and Systems*. Saunders College Publishing, Fort Worth, TX, fourth edition, 1988.

A. I. Markushevich. *Theory of Functions*. Chelsea Publishing Company, New York, second edition, 1977.

J. E. Marsden and A. J. Tromba. *Vector Calculus*. W. H. Freeman and Company, New York, fourth edition, 1996.

P. M. Morse and H. Feshbach. *Methods of Theoretical Physics*. McGraw Hill, New York, 1953.

D. Richards. *Advanced Mathematical Methods with Maple*. Cambridge University Press, Cambridge, U.K. ; New York, 2002.

B. Ryden. *An Introduction to Cosmology*. Addison Wesley Publishing Company, San Fransisco, CA, 2002.

J. Stewart. *Calculus: Early Transcendentals*. Brooks Cole, sixth edition, 2007.

S. H. Strogatz. *Nonlinear Dynamics and Chaos: With Applications to Physics, Biology, Chemistry, and Engineering*. Perseus Publishing, Cambridge, MA, 1994.

J. R. Taylor. *Classical Mechanics*. University Science Books, Sausalito, CA, 2005.

G. B. Thomas and R. L. Finney. *Calculus and Analytic Geometry*. Addison-Wesley Press, Cambridge, MA, ninth edition, 1995.

H. F. Weinberger. *A First Course in Partial Differential Equations: With Complex Variables and Transform Methods*. Dover Publications, New York, 1995.

Index